Sigeru Torii

Electroorganic Reduction Synthesis

Volume 1

Related Titles

Bard, A. J., Stratmann, M., Schäfer, H. J. (Eds.)

Encyclopedia of Electrochemistry

Volume 8: Organic Electrochemistry

2004
ISBN 3-527-30400-2

Alkire, R. C., Kolb, D. M. (Eds.)

Advances in Electrochemical Science and Engineering

Volume 8

2003
ISBN 3-527-30211-5

Izutsu, K.

Electrochemistry in Nonaqueous Solutions

2002
ISBN 3-527-30516-5

Hodes, G. (Ed.)

Electrochemistry of Nanomaterials

2001
ISBN 3-527-29836-3

Sigeru Torii

Electroorganic Reduction Synthesis

Volume 1

KODANSHA

WILEY-VCH Verlag GmbH & Co. KGaA

Sigeru Torii
Professor Emeritus, Okayama University, Japan

This book was carefully produced. Nevertheless, authors and publisher do not warrant the information contained therein to be free of errors. Readers are advised to keep in mind that statements, data, illustrations, procedural detailes or other items may inadvertently be inaccurate.

Published jointly by
Kodansha Ltd., Tokyo (Japan),
WILEY-VCH Verlag GmbH & Co. KGaA, Weinheim (Federal Republic of Germany)

Library of Congress Card No. : applied for.

British Library Cataloguing-in-Publication Data
A catalogue record for this book is available from the British Library.

Bibliographic information published by
Die Deutsche Bibliothek
Die Deutsche Bibliothek lists this publication in the Deutche Nationalbibliografie; detailed bibliographic data is available in the Internet at <http://dnb.ddb.de>.

ISBN 4-06-208193-8 (KODANSHA)
ISBN 3-527-31539-X (WILEY-VCH)
ISBN 978-3-527-31539-0 (WILEY-VCH)

Printed in Japan

Preface

Awareness of the versatility and uniqueness of electrolytic procedures in organic synthesis is increasing in recent years as a result of various applications being spread gradually to cover many areas of academic and industrial organic chemistry. Such advances in methodology brought about a general interest among organic chemists with versatile synthetic devices of great promise, and basic concepts for the design of conditions for electrolysis are now becoming familiar to them by considering the complex interaction between components and reactive species in electrolysis media as the reaction site.

Electroorganic synthesis and its applications have been developed largely over the past two decades with a clear structure and its focus on synthetic methods. Most efforts of electroorganic synthetic chemists have been devoted especially to the investigation of electrolysis conditions for obtaining the desired products. Consequently, product-selectivity has been found to be derived from the most appropriate choice of a combination of influential factors such as solvents, electrolytes, electrode materials, additives, electrolysis methods (direct or indirect) together with electrolysis equipment and applied voltages (electrode potential).

This is a companion volume to *ELECTROORGANIC SYNTHESES Methods and Applications Part I: Oxidations* published by VCH Weinheim and Kodansha Tokyo in 1985. The present two-volume monograph consists of 13 chapters on electrochemical reduction and its product selectivity and covers all important organic substance classes such as aldehydes, ketones, acids, esters, acids anhydride, olefins, aromatics, and nitrogen-, sulfur-, selenium- and tellurium-containing compounds, together with halogenated compounds, alcohol and related derivatives, organic compounds involving group IIIA, IVA, VA, IB and IIB elements, organometallic compounds as well as the methodology on indirect reduction redox mediators, electrogenerated base-assisted reactions, and electropolymerization. The last chapter (13) deals with electroreduction together with electrooxidation in the course of polymerization reactions.

Like the preceding volume on oxidation, this work also provides a survey of synthetically interesting references on electroreduction reactions of organic compounds in terms of conditions for electrolysis and product selectivity, and aims to present the results in a form useful for synthetic applications. Although over 3400 references are cited, the author has not attempted to provide an exhaustive review, and articles which appeared before 1965 have generally not been included. The results presented here can be readily understood on the basis of organic chemical considerations, and detailed discussion of electrode reactions and reaction mechanisms is not included, since the major aim of this book is to aid electrosynthetic studies. Appendix 1, which indicates the relation between partial structures before and after electrolysis together with typical conditions for electrolysis, should assist readers in gaining easy access to the appropriate conditions for electrolysis for a desired functionalization.

The author is particularly indebted to Professor Dr. Hideo Tanaka for his invaluable assistance in checking the references, discussing the contents and commenting on the manuscript. I acknowledge the careful secretarial work of Ms. Yuuko Fukushima, Ms. Noriko Sera, Ms. Hiroko Nakanishi and Ms. Naomi Miyake, who arranged the data concisely and typed the manuscript. The author sincerely thanks Ms. Fukushima for her devoted assistance in arranging the manuscript in a most careful manner over a long period of time. Mr. Ippei Ohta of Kodansha were always helpful and cooperative, and I am grateful to Ms. Cecilia M. Hamagami for assistance in finalizing the English manuscript. Finally, special thanks to my wife, Hiroko, for her constant support and encouragement.

Sigeru Torii
Professor Emeritus, Okayama University
Director, The Institute of Creative Chemistry, Okayama, Japan
November 2005

Contents

Contents for Volume 2

1. Electrochemical Reduction and Product Selectivity

1.1 Introduction

The basic concept elucidating the influential factors for product-selective electrosynthesis has not yet been established [1], although the factors themselves have been discussed in reference [1e]. This may retard the prevalence of electrosynthetic methods in the organic chemist community. On the other hand, keen demand arising from the manufacturing processes to solve enviromental pollution problems pay much attention to synthetic methods which are operative as non-polluting processes. In this sense, modern electrosynthetic methods are facing a good opportunity to provide the solution of the enviromental problems due to their high product-selectivities under mild and easy reaction conditions as well as to save total amounts of materials being used for synthetic operations. The aim of the present monograph is to introduce versatile features of electrolysis methods as a potent tool in synthetic organic chemistry, as examplified by typical reactions clarified in each chapter.

In the last decade, a wide range of electroreduction reactions have been developed intensively so that organic compounds are readily functionalized by selecting suitable conditions. Today, different compounds can be synthesized selectively from the same starting compound by choosing the proper electroreduction conditions. These facts recall attention to the phenomena in which the fate of reactive spieces formed in electrolysis media can be controlled by choosing different chemical potential surface (CPS) under changing influential factors in electrolysis media (Fig. 1.1). For example, the substrate which undergoes

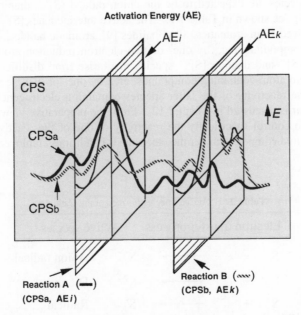

Fig. 1. 1 Different compounds from the same compound based on choosing different chemical potential surface (CPS) under changing influential factors in electrolysis media.

electron-transfer under electrolysis conditions shows different reduction potentials depending on not only the intrinsic nature of its functional groups but also on the interaction with the surroundings so that one may expect to have a variety of potential diagrams as CPS whose largest peak probably plays an important role for the determination of the reaction fate.   Thus, the selective synthesis of different compounds arising from the same compound can be realized by choosing the proper CPS from the CPS diagrams which would be affected by numerous parameters, e.g., effects of solvents, electrolytes, additives, electrode materials, pH values, etc. [2].   In this monograph, the author wishes to highlight the potentiality of recent electrosynthetic reactions.   For instance, the characteristic features of the electroreduction reactions involving activated olefins hydrodimerization, reductive addition of halides to carbonyl compounds, metal-complex systhesis [3], metal-complex catalyzed reactions [4], electrosynthesis of hypervalent compounds [5], intermolecular carbon-carbon bond formations [6], electroreductive cyclization [7], stereocontrol of electroreductive coupling reactions, radical induced reactions by indirect electroreductions, electrogenerated base-assisted reaction, and reaction of electrogenerated nucleophiles are demonstrated.   A new challenge for workers in the field is asymmetric induction in electrolysis systems.

The important results obtained over the past decade have been reviewed in several monographs [1].

1.1.1   Active Reaction Species Produced at Cathode

In the last decade, the utilization of electrochemically generated reactive species for the product-selective as well as stereo-selective [8,9] organic syntheses have been investigated extensively.   Actually, electroreduction procedures provide a variety of active reaction species by the electron-transfer from cathode to substrate.

The active species are expected to be the anion radical $[S]^{-\cdot}$, dianion $[S]^{2-}$, radical $[S]^{\cdot}$, and anion $[S]^{-}$ as shown in Table 1.1.   The active intermediates $[S]^{-\cdot}$ and radical $[S]^{\cdot}$ are generally derived from aromatics, alkyl halides [9], aromatic halides, carbonyl and active methylene compounds, etc., by one- or two-electron reduction around the cathode, while the anion $[S]^{-}$ and dianion $[S]^{2-}$ species may arise from disulfides, alkyl halides, haloketones, hetero atom-containing compounds, etc., by one- or two-electron additions at the cathode.   The reactivity of the latter species containing electrogenerated bases (EG bases) in particular is discussed in Chapter 12.   From the preparative viewpoint, one major concern is how to control specifically the reactivity of one of these active intermediates, leading to the desired compound under the electrolysis conditions employed.

Table 1.1 Active Species Formed at the Cathode

Electron transfer process			Active species	
S	$+ e^{-}$	$S^{-\cdot}$	Anion radical
$S^{-\cdot}$	$+ e^{-}$	S^{2-}	Dianion
S^{+}	$+ e^{-}$	$S^{\cdot}$	Radical
$S^{\cdot}$	$+ e^{-}$	S^{-}	Anion

1.1.2   Electron-transfer Processes and Electron Transmission Segments

Electroreduction of organic compounds proceeds in a stepwise manner through electron up-taking at the cathode (E process) and subsequent chemical reaction (C process) (eq. (1.1)).

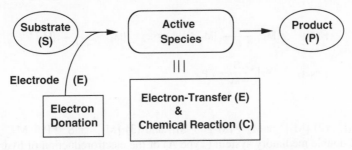

$$(1.1)$$

Organic compounds (substrates) are converted firstly into electron-enriched active species by electron-uptaking (an electron donation process), which then undergo a series of electron-transter and chemical reactions to yield products (Fig. 1.2).   The electron donating processes to the substrate (S) involves either (1) a direct electron-transfer from the cathode or (2) an indirect electron-transfer from electroreductively activated materials (mediator), *e.g.*, electrogenerated anion radicals, anions, radicals, and low-valent metal complexes.

Fig. 1. 2   Electroreductive conversion processes of organic compounds

The fate of the active species, *i.e.*, [S]⁻ˑ, [SH]ˑ, [SH]⁻ etc., is always affected by solvents, electrolytes, additives, electrode materials, etc., in which their interactions in electrolysis media provide a variety of chemical potential surfaces.

Electroorganic reactions involve electron-transfer (ET) processes which comprise a set of electron transmission segments, i.e., [S → AIa], [AIa → AIb], [AIb → AIn], [AIn → P], as shown in Fig. 1.3.   The substrate (S) is converted into an active intermediate (AIa) by a one-electron uptaking at the cathode.   This is the initial segment [S → AIa] of the electron transfer in electrolysis media.   Generally, the reaction process [S → P] consists of a series of multi-segments as an electron transmission unit before arriving at the product (P).

Let us consider the electron-transfer process regarded as an accumulation of electron transmission segments [1e].   As shown in Fig. 1.4, the reaction Type A is a typical double mediatory system which contains a set of three electron transmission segments as follows:

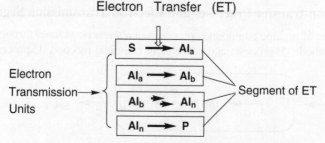

Fig. 1. 3 Active intermediate in electroorganic synthesis [1n];
S = Substrate;   AI = Active Intermediate;   P = Product

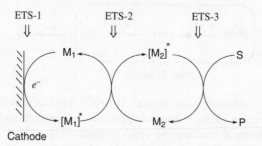

Fig. 1. 4 Electron-transmission unit (Type A) [2a]; M = Mediator,
S = Substrate, P = Products,
⇓ = Electron-transfer segment (ETS)

$$\text{ETS-1} \; : \; M_1 \xrightarrow{e^-} [M_1]^* \; ; \quad \text{ETS-2} \; : \; [M_1]^* \xrightarrow[-M_1]{M_2} [M_2]^*$$

$$\text{ETS-3} \; : \; [M_2]^* \xrightarrow[-M_2]{S} P$$

(1) $M_1 \rightarrow [M_1]^*$, (2) $[M_1]^* + M_2 \rightarrow [M_2]^* + M_1$, and (3) $[M_2]^* + S \rightarrow P + M_2$.   For example, a typical double mediatory system (Type A) of the electroreduction of hydroxyacetone **1** in the presence of yeast is shown in eq. (1.2) [10].   Apparently, the reduction system involves three transmission segments as an electron-transfer unit.   The presence of NAD as a mediatory catalyst results in the efficient formation of (R)-1,2-propanediol **2**.

(1.2)

A more complex electron transmission unit is shown in Type B, in which four electron transmission segments marked with arrows (Fig. 1.5) are included as follows: (1) $[M]^* + M \rightarrow [M]^* + M$, (2) $[M]^* + S_1 \rightarrow [M\text{-}S_1]^*$, and (3) $[M\text{-}S_1]^* + S_2 \rightarrow P + M$.   The type B unit contains not only four electron transmission segments, but also a process consisting of the formation of an *electron-transferring complex* $[M\text{-}S_1]^*$.

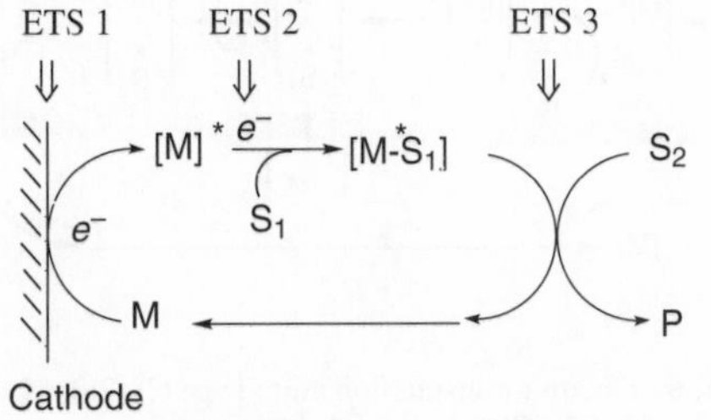

Fig. 1. 5 Electron-transmission unit (Type B) [2a]; M = Mediator, S = Substrate, P = Products, $\Downarrow$ = Electron-transfer segment (ETS)

ETS 1: $[M]^* + M \longrightarrow [M]^* + M_1$

ETS 2: $[M]^* + S_1 \xrightarrow{e^-} [M\text{-}S_1]^*$

ETS 3: $[M\text{-}S_1]^* \xrightarrow{S_2} P + M$

Type A involves a simple indirect electroreaction without the formation of any *electron-transferring complex*. In contrast, the *electron-transferring complexes* in type B cannot be fully characterized in terms of a simple electron transfer process. The actual reaction of the type B system is examplified by the Barbier-type allylation of carbonyl compounds **3** with electrogenerated allyltin(IV) reagents **7** [11]. Diallyltin(IV) dibromide **7** form allyl tin **6** and allyl bromide **8** in methanol reacts with **3** to give the product **4** (eq.(1.3)). Accordingly, the formation of the electron transferring complex may be well defined by assuming an electron transmission segment in the electron-transfer process.

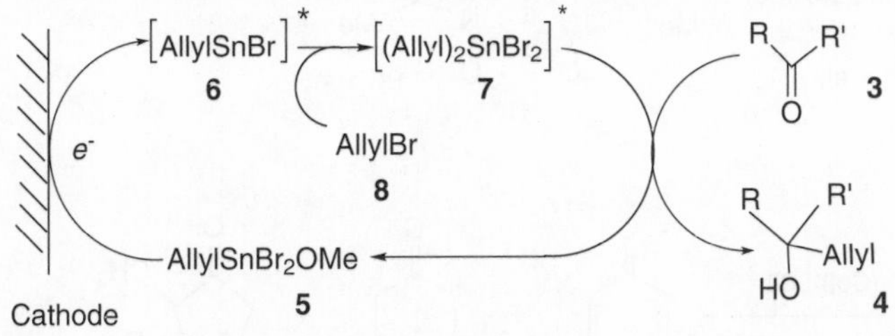

(1.3)

A number of transition metal redox complexes have been used as a mediatory system. Such metal complexes often provide unstable electron-transferring complexes as electron-transmission intermediates. The type C units indicates a mediatory system which involves electron-transfer segments from ETS-1 to ETS-5 (Fig 1.6). In particular, the ETS-3 and ETS-4 segments as carbon radical intermediates exist in a reversible process with alkyl-cobalt complexes in a matrix. The type C reaction is assigned to be the ring enlargement of α-(bromomethyl)cycloalkanones **9** *via* an acyl rearrangement of a radical intermediate **11** into **17, 18,** and **19**. The reaction is catalyzed by the electroreductively produced low-valent cobalt complexes **14** derived from the chloropyridine cobaloxime **15** (eqs. (1.4), (1.5)). The electrocatalytic rearrangement of **16** proceeds in an MeOH-Et4NOTs/KOH-(Pt) system in the presence of **15** to give the ring-enlarged products **17** (74%) and **18** (4%) together with **19** (17%) (eq. (1.5)) [12].

ETS-1   ETS-2   ETS-3   ETS-4   ETS-5

Fig. 1. 6 Electron-transmission unit (Type C) [2a]; M = Mediator,
S = Substrate, P = Products,
⇓ = Electron-transfer segment (EST)

(1.4)

(1.5)

The type D system demonstrates a set of cathode and anode reactions, which consists of more than five electron-transfer segments (Fig. 1.7).   The most complex electron-transmission unit (Type D) has to be the electroreductive allylation of the imine **20** in a metal redox mediatory system (a paired reaction system) (eq. (1.6)) [13].   The allylation of **20** proceeds in a THF-Bu₄NBr(0.1M)-(Al/Pt) system in the presence of lead (II)bromide to give the product **23**, in which the transmission unit, at least, comprises a set of five electron-transfer segments.

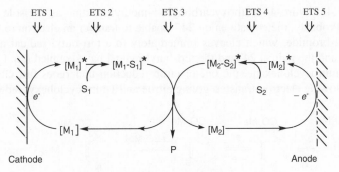

Fig. 1. 7 Electron-transmission unit (Type D) [2a]; M = Mediator,
S = Substrate, P = Products,
⇓ = Electron-transfer segment (EST)

$$(1.6)$$

The importance of the present concept which refers to an electron transmission unit is noted on judging the relative facility of an electron-transfer in the employed system. Namely, efforts at analyzing complex electrochemical systems in order to discover why the system chosen facilitates an electron-transfer in a suitable medium, could be greatly assisted by taking an electron transmission unit into consideration.   The net processes for indirect electrochemical reactions are hidden behind the veil of a multi electron-transmission unit; hence the segmentation of an electron transfer process into several electron transmission segments provides a helpful tool for elucidating a complex electrochemical reaction.

1.1.3   Electron-transfer Catalysts

Most electrolysis reactions in electroorganic synthesis consist of complex electron-transfer processes in which electrons-transfer from the cathode to the organic substrate by the aids of electrochemically activated intermediates, mediators, redox catalysts, and/or redox polymer-coated electrodes.   Synthetic features of electron-transfer to (or from) neutral molecules [14], anion radicals [15], anions [16], dianions [17], photochemically excited anion radicals [18,19], and atom transfer reactions have been reviewed [19,20].

An electron-transfer mechanism has been proposed for electrochemically induced aliphatic nucleophilic substitution reaction [21].   For instance, the enolate ion **24**$^-$ obtained by two successive one-electron reductions of 1-methyl-4-(methoxycarbonyl)-pyridinium iodide **24**$^+$ at the cathode has been found to be an excellent electron donor and acts with tert-butyldimethylsulfonium iodide (or tert-butyl bromide) in DMF with formation of

4-tert-butyl-1,4-dihydro-4-(methoxycarbonyl)-1-methylpyridine as the sole product (eq. (1.7)) [16]. Probably, the enolate anion 24^- is able to transfer an electron to the substrate, *e.g. tert*-butyl bromide, which cleaves immediately to a tert-butyl radical and a leaving group (Br$^-$), and the radical $24^{\bullet}$ and the tert-butyl radical have coupled (eq. (1.8)). In the case of 1,1-dinitrocyclohexane, the one-electron reduction undergoes fast cleavage of the C-N bond following electron-transfer, giving nitrile and 1-nitrocyclohexyl radical [22].

$$24^- + t\text{-BuBr} \longrightarrow \left[24^{\bullet} + t\text{-Bu}^{\bullet} + Br^- \right] \longrightarrow 25 + Br^- \quad \text{-------} \quad (1.8)$$

α-Aryloxyacetophenones can be used as an electron-transfer probe on the fragmentation of their electrochemically produced anion radical, in which the fragmentation rate constants are shown to correlate with the pK_a of the corresponding phenols [23]. By using a redox mediatory process of electron-transfer initiated radical chain reactions, the cross-coupling between aryl halides and phenoxide ions has been attained [24].

Among metal complexes as redox catalysts, activation of aromatic halides by the nickel-catalyzed electron-transfer, leading to biphenyl [25] and electrocarboxylation [26] has been intensively investigated. It has been demonstrated that the electrogenerated anion radicals tend to transfer electrons to aromatic halides [27a], resulting in a reduction of the halide and the generation of aromatic compounds [27].

A concerted electron-transfer-bond-breaking reaction has been suggested for the reduction of perfluoroalkyl bromides, arylmethyl halides, and aryldialkylsulfonium cations [28].

The fundamentals of catalysis at redox polymer-coated electrodes have been developed since the late 1970s [29,30].

To know electron-transfer pathways in the living cell is a current topic referring to the electrochemistry of biological molecules, e.g. enzymes and cofactors [31]. The electrochemistry has contributed to the elucidation of the midpoint potentials of different redox proteins and to finding a way to construct a variety of biosensors. A stable monolayer of flavin adenine dinucleotide (FAD) absorbed on a glassy carbon electrode can act as a mediator through which electron-transfer from FAD molecule to the electrode can occur [31]. The electrical communication between FAD redox centers of glucose oxidase and a conventional carbon paste electrode has been attained by using electron-transfer relay systems based on polysiloxanes [32]. Direct electron-transfer between an immobilized mediator and an immobilized viologen-accepting pyridine nucleotide oxidoreductase on carbon electrode has been devised [33]. Without using electron-transfer mediators, a poly(indole-5-carboxylic acid)-modified glassy carbon electrode shows a catalytic effect to ascorbate and NADH [34]. The electron-transfer reactions of high-spin metmyoglobin and low-spin cyanometmyoglobin have been found to be implications for long-range electron-transfer in proteins [35]. The direct electron-transfer of ferredoxins at an indium oxide electrode modified with poly-L-lysine [36], poly-L-ornithine [37], or aminosilane [38] has been investigated. Further details of electron-transfer catalysis are discussed in each chapter.

Viologens (1,1′-dialkyl-4,4′-bipyridinium, V^{2+}) have received much attention as an electron-transfer mediator especially in conversion of light energy systems, [39]. The viologen modified electrodes show a reversible wave due to a redox reaction between V^{2+} and viologen cation radical, $V^{+\bullet}$ [40,41]. An ordered electron-transfer has been observed in the cathodic redox reaction of flavin mononucleotide by using the oriented viologen pendant-modified electrode [42].

1.2 Design of Solvent-Electrolyte-Electrode Systems

1.2.1 Effects of Solvent Systems

The nature of the solvent used has a profound influence on the interaction between the solvent and the dissolved substrate and other species as well as their ability to donate or accept electrons. The solvent parameters investigated are Gutman donor number (DN), Gutman acceptor number (AcN), Dimroth Reichardt parameter (Ec), and solvent dielectric constant (ε) (Table 1.2) [43]. The effect of solvent on the half-wave potential is found to be correlated with the above solvent parameters. The correlation coefficient values, Stokes radii values, and heterogenous rate constant values reveal that Gutman donor number (DN) can act as a good index for the degree of reversibility of the charge transfer process [43].

In electroreduction of picolinic acid in an aqua-organic (1/1) mixture of various organic solvents, $e.g.$, MeOH, EtOH, PrOH, dioxane, DMF, DMSO, and ethylene glycol at pH 1.15 and 3.05, the limiting current is found to decrease, while half-wave potential is found to shift to a more negative value in the presence of an organic solvent [44].

It has been found that the interaction of the solvent and ferric perchlorate (Fe^{3+} ion) affects the half-wave potential ($E_{1/2}$) and heterogenous rate constant (k_s). The half-wave potential of the ferric-ferrous system has been reported in alcohols [45], nitriles [46,47], and DMF [48]. The kinetic parameters (k_s = rate constant, α = electron-transfer coefficient, and D = diffusion coefficient) for ferric perchlorate have been reported in DMF [49] and MeOH [50]. It is found that, for the reduction of the charged $[CoCp_2]^+$ ion ($[CoCp_2]^+ = e^- \; CoCp_2$), changes in the migration current as a function of solvent in the absence of electrolyte are related to the dependence of ionic mobility on solvent viscosity [51].

It is found that, in the case of the indirect reduction of aromatic halides in benzonitrile, the solvent itself plays the role of an electron carrier [52]. The solvent effect of the redox potential for the oxidation of ferrocene derivatives has been observed [53]. Interfacial solvation and double-layer effects on redox reactions in a self-assembled monolayer of N-(7-mercaptoheptyl)ferrocene carboxamide coadsorbed with n-alkanethiol derivatives have been observed [54].

Table 1. 2 Parameters of the Solvents Investigated

Solvent	AcN	DN	Eτ	ε	η (poise x 10^2)
MeCN	18.9	14.1	46.0	36.0	0.34
DMF	16.0	26.6	43.8	36.7	0.80
DMSO	19.3	29.8	45.0	46.7	2.00

AcN = Gutman acceptor number   DN = Gutman donor number
Eτ = Dimroth Reichardt parameter   ε = Dielectric constant   η = Viscosity
[Reproduced with permission from S. Kumbhat, *J. Indian Chem. Sect. A*, **33A**, 162 (1994)]

1.2.2   Behaviors of Electrolytes

The suitable choice of a solvent and an electrolyte is essential for product-selective electrolysis [1e].   The behavior of cationic species in a catholyte is directly related to the product-selectivity in electroreduction.   Tetraalkylammonium ions play an especially important role in a variety of electroreduction reactions.

The cathodic formation of tetraalkylammonium amalgams has been recognized at mercury electrodes [55].   The formation of tetraalkylammonium-graphite lamellar compounds are known to form electrochemically [56].   The electrolysis of dimethylpyrrolidinium salts in a DMF-Electrolyte-(Hg) system affords an ordered amalgam as a film on the surface of the mercury cathode.   At solid metal electrodes, a metal-dimethylpyrrolidinium composite material is produced [57].   Under electroreduction conditions, aliphatic tetraalkylammonium species tend to form amalgams ($R_4N(Hg)_x$) [58].   Tetraethylammonium tetrafluoroborate can be electrochemically reduced to ethane and ethylene at -1.44 V (SCE) in a DMF-Bu_4NBF_4(0.1M)-(C) system [59].   Similarly, tetrabutylammonium tetrafluoroborate is reduced into a mixture of n-butane and 1-butene at -1.44 V (SCE).

In the course of electrochemical copolymerization of acrylonitrile (AN) with methacrylic acid (MAA), the effect of supporting electrolytes and cathode materials is prominent.   For instance, when tetraalkylammonium salts are used as supporting electrolytes, the resulting copolymers are found to contain 90% by weight AN, whereas the AN content is 30% or 73% for $LiClO_4$ or $NaNO_3$, respectively [60].   Both tetrabutylammonium perchlorate and bromide are found to be a good choice for the electropolymerization of N-substituted acrylamide [61a].   Ferric chloride does not give any polymer.

Adsorption behavior of o-aminobenzoic acid on the mercury surface from three different electrolytes, $i.e.$, NaCl (0.1M), $NaNO_3$ (0.1M), and Na_2SO_4 (0.05M), has been investigated in aqueous media [62].

A solid polymer electrolyte (SPE)-furnished electrolyzer has been developed for electroorganic synthesis [63a-d].   SPE composite electrodes are prepared by deposition of platinum on Nafion; these can be possibly modified by electrochemical deposition of nickel and copper.   The reduction of nitrobenzene has been attempted on SPE composite electrodes [63e-h].   Copper-modified SPE electrodes have been used for the electroreduction of p-nitrophenol and p-nitrotoluene [63i].

Electroreduction of benzaldehyde using reduced silver bound on SPE electrode brings about very good selectivity in formation of benzyl alcohol [64].   The current efficiency of the cathodic reduction depends on cell voltage, type of solvent, and temperature employed [64].

1.2.3   Behaviors of Cathode Materials

1.2.3.1   Proper Choice of Electrode

The role of the cathodic material in the electrosynthesis of organic compounds has been well discussed [65].   The direct electron transfer from the cathode to the organic compound takes place predominantly as sp metals such as Pb, Hg, Sn, Ga, Tl, Zn, Cd, Bi, Al, In etc., and it is known that the highest hydrogen overpotential is exhibited by these sp metals.   In general, the absorption of hydrogen on the sp metals does not occur (or occurs little) and the controlling stage in the cathodic electrosynthesis is the transfer of charge to the hydroxonium ions, which cause hydrogen to evolve at the cathode.   The use of the sp metals also favors the production of anion radicals and radicals, and their secondary chemical reactions lead to hydrodimerization, cathodic coupling, and the formation of organometallic compounds.

In contrast, the lowest hydrogen overpotential is observed on d metals with the maximum degree of occupancy of the outer d orbitals of metals such as Pt, Ru, Ni, Pd, Rh, Fe, Co, Re, etc., in which hydrogen atoms are readily absorbed.   Such hydrogen atoms sorbed on the d metals cathode surface act as a hydrogenating agent.

The electroreduction of a ketone to the corresponding carbinol without forming pinacole as a by-product can be conducted at platinized platinum, Raney nickel, or other electrodes having electrocatalysts.   Purely catalytic reactions that either do not take place electrochemically or are realized with great difficulty on account of the need to reach potentials that are too negative are possible at these d metals.   Some typical reactions taking place at d metal catalysts are listed in Table 1.3 [66-84].   As shown in Table 1.4 [85-87], iron and nickel in the electrochemical cleavage of a carbon-hydrogen bond exhibit activity close to that of platinum.   For example, during the electroreduction of cyclopentadiene at iron or platinum cathodes, hydrogen is released, and the cyclopentadiene anion is formed [85,88].

The product-selectivity of the reduction of acetone varies depending on the cathode material as exemplified in Table 1.5 [65].   A high yield of isopropyl alcohol is obtained only at mercury and lead cathodes.   The formation of propane is observed at cadmium and aluminum cathodes [89].

The choice of a hydrophobic electrode is effective for the dimerization of deactivated olefins, $e.g.$, fumarodinitrile.   A hydrophobic environment can be provided effectively by the self-association of adsorbed hydrophobic molecules, which afford a pinhole free, self-repairing monolayer as long as the bulk concentration of the adsorbate is high enough to support the condensed film.   Since the adsorbate is not consumed in the electrode processes, the above requirement is readily satisfied.   The electroreduction of fumarodinitrile in an aqueous potassium $(1S)$-$(+)$-10-camphorsulfonate at mercury cathode tends to form the corresponding dimer, because the supporting electrolyte strongly adsorbed on the surface can provide a hydrophobic micro-environment around the electrode [90].   Specifically, the

Table 1.3 Types of Reactions Taking Place at Metal Catalysts (Group 1)

| Substrate | Cathode Material | | | | | | Product |
| | Ni | | Pd | | Pt | | |
	Yield, %	Ref.	Yield, %	Ref.	Yield, %	Ref.	
$RC{\equiv}CH$	~11	[67, 68, 117]	55-60	[69, 70]	90	[69, 70]	$RCH{=}CH_2$
$RCH{=}CH_2$	75	[117]	35-40	[69]	–	–	RCH_2CH_3
	70	[71,73]	64	[71]			
RCHO	70-83	[71, 72, 73, 74, 117]	–	–	96-98	[71]	RCH_2OH
RCOR	65-75	[71, 117]	30-38	[69, 71]	60	[69]	R_2CHOH
	97	[73, 74]			98-99	[71]	
$RC{\equiv}N$	72-76	[117]	80-85	[75a]	~3	[71]	RCH_2NH_2
	80-83	[73]	75	[75b, 76, 77]	5-10	[78, 79]	
$RCH{=}NOH$	62-76	[73, 117]	–	–	–	–	RCH_2NH_2
	92	[80, 81]					
Pyridine	68-83	[82]	–	–	–	–	Piperidine
Sugar	90	[83]	–	–	–	–	Alcohols
$ArNO_2$	58-63	[84, 117]	–	–	80	[71]	$ArNH_2$

Table 1. 4 Reduction Potentials (Ep)[a] of Substrates
in Different Cathode Materials

$$\text{H-CXYZ} \xrightarrow{e^-} [\text{H-CXYZ}]^{-\bullet} \longrightarrow H^{\bullet} + [\text{CXYZ}]^-$$

Substrate, H-CXYZ			Ep, V (Ag/AgCl)			Ref.
X	Y	Z	Pt	Fe	Ni	
H	R[b]	R	2.06	2.20	2.30	[85]
H	Ac	CO_2Et	–	1.90	1.70	[85]
H	CN	CN	–	1.28	–	[85]
H	H	NO_2	1.45	1.43	1.28	[86]
H	R	NO_2	1.53	1.64	1.35	[86]
Cl	Cl	Cl	2.32	2.18	2.04	[87]
Br	Br	Br	1.30; 2.04	1.04; 1.32	1.20; 1.84	[87]

[a] MeCN-Et$_4$NClO$_4$ (0.1 M)   [b] R = alkyl

Table 1. 5 Effect of Cathode Material in the
Reduction of Acetone

Cathode material	Isopropyl alcohol, %	Pinacol, %	Propane, %
Hg	95	3	2
Pb	68	7	25
Zn	3	0	97
Cd	0	0	100
Al	0	0	100

[Reproduced with permission from A. P. Tomilov, *Zh. Org. Khim.*, **29**, 657 (1993)]

absence of interfacial water may prevent protonation of anion radicals formed by electron transfer, even when the electrode is immersed in an aqueous solution.

The electrocatalytic hydrogenation of conjugated enones to the corresponding saturated carbonyl compounds has been intensively investigated in aqueous methanol at nickel boride, fractal nickel, and Raney nickel electrodes made of pressed powder [91].   Under given electrolysis conditions, fractal nickel electrodes are found to give the highest product-selectivity.

The preparative electroreduction of nitrobenzene, phenyl-hydroxylamine, azoxyben-zene, azobenzene, and hydrazobenzene at Devarda copper and Raney nickel electrodes in neutral and basic aqueous methanolic solutions affords aniline in 80~100% yields (80~100% current efficiency) [73,92a].   The electrocatalytic hydrogenation of *o*- and *p*-ni-trobenzoic acids and *m*- and *p*-nitrobenzenesulfonic acids at Raney-Ni and Devarda copper electrodes in an aqueous NaOH system gives the corresponding amino acids in 80~100% yields (current efficiency, 75~100%) [92b].   Other reaction examples are discussed in Chapter 3 [92c].   The effect of Ni, Cu, Pb, Cu-Hg, and Pb-Hg electrodes for the reduction of *m*-chloronitrobenzene, yielding 3,3′-dichloroazobenzene in around 90% yields, has been well investigated [93].

The platinized platinum electrode has been used for a radiotracer study of the electrocatalytic reduction of nicotine acid [94].   A highly dispersed platinum on graphite particles is found to be effective for the electrocatalytic hydrogenation of phenol, leading to cyclo-hexanol (current efficiency, 85%) [95].   Electrochemical reactions of [Ni(II)(RNC)$_2$]I$_2$ (R = 2,6-Me$_2$C$_6$H$_3$) have been found to proceed in different ways according to the electrode material.   The reaction at a platinum electrode proceeds in the usual electrochemical man-ner, but at a mercury electrode, the reaction undergoes a CE mechanism [96].   A nickel foam cathode has been used for the electroreductive coupling of aroyl and arylacetyl chlo-rides in an undivided cell, leading to the corresponding symmetrical ketones [97].

Hydrophobic zinc and lead electrodes are examined for the electroreduction of car-boxylic acids, halides, enolates, and heteroaromatic compounds [98].   For the dechlorina-tion of α,α-dichloropropionic acid, lead, copper, and compact pyrographite are found to be the most applicable cathode materials, leading to monochloropropionic acid [99].

Preparative scale of electrosynthesis of toluidines using thermally coated Ti/TiO$_2$ elec-trode in an aqueous H$_2$SO$_4$(1M) system has been attained in 90% yield [100a].   The ther-mally coated Ti/TiO$_2$ electrode can be reused at least five times without any loss in yield and current efficiency.   The reduction of nitrobenzene and m-dinitrobenzene proceeds to give amino compounds in 83~87% product yields [100b,c,e].   1-Chloro-2,4-diaminoben-zene has been also synthesized by use of Ti/TiO$_2$ electrodes [100d].

The electroreduction of aromatic diazonium salts in an aqueous solution at mercury mi-cro- and macroelectrodes has been investigated [101,102].   Electrodeposition of di-methylpyrrolidinium-mercury and tetramethylphosphonium-mercury as an initial molecular layer has been reported [103].   The formation of micelles on the mercury electrode surface in an aqueous Na$_2$SO$_4$(0.1M) system containing cetylpyridinium cations has been electrore-duced, leading to a film of uncharged monomer species of the reduction product [104]. The use of a mercury bubble electrode for cathodic reductions of p-hydroxybenzaldehyde, 1,2,4,6-tetramethylpyridinium methylsulfate, nitrobenzene, and acetophenone has been in-vestigated [105].

The role of the thiol group of 11-ferrocenylundecanethiol modified on indium-tin oxide (In/SnO) electrode has been studied [106].

Thermodynamically controlled electrochemical formation of thiolate monolayer self-assembled from dilute solutions of X(CH$_2$)$_n$SH on gold electrode has been reported [107]. Potential-dependent stability of self-assembled organothiols on gold electrodes in methyl-ene dichloride has been determined by electrochemical measurement [108].

Surface effects of glassy carbon electrode (GCE) have been observed on responding aromatic carbonyl compounds [109b].   The interaction between acidic surface functional groups of GCE and the aromatic carbonyl group such as benzil has been noticed [109a]. The reactant-incorporated graphite powder electrodes has been used for the reduction of p-dinitrobenzene in a dilute aqueous acid solution [110].

The glassy carbon electrode modified with iron-sulfur and iron-selenium clusters (Bu$_4$N)$_2$ [Fe$_4$X$_4$(YC$_6$H$_4$-p-t-Bu)$_4$] (X, Y= S, Se) demonstrates the stable (2$^-$ / 3$^-$) redox couple in CV measurements and the electron transfer between the cluster and glassy carbon plate is influenced by the protonation equilibrium [111].   Alkylazides can be electrore-duced by a [Bu$_4$N]$_3$ [Mo$_2$Fe$_6$S$_8$(SPh)$_9$]-modified glassy carbon electrode, leading to alky-lamines [112].

Platinum-base modified electrodes from organic-inorganic hybrid gels containing fer-rocene units covalently bonded inside a silicon network have been reported [113].   The coated films are very stable and possess the unique feature of very tight bonding of the electroactive moieties to the polymer, which offers an interesting scope for the analysis of electron-transfer phenomena.   Electrochemically-produced hydrogen atoms diffused into a Pd sheet electrode have been used for successive hydrogenation of styrene on the reverse side [114].   The hydrogenation of styrene proceeds in an aq. KOH-(Pt/Pd) system to give ethylbenzene in 93% current efficienty.

The use of tin cathode is found to be the best choice for the intramolecular electroreductive cyclization of non-conjugated olefinic and acetylenic ketones yielding cyclic tertiary alcohols [115].

Electroreduction of molecular oxygen on platinum single crystals in an aqueous acid medium has been investigated [116].

Glassy carbon-supported platinum-modified cobalt oxides (Pt + Co_3O_4) electrodes exhibit good electro-catalytic activity towards the dioxygen in oxygen-saturated sodium hydroxide solution [117].

Catalytic electroreduction of dioxygen by an iron porphyrin ion-complex modified glassy carbon electrode, which can provide the effective sites for the reduction, has been studied in aqueous solutions [118].

A new kind of electrode modified by hydrophobic Nafion gels loaded with 9-phenylacridinium species is found to exhibit fairly good catalytic activity due to the affinity of the perfluoronated chains for oxygen [119].   The electroreduction of dioxygen to hydrogen peroxide by different substituted acridinium has been attained in homogeneous solutions with Nafion gel-modified electrodes [120].   Nafion-modified stainless steel-carbon composite electrodes has been also proposed for the electroreduction of dioxygen [121].

In fuel cell operations, wet-proofed electrodes play an important role to improve the intensity of current-generating processes, simplifying the design, increasing the reliability, and lowering the consumption of electrocatalysts.   The wet-proofed porous electrodes are used for the electrosynthesis of organic compounds from gaseous and liquid materials which have limited solubility in water.   Recently, the wet-proofed porous electrodes have proven to be useful for electroreduction of nitromethane, chlorination of alkanes, and dehalogenation of perhalogenated compounds [122].

A particular sulfurized cathode has been prepared by heating a mixture of sulfur and carbon powders (2 to 1 ratio) at 130 °C.   The sulfur-carbon composite electrode works as a sulfur source for the sulfurization of propionitriles bearing a leaving group at the _-position [123].   The direct synthesis of diaryl diselenides and ditellurides has been performed using Se and Te electrodes in an MeCN-Bu_4NPF_6 (0.1M)-(Se or Te) system in 14~40% yields [124].

The feasibility of a solid polymer electrolyte (SPE) method for Kolbe type reaction has been investigated by using Pt-SPE composed of Nafion 415 and platinum [125].   The electrohydrogenation of olefinic double bonds at the SPE electrode also takes place [63b].

The merits of channel electrodes for the chronoamperometry study using double-channel electrodes have been reported [126].

Various quinones anchored thiol-monolayered gold electrodes have been intensively investigated for the development of molecular electronic and optoelectronic systems [127]. 2,3-Dichloro-1,4-naphtoquinone has been anchored to a gold electrode surface through the self-assembled monolayers of aminoalkanethiols [128].   It has been found that the tailored monolayer assemblies composed of redox protein linked to thiol-derivatized Au electrodes provide novel configurations for mediated electron transfer in enzymes [129].   As shown in Scheme 1.1 [129], the enzyme glutathione reductase **28** has been covalently attached by forming amide **29** with monolayer chemisorbed to an Au electrode.   The resulting electrode-immobilized protein **29** is treated by *N*-methyl-*N*´-(carboxyalkyl)-4,4´-bipyridinium in the presence of urea to yield an electron relay modified enzyme **30** exhibiting electrical communication with the electrode.   The bipyridinium-modified enzyme monolayer electrodes 29 play a role as biocatalytic redox assemblies for electroreduction of oxidized glutathione, GSSG.

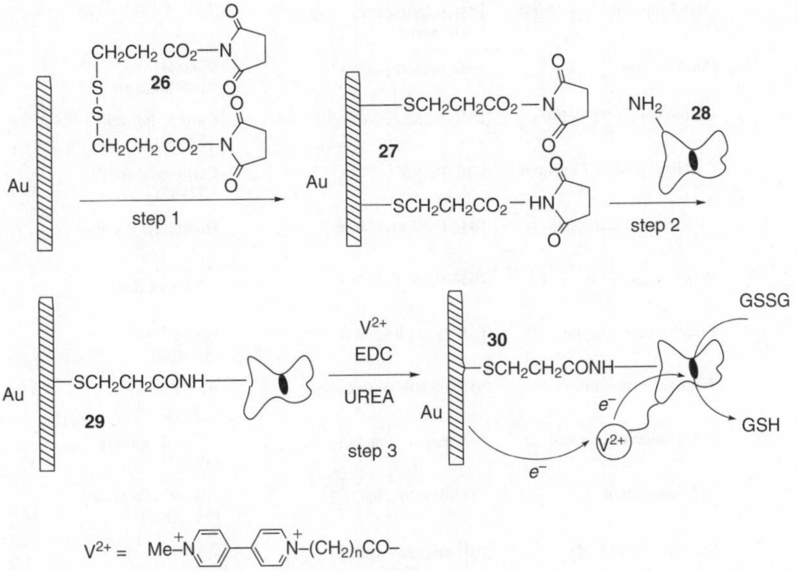

Scheme 1. 1 Sequence for assembling the monolayer of glutathione reductase exhibiting electrical communication.
[Reproduced with permission from I. Willner et al., *J. Am. Chem. Soc.*, **114**, 10965 (1992)]

1.2.3.2 Metal Effects of Sacrificial (Consumable) Anodes

Potentialities of the preparative electrolysis in a diaphragmless electrolyzer equipped with sacrificial (consumable) anodes are known as an anodic reduction [130,131].   A wide range of usages of sacrificial anodes have been recorded in the literature, indicating that they are quite effective for homo-couplings (C-C bond formation), carboxylations, heterocouplings, and formation of C-P, C-Si, O-Si, and C-S bonds [130].

The carbon-carbon bond formation of arylalkenes and alkyl bromides has been attained using a sacrificial anode [132].   The combination of a nickel catalyst and a sacrificial aluminum anode in an undivided cell can provide a novel electroreductive cross-coupling procedure [133].   Indium and zinc metals play an important role in Reformatsky reactions [134].

The electroreductive silylation of activated olefins such as α,β-unsaturated esters, ketones, and nitriles takes place by use of a reactive metal anode (Mg, Al, Zn, etc.) in an undivided cell [135a].   The stereoselective synthesis of Z-silyl enol ethers occurs via deprotonation of enolizable ketones with a sacrificial magnesium anode [136].

Consumable zinc, magnesium and aluminum anodes have also been found to be feasible for use in the electroreductive acylation of activated olefins [135b].   It is suggested that some stabilization effect for anionic species by metal ions generated from a sacrificial anode facilitates the subsequent acylation [135b].   The electroreductive radical cyclization of 2-bromoethyl 2-alkynyl ethers has been improved by the combined use of Co(III)Cl(py)

Table 1. 6 Sacrificial Anode Used in Electrolysis Systems

Sacrificial Anode	Type of Reaction	Electrolysis Condition	Product (yield %)	Ref.
Al	Alkylation (ring formation)	DMF-Bu$_4$NBr-(Al/Ni)	Cyclic products (20~50)	[132]
Al	Arylation of α-chloroesters	DMF-Bu$_4$NBr-(Al/C) [Ni(II)(bpy)]Br$_2$		[133]
Al	Pinacolization	DMF-Bu$_4$NBF$_4$-(Al/C)	Pinacol (quantitative)	[142]
Al	Carboxylation of Halides	DMF-Bu$_4$NBr-(Al/C)	Carboxylic Acids (57~88)	[151]
Al	Carboxylation of Carbonyl	DMF-Bu$_4$NBr-(Al/C)	Carboxylic Acids (32~85)	[151]
Cd	Allylation of Carbonyl	DMF-Et$_4$NClO$_4$-(Cd-Pt)	Homoallyl Alcohols	[150]
Cd	Pinacolization	DMF-Me$_4$NCl-(Cd/Pt)	1.2-Diol Cd salt	[147]
Cu	Coupling dimerization	DMF-Bu$_4$NClO$_4$/NaI-(Cu/Pt)	Dimer (53~100)	[149]
Cu	Trifluoromrthylation	DMF-Bu$_4$NBr/Ph$_3$P-(Cu/SUS)	Ar-CF$_3$ (23~98)	[148b]
Mg, Al, Zn	β-Acylation of enoates	DMF-Bu$_4$NBr-(Mg/SUS)	4-Oxoalkanoates (49~88)	[135b]
Mg	Dehalogenation	DMF-Bu$_4$NBF$_4$/SmCl$_3$-(Mg)	Alkane, alkene, ester (58~100)	[143]
Mg	Degradation of DMF	DMF-Bu$_4$NBr-(Mg/SUS)	Glyoxal	[138]
Mg	Trimethylsilylation of ketones	DMF/HMPA-Bu$_4$NBr-(Mg/SUS) 2-pyrrolidone	Silyl enol ethers (80~88)	[137]
Mg	Trimethylsilylation of enoates	DMF-Bu$_4$NBr-(Mg/SUS)	β-Silylated esters (14~79)	[135]
Mg	Deprotection of allyl ether	DMF-Bu$_4$NBF$_4$-(Mg/C) [Ni(bipy)$_3$](BF$_4$)$_2$	Alcohols (40~99)	[146]
Mg	Metallation with Zn	DMF-ZnBr$_2$/NiBr$_2$(bpy)-(MgNi)	3-Thienylzine bromide (80)	[148a]
Ni	Polymerization of olefins	DMF-Bu$_4$NClO$_4$-(Ni/Ni)	Polymer (89)	[61a]
Ni	Carbonation of epoxides	DMF-KBr/Ni(cyclam)-(Mg/Ni)	Cyclic carbonstes (51~91)	[61b]
Zn, Sn, Al, In, or Fe	Reformatsky reaction	DMF-Bu$_4$NBr-(Zn/Ni)	β-Keto esters (34~80)	[134(c)]
Zn	Reformatsky reaction	DMF/THF-Bu$_4$NBr-(Zn/Ni)	β-Lactones (59~88)	[134a, 134d, 134d]
Zn, Al	Acyl of haloketones	DMF-Bu$_4$NBF$_4$-(Mg/C) Ni(II)(bpy)	Benzyl ketones (54~79)	[145]
Zn (powder)	Carboxylation	THF/HMPA-Bu$_4$NBF$_4$/NiCl$_2$(dppe)-(Zn)	Aryl-2-propanoic acids (62~89)	[144]
Zn	Radical cyclization	MeOH-Et$_4$NOTs/Cobaloxime-(Zn)	2,9-Dioxabicyclo-[4.3.0]nonane (54~77)	[137]

Table 1.6 Sacrificial Anode Used in Electrolysis Systems (continued)

Sacrificial Anode	Type of Reaction	Electrolysis Condition	Products (yield, %)	Ref.
Zn (powder)	Carboxylation	THF/HMPA-Bu$_4$NBF$_4$/NiCl$_2$(dppe)-(Zn)	Aryl-2-propanoic acids (62~89)	[144]
Zn	Radical cyclization	MeOH-Et$_4$NOTs/Cobaloxime-(Zn)	2,9-Dioxabicyclo[4.3.0]nonane (54~77)	[137]

complex and a zinc plate as a sacrificial anode [137]. The use of a sacrificial magnesium anode causes reductive degradation of DMF as a solvent, depending mainly on the temperature and current density [138].

Hexafluoroacetone undergoes pinacolization when it reacts with anode metals [139-142]. The use of a consumable magnesium anode with a catalytic amount of SmCl$_3$ is a potent system for the dehalogenation of organic halides [143].

Zn powder plays a role as a good alternative to the use of sacrificial anodes [144]. A variety of sacrificial anodes used in electrolysis systems are shown in Table 1.6 [61, 132-138, 142, 146-151].

1.2.3.3 Electrochemically Functioning Electrodes

Quinones promote electron transfer process from the excited state of fluorophores and quench fluorescence. Chemical and electrochemical reduction of quinones to hydroquinones prevents electron transfer and revives fluorescence in a reversible way. Such a system is the prototype of a molecular photoswitch which can be operated from outside by varying the redox potential. A typical device should combine (1) a luminescent fragment and (2) a control unit, capable of modifying the light-emitting properties of the adjacent subunit. A redox switching of anthracene fluorescence through a Cu(II)/Cu(I) couple has been proposed [152].

Electrochemically switched systems based on anthraquinone-containing cation complexing subunits which are potentially useful in enhanced cation binding and transport across membranes have been intensively investigated [153-155]. Anthraquinones bearing phtochemically isomerizable 4′-substituted stilbene moieties can be interconverted electrochemically between four states as *cis* or *trans* isomers as a molecular switching [156].

Recently, electrochemical switching enhanced sodium binding capabilities of novel bis(anthraquinone) diazacrown ethers are being evaluated, indicating that both compounds exhibit enhanced sodium binding properties upon electroreduction [157]. 1,4-Dithiafulvenyl substituted bianthrones have been prepared with the aim of making a light-activated molecular switch [158]. The redox potentials of the donor part of the bianthrones depends on the conformation of the bianthrone moiety.

Poly[*trans*-bis(3,4-ethylenedioxythiophene)vinylene] obtained by electropolymerization of *trans*-bis(3,4-ethylenedioxythiophene)vinylene exhibits a band gap of 1.4 eV and can be rapidly switched between insulating deep purple absorptive and highly conducting light blue transmissive states [159].

A light-triggered optical and electrochemical switching function has been found in diarylethene modified electrode devices [160].

The realization of nanoscale switching devices may have far-reaching implications for computing and biomimetic engineering. A rotaxane exemplified as a class of molecular devices in terms of a chemically and electrochemically switchable molecular shuttle has been developed [161].

The electrochemical communication of enzyme redox sites and electrodes is the basis for various amperometric biosensor devices [162,163].   Immobilization of redox enzymes in functionalized redox polymers [164,165] or chemically modified proteins with electron-transfer mediators [166] provides general procedures to establish electrical interactions between insulated enzyme redox centers and electrodes.   Recently, mediated electron-transfer biosystems in glutathione reductase organized in self-assembled monolayers on Au electrodes have been used for electrical communication with the electrode [129]. Spiroperimidine shows a reversible rearrangement with electrochemical activation and also photochromic isomerization, which play a switching role by light irradiation [167].   A biosensor based on direct communication between glucose oxidase and polypyrrole, incorporated in the pores of track-etch membranes, has been used for the detection of glucose in the concentration range 1~25 mmol/dm³ [168].   Recently, numerous switching devices have been proposed as follows: electrochemically switchable diffraction grating of poly(bpy)$_2$Ru(vpy)$_2^{2+}$ (vpy = 4-vinylpridine) [169,170], switchable molecular shuttle [161], revesible redox conductivity state switching [171], and proton-induced switching of electron-transfer pathways in Dendrimer-type tetranuclear RuOs$_3$ complexes [172].

The direct transfer of electrons between redox active prosthetic groups of oxidoreductases and electrodes is far to slow for preparative conversions.   In order to improve the electron transfer processes, the binding of the mediator as well as the enzyme covalently to a functionalized carbon surface in order to check whether electron flow from the cathode *via* the mediator acting as "molecular wire" has been investigated (see Scheme 1.2). [33]. A poly(indole-5-carboxylic acid)-modified glassy carbon electrode plays the role of a mediatorless electrocatalysis to ascorbate and NADH [173].

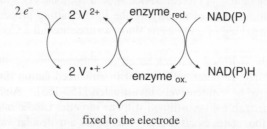

fixed to the electrode

Scheme 1. 2 Flow of electrons via immobilized viologens to an immobilized enzyme that reduces soluble NAD.

Antibody-antigen interactions play a major role in biosensor technology.   The potentiality of the antigen-antibody interactions limits the applicability of antigen (or antibody) sensing surfaces to single-cycle analytic devices.   The development of amperometric immunosensors by the organization of antigenic self-assembled monolayer electrodes has been attempted as an approach to develop reversible amperometric immunosensors utilizing a photoisomerizable monolayer electrode [174].

The principle of reversible amperometric immunosensors is shown in Fig. 1.8 [174]. The antigen monolayer consists of a photoisomerizable component.   In one photoisomer state (state A), the antigen recognizes the Ab (antibody), and the difference in the amperometric responses of the Ab-linked electrode and the antigen monolayer electrode, lacking the Ab, towards the solubilized redox couple provides a quantitative measure for the Ab concentration.   Upon completion of the measuring cycle the antigen monolayer is photoi-

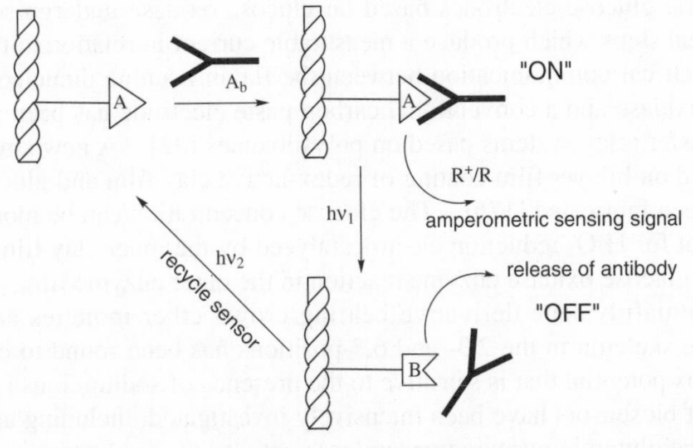

Fig. 1. 8 Schematic configuration of a reversible amperometric immunosensor utilizing a photoisomerizable monolayer electrodes.
[Reproduced with permission from I. Willner, *J. Am. Chem. Soc.*, **116**, 9365 (1994)]

somerized to state B.   This leads to distortion of the antigen monolayer, and the perturbed surface lacks affinity towards the Ab.   The Ab is washed off and the monolayer electrode is further illuminated ($h\nu_2$) to restore the active antigen monolayer electrode.   Thus, a two-step illumination procedure regenerates the active antigen electrode.

Transformation of the DNP(dinitrophenyl)-Ab amperometric immunosensor into a reversible multicycle sensing device was accomplished by organization of a photoisomerizable dinitrophenyl spiropyran antigen monolayer electrode (Scheme 1.3) [174].   The resulting dinitro spiropyran monolayer electrode **34** exhibits reversible photoisomerizable properties.   Illumination of the **34** monolayer electrode, $\lambda = 360\sim380$ nm, results in the **35** monolayer electrode, and further irradiation of the **35** monolayer electrode, $\lambda > 495$ nm, restores the **34** monolayer.

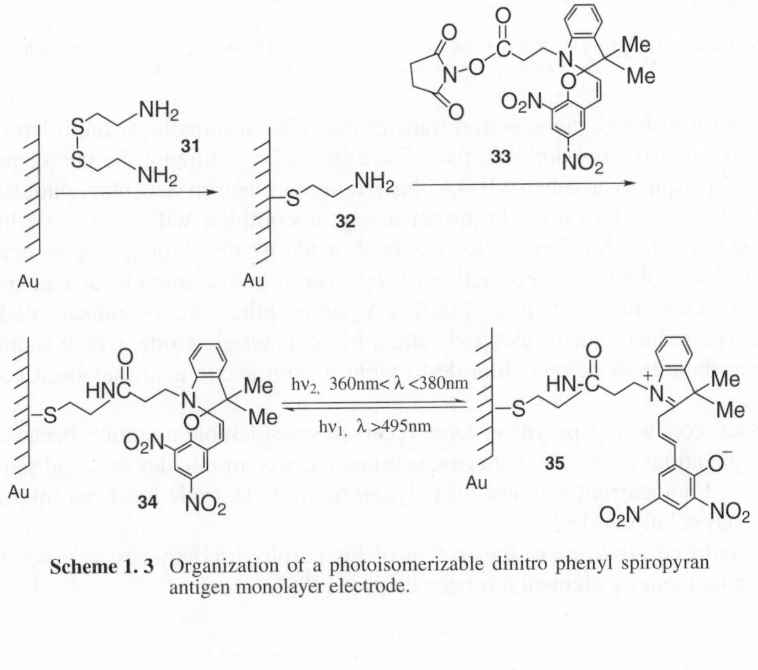

Scheme 1. 3 Organization of a photoisomerizable dinitro phenyl spiropyran antigen monolayer electrode.

Amperometric glucose electrodes based on glucose oxidase undergo several chemical or electrochemical steps which produce a measurable current in relation to the glucose concentration.   Electrical communication between the flavin adenine dinucleotide redox centers of glucose oxidase and a conventional carbon paste electrode has been achieved by using electron-transfer relay systems based on polysiloxanes [32].   A new amperometric glucose sensor based on bilayer film coating of redox-active clay film and glucose oxidase enzyme film has been fabricated [175].   The glucose concentration can be monitored by measuring the current for H_2O_2 reduction electrocatalyzed by the inner clay film where H_2O_2 is produced by the glucose oxidase enzyme reaction in the outer enzyme film.

A new tetrathiafulvalene derivative bearing crown ether moieties annulated to the tetrathiafulvalene skeleton in the 2,3- and 6,7-positions has been found to be unique in the nature of its redox potential that is sensitive to the presence of sodium ions [176].

A number of biosensors have been intensively investigated, including an ISFET sensor for monitoring the glutamic-pyruvic transaminase activity using glutamine synthetase from a thermophilic bacterium [177];

A new type of enzyme electrode, "the dialysis electrode," based on the enzyme glutamate oxidase, is used for continuous measurement of levels of the neurotransmitter glutamate **36** in the brain of a freely moving rat (Scheme 1.4) [178].

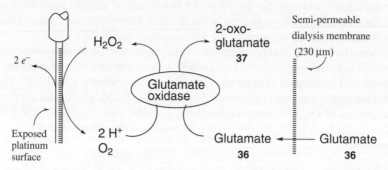

Scheme 1. 4 Dialysis electrode for measuring glutamate.   [Reproduced with permission from W. J. Albery et al., *J. Chem. Soc., Chem. Commun.*, **1992**, 900,]

The control of interfacial electron transfer reactions at monolayer-modified electrodes has attracted extensive research efforts.   For example, by utilizing the photoisomerizable *N*-methyl-*N*′-[4-(phenylazo)benzyl]-4,4′-bipyridinium electron acceptor, photostimulated formation and dissociation of the donor-acceptor complex with eosin dye have been demonstrated [179].   Another device has been made by developing electrochemical and quartz-crystal-microbalance transduction of light-controlled supramolecular interactions at monolayer-functionalized electrodes [180].   A photosynthetic semiconductor cell, consisting of a *p*-type semiconductor electrode and a Mg counter-electrode, which spontaneously fixes carbon dioxide in benzyl chloride to yield phenyl acetic acid, has been constructed [181].

Electrical conducting polymers have received considerable attention because of their possible applications to organic batteries, sensors, electrochromic devices, and semiconductor devices.   Photocurrent response of polyaniline film electrode has been investigated in acidic aqueous solutions [182].

Redox-induced *cis-trans* isomerization of bis(porphyrinyl)ethenes as a possible basis for a molecular memory element has been devised [183].

1.2.3.4 Polymer and Chemicals Modified Electrodes

Catalytic electrochemical reactions is one of the desired goals in the use of chemically modified electrodes [184].

Techniques for binding redox-active polymers and block polymers for covalent attachment to electrode surfaces have been investigated. For example, ferrocene- and phenothiazine-based redox-active polymers bearing $Si(OEt)_3$, pyridyl, bromobenzyl, and pyrenyl derivatives as the terminal group for covalent attachment to Pt, $In_2(Sn)O_3$, and n-Si electrode surfaces have been reported [185]. The evaporative deposition of films on the electrode surface is also a versatile procedure [186].

α-Methoxy-ω-mercaptopoly(ethylene glycol) has been used for complex formation of the monolayer with α-cyclodextrin on the surface of gold electrodes [187].

Absorbed polymeric nickel phosphine complexes are found to be more electroactive than covalently anchored monomeric nickel analogues [188]. The electrochemical activities of the nickel complexes provide useful probes for the characterization of the chemically modified electrode. The Ni-based chemically modified electrode exhibits high electrocatalytic activity towards the oxidation of carbohydrates in alkaline solution [189]. An electroreductive intercalation method for MoO_3 thin film opens a route for the preparation of various classes of useful thin layer materials [190].

Three electron-hopping conductive polymer films on electrodes for electron-transfer studies have been prepared by electrochemical copolymerization of osmium with ruthenium and zinc complexes [191]. An electrocatalytic reaction for olefin epoxidation has been attained by reduction of dioxygen in methylene dichloride containing benzoic acid as a proton source at a poly-$[Ru(vbpy)_3]^{2+}$ modified carbon electrode in the presence of manganese *meso*-tetraphenylprophyrin catalyst [192]. The polymer film assists the reduction of dioxygen leading to hydrogen peroxide and prevents electroreduction of high-valent metal oxoporphyrin.

Electroreduction on polymer modified electrodes is listed in Table 1.7 [193-198].

Table 1. 7 Electroreduction of Organic Compounds on Functional Polymer-modified Electrodes

Substrate	Coating (Modification)	Electrode	Products Conv. Yield (%)	Ref.
Anthracene	Poly(vinylferrocenium)	Pt	Dihydroanthracene	[193]
Acetophenone	Pt-Nafion Composite	Sb	1-Phenylethanol (76), Pinacol (9)	[194]
Diphenylacetylene	Pd-Incorporated Polypyrrole	C	1,2-Diphenylethane (91), *cis*-Stilbene (1)	[195]
2-Methylcyclohexanone	Nafion and Poly(acrylic acid)	Graphite Felt	2-Methylcyclohexanol(49.8) (1S,3S)-isomer(%ee, 100)	[196]
H_2O	Poly(2,2'-bipyridine-5,5'-diyl(Ru complex)	Pt	H_2	[197]
Alkyl azides	Poly(pyrrole-alkylammonium) films	Carbon Felt	Alkylamines	[198]

The overpotential required for the electroreduction of carbon dioxide has been diminished due to the effect of mediated activation by a metal complex fixed on a electrode. For this purpose, metal complex-fixed polyaniline-prussian blue-modified electrodes have been used [199a]. The two laminated films and a fixed metal complex constitutes the functional part of the modified electrode. The electroreduction of carbon dioxide to lactic acid has been realized by use of the above modified electrode [199b-d]. The poly[N-(5-hydroxypentyl)pyrrole] film-coated electrode incorporating Pd microparticles exhibits catalytic activity with regard to the hydrogenation of diphenylacetylene, giving 1,2-diphenylethane (91%) and cis-stilbene (1%) [195]. Cathodic cleavage of the sulfur-carbon bond of aromatic sulfones at electrodes modified by polyfluorenes or polydibenzofuran has been attained [200]. An electrochemical response towards urea has been recorded using a poly(L-glutamate)-immobilized Pt electrode (an enzyme-free electrode) [201]. The $(SN)_x$ modified electrode can be prepared by electroreduction of cyclopenta-azathienium chloride, S_5N_5Cl, in a SO_2-S_5N_5Cl-(Pt) system [202].

A combination of the "electro-coated" Nafion-modified stainless steel carbon composite and platinum electrodes has been used for the electrochemical reduction of dioxygen [121]. The electrocatalytic oxygen reduction has been performed on a (tetrasulfonatophthalocyaninato)cobalt-containing incorporated polypyrrole film electrode [203]. Small particles (*e.g.* platinum black) or neutral molecules [*e.g.* cobalt(II) meso-tetraphenylporphine] have been immobilized as catalytic sites in a conducting polymer film electrode, consisting of polypyrrole-poly(chloride) alloy [204].

Electro-copolymerization of acrylonitrile and methyl acrylate onto graphite fibers has been intensively investigated [205]. Their properties can be systematically varied by controlling the monomer ratio in the electrolysis solution. The preparative electroreduction of 1,2-dibromo-1,2-diphenylethane has been carried out using electrodes coated by a polypyrrole film containing viologen systems [206]. The electrode surfaces appear to be sufficiently stable to work out preparative-scale electroreduction.

Glassy carbon felt electrodes are modified by electrodeposition of poly(pyrrole-viologen) films, followed by electroprecipitation of precious metal (Pt, Pd, Rh, or Ru) microparticles. The resulting electrodes have been proven to be active for the electrocatalytic hydrogenation of conjugated enones in aqueous media (pH 1.0) [207]. Poly(2,3-diaminonaphthalene)-filmed glassy carbon electrode shows excellent electrocatalytic activity for dioxygen reduction to hydrogen peroxide in an aqueous solution (pH 1.0) [208].

Remarkably high catalytic currents and turn over numbers are found for the electroreduction of halides at redox modified electrodes coated with thin films of polysiloxanes, bearing viologens or nitroxides as redox centers [209].

The electrochemically produced $DMP(Hg)_5$ derived from the reduction of the dimethylpyrrolidinium cation $(DMP)^+$ at mercury film cathodes forms as an insoluble conducting phase which grows with increasing charge-transfer [210].

1.2.3.5 Viologen, Enzyme and Protein Modified Electrodes

Much attention has been paid to the electron-transfer between an electrode and enzymes. Oxidoreductase with cofactors at their active centers is especially well investigated, because the enzymes can be used for new types of biosensors and activity-controllable bioreactors.

Flavoenzymes such as glucose oxidase and L-amino acid oxidase have been immobilized on electrode surfaces to study the possibility of direct electron-transfer. However, little is known about the direct electron- transfer in the absence of a mediator [211]. A method for the direct electron-transfer between a monolayer of quinoprotein oxidoreductase, fructose dehydrogenase, and various electrodes (Pt, Au, and GC) has been demonstrated [211c].

The electroreduction of methylviologen (MV^{2+}) has been investigated in the presence of sodium decyl sulfate (SDS) at concentration levels below and above the critical micelle concentration of this anionic surfactant [212].   The dimerization of the cation radical ($MV^{\cdot+}$) is remarkably enhanced by the surfactant system at low micelle concentration. The electro-enzymatic viologen-mediated stereospecific reduction of 2-enoates has been performed using free and immobilized enoate reductase on cellulose filters or modified carbon electrodes.   A preparative electro-enzymatic reduction of (E)-2-methyl-3-phenyl-butenoate can be performed in a long time experiment in an aqueous MeOH(30%)-tris-acetate(0.1M)/Ca(OAc)$_2$(0.7%w/v)Buffer(pH7)-(C) system to give the corresponding R-enantiomer of 2-methyl3-phenylpropionic acid in 98% e.e. [213].   The electrochemistry of 1,1′-diheptyl-4,4′-bipyridinium ion (heptylviologen, HV^{2+}) reveals that the heptylviologen plays a role as an electron transfer mediator in the presence of 1-phosphatidylcholine liposomes [214].

A poly(phenosafranine)-modified electrode prepared by a potential-sweep oxidative electrolysis of phenosafranine shows excellent electrocatalytic activity for NADH oxidation in phosphate buffer solution (pH 7) [215].   A preparative scale conversion of pyruvate to lactate by NADH using an enzyme-viologen modified electrode has been attained [216]. The use of the electrode for lactate dehydrogenase-mediated reduction on a gram scale operation yields the lactate in 52% yield.

Glucose electrodes made with the redox polymer, i.e., poly(4-vinylpyridine) complexes of Os(bpy)$_2$Cl, modified enzyme are relatively stable and sensitive.   When a polycationic polymer is covalently bound to the enzyme, the electrooxidation of glucose occurs even at high electrolyte concentration [217].

A gold electrode immobilizing alcohol dehydrogenase from gluconobacter suboxydans, a membrane-bound quinohemoprotein, on the surface by adsorption produces an anodic current due to the enzyme-catalyzed oxidation of ethanol [218].   An enzyme electrode immobilizing lactate oxidase and lactate dehydrogenase has been prepared for the catalytic assay of L-lactate or pyruvate monitored by an oxygen electrode [219].

Viologen-linked N-substituted pyrroles are polymerized by electrooxidation on glassy carbon, giving viologen-modified electrode [220].   The electropolymerization of N,N′-1,1′-dialkylvilogen-linked pyrroles on platinum electrode is carried out by a potential sweep method in an MeCN-Bu$_4$NClO$_4$(0.1M)-(Pt) system at 0~1.8 V (SCE) [42].   The electrons are transferred from the modified electrode to dioxygen through the viologen moiety.   The modified electrode works as a reactive electrodes for the selective electrocatalytic reduction of dioxygen to hydrogenperoxide.   A Langmuir-Blodgett viologen film is prepared by electrostatically fixing the viologen monolayer containing the alkyl chain moiety on the poly(potassium 1-sulfactoethylene) [221].   The presence of poly-L-lysine promotes the electron-transfer between the electrode and the ferredoxin molecules [222].   The electron-transfer reactions of four redox proteins, cytochrome c, ferredoxin, plastocyanin, and azurin have been investigated at peptide-modified gold electrodes [223].   Cytochrome c is found to undergo electron-transfer reactions at a gold electrode modified with a Langmuir-Blodgett monolayer of 20-mercaptoeicosane-1-ol/dioleoyl-L-α-phosphatidylcholine [224]. The enhancement of the electrocatalytic response due to heterogeneous biomolecular recognition between cytochrome c peroxidase and cytochrome c at the protein-modified gold electrode has been observed [225].   Polypeptide-modified indium oxide electrodes for direct electron transfer has been devised [36,37].   Good, quasi-reversible electron transfer has been attained at different peptide-protein configurations by changing the pH or the ionic strength of the solution.   A variety of electron transfer systems on viologen, enzyme, and protein modified electrodes is listed in Table 1.8 [36-38, 42, 211c, 215-218a, 223 224, 226-232].

Table 1.8 Viologen-, Enzyme-, and Protein- modified Electrodes

Enzyme, Protein	Electrode Material	Redox Potential mV (reference)	pH	Ref.
Quinoprotein oxidoreductase	Pt	80 (Ag/AgCl)	4.5	[211c]
fructose dehydrogenase	Au	80 (Ag/AgCl)	4.5	
	GC	40 (Ag/AgCl)	4.5	
Glucose oxidase (antigen, gelatin)	GC	250 (SCE) 180	8.0	[226]
Glucose oxidase (copolymer of vinylferrocene and acrylamide)	GC	330 (SCE)	7.0	[217]
NADH (poly(phenosafranine))	GC (MeCN-NaClO$_4$)	515 (Ag/AgCl)	7.0	[215]
Viologen/NADH dehydrogenase (poly(o-nitroaniline)modified)	GC	−500 (SCE)	7.5	[227]
NADH (arrageenan, [Rh(bpy)$_2$]$^{3+}$ (Rh complex immobilized)	GC	−1.050 (SCE)	8	[228a, 228b]
Viologen/flavin mononucleotide (polypyrrole modified)	Pt (MeCN-Bu$_4$NClO$_4$)	− 450 (SCE)	–	[42]
Alcohol dehydrogenase	Au	200 (Ag/AgCl) –	6.0	[218a]
Lys-Cys	Au	90 (SCE)	7.0	[223]
Cys-glu	Au	140 (SCE)	7.0	
Cys-phe	Au	–	7.0	
Cys-Tyr	Au	200 (SCE)	7.0	
Cytochrome c (20-mercaptoeicosane-1-ol/ dioleoyl-L-α-phosphatidylcholine)	Au	−70 (SCE) 400	7.0	[224]
Ferredoxins (poly-L-lysine)	ITO (aq. NaCl buffer)	−570 (Ag/AgCl)	7.5	[36]

Table 1. 8 Viologen-, Enzyme-, and Protein- modified Electrodes (continued)

Enzyme, Protein	Electrode Material	Redox Potential mV (reference)	pH	Ref.
Polylysine (ferredoxin)	ITO	−350 (Ag/AgCl)	7.2	[37]
Polyornithine (ferredoxin)	ITO	(Ag/AgCl)	7.2	
Ferredoxins (aminosilane modified)	ITO (aq. NaCl/HCl buffer)	− 650 (Ag/AgCl)	7.2	[38]
Horseradish peroxidase (PhNMe$_2$ → PhNHMe)	GlC	− 0.30 (SCE)	5.5	[229a, 97b]
NAD$^+$ → NADH Hydrogenase (*Alcaligenes eutrophus H16*)	Pt	− 0.70 (SCE)	7.3	[230]
NAD$^+$ → NADH Lipoamide dehydrogenase (pyruvate → lactate)	GlC	− 0.90 (SCE)	7.0	[216a, 216b]
NAD$^+$ → NADH Diaphorase Methyl viologen (mediator)	GlC	− 0.70 (SCE)	7.5	[232a, 232b]
Succinate dehydrogenase	GC	− 0.40 (SCE)	7.7>	[231]
Spinach Ferredoxins Poly-*L*-lysine	GC	− 0.82 (Ag/Ag$^+$)	5.8	[222]

GC: graphite carbon ;   GlC: glassy carbon

1.2.4   Effects of Additives

1.2.4.1   Choice of Proton Donors

A trace amount of water in electrolysis medium also plays an important role as a proton donor. For example, the reduction of carbon monoxide proceeds on a bright and platinized platinum electrode in acetonitrile containing a trace amount of water [233].   The electrodimerization of diethyl fumarate in an MeCN-Bu$_4$NBF$_4$-(Au) system is found to be accelerated by the presence of water in the electrolysis medium.   The role of water is to specifically solvate the anion radicals as a proton donor and hence increase the dimerization rate [234].

The influence of the water content in the solvent on the product distribution for the electrohydrodimerization of 2-cyclohexenones has been studied [235].   The reduction potential on the electroreduction of nitrobenzene derivatives is found to be affected by the concentration of water in DMF [236].

The cis/trans isomer ratio of cathodic hydrogenation of methyl 4-tert-butylcyclohex-1-enecarboxylate has been found to depend slightly on the choice of proton source in a DMF(THF)-Et4NBr(0.1M)-(Hg) system [237]. DMF-Hydroquinone system affords the corresponding cyclohexane in 95% yield (*cis/trans* = 62/38).

Electroreductive removal of the methylsulfinyl group from *p*-substituted α-(methyl-sulfinyl)-α-(methylthio)acetophenone proceeds in an MeCN-Bu₄NClO₄(0.1M)-(Hg) system in the presence of benzoic acid as a proton donor in good yield [238]. Under similar electrolysis conditions, phenol is somewhat inferior to benzoic acid in conversion yield.

The selective cathodic cleavage of one of the alkoxycarbonyl or acyl groups from various imidodicarbonates, acylamides, and diacylamides takes place in an MeCN-Et₄NCl-(Pt/C) system in the presence of Et₃NHCl as a proton donor to give cleavage products in 89~100% yields [239].

Examples of proton donor in electrolysis systems are listed in Table 1.9 [237-245].

Table 1. 9 Role of Proton Donor in Electrolysis Systems

Proton Donor	Type of Reaction	Electrolysis Condition	Products	Ref.
Hydroquinone	*cis/trans* Isomer ratio	DMF(THF)-Et₄NBr(0.1M)-(Hg)	Hydrogenated products	[237a, 237b]
PhCO₂H	Removal of the methylsulfinyl group	MeCN-Bu₄NClO₄(0.1M)-(Hg)	Demethylsulfinylation products	[238a, 238b]
Et₃NHCl	Cathodic cleavage of C-S bonds	MeCN-Et₄NCl-(Pt/C)	Dealkoxylation	[239]
Phenol, (CF₃)₂CHOH	Reduction of 1,4-dihalobutanes	DMF-Me₄NClO₄-(C)	1-Butene, 2-butene, butane, cyclobutane, etc.,	[240]
Phenol	Reduction of double bonds	DMF-Bu₄NClO₄-(Hg)	Hydrogenated products	[241]
MeOH	Reductive hydrogenation	DMF-Bu₄NClO₄-(Pt)	N-H Cleavage products	[242]
NH₄Cl	Hydrogenation of double bond	DMF-Et₄NClO₄-(Hg)	Hydrogenated products	[243] [244]
p-MeC₆H₄SO₃H	Shift of redox potentials	MeCN-Et₄NPF₆-(C)	Dealkoxylation	[245]

[Reproduced with permission from C. Moinet et al., *Electrochim. Acta*, **38**, 325 (1993)]

1.2.4.2   Mediated Activation with Additives

Mediated activation with additives is a common feature in electroorganic reactions. For example, the addition of methyl formate or potassium iodide enhances the formation of methanol, when carbon monoxide is electrochemically reduced into methanol using Everitt's salt in the presence of pentacyanoferrate(II) [199c]. Such phenomena have been well recognized in a variety of electrochemical reactions and their details are discussed in each chapter.

1.2.5   Effects of pH on Electrolysis Media

α-β-Unsaturated ketones are reduced on a mercury or glassy carbon electrode, leading to various products such as saturated ketones, unsaturated alcohol, hydrodimers (diketones), organomercuric derivatives and others, depending on the type of substrate, reaction medium, pH, the applied potential etc.

It has been suggested that the electroreductive hydrodimerization of 2-cyclopenten-1-

one follows two different radical-radical coupling pathways to form its hydrodimer in basic media [246].   In the pH range 6.3~7.5, the final hydrodimer is formed *via* protonation of the radical anion to generate the neutral radical, followed by coupling of the radical anion with neutral radical and further protonation of the resulting dimer anion.   At pH zone 8.0~11.0, the same final hydrodimer is produced *via* dimerization of the radical anion and subsequent protonation of the dimer dianion formed.   In the range at pH 8.0~8.5, the proton donor is either boric acid of the buffer or H^+ ion, whereas at pH 9.0~11.0, it is only the H^+ ion [246b].   Under the different pH values, electroreductive dimerization of 3-buten-2-one [246c], benzalacetone [247], and benzal acetophenone [247], has been investigated.

The polarographic behavior of dichloromaleic acid, chloromaleic acid, dichlorofumaric acid, and chlorofumaric acid as a function of pH has been discussed in details [248].

Electrochemical modulation of self-doped sulfonated polyaniline is shown to provide electronic control of pH and enzyme activity in the vicinity of the electrodes [249].

One approach of producing pH-sensing devices involves the electrode-deposition of polymer films on electrode surface.   It is suggested that the development of potentiometric sensors take into account the properties of the substrate material employed in their construction [250].   Voltammetric pH measurements with a surface-modified Pt/Nafion, $Ru(bpy)_3^{2+}$ electrode *vs.* a Pt/poly(OPD) (OPD = orthophenylenediamine) reference as pH probes have been attained.   Changes in the pH translate to shifts in voltammetric peak potentials [251].

1.2.6   Effects of Ultrasound

The beneficial effects of ultrasound on electrochemical reactions have been widely surveyed since 1963 [252].   Particularly, the effects of ultrasound have been investigated on the product-selectivity of an electrosynthetic reaction, depassivation and erosion of electrode surfaces, electrochemiluminescence, and mass transport to a microelectrode [253,254].

The effect on product-distribution in Kolbe electrolysis by ultrasonic irradiation has been observed [255].   The hydrodimerization has been found to be affected by ultrasonic irradiation consuming less electrons [256b].   For example, the direct use of ultrasound-vibrators as cathode for the electroreduction of benzaldehydes has been attempted [256].   The product-selectivity for the hydrodimeric product can be improved to some extent.   The ultrasonic effect for the formation of methyltin derivatives has been studied in detail [257].

1.3   Design of Electrolysis Cells and Reactors

1.3.1   Laboratory Electrolysis Cells

The cells used for electroorganic synthesis look quite different, depending on the type of reaction and on the handling substrates. In particular, separation or extraction of the product from the running electrolytic solution must be considered for the cell design.   Several typical cell designs are discussed in detail in the literature [258].

A "redox cell" has been used for the electroreduction of porphyrin with a 4-nitrophenyl group in the meso position[259].   The electrosyntheses of the corresponding amino, nitroso, and dinitroso derivatives are carried out in the redox cell with two consecutive porous electrodes of opposite polarities (Fig.1.9) [259c].   Both reduction of the nitro group and reduction of the porphyrin ring occur at the first porous electrode, and the immediate reoxidation of the reduced porphyrin at the second porous electrode prevents the chemical re-

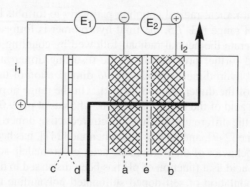

Fig. 1. 9 Schematic diagram of the "redox" cell with two electrical circuits and two working electrodes. E1, E2 = power supplies; i1, i2 = current intensities. a and b = porous cathode and porous anode, c = auxiliary counter electrode, d = separator, e = porous insulator.
→ electrode flow circuit.
[Reproduced with permission from C. Moinet et al., *Electrochim. Acta*, **38**, 325 (1993)]

arrangement of the macrocycle.   The procedure of continuous electrolysis through the cell fitted with porous electrodes has been used for the preparation of *N*-amino-2-methylindoline by electroreduction of *N*-nitroso-2-methylindoline [260].

Efficient carboxylation of organic halides can be achieved by electroreduction in aprotic solvents under at a carbon dioxide atmosphere.   For this purpose, large scale stainless steel pressurized production cells with a consumable magnesium anode are successfully used for the synthesis of various carboxylic acids such as naphthylacetic acid, diphenylacetic acid, trifluoromethylbenzoic acid, and isobutylhydratropic acid [261].

A bipolar fluidized bed cell constructed from vitreous carbon microspheres between two feeder electrodes has been used for the cathodic coupling of carbon dioxide and butadiene [262].   The cell gives a high space time yield and the products are a mixture of two isomeric pentenoic acids, two hexenedioic acids, and three decadienedioic acids, which can form in reasonable current yields.

An electrochemical membrane reactor using $(CeO_2)_{0.8}(SmO_{1.5})_{0.2}$ as a solid electrolyte has been studied at 350 °C to 450 °C for the partial oxidation of propene to acrylaldehyde [263].

The design and evaluation of an electrolysis cell which can be used at increased gas pressure have been proposed [264].

A successful example of electrochemical deionization cell using ion adsorption electrodes based on organic redox polymers has been proposed [265].

Carbon dioxide and monoxide can be converted into the corresponding reduced products, *i.e.*, methanol, with an electrochemical photocell in which *n*-CdS and Everitt's salts-modified platinum electrodes operate as the photoanode and the cathode [266].   Gas diffusion electrode cells, which promote carbon dioxide reduction to alcohols, are fabricated. Electrocatalysts containing $La_{1.8}Sr_{0.2}CuO_4$ Teflon-bonded on carbon gas diffusion electrodes appears active for the above purpose [267].

A ZnS microcrystalline solid has been used for a photochemical reduction of carbon dioxide and formate to methanol in the presence of methanol dehydrogenase, 2-propanol, and pyrroloquinoline quinone as an electron mediator [268].

1.3.2 Electrolysis Reactors (Electrolyzers)

Recent trends concerning the commercial processes on industrial electroorganic synthesis in the USA and Canada [269], in Europe [270], and in Japan [271] have been reviewed [272].

The solid oxide fuel cell technology has been used for the partial oxidation of methane to produce ethane and ethylene. The electrocatalytic cell consists of a solid electrolyte (yttria-stabilized zirconia) coated on either side with a conductive metal to from electrodes. Air is passed over one side of the cell where it reacts with the cathode to form oxygen anion radicals. The oxygen anion radicals are transported through the zirconia to the anode where they oxidize the substrate. The cathode and anode are connected by an external circuit so that a current is generated. External circuit reactions are carried out in either a tubular or disc reactor. The choice of electrocatalyst on the anode affects the product-selectivity [273].

The optimum conditions for the generation of hydrogen peroxide from humidified oxygen in a proton exchange membrane electrochemical flow reactor have been investigated as a function of the applied voltage, electrode materials, catalyst loadings, reactant flow-rates, pressure, etc. [274]. A new electrolyzer for the oxidation of metal powder as an alternative to sacrificial anode in organic solvents has been devised as a tool for the nickel-catalyzed electrosynthesis of anti-inflammatory agents [144].

A FM01-LC parallel plate, a laboratory electrolyzer, has been modified by the incorporation of stationary, flow-by, three-dimensional electrodes [275]. A reactor fitted with a porous cathode for reducing oxalic acid to either glyoxylic acid or ethylene glycol has been patented [276].

A pilot plant for the electroreduction of nitrobenzene to p-aminophenol has been examined by a packed-bed electrode reactor [277]. The pilot scale production of 4-aminomethylpyridine by the electrode in 89% isolated yield (current efficiency) [278].

A flow reactor for the N-demethylation of N,N'-dimethylaniline with hydrogen peroxide generated by horseradish peroxidase immobilized on graphite felt has been investigated on [229b]. A two-compartment packed-bed flow reactor is constructed (Fig. 1.10) [229b],

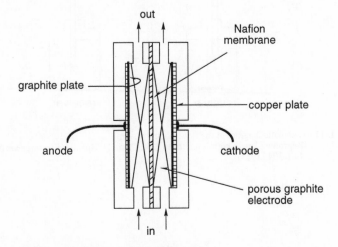

Fig. 1. 10 Two-compartment packed-bed flow reactor.
[Reprodcued with permission from J. K. Chen, K. Nobe, *J. Electrochem. Soc.*, **140**, 304 (1993)

and a nafion membrane is positioned between the two halves of the reactor. The flow reactor can be operated in the recycle mode (Fig. 1.11) [229b] in which the electrolyte solution is returned to a receiver. Bioelectrosynthesis of 5-chlorobarbituric acid using chloroperoxidase (CPO) has been achieved by use of a system which consists of an electrolytic cell, a hollow-fiber membrame reactor and an anion exchanger as illustrated in Fig. 1.12 [229b]. The hydrogen peroxide produced in the electrolytic cell is consumed by chloroperoxidase in the membrane reactor and the product thus formed is subsequently scavanged by an anion exchanger. Besides producing hydrogen peroxide, the electrolytic cell reverses the further enzymatic halogenation of 5-chlorobarbituric acid by dechlorination in an electro-reduction fashion [279].

For the study of the kinetics and mechanism of photo-induced electron-transfer processes, the use of microelectrodes has been demonstrated [280]. The rate of light induced electron transfer between anthraquinone and the tetraphenylborate anion is quantified using phototransient experiments at microband electrodes.

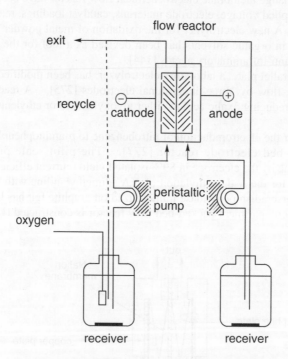

Fig. 1. 11 Schematic diagram of the flow reactor system.
[Reproduced with permission from J. K. Chen, N. Nobe, *J. Electrochem. Soc.*, **140**, 304 (1973)]

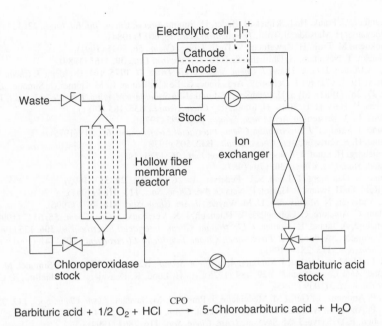

Electrolytic cell

$$\text{Barbituric acid} + 1/2\,O_2 + HCl \xrightarrow{\text{CPO}} \text{5-Chlorobarbituric acid} + H_2O$$

Fig. 1. 12 Schematic representation of the complete chloroperoxidase (CPO) containing bioelectrolytic system. [Reproduced with permission from C. Laane et al., *Enzyme Mirob. Technol.*, **8**. 345 (1986)]

References

[1] (a) M. R. Rifi, F. H. Covitz, *Introduction to Organic Electrochemistry*, Marcel Dekker Inc., New York (1974); (b) H. R. Thirsk, *Electrochemistry*, **6**, The Chemical Society, Burlington House and London (1978); (c) T. Shono, *Electroorganic Chemistry as a New Tool in Organic Synthesis*, Springer-Verlag, Berlin, Heidelberg, New York, Tokyo (1984); (d) A. Tallec, *Électrochimie Organique*, Masson, Paris (1985); (e) S. Torii, *Electroorganic Syntheses Part1 : Oxidations Methods and Applications Monographs in Modern Chemistry*, Kodansha and VCH, Weinheim (1985); (f) A. J. Fry, W. E. Britton (eds.), *Topics in Organic Electrochemistry*, Plenum Press, New York (1986); (g) S. Torii (ed.), *Recent Advances in Electroorganic Synthesis*, Kodansha and Elsevier, Amsterdam (1987); (h) E. Steckhan (ed.), *Topics in Current Chemistry*, **142**, Springer-Verlag, Berlin, Heidelberg (1987); (i) E. Steckhan (ed.), *Topics in Current Chemistry*, **143**, Springer-Verlag, Berlin, Heidelberg (1988); (j) E. Steckhan (ed.), *Topics in Current Chemistry*, **148**, Springer-Verlag, Berlin, Heidelberg (1988); (k) A. J. Fry, *Synthetic Organic Electrochemistry*, Wiley Interscience Publication, New York, Chichester, Brisbane, Toronto, Singapore (1989); (l) H. Lund, M. M. Baizer (eds.), *Organic Electrochemistry*, Marcel Dekker, Inc., New York, Basel, Hong Kong (1991); (m) T. Shono (ed.), *Modern Methodology in Organic Synthesis*, Kodansha and VCH, Weinheim, New York (1992); (n) S. Torii, *Organic Synthesis in Japan Past, Present, and Future*, **1992**, 481; (o) S. Torii (ed.), *Novel Trends in Electroorganic Synthesis*, Kodansha (1995).

[2] (a) S. Torii, *Frontiers in Organic Synthesis-Design and Control of ReactionsÉ~ Role of Electron-Transfer Segments in Indirect Electroreaction*, T. Mukaiyama (ed.) Gendaikagaku, Zokan, **21**, p.21, Tokyo Kagaku Douzin, Tokyo (1992); (b) S. Torii, *New Trends in Electroorganic Reaction: Role of Multi Mediatory Systems in Electroorganic Reactions*, Organic Synthesis in Japan-Past, Present, and Future, R. Noyori (ed.) p. 481, Tokyo Kagaku Dozin, Tokyo (1992).

[3] Y. Yamamoto, *Yuki Gosei Kagaku Kyokaishi*, **51**, 804 (1993).

[4] T. Fuchigami, *Yukagaku*, **39**, 888 (1990).

[5] T. Fuchigami, T. Fujita, *Kagaku Kogyo*, **52**, 164 (1994).

[6] H. J. Schäfer, *Angew. Chem., Int. Ed. Engl.*, **20**, 911 (1981).

[7] J. Simonet, G. Le Guillanton, *Bull. Soc. Chim. Fr.*, **1986**, 221.

[8] (a) T. Shono, Y. Morishima, N. Moriyoshi, M. Ishifune, *J. Org. Chem.* **59**, 273 (1994); (b) S. Torii, P. Liu, N. Bhuvaneswari, C. Amatore, A. Jutand, *J. Org. Chem.*, **61**, 3055 (1996).

[9] S. U. Pedersen, T. Lund, *Acta Chem. Scand.*, **45**, 397 (1991).

[10] H. Gunther, C. Frank, H.-J. Schuets, J. Bader, H. Simon, *Angew. Chem., Int. Ed. Engl.*, **22**, 322 (1983).
[11] K. Uneyama, H. Matsuda, S. Torii, *Tetrahedron Lett.*, **25**, 6017 (1984).
[12] T. Inokuchi, M. Tsuji, H. Kawafuchi, S. Torii, *J. Org. Chem.*, **56**, 5945 (1991).
[13] H. Tanaka, T. Nakahara, H. Dhimane, S. Torii, *Tetrahedron Lett.*, **30**, 4161 (1989).
[14] (a) L. A. Avaca, J. H. P. Utley, *J. Chem. Soc., Perkin Trans. II*, **1975**, 161; (b) idem, *J. Chem. Soc., Perkin Trans. I*, 1975, 971; (c) J. Delaunay, A. Lebouc, G. Le Guillanton, L. M. Gomes, J. Simonet, *Electrochim. Acta*, **27**, 287 (1982); (d) S. U. Pedersen, H. Lund, *12th Sandbjerg Meeting, Extended Abstract*, **1985**, 93.
[15] P. Fuchs, U. Hess, H. H. Holst, H. Lund, *Acta Chem. Scand.*, **B35**, 185 (1981).
[16] H. Lund, L. A. Kristensen, *Acta Chem. Scand.*, **B33**, 495 (1979).
[17] H. Lund, J. Simonet, *J. Electroanal. Chem., Interfacial Electrochem.*, **65**, 205 (1975).
[18] H. Lund, H. S. Carlsson, *Acta Chem. Scand.*, **B32**, 505 (1978).
[19] P. Nelleborg, H. Lund, J. Eriksen, *Tetrahedron Lett.*, **26**, 1773 (1985).
[20] H. Lund, *Stud. Org. Chem.*, **30**, 179 (1987).
[21] H. Lund, K. Daasbjerg, T. Lund, S. U. Pedersen, *Acc. Chem. Res.*, **28**, 313 (1995).
[22] J. C. Ruhl, D. H. Evans, P. Hapiot, P. Neta, *J. Am. Chem. Soc.*, **113**, 5188 (1991).
[23] M. L. Andersen, N. Mathivanan, D. M. Wayner, *J. Am. Chem. Soc.*, **118**, 4871 (1996).
[24] N. Alam, C. Amatore, C. Combellas, A. Thiebault, J. N. Verpeaux, *J. Org. Chem.*, **55**, 6347 (1990).
[25] C. Amatore, A. Jutand, L. Mottier, *J. Electroanal. Chem., Interfacial Electrochem.*, **306**, 125 (1991).
[26] (a) C. Amatore, A. Jutand, *J. Electroanal. Chem., Interfacial Electrochem.*, **306**, 141 (1991); (b) idem, *J. Am. Chem. Soc.*, **113**, 2819 (1991).
[27] (a) H. Lund, M.-A. Michel, J. Simonet, *Acta Chem. Scand.*, **B28**, 900 (1974); (b) J. Simonet, M.-A. Michel, H. Lund, *Acta Chem. Scand.*, **B29**, 489 (1975); (c) H. Lund, K. Daasbjerg, D. Ochiallini, S. U. Pedersen, *Elektrokhimiya*, **31**, 939 (1995).
[28] (a) C. P. Andrieux, L. Gelis, M. Medebielle, J. Pinson, J.-M. Savéant, *J. Am. Chem. Soc.*, **112**, 3509 (1990); (b) C. P. Andrieux, A. Le Gorande, J.-M. Savéant, *J. Am. Chem. Soc.*, **114**, 6892 (1992); (c) C. P. Andrieux, M. Robert, F. D. Saeva, J.-M. Savéant, *J. Am. Chem. Soc.*, **116**, 7864 (1994).
[29] (a) C. P. Andrieux, J.-M. Savéant, *Techniques of Chemistry; Molecular Design of Electrode Surfaces* R. W. Murray (ed.), Vol. 22, Wiley Interscience, New York, 1992 and references cited therein; (b) R. W. Murray, *Electroanalytical Chemistry* A. J. Bard (ed.), Vol. 13, pp. 191, Marcel Dekker, New York (1984); (c) C. P. Andrieux, J.-M. Savvant, *J. Electroanal. Chem., Interfacial Electrochem.*, **142**, 1 (1982).
[30] (a) E. Laviron, *J. Electroanal. Chem., Interfacial Electrochem.*, **131**, 61 (1982); (b) W. J. Albery, A. R. Hillman, *Annual Report 1981*, **1983**, 317; (c) A. F. Diaz, K. K. Kanazawa, G. P. Gardini, *J. Chem. Soc., Chem. Commun.*, **1979**, 635; (d) M. D. Levi, M. Lapkowski, *Electrochim. Acta*, **38**, 271 (1993).
[31] M. F. J. M. Verhagen, W. R. Hagen, *J. Electroanal. Chem.*, **334**, 339 (1992).
[32] P. D. Hale, L. I. Boguslavsky, T. Inagaki, H. I. Karan, H. S. Lee, T. A. Skotheim, Y. Okamoto, *Anal. Chem.*, **63**, 677 (1991).
[33] H. Gunther, A. S. Paxinos, M. Schultz, C. van Dijk, H. Simon, *Angew. Chem., Int. Ed. Engl.*, **29**, 1053 (1990).
[34] M. Somasundrum, J. V. Bannister, *J. Chem. Soc., Chem. Commun.*, **1993**, 1629.
[35] B. C. King, F. M. Hawkridge, B. M. Hoffman, *J. Am. Chem. Soc.*, **114**, 10603 (1992).
[36] I. Taniguchi, K. Hayashi, M. Tominaga, R. Muraguchi, A. Hirose, *Denki Kagaku*, **61**, 774 (1993).
[37] I. Taniguchi, Y. Hirakawa, K.-I. Iwakiri, M. Tominaga, K. Nishiyama, *J. Chem. Soc., Chem. Commun.*, **1994**, 953.
[38] K. Nishiyama, H. Ishida, I. Taniguchi, *J. Electroanal. Chem.*, **373**, 255 (1994).
[39] C. L. Bird, A. T. Kuhn, *Chem. Soc. Rev.*, **10**, 49 (1981).
[40] (a) J. Moutet, C. J. Pickett, *J. Chem. Soc., Chem. Commun.*, **1989**, 188; (b) G. Bidan, A. Deronizier, J. Moutet, *ibid.*, **1984**, 1185; (c) A. Deronizier, J. Moutet, *Acc. Chem. Res.*, **22**, 249 (1989); (d) A. Bettelheim, B. A. White, S. A. Raybuck, R. W. Murray, *Inorg. Chem.*, **26**, 1009 (1987).
[41] (a) H. Akahoshi, S. Toshima, K. Itaya, *J. Phys. Chem.*, **85**, 818 (1981); (b) N. Oyama, N. Oki, H. Ohno, Y. Ohnuki, H. Matsuda, E. Tsuchida, *ibid.*, **87**, 3642 (1983); (c) R. J. Mortimer, F. C. Anson, *J. Electroanal. Chem., Interfacial Electrochem.*, **138**, 325 (1982).
[42] M. Kijima, A. Sakawaki, T. Sato, *Chem. Lett.*, **1991**, 499.
[43] S. Kumbhat, *J. Indian Chem., Sect. A*, **33A**, 162 (1994).
[44] G. Veerabhadram, K. S. Sastry, *J. Electrochem. Soc. India*, **34**, 30 (1985).
[45] D. B. Bruss, T. Devries, *J. Am. Chem. Soc.*, **78**, 733 (1956).
[46] I. M. Kolthoff, J. G. Coetzee, *J. Am. Chem. Soc.*, **79**, 1852 (1957).
[47] R. C. Larson, R. T. Iwamoto, *J. Am. Chem. Soc.*, **82**, 3526 (1960).
[48] G. P. Kumar, D. A. Pantony, *J. Polarog. Soc.*, **14**, 84 (1968).
[49] P. Kanathanana, M. S. Spritzer, *Anal. Chem.*, **56**, 958 (1974).
[50] S. Kumbhat, P. Sharma, *Bull. Electrochem.*, **6**, 737 (1990).
[51] J. B. Cooper, A. M. Bond, *J. Electroanal. Chem., Interfacial Electrochem.*, **315**, 143 (1991).

[52] J. Simonet, M. Chaquiq, C. Bedre, G. Mabon, *J. Elcectroanal. Chem., Interfacial Electrochem.*, **281**, 289 (1996).

[53] Z.-H. Zhong, T. M.-Inoue, A. Ichimura, *Anal. Sci.*, **8**, 877 (1992).

[54] G. K. Rowe, S. E. Creager, *J. Phys. Chem.*, **98**, 5500 (1994).

[55] H. N. McCoy, W. C. Moore, *J. Am. Chem. Soc.*, **33**, 273 (1911).

[56] (a) J. Simonet, H. Lund, *J. Electroanal. Chem., Interfacial Electrochem.*, **75**, 719 (1977); (b) G. Bernard, J. Simonet, *J. Electroanal. Chem., Interfacial Electrochem.*, **96**, 249 (1979); (c) J. Berthelot, M. Jubault, J. Simonet, *J. Chem. Soc., Chem. Commun.*, **1982**, 759.

[57] (a) E. Kariv-Miller, C. Nanjundiah, J. Eaton, K. E. Swenson, *J. Electroanal. Chem., Interfacial Electrochem.*, **167**, 141 (1984); (b) E. Kariv-Miller, V. Svetlicic, *ibid.*, **205**, 319 (1986); (c) V. Svetlicic, E. Kariv-Miller, *ibid.*, **209**, 319 (1986); (d) E. Kariv-Miller, P. B. Lawin, Z. Vajtner, *ibid.*, **195**, 435 (1985).

[58] B. C. Southworth, R. Osteryoung, K. D. Fleischer, F. C. Nachod, *Anal. Chem.*, **33**, 208 (1961).

[59] C. E. Dahm, D. G. Peters, *J. Electroanal. Chem.*, **402**, 91 (1996).

[60] V. Yuayanathan, S. Venkatachalam, V. N. Drishnamurthy, *Eur. Polym. J.*, **29**, 1373 (1993).

[61] (a) S. H. El-Hamouly, M. A. Azzem, U. S. Yousef, *Eur. Polym. J.*, **29**, 1271 (1993); (b) ER95003 P. Tascedda, E. Dunach, *J. Chem. Soc., Chem. Commun.*, **1995**, 43.

[62] S. Alwarappan, A. S. Mideen, M. A. Kulandainathan, P. V. Krishnamoorthy, P. Manisankar, T. Vasudevan, S. V. Iyer, *J. Electrochem. Soc. India*, **39**, 169 (1990).

[63] (a) Z. Ogumi, K. Nishio, S. Yoshizawa, *Denki Kagaku*, **49**, 212 (1981) ; (b) *idem, Electrochim. Acta*, **26**, 1779 (1981); (c) Z. Ogumi, H. Yamashita, K. Nishio, Z.-I. Takehara, S. Yoshizawa, *Electrochim. Acta*, **28**, 1687 (1983); (d) Z. Ogumi, H. Yamashita, K. Nishio, Z.-I. Takehara, *Denki Kagaku*, **52**, 180 (1984) ; (e) Z. Ogumi, M. Inaba, S.-I. Ohashi, M. Uchida, Z.-I. Takehara, *Electrochim. Acta*, **33**, 365 (1988); (f) M. Inaba, K. Fukuta, Z. Ogumi, Z.-I. Takehara, *Chem. Lett.*, **1993**, 1779; (g) M. Inaba, Z. Ogumi, Z.-I. Takehara, *J. Electrochem. Soc.*, **140**, 19 (1993); (h) M Inaba, S. Kinoshita, Z. Ogumi, Z.-I. Takehara, *Denki Kagaku*, **62**, 1188 (1994); (i) M. Inaba, J. T. Hinatsu, Z. Ogumi, Z.-I. Takehara, *J. Electrochem. Soc.*, **140**, 706 (1993).

[64] Y.-L. Chen, T.-C. Chou, *Ind. Eng. Chem. Res.*, **33**, 676 (1994).

[65] A. P. Tomilov, *Zh. Org. Khim.*, **29**, 657 (1993).

[66] T. Chiba, M. Okimoto, H. Nagai, Y. Takata, *Bull. Chem. Soc. Jpn.*, **56**, 719 (1983).

[67] K. N. Campbell, E. E. Young, *J. Am. Chem. Soc.*, **65**, 965 (1943).

[68] D. V. Sokol'skii, I. V. Kirilyus, U. B. Bekenova, Z. A. Malikov, *Elektrokhimiya*, **20**, 557 (1984).

[69] T. Nonaka, M. Takahashi, T. Fuchigami, *Bull. Chem. Soc. Jpn.*, **56**, 2584 (1983).

[70] M. Sakuma, *J. Electrochem. Soc. Jpn.*, **28**, 164 (1960).

[71] M. A. Casadei, D. Pletcher, *Electrochim. Acta*, **33**, 117 (1988).

[72] X. Hemptinne, *Acad. R. Belg. Cl. Sci. Mem.*, **32**, 7 (1961).

[73] G. Belot, S. Desjardins, J. Lessard, *Tetrahedron Lett.*, **25**, 5347 (1984).

[74] M. J. Lain, D. Pletcher, *Electrochim. Acta*, **32**, 109 (1987).

[75] (a) V. Krishnan, K. Ragupathy, H. V. K. Udupa, *J. Appl. Electrochem.*, **5**, 125 (1975); (b) *idem, J. Appl. Electrochem.*, **8**, 169 (1978).

[76] G. Horanyi, *Electrochim. Acta*, **31**, 1095 (1986).

[77] T. Osa, T. Matsue, A. Yokozawa, T. Yamada, *Denki Kagaku*, **54**, 484 (1986).

[78] A. P. Tomilov, I. V. Kirilyus, I. P. Andriyanova, *Elektrokhimiya*, **8**, 1050 (1972).

[79] V. Krishnan, A. Muthukumaran, H. V. K. Udupa, *J. Appl. Electrochem.*, **9**, 657 (1979).

[80] N. Ayyaswami, V. Krishnan, *J. Appl. Electrochem.*, **13**, 731 (1983).

[81] N. Ayyaswami, V. Krishnan, *J. Appl. Electrochem.*, **14**, 557 (1984).

[82] I. V. Kirilyus, G. K. Murzatova, D. V. Sokol'skii, *Elektrokhimiya*, **15**, 1543 (1979).

[83] P. N. Pintauro, N. Phan, M. M. Baizer, N. Nobe, *AlchE Symp. Ser.*, **83**, 34 (1987).

[84] A. Velin, J. Lessard, *Proc. Electrochem. Soc.*, **90**, 10 (1990).

[85] A. V. Bukhtiarov, E. A. Chernyshev, O. V. Kuz'min, B. K. Kabanov, V. N. Golyshin, A. P. Tomilov, *Dokl. Akad. Nauk SSSR*, **278**, 1127 (1984).

[86] A. V. Bukhtiarov, V. V. Mikheev, Yu. G. Kudryavtsev, A. V. Lebedev, *Dokl. Akad. Nauk SSSR*, **304**, 620 (1989).

[87] A. V. Bukhtiarov, V. N. Golyshin, A. P. Tomilov, O. V. Kuz'min, *Zh. Obshch. Khim.*, **58**, 1398 (1987).

[88] A. V. Bukhtiarov, V. N. Golyshin, I. A. Rodnikov, O. V. Kuz'min, A. P. Tomilov, L. N. Nekrasov, T. N. Khomchenko, *Elektrokhimiya*, **22**, 663 (1986).

[89] T. Nonaka, T. Sekine, K. Sugino, *J. Electrochem. Soc.*, **115**, 242 (1968).

[90] B. Kurtyka, R. de Levie, *J. Electroanal. Chem.*, **397**, 311 (1995).

[91] B. Mahdavi, P. Chambrion, J. Binett, E. Martel, J. Lessard, *Can. J. Chem.*, **73**, 846 (1995).

[92] (a) A. Cyr, P. Huot, G. Belot, J. Lessard, *Electrochim. Acta*, **35**, 147 (1990); (b) J. Lessard, G. Belot, Y. Couture, S. Desjardins, C. Roy, *Int. J. Hydrogen Energy*, **18**, 681 (1993); (c) E. N. Pervii, A. N. Sofronkov, N. M. Fedyshina, *Zh. Prikl. Khim.*, **58**, 1905 (1985).

[93] V. K. Sharma, Madhu, D. K. Sharma, *Bull. Electrochem.*, **9**, 117 (1993).

34   1. Electrochemical Reduction and Product Selectivity

[94] G. Horanyi, *J. Electroanal. Chem., Interfacial Electrochem.*, **284**, 481 (1990).
[95] K. Amouzegar, O. Savadogo, *Electrochim. Acta*, **39**, 557 (1994).
[96] K. Ehara, K. Kumagaya, Y. Yamamoto, K. Takahashi, H. Yamazaki, *J. Organomet. Chem.*, **410**, C49 (1991).
[97] J.-C. Folest, E. Pereira-Martins, M. Troupel, J. Périchon, *Tetrahedron Lett.*, **34**, 7571 (1993).
[98] Y. Kunugi, T. Nonaka, Y.-B. Chong, N. Watanabe, *J. Electroanal. Chem.*, **356**, 163 (1993).
[99] G. V. Itov, I. A. Avrutskaya, *Elektrokhimiya*, **31**, 1245 (1995).
[100] (a) C. Ravichandran, M. Noel, P. N. Anantharaman, *J. Appl. Electrochem.*, **24**, 965 (1994); (b) idem, *J. Appl. Electrochem.*, **24**, 1256 (1994); (c) V. P. Gul'tyai, V. N. Leibzon, *Russ. J. Electrochem.*, **32**, 59 (1996); (d) C. Ravichandran, M. Noel, P. N. Anantharaman, *Bull. Electrochem.*, **10**, 283 (1994); (e) F. Beck, W. Gabriel, *Angew. Chem., Int. Ed. Engl.*, **24**, 771 (1985).
[101] O. Orange, C. Elfakir-Hamet, C. Caullet, *J. Electrochem. Soc.*, **128**, 1889 (1981).
[102] C. Hamet-Elfakir, C. Caullet, *Bull. Soc. Chim. Fr.*, **1986**, 687.
[103] E. Kariv-Miller, D. M. Loffredo, V. Svetlicic, *Denki Kagaku*, **62**, 1276 (1994).
[104] P. Nikitas, A. Pappa-Louisi, S. Antoniou, *J. Electroanal. Chem.*, **367**, 239 (1994).
[105] F. Pragst, F. Scholz, P. Woitke, V. Kollek, G. Henrion, *J. Prakt. Chem.*, **327**, 1028 (1985).
[106] Y. Sato, K. Uosaki, *Denki Kagaku*, **62**, 1269 (1994).
[107] D. E. Weisshaar, B. D. Lamp, M. D. Porter, *J. Am. Chem. Soc.*, **114**, 5860 (1992).
[108] W. R. Everett, T. L. Welch, L. Reed, I. Fritsch-Faules, *Anal. Chem.*, **67**, 292 (1995).
[109] (a) M. Chandrasekaran, M. Noel, V. Krishnan, *J. Chem. Soc., Perkin Trans. II*, **1992**, 979; (b) M. Chandrasekaran, Ph. D. Thesis, Madurai-Kamaraj University, Madurai, 1989.
[110] R. H. Dahm, R. J. Latham, S. E. Mosley, *J. Appl. Electrochem.*, **16**, 213 (1986).
[111] K. Tanaka, M. Nakamoto, Y. Taru, T. Tanaka, *Bull. Chem. Soc. Jpn.*, **62**, 2830 (1989).
[112] S. Kuwabata, Y. Hozumi, K. Tanaka, T. Tanaka, *Chem. Lett.*, **1985**, 401.
[113] P. Audebert, P. Calas, G. Cerveau, R. J. P. Corriu, N. Costa, *J. Electroanal. Chem.*, **372**, 275 (1994).
[114] (a) H. Inoue, T. Abe and C. Iwakura, *J. Chem. Soc., Chem. Commun.*, **1996**, 55; (b) C. Iwakura, *Yuasa Jiho*, **80**, 1 (1996) (in Japanese); (c) C. Iwakura, T. Abe, H. Inoue, *J. Electrochem. Soc.*, **143**, 71 (1996); (d) C. Iwakura, Y. Yoshida, H. Inoue, *J. Electroanal. Chem., in press*; (e) Y. Matsuda, C. Iwakura, *Denki Kagaku Gairon*, p. 93-102, Maruzen, Tokyo.
[115] T. Shono, N. Kise, T. Suzumoto, T. Morimoto, *J. Am. Chem. Soc.*, **108**, 4676 (1986).
[116] F. El Kadiri, R. Faure, R. Durand, *J. Electroanal. Chem., Interfacial Electrochem.*, **301**, 177 (1991).
[117] S. Wada, K. Nishimura, K. Yahikozawa, M. Enyo, Y. Takasu, *Chem. Lett.*, 1993, **45**.
[118] T. Sawaguchi, T. Matsue, K. Itaya, I. Uchida, *Electrochim. Acta*, **36**, 703 (1993).
[119] C. P. Andrieux, P. Audebert, P. Hapiot, B. Divisia-Blohorn, P. Aldebert, *J. Electroanal. Chem., Interfacial Electrochem.*, **296**, 129 (1990).
[120] P. Audebert, P. Hapiot, *J. Electroanal. Chem.*, **361**, 177 (1993).
[121] G. M. Swain, B. J. Tatarchuk, *J. Electrochem. Soc.*, **140**, 1026 (1993).
[122] V. L. Kornienko, G. A. Kolyagin, G. V. Kornienko, Yu. V. Saltykov, *Elektrokhimiya*, **28**, 507 (1992).
[123] G. L. Guillanton, Q. T. Do, J. Simonet, *Bull. Soc. Chem. Fr.*, **1989**, 433.
[124] C. Degrand, R. Prest, *J. Electroanal. Chem., Interfacial Electrochem.*, **282**, 281 (1991).
[125] Z. Ogumi, H. Yamashita, K. Nishio, Z. Takehara, S. Yoshizawa, *Electrochim. Acta*, **28**, 1687 (1983).
[126] R. G. Compton, B. A. Coles, J. J. Gooding, *J. Phys. Chem.* **98**, 2446 (1994).
[127] A. Doron, M. Portnoy, M. L.-Dagan, E. Katz, I. Willner, *J. Am. Chem. Soc.*, **118**, 8937 (1996).
[128] F. Mukae, H. Takemura, K. Takehara, *Bull. Chem. Soc. Jpn.*, **69**, 2461 (1996).
[129] I. Willner, E. Katz, A. Riklin, R. Kasher, *J. Am. Chem. Soc.*, **114**, 10965 (1992).
[130] A. P. Tomilov, *Elektrokhimiya*, **32**, 30 (1996).
[131] J. Shoussard, J.-C. Folest, J.-Y. Nédélec, J. Périchon, S. Sibille, M. Troupel, *Synthesis*, **1990**, 369.
[132] E. Léonel, J. P. Paugam, J.-Y. Nédélec, J. Périchon, *J. Chem. Res. (S)*, **1995**, 278.
[133] A. Conan, S. Sibille, E. d'Incan, J. Périchon, *J. Chem. Soc., Chem. Commun.*, **1990**, 48.
[134] (a) H. Schick, R. Ludwig, K.-H. Schwarz, K. Kleiner, A. Kunath, *J. Org. Chem.*, **59**, 3161 (1994); (b) idem, *Angew. Chem., Int. Ed. Engl.*, **32**, 1191 (1993); (c) K.-H. Schwarz, K. Kleiner, R. Ludwig, H. Schick, *Chem. Ber.*, **126**, 1247 (1993); (d) H. Schick, R. Ludwig, K.-H. Schwarz, K. Kleiner, A. Kunath, *Angew. Chem.*, **105**, 1218 (1993).
[135] (a) T. Ohno, H. Nakahiro, K. Sanemitsu, T. Hirashima, I. Nishiguchi, *Tetrahedron Lett.*, **33**, 5515 (1992); (b) T. Ohno, H. Aramaki, H. Nakahiro, I. Nishiguchi, *Tetrahedron*, **52**, 1943 (1996).
[136] (a) D. Bonafoux, M. Bordeau, C. Biran, J. Dunogués, *J. Organomet. Chem.*, **493**, 27 (1995); (b) C. Biran, M. Bordeau, D. Bonafoux, D. Deffieux, C. Duprat, V. Jouikov, MP Léger-Lambert, C. Moreau, F. Serein-Spirau, *J. Chem. Phys.*, **93**, 591 (1996).
[137] T. Inokuchi, H. Kawafuchi, K. Aoki, A. Yoshida, S. Torii, *Bull. Chem. Soc. Jpn.*, **67**, 595 (1994).
[138] C. Saboureau, M. Troupel, J. Périchon, *J. Appl. Electrochem.*, **20**, 97 (1990).
[139] M. Allan, A. F. Janzen, C. J. Willis, *Can. J. Chem.*, **46**, 3671 (1968).

[140] I. L. Knunyants, S. A. Postovoi, N. I. Delyagina, Yu. V. Zeifman, *Izv. Akad. Nauk SSSR, Ser. Khim.*, **1987**, 2256.
[141] S. A. Postovoi, I. M. Vol'pin, E. I. Mysov, Yu. V. Zeifman, L. S. German, *Izv. Akad. Nauk SSSR, Ser. Khim.*, **1989**, 1173.
[142] V. A. Grinberg, V. R. Polishchuk, S. R. Sterlin, *Elektrokhimiya*, **30**, 1068 (1994).
[143] H. Hebri, E. Dunach, J. Périchon, *Synth. Commun.*, **21**, 2377 (1991).
[144] J. F. Fauvarque, Y. De Zelicourt, C. Amatore, A. Jutand, *J. Appl. Electrochem.*, **20**, 338 (1990).
[145] M. Durandetti, S. Sibille, J.-Y. Nédélec, J. Périchon, *Synth. Commun.*, **24**, 145 (1994).
[146] S. Olivero, E. Duñach, *J. Chem. Soc., Chem. Commun.*, **1995**, 2497.
[147] P. Kaur, J. S. Banait, S. S. Pahil, *Bull. Electrochem.*, **7**, 562 (1991).
[148] (a) C. Gosmini, J. Y. Nédélec, J. Périchon, *Tetrahedron Lett.*, **38**, 1941 (1997); (b) J. M. Paratian, S. Sibille, J. Périchon, *J. Chem. Soc., Chem. Commun.*, **1992**, 53.
[149] M. Tokuda, K. Endate, H. Suginome, *Chem. Lett.*, **1988**, 945.
[150] M. Tokuda, M. Uchida, Y. Katoh, H. Suginome, *Chem. Lett.*, **1990**, 461.
[151] G. Silvestri, S. Gambino, G. Filardo, *Tetrahedron Lett.*, **27**, 3429 (1986).
[152] G. De Santis, L. Fabbrizzi, M. Licchelli, C. Mangano, D. Sacchi, *Inorg. Chem.*, **34**, 3581 (1995).
[153] L. Echegoyen, Y. Hafez, R. C. Lawson, J. De Mendoza, T. Torres, *J. Org. Chem.*, **58**, 2009 (1993).
[154] L. Echegoyen, R. C. Lawson, C. Lopez, Y. Hafez, J. De Mendoza, T. Torres, *J. Org. Chem.*, **59**, 3814 (1994).
[155] J. R. Allen, T. Gynkowski, J. Desai, L. G. Bachas, *Electroanalysis*, **4**, 533 (1992).
[156] A. K. Newell, J. H. P. Utley, *J. Chem. Soc., Chem. Commun.*, **1992**, 800.
[157] L. Echegoyen, Y. Hafez, R. C. Lawson, J. De Mendoza, T. Torres, *Tetrahedron Lett.*, **35**, 6383 (1994).
[158] M. Jorgensen, K. Lerstrup, P. Frederiksen, T. Bjornholm, P. Sommer-Larsen, K. Schaumburg, K. Brunfeldt, K. Bechgaard, *J. Org. Chem.* **58**, 2785 (1993).
[159] G. A. Sotzing, J. R. Reynolds, *J. Chem. Soc., Chem. Commun.*, **1995**, 703.
[160] N. Nakashima, Y. Deguchi, T. Nakanishi, K. Uchida, M Irie, *Chem. Lett.*, **1996**, 817.
[161] R. A. Bissell, E. Cordova, A. E. Kaifer, J. F. Stoddart, *Nature*, **369**, 133 (1994).
[162] A. Heller, *Acc. Chem. Res.*, **23**, 128 (1990).
[163] G. Wilson, In *Biosensors: Fundamentals and Applications*, A. Turner, I. Karube, G. Wilson (eds.), Oxford University Press, New York (1987).
[164] Y. Degani, A Heller, *J. Am. Chem. Soc.*, **111**, 2357 (1989).
[165] L. Gorton, H. Karan, P. Hale, T. Inagaki, Y. Okamoto, T. Skotheim, *Anal. Chem. Acta*, **228**, 23 (1990).
[166] Y. Degani, A. Heller, *J. Am. Chem. Soc.*, **110**, 2615 (1988).
[167] J. Salbeck, V. N. Komissarov, V. I. Minkin, J. Daub, *Angew. Chem., Int. Ed. Engl.*, **31**, 1498 (1992).
[168] C. G. J. Koopal, B. D. Ruiter, R. J. M. Nolte, *J. Chem. Soc., Chem. Commun.*, **1991**, 1691.
[169] B. T. Hauser, T. S. Bergstedt, K. S. Schanze, *J. Chem. Soc., Chem. Commun.*, **1995**, 1945.
[170] T. S. Bergstedt, B. T. Hauser, K. S. Schanze, *J. Am. Chem. Soc.*, **116**, 8380 (1994).
[171] J. Bobacka, A. Ivaska, M. Grzeszczuk, *Synth. Met.*, **44**, 21 (1991).
[172] M. Haga, M. M. Ali, R. Arakawa, *Angew. Chem., Int. Ed. Engl.*, **35**, 76 (1996).
[173] M. Somasundrum, J. V. Vannister, *J. Chem. Soc., Chem. Commun.*, **1993**, 1629.
[174] I. Willner, R. Blonder, A. Dagan, *J. Am. Chem. Soc.*, **116**, 9365 (1994).
[175] T. Ohsaka, Y. Yamaguchi, N. Oyama, *Bull. Chem. Soc. Jpn.*, **63**, 2646 (1990).
[176] T. K. Hansen, T. Jorgensen, J. Becher, *J. Chem. Soc., Chem. Commun.*, **1992**, 1550.
[177] T. Iida, Y. Ogura, H. Kobayashi, T. Mitamura, K. Nagata, K. Tomita, *Denki Kagaku*, **56**, 1118 (1988) (in Japanese).
[178] W. J. Albery, M. G. Boutelle, P. T. Galley, *J. Chem. Soc., Chem. Commun.*, **1992**, 900.
[179] I. Willner, S. Marx, Y. Eichen, *Angew. Chem., Int. Ed. Engl.*, **31**, 1243 (1992).
[180] S. Marx-Tibbon, I. Ben-Dov, I, Sillner, *J. Am. Chem. Soc.*, **118**, 4717 (1996).
[181] S. Nakabayashi, J.-I. Ushizaki, K. Uosaki, *J. Electroanal. Chem.*, **371**, 111 (1994).
[182] T. Komura, H. Sakabayashi, K. Takahashi, *Bull. Chem. Soc. Jpn.*, **67**, 1269 (1994).
[183] D. P. Arnold, V. V. Borovkov, G. V. Ponomarev, *Chem. Lett.*, **1996**, 485.
[184] (a) R. W. Murray, *ElectroAnal. Chem., Interfacial Electrochem.*, **13**, 191 (1984); (b) A. Merz, *Topics Curr. Chem.*, **152**, 49, (1990).
[185] D. Albagli, G. C. Bazan, R. R. Schrock, M. S. Wrighton, *J. Am. Chem. Soc.*, **115**, 7328 (1993).
[186] A. L. Crumbliss, D. Cooke, J. Castillo, P. W.-Neilson, *Inorg. Chem.*, **32**, 6088 (1993).
[187] N. Nakashima, Y. Miyata, M. Tominaga, *Chem. Lett.*, **1996**, 731.
[188] M. A. Fox, D. A. Chandler, P.-W. Wang, *Macromolecules*, **24**, 4626 (1991).
[189] (a) T. R. I. Cataldi, D. Centonze, G. Ricciardi, *Electroanalysis*, **7**, 312 (1995); (b) T. R. I. Cataldi, E. Desimoni, G. Ricciardi, F. Lelj, *Electroanalysis*, **7**, 435 (1995).
[190] H. Tagaya, K. Ara, T. Ogata, K. Matsushita, J. Kadokawa, M. Karasu, K. Chiba, *Chem. Lett.*, **1994**, 2439.
[191] N. A. Surridge, F. R. Keene, B. A. White, J. S. Facci, M. Silver, R. W. Murray, *Inorg. Chem.*, **29**, 4950

(1990).

[192] H. Nishihara, K. Pressprich, R. W. Murray, J. P. Collman, *Inorg. Chem.*, **29**, 1000 (1990).

[193] H. Gülce, H. Özyörük, A. Yildiz, *Ber. Bunsenges. Phys. Chem.*, **98**, 228 (1994).

[194] M. Yasuzawa, A. Kunugi, M. Inaba, Z. Ogumi, *Denki Kagaku*, **62**, 1183 (1994).

[195] N. Takano, M. Nakabayashi, N. Takeno, *Chem. Lett.*, **1995**, 219.

[196] Y. Kashiwagi, Y. Yanagisawa, N. Shibayama, K. Nakahara, F. Kurashima, J. Anzai, T. Osa, *Electrochim. Acta*, **42**, 2267 (1997).

[197] T. Yamamoto, Y. Yoneda, T. Maruyama, *J. Chem. Soc., Chem. Commun.*, **1992**, 1652.

[198] J.-C. Moutet, A. Ourari, A. Zouaoui, *Electrochim. Acta*, **37**, 1261 (1992).

[199] (a) K. Ogura, N. Endo, M. Nakayama, H. Ootsuka, *J. Electrochem. Soc.*, **142**, 4026 (1995); (b) K. Ogura, M. Higasa, J. Yano, N. Endo, *J. Electroanal. Chem.*, **379**, 373 (1994); (c) K. Ogura, S. Yamasaki, *J. Chem. Soc., Faraday Trans. I*, **81**, 267 (1985); (d) K. Ogura, H. Sugihara, J. Yano, M. Higasa, *J. Electrochem. Soc.*, **141**, 419 (1994).

[200] J. Simonet, J. R-Berthelot, M. M. Granger, H. Le Deit, *J. Electroanal. Chem.*, **372**, 185 (1994).

[201] M. Maeda, Y. Tsuzaki, K. Nakano, M. Takagi, *J. Chem. Soc., Chem. Commun.*, **1990**, 1529; (b) M. Maeda, Y. Fujita, K. Nakano, M. Takagi, *J. Chem. Soc., Chem. Commun.*, **1991**, 1724.

[202] A J. Banister, Z. V. Hauptman, A. G. Kendrick, *J. Chem. Soc., Chem. Commun.*, **1983**, 1016.

[203] T. Osaka, K. Naoi, T. Hirabayashi, S. Nakamura, *Bull. Chem. Soc. Jpn.*, **59**, 2717 (1986).

[204] F. Mizutani, S. Iijima, Y. Tanabe, K. Tsuda, *J. Chem. Soc., Chem. Commun.*, **1985**, 1728.

[205] J. Chang, J. P. Bell, S. Shkolnik, *J. Appl. Polym. Sci.*, **34**, 2105 (1987).

[206] L. Coche, A. Deronzier, J.-C. Moutet, *J. Electroanal. Chem., Interfacial Electrochem.*, **198**, 187 (1986).

[207] L. Coche, B. Ehui, D. Limosin, J.-C. Moutet, *J. Org. Chem.*, **55**, 5905 (1990).

[208] T. Ohsaka, T. Watanabe, F. Kitamura, N. Oyama, K. Tokuda, *J. Chem. Soc., Chem. Commun.*, **1991**, 487.

[209] A. Merz, S. Reitmeier, *J. Chem. Soc., Chem. Commun.*, **1990**, 1054.

[210] E. Kariv-Miller, V. Svetlicic, P. B. Lawin, *J. Chem. Soc., Faraday Trans. I*, **83**, 1169 (1987).

[211] (a) C. Bourdillon, J. P. Bourgeois, D. Thomas, *J. Am. Chem. Soc.*, **102**, 4231 (1980); (b) T. Ikeda, I. Katasho, M. Kamei, M. Senda, *Agric. Biol. Chem.*, **48**, 1969 (1984); (c) G. F. Khan, H. Shinohara, Y. Ikariyama, M. Aizawa, *J. Electroanal. Chem., Interficial Electrochem.*, **315**, 263 (1991).

[212] P. A. Quintela, A. E. Kaifer, *Langmuir*, **3**, 769 (1987).

[213] I. C. G. Thanos, H. Simon, *J. Biotechnol.*, **6**, 13 (1987).

[214] A. E. Kaifer, *J. Am. Chem. Soc.*, **108**, 6837 (1986).

[215] K. Tanaka, K. Tokuda, T. Ohsaka, *J. Chem. Soc., Chem. Commun.*, **1993**, 1770.

[216] (a) A. J. Fry, S. B. Sobolov, M. D. Leonida, K. I. Voivodov, *Denki Kagaku*, **62**, 1260 (1994); (b) *idem, Tetrahedron Lett.*, **35**, 5607 (1994).

[217] M. V. Pishko, I. Katakis, S.-E. Lindquist, A Heller, Y. Degani, *Mol. Cryst. Liq. Cryst.*, **190**, 221 (1990).

[218] (a) H. Yanai, K. Miki, T. Ikeda, K. Matsushita, *Denki Kagaku*, **62**, 1247 (1994); (b) T. Ikeda, D. Kobayashi, F. Matsushita, T. Sagara, K. Niki, *J. Electroanal. Chem.*, **361**, 221 (1993).

[219] F. Mizutani, Y. Shimura, K. Tsuda, *Chem. Lett.*, **1984**, 199.

[220] M. Kijima, A. Sakawaki, S. Sato, *Bull. Chem. Soc. Jpn.*, **67**, 2323 (1994).

[221] N. Oyama, S. Ikeda, O. Hatozaki, M. Shimomura, K. Mishima, S. Nakamura, *Bull. Chem. Soc. Jpn.*, **66**, 1091 (1993).

[222] J. Haladjian, P. Bianco, L. Asso, F. Guerlesquin, M. Bruschi, *Electrochim. Acta*, **31**, 1513 (1986).

[223] P. D. Barker, K. Di Gleria, H. A. O. Hill, V. J. Lowe, *Eur. J. Biochem.*, **190**, 171 (1990).

[224] N. Nakashima, K. Abe, T. Hirohashi, K. Hamada, M. Kunitake, O. Manabe, *Chem. Lett.*, **1993**, 1021.

[225] L. Jiang, C. J. McNeil, J. M. Cooper, *Angew. Chem., Int. Ed. Engl.*, **34**, 2409 (1995).

[226] C. Bourdillon, C. Demaille, J. Moiroux, J.-M. Savéant, *J. Am. Chem. Soc.*, **117**, 11499 (1995).

[227] Y. Nakamura, J.-Y. Cheng, I. Tabata, S. Suye, M. Senda, *Denki Kagaku*, **62**, 1235 (1994).

[228] (a) Y. Shimizu, A. Kitani, S. Ito, K. Sasaki, *Denki Kagaku*, **61**, 872 (1993); (b) *idem, ibid.*, **62**, 1233 (1994).

[229] (a) J. K. Chen, K. Nobe, *J. Electrochem. Soc.*, **140**, 299 (1993); (b) *idem, ibid.*, **140**, 304 (1993).

[230] J. Cantet, A. Bergel, M. Comtat, *Bioelectrochem. Bioenergetics*, **27**, 475 (1992).

[231] J. Hirst, A. Sucheta, B. A. C. Ackrell, F. A. Armstrong, *J. Am. Chem. Soc.*, **118**, 5031 (1996).

[232] (a) T. Matsue, H.-C. Chang, I. Uchida, T. Osa, *Tetrahedron Lett.*, **29**, 1551 (1988); (b) H.-C. Chang, T. Matsue, I. Uchida, T. Osa, *Chem. Lett.*, **1989**, 1119.

[233] G. Pierre, A. Ziade, Nouv. *J. Chem.*, **10**, 233 (1986).

[234] (a) V. D. Parker, *Acta Chem. Scand.*, **B35**, 147 (1981); (b) *idem, ibid.*, **B37**, 393 (1983).

[235] P. Tissot, J.-P. Surbeck, F. O. Gülacar, *Helv. Chim. Acta*, **64**, 1570 (1981).

[236] S. Fujii, R. Inaba, *Denki Kagaku*, **53**, 333 (1985) (in Japanese).

[237] (a) C. I. De Matteis, J. H. P. Utley, *J. Chem. Soc., Perkin Trans. II*, **1992**, 879; (b) J. H. P. Utley, M. Güllü, C. I. De Matteis, M. Motevalli, M. F. Nielsen, *Tetrahedron*, **51**, 11873 (1995).

[238] (a) A. Kunugi, K. Minami, M. Yasuzawa, K. Abe, T. Hirai, *Chem. Express*, **4**, 189 (1989); (b) A. Kunugi, N. Takahashi, K. Abe, T. Hirai, *Bull. Chem. Soc. Jpn.*, **62**, 2055 (1989).

[239] H. L. S. Maia, L. S. Monteiro, F. Degerbeck, L. Grehn, U. Ragnarsson, *J. Chem. Soc., Perkin Trans. II*, **1993**, 495.

[240] W. A. Pritts, D. G. Peters, *J. Electroanal. Chem.*, **380**, 147 (1995).

[241] G. Farnia, F. Marcuzzi, G. Melloni, G. Sandona, M. V. Zucca, *J. Am. Chem. Soc.*, **111**, 918 (1989).

[242] D. A. Van Galen, J. H. Barnes, M. D. Hawley, *J. Org. Chem.*, **51**, 2544 (1986).

[243] W. Jugelt, U. Dünnbier, *Z. Chem.*, **30**, 173 (1990).

[244] U. Dünnbier, W. Jugelt, *Pharmazie*, **46**, 512 (1991).

[245] A. Kikuchi, T. Fukumoto, K. Umakoshi, Y. Sasaki, A. Ichimura, *J. Chem. Soc., Chem. Commun.*, **1995**, 2125.

[246] (a) R. M. Bastida, E. Brillas, J. M. Costa, *J. Electroanal. Chem., Interficial Electrochem.*, **227**, 55 (1987); (b) idem, *J. Electrochem. Soc.*, **138**, 2289 (1991); (c) *idem, ibid.*, **138**, 2296 (1991).

[247] P. Manisankar, A. Gomathi, T. Vasudevan, D. Yelayutham, R. K. Srinivasan, S. Chidambaram, *Trans SAEST*, **28**, 152 (1993).

[248] R. Annino, R. J. Boczkowski, D. J. Bolton, W. E. Geiger Jr., D. T. Jackson Jr., J. Mahler, *J. Electroanal. Chem., Interfacial Electrochem.*, **38**, 403 (1972).

[249] J. Yue, A. J. Epstein, *J. Chem. Soc., Chem. Commun.*, **1992**, 1540.

[250] G. Cheek, C. P. Wales, R. J. Nowak, *Anal. Chem.*, **55**, 380 (1983).

[251] I. Rubinstein, *Anal. Chem.*, **56**, 1135 (1984).

[252] A. J. Bard, *Anal. Chem.*, **35**, 1125 (1963).

[253] (a) A. Chyla, J. P. Lorimer, T. J. Mason, G. Smith, D. J. Walton, *J. Chem. Soc., Chem. Commun.*, **1989**, 603; (b) R. G. Compton, J. C. Eklund, S. D. Page, G. H. W. Sanders, J. Booth, *J. Phys. Chem.*, **98**, 12410 (1994); (c) H. Zhang, L. A. Coury, *Anal. Chem.*, **65**, 1552 (1993); (d) D. J. Walton, S. S. Phull, D. M. Bates, J. P. Lorimer, T, J. Mason, *Electrochim. Acta*, **38**, 307 (1993).

[254] P. R. Birkin, S. S.-M. Martinez, *J. Chem. Soc., Chem. Commun.*, **1995**, 1807.

[255] T. J. Mason, J. P. Lorimer, D. J. Walton, *Ultrasonics*, **28**, 333 (1990) and references cited therein.

[256] (a) K. Matsuda, M. Atobe, T. Nonaka, *Chem. Lett.*, **1994**, 1619; (b) M. Atobe, T. Nonaka, *Chem. Lett.*, **1995**, 669.

[257] M. Atobe, K. Matsuda, T. Nonaka, *Denki Kagaku*, **62**, 1298 (1994).

[258] (a) R. E. White (ed.), Electrochemical *Cell Design*, Plenum Press, New York, London (1983); (b) D. Pletcher, F. C. Walsh (eds.), *Industrial Electrochemistry*, Blackie Academic & Professional, London, Glasgow, New York, Tokyo, Melbourne, Madras (1993).

[259] (a) C. Lamoureux, C. Moinet, *Bull. Soc. Chim. Fr.*, **1988**, 59; (b) C. Gault, C. Moinet, *Tetrahedron*, **45**, 3429 (1989); (c) C. Moinet, G. Simonneaux, M. Autret, F. Hindre, M. Le Plouzennec, *Electrochim. Acta*, **38**, 325 (1993).

[260] G. Jacob, C. Moinet, A. Tallec, *Electrochim. Acta*, **28**, 635 (1983).

[261] J. Chaussard, M. Troupel, Y. Robin, G. Jacob, J. P. Juhasz, *J. Appl. Electrochem.*, **19**, 345 (1989).

[262] D. Pletcher, J. T. Girault, *Inst. Chem. Eng. Symp. Ser.*, **98**, 13 (1986).

[263] S. Hamakawa, T. Hayakawa, A. P. E. York, T. Tsunoda, Y. S. Yoon, K. Suzuki, M. Shimizu, K. Takehira, *J. Electrochem. Soc.*, **143**, 1264 (1996).

[264] J. E. Anderson, E. T. Maher, *Anal. Chem.*, **63**, 2073 (1991).

[265] A. Factor, T. O. Rouse, *J. Electrochem. Soc.*, **127**, 1313 (1980).

[266] (a) K. Ogura, M. Takagi, *J. Electroanal. Chem., Interfacial Electrochem.*, **201**, 359 (1986); (b) *idem, ibid.*, **195**, 357 (1985).

[267] M. Schwartz, R. L. Cook, V. M. Kehoe, R. C. MacDuff, J. Patel, A. F. Sammells, *J. Electrochem. Soc.*, **140**, 614 (1993).

[268] S. Kuwabata, K. Nishida, R. Tsuda, H. Inoue, H. Yoneyama, *J. Electrochem. Soc.*, **141**, 1498 (1994).

[269] N. L. Weinberg, D. J. Mazur, *Kagaku to Kogyo*, **43**, 2002 (1990).

[270] F. Beck, *Kagaku to Kogyo*, **43**, 1997 (1990).

[271] I. Nishiguchi, K. Yamataka, M. Taniguchi, S. Takenaka, *Kagaku to Kogyo*, **43**, 1992 (1990) (in Japanese).

[272] D. Petcher, F. C. Walsh, *Industrial Electrochemistry*, Chapman & Hall, New York (1990).

[273] T. J. Mazanec, T. L. Cable, J. G. Fry Jr., *Solid State Ionics*, **53**, 111 (1992).

[274] P. Tatapudi, J. M. Fenton, *J. Electrochem. Soc.*, **140**, L55 (1993).

[275] C. J. Brown, D. Pletcher, F. C. Walsh, J. K. Hammond, D. Robinson, *J. Appl. Electrochem.*, **24**, 95 (1994).

[276] A. Morduchowitz, A. F. Sammells, *U. S. Pat.* 4560450 (1985).

[277] W. L. Xu, P. Ding, W. K. Yuan, *Chem. Eng. Sci.*, **47**, 2307 (1992).

[278] J. E. Toomey Jr., G. A. Chaney, M. Wilcox, *Stud. Org. Chem.*, **30**, 245 (1987).

[279] C. Laane, A. Weyland, M. Franssen, *Enzyme Microb. Technol.*, **8**, 345 (1986).

[280] R. G. Compton, J. C. Eklund, L. Nei, *J. Electroanal. Chem.*, **381**, 87 (1995).

2. Electroreduction of Aldehydes, Ketones, Acids, Esters and Acids Anhydride

2.1. Electroreduction of Carbon Oxides

2.1.1 Electroreduction of Carbon Dioxide

Carbon dioxide (CO_2) can be electroreduced to various organic compounds, i.e., carbon monoxide, formic acid, oxalic acid, glyoxalate, formaldehyde, methanol, ethanol, methane, ethane, etc. The product selectivity depends on the reaction conditions such as electrode materials, solvent systems, operational parameters, so on [1, 2]. The proper catalyst activity of the electrode materials in a solvent is especially important for product selectivity in the electroreduction of CO_2. Recently, light has been shed on the product distribution with regard to the mechanism. [3].

The carbon dioxide is equivalent to an HCO_3^- species in an aqueous medium, as shown in eq. (2.1) [4]. The reduction of HCO_3^- to HCO_2^- in aqueous media is known to

$$CO_2 + H_2O \; \xrightleftharpoons{} \; H_2CO_3 \; \underset{-H^+}{\overset{H^+}{\rightleftharpoons}} \; HCO_3^- \; \underset{-H^+}{\overset{H^+}{\rightleftharpoons}} \; CO_3^{2-} \qquad (2.1)$$

occur at electrode surfaces under high current efficiency [2b, 5]. The catalytic electrodes coated with a Pd-impregnated polymer has been found to be effective for the reduction of HCO_3^- to HCO_2^- [4]. Electrocatalytic reduction of CO_2 proceeds in a stepwise fashion according to (eq. (2.2)) [6]. The difficulties in the overall electrochemical reaction are the

$$CO_2 + 2e^- + 2H^+ \longrightarrow HCO_2H \qquad \cdots\cdots(a)$$

$$HCO_2H + 2e^- + 2H^+ \longrightarrow CH_2O + H_2O \qquad \cdots\cdots(b)$$

$$CH_2O + 2e^- + 2H^+ \longrightarrow CH_3OH \qquad \cdots\cdots(c) \qquad (2.2)$$

$$CH_3OH + 2e^- + 2H^+ \longrightarrow CH_4 + H_2O \qquad \cdots\cdots(d)$$

further reduction of formic acid or formate ions to formaldehyde or methanol. The reduction of formic acid at metal electrodes occurs only in a narrow potential range and at impractically small current densities [2b, 5b, 7]. In mildly acidic solutions, the reduction of H_3O^+ rather than formic acid proceeds favorably [8]. The characteristic features of electroreduction efficiencies on various metal electrodes and/or additions are shown in Table 2.1 [9~48].

Various investigations have shown that the electroreduction of CO_2 in aqueous weakly acidic and basic solutions using metal electrodes (Hg, Au, Pb, Zn, Cd, Sn, and In) having rather high overpotential for hydrogen evolution yields formic acid or formate ions as major products [49]. For the reduction of formic acid or formate to methanol, the use of neutral or alkaline media may minimize the reduction of H_3O^+ [8]. Metal electrodes (Zn, Cd, Hg(IIB), Tl, In (IIIA), Sn, Pb (IVA)) with aqueous tetraalkylammonium salt solutions result in the formation of carbon monoxide, formic, oxalic, and other carboxylic acids [50]. The

Table 2.1 Current (Faradaic, %)-Efficiency for Electroreduction Products of CO_2 at Various Metallic Electrodes

Group	Metal	HCOOH	(CO2H)2	CHO CO2H	OH MeCHCO2H	CO	CH4	C2H4	MeOH	EtOH	H2
IB	Cu	10.2[10] 32[39d] 80.4[34d]	57[25l] 70.7[34a] 34.6[34d]			16.5[10] 48[25l] 80[25l] 40[29] 33[39d] 46.6[34d] 64*[35l] 64.7[9] 40.7[10] 89.9[32a] 80[37] 99[47]	23[9] 35[10] 55[31] 55[32c] 39[39d] 73[35d] 40[32a,46]	40[9] 12.7[10] 25[29] 10[31] 30.5[32c] 41[39d] 25[35d] 40[46]	3[27l] 72[29]	10[25l] 10.9[32c]	52.0[10] 100[25l] 54.1[26]
	Ag	34.8[39d]									28[10] 50.0[10]
	*(Ag 0.05%)	20.5[10]									
	Au	10.3[32b,38]	10[32b,38]			81.5[9] 16.9[10] 91[32b,38] 93[35l] 82[47]					86[9] 23[9] 73.4[10] 9[32b,38]
IIB	Zn	20[9]	19.5[10] 50[39] 80[40]	48[39]		39.6[9] 80[25l] 65[39d] 90[47]	—	—			40[9]
	Cd	39[9]	55.9[10] 90.2[41a] 100[41a] 83[42]	35[41b]		97-98[39d]	—				39[9] 63.2[10]
	Hg	94[9] 100[15l] 65[44]	93[43] 73[43a]	58[41b]		69[43a]					68.1[10]
IIIA	Al	—	—			—	—				99[9] 95.7[10]
	Ga	—	—			—	—				91[9]
	Tl	53.4[10]									46.2[10]
	In	69[9] 70.0[10] 100[15l] 95[39d]	65.2[39a]			14.7[10]	—				25[9] 56.5[10]
IVA	Sn	63[9] 50.5[39a] 100[39c]	28.5[10] 100[39e]			99[47]	—				26[9] 94.9[10]
	Pb	50[9] 16.5[10]	76[39a] 86[39a]		54[41b]	—	—				41[9] 93.3[10]
	Si										102.2[10]
	C				63[41b]						

Table 2.1 Current (Faradaic, %)-Efficiency for Electroreduction Products of CO_2 at Various Metallic Electrodes (continued)

Group	Metal	HCOOH	(CO2H)2	CHO CO2H	OH MeCHCO2H	CO	CH4	C2H4	MeOH	EtOH	H2
IV B	Ti					13.5[10]	—	—			69.4[10]
	Zr										99.9[10]
	Hf										99.2[10]
V B	V	—					—	—			86[9] 91.9[10]
	Nb	—					—	—			97[9] 97.3[10]
	Ta	—					—	—			90[9] 102.2[10]
VI B	Cr	87[19b]					—	—	96.8[11]		99[9] 92.2[10]
	Mo	—					—	—	85[12]		103[9] 99.9[10]
	W	81[4]					—	—			102[9] 96.9[10]
VII B	Re	48[45a]				92.3[13]	—	—			98[9] 99.0[10]
	Mn										90.9[10]
VIII	Fe	100[19a] 51.4[45b]		72.2[11f]	59.5[11g]	40[14]	—	—	86.6[11]		104[9] 89.8[10]
	Co	64[13c, f] 60[18c]				20.1[15, 48]	—	—			102[9] 92.9[10]
	Rh		60[18c]				—	—			99.3[10]
	Ni	13.7[10] 23.2[15] 90[16]	70-90[22]			37.5[15] 21.0[10]	—	—		79[29b]	93[9] 98.8[10]
	Ru	84.3[20] 57[17] 76[20] 78[27b]				19.9[18a] 87.8[21]	—	—	84[29a]		111[9] 99.1[10] 50.7[18a]
	Os	12[22]				80[13c] 90[13d]	—	—			29.8[13c]
	Pd	16.1[10] 100[15b, 28] 21[30] 70.7[34a]				62.3[15] 85[24] 11.6[10] 37[36]	—	—	30[23b]		29[9] 90.3[10] 100[24] 68[36]
	Ir	99[33]					—	—			99[9]
	Pt	85[4] 24.1[15]				33.9[15]	—	—	50[23a]		99[9] 92.6[10]

Table 2.2 Correlation of Reaction Path with Electrode Material and Solvent System

Reaction Path	Electrode Materials	Solvent * System

H^+, e^- → $HCOO^-$

W, Re, Rh, Ni
Ru, Os, Pd, Pt
Cu, Ag, Au, Zn
Ga, Cd, Hg, Tl, In
Sn, Pb

aqueous
solutions
~pH 9.5

H^+, e^- → $CO + OH^-$

Ti, Re, Fe, Co, Ni
Ru, Os, Pd, Pt, Cu
Ag, Au, Zn, Hg, In

aqueous
solutions
~pH 6.0

$4H^+, 4e^-$ → $\bullet CH_2 + H_2O$

CO_2

e^-

$CO_2^{-\bullet}$

$2H^+, 2e^- \rightarrow CH_4$

$\bullet CH_2 \rightarrow C_2H_4$

$\bullet CH_2, H_2O \rightarrow CH_3OH$

Cr, Mo, Fe
Co, Cu
Cu/Ag
Enzyme

aqueous
solutions

- -

CO_2, e^- → $CO + CO_3^{2-}$

$CO_2^{-\bullet}$ → $(COO)_2^{2-}$

Re, Fe, Os
Au, Zn, In
Sn

DMF
MeCN
EtOH

$2H^+, 2e^-$ →

$HCO + OH^-$
|
COO^-

$2H^+, 2e^-$ →

H_2COH
|
COO^-

Rh, Zn, Hg
Pb, Tl

aq. NH_4^+

DMF
DMSO
PC*

Acetonitrile = MeCN ; *N,N*-Dimethylformamide = DMF; Dimethyl Sulfoxide = DMSO;
Propylene Carbonate = PC*
[Reproduced with permission from K. Ito, *Denki Kagaku*, **58**, 984]

correlation of reaction paths with electrode materials and solvent systems is outlined in Table 2.2 [9]. The electrode metal-catalyzed reduction of CO_2 can be classified into three major groups based on their products from CO_2. The In and Pb together with Hg electrodes are useful for the selective formation of HCO_2H in aqueous media. In non-aqueous media, Pb, Hg, and Tl tend to give oxalic acid preferentially and In, Zn, Au, and Sn electrodes can be used for CO production [51]. The efficient conversion of CO_2 into formic acid at a mercury cathode has been accomplished in an aqueous sodium hydrogen carbonate solution in 65% current efficiency (current eff.) [44b]. Using platinum electrodes, CO_2 can be electroreduced to formic acid in aqueous solutions containing various kinds of viologen derivatives [44a]. Metal phthalocyanines (metal = Co, Ni) deposited on carbon electrodes are found to catalyze the electroreduction of carbon dioxide to formic acid at pH 3~7 [52]. The complex *cis*-[Os(bpy)₂(CO)H]⁺[PF₆]⁻ has been found to be an electrocatalyst for CO_2 reduction in an MeCN-Bu₄NPF₆(0.1M)-(Pt) system [23]. Gallium arsenide electrodes have been shown to be efficient for the selective reduction of CO_2 into formic acid [6]. The Cd, In, Sn, and Pb electrodes preferentially yield formate (65~97% current eff.) together with a small amount of CO, methane, and hydrogen. However, formic acid is the only product in the photoelectrolytic reduction of CO_2 at a p-type gallium phosphide (GaP) photoelectrode in aqueous electrolytes at -1.2 V (Ag/AgCl) [53]. Polypyridyl complexes of Rh and Ru are used for the CO_2 reduction, leading to formate (64% current eff.) [17]. Under anhydrous conditions, carbon monoxide (CO) is the dominant product, but addition of water results in up to a 25% ratio of formate. Addition of weak Brönsted acids such as 1-propanol, 2-pyrrolidone, and CF_3CH_2OH triggers a considerable improvement of the catalyst of CO_2 reduction by iron(0)tetraphenylporphyrins. Both the catalytic currents and the life time of the catalyst increase without significant formation of hydrogen [54].

Oxalate and glyoxylate are electrosynthesized in an aqueous solution containing tetramethylammonium ions at pH 9. The reaction proceeds in two steps to yield oxalic acid (100% yield) on carbon at -0.9 V (Ag/AgCl) and glyoxylic acid (35% yield) on mercury electrode at -1.8 V (eq. (2.3)) [41]. Electrosynthesis of oxalic acid on a preparative

$$2\,CO_2 \xrightarrow{2e^-} \begin{matrix} CO_2^- \\ | \\ CO_2^- \end{matrix} \xrightarrow[3H^+]{2e^-} \begin{matrix} CH(OH)_2 \\ | \\ CO_2^- \end{matrix} \rightleftharpoons \begin{matrix} CHO \\ | \\ CO_2^- \end{matrix} + H_2O \qquad (2.3)$$

$$\text{oxalate} \qquad\qquad\qquad\qquad \text{glyoxylate}$$

scale has been performed in a DMF-Bu_4ClO_4-(Pb) system in the presence of zinc salts [40]. The conversion yield is ca. 85% with an 80% current eff. Oxalate formation in electrochemical CO_2 reduction can be attained in 60% current eff. by use of rhodium-sufur cluster catalysts at -1.50 V (SCE) [18]. Metal electrodes in non-aqueous media yield carbon monoxide, formic, and oxalic acids [2a, 39a, 55]. The electroreduction of CO_2 in a DMSO (or propylene carbonate)-Et_4NClO_4-(Pb) system yields oxalic acid as a major product [39a]. Electrolysis of CO_2 in a DMF-Et_4NBr(0.2 M)-(Hg) system at $-2.15 \sim -2.20$ V (SCE) gives oxalate as a major product (83% current eff.) [42]. Triangular metal-sulfide clusters, $[\{Ir(C_5Me_5)\}_3(\mu3\text{-}S)_2]^{2+}$ and $[\{Co(C_5H_4Me)\}_3(\mu3\text{-}S)_2]^{2+}$, catalyze the electrochemical CO_2 reduction to selectively produce oxalate at -1.30 and -0.70 V (Ag/AgCl), respectively, in MeCN [18b]. A process and apparatus for converting metallic elements to metallic oxalates in the prsence of CO_2 in a two-redox couple electrolyte solution separated by the membrane having photosensitizers are patented [56]. (2,2'-Bipyridine)tricarbonylchlororhenium(I) has been used as a homogeneous mediator catalyst for the conversion of CO_2 to CO [57]. Also, $[Re(bpy)(CO)_3Br]$ (bpy = 2,2'-bipyridine) and $[Re(terpy)(CO)_3Br]$ (terpy = 2,2':6',2''-terpyridine) complexes are found to be effective as efficient catalysts for CO_2 electroreduction to form formic acid and CO in an aqueous medium when incorporated into a coated Nafion membrane [45].

Electrocatalytic reduction of CO_2 to methanol has been intensively investigated [58a]. Electrolysis using molybdenum electrodes in an H_2O-Na_2SO_4(0.2M)-CO_2 system at pH 4.2 produces methanol as a major product (85% faradaic efficiency) at room temperature and -0.7 to -0.8 V (SCE) [12]. The reduction of CO_2 on Cu-Ni alloys gives rise to the formation of methanol [28]. The catalytic reductions of CO and CO_2 to methanol on a mediated electrode by surface-confined iron metal complexes [58b], by quinone immobilized electrodes [58c], and by dual-film electrodes modified with and without cobalt(II) and iron(II) complexes [59] have been demonstrated. A fuel system consisting of hydrogen and CO_2 as an oxidant provides methanol as a reduction product. The carbon dioxide is reduced catalytically with an electrode mediator and homogeneous catalysts. The platinized platinum gauze anode and a platinum gause cathode modified with Everitt's salt, $K_2Fe(II)$ [Fe(II)(CN)$_6$], are used for the electroreduction of CO_2 in an MeOH-KCl (0.1M) system in the presence of 1-nitroso-2-naphthol-3,6-disulfonatocobalt(II) and/or aquapentacyanoferrate(II) to give methanol [60]. Carbon dioxide has been reduced, giving methanol at the Everitt's salt-mediated electrode in the presence of 1,2-dihydroxybenzene-3,5-disulphonato-ferrate(III) complex in ethanol [61]. Electroreductive conversion of CO_2 to methanol with the aid of formate dehydrogenase and methanol dehydrogenase as biocatalysts has been studied [62]. The current efficiencies of methanol formation with an electrochemical pho-tocell in the presence of pentacyanoferrate(II), pentachlorochromate(III), bis(oxalato)chro-mate(III) and aminepentacyanoferrate(II) are 86.6, 83.3, 96.8, and 61.7% (current eff.), re-spectively [11a, b]. Photoinduced reduction of formate to methanol is achieved using ZnS microcrystalline colloid as a photocatalyst in the presence of methanol dehydrogenase [63]. Electrochemical fixation of CO_2 in pyruvic acid to yield malic acid using malic enzyme as an electrocatalyst has also been reported [64].

The Ag and Au cathodes give principally carbon monoxide in 61~93% current efficien-cies [36]. The gaseous products at the Cu electrode contain significant amounts of methane (40% current eff.) and hydrogen (33% current eff.). The Ni and Fe electrodes provide mostly hydrogen gas in 96~97% current efficiencies [36]. Mechanistic and kineti-cal studies of the electroreduction of CO_2 in an aqueous solution on metal electrodes with high and moderate hydrogen overvoltages (Sn, In, Bi, Sb, Cd, Zn, Cu, Pb, Ga, Ag, Au, Ni, Fe, W, Mo, and glassy carbon) reveal that the optimal electrodes for the reduction are those with moderate hydrogen overvoltages [65]. Chemisorbed species of CO_2 in aqueous solu-tions are suggested to be produced preferentially on Pt and Rh electrodes, and the interac-tion of the chemisorbed CO_2 and hydrogen may provide hydrocarbon products. Such inter-action has not been observed for Ir, Pd, Os, or Ru [66].

Carbon monoxide is produced at Sn and In electrodes as major products in the above electrolysis system [39b]. A polymer film of Re(bpy)(CO$_3$)Cl complexes on a Pt electrode can reduce CO_2 to CO monoxide with greatly improved turnover numbers [13b] in compar-ison with those observed for the analogous reduction with Re(bpy)(CO)$_3$Cl complexes [13a]. In aprotic solvents with either Hg, Pb, Sn, In or Pt, the rate-determining step is the transfer of the second electron to the (CO$_2$)$_2^{-\cdot}$ anion radical formed as a result of the inter-action of the initially generated $CO_2^{-\cdot}$ anion radical with the absorbed CO_2 molecule [67]. An irreversible absorption on the platinized titanium dioxide film electrodes is observed in aqueous media [68]. Both electroplated and teflon-supported Ru electrodes are used for CO_2 electrolysis, leading to CH_4, MeOH, and CO [69]. The controlled potential electroly-sis of CO_2 in an aqueous EtOH (80%)-[Ru(bpy)(trpy)(CO)]$^{2+}$–(C) system at −1.70 V (Ag/Ag$^+$) affords a mixture of HCO$_2$H, CH$_2$(OH)CO$_2$H, CH$_2$O, CHO-CO$_2$H, MeOH, and CO [70a]. The electroreduction of CO_2 in the presence of [Ru(bpy)$_2$(CO)$_2$](PF$_6$)$_2$ affords CO and HCO$_2$H under the similar conditions [70b]. Methanol as solvent is found to be ef-fective for the CO production in the electroreduction of CO_2 catalyzed [RuL^1(L^2)(CO)$_2$]$^{2+}$ [L^1, L^2 = (bpy)$_2$ (bpy = 2,2'-bipyridine), (bpy)(dmbpy) (dmbpy = 4,4'-dimethyl-2,2'-bipyri-dine), (dmbpy)$_2$, or (phen)$_2$ (phen = 1,10-phenanthroline)], [Ru(phen)$_2$(CO)Cl]$^+$, and [RuL(CO)$_2$Cl$_2$] at −1.30 V (SCE) [21] (see Section 2.1.2). Iron(0) porphyrin complexes catalyze the electroreduction of CO_2 to give CO as a major product [71]. The presence of a hard electrophile such as Mg^{2+} ion dramatically improves the rate of the reaction. The electroreduction of CO_2 to CO at Ni electrodes modified with Cd has been effectively at-tained in an aqueous solution, although almost no CO_2 is reduced at Ni electrodes [72].

Carbon dioxide electroreduction mediated by electropolymerized electrodes of a nickel tetraazaannulene complex, Ni[Me$_4$Bzo$_2$[14]tetraeneN$_4$], has been also investigated [16].

Copper electrodes tend to form hydrocarbons such as CH$_4$ and C$_2$H$_4$ with high current efficiencies [32, 36, 72, 74]. The Cu/Ag alloy (2/3 atomic ratio) electrode gives a high current efficiency for ethylene and a lower one for methane [29a]. A detail comparison of various Cu alloy electrodes for CO$_2$ electroreduction has been investigated [25]. Electrocatalytic activities of Cu-Sn and Cu-Zn alloys for CO$_2$ reduction depend on their microcrystalline phases for the selective formation of CO with a high reaction rate (ca. 80% current eff.) [30]. The electroreduction of CO$_2$ on Cu electrodes by the pulsed method proceeds with high faradaic efficiency for the generation of CH$_4$ and C$_2$H$_4$ at –2.6 V (SCE) [31]. The electroreduction of CO$_2$ to methane and ethylene has been performed using a copper/Nafion electrode (solid polymer electrolyte structures) [75]. The photoelectrochemical and electrochemical reduction of CO$_2$ at various semiconductor electrodes results in the formation of methanol, formaldehyde, formic acid, methane, and carbon monoxide [4a, 76a~f]. Liquid CO$_2$ is reduced electrochemically at a Cu electrode producing CO, CH$_4$, C$_2$H$_4$, and HCO$_2$Me [76g].

One of the problems in the CO$_2$ reduction comes from its large overpotential. Indeed, potentials farther negative than –2 V (SCE) are required in preparative-scale electrolyses. A great deal of effort has, therefore, been devoted to searching for the appropriate catalysis to reduce the potential. Transition metal complexes are reported to catalyze the electroreduction of CO$_2$ to produce carbon monoxide efficiently [43, 77, 78]. The catalytic electroreduction of CO$_2$ with [Ru(bpy)$_2$(CO)$_2$](PF$_6$)$_2$ in an aqueous DMF solution is influenced by pH value in the media. For example, formic acid, carbon monoxide, and hydrogen are formed in an alkaline solution (pH 9.5) while in acidic solution (pH 6.0) carbon monoxide and hydrogen are only evolved [79]. The transition-metal complexes of tetraazamacrocycles including phthalocyanines or tetraphenylporphyrins have been found to act as electrocatalysts for CO$_2$ reduction in aqueous systems [77a, 80]. Quite a different type of complex, the tetranuclear ion-sulfur cluster, [Fe$_4$S$_4$(SR)$_4$]$^{2-}$, catalyzes the electroreduction of CO$_2$ in a non-aqueous DMF solution [43].

2.1.2 Electroreductive Incorporation of Carbon Dioxide

The electrochemical carboxylation for producing carboxylic acids under estimating the preparative possibilities of this method has been well documented [81].

The cathodic coupling of carbon dioxide and butadiene in acetonitrile in undivided parallel plate cells has been discussed [82a]. The cathodic coupling of carbon dioxide and butadiene in a bipolar fluidized bed cell constructed from vitreous carbon microspheres between two feeder electrodes has been attempted [82b]. The cell gives a high space time yield and the product is a mixture of two isomeric pentenoic acids, two hexenedioic acids and three decadienedioic acids which can be formed in a reasonable current yield.

The incorporation of CO$_2$ into non-activated alkynes is found to be catalyzed by electrogenerated Ni(0)(bpy)$_2$ complexes in the presence of a sacrificial magnesium anode, affording α,β-unsaturated acids in moderate to good yields. The formation of acids from alkynes is stoichiometric with respect to the nickel complex if performed in a two-compartment cell, but can be made catalytic in a single-compartment cell. An intermediate nickelacycle **3** can be isolated from the reaction with 4-octyne **1** (R = C$_3$H$_7$). The cleavage of this metallacycle **3** by magnesium ion is the key step to explaining the catalytic cycle, as shown in a mechanistic cycle presented in eq. (2.4) [83a]. The activation of CO$_2$ and diynes by electrogenerated Ni(0)L complexes (L = bpy, PMDTA(= pentamethyldiethylenetriamine)) leads to the selective incorporation of one molecule of CO_2 into unsaturated systems. A series of non-conjugated diynes affords selectively linear or cyclic adducts depending on the ligand [83b]. For instance, the carboxylation of 1,6-diyne **4** using complex having bpy as the ligand gives selectively cyclic carboxylic acid **5** (eq. (2.5a)), whereas with complex having PMDTA, linear unsaturated acid **7** is formed preferentially (eq. (2.5b))[83b]. The Ni(II) complex associated to PMDTA is proved to be the catalyst precursor for the fixation of CO_2 into alkynes in electrolyses using a magnesium anode [83c]. The electrosynthesis of 2-vinylidene-3-yne carboxylic acids **9** from CO_2 and substituted 1,3-diyne **8** is attained in a DMF-Bu$_4$NBF$_4$-(Mg) system by use of a nickel-triamine complex (eq. (2.6))[83d]. The hydrocarboxylation of the triple bond occurs through stereoselective cis addition. A large-scale stainless steel pressurized production cell for the CO_2-incorporation procedure has been proposed [84]. The CO_2-incorporation method with the electrogenerated Ni(0)(bipy)$_2$ complexes can be extended to a series of 1,3-enynes [85].

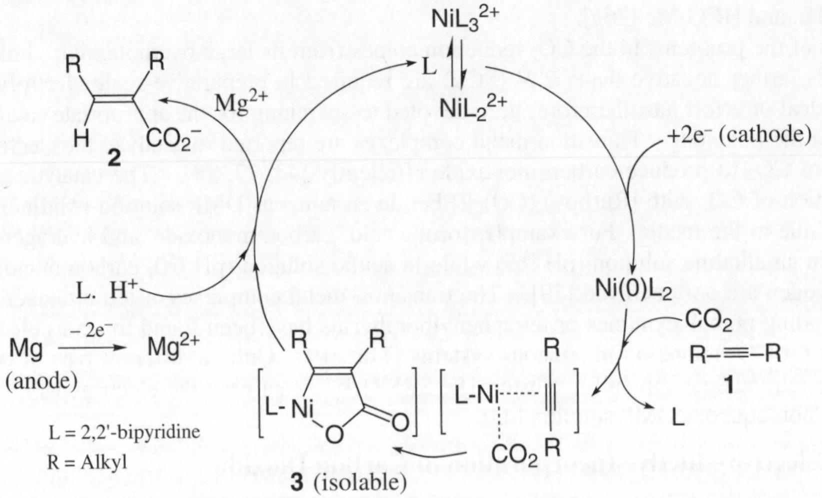

(2.4)

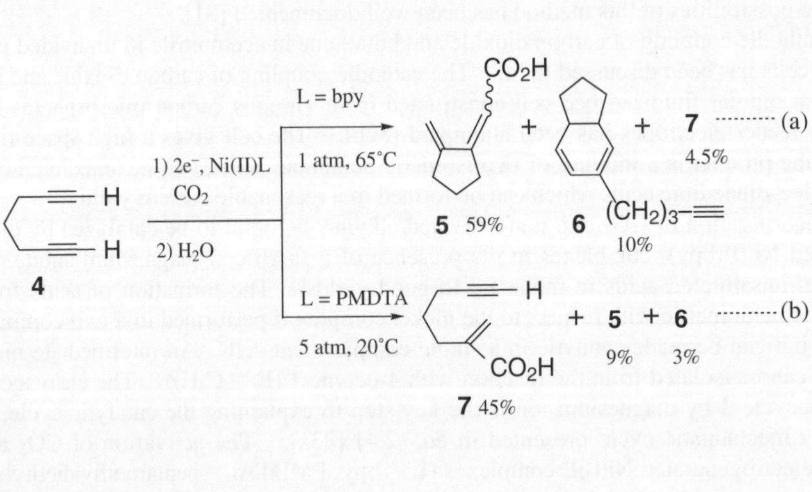

(2.5)

$$\text{(2.6)}$$

Unsaturated carboxylic acids **11** and **12** are obtained by reductive electrocarboxylation of allenes **10** with nickel(II) complexes as the catalyst in the presence of a magnesium anode (eq. (2.7a)) [86a]. Electrocarboxylation of styrene **13** is carried out in a DMF (40 ml)-Bu$_4$NBF$_4$ (0.3 mmol)-(Mg) system in the presence of NiBr$_2$(dme) (dme = dimethoxyethane) and PMDTA under CO$_2$ (5 atm) to give dicarboxylic acid **14** in 85% yield (eq. (2.7b)) [86b]. Cyclic carbonates **16** are electrosynthesized in good yields from terminal epoxides **15** by a Ni(cyclam)Br$_2$ catalyzed CO$_2$-fixation (eq. (2.8)) [87]. Carbon dioxide fixation to methyl acrylate can be performed in an MeCN-Bu$_4$NBF$_4$-(C) system in the presence of [Bu$_4$N]$_3$[Mo$_2$Fe$_6$S$_8$(SEt)$_9$] complex as a catalyst [88]. The above complex catalyst has been shown to be effective for electroreductive CO$_2$-fixation to thioester catalyzed by [Mo$_2$Fe$_2$S$_8$(SEt)$_9$]$^{3-}$ [89].

$$\text{(2.7)}$$

Methyl 2-thiophenecarboxylate **17** can be carboxylated in a DMF-Bu$_4$NI(0.2M)-(Hg) system at −1.95 V (SCE) to give the acid **18** in 78~80% yields (eq. (2.9)) [90]. Electrochemically activated CO$_2$ reacts with amines (RR'NH) and ethyl iodide in an MeCN-Et$_4$NClO$_4$(0.1M)-(Pt-Cu) system to afford the corresponding carbamate (RR'NCO$_2$Et) in 63-92% yields (eq. (2.10)) [91].

$$\text{(2.9)}$$

$$CO_2 \xrightarrow[\text{63-92\% yields}]{+e^-,\ \text{MeCN-Bt}_4\text{NClO}_4(0.1M)\text{-(Pt/Cu)}} RR'NCO_2Et \tag{2.10}$$

R = H, Me; R' = alkyl, Ph, Bn

Catalytic formation of ketones via double alkylation of CO derived from the electroreductive disproportionation of CO_2 occurs in a DMSO/MeCN-Me$_4$NBF$_4$-(C) system in the presence of [Ru(bpy)$_2$(qu)(CO)]$^{2+}$ (bpy = 2,2'-bipyridine, qu = quinoline) [92].

The carbon dioxide anion radical (eq. (2.11a)) catalyzing coupling reaction of pyridine **19** has been shown to proceed in an MeCN-Et$_4$NBr-(Pb) system in the presence of pyridine (0.1M) under continuously bubbling of CO_2 to give the dimer **21** (eq. (2.11)) [93]. The nickel-catalyzed electrocarboxylation of aromatic halides is discussed in Chapter 7 [94]. The reductive addition of CO_2 to quinones has been investigated in acetonitrile [95] and in dimethyl sulfoxide [96].

$$CO_2 + e^- \longrightarrow CO_2^{-\bullet} \qquad \text{············ (a)}$$

$$\text{(b)}$$

$$\text{(c)}$$

$$\tag{2.11}$$

The [Ru(bpy)$_2$(qu)(CO)]$^{2+}$ complex catalyzes electroreductive disproportionation of CO_2 to give CO (78% current eff.) and CO_3^{2-} in an MeCN-LiBF$_4$-(C) system (eq. (2.12a)), while the same reduction in a DMSO/MeCN-Me$_4$NBF$_4$-(C) system affords CH_3COCH_3 (16% current eff.), $CH_3COCH_2CO_2^-$ (5.8%), HCO_2^- (6.7%), and CO (42%) (eq. (2.12b,c)) , respectively [97].

$$2CO_2 \xrightarrow{2e^-} CO + CO_3^{2-} \qquad \text{·········(a)}$$

$$2CO_2 + 2(CH_3)_4N^+ \xrightarrow{4e^-} CH_3C(O)CH_3 + CO_3^{2-} \text{·········(b)}$$
$$+ 2(CH_3)_3N$$

$$\tag{2.12}$$

$$CH_3C(O)CH_3 + 2CO_2 \xrightarrow{2e^-} CH_3C(O)CH_2COO^- \qquad \text{·········(c)}$$
$$+ HCOO^-$$

2.1.3   Electroreduction of Carbon Monoxide

Carbon monoxide (CO) is a material of primary importance as C$_1$ chemistry from the standpoint of utilization of carbon resources.   The electroreduction of CO in an MeCN-LiClO$_4$

(0.1 M)/HClO$_4$-(Pt) system containing a trace amount of water on a bright and platinized platinum electrode at -2.5 V vs. Ag/Ag$^+$ can lead to the formation of methanol (17% current eff.) in an undivided cell [98]. Carbon dioxide was not electroreduced at Fe electrode, whereas CO can be reduced to CH$_4$, C$_2$H$_4$, and C$_2$H$_6$ [99]. The current efficiency of the hydrocarbon formation amounts to 10%. The carbon monoxide is adsorbed on the electrode, which prevents oxidation of iron. An effective cathodic reduction of CO to afford hydrocarbons, i.e., CH$_4$ and C$_2$H$_4$, and alcohols has been attained at a Cu cathode in aqueous solutions [100]. The solid-state electrochemical technique can significantly enhance the rate of hydrogenation of CO, leading to CH$_4$ over Fe, Ni, and Co catalyst [101].

Carbon monoxide can be converted into methanol by using an electrochemical photocell composed of an n-CdS photoanode and Everitt's salt-modified platinum cathode. The current efficiency of the methanol formation is almost 100% [102]. Everitt's salt [ES, K$_2$Fe(II)Fe(II)(CN)$_6$] can be provided by the electrochemical reduction of Prussian blue [PB, KFe(III)Fe(II)(CN)$_6$] (eq. (2.13)), which is electroplated on a platinum plate from a fresh solution of FeCl$_3$ and K$_3$Fe(CN)$_6$ [103, 104]. Pentachlorochromate(III),

$$CO + 4H^+ + 4ES \xrightleftharpoons[4e^-]{} MeOH + 4PB + 4K^+ \tag{2.13}$$

bis(oxalato)chromate(III) and iron(III)-tiron complexes in ethanol are also used in the photocell [105]. The net process involving a recycling system of the iron complexes ($22 \rightarrow 23 \rightarrow 24 \rightarrow 25$) for the reduction of CO to methanol is shown in eq. (2.14). This reaction can be activated by a homogeneous catalyst system consisting of metal complex and methanol. A combination of methanol and [Fe(CN)$_5$]$^{3-}$ operates as a homogeneous catalyst, and the formation of methyl formate is assumed as a catalytic complex. As shown in eq. (2.14),

(2.14)

the methanol formation terminates when the oxidation of ES reaches equilibrium. However, the conversion continues under the consumption of a proton in the solution under controlled potential conditions. This external energy is consumed by the reduction of PB to ES [106]. A 5,10,15,20-tetraphenylporphyrinatoiron(II)⁻ coated *p*-gap-glassy carbon electrode can be used for the reduction of CO to methanol in aqueous media [107]. Photoelectrochemical reduction of CO using a *p*-Si semiconductor electrode in an aqueous sulfuric acid solution affords formaldehyde (eq. (2.15)) [108].

$$
CO
\begin{cases}
\xrightarrow[\text{100\% c.e.}]{\text{n-CdS Photocell}} MeOH \quad \text{------------} \quad (a) \\
\\
\xrightarrow{\text{aq. H}_2\text{SO}_4\text{–p–Si Semiconductor}} HCHO \quad \text{------------} \quad (b)
\end{cases}
\qquad (2.15)
$$

2.2   Alcohols from Aldehydes and Ketones by Electroreduction

2.2.1   Alcohols from Aliphatic Aldehydes and Ketones

The electroreduction of carbonyl compounds has been investigated under a wide range of conditions [109].

The electroreduction of formaldehyde, acetaldehyde and acetone into the corresponding alcohols has been attained using alcohol dehydrogenase as an electrocatalyst [110]. In all cases, the current efficiency and the conversion yield reach *ca.* 100%. 1,5-Dihydroflavin analogues can be used as a mediator for the electroreduction of acetaldehyde in an MeCN-HClO₄-(Au) system in 100% yield [111]. The cathodic reduction of oxalic acid, leading to glyoxylic acid has been intensively studied [112]. Suitable cathodes for the above reduction are high overvoltage metals such as lead, and graphite is also a preferred material [113]. Favorable conditions of the operation which give high yields and current efficiencies are temperatures of 20 °C and less acidic electrolyte (usually sulfate), and electrode potentials between –0.75 and –1.3 V (SCE). The preparative synthesis of glyoxylic acid by electroreduction of oxalic acid has been reported [113]. Electroreduction of 2-propenal leading to propanol has also been discussed [114].

Electroreduction of cyclohexanone **25** in a DMF-Bu₄NBF₄-(Hg) system at –3.1 V (SCE) affords cyclohexanol **26** in 75% yield (eq. (2.16)) [115]. Cyclohexanol can be also obtained in aqueous diglyme at –2.9 V in 73% yield. Cyclohexanone is preferentially

$$
\text{25} \quad \underset{\text{Conditions A or B}}{\overset{2e^-}{\longrightarrow}} \quad \text{26} \qquad (2.16)
$$

Conditions A: –3.1 V *vs.* SCE, DMF-Bu₄NBF₄-(Hg) :   75%
B: –2.9 V *vs.* SCE, H₂O/Diglyme-Bu₄NBF₄-(Hg): 73%

and more readily reduced than cyclopentanone under acidic conditions. Use of cadmium cathode in an aqueous EtOH(30%)-H₂SO₄ (0.13M) system improves the current efficiency up to 90% [116]. The reduction of 6-hepten-2-one **27** to the alcohol **28** has been performed under varying electrolysis conditions (eq. (2.17)) (a) [115b, 117, 118]. Lead cathode is found to be effective for the electroreduction of 6-hepten-2-one, 6-heptyn-2-one, and

5-phenylpentan-2-one but not for other ketones [119].   The *i*-PrOH-LiClO₄-(Sn) system tends to give simple carbonyl reduction products, whereas tetraalkylammonium salts may provide intramolecular cyclization products (see section 2.4.2) [120].   Electroreduction of alicyclic ketones in an aqueous H₂SO₄-(Pt) system yields the corresponding alcohols and hydrocarbons [121].   Product-selective reduction of ketones and aldehydes leading to the corresponding pinacol is discussed in section 2.3.1.

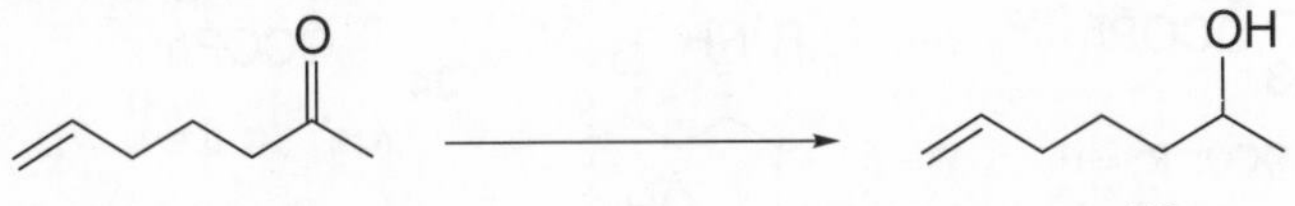

$$ \text{(2.17)} $$

	27	**28**
DMF-Bu₄NBF₄-(Hg), −3.1 V (SCE)		85%
H₂O/Diglyme-Bu₄NBF₄-(Hg), −3.2 V		86%
DMF-Bu₄NBF₄-(C), −3.1 V		85%

    The hydrogenation of aldehydes and ketones with the electrogenerated hydrogen has been investigated using Raney nickel modified electrodes to avoid the use of autoclave under high pressure conditions.   Ketones can be reduced in an MeOH-MeONa system to give the corresponding secondary alcohols in 70~82% yields.   Similar results are obtained with aromatic aldehydes (70~83% yields) but aliphatic aldehydes form aldol adducts [122].   Reduction of 4-*tert*-butylcyclohexanone **29** by using an electrochemically deposited platinum black cathode affords the corresponding alcohol **30** in 63% yield (eq. (2.18)) [123].   Electrohydrogenation of 4-*tert*-butylcyclohexanone with catalytically active powder electrodes has also been investigated [124].   The *C*- and *O*-alkylation of cyclohexanone derivatives has been attained in a DMF-Et₄NClO₄-(Hg) system [125].   The electroreduction of unhydrated 1,2-cyclohexanedione on mercury electrode has been examined at pH < 10, leading to α-hydroxycyclohexanone [126].   The practical electroreduction of 7-ketolithocholic acid **31** to ursodeoxychlolic acid **32** has been attained in a HMPA/EtOH-LiCl(0.2M)-(Ru or Ti) system (eq. (2.19)) [127].   The electroreduction of the carbonyl group of docetaxel **33** (Taxotere®) is performed in an MeOH-NH₄Cl-(Hg) system to give the corresponding 9α- and 9β-dihydrodocetaxels **34** in 64% yield (α/β ratio: ca. 2/1)(eq. (2.20)) [128].

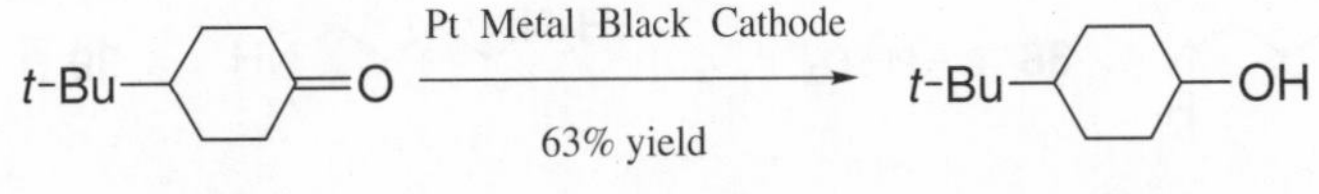

$$ \text{(2.18)} $$

29   *cis*/*trans* = 1 / 1   **30**

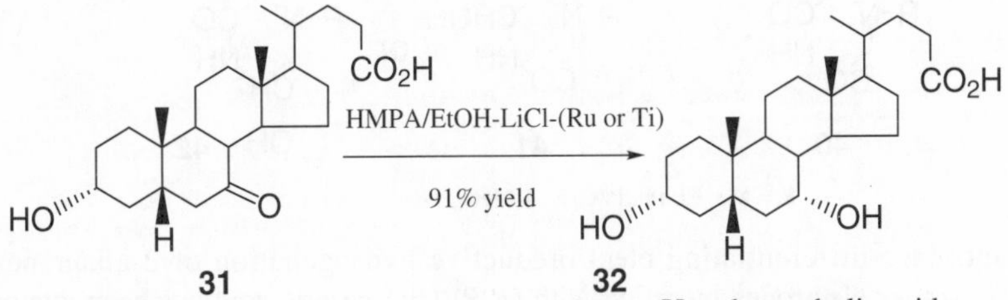

$$ \text{(2.19)} $$

31   **32**

Ursodeoxycholic acid

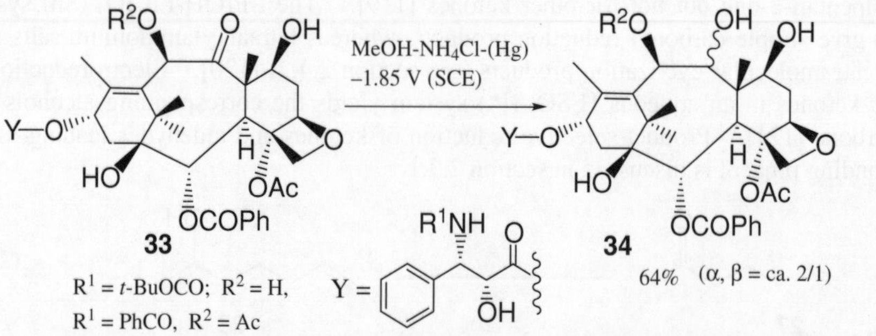

$$(2.20)$$

33

R[1] = *t*-BuOCO; R[2] = H,   Y =

R[1] = PhCO, R[2] = Ac

34

64%   (α, β = ca. 2/1)

The electroreduction of isatin **35** in acidic and neutral solutions in an aqueous NaHPO$_4$-(Hg) system at -1.5 V (SCE) affords isatide **37** in good yields (eq. (2.21)) [129].   The large ring hydroxy lactam **39** can be prepared by the electroreduction of **38** in an EtOH-Me$_4$NCl-(C/Hg) in 98% yield (eq. (2.22)) [130].   The amido carbonyl can be electroreduced to the corresponding alcohol.   For instance, the reduction of 1-substituted 1,4-diazaspiro[5.5]un-decane-3,5-diones **40** in an MeCN-LiClO$_4$-(Pt) system at -1.6 V (Ag/AgCl/Cl$^-$) gives the alcohols **41** and **42** in 60% yield (eq. (2.23)) [131].

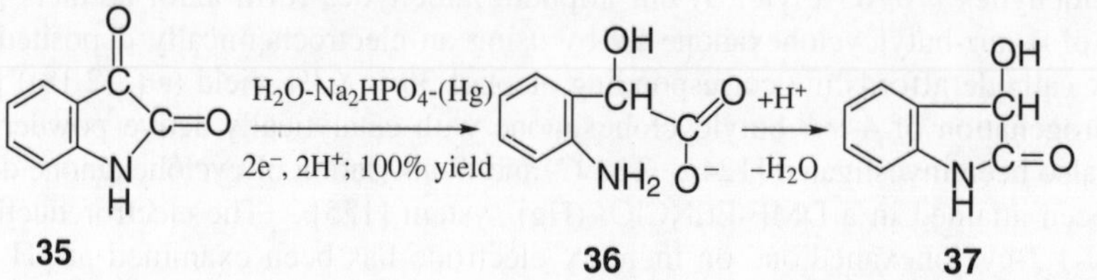

$$(2.21)$$

35   **36**   **37**

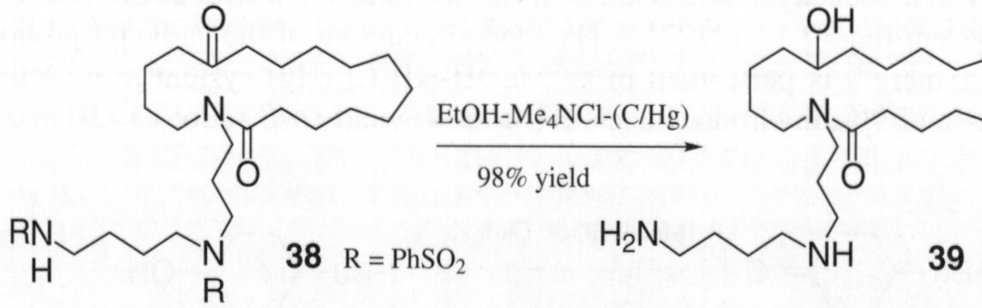

$$(2.22)$$

38 R = PhSO$_2$ **39**

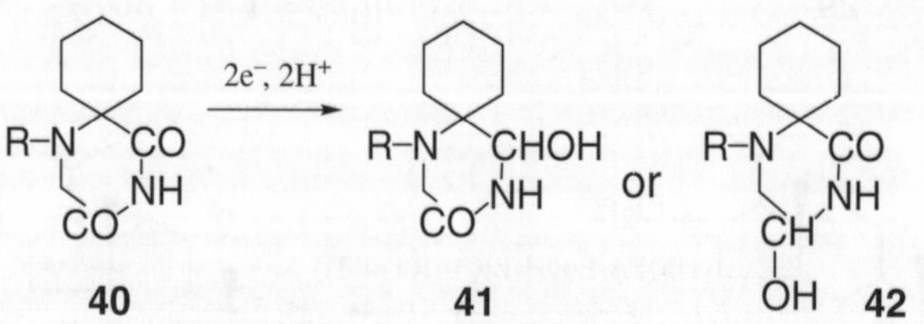

$$(2.23)$$

40   **41**   **42**

R = Me, Et, Ph, PhCH$_2$, 4-MeC$_6$H$_4$

The enantioface-differentiating electroreductive hydrogenation of 2-alkanones on Raney nickel powder electrodes modified with (*R,R*)-(+)-tartaric acid has been attempted [132].   (S)-(+)-2-Hexanol is synthesized in 2~6% optical purities.   The asymmetric reduction of α-keto acids in the magnetic fields of 980-1680 Gauss on mercury electrode has been attempted.   Phenylglyoxylic acid is reduced under the conditions of pH 3.8, 10 °C, 1680 Gauss, −1.25 V (SCE) to give the α-hydroxy acid in 21% maximum optical yield [133].   However, a recent report reveals that only optically inactive products are isolated [134].

Preparative electroreductions of dl-glyceraldehyde derivative 43 is carried out in an aqueous KCl (0.1M)-(Hg) system at −1.5 V (SCE) at pH 7 to afford a mixture of 44 and 45 (eq. (2.24)) [135].   Electroreductive conversion of glucose to sorbitol has been attained under 70~100% current eff. [136].   In order to maintain high sorbitol current eff. as glucose is consumed, a systematic procedure for lowering the applied current is devised in which the current is adjusted from 500 mA/10 g of Raney nickel at 1.6 M glucose to <100 mA/10 g at 0.2 M glucose.   Simultaneous electrolytic production of xylitol 47 and xylonic acid 48 from xylose 46 has been reported (eq. (2.25)) [137].   An electrocatalytically active cathode having hydridic features has been used to enhance Faradaic yields in xylitol production. Two types of cathode are used: amalgamated zinc and a catalytically active nickel or titanium coated electrode.   The electrolysis is carried out in an H_2O-Na_2SO_4(0.8M)-(MoFe coated Pt) system at pH 8.5 under a current density of 2 mA/cm^2.

$$
\begin{array}{c}
\text{CHO} \\
|\\
\text{CH–OMe} \\
|\\
\text{CH–OMe}
\end{array}
\quad
\xrightarrow[\text{−1.5 V (SCE)}, p\text{H }7]{\text{H}_2\text{O-KCl(0.1M)-(Hg)}}
\quad
\begin{array}{c}
\text{CHO} \\
|\\
\text{CH}_2 \\
|\\
\text{CH}_2\text{OMe}
\end{array}
\quad + \quad
\begin{array}{c}
\text{CH}_2\text{–OH} \\
|\\
\text{CH–OMe} \\
|\\
\text{CH}_2\text{–OMe}
\end{array}
\qquad (2.24)
$$

43   **44** 64% yield   **45** 36% yield

$$
\begin{array}{c}
\text{CHO} \\
\text{H}\!-\!\!\!-\!\text{OH} \\
\text{HO}\!-\!\!\!-\!\text{H} \\
\text{H}\!-\!\!\!-\!\text{OH} \\
\text{CH}_2\text{OH}
\end{array}
\quad
\xrightarrow[2e^-]{\text{2 H-adatom}}
\quad
\begin{array}{c}
\text{CH}_2\text{OH} \\
\text{H}\!-\!\!\!-\!\text{OH} \\
\text{HO}\!-\!\!\!-\!\text{H} \\
\text{H}\!-\!\!\!-\!\text{OH} \\
\text{CH}_2\text{OH}
\end{array}
\quad + \quad
\begin{array}{c}
\text{CO}_2\text{H} \\
\text{H}\!-\!\!\!-\!\text{OH} \\
\text{HO}\!-\!\!\!-\!\text{H} \\
\text{H}\!-\!\!\!-\!\text{OH} \\
\text{CH}_2\text{OH}
\end{array}
\qquad (2.25)
$$

46 d–Xylose   **47** d–Xylitol   **48** Xylomic Acid

2.2.2   Alcohols from Alkyl Aryl Ketones

The effect of pH on the electroreduction of alkyl aryl ketones has been studied [138].   The half-wave potentials of several alkyl arenyl ketones in a DMF-Bu$_4$NI-(Hg) system are listed in Table 2.3 [139].   Acetophenone can be electroreduced to a mixture of 1-phenylethanol, 2,3-diphenyl-2,3-butanediol, and ethylbenzene, and the product distribution depends on the reaction conditions [140, 141].   Selective conversion of acetophenone to 1-phenylethanol has been achieved by indirect electroreduction in an aqueous EtOH/BuOH(1/1.5v/v)-HCl(2M)-(Pt/Pb) system in the presence of a catalytic amount of antimony(III) chloride [142].   Electrolysis is carried out at a constant current density (~3.7 mA/cm^2) corresponding to the reduction potential of SbCl3 (−0.25 V (Ag/AgCl)) with stirring at room temperature, affording 1-phenylethanol in 99% yield.   Use of zinc chloride in an EtOH/BuOH(1/1.5)-HCl(2M)-(Pt) system selectively gives 1-phenylethanol [143]. Similarly, 4-haloacetophenone can lead to the corresponding 1-(4 halophenyl)ethanol in 91~98% yields [144].   The influence of cationic surfactants for the electroreduction of acetophenone has been investigated [145].   The specific interactions between the surfactants

Table 2.3 Half-Wave Potentials* of Substituted
Alkyl Arenyl Ketons

C$_6$H$_4$	X—C$_6$H$_4$—CO—R		
X \ R	H	Me	t-Bu
H	−1.79	−1.89	−2.05
4-Me	−1.89	−2.02	−2.10
4-MeO	−1.97	−2.08	−2.20
4-Me$_2$N	−2.13	−2.22	−

* vs. SCE in DMF-Et4NI
[Reproduced with permission from J.-P. Seguin et al.,
C. R. Acad. Sci. Paris, Sér.C, **278**, 129 (1974)]

and acetophenone induce the product-selectivity on the product ratio of pinacol and carbinol. Electroreduction of benzaldehyde and acetophenone leading to their corresponding alcohols and hydrocarbons has been performed by use of a nickel on graphite or vitreous carbon cathode formed in situ in a catholyte containing Ni^{2+} [146].

Electroreduction of benzaldehyde and acetophenone at a nickel-poly(tetrafluoroethylene) composite-plated electrodes has also been examined [147]. The trialkoxysilane-assisted electroreduction method for carbonyl compounds can be successfully extended to alkyl aryl ketones [148]. By this procedure, acetophenone can be reduced into 1-phenylethanol in 67~80% yields. Three head-to-tail dimers **50**, **52**, and **53** (eqs. (2.26, 2.28) are formed together with pinacol **51** (eq. (2.27)), when acetophene is cathodically reduced in a DMF-Bu$_4$NBr$_4$-(Hg) system in the presence of cyclodextrins [149] (see section 2.3.2). In the course of the reaction, some asymmetric induction has been observed. Electroreduction of β-chloropropiophenone in a DMF-KClO$_4$(0.1M)-(Pt) system affords two different products such as 1,5-diphenyl-2,6-dioxabicyclo[3.3.0]octane **55** and 1,6-diphenylhexane-1,6-dione **56** (eq. (2.29))[150]. Cathodically formed 1,4-diacetylbenzene dianion may react as a reducing agent toward aromatic halides and also as an electrogenerated base with 1,4-diacetylbenzene as a pro-nucleophile[151]. Controlled-potential coulometry of 1-benzoyl-2-phenylcyclopropane **57** yields phenyl 3-phenylpropyl ketone **58** in 66% yield *via* the cyclopropylcarbinyl rearrangement induced by an aryl cyclopropyl ketyl radical (eq. (2.30)) [152].

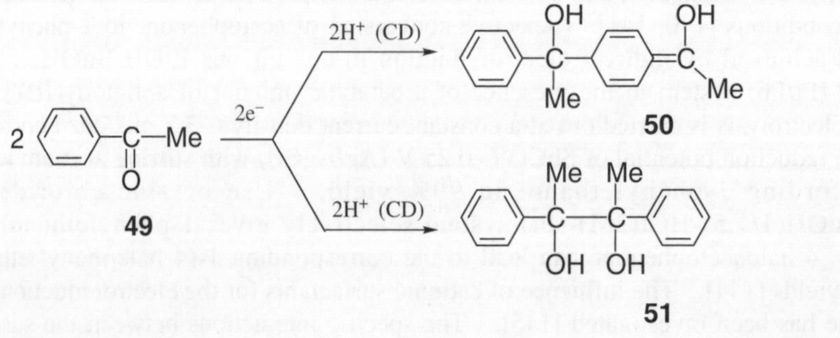

(2.26)

(2.27)

$$(2.28)$$

$$(2.29)$$

$$(2.30)$$

The effect of cathode materials for the electroreduction of acetophenone, 4-acetylpyridine, and 2-acetylthiophene has been investigated [153].   Cadmium and lead cathodes are found to be more effective rather than copper cathodes.   Reduction of phenyl 1-phenylethyl ketone 59 by using a platinum black modified cathode gives the alcohol 60 in quantitative yield (eq. (2.31))[123a].   Phenyl *tert*-butyl ketone has been reduced on chemically modified electrodes bearing methylstrychninium or methylbrucinium cations for asymmetric induction [154].   The electroreduction of rotenone 61 leading to retenol 62 has been attained in a DMF-Bu₄NClO₄-(Hg) system in good yield (eq. (2. 32))[155].   This may be an alternative to the chemical preparation of rotenol.

$$(2.31)$$

$$(2.32)$$

2.2.3   Alcohols from Aryl Aldehydes and Diaryl Ketones

Electroreduction of aromatic carbonyl compounds tends to give the corresponding alcohols along with dimerization products.   Low concentration of substrate, high current density, and low pH value suppress the formation of dimerized products [156]. The electroreduction of benzaldehyde on an silver deposited polymer electrode provides benzyl alcohol in 100% selectivity with 9.9% current efficiency [157]. The reduction of aromatic aldehydes in aprotic solvents has been studied.   All the aldehydes possess the highest reducibility in acidic media and the lowest in basic media [158]   In the reduction of benzaldehyde, the first reduction step is reversible in cyclic voltammetric experiments. The α-hydroxybenzyl radicals [PhCHOH]$^{\cdot}$ formed upon protonation of the initial ketyl radicals [PhCHO$^{-\cdot}$] are more difficult to reduce than the starting substrate [159].   The electroreduction of benzaldehyde results in the formation of benzyl alcohol and pinacols, and the formation of benzyl alcohol is favored under acidic conditions and pinacol formation under alkaline conditions [156].   The electroreduction using Pt-Pb/Nafion® without supporting electrolyte in THF (or benzene) affords benzyl alcohol in almost 100% selectivity [157b]. The electroreduction of benzaldehyde and acetophenone at the Pd-deposited carbon felt electrode in an aqueous EtOH(90%)-H$_2$SO$_4$(0.1M) system proceeds under 50~55% current efficiency [157c].   The addition of α-cyclodextrin to the electroreduction system of benzaldehyde suppresses the yield of hydrobenzoin and dimerization products [160].   Ultrasonic irradiation is effective in the electroreduction of benzaldehyde [161a,b].   Mass transfer coefficients derived from the formation ratio of the hydrodimeric to hydromonomeric products have been discussed in electroreduction of p-methylbenzaldehyde in a parallel plate electrode flow cell and a rotating cylinder electrode-beaker cell. These coefficients have been shown in good agreement with those obtained by ordinary current-potential and concentration-time methods [162].   The electroreduction of vanillin **63** in an H$_2$O-NaHCO$_3$(0.5M)-(Hg) system as on around neutral condition preferentially tends to give vanillyl alcohol **64**, whereas hydrovanilloin **65** is formed exclusively under strong basic conditions (eqs. (2.33, 2.34)) [163, 164].   Constant potential electroreduction of vanillin in an acidic medium also yields vanillyl alcohol **64** [165].   Quinoline-4-carbaldehyde can be electroreduced through

two one-electron irreversible steps in both acid and alkaline media [166].   The anion radicals of p-substituted benzaldehydes form hydrogen bonded complexes with water and methanol in DMF, depending on the electron donating ability of the substituents [167].   *Ortho*-nitrobenzaldehyde is converted into o-aminobenzyl alcohol in a H$_2$O/EtOH(1/1)-H$_2$SO$_4$(0.1M)-(Pt/Pb) system in 61% yield (current efficiency 30.1%) [168].   The electron-withdrawing nitro group shifts the reduction potential of the carbonyl group present in the nitrobenzaldehydes to the positive side relative to unsubstituted benzaldehydes.

A versatile procedure for the electroreduction of aromatic aldehydes and aryl ketones, leading to the corresponding alcohols, has been developed [148]. The method involves the cathodic reduction of the carbonyl compound **66** in the presence of trialkoxysilanes in a DMF-Bu$_4$NBr-(Pt) system, giving the alcohol **68** in 76% yield via the silyl ether **67** (eq. (2.35)) [148].

$$\text{(2.35)}$$

The reduction of benzophenone proceeds by two reversible one-electron steps [109c]. The first step produces an anion radical which does not form ion pairs with a tetraethylammonium cation [169]. In the presence of acid as a proton donor, the reduction of benzophenone proceeds *via* an acid conjugated complex which causes to give a new wave appearing at more positive potentials than the original reduction one [170]. The peak potentials (Ep) of benzophenone in anhydrous liquid ammonia at -50 °C have been found to be -1.25 and -1.78 V (Ag/AgNO$_3$) [171]. The half wave potentials observed in DMF are -1.21 and -1.78 V (Hg pool), respectively [172]. Generally, biaryl ketone anion radicals undergo a dimerization or a disproportionation reaction. It has been found that the disproportionation of electrogenerated benzophenone anion radicals is accelerated by the presence of cationic species, i.e., Li$^+$, in DMF [173].

Electrochemical reduction of diaryl ketone **69** (R = Ar or Ph) and its substituted homologues in an aqueous solution yields various products **70**, **71**, and **72**, depending on the pH conditions and consumed electricities (eq. (2.36)) [138, 174]. Solvent effects on the Hammett ρ value for the electroreduction of substituted benzophenones are observed [175].

$$\text{(2.36)}$$

Benzophenones undergo a one-step two-electron reduction in molten acetamide at 85°C in a AcONa-(Hg or C) system [176]. Chemoselective electroreduction of benzophenone **73** leading to the corresponding alcohol **74** has been achieved in a DMF-Bu$_4$NBr-(Hg) system in the presence of manganese metal ions (eq. (2.37)) [177]. Complete electroreduction of the carbonyl group of benzophenone has been attained using nickel, copper, amalgamated copper, and amalgamated lead electrodes [178a].   The electroreduction of benzophenone in an aqueous H_2SO_4(10%)-(Ni) system affords diphenylmethane in 85% yield.   The adsorption processes on the glassy carbon electrode as well as their surface protonation effect to the aromatic carbonyl compound, *e.g.*, *p*-chlorobenzophenone, are discussed in comparison with the behavior on mercury electrode [178b].

$$\text{73} \xrightarrow[\text{−1.57 V, 87\% yield}]{\text{DMF-Bu}_4\text{NBr/MnCl}_2\text{-(Hg)}} \text{74} \qquad (2.37)$$

Electrolysis of biphenyl phenyl ketone in a DMF-Et$_4$NI/LiI-(Pt/Hg) system at −1.80 V (SCE) yields α-(4-biphenyl)benzyl alcohol in 70% yield [173a].   Benzoin can be electroreduced to the corresponding alcohol in an aqueous pyridine (90%) solution [179]. Controlled-potential electrolysis of fluorenone **75** (Y = H) yields fluorenol **76** (Y = H) in 95~99% yields in a DMF-Bu$_4$NBF$_4$-(Hg) system in the presence of phenol as a proton donor at −1.40 V (SCE) (eq. (2.38)) [180].   By using benzoic acid as a strong proton donor, **75** gives **76** (62%) together with pinacol (27%).   Fluorenones bearing substituents at positions β and γ can be electroreduced in a DMF-Bu$_4$NClO$_4$-(Hg) system to form paramagnetic radical anions by a one-electron reduction [181].   The macroscale electrolysis of

$$\text{75} \xrightarrow[\substack{\text{−1.40V (SCE), 95~99\% yields}\\ \text{Y = H, 2–NH}_2\text{, 2–Br, 3–Ph}}]{\text{DMF-Bu}_4\text{NBF}_4\text{-(Hg)}} \text{76} \qquad (2.38)$$

indolyl heteroaryl ketones **77** under protic conditions always affords the corresponding alcohols **78** in high yields (eq. (2.39)) [182].   The electrochemical conversion of phenalenone **79** into the corresponding peroyrene **80** and its derivatives **81** and **82** has been performed in a DMF-Bu$_4$NClO$_4$-(Pt/Hg) system in the presence of excess acetic anhydride

$$\text{77} \xrightarrow[\text{~100\% yield}]{\text{MeCN-Bu}_4\text{NClO}_4\text{(0.1M)-(Pt)}} \text{78} \qquad (2.39)$$

Het = 2–benzothiazolyl
2–benzoxazolyl
1–methyl–2–benzimidazolyl
4–pyridyl
2–pyridyl

(eq. (2. 40)) [183].   The phenalenone radical anion formed by one-electron reduction of **79** may lead to a neutral acetoxyphenalenium radical by acetylation with acetic anhydride, which enters into a dimerization reaction, giving rise to the hydro carbon peropyrene **80** and it derivatives **81** and **82**.   The controlled potential reduction of acecyclone yields the corresponding *cis*- and *trans*-isomeric reduction products [184].   The attack of electrogenerated

$$(2.40)$$

intermediates on the electrophilic reagents can take place instead of simple protonation. Thus, the electroreduction of diaryl-1,2-diketones **83** in a DMF-LiClO$_4$(0.2M)-(Hg) system in the presence of *N*-arylcarbonimidoyl dichlorides leads to 2-arylimino-4,5-diaryl-1,3-dioxoles **84** in almost quantitative yields (eq. (2.41)) [185]. Electroreductive cleavage of the C-C bond of aryl trityl ketones **85** in a DMF-Et$_4$NI-(Hg) system has been observed *via* an anion radical intermediate to form triphenylmethane **86**, triphenylmethanol **87**, and aryl carboxylates **88** as major products (eq. (2.42)[186].

$$(2.41)$$

| | Ar, yield, % | |
	Phenyl	2-Naphthyl
86	85	78
87	15	16
88	87	84

$$(2.42)$$

* DMF-Et$_4$NI-(Hg)

2.2.4   Complete Reduction of Aldehydes and Ketones Alkanes

The formation of hydrocarbons from alkanones and alkanediones has been investigated. Acetone, 2-butanone, and 2-pentanone are electrochemically reduced to their corresponding hydrocarbons [187][188][189]. Simple dioxo compounds such as glyoxal and biacetyl are also reducible [190]. Electrohydrogenation of 2,4-pentanedione **89** (n = 1) and 2,5-hexanedione **89** (n = 2) has been carried out in a H$_2$O-H$_2$SO$_4$-(Pt) system to give hydrocarbons **90** in 92~99% yields (eq. (2.43)) [191].

The electroreduction of acetophenone to ethylbenzene has been partially attained at a platinized platinum cathode in acidic ethanol. The major by-product is generally the alcohol, and this may be minimized by a periodic current reversal to reactivate the electrode [141].

$$\underset{\substack{\displaystyle \text{Me-C-(CH}_2)_n\text{-C-Me} \\ \parallel \qquad\quad \parallel \\ O \qquad\qquad O \\ \textbf{89} \;\; n = 1, 2}}{} \quad \xrightarrow{\text{H}_2\text{O-H}_2\text{SO}_4\text{-(Pt)}} \quad \text{MeCH}_2\text{-(CH}_2)_n\text{-CH}_2\text{Me}$$

(2.43)

90

n	Yield
1	95 ~ 99
2	92.4

2.3   Pinacol Type Dimerization

2.3.1   Pinacols from Aliphatic Aldehydes and Ketones

Electrochemical hydrodimerization of formaldehyde to ethylene glycol has been attained by the proper choice of electrolysis conditions.   The reaction preferentially proceeds by employing the graphite cathode bearing a specified surface oxide structure and certain quaternary ammonium salts, and by holding the pH value in the range of 5 to 7 [192].   On electrohydrodimerization of formaldehyde to ethylene glycol, the experimental data regarding the influence of the major parameters on the current yield for the manufacturing are discussed [193].   The optimum conditions (current and conversion yields: 92 and 95%) for ethylene glycol production by cathodic hydrodimerization of formaldehyde are found to be as follows: current density 1.0 kA/m^2; pH 5.5; temperature 40~45°C; initial formaldehyde concentration 5 M [194].   The presence of a quaternary ammonium salt brings about improved results [195].   A commercial process for the preparation of ethylene glycol from formaldehyde in neutral or acidic solutions has been patented [196].

The electrohydrodimerization of acetone at different electrode materials in alkaline media has been intensively investigated.   The electrode's activity rises dramatically when the electrochemical alloys of Cu/Zn coatings on the nickel cathode are used [197].   Photoexcited electrogenerated quinone anion radicals reduces acetaldehyde, acetone, acetophenone, benzaldehyde, and benzophenone, giving the corresponding pinacols as major products [198].   Dramatic effects of a small amount of some kinds of tetraalkylammonium ions, e.g., DMP$^+$ (DMP = dimethylpyrrolidinium) and TEA$^+$ (TEA = tetraethylammonium), on the electroreductive pinacolization of 6-hepten-2-one **91** or cyclohexanone have been observed.   The electroreduction of **91** in an aqueous diglyme-DMPBF$_4$-(Hg) system at –2.7 V (SCE) affords the corresponding pinacol **92** in 75% yield (eq. (2.44)) [115].   As mentioned in section 2.2.1, the electrolysis at –3.1 V in an aqueous diglyme-Bu$_4$NBF$_4$-(Hg) system, 6-hepten-2-one **91** consuming 2 F/mol of electricities gives the alcohol **94** in 86% yield (eq. (2.45b)) [115, 118]. The electroreductive cyclization of **91** is discussed in section 2.4.2.   The DMP$^+$-mediated electroreduction of cyclohexanones consumes only 1 F/mol to form pinacols, and the addition of water results in an increase in the catalytic current up to a limiting plateau [118]. In contrast to the formation of the alcohol **97** and 2,2,6,6-tetramethylpiperidine **98** from 4-oxopiperidine **95** under conventional electrolysis conditions, the pinacol **96** is formed at a zinc cathode in up to 10% yield together with alcohol **97** as a major product (eq. (2.46)) [199].   Evidently, the relatively low yield of pinacol **96** is due to a steric hindrance.   Electroreductive intramolecular pinacolization of the cyclohexanetrione **99** in an MeCN-Et$_4$NBr-(Hg) system in the presence of acetic anhydride can lead to the corresponding cyclopropane derivative **100** in 71% yield (eq. (2.47)) [200].

$$\text{(2.44)}$$

$$\text{aq. Diglyme-DMPBF}_4\text{-[Hg]}$$
$$-2.7 \text{ V } vs \text{ SCE, } 75\% \text{ yield}$$

$$\text{DMF-Bu}_4\text{NBF}_4\text{-(Hg)}$$
$$-2.8 \text{ V, } 94\% \text{ yield} \quad \cdots\cdots(a)$$

$$\text{aq. Diglyme-DMPBF}_4\text{-(Hg)}$$
$$-3.2 \text{ V, } 86\% \text{ yield} \quad \cdots\cdots(b) \quad \text{(2.45)}$$

DMP$^+$: dimethylpyrrolidinium

$$2e^-, 2H^+$$
$$10\%$$

$$4e^-, 4H^+$$

$$2e^-, 2H^+$$

$$\text{(2.46)}$$

$$\text{MeCN-Et}_4\text{NBr/Ac}_2\text{O-(Hg)}$$
$$-2.8 \text{ V (SCE), } 71\% \text{ yield}$$

$$\text{(2.47)}$$

The samarium (II)-catalyzed pinacolization of alphalic aldehydes has been attained in good yields [201]. The electrolysis procedure based on the use of sacrificial anodes of magnesium or aluminum allows the use of SmCl$_3$ (5~10%) as a catalyst precursor. The electrolysis is carried out in an N-methylpyrrolidone-SmCl$_3$-(Al/Ni) system: substrate (pinacole yield%), Me$_3$CCHO (72%); n-C$_6$H$_{13}$CHO (75%); 2-heptanone (40%).

2.3.2 Pinacols from Alkyl Aryl Ketones

It has been found that the pinacolization of alkyl aryl ketones, *i.e.* indanone, tetralone, chromanone, etc., is enhanced when a divalent transition metal cation is present. This phenomenon is general and occurs with Cr^{2+}, Mn^{2+}, Fe^{2+}, CO^{2+}, Zn^{2+}, but not with Ni^{2+} [202, 203]. The greatest specificity is generally observed when Fe^{2+} is present. It is interesting to note that antimony(III) chloride is found to have electrocatalytic property for the electro-chemical formation of 1-phenylethanol [204].   The presence of $EuCl_3\cdot6H_2O$ in DMF (or aqueous DMF)-(Hg) system for the electroreduction of alkyl aryl ketones enhances the for-mation of pinacol dimers in contrast to the results from $CrCl_3\cdot6H_2O$ [205]. The electrogen-erated Eu(II) and the ketones may occur via the Eu(II)-carbonyl complexes. Pinacols are obtained as follows: acetophenone (97%), benzal-acetophenone (94%), benzophenone (83%), fluorenone (94%).

At higher substrate concentration of pH 2.8~3.2 or dilute substrate concentration under highly basic media, each electrolysis condition provides the pinacol in a good yield [206]. Electroreduction of acetophenone using a mercury bubble electrode has been attempted [207].   An amalgamated copper cathode provides pinacol from acetophenone in 70% yield [208].   The electroreduction of acetophenone in aqueous micellar media brings about im-proved results [206]. Product-selectivity for the electroreduction of acetophenone to the corresponding hydrodimeric product is found to be affected by temperature [209]. In aque-ous alkaline media, the macroelectrolysis of acetophenone **101** and p-methoxyacetophe-none 103 preferentially undergoes pinacolization, giving **102** and **104** (eq. (2.48), (2.49)), in the case of acetophenone, also head-to-tail dimers (27%) (eq. (2.49)) [210].   In the pres-ence of β-cyclodextrin, the electrochemically produced dimers show some optical activity (% ee, ca. 24 for each diastereoisomer).

$$\text{Acetophenone} \quad \xrightarrow[\substack{-1.60 \sim -1.63 \text{ V (SCE), 73\% yield}}]{\text{aq. MeOH(30\%)-KOH(0.5M)-(Hg)}} \quad \text{102} \qquad (2.48)$$

101

$$\text{MeOC}_6\text{H}_4\text{COMe} \quad \xrightarrow[\substack{1.70\sim1.73\text{V}}]{\text{aq. MeOH(30\%)-KOH(0.5M)-(Hg)}} \quad \text{104} \qquad (2.49)$$

103

The level of water in the medium is known to affect the rate of formation of pinacols, decreasing with decreasing water concentration in the range of 0~5% [211].   A high cur-rent density is favorable in obtaining pinacols [212].   Stereochemical factors in the cathod-ic pinacolization of acetophenone at a mercury pool have been studied.   The dl/meso ratio is as high as 12.5/1 using a dry DMF-LiClO$_4$ system [213].   In a DMF-Bu$_4$NI-(Pt/Hg) sys-tem, acetophenone forms mainly pinacol [172, 214].   The pinacol **106** of α-trifluoroace-tophenone **105** has been obtained in 35% yield by electroreduction in an MeCN-Et$_4$NCl/LiClO$_4$-(?) system (eq. (2.50)) [215a].   The electrode process of substituted α,α,α-trifluoroacetophenones probably proceeds via a single transfer to form a pinacol [215b].

$$\text{Ph}-\text{CO}-\text{CF}_3 \quad \xrightarrow[\substack{-1.45 \text{ V (SCE), 35\% yield}}]{\text{MeCN-Et}_4\text{NClO}_4/\text{LiClO}_4\text{-(?)}} \quad \text{Ph}-\overset{\overset{\displaystyle \text{OH}}{|}}{\underset{\underset{\displaystyle \text{CF}_3}{|}}{\text{C}}}-\overset{\overset{\displaystyle \text{CF}_3}{|}}{\underset{\underset{\displaystyle \text{OH}}{|}}{\text{C}}}-\text{Ph} \qquad (2.50)$$

105   **106**

Alkyl aryl ketones can be reduced in a DMF-Bu$_4$NBr-(Hg) system in the presence of manganese dichloride or chromium trichloride, leading to the corresponding pinacols in

63~85% yields, *dl* diastereoisomers are produced preferentially [177]. It has also been found that the electroreduction in the presence of the metal salts always proceeds at less negative potentials than the electroreduction of ketones [216-220]. The pinacolization of various carbonyl compounds using a zinc or magnesium sacrificial anode in a DMF-Bu_4NBF_4-(Zn (or Mg)/SUS) system affords the corresponding pinacols in 90~95% yields [221]. The samarium(II)-catalyzed pinacolization of aryl carbonyl compounds proceeds in a DMF-$SmCl_3$(5~10%)-(Mg/Ni) system to give the pinacols in 80~90% yields [201].

Electroreduction of 1,8-diacetylnaphthalene **107** in a DMF-Bu_4NClO_4-(Hg) system undergoes an aldol type cyclization to give 3-methyl-1H-phenalen-1-one **108** in 50% yield (eq. (2.51)) [222]. Large-scale electrolysis of 4-acetylphenoxylalkyl trimethyl ammomium salts **109** has been performed in an aqueous DMF-Bu_4NBF_4-(Hg) system to afford the corresponding pinacol **110** in 85% yield (eq. (2.52)) [223]. The electroreduction of dibenzoylmethane **111** in an aqueous EtOH(40%)-LiCl(0.1M) system at pH 2.2~5.0 affords preferentially the corresponding pinacol **112** (eq. (2.53)), whereas at pH 6.0~7.6, the electrolysis gives the monomeric carbinol [224]. Intramolecular cyclization of 1,5-diketones proceeds through anion radical intermediates. Electroreduction of 1,3-dibenzoyl-1,3-diphenyl-propane **113** in an MeCN-Me_4NBF_4-(Pt) system gives 1,2,3,5-tetraphenyl-2-hydroxycyclopentanol **114** in 50% yield (eq. (2.54)) [225].

$$\text{(2.51)}$$

$$\text{(2.52)}$$

$$Y = (CH_2)_n \overset{+}{N}Me_3 X^-$$
$$(X = Br, ClO_4; \ n = 3, 4, 6, 8, 10)$$

$$\text{(2.53)}$$

$$\text{(2.54)}$$

The effect of proton donors on the electroreduction of rotenone **115** in acetonitrile has been investigated. In the presence of acetic acid, one-electron reduction of the carbonyl group of **115** leads to the formation of the neutral ketyl radical, which dimerizes to the pinacol **116** (eq. (2.55)). The electrolysis in water containing acetonitrile, however, gives rotenol **117** as the sole product (eq. (2.56)) [226]. The reduction of the Wieland-Miescher ketone **118** in an MeOH/H_2O-KCl-(Ag/Hg) system at –1.7 V (SCE) results in the corresponding pinacol **120** as a major product together with non-conjugated ketone **119** (eq. (2.57)) [227]. The electroreduction of 4*β*,9*β*-dihydroindeno[2,1-a]indene-5,10-dione in a DMF-Bu_4NBr-(Hg) system gives the corresponding pinacol dimer in 24.3% yield [228].

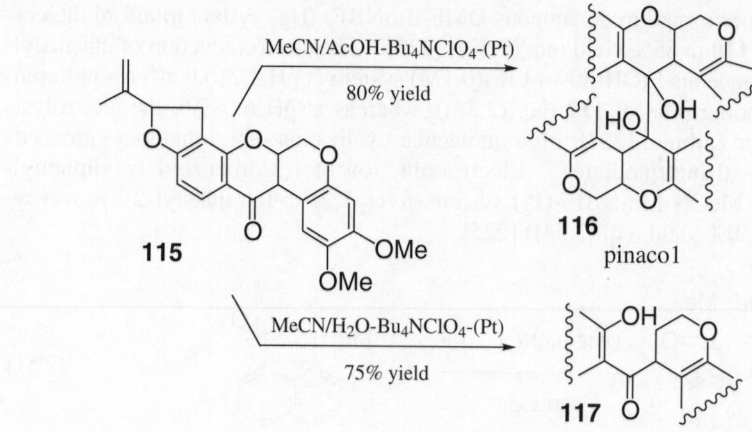
MeCN/AcOH-Bu4NClO4-(Pt)
80% yield
HO
OH
115
OMe
OMe
116
pinacol
MeCN/H2O-Bu4NClO4-(Pt)
75% yield
OH
117

(2.55)

(2.56)

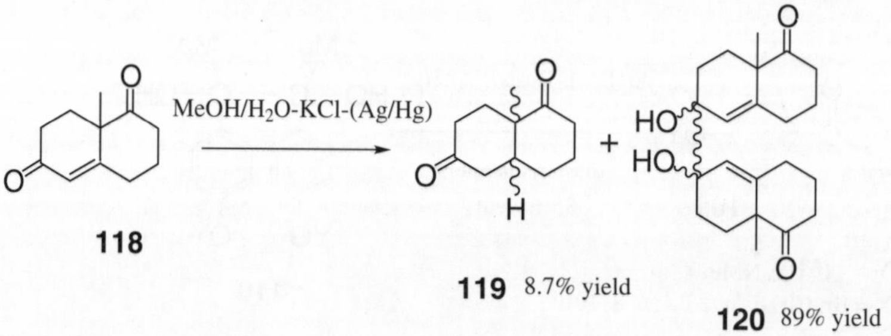
MeOH/H2O-KCl-(Ag/Hg)
H
HO
HO
118
119 8.7% yield
120 89% yield

(2.57)

A highly diastereospecific dimerization has been performed on electroreduction of 6-acetyl-1,3,7-trimethyllumazine **121** in a PrOH-KCl system, leading to the formation of the corresponding pinacol **122** in 84% yield (eq. (2.58)) [229].

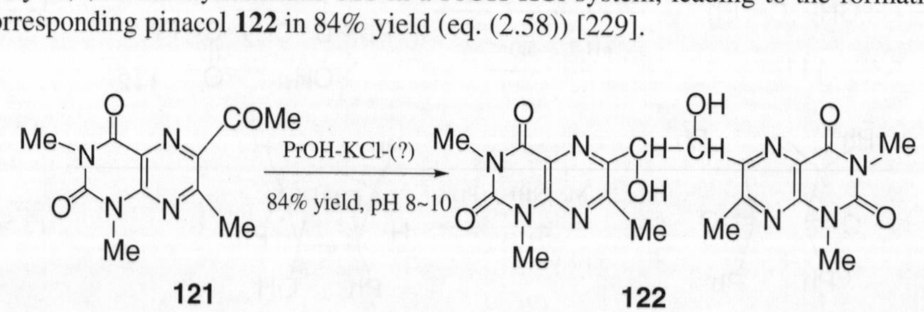
COMe
Me
PrOH-KCl-(?)
84% yield, pH 8~10
OH
CH–CH
OH
121
122

(2.58)

2.3.3   Pinacols from Aromatic Aldehydes and Biaryl Ketones

Benzaldehyde anion radical readily undergoes dimerization in protic solvents. Two mechanisms have been proposed to account for the dimerization of benzaldehyde anion radical in ethanol [230]. In the presence of water and hydroxide ion, the predominant mechanism is shown in eq. (2.59). In ethanol containing a low concentration of acetic acid, the mechanism would be as shown in eq. (2.60). In a DMF-Bu$_4$NBr-(Hg) system, benzaldehyde and anisaldehyde form mainly pinacols [172]. The reactivity of the anion radicals of aromatic aldehydes initially formed decreases in the following order: 9-anthraldehyde > benzaldehyde > 2-naphthaldehyde > 1-naphthaldehyde [231a]. The sacrificial cadmium anode has been found to be effective for the pinacolization of aryl aldehydes [231b].

$$Ph{-}CHO \rightleftharpoons \left(Ph{-}CHO\right)^{-\bullet} \qquad \text{-------(a)}$$

$$2\left[Ph{-}CHO\right]^{-\bullet} \rightleftharpoons \underset{123}{} \; Ph{-}\underset{O^-}{CH}{-}\underset{}{CH}{-}Ph \qquad \text{------(b)} \qquad (2.59)$$

$$\underset{124}{124} + H_2O \longrightarrow Ph{-}\underset{OH}{CH}{-}\underset{125}{CH}{-}Ph + OH^- \; \text{----(c)}$$

$$\underset{123}{123} + AcOH \rightleftharpoons Ph{-}\overset{\bullet}{C}HOH + AcO^- \; \text{-----(a)} \qquad (2.60)$$

$$\underset{126}{126} + Ph{-}CHO \longrightarrow Ph{-}\underset{\bullet O}{CH}{-}\underset{127}{CH}{-}Ph \qquad \text{-----(b)}$$

β-Cyclodextrin complex can alter the course of electroreduction of benzaldehyde. Electrolysis of benzaldehyde **128** in a DMF-Bu$_4$NBF$_4$-(Hg) system gives hydrobenzoin **129** in 60% yield, but β-cyclodextrin complex of **128** produces **129** in 80% yield (eq. (2.61)) [232]. The stereochemistry of the product is remarkably affected. The nature of the proton source causes variations in the *dl* to meso ratio but the *dl* isomer always predominates except for the reduction at –2.4 V of the β-cyclodextrin complex, which gives 75% meso-rich pinacols in 80% yield.

$$
\begin{array}{c}
\text{DMF-Bu}_4\text{NBF}_4\text{-(Hg)} \\
60\% \;(dl\,/meso = 14\,/\,1) \\
Ph{-}CHO \\
\mathbf{128} \\
\text{DMF-Bu}_4\text{NBF}_4/\beta\text{-CD-(Hg)} \\
80\% \;(dl/meso = 1\,/\,3)
\end{array}
\qquad
\begin{array}{c}
Ph{-}\underset{OH}{CH}{-}\underset{OH}{CH}{-}Ph \\
\mathbf{129}
\end{array}
\qquad (2.61)
$$

β–CD : β–cyclodextrin

2,2'-Dihydroxyhydrobenzoin **132** has been synthesized by the electroreduction of sali-cylaldehyde **130** *via* **131** under alkaline conditions at different metal electrodes (eq. (2.62)) [233].   The optimum conditions are shown to be as follows: electrode (current density A/cm^2), zinc (0.074 to 0.144); lead (0.078 to 0.122); amalgamated copper (0.111 to 0.29); amalgamated lead (0.105 to 0.150), temperature 20°C, concentration of NaOH (1.5~2.0 M), yields 80~85%.   The electroreduction of 9-anthraldehyde in a DMF-Et$_4$NClO$_4$-(Hg) system affords dimerization products via an anion radical intermediate [234].

$$ (2.62) $$

130   **131**   **132**

The pinacol **134** can be obtained in 85% yield by electroreduction of 5-methylfurfural **133** in an H$_2$O-NaOH-(C/Cu) system at –1.6 V (SCE) (eq. (2.63)) [235].   Use of a mer-cury-bubble-electrode for the preparative pinacolization of 4-hydroxybenzaldehyde and acetophenone is found to be effective [236].

$$ (2.63) $$

133   **134**

The electroreduction of various aromatic ketones has been intensively studied in an aqueous, organic, and mixed solvent systems [216].   The first reduction peaks (a) of aro-matic ketones in an aprotic solvent in the presence of metal salts are always proceeded by other peaks (c) which appear at a less negative potential.   Actually, electroreduction of benzophenone in a DMF-Bu$_4$NBF$_4$/FeCl$_2$ system provides the peak (a) at –1.40 V (Ag/Ag$^+$) and the peak (c) at –1.12 V, respectively [237].   The latter peaks (c) are ascrib-able to a fast chemical reaction between low-valent metals and anion radicals of the ke-tones.   Electroreduction of benzophenone in a DMF-Et$_4$NClO$_4$-(Hg) system by an imped-ance technique has been studied [238].   The effect of magnesium and lithium salts for the reduction of benzophenone has been elucidated [239].   3,5-Dibromo-4-hydroxybenzophe-none undergoes the same reduction as that of the unsubstituted benzophenone forming the corresponding pinacol in acidic medium and carbinol in neutral and basic media [240].

The reduction of aromatic ketones in a DMF-Bu$_4$NI-(Hg) system in the presence of chromium(III) salts leads to the formation of a ketone-chromiun(II) complex or to the for-mation of a ketyl-chromium(II) complex which is easily reducible even when the ketone is intact [216].   The formation of the complexes is proven by the presence of a specific peak on voltammogramms.

The polarographic and kinetic studies of the electroreduction of benzophenone demon-strate that 1) the electron transfer is reversible, 2) a protonation step proceeds the electron transfer, and 3) the process takes place either at the interface or in the reaction layer, de-pending on the experimental conditions [241].   Benzophenone and its methylated homo-logues **135** are electroreductively dimerized in an aqueous EtOH(50%)-NaOH(3M)-(Au) system to form the corresponding glycols **136** in 48~87% yields (eq. (2.64)) [242].   The controlled potential electrolysis of the diketone **137** in an aqueous EtOH-KOH-(Pb) system at −2.0 V (SCE) undergoes intramolecular pinacolization to give the diol **138** in 27.3% yield (eq. (2.65)) [243].   A preparative electroreduction of fluorenone **139** in an aluminum chloride −1-methyl-3-ethylimidazolium chloride (MEIC) molten salt system as the basic melt (−1.00 V (reference melt)) results in the formation of fluorenone pinacol **140** in 80% yield together with pinacolone **141** (10%) (eq. (2.66)) [244].   However, the electroreduction of **139** in acidic melt (+ 0.10 V) affords the fluorenone pinacolone **141** in 75% yield (eq. (2.67)) [245].   The lifetime of the electrogenerated ketyl depends on melt acidity, becoming longer (more stable) as acidity increases. (Melt acidity is usually reported as the ratio of AlCl$_3$ to MEIC, with acidic melts having ratios greater than one and basic melts having raios less than one.)

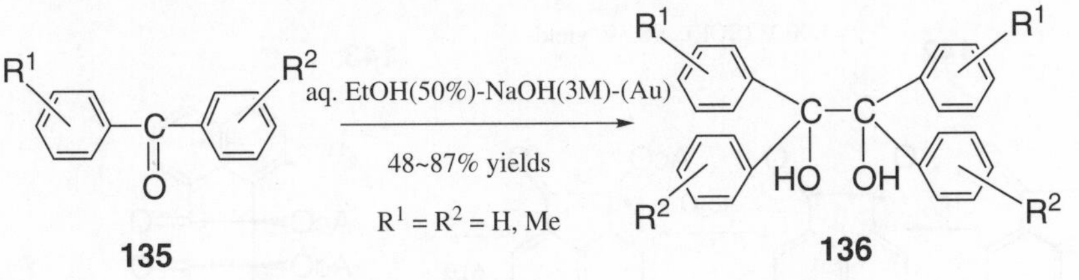

(2.64)

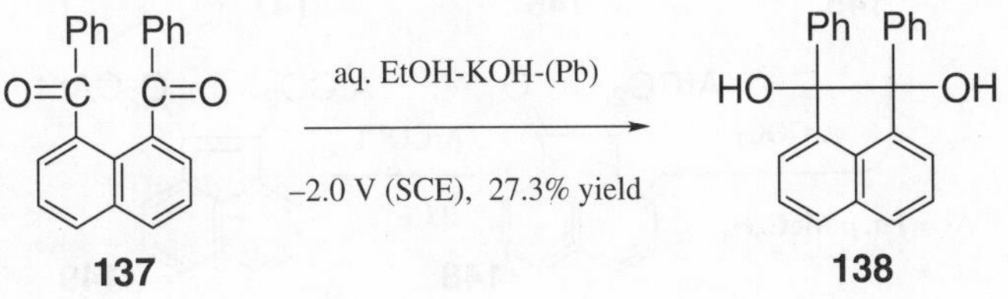

135   **136**

(2.65)

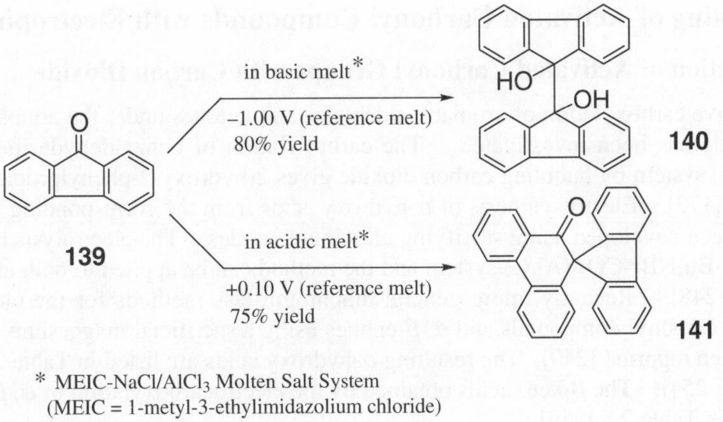

137   **138**

(2.66)

(2.67)

139   **140**

141

* MEIC-NaCl/AlCl$_3$ Molten Salt System
(MEIC = 1-metyl-3-ethylimidazolium chloride)

2.3.4   Miscellaneous Pinacol Type Dimerization

Carbonyl compounds can lead to the olefins by electrochemically reduced tungsten species. The electroreduction of benzaldehyde **142** in a THF-Bu$_4$NClO$_4$-(Al/Pt) system in the presence of tungsten hexachloride results in the formation of stilbene **143** in 98.5% yield (eq. (2.68)) [246].   Cathodic acylation of 1,2-acenaphthenedione **144** in an Acetone-LiClO$_4$-(Pt/Hg) system in the presence of acetic anhydride gives rise to the formation of the corresponding pinacol type product **147** (61% yield), but the similar reaction with aroyl chloride instead of acetic anhydride undergoes a different reaction process to produce the corresponding 1,2-bis(aroyloxy)-acenaphthylene **149** (73%) (eq. (2.69)) [247].

$$2 \ Ph-CHO \xrightarrow[\text{−1.90 V (SCE), 98.5\% yield}]{\text{THF-Bu}_4\text{NClO}_4\text{-(Al/Pt), WCl}_6} Ph-CH=CH-Ph \qquad (2.68)$$

142   **143**

144   **145**   **146**   **147**   (2.69)

Ar = Ph, p-MeC$_6$H$_4$

148   **149**

2.4   Coupling of Activated Carbonyl Compounds with Electrophiles

2.4.1   Reaction of Activated Carbonyl Groups with Carbon Dioxide

Electroreductive carboxylation of aromatic aldehydes and ketones under the atmosphere of carbon dioxide has been investigated.   The carboxylation of benzaldehyde in a DMF-Bu$_4$NI-(Pt/Hg) system by bubbling carbon dioxide gives 2-hydroxy-2-phenylpropionic acid in 27% yield [172].   Electrosynthesis of α-hydroxy acids from the corresponding aromatic ketones has been developed using sacrifying aluminum anodes.   The electrolysis is carried out in a DMF-Bu$_4$NBr-CO$_2$-(Al/C) system and the method can be applied to both aldehydes and ketones [248].   Recently, more general diaphragm-less methods for the electrocarboxylation of carbonyl compounds and α, β-enones using a sacrificial magnesium anode in DMF have been reported [249].   The resulting α-hydroxy acids are listed in Table 2.4 [172, 249, 250, 252, 254].   The β-keto acids obtained by the electrocarboxylation of α, β-enones are indicated in Table 2.5 [249].

Electrosynthesis of 2-hydroxy-2-(p-isobutylphenyl)propionic acid **151** from the ketone **150** is performed in a DMF-Bu$_4$NI-(Pt/Hg) system by bubbling carbon dioxide during the electrolysis (eq. (2.70)) [250, 251].   This procedure can be successfully applied to the synthesis of benzylic acids starting from benzophenones [252].   Independently, the electrolysis is performed in an MeCN-Bu$_4$NBr-(C/Hg) system in the presence of carbon dioxide to yield **151** in 83% yield (eq. (2.70)) [253a].   The improved procedure for the preparation of 2-aryllactic acids has been recently reported [253b]. The soluble metal anodes in diaphragm-less cells are also effective for the electrocarboxylation of a variety of carbonyl compounds [254].

$$iso\text{-}Bu\text{—}\bigcirc\text{—}\overset{\overset{\displaystyle O}{\|}}{C}\text{—}Me \xrightarrow[\text{CO}_2,\ 0\ °C]{\substack{\text{DMF-Bu}_4\text{NI-(Pt/Hg)}\\ \text{or MeCN-Bu}_4\text{NBr-(C/Hg)}}} iso\text{-}Bu\text{—}\bigcirc\text{—}\underset{\underset{\displaystyle CO_2R}{|}}{\overset{\overset{\displaystyle OH}{|}}{C}}\text{—}Me \qquad (2.70)$$

150   **151**

Table 2.4 Electrocarboxylation of Carbonyl Derivatives and Activated Olefins*

R^1	R^2	Electrolysis Conditions	Products, Yield, % $R^1 \diagdown R^2$ $HO \diagup CO_2H$ Conv. (Current Eff.)		Ref.
			Conv.	(Current Eff.)	
H	Me	DMF-Bu$_4$NI/CO$_2$-(Zn/C)	9	(8)	[254]
H	C$_7$H$_{15}$	DMF-Bu$_4$NBr/CO$_2$-(Mg/SUS)	20	(–)	[249]
H	Ph	DMF-Bu$_4$NBr/CO$_2$-(Mg/SUS)	60	(–)	[249]
		DMF-Bu$_4$NI/CO$_2$-(Zn/C)	24	(20)	[254]
Me	Ph	DMF-Bu$_4$NBr/CO$_2$-(Mg/SUS)	75	(–)	[249]
		DMF-Bu$_4$NI/CO$_2$-(Zn/C)	62	(55)	[254]
Me	4-*i*-Bu-Ph	DMF-Bu$_4$NI/CO$_2$-(Zn/C)	87	(70)	[254]
		DMF-Bu$_4$NI/CO$_2$-(Pt/Hg)	85	(–)	[250] [172]
Me	2-Naphthyl	DMF-Bu$_4$NI/CO$_2$-(Zn/C)	81	(57)	[254]
Me	6-MeO-2-Naphtyl	DMF-Bu$_4$NI/CO$_2$-(Zn/C)	90	(83)	[254]
Ph	Ph	DMF-Bu$_4$NBr/CO$_2$-(Mg/SUS)	70	(–)	[249]
		DMF-Bu$_4$NI/CO$_2$-(Zn/C)	75	(57)	[254]
		DMF-KI-(Hg)/CO$_2$	86	(–)	[252]
		DMF-Bu$_4$NBr/CO$_2$-(Mg/SUS)	70	(–)	[249]

* Carried out in a DMF (300 ml)-Bu$_4$NX (X = Br, I or BF$_4$, 0.04 M)-(Mg) system in the presence of carbon dioxide
[Reproduced with permission from S. Mcharek et al., *Bull. Soc. Chim. Fr.*, **1989**, 95]

Table 2.5 Electrocarboxylation of Carbonyl Derivatives and α,β-Enones[a]

R^1 R^2 (C=O)		Products, Yield, % Conv. (Current Eff.)
R^1	R^2	
Me	MeCH=CH	(structure with O and CO_2H) 35 (–)
Me	PhCH$_2$	Ph—(O)—CO_2H 82 (–)
Me	PhCH=CH	CO_2H O, Ph 85 (–)
Me$_2$C=CH	Me$_2$C=CH	(structure with CO_2H) ca. 60[b] (–)
–(CH$_2$)$_3$CH=CH–		O=(cyclohexyl)CO_2H 40 (–)
–CH$_2$CCH$_2$C=CH– (Me Me, Me)		O=(cyclohexyl)CO_2H 42 (–)

[a] Carried out in a DMF (300 ml)-Bu$_4$ NX (X = Br, I or BF$_4$, 0.04 M)-(Mg/SUS) system in the presence of carbon dioxide.
[b] Unstable compound contaminated with isomers.

2.4.2   Intramolecular Cyclization of Activated Carbonyl Groups with Electrophiles

Intramolecular cyclization of non-conjugated olefinic ketones initiated by the electroreduction of the carbonyl group has been developed [255].   The electroreduction of 6-hepten-2-one **152** in a dioxane/MeOH-Et$_4$NOTs-(C) system at the potential of –2.7 V (SCE) gives 1,2-dimethylcyclopentanol **153** in 98% yield (eq. (2.71)) [117, 256a].   Similarly, the electroreduction of 6-trimethylsilyl-6-hepten-2-one affords *cis*-1-methyl-3-trimethylsilyl cyclohexanol [256b].   The cyclization of **152** is effected by homogeneous redox catalysts, which include biphenyl, 2-methoxybiphenyl, naphthalene, and phenanthrene [257].   The yield of the cyclized alcohol **153** increases with redox catalyst concentration, increasingly positive reduction potential of the catalyst and amount of water in the medium [257]. Investigation of the highly stereoselective cathodic cyclization of 6-alken-2-one systems provides valuable information on a delicate balance of kinetic factors which allows stereospecific ketyl addition to an inactivated alkene to succeed in competition with protonation and further electron transfer [257].   Bicyclo[3.3.0]octanols **155** are synthesized by electroreduction of the allyl pentenyl ketone **154** in a DMF-Bu$_4$NBF$_4$(0.1M)-(Hg) system *via* a tandem cyclization process (eq. (2.72)) [258].

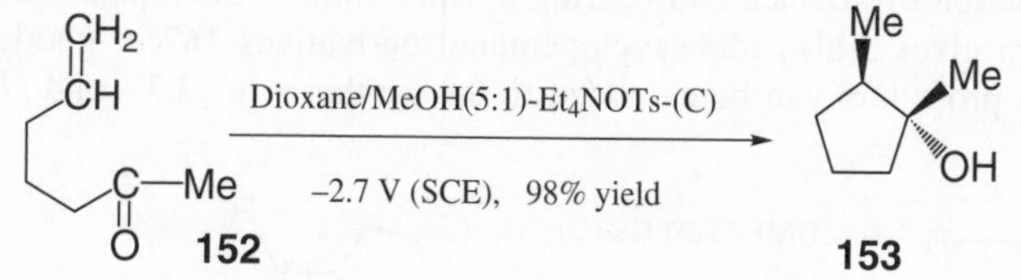

$$\text{Dioxane/MeOH(5:1)-Et}_4\text{NOTs-(C)} \qquad (2.71)$$
$$-2.7 \text{ V (SCE)}, \quad 98\% \text{ yield}$$ 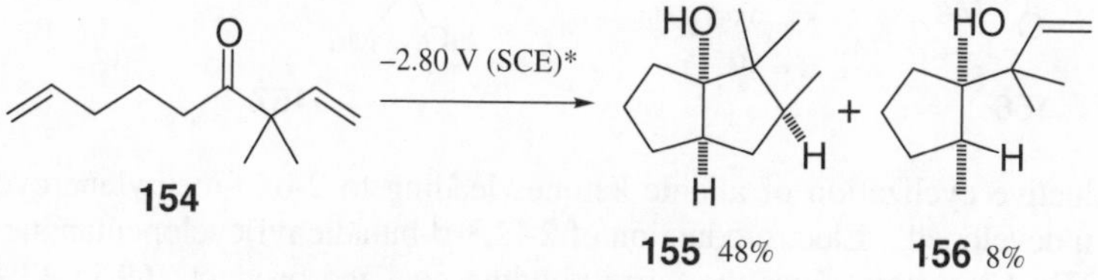

152 → **153**

$$-2.80 \text{ V (SCE)}^* \qquad (2.72)$$

154 → **155** 48% + **156** 8%

*DMF/i-PrOH(0.2M)-Bu₄NBF₄-(Hg)

Electroreductive cyclization of γ- and δ-cyano alkanones **157** proceeds in an i-PrOH-Et₄NOTs-(C/Sn) system at –2.8 V (SCE) to give the cyclized products **158** and **159** in 58~84% yields (eq. (2.73)) [259]. The electroreductive transannular cyclization of (Z,E)-4,8-cyclododecadien-1-one **160** proceeds in a DMF-Bu₄NBF₄(0.1M)-(Pt/Hg) system to afford a mixture of two bicyclic alcohols **161** and **162** via a ketyl anion radical as an intermediate (eq. (2.74)) [260]. The electroreductive cyclization of the olefinic aldehyde **163** proceeds in 89% yield to afford **164** and **165** (eq. (2.75)) [261]. Both products can be used for the total synthesis of quadrone. The electrochemical cyclization proceeds through regio-and stereoselective manners to give *cis*-isomers, and the procedure does not give any cyclic tertiary alcohols other than five and six-membered ring systems [117].

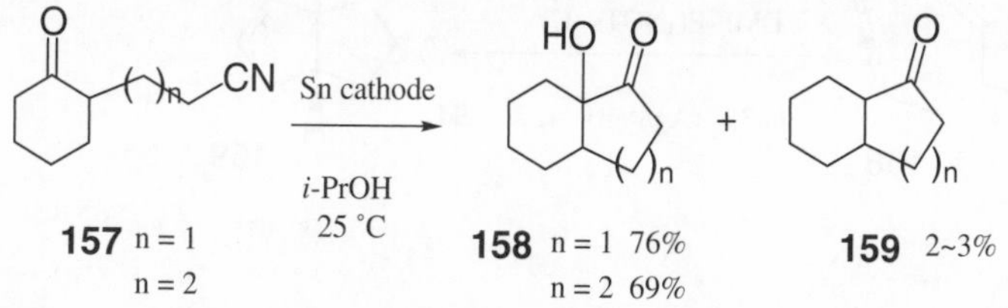

$$\text{Sn cathode} \qquad (2.73)$$
$$i\text{-PrOH}$$
$$25 \text{ °C}$$

157 n = 1   n = 2   **158** n = 1 76%   n = 2 69%   **159** 2~3%

$$\text{DMF-Bu}_4\text{NBF}_4(0.1\text{M})-(\text{Pt/Hg}) \qquad (2.74)$$
$$-2.94 \text{ V (SCE)}$$

160 → **161** 28.6% + **162** 19.8%

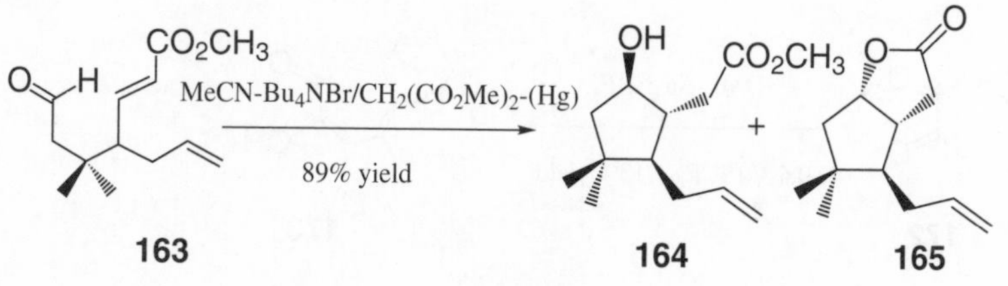

$$\text{MeCN-Bu}_4\text{NBr/CH}_2(\text{CO}_2\text{Me})_2\text{-(Hg)} \qquad (2.75)$$
$$89\% \text{ yield}$$

163 → **164** + **165**

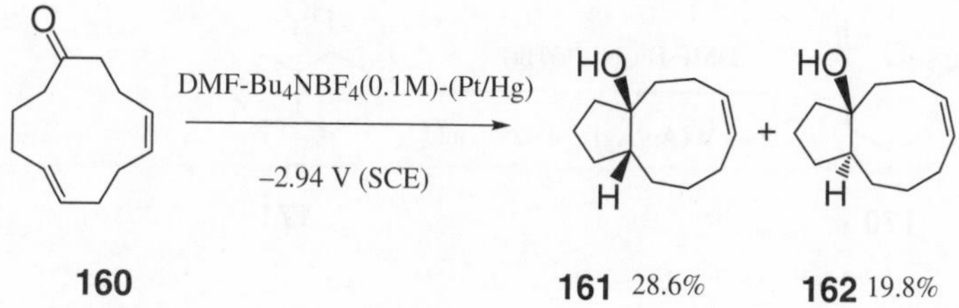

The electroreduction of ketones **166** bearing a triple bond at the δ position in a DMF-Et$_4$NOTs-(C) system gives 2-alkylidenecyclopentanol derivatives **167** in good yields (eq. (2.76)) [262].   The procedure can be extended to the synthesis of [3.3]- and [4.3]bicyclic systems.

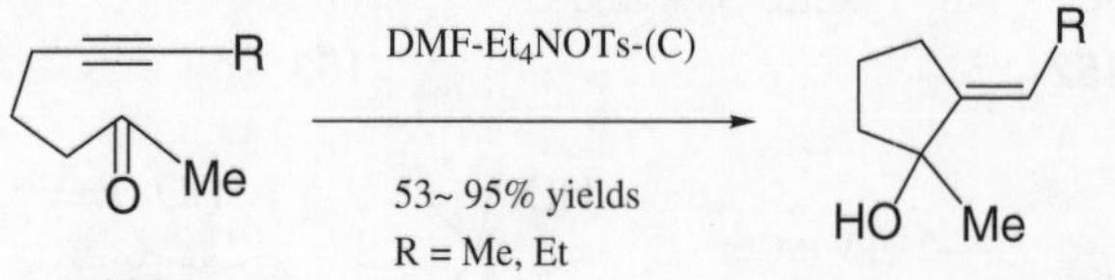

$$\text{DMF-Et}_4\text{NOTs-(C)}$$

53~ 95% yields
R = Me, Et

166   **167**

(2.76)

Electroreductive cyclization of allenic ketones leading to 2-oxo-methylene cyclopentenols has been developed.   Electroreduction of 2-(2,3-β-butadienyl)cyclopentanone **168** in a DMF-Et$_4$NOTs-(C) system gives the corresponding cyclized product **169** in 42% yield (eq. (2.77)) [263a].   The same compound can be obtained in 30% yield when the allenic ketone 168 is treated with sodium naphthalenide in THF at 25°C.   The intramolecular electroreductive cyclization of terminal allenic ketones **170** has been attained in a DMF-Bu$_4$NI-(C/Hg) system (eq. (2.78)) [263b].   The electrolysis of the allenyl ketones **170** is carried out at –2.43 V (Ag/AgI) at 10 °C to give the cyclized products **171** in 43.2% yield.   The electroreductive cyclization of **172** is performed in a DMF-Bu$_4$NBF$_4$ system at –0.4 V to give the cyclized product **173** in 13% yield (eq. (2.79)) [264].   In the course of the electroreduction of ethyl 3,7-dimethyl-10-oxo-2,6-decadienoate, an intramolecular electron transfer from the enoate group to the terminal aldehyde has been observed [265].

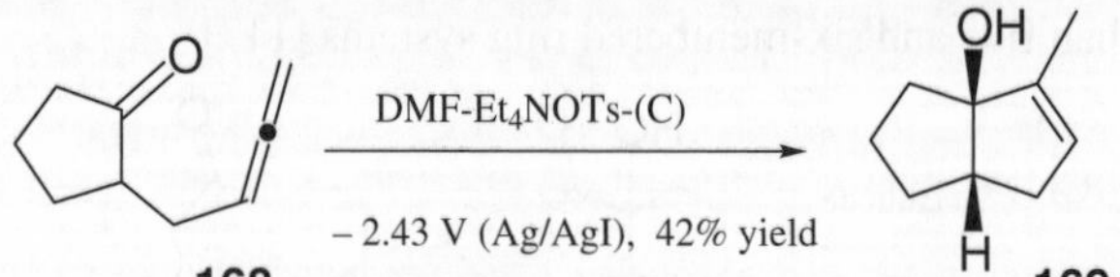

$$\text{DMF-Et}_4\text{NOTs-(C)}$$

– 2.43 V (Ag/AgI), 42% yield

168   **169**

(2.77)

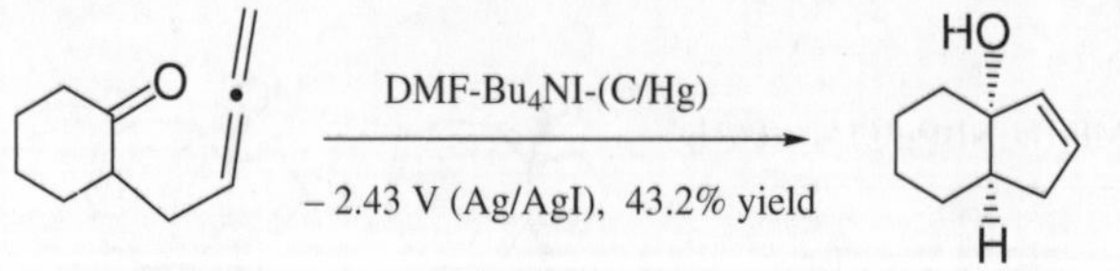

$$\text{DMF-Bu}_4\text{NI-(C/Hg)}$$

– 2.43 V (Ag/AgI), 43.2% yield

170   **171**

(2.78)

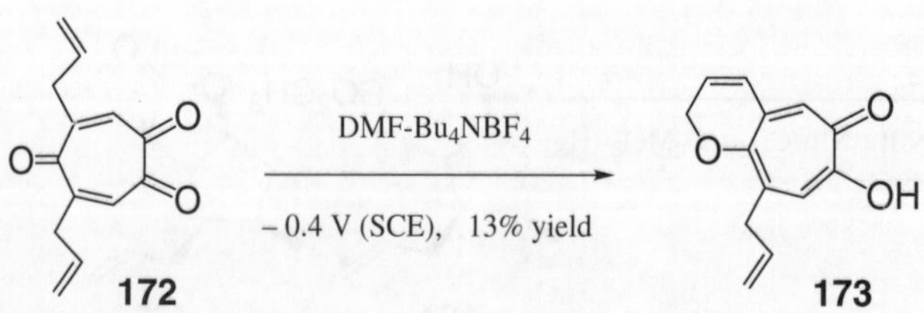

$$\text{DMF-Bu}_4\text{NBF}_4$$

– 0.4 V (SCE),   13% yield

172   **173**

(2.79)

Electroreductive intramolecular addition of a carbonyl group to an aromatic ring leading to six-membered ring formation has been developed. Electrolysis of 5-phenyl-2-pentanone **174** in an Me_2CHOH-Et_4NOTs-(C/Sn) system gives the cyclized product **175** (70%) and 5-phenyl-2-pentanol **176** (7%) (eq. (2.80)) [266]. Tetrabutylammonium bromide can be used for this purpose, but lithium perchlorate does not give any cyclized product. Tin cathode is the best choice. Electrochemically induced pericyclic reactions of triene systems which are conjugated with carbonyl groups have been reported. The controlled potential one-electron reduction of the compounds **177** in a THF-Bu_4NClO_4-(Pt) system at –2.1 V gives the cyclized products **178** in good yields (eq. (2.81)) [267].

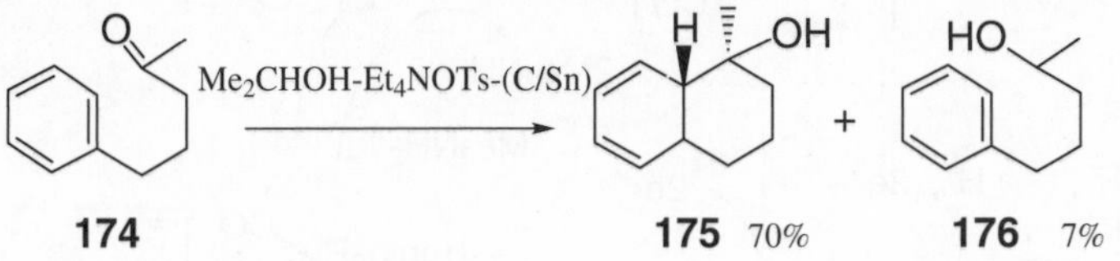
Me2CHOH-Et4NOTs-(C/Sn)
H
OH
HO
+
174
175 70%
176 7%

(2.80)

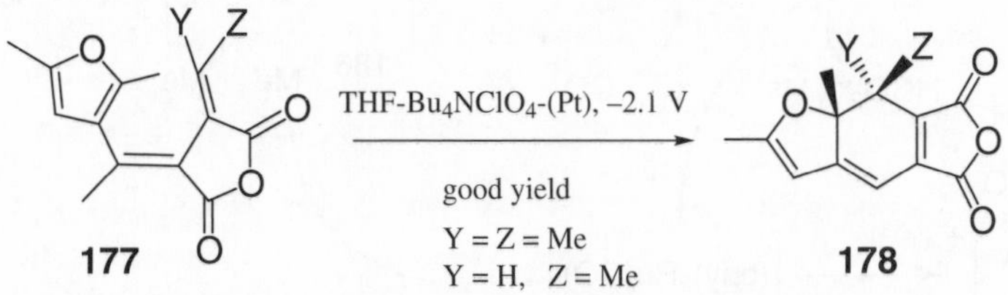
Y Z
O
THF-Bu4NClO4-(Pt), –2.1 V
good yield
Y = Z = Me
Y = H, Z = Me
177
178

(2.81)

Electroreductive ring opening of *N*-acylaziridines **179** leading to the corresponding oxazolines 180 has been observed in a THF-Bu_4NClO_4-(Hg) system (eq. (2.82)) [268].

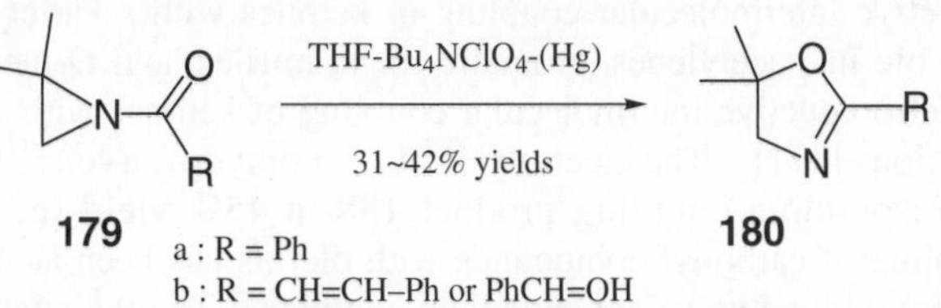
THF-Bu4NClO4-(Hg)
31~42% yields
N
O
R
179
180
a : R = Ph
b : R = CH=CH–Ph or PhCH=OH

(2.82)

Cathodic coupling of ketones with ethoxydimethylvinylsilanes takes place regioselectively at the β position to the ethoxydimethylsilyl group to give 1-oxa-2-silacyclopentane derivatives **181** in 62-79% yields (eq. (2.83)) [269].

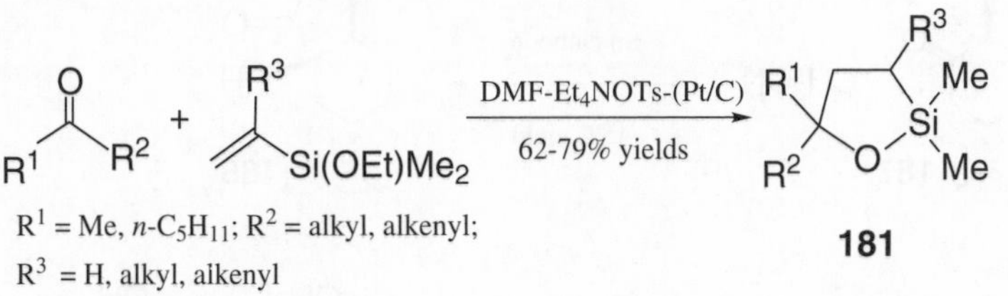
R1
R2
+
R3
Si(OEt)Me2
DMF-Et4NOTs-(Pt/C)
62-79% yields
Me
Si
Me
R1 = Me, n-C5H11; R2 = alkyl, alkenyl;
R3 = H, alkyl, alkenyl
181

(2.83)

2.4.3 Intermolecular Coupling of Carbonyl Groups with Electrophiles

Electroreduction of carbon dioxide catalyzed by $[Ru(bpy)_2(CO)_2]^{2+}$ in the presence of Me_2NH and $Me_2NH{\cdot}HCl$ in anhydrous acetonitrile produces HCO_2^- and *N,N*-dimethylformamide in good current efficiencies [270]. A possible catalytic cycle for the reaction is shown in Scheme 2.1.   The penta-coordinated Ru(0) complex **182** reacts with carbon dioxide to yield $[Ru(bpy)_2(CO)(CO_2^-)]^+$ **183**, which exists as an equilibrium mixture with $[Ru(bpy)_2(CO)C(O)OH]^+$ **184** and $[Ru(bpy)_2(CO)_2]^{2+}$ **185**.   In the presence of Me_2NH, $[Ru(bpy)_2(CO)_2]^{2+}$ **185** effectively reacts with the amine to produce $[Ru(bpy)_2(CO)C(O)NMe_2]^+$ **186**, which undergoes two-electron reduction to afford DMF with generating Ru(0) complex **182**.

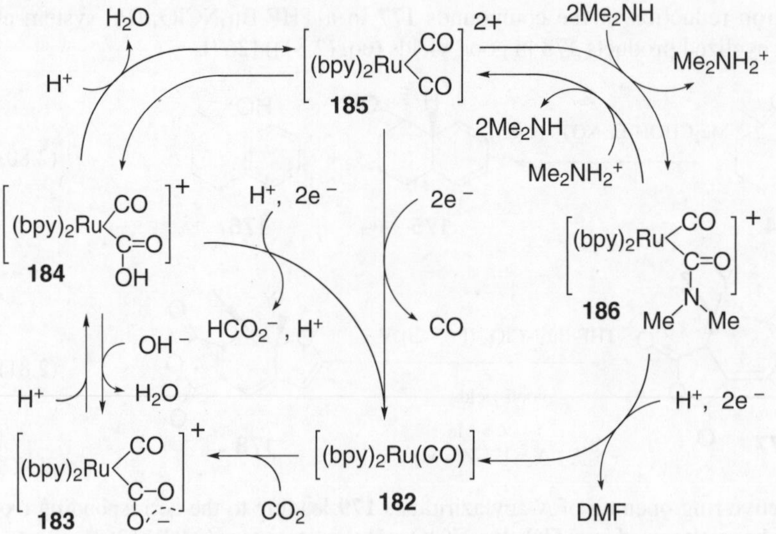

Scheme 2.1

The electroreductive intermolecular coupling of ketones with a variety of unsaturated compounds such as olefins, acetylenes, aromatic rings, nitriles, and O-methyl oximes has been reported.   Electroreductive intermolecular coupling of ketones with nitriles proceeds under similar conditions [271].   The electrolysis of (+)-dihydrocarvone **187** with acetonitrile gives the corresponding coupling product **188** in 45% yield (eq. (2.84)) [271]. Intermolecular coupling of carbonyl compounds with olefins has been achieved by activation of olefinic moiety with a low valent palladium complex.   The Pd-catalyzed reaction of allylic acetates **189** with benzaldehyde proceeds in a $DMF-Et_4NOTs/PdCl_2(PPh_3)_2$-(Pt) system in the presence of $ZnCl_2$ as an additive, giving **190** and **191** (eq. (2.85)) [272].

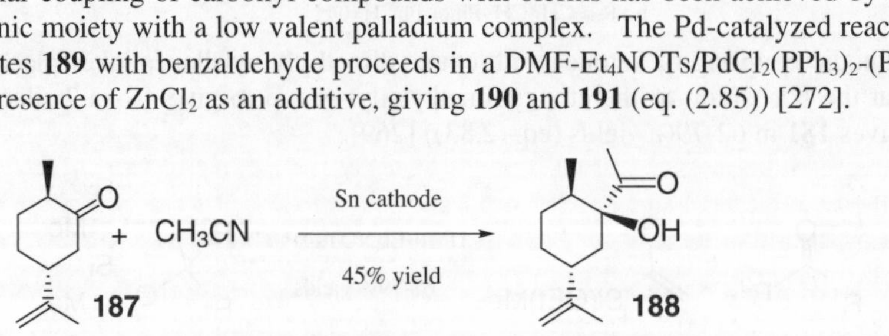

(2.84)

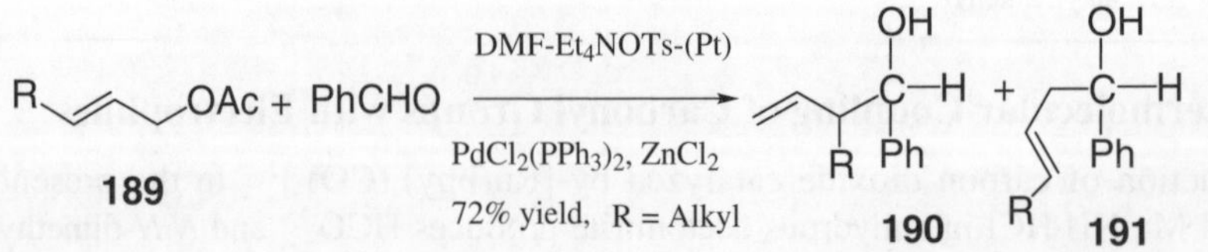

(2.85)

Electrochemically induced chain reaction of 2,6-dichlorobenzaldehyde **192** with fluorene **193** takes place in a THF-Bu$_4$NClO$_4$(0.2M)-(C) system at -30 °C to give the coupling product **194** in 93% yield (eq. (2.86a)) [273].

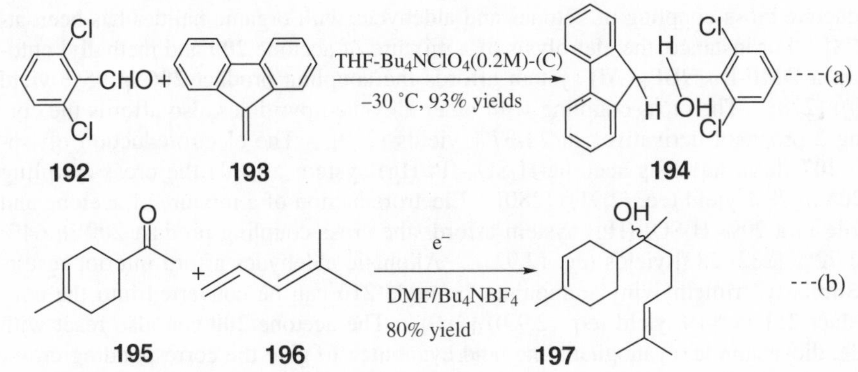

(2.86)

The cathodic coupling of a carbonyl group with olefins has been reported. The electroreduction of non-conjugated olefinic ketones can lead to the corresponding intramolecularly coupled alcohols with good regio- and stereoselectivities (see Section 2.4.2). [117, 256, 262, 267]. Recently, the electroreductive intermolecular coupling of the ketone **195** with the olefin **196** has been attained under similar conditions [274]. The electrolysis of **195** is performed in a DMF-Bu$_4$NBF$_4$-(C) system to give **197** in 80% yield (eq. (2.86b)).

The electroreductive intermolecular coupling of cyclohexanone with O-methyl oximes **198** proceeds in an i-PrOH-Et$_4$NOTs-(C/Sn) system in a divided cell to afford 2-methoxyamino alcohol **199** in 90% yield (eq. (2.87)) [274b].

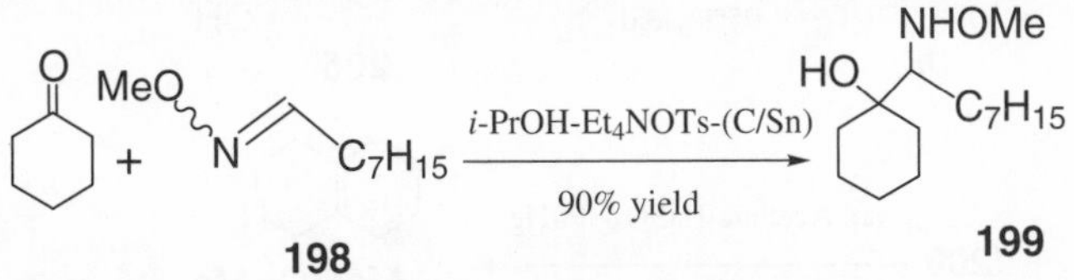

(2.87)

A radical addition process has been proposed for the cross-coupling of acetone **200** with electron acceptors. The reaction proceeds with various electron acceptors, *e.g.*, pyridium salts, isoquinoline, vinylphosphonates, trimethylvinylammonium halide, crotonate, etc., which are reduced at a more negative potential than that of acetone **200** in acidic media. For example, the cross-coupling of **200** with pyridine gives 90.6% yield of 2-(2-tetrahydropyridyl)-2-propanols **201** and **202** together with 2-(2-pyperidyl)-2-propanol (4.8%) (eq. (2.88)) [275].

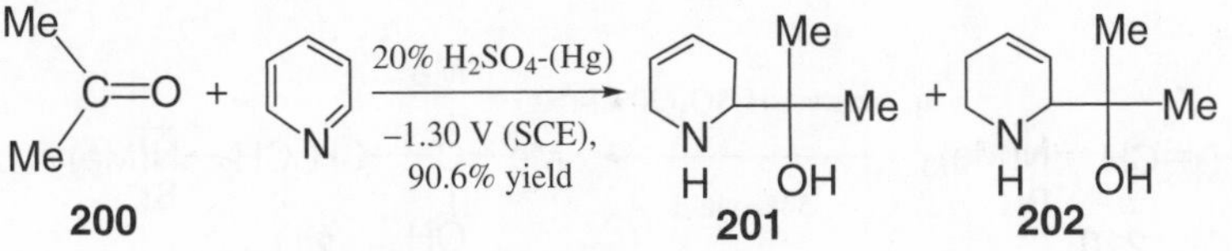

(2.88)

The electroreductive cross-hydrocoupling of ketones and alkanals with pyridine has been investigated [276].   The electroreductive cross-coupling of acetone **200** with *trans-*(*S*)-1-phenyl-2-buten-1-ol **203** in a DMF-Bu$_4$NBF$_4$-(Pt/C) system affords (*S*)-2,3-dimethyl-5-phenyl-2-pentanol **204** in 90% yield with higher than 90% ee (eq. (2.89)) [277].   The electroreductive cross-coupling of ketones and aldehydes with organic halides has been attained [278].   For instance, the eletrolysis of a mixture of acetone **200** and methallyl chloride **205** in a DMF-Bu$_4$NBF$_4$-(Al) system affords the coupling product **206** in 95% yield (eq. (2.90)) [278].   The cross-coupling with other alkylated pyridines also affords the corresponding 2-propanol derivatives in 23-37% yields [279].   The electroreduction of isoquinoline **207** in an aqueous acetone-H$_2$SO$_4$-(Pt/Hg) system affords the cross-coupling product **208** in 78% yield (eq. (2.91)) [280].   Electroreduction of a mixture of acetone and acrylonitrile in a 20% H$_2$SO$_4$-(Hg) system affords the cross-coupling product **209** in 64% [281] and 72% [282, 283] yields (eq. (2.92)).   Aliphatic aldehydes afford inferior results [284].   Similarly, trimethylvinylammonium bromide **210** can be converted into the coupling product **211** in 54% yield (eq. (2.93)) [280].   The acetone **200** can also react with cyanamide, dicyanamide, cyanoguanidine, and cyanourea to give the corresponding cross-coupling products.   For example, the cathodic cross-coupling of **200** with cyanamide is

$$ (2.89) $$

$$ (2.90) $$

$$ (2.91) $$

$$ (2.92) $$

$$ (2.93) $$

carried out in an aqueous acetone-H_2SO_4(20%)-(Hg) system to give 2-imino-4,4,5,5-tetramethyloxazolidine **212** in 90.8% yield (eq. (2.94)) [285].   Electroreductively activated carbonyl groups arising from tetralones and indanones undergo alkylation with alkyl halides at their aromatic nuclear and carbonyl groups.   Electrolysis of 2-methyl-2-phenyltetralone **213** in a DMF-Bu$_4$NI-(Hg) system in the presence of alkyl halide affords 6-alkylated products **214** in moderate yields (eq. (2.95)) [286].

200 + H$_2$N—CN   →   aq. Acetone-H$_2$SO$_4$(20%)-(Hg)   90.8% yield   **212**   (2.94)

213   1) DMF-Bu$_4$NI/RX-(Hg)   2) O$_2$ **214** R = t–Bu, 60%   Bu, 40%   (2.95)

Electroreductive acetylation of benzophenone is performed in an MeCN-Et$_4$NBr-(Cu/Hg) system in the presence of acetic anhydride to provide the acetylated product **215** in 66% yield (eq. (2.96)) [287].   The ratio between 1,2- and 1,6-addition (**217/218**) in the reactions of electrogenerated fluorenone anion radical derived from **216** with RX in THF, leading to the compound **219** (1,2-adduct), has been found to be similar to the ratio obtained in the Grignard reaction of fluorenone with RMgX in THF (eq. (2.97)) [288]. Electroreduction of fluorenone **216** in an MeCN-Bu$_4$NPF$_6$-(Pt/Hg) system in the presence of 4-chlorobutanoyl chloride gives O-acylated products **221**, **222**, and **223** along with **224** (eq. (2.98)) [289].   The anion radical intermediate, which may form by a one-electron reduction, can react with acyl chloride to form a neutral radical and subsequent one-electron reduction affording an anion intermediate **220**.   The cathodic reduction of 2,2,4,4-tetramethylcyclobutan-1,3-dione **225** in a THF-Bu$_4$NBF$_4$-(Mg/Pt) system in the presence of acetic anhydride affords a mixture of 3-acetoxycyclobutan-1-one **226** and 1,4-diacetoxy-5,5,6,6-tetramethylbicyclo[2.1.1]hexan-2-one **227** (eq. (2.99)) [290].   Allylation takes place preferentially at the less substituted carbon side when the carbonyl group is more easily reduced than allyl halides [291].

Ph₂C=O → MeCN-Et$_4$NBr-(Cu/Hg)   66% yield   **215**   (2.96)

217 1,2-addition   **218** 1,6-addition   → RMgX / **216** / e⁻ / RX   **219**   (2.97)

61~94% yields   R = Me, Bu, PhCH$_2$

$$(2.98)$$

221 3.5%   **222** 17%   **223** 8%   **224** 11%

$$(2.99)$$

225 $R^1 = R^2 = Me, -(CH_2)_5-$

	226	**227**
Yield	35%	12%

Fluorenone-Cr(CO)$_3$ complexes **228** are electroreduced in a DMF-Bu$_4$NPF$_6$-(Pt/Hg) system in the presence of 1,3-dibromopropane as an electrophile, affording the spiro ether product **229** in 64% yield (eq. (2.100)) [292].   Unusual homocoupling of 4-chloro-3-formyl-2H(I)-benzopyran **230** proceeds in an EtOH-H$_2$O system, giving **231** (eq. (2.101)) [293a].   The diastereoselective electrosynthesis of d,l-(2R, 4S, 6R)-6-[(Z)-1'-bromo-2'-phenylethenyl]-2,4-dimethyltetrahydropyran-2,4-diol (41% yield) has been accomplished by electrolysis of α-bromo-cis-cinnamaldehyde in an Acetone-LiClO$_4$-(Hg) system [293b].

Ar^1, Ar^2 = C$_6$H$_4$-(C$_6$H$_4$)Cr(CO)$_3$

$$(2.100)$$

$$(2.101)$$

2.5   Electroreduction of Enones and Enoates

2.5.1   Electroreduction of Enones

2.5.1.1   Electroreductive Hydrogenation and Carboxylation of Enones

The achievement of selective and electrocatalytic hydrogenation of enone systems is an important goal in electrosynthesis. An early communication demonstrates some attempts on the electroreduction of α, β-unsaturated aldehydes and ketones, which involve acrolein, crotonaldehyde, cinnamaldehyde etc., leading to the corresponding either unsaturated alcohols or saturated aldehydes and ketones [294]. Electrocatalytic hydrogenation of conjugated enones 232 takes place in a MeOH/H$_2$O(1/1)-NaCl(0.1M)/H$_3$BO$_3$(0.1M) system by use of a fractal nickel electrode to give the corresponding saturated ketones 233 in 80-100% yields (current eff. 25-75%) (eq.(2.102a)) [295]. Partial electrochemical reduction of the carbonyl moieties of acetyl and aroyl functional groups attached to chromones at position _ proceeds in a DMF-Bu$_4$NBF$_4$(0.1M)-(Hg) system to give either the alcohol 235a (Y = CHOH-Ph) from the 2-benzoyl derivative 234 (R = Ph) (eq. (2.102b)) or the saturated ketone 235b (Y = CO-Me) (eq. (2.102c)) from the 2-acetyl derivative 234 (R = Me), respectively [296]. The quantitative hydrogenation of the double bond bearing electron-withdrawing groups in an ROH/H$_2$O-(Hg) system has been realized. Under the conditions, 4-(ethoxycarbonyl)coumarin can be hydrogenated to give the corresponding dihydrocoumarin in quantitative yield [297]. The regioselective electrohydrogenation of the C = C bond of 2-substituted 6-methoxycarbonyl-4H-1,3-thiazine-4-ones 236 has been attained in an AcOH/EtOH(1/1v/v)-AcONa(0.5M)-(Hg) system to give the corresponding 2,3,5,6-tetrahydro-4H-1,3-thiazine-4-ones 237 in 86~95% yields (eq. (2.103)) [298].

$$R^1, R^3 = H, alkyl; R^2 = alkyl, Ph$$

$$R \underset{N_3}{\overset{S}{\underset{4}{\bigcirc}}}{}^{1}_{5}{}^{6}CO_2Me \xrightarrow[-1.5 \text{ V (SCE), } 86\sim95\% \text{ yields}]{\text{AcOH/EtOH(1/1v/v)-AcONa(0.5M)-(Hg)}} R \underset{N}{\overset{S}{\bigcirc}}CO_2Me \tag{2.103}$$

236   R = EtO, Ph, PhCH$_2$S   **237**

Electroreduction of tetramethyl bicyclo[3.3.0]oct-1-en-3,7-dione-2,4,6,8-tetracarboxylate **238** in an H$_2$O-NaClO$_4$(0.25M)/HCl-(Pt/Hg) system affords the corresponding saturated bicyclo product **239** in 77% yield (eq. (2.104)) [299]. Electrohydrogenation of (E)-4-(2-furyl)but-3-enone **240** has been performed in an EtOH/H$_2$O(1/2)-NaHCO$_3$-(Pb) system to give the corresponding saturated ketone **241** (eq. (2.105)) [300]. Asymmetric electro-hydrogenation is carried out in an EtOH/H$_2$O(1/1)-AcOH/AcONa-(Hg) system in the presence of brucine acetate at pH 4.7 to afford ethyl 3,4-dihydrocoumarin-4-carboxylate in a ca. 20% optical yield. The catalytic powder electrodes (Raney-Ni and Pd (or Pt))-carbon systems are found to be effective on hydrogenation of 2-cyclohexen-1-one, leading to cyclohexanone (conv. 88%, select. 100%) [301a]. Mesityl oxide and trans-cinnamaldehyde was also hydrogenated under similar conditions [301b].

$$\text{238} \xrightarrow[77\% \text{ yield}]{\text{H}_2\text{O-NaClO}_4(0.25\text{M})/\text{HCl-(Pt/Hg)}} \text{239} \tag{2.104}$$

238   **239**

$$\text{240} \xrightarrow[90\% \text{ yield}]{\text{EtOH/H}_2\text{O}(1/2)\text{-NaHCO}_3\text{-(Pb)}} \text{241} \tag{2.105}$$

240   **241**

The enone system of acecyclone is electrochemically reduced in the presence of proton donors, yielding cis- and trans-isomers [184]. The electroreductive hydrogenation of **242** is carried out in a DMF-Bu$_4$NBF$_4$-(Pt/Hg) system in the presence of benzoic acid (or acetic acid) as a proton donor to give the partially hydrogenated product **243** in 88~89% yields (eq. (2.106)). The double bond of 4,4'-dimethoxybenzalacetophenone is also hydrogenated in an aqueous EtOH(40%v/v)-Britton-Robinson Buffer system [302]. The electrohydrogenation of the carbon-carbon double bond of enone systems such as 2-benzylidene-1-acenaphthenones proceeds through the reduction of the carbonyl group at the initial stage in an aqueous-alcoholic media, leading to the corresponding hydrogenated product **245** (eq. (2.107)) [303]. The influence of the substituents attached to the phenyl group on the reactivity of the enone group is discussed. The nitro group as a strong electron-withdrawing group attached to the benzene ring of aurones **246** causes the formation of hydrogenated products **248** under electrolysis in a DMF-Bu$_4$NBF$_4$(0.1M)-(Hg) system (eq. (2.108)) [304]. In contrast, the methoxy group as an electron-donating group tends to give the hydrodimerization product (**246**→**247**). The electroreductive conversion of the 1,4-benzoxyzine-8-one derivative **249** has been performed in an MeOH-Et$_4$NClO$_4$-(Pt) system to afford the phenolic product **250** in 80% yield (eq. (2.109)) [305].

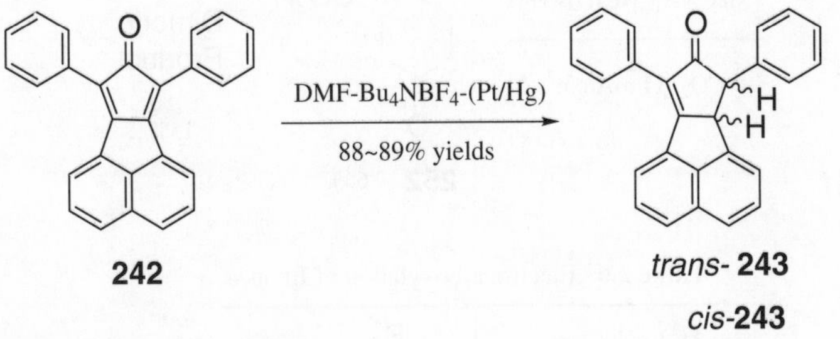

$$\text{(2.106)}$$

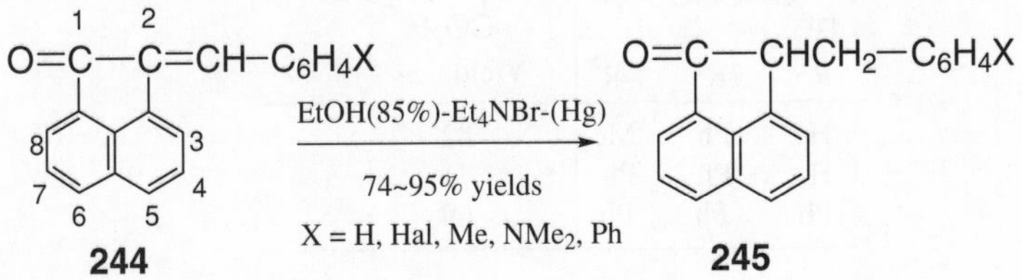

242   *trans*- **243**

        cis-**243**

$$\text{(2.107)}$$

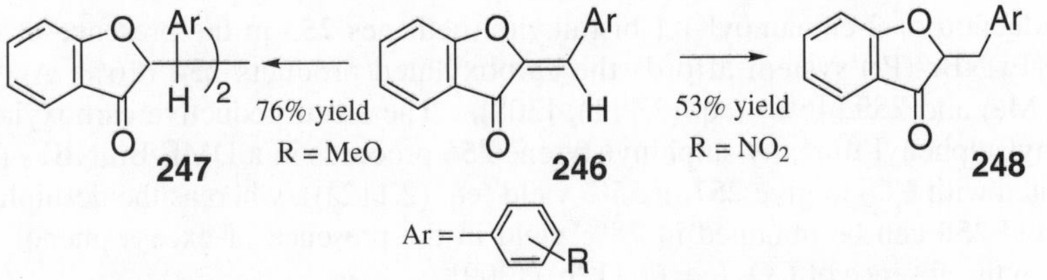

244   **245**

DMF-Bu₄NBF₄-(Pt/Hg)

88~89% yields

EtOH(85%)-Et₄NBr-(Hg)

74~95% yields

X = H, Hal, Me, NMe₂, Ph

$$\text{(2.108)}$$

247   76% yield   **246**   53% yield   **248**

R = MeO   R = NO₂

Ar = ⟨⟩—R

$$\text{(2.109)}$$

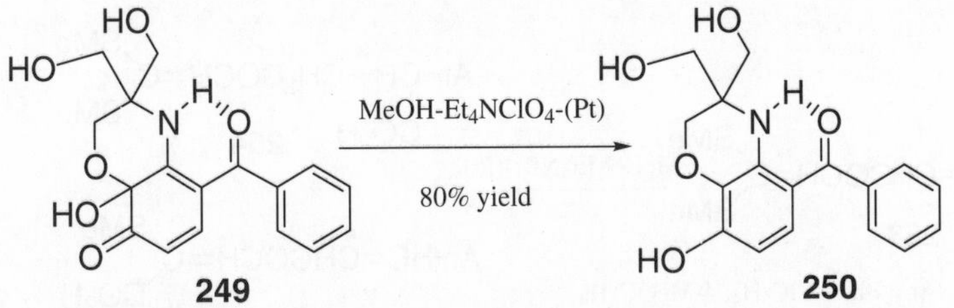

MeOH-Et₄NClO₄-(Pt)

80% yield

249   **250**

The reductive cross-coupling reaction of active enones with carbon dioxide has been carried out in an MeCN-Et₄NOTs-(Hg) system to give carboxylated products in 22~52% current efficiencies [42].   Initial large-scale electroreductive carboxylation of benzalacetophenone has been investigated in a DMF-KI-(Hg) system, affording the carboxylated products in 20~30% yields [306]. Carboxylation of benzalacetone also proceeds to give α-phenyl levulinic acid under conditions similar to those above.   An improved method is performed by lowering the current density in an MeCN-Et₄NClO₄-(Hg) system and the desired γ-keto acids are obtained in 35~82% yields [307].   For example, the electroreduction of 2-cyclohexen-1-one **251** under the lower current density conditions in an MeCN-Et₄NClO₄-(Hg) system by bubbling carbon dioxide gives the corresponding carboxylated product **252** as a major product (64%) together with a dimeric product (15%) (eq. (2.110)) [307].   Some results are listed in Table 2.6.

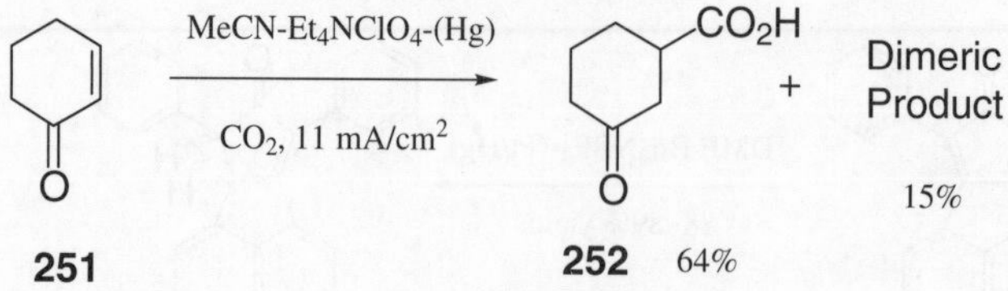

$$\text{251} \xrightarrow[\text{CO}_2,\ 11\ \text{mA/cm}^2]{\text{MeCN-Et}_4\text{NClO}_4\text{-(Hg)}} \text{252} \quad + \quad \text{Dimeric Product} \qquad (2.110)$$

251   **252** 64%   15%

Table 2.6 Electrocarboxylation of Enones

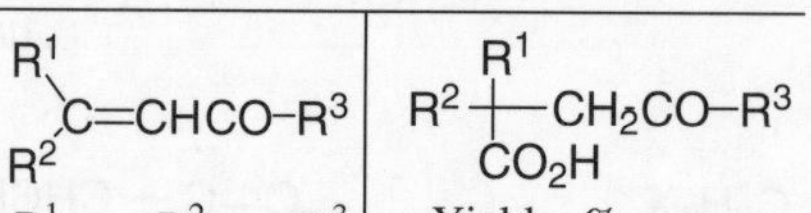

R^1 $C=CHCO-R^3$ R^2		R^1 $R^2-C-CH_2CO-R^3$ CO_2H	
R^1	R^2	R^3	Yield , %
H	Ph	Me	82
H	Ph	Ph	71
Ph	Ph	Ph	69

Electroreduction of 2-cinnamoyl-1,1-bis(alkylthio)ethenes **253** in the presence of CO_2 in an MeCN-Et$_4$NBr-(Pt) system affords the carboxylated products **254** (26%; Ar = p-ClC$_6$H$_4$; R = Me) and **255** (46%) (eq. (2.111)) [308].   The electroreductive carboxylation of (E)-ß-methylsulphenyl-ß-methylsulphinylstyrene **256** proceeds in a DMF-Bu$_4$NBF$_4$-(Hg) system saturated with CO_2 to give **257** in 55% yield (eq. (2.112a)), whereas the desulphinylated compound **258** can be obtained in 78% yield in the presence of excess phenol as a proton donor in the absence of CO_2 (eq. (2.112b)) [309].

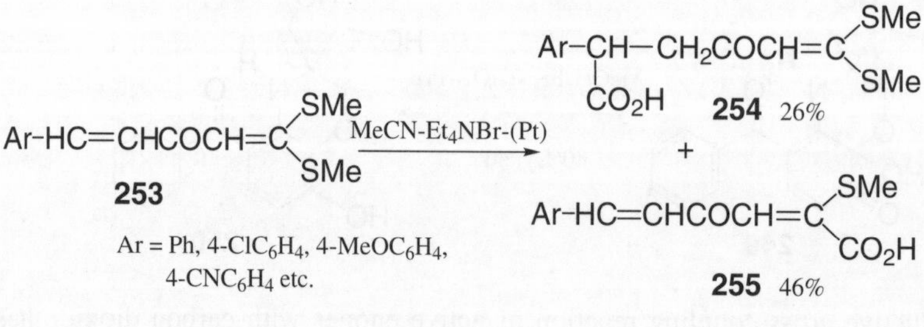

$$\text{Ar-HC=CHCOCH=C}\begin{smallmatrix}\text{SMe}\\\text{SMe}\end{smallmatrix} \xrightarrow{\text{MeCN-Et}_4\text{NBr-(Pt)}}$$

253

Ar = Ph, 4-ClC$_6$H$_4$, 4-MeOC$_6$H$_4$,
4-CNC$_6$H$_4$ etc.

$$\text{Ar-CH-CH}_2\text{COCH=C}\begin{smallmatrix}\text{SMe}\\\text{SMe}\end{smallmatrix}$$
$$\overset{|}{\text{CO}_2\text{H}}$$
254 26%

+

$$\text{Ar-HC=CHCOCH=C}\begin{smallmatrix}\text{SMe}\\\text{CO}_2\text{H}\end{smallmatrix}$$
255 46%

(2.111)

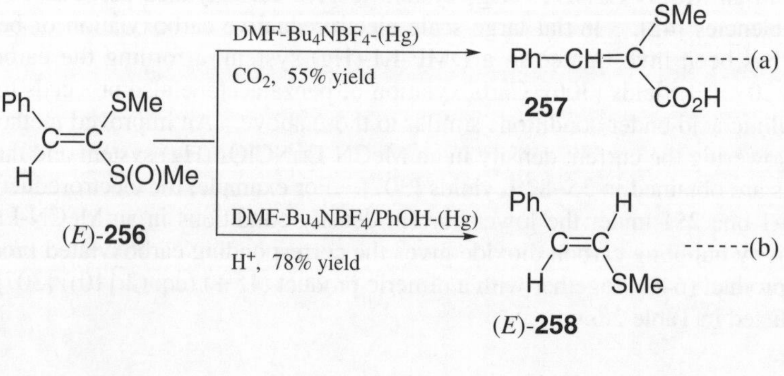

$$(E)\text{-}\mathbf{256} \quad \begin{matrix}\text{Ph}\\\text{H}\end{matrix}\text{C=C}\begin{matrix}\text{SMe}\\\text{S(O)Me}\end{matrix}$$

$$\xrightarrow[\text{CO}_2,\ 55\%\ \text{yield}]{\text{DMF-Bu}_4\text{NBF}_4\text{-(Hg)}} \text{Ph-CH=C}\begin{smallmatrix}\text{SMe}\\\text{CO}_2\text{H}\end{smallmatrix} \text{-----(a)}$$
257

(2.112)

$$\xrightarrow[\text{H}^+,\ 78\%\ \text{yield}]{\text{DMF-Bu}_4\text{NBF}_4\text{/PhOH-(Hg)}} \begin{matrix}\text{Ph}\quad\text{H}\\\text{C=C}\\\text{H}\quad\text{SMe}\end{matrix} \text{-----(b)}$$

(E)-**258**

2.5.1.2   Electrochemical Hydrodimerization of Enones

The electroreduction of conjugated enones has been shown to yield hydrodimers in acidic solutions and saturated ketones in alkaline solutions [310]. Electroinitiated polymerization of acrolein has been achieved by controlled potential electrolysis at the reduction peak potential of the monomer [311]. Hydrodimerization of activated olefins are discussed in Chapter 3.3.1. The mechanism of electrohydrodimerization of 2-cyclohexen-1-ones in an aqueous systems is well discussed in the literature [294, 312]. A more comprehensive study on the electroreduction of the enone system in an aqueous media over a wide pH range has been reported [313]. Detailed cyclic voltammetry investigations on the enone systems reveal characteristic behaviors of reactive species in non-aqueous media [314, 315, 316, 317].

Investigation on the electrohydrodimerization of 2-cyclohexen-1-one **259** in buffered aqueous ethanolic solutions (20% EtOH, v/v) over the pH ranges of 1.0~12.1 reveals that two kinds of one-electron reductions proceed depending on the solution pH: at pH > 8, the hydrodimers **260** and **261** are formed by coupling of the neutral radical generated *via* a one-electron reduction of the protonated form of **259**, and at pH < 5, **260** and **261** are formed by an anion radical formation-protonation-coupling sequence (eq. (2.113)) [316]. The hydrodimerization of 2-cyclohexen-1-one, 4,4-dimethyl-2-cyclohexen-1-one, and 4,4-diphenyl-2-cyclohexen-1-one has been performed in buffered aqueous ethanolic solutions (50% v/v). The cyclohexenones undergo two one-electron processes corresponding to the reduction of their respective protonated and unprotonated forms. In the pH range *ca.* 5.5~8.5, the two processes of each substrate compete and their relative contributions are given by the rate of establishment of the equilibrium between their unprotonated and protonated forms [318]. The product distribution of dimers **263**, **264**, and **265** formed in the electrohydrodimerization of 2-cyclohexen-1-ones **262** is shown as function of the water content in MeCN (eq. (2.114)) [319]. The presence of water favors the formation of the glycols **265** (Table 2.7, Fig. 2.1). The influence of water content of the solvent on the product distribution in the hydrodimerization of 2-cyclohexen-1-ones has been elucidated [314c, 319]. The mechanism in non-aqueous media has also been investigated [314a]. The electrolysis conditions of the hydrodimerization of cyclohexenone and the composition of the products are summarized in Table 2.8 [317].

(2.113)

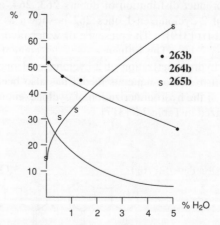

(2.114)

262a: $R^1=R^2=R^3=H$

b: $R^1=R^2=H$, $R^3=Me$

c: $R^2=H$, $R^1=R^3=Me$

d: $R^1=H$, $R^2=R^3=Me$

Table 2.7 Relative Product Distribution of Hydrodimers **263**, **264**, and **265** in the Electrolysis of **262** (2.10^{-1}M in CH$_3$CN)

Substrate	262a			262b			262c			262d		
Product	263a	264a	265a	263b	264b	265b	263c	264c	265c	263d	264d	265d
0% H$_2$O	97	–	–	52	31	16	10	40	50	10	60	30
5% H$_2$O	95	–	–	28	4	67	–	10	90	–	45	50

vs. water content (error: ± 5%)

Fig. 2.1 Variation of product distribution of hydrodimers **263b–265b** with the water content (v/v) of the solvent.

Effect of additives on electroreduction of 3-phenyl-2-cyclohexen-1-one **268** has been studied. In contrast to the non-substituted 2-cyclohexen-1-one **259**, the major coupling occurs at the 1,3- and 1,1-positions, leading to **270** and **271** in the presence of additives at the potential of –1.80 V (SCE). The electrolysis at the potential of –2.40 V provides 2,3-hydrogenation product **272**, which may be the result of the reduction with lithium metal amalgam (eq. (2.115)) [320].

Table 2.8 Hydrodimerization of Cyclohexenone

Electrolysis Media	Reduction Potential, V	**266**	Yield, % **267**
MeCN–HCl/KCl	−1.00 −1.05	59	21
MeCN/AcOH–AcONa	−1.45 −1.54	64	18
MeCN–NaOH/KCl	−1.65 −1.70	72	8

[Reproduced with permission from E. Touboul et al., *C. R. Acad. Sci. Paris, Sér. C*, **268**, 1170 (1969)]

$$\text{268} \xrightarrow{\text{DMF-Bu}_4\text{NI-(Pt/Hg)}} \textbf{269} + \textbf{270} + \textbf{271} + \textbf{272} \qquad (2.115)$$

Additive	Potential (SCE)	**269**	**270**	**271**	**272**
			Yield, %		
None	−1.80	–	–	2.9	–
Na⁺ (200 mM)	−1.80	–	72	–	–
H₂O (10%)	−1.80	2.5	1.7	80	–
Li⁺ (20 mM)	−2.40	1.8	18	1.8	8.4

Preparative electrolysis of **273** in a DMF-Bu$_4$NBF$_4$-(Hg) system at the potential −2.3 V affords the tetracyclic compound **274** (eq. (2.116) [321].

$$\textbf{273} \xrightarrow[-2.3 \text{ V (Ag/Ag}^+)]{\text{DMF-Bu}_4\text{NBF}_4\text{-(Hg)}} \textbf{274} \text{ major product} \qquad (2.116)$$

The effect of pH on the electrohydrodimerization of 3-buten-2-one is also discussed under similar electrolysis conditions [322].

The electroreductive coupling of the α, β-enone **275** in a DMF-Et$_4$NCl-(Hg) system in the presence of dimethyl sulfate as an alkylating agent affords a mixture of **276** (80%) and **277** (5~15%), respectively (eq. (2.117)) [323].

(2.117)

2.5.2   Electroreduction of Dienones and Quinones

Quinones demonstrate a variety of redox reactions accompanying protonation reactions. The thermodynamic parameters of these reactions (standard redox potential $E°$ and pK values) depend on the substituents of the quinone ring as well as on the environment (*i.e.*, solvent) which surrounds the molecule [324].   Prominent features of quinone homologues which involve investigations on cation binding properties of calixquinones [325, 326], four-electron redox properties [327], and standard electrochemical parameters have been unveiled [324].

The electroreduction of *p*-benzoquinone in DMF or MeCN exhibits two well-defined one-electron steps separated by *ca.* 0.6 V (SCE) (eq. (2.118) [328].   In an aqueous solution, *p*-benzoquinone is electroreduced in aqueous perchloric acid, first to produce the monoprotonated cation, HQ^+, which is found to be reducible at more positive potentials than neutral quinone.   The percent yields of hydroquinone formation under different electrolysis conditions with two equivalents of acid present during electrolysis are shown in Table 2.9. [330, 331, 332].   The reversibility of the electroreduction of quinones increases in the order benzoquinone < naphthoquinone $\cong$ phenanthraquinone < anthraquinone [329].   Quinones are initially reduced to the quinol ions [332].

Table 2.9 Hydroquinones from Quinones under Different Conditions

Compound	Hydroquinone Conditions, Yield, (%)		
	A	B	C
1,4-benzoquinone	90	90	90
9,10-phenantraquinone	100	78	55
1,4-naphthaquinone	80	80	80
9,10-anthraquinone	33	75	60

Conditions : (A) Cathodic two electrons per mole electrolysis just before
the first reduction potential.
(B) Beyond the first reduction potential.
(C) Beyond the second reduction potential of the quinone.

$$Q + e^- \longrightarrow Q^{-\bullet} \quad \text{............ (a)}$$
$$Q^{-\bullet} + H^+ \text{(solvent)} + e^- \longrightarrow HQ^- \quad \text{............ (b)}$$
$$HQ^- + H^+ \longrightarrow H_2Q \quad \text{............ (c)}$$
$$Q + 2e^- \longrightarrow Q^{2-} \quad \text{............ (d)} \quad (2.118)$$
$$Q^{2-} + H^+ \text{(solvent)} \longrightarrow HQ^- \quad \text{............ (e)}$$
$$HQ^- + H^+ \longrightarrow H_2Q \quad \text{............ (f)}$$

$$Q = \text{quinone}$$

Hydroquinones work as a reducing agent for oxygen. The 8-hydroxy-1,2-naphthohy-droquinone-3,6-disulfonic acid derived from its quinone by electroreduction can be used for the above purpose [333]. The electroreductive addition of ClCO$_2$Me to 9,10-phenanthrene quinone has been attained in an MeCN-Et$_4$NBF$_4$(0.2M)-(Pt) system, and carbon dioxide also can be trapped under similar conditions [334]. The reduction of indigo **278** to the disodium salt of leuco-indigo **281**, which is soluble in alkaline solution, is crucial to the application of indigo as a textile dye. An indirect electroreduction procedure, involving iron(II)/iron(III) redox complexes **279** and triethanolamine (TEA), stable enough in alkaline medium, has been developed [335]. The electrolysis is carried out in 1 L of an aqueous TEA(40g)/Fe$_2$(SO$_4$)$_3$(6g)/NaOH(10g)-(SUS/Cu) system in the presence of **278** (5g) in a cathode cell (eq. (2.119)).

(2.119)

Electroreduction of 2,3-disubstituted 1-indenones **282** in a DMF-Bu₄NBF₄-(Pt/Hg) system in the presence of acetic anhydride affords exclusively the *O*-acetylated products **283** and **284** in good yields (eq. (2.120)) [336]. 2,3-Dimethylindenone **282** also gives the corresponding acetate dimers **286** in 83~95% yield. The reduction of **282** in the presence of hydroquinone as a proton donor proceeds through an ECE mechanism [337]. In contrast, a new type of electrochemical dimerization of **282** has been reported. The dimerization proceeds in a DMF-Bu₄NI-(Hg or C) system in the presence of a proton donor at − 0.8 V (Ag/AgI) to give a mixture of isomeric dimers **286** in good yields (eq. (2.121)) [338].

$$(2.120)$$

$$(2.121)$$

2.5.2ER84162 ver.0.1/93.2.5

The electrochemistry of a cyclopentadienone system has been intensively investigated in order to accumulate theoretical knowledge on aromaticity [339]. The electrochemical behavior of tetraphenylcyclopentadienone **287** has been well elucidated. The controlled-potential electrolysis of **287** in a DMF-Bu₄NBF₄-(Pt/Hg) system provides first a relatively stable anion radical after one-electron reduction. In the presence of a proton donor, the anion radical yields a trans-reduction product **288** *via* prototropy of the enolic form (eq. (2.122)). When the anion radical is acetylated with acetic anhydride, the enol acetate **289** is also isolated as a major constituent (eq. (2.123)) [340]. Kinetics and effects for the protonation of various 2,3,4,5-substituted cyclopentadienone anion radicals have also been investigated [341].

$$(2.122)$$

$$(2.123)$$

The cathodic reaction of 4-pyranones **290** has been studied in protic and aprotic media. In protic mixed solvent systems, *e.g.*, H$_2$O/EtOH(1/1) and MeCN/H$_2$O(4/1), the electrolysis of 2,6-dimethyl-4-pyranone **290** in the presence of lithium chloride at -2.1 V (SCE) gives a dimer **291** (eq. (2.124)), whereas in an aprotic solvent, the electrolysis provides a ring-opening dimer **292** by the electroreductive cleavage of carbon-oxygen bonds (eq. (2.125)) [342].

$$\text{(2.124)}$$

$$\text{(2.125)}$$

The one-electron reduction of phenalen-1-one **293** in a DMF-Bu$_4$NClO$_4$-(Hg) system provides a very stable anion radical. On addition of acetic anhydride to the electrolysis system, the anion radicals are converted into a neutral acetoxyphenalene radical by the acetylation of the alcoholate. The latter intermediate is much less stable and undergoes an intermolecular coupling reaction to give a mixture of perylene **294** a (R^1 = R^2 = H) and its mono- and di-acetoxy derivatives **294** b (R^1 = OAc, R^2 = H) and **294** c (R^1 = R^2 = OAc) (eq. (2.126)) [343].

$$\text{(2.126)}$$

294 a. R^1 = R^2 = H   10%
b. R^1 = OAc, R^2 = H   15%
c. R^1 = R^2 = OAc   15%

As a very special case, 4,4-diphenyl-2,5-cyclohexadienone **295** can lead to the corresponding pinacol **296** by the electroreduction in a DMF-KI system (eq. (2.127))[344]. The electroreductive coupling reaction of 3,4-estrone-o-quinone **297** and adenine **298** occurs in a DMF-LiClO$_4$(0.1M)-(Pt) system to give the coupling product **299** (eq. (2.128)) [345].

$$\text{(2.127)}$$

$$(2.128)$$

Terpenoids bearing a 1,4-benzoquinone moiety often occur in plants and animals, and they play an important role in electron transfer processes in biological systems. Details of the avarol-avarone redox behavior (**300↔301**) in an MeCN-Et$_4$NClO$_4$-(Pt) system have been investigated (eq. (2.129)) [346]. The reduction of avarone **301** takes place in two steps by the formation of the stable radical anion (–0.55 V (SCE)), which can be reduced to the dianion at a more negative potential (–1.18V). The quinones **302** bridged with a polyether chain as a crowned redox system are known to behave as ligands for complexations with lithium, sodium, and potassium ions (eq. (2.130)) [347, 348]. Cyclic voltammetric studies of **302** in DMF provide evidence for the coupling between the complexation and redox reactions [348]. The twin benzoquinone type compound **304** as a novel crowned redox system shows highly enhanced cation-binding properties for Li$^+$ cation upon electroreduction [349]. A variety of hydroxy and methoxy-substituted 2-(acetoxymethyl)-9,10-anthraquinones **305** are electroreduced in an aqueous MeCN(50%)-KCl-(C) system under buffering with KH$_2$PO$_4$ and KOH (pH 7,8, and 9) in the presence of dithionite (S$_2$O$_4^{2-}$) (eq. (2.131)) [350, 351]. The substituted 2-methyl-9,10-anthraquinones **310** are formed from the anthraquinone acetates **306** or **307** (16~82% yields) via **308** or **309** at higher pH ranges. The structures of semiquinoid radicals of 1,2-dihydroanthraquinone (alizarin) and 1,2,4-trihydroxyanthraquinone (purpurin) generated by electroreduction are elucidated by EPR spectra [352]. A new conjugated function of photochromism and electrochromism of an azo-quinone system has been demonstrated [353].

$$(2.129)$$

$$(2.130)$$

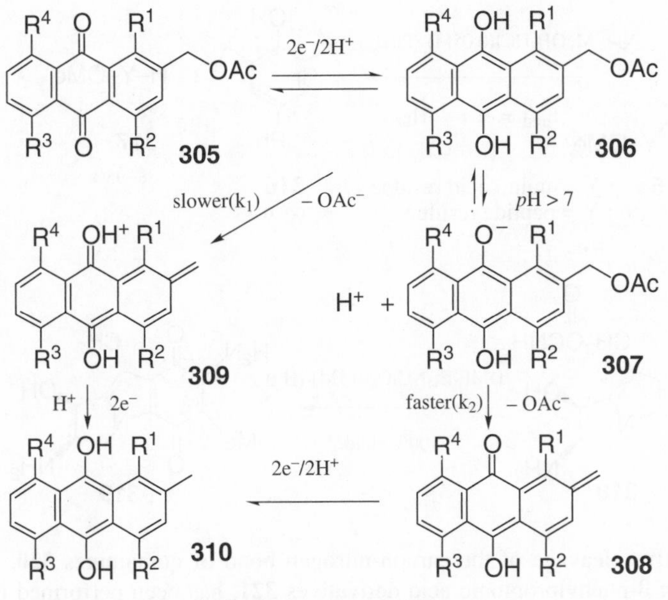

$$R^1, R^2, R^3, R^4 = H, OH, OMe, OAc$$

(2.131)

The electroreduction of anthraquinone **311** to anthrahydroquinone proceeds in an aqueous EtOH(50%)-H$_2$SO$_4$(50vol%)-(Hg) system, [354] and the second step is a two-electron reduction of protonated anthrahydroquine to anthrone **312** (eq. (2.132)). The optimun reduction conditions for the anthrone formation have been clarified [355]. Primary alcohols protected with the tritylone group **313** can be deprotected cathodically to give **314** in an MeOH-LiBr(0.1M)-(Pt/Hg) system at −1.4 V (Ag/AgCl) in good yields (eq. (2.133)) [356]. Deprotection of amino acids and peptide esters bearing di-*tert*-butylated 1-phenyl-4-oxo-2,5-cyclohexadienyl group **315** proceeds in an MeOH-HCl(0.05M)-(Pt/Hg) system to give the desired amino acids and/or peptide ester hydrochlorides **317** in 88~93% yields together with phenol **316** (90~94% yields) (eq. (2.134)) [357]. Flow cell reduction of 1,2-*cis*-2,7-diamino-1-hydroxymitosene **318** in a DMF-Et$_4$NClO$_4$(0.1M)-(Hg) system at 1.2~1.5 V (Ag/AgCl) affords the C-O bond cleavage product **319** as the sole product by the initiation of uptaking an electron at the quinone system (eq. (2.135)) [358, 359].

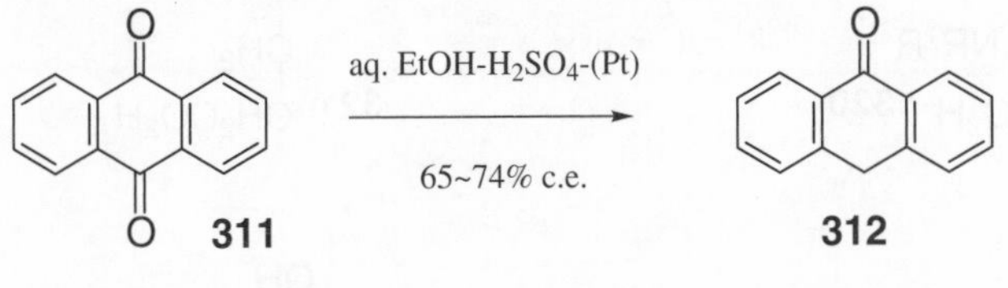

(2.132)

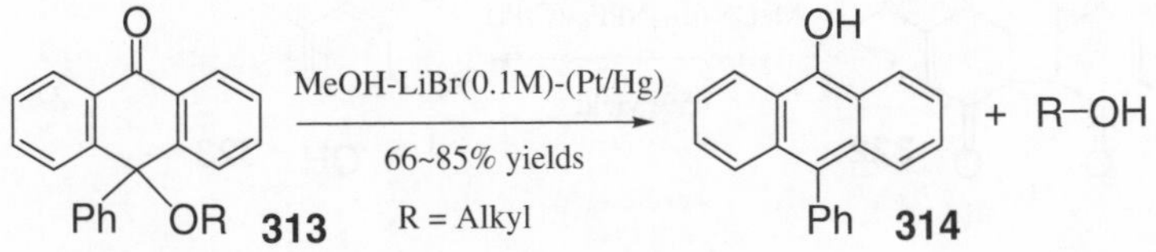

(2.133)

$$\text{(2.134)}$$

315 a : Y = amino acid residue   **316**
        b : Y = peptide residue   90~94%

317
88~93%

$$\text{(2.135)}$$

318   **319**

Electroreductive cleavage of the carbon-nitrogen bond of quinamines **320**, leading to the corresponding ß-phenylpropionic acid derivatives **321**, has been performed in a DMF-Bu$_4$NClO$_4$-(Hg) system at –2.0 V (SCE) (eq. (2.136)) [360].   The one-electron reduction of the anthracycline glycosides and daunomycin produces the corresponding semiquinone radical anion, and the two-electron reduction is followed by glycoside elimination. The rate of this elimination decreases markedly with increasing acidity of the medium [361]. Controlled potential electrolysis (–1.67 V *vs.* ferrocene) of dithymoquinone **322** in an MeCN-Bu$_4$NPF$_6$-(C/Pt) system at 0 °C affords thymohydroquinone **323** in 65% yield (eq. (2.137)) [362].   The antitumour anthrapyrazole CI-941 **324** undergoes two successive one-electron reductions in an aprotic solvent (DMF) and a one-step two-electron reduction in the presence of proton donors.   As a result, the corresponding leuco form **325** has been isolated in 97% yield (eq. (2.138)) [363].

$$\text{(2.136)}$$

320   **321**

$$\text{(2.137)}$$

322   **323**

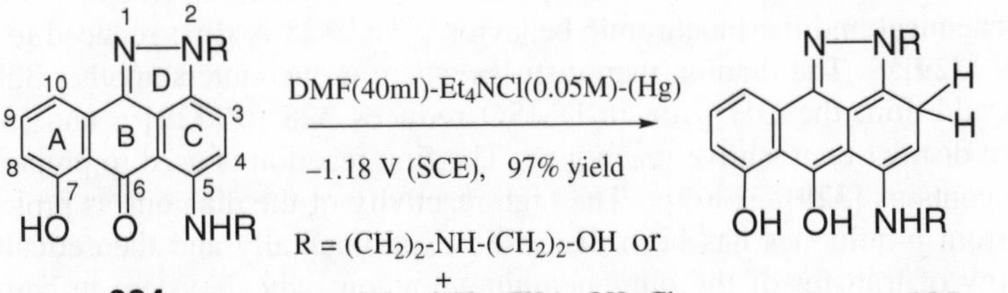

$$\text{(2.138)}$$

324

R = $(CH_2)_2$-NH-$(CH_2)_2$-OH or
R = $(CH_2)_2$-$\overset{+}{NH_2}$-$(CH_2)_2$-OH, Cl$^-$

325

Relatively stable quinonemethides have been subjected to cathodic reduction [364, 365, 366]. The lifetime of the electrogenerated radical-anions and dianions is governed by steric hindrance in the absence of an added electrophile. Hindered intermediates are relatively long-lived yet hydrogenated in the presence of a proton donor and alkylated in the presence of methyl iodide. Less hindered analogues rapidly dimerize at carbon, with concomitant protonation or O-methylation depending on the added electrophile. The electroreduction of the quinonemethide **326** in a DMF-Bu$_4$NCl$_4$(0.1M)-(Hg) system in the presence of methyl iodide gives the dimer **327** in 80% yield (eq. (2.139)) [366].

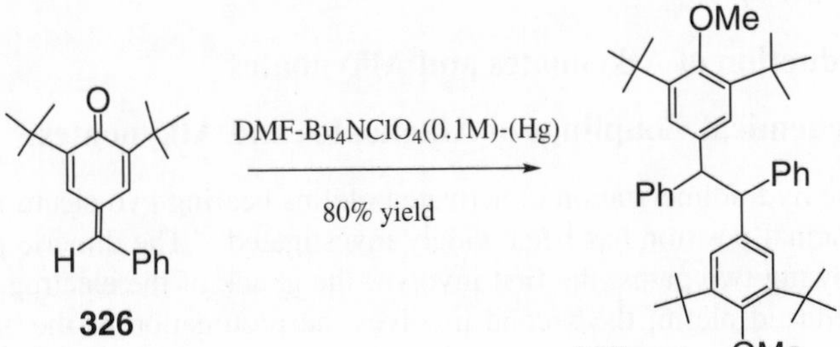

326

DMF-Bu$_4$NClO$_4$(0.1M)-(Hg)

80% yield

327

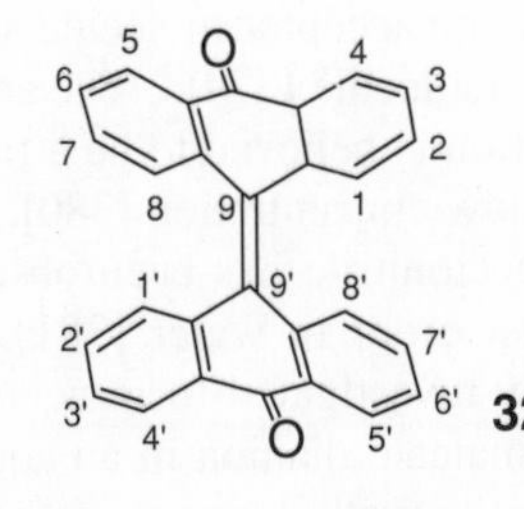

328

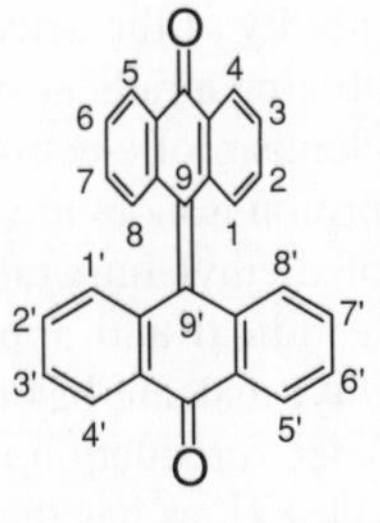

329

$$\text{(2.139)}$$

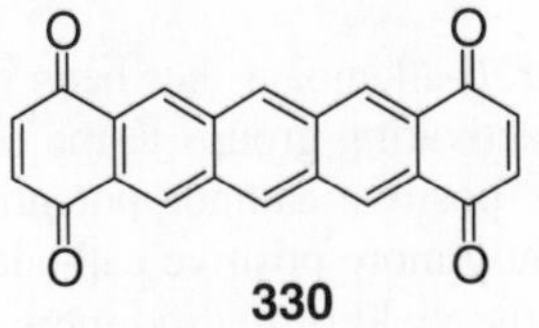

330

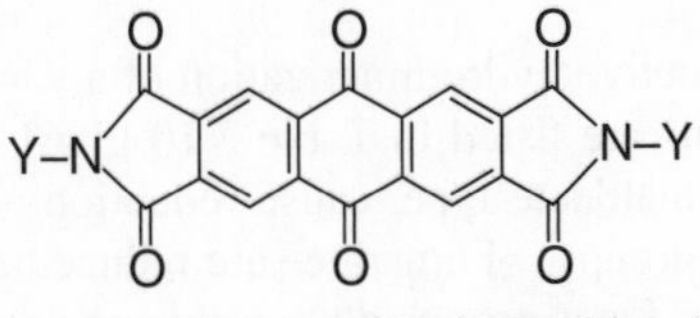

331   Y = 4-t-butylphenyl;
2,5-di-t-butylphenyl

Biathrone **328** is a sterically hindered quinone-type ethylene derivative which shows unique photochemical and thermochromic behavior. The 328 is first reduced to generate the dianion of $[329]^{2-}$. The dianion then diffuses where it encounters another 328 to give $[329]^{-\cdot}$. In addition, the OH$^-$ ion in DMSO reduces **328** to $[329]^{-\cdot}$ and $[329]^{-\cdot}$ to $[329]^{2-}$ in two distinct consecutive reactions. The first reaction is used to prepare a solution that only contains $[329]^{-\cdot}$ [367]. The high reactivity of the dianions as proton acceptors derived from p-quinones has been discussed experimentally and theoretically [368]. The free energy of transfer of the quinone/quinone anion radical system in benzonitrile, acetonitrile, propylene carbonate, DMF, and/or DMSO has been shown not to exceed 1.01 ± 0.3 kcal/mol and is almost negligible [369]. The linear aromatic compounds **330** and **331** possessing imide and quinone electron-acceptor groups are electroreduced to produce a solution of the corresponding anion radicals, and their near-infrared spectra have been recorded [370].

The polarographic reduction of chalcone derivatives has been investigated [371~376]. In acidic media, chalcone undergoes two-electron step reaction to yield a saturated ketone. But in neutral or alkaline solution, the reduction proceeds fast through the saturated ketone stage to yield the corresponding secondary alcohol. The electroreduction of 2,6-dibenzylidenecyclohexanone proceeds through a defined one-electron wave in acidic solutions [377]. The cyclic voltammograms of some aryldienone derivatives demonstrate two irreversible reduction waves [378].

2.5.3   Electroreduction of Alkenoates and Alkynoates

2.5.3.1   Electrochemical Couplings of Alkenoates and Alkynoates

The electroreductive hydrodimerization of activated olefins bearing two electron-withdrawing groups at the vicinal position has been widely investigated. The dimeric products are formed by the following two paths: the first involves the attack of the electrogenerated anion radical on unreduced olefin; the second involves the protonation of the anion radical followed by reduction to an anion and subsequent attack on the olefin. The isomer distribution in the dimeric product obtained by the first route is rationalized on the basis of the relative anion stabilizing ability of the activating groups in the acceptor molecule and the relative ability toward stabilizing a radical site in the donor molecule [379]. The anion-radical substrate coupling mechanism of electrohydrodimerization is believed to be a favorable reaction pathway when proton donors are kept at a very low concentration [380]. The electrohydrodimerization of diethyl fumarate in aqueous acetonitrile has been observed to be second order in anion radical and approximately first order in water [381]. The hydrodimerization of diethyl maleate has been intensively investigated for practical preparation [382, 383]. The electroreduction mechanism of maleate dianion in an aqueous solution has been discussed [384], as has the protonation of the methyl cinnamate anion radical [385].

Electroreductive hydrodimerization of a variety of α, β-alkenoates has been investigated. The results are listed in Table 2.10 [386]. Two activating groups at the α-position, *e.g.* ethylidenemalonate type, cause reduction at a more positive cathode potential than at one activating groups. Fumarates are reduced at a slightly more positive cathode voltages than maleates. Ester groups do not change drastically the yield of hydrodimers. Two β-alkyl groups do not nullify the hydrodimerization process but additional alkyl groups at the α position cause hydrocoupling to cease .

Hydrocoupling of ethyl cinnamate in mixed solvent systems, *e.g.*, DMSO/EtOH, is

Table 2.10 Hydrodimerization of Enoates

W	X	Y	Z	V(SCE)	Conditions	Hydrodimer Yield, %
H	H	H	OEt	−1.87	DMF–MeEt$_3$NOTs–(Hg)	74–87
Me	Me	H	OEt	−2.10~−2.18	MeCN–Et$_3$NOSO$_2$Ph–(Hg)	66.2
CO$_2$Et	H	H	OEt	−1.32~−1.40	DMF–Et$_3$NOTs–(Hg)	61.5
H	CO$_2$Et	H	OEt	−1.19~−1.30	DMF–Et$_4$NOTs–(Hg)	28
H	Ph	H	OEt	−1.57~−−1.61	H$_2$O–MeBu$_3$NOTs–(Hg)	28
H	Me	H	NEt$_2$	−2.03~−2.12	H$_2$O–Et$_4$NOTs–(Hg)	61.4
H	H	H	NEt$_2$	−1.91~−1.95	H$_2$O–Et$_4$NOTs–(Hg)	73.3
H	H	H	NH$_2$	−1.82~−2.00	H$_2$O–Et$_4$NOTs–(Hg)	39.6
EtO	H	H	OEt	−2.22	DMF–MeBu$_3$NOTs–(Hg)	57
CO$_2$(2-Et)Hex	H	H	O(2–Et)Hex	−1.41	EtOH–MeBu$_3$NOTs–(Hg)	66
Bu	H	H	OBu	−1.30	None–MeBu$_3$NOTs–(Hg)	70
H	CO$_2$(2-Et)Hex	H	O(2–Et)Hex	−1.22	EtOH–MeBu$_3$NOTs–(Hg)	80
H	Ph	CO$_2$Et	OEt	−1.38~−1.47	DMF–Et$_4$NOTs–(Hg)	45
H	Me	CO$_2$Et	OEt	−1.41~−1.68	DMF–Et$_4$NOTs–(Hg)	90

found to depend on the relative amounts of protic and aprotic components in the mixture. The logarithm of the dimerization rate constant of the ethyl cinnamate radical anions is found to depend linearly on acidity function [387]. Electrodes coated with LB-films of dimethyldioctadecyl-ammonium poly(vinyl sulfonate) are used for hydrodimerization of enoates such as ethyl cinnamate, dimethyl maleate, and acrylonitrile [388]. The effect of the mass-transfer rate on the electroreduction of methyl cinnamate has been reported [389].

In the prepilot scale synthesis of ethyl dimalonate **337** and dimethyl adipate **334**, the effect of mixing on paired dimerizations has been studied in a variety of flow cells operated with batch recycling. The cross-coupling of methyl acrylate **332** with malonate anion **335** affords Michael adduct **336**. The yields of normal Michael adducts **336** and ethyl malonate dimer **337** can be improved up to 30% and 20%, respectively (eq. (2.140)) [382]. The electrolysis of methyl benzylideneacetoacetate **338** under similar conditions leads to the formation of only *trans-trans* isomer of cyclopentene derivative **339** (eq. (2.141)) [390].

The electroreduction of substituted 3-(2-furyl)-acrylonitriles **340** in an aqueous EtOH/DMF-Et$_4$NI(0.2M)-(Hg) system gives a mixture of the *dl*-hydrodimer **341** and the enamine **342** (*cis*-configuration) in 80% yield (eq. (2.142)) [391].

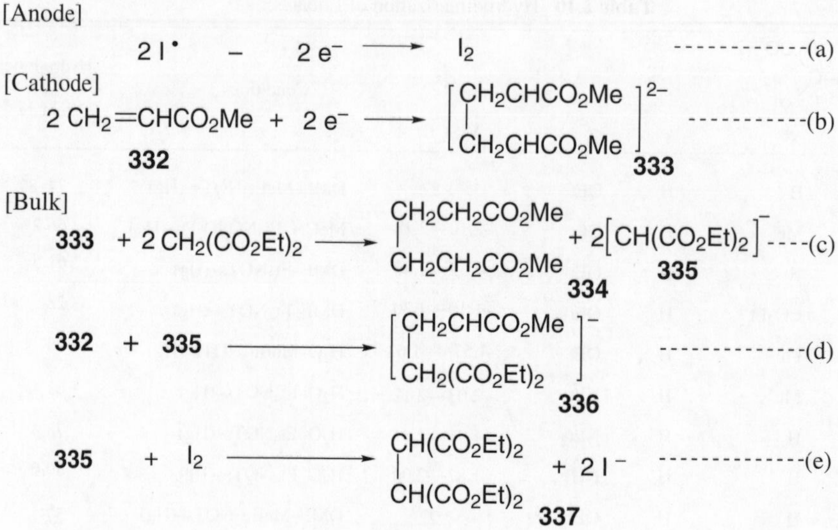

[Anode]

$$2\,I^{\bullet} \;-\; 2\,e^- \;\longrightarrow\; I_2 \qquad\qquad\qquad \text{----------(a)}$$

[Cathode]

$$2\,CH_2{=}CHCO_2Me \;+\; 2\,e^- \;\longrightarrow\; \begin{bmatrix} CH_2CHCO_2Me \\ | \\ CH_2CHCO_2Me \end{bmatrix}^{2-} \quad \text{----------(b)}$$

$$\underset{\mathbf{332}}{} \qquad\qquad\qquad\qquad\qquad\qquad\qquad\qquad \underset{\mathbf{333}}{}$$

[Bulk]

$$\mathbf{333} \;+\; 2\,CH_2(CO_2Et)_2 \;\longrightarrow\; \begin{matrix} CH_2CH_2CO_2Me \\ | \\ CH_2CH_2CO_2Me \end{matrix} \;+\; 2\Big[CH(CO_2Et)_2\Big]^- \text{----(c)} \quad (2.140)$$

$$\underset{\mathbf{334}}{}\qquad\qquad\qquad \underset{\mathbf{335}}{}$$

$$\mathbf{332} \;+\; \mathbf{335} \;\longrightarrow\; \begin{bmatrix} CH_2CHCO_2Me \\ | \\ CH_2(CO_2Et)_2 \end{bmatrix}^- \qquad \text{----------(d)}$$

$$\underset{\mathbf{336}}{}$$

$$\mathbf{335} \;+\; I_2 \;\longrightarrow\; \begin{matrix} CH(CO_2Et)_2 \\ | \\ CH(CO_2Et)_2 \end{matrix} \;+\; 2\,I^- \qquad \text{----------(e)}$$

$$\underset{\mathbf{337}}{}$$

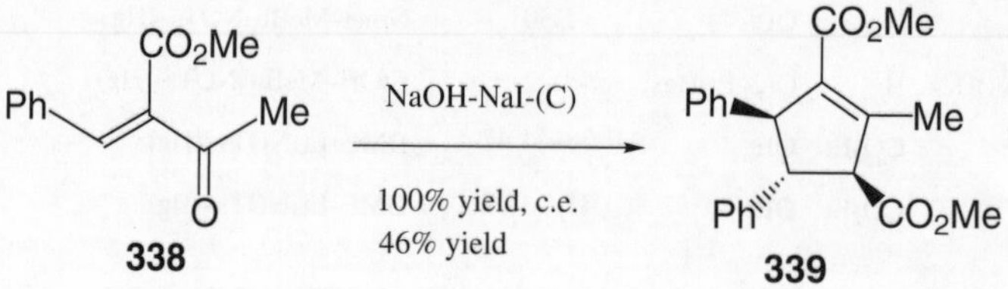

$$(2.141)$$

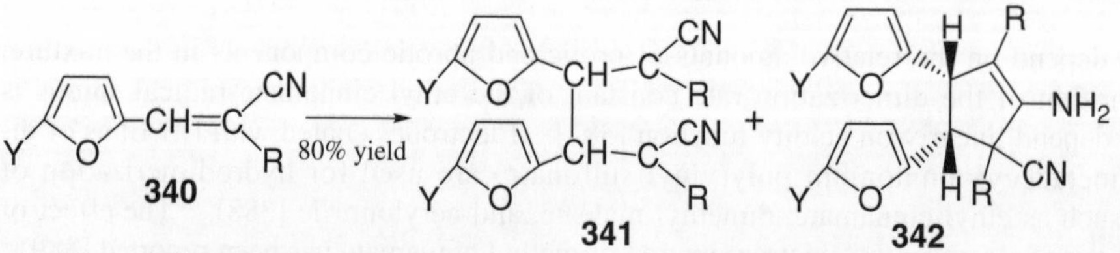

Conditions: $H_2O/EtOH/DMF/-Et_4NI(0.2M)-(Hg)$
        -1.7 V (SCE)   **341/ 342** = 60/40
    $Y = H$, Me, $PhCH_2$, $Ph(CH_2)_2$, $R = H$, Ph

The controlled potential electrolysis of methyl 2(*E*)-4-(3-butenyl)-7-oxo-2-heptenoate **343** has been performed in a DMF-Bu$_4$NPF$_6$ (0.1 M)-(Hg) system in the presence of 3,5-dimethylphenol as a proton donor to give the cyclized compounds **344** and **345** (eq. (2.143)) [392]. The double cyclization of the en-yne system of β-dicarbonyl enol phosphate **347** derived from **346** has been realized by electroreduction in a DMF-Bu$_4$NClO$_4$(0.2M)-(Hg) system to give bicyclo[3.1.0] derivation **348** (eq. (2.144)) [393].

$$ \text{(2.143)} $$

$$ \text{(2.144)} $$

*DMF-Bu₄NClO₄(0.2M)-(Hg)

2.5.3.2 Electroreductive Couplings of Enoates with Electron Acceptors

An efficient intramolecular electroreductive cyclization method for the preparation of five- and six-membered ring systems starting from enoate and electron-accepter has recently been developed. The isolated yields of cyclized products range from 70% to 80% and both formyl and keto-enoates can be used as starting materials. Electroreduction of methyl 8-oxo-2-octenoate **349** in an MeCN-Et₄NOTs-(Hg) system in the presence of diethyl malonate as a proton donor at −2.25 V (SCE) gives the cyclized product **351** in 70% of the isolated yield (eq. (2.145) [394]. Electroreductive cyclization of the keto enoate **352** can lead to two products, **353** and **354**, in 70% yield. The stereochemistry of the products is proposed to be an ester, and the hydroxyl groups are predominantly *cis* rather than *trans* (eq. (2.146)) [395].

$$ \text{(2.145)} $$

$$ \text{(2.146)} $$

A variety of γ-keto ester enol phosphates can be reduced to afford reactive intermediates which, *via* a double cyclization process, are converted into bicyclo[n.1.0]alkanes (n = 3 or 4). For example, the electroreduction of the enol phosphate **355** in an DMF-Bu$_4$NClO$_4$(0.2M)-(Pt/Hg) system at –2.30 V (SCE) gives the bicyclo[3.1.0]alkane **356** in 58% yield (eq. (2.147)) [396].

$$(2.147)$$

The electrolysis is carried out with a ten-fold molar excess of methyl vinyl ketone at a potential of –1.2 ~ –1.3 V (SCE). The electroreductive coupling of methyl vinyl ketone with 4-vinylpyridine (–1.5 V (SCE)) in a DMF/H$_2$O(15/1)-Et$_4$NOTs-(Hg) system gives 4-(5-oxohexyl)pyridine in 25% yield [397]. Electroreductive coupling of diethyl maleate with Michael acceptors, *i.e.*, ethyl acrylate and acrylonitrile, has been examined in an aqueous DMF-Et$_4$NOTs-(Hg) system, giving some cross-coupling products [398].

The cross-hydrocoupling of acrylonitrile **358** and/or maleic acid with alkanones **357** proceeds to afford a mixture of the corresponding γ-hydroxynitriles **359** and γ-lactones **360** (eq. (2.148)) [282, 283, 399]. The electrolysis of the ketones 357 (0.05 mol) is carried out in an aqueous EtOH-H$_2$SO$_4$(20%)-(Pt/Hg) system in the presence of acrylonitrile **358** at 30~35 °C to give a mixture of **359** and **360**, respectively.

$$(2.148)$$

* aq-EtOH-H$_2$SO$_4$(20%)-(Pt/Hg)
a. R = Et;   R' = Me
b. R = *i*-Bu;   R' = Me
c. R = R' = Bu
d. R, R' = –(CH$_2$)$_5$–
e. R, R' = –(CH$_2$)$_6$–

2.6   Electroreduction of Aldehydes, Carboxylic Acids, Esters, Amides and Acid Anhydrides

2.6.1   Electroreduction of Aldehydes, Ketones and Carboxylic Acids

Electroreductive removal of the carbon-oxygen atom of 15-oxo-lanosterol **361** has been attained in an aqueous Dioxane(90%)-H$_2$SO$_4$(10%)-(Pb) system to yield the deoxygenated product **362** in 67% yield (eq. (2.149)) [400].

361   **362**   67%   **363**   17%

R =

$$(2.149)$$

Electroreduction of pyridine 2,6-dialdehyde **364**, leading to the corresponding alcohol **365** is found to proceed quantitatively in a neutral phosphate buffer solution but non-quantitatively in acidic phosphate and acetate buffer solutions (eq. (2.150)) [401, 402].

$$(2.150)$$

The reductive amination of alkenones in an aqueous solution of primary amines can be realized by an electrochemical method.   The best yields of the secondary amines are attained at lead and cadmium cathodes in an aqueous electrolytic solution at pH 11~12 [403].

The electroreduction of aliphatic and aromatic carboxylic acids seems to be a promising tool for the preparation of the corresponding aldehydes and alcohols.   However, ordinary electroreduction of the acids faces difficulties in obtaining either aldehydes or alcohols, selectively.   Aromatic aldehydes may be produced under buffering conditions with boric acid or phosphate [404, 405].   The straightforward conversion of carboxylic acids to the corresponding aldehydes has been achieved in a CH_2Cl_2-Bu_4NBr-(C/Sn) system in the presence of a mixture of Bu_3P and $MeSO_3H$ without over-reduction [406].

The electrochemical transformation of aliphatic and aromatic carboxylic acids into the corresponding primary alcohols is a key technology in organic synthesis [407, 408, 409]. The reduction of carboxylic acids with diborane derived from sodium borohydrate by electroreduction has been proposed [410].

The electrosynthesis of glyoxylic acid from the electroreduction of oxalic acid has been actively investigated and the commercialization attempted.   Glyoxylic acid can be produced in an aqueous sulfuric acid system with current efficiencies better than 70% and chemical yields better than 90% [411].   The preliminary scale-up studies of the electroreduction of oxalic acid to glyoxylic acid have been investigated in filter press type cells varying form a small bench scale to a pilot size [112a,412~414].   Titanium grills coated with lead dioxide electrochemically deposited have been shown to be appropriate anode materials for oxalic acid reduction because of their low oxygen overvoltage and high corrosion resistance [112b].   For preparing glyoxylic acid, stainless chromium-nickel steel or graphite electrodes have been patented [415].   A pilot-scale electrosynthesis (20 t/year) of glyoxylic acid has been performed in an H_2O-H_2SO_4(10%)-(Pb) system and Nafion 425 cation-exchange membrane has been used as a diaphragm.   At current density of 20 A/dm^2 and current efficiency of 60~65%, the yields of glyoxylic acid are 94~96% (eq. (2.151)) [416,416].   A process for overcoming rapid deactivation of the Pb cathode in the course of electroreduction of oxalic acid has been proposed by adding small amounts of a Pb(II) salt in both anolyte and catholyte [417]. The electroreductive sythesis of homoserine **367** from aspartic acid **366** has been attained in an aqueous EtOH(85%)-LiCl-(Pt/Pb) system in 71% yield (eq. (2.152)) [418]. Electrocatalytic reduction of phenylglyoxylic acid is performed qualitatively using methyl viologen as a mediator in an aqueous solution [419], and the presence of β-CD induces enantioselectivity (19% ee(S)).

$$\underset{CO_2H}{\overset{CO_2H}{|}} + 2H^+ \xrightarrow[\text{94~96\% yields}]{2e^-} \underset{CO_2H}{\overset{CHO}{|}} + H_2O \tag{2.151}$$

$$\begin{array}{c} \text{(366)} \end{array} \xrightarrow[\text{71\% yield}]{\text{aq. EtOH(85\%)-LiCl-(Pt/Pb)}} \begin{array}{c} \text{(367)} \end{array} \tag{2.152}$$

366   **367**

Benzoic acid has been led to benzaldehyde and benzyl alcohol and their product ratios are found to be a function of potentials [404]. Benzaldehyde can be reduced much more easily than benzoic acid. Electroreduction of 2,3,4,5,6-pentafluorobenzoic acid **368** is performed in an aqueous H_2SO_4(5wt%)-(Ti/Zn) system to form 2,3,5,6-tetrafluorobenzyl alcohol **369** (eq. (2.153)) [420a]. 2,3,5,6-Tetrafluorobenzaldehyde has been obtained in 75% selectivity from **368** under controlling electrolysis conditions [420b].

$$\begin{array}{c} \text{(368)} \end{array} \xrightarrow[\text{ca. 50\% yield}]{\text{aq.}H_2SO_4(5wt\%)-(Zn)} \begin{array}{c} \text{(369)} \end{array} \tag{2.153}$$

368   **369**

3-Phenoxybenzyl alcohol **371**, a potential key intermediate for the pyrethroid insecticides, has been prepared by the electroreduction of 3-phenoxybenzoic acid **370**. The electroreduction of **370** in an aqueous EtOH(56%)-H_2SO_4(22%)-(Pb) system gives the corresponding benzyl alcohol **371** in 93% yield (eq. (2.154)) [421, 422]. Under almost similar conditions, the large scale electroreduction of m-hydroxy benzoic acid **372** into the corresponding benzyl alcohol **373** *via* the aldehyde intermediate has been successfully performed in an aqueous H_2SO_4(17.5%)-(Pb) system at 50 °C (eq. (2.155)) [423, 424~426]. 1,4-bis(Hydroxymethyl)benzene has been synthesized from 4-hydroxymethyl benzoic acid in an aqueous H_2SO_4(10%)-(Pb) system in 68% yield [427].

$$\begin{array}{c} \text{(370)} \end{array} \xrightarrow[\text{93\% yield}]{\text{EtOH/H}_2\text{O(56/22v/v)-H}_2\text{SO}_4\text{-(Pb)}} \begin{array}{c} \text{(371)} \end{array} \tag{2.154}$$

370   **371**

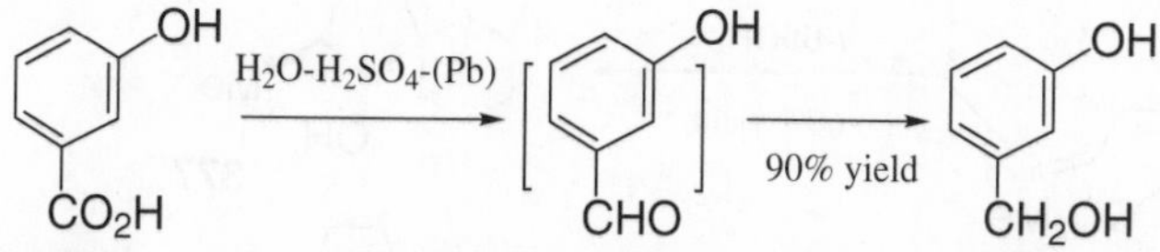

$$ (2.155) $$

372 → **373**

The electroreduction of pyridine 2- and 4-carboxylic acids has been examined in an aqueous sulfuric acid solution [428].

A cathodic esterification of carboxylic acids, leading to the corresponding esters, has been performed in a DMF-Et$_4$NOTs-(Pt) system in the presence of alkyl iodides (eq. (2.156)) [429]. An improved method for esterification is the addition of alkylating agents, *i.e.* alkyl halides and alkyl tosylates, to the catholyte after the electrolysis [430].

$$ R^1{-}CO_2H \xrightarrow[\substack{R^2I\ (or\ R^2OTs) \\ 42\sim96\%}]{DMF\text{-}Et_4NOTs\text{-}(Pt)} R^1CO_2R^2 \qquad (2.156) $$

$$ R^1 = Me,\ Pr,\ Ph,\ CH_2Cl $$
$$ CH{=}CHMe,\ C{\equiv}CH,\ Py $$
$$ R^2 = Me,\ Et,\ Bu $$

2.6.2   Electroreduction of Esters and Amides

The theoretical and preparative aspects of the electrochemical method for the deprotection of various esters have been well documented [431]. The reduction of esters and amides has been attempted in liquid ammonia and methylamine [432,433].

The electroreduction of aliphatic carboxylates **374** to primary alcohols **375** has been attained in good yield in a THF/t-BuOH-LiClO$_4$-(Mg) system in an undivided cell (eq. (2.157)) [434a]. The electrolysis system employed may overcome high negative reduction potentials ($\sim{-}3.0$V *vs* SCE). The electroreduction of the unsaturated ester **376** provides cyclized compounds **377** (eq. (2.158) and **378** (eq. (2.159)), depending on the presence of t-BuOH or Me$_3$SiCl [434a]. The cathodic cyclo-coupling of 1,3-dienes with aliphatic esters can be promoted by a magnesium electrode and yields homologues of cyclo-3-pentenols [434b]. The electrolysis of a mixture of the diene **379** and the ester **380** in a THF-LiClO$_4$-(Mg) system affords the corresponding cyclopentenol **381** in good yields (eq. (2.160)).

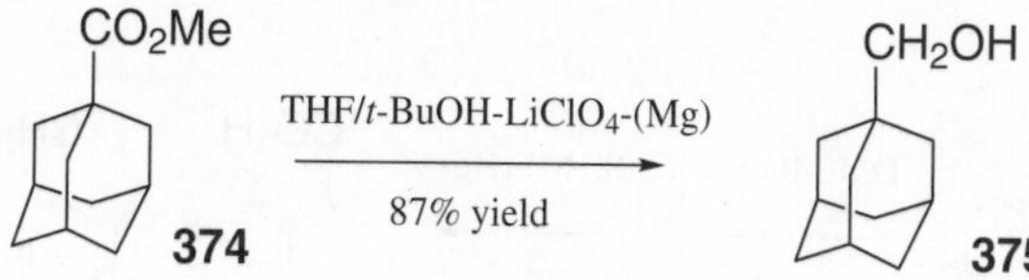

$$ (2.157) $$

374 → **375**

$$t\text{-BuOH} \quad \xrightarrow{} \quad \text{377} \tag{2.158}$$
60% yield

$$\xrightarrow{\text{Me}_3\text{SiCl}} \quad \text{378} \tag{2.159}$$
67% yield   TMSO

376

$$\text{379} \quad + \quad i\text{-PrCO}_2\text{Me} \quad \xrightarrow[\substack{62\sim72\% \text{ yields} \\ n=1,2}]{\text{THF-LiClO}_4\text{-(Mg)}} \quad \text{381} \tag{2.160}$$

380

The electroreduction of aromatic carboxylates to benzyl alcohols has been well documented in the literature [1,435]. Electroreductive cleavage of aromatic esters has been studied. Benzoate esters are reasonably easy to reductively cleave the C-O bond under aprotic conditions and the anion radicals formed after initial uptake of an electron lead to give benzoate and the more readily reducible benzyl radical. Large-scale controlled-potential electrolysis of benzyl benzoate **382** in a DMF-Bu$_4$NClO$_4$(0.1M)-(Hg) system affords benzoic acid **383** (50% yield) and toluene **384** (47% yield) (eq. (2.161)) [436]. The mechanistic features of electrolytic cleavage of acyclic and cyclic aromatic esters have been discussed based on cyclic voltammograms [436]. The electroreduction of a variety of benzoates in an EtOH-R$_4$NCl-(Hg) system (R = Me, Et, Octyl) results in the formation of the benzyl alcohol **386** along with a mixture of the ring-hydrogenated products **387**, **388**, and **389** (eq. (2.162)) [409]. The ring-hydrogenation is, however, suppressed drastically, when small amount of acetic acid is added to the electrolysis medium [409]. The yield of benzyl alcohol is more than 70% (Table 2.11). The combination of the solvent and the electrode material seems to be essential. Phenyl benzoate can be reduced to benzyl alcohol in quantitative yield. Ring substituents in the benzoic acid do not alter the yield (63~81%) when acetic acid is present. Electrolysis with mercury, cadmium, or lead electrode as a cathode proceeds smoothly to give benzyl alcohol in 60~75% yields. Phenyl aliphatic carboxylates can be reduced to the corresponding alcohols in 30~40% yields under controlled conditions. *p*-Methoxy-carbonylbenzyl acetate is known to be cleaned electrochemically [437a]. The deprotection technique has been used in antibiotics synthesis [437b].

$$\xrightarrow[-1.5 \text{ V (SCE)}]{\text{DMF-Bu}_4\text{NClO}_4(0.1\text{M})\text{-(Hg)}} \quad \text{383}\ 50\% \quad + \quad \text{384}\ 47\% \tag{2.161}$$

382

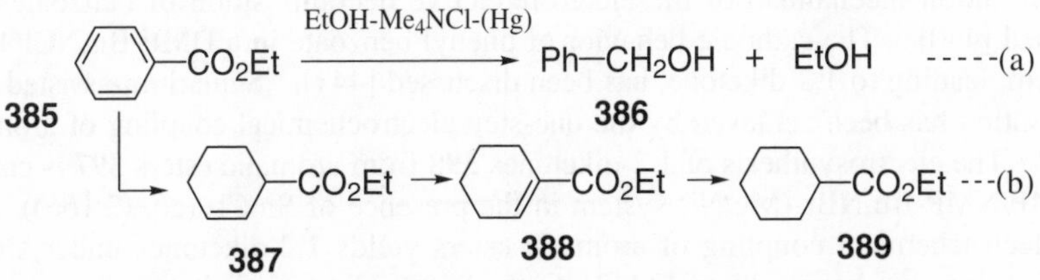

$$\text{385} \xrightarrow{\text{EtOH-Me}_4\text{NCl-(Hg)}} \text{Ph–CH}_2\text{OH} + \text{EtOH} \quad \text{-----(a)}$$
$$\text{386}$$

$$\text{387} \rightarrow \text{388} \rightarrow \text{389} \quad \text{---(b)}$$

(2.162)

Table 2.11 Product Distribution of the Electroreduction of **385**

AcOH, mol/l	386	387	388	389	Others
0	26.	20.5	3.1	12.1	27.7
0.05	78.0	–	–	–	5.0
~ 0.35					

Benzyl methyl ether **391** has been synthesized by electroreduction of methyl benzoate **390** in 30% aqueous sulfuric acid at an amalgamated lead cathode (eq. (2.163)) [438]. The unusual head-to-tail coupling of methyl benzoate **390** is found by the electrolysis in an *i*-PrOH-Et$_4$NOTs-(Sn) system to give **392** (eq. (2.164)) [439]. Attempts to stop the electroreduction of esters at the aldehydes have been unsuccessful on a preparative scale [409]. The dimethyl tele- and isophthalates are chemoselectively reduced in the presence of acetic acid to the corresponding methyl hydroxymethylbenzoates.

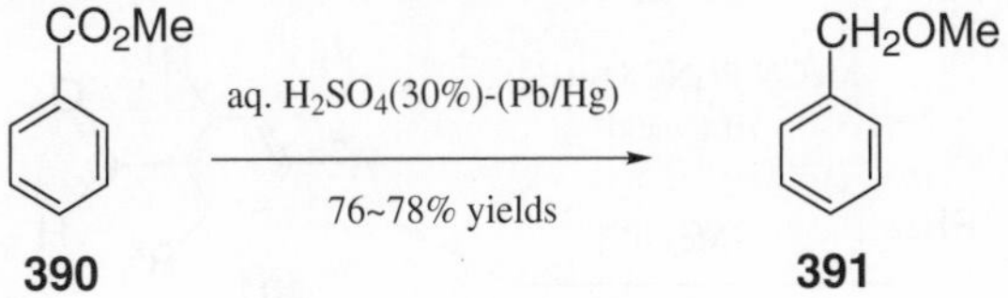

$$\text{390} \xrightarrow[\text{76~78\% yields}]{\text{aq. H}_2\text{SO}_4(30\%)\text{-(Pb/Hg)}} \text{391}$$

(2.163)

$$2 \; \text{390} \xrightarrow[\text{70\% yield}]{\textit{i}\text{-PrOH-Et}_4\text{NOTs-(Sn)}} \text{392}$$

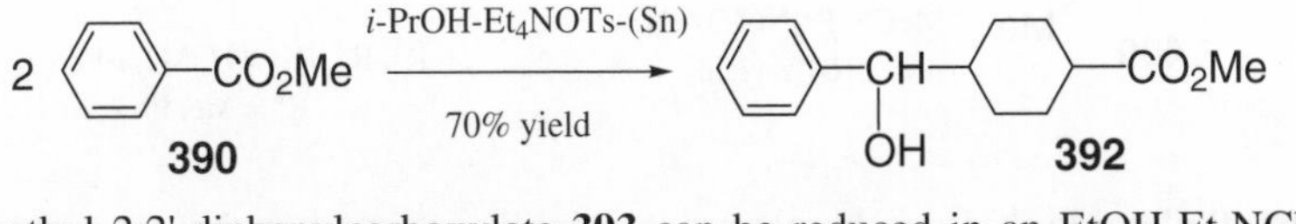

(2.164)

Dimethyl 2,2'-diphenylcarboxylate **393** can be reduced in an EtOH-Et$_4$NCl-(Hg) system to 2,2'-biphenyldiyldimethanol **394** (58%), 9,10-dihydrophenanthrene-9,10-diol **395** (35%), and 9,10-phenanthrendiol **396** (10%) (eq. (2.165)) [409].

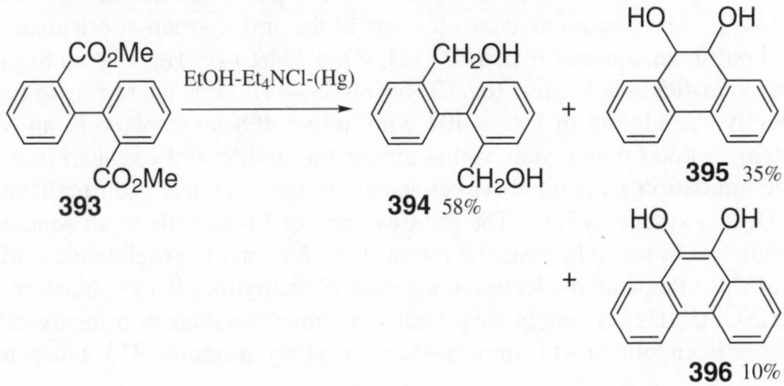

$$\text{393} \xrightarrow{\text{EtOH-Et}_4\text{NCl-(Hg)}} \text{394 } 58\% + \text{395 } 35\% + \text{396 } 10\%$$

(2.165)

A radical anion mechanism of the electroreductive decomposition of benzoates has been proposed [440]. The cathodic behavior of phenyl benzoate in a DMF-Bu$_4$NClO$_4$-(Pt or Hg) system, leading to 1,2-diketone, has been discussed [441]. Samarium-assisted acyloin condensation has been achieved by the one-step electrochemical coupling of aromatic esters [442]. The electrosynthesis of 1,2-diketones 398 from aromatic esters 397 is carried out in a DMF/NMP-Bu$_4$NBr-(Mg/Ni) system in the presence of SmCl$_3$ (eq. (2.166)). An improved electrochemical coupling of aromatic esters yields 1,2-diketones under simple and mild conditions has been achieved [443]. Cathodic coupling of aliphatic esters, leading to 1,2-diketone and acyloin, with magnesium electrode in the presence of chlorotrimethylsilane has been attained [444]. Electroreduction of alkyl benzoates in dry acetonitrile yields benzoylacetonitrile in 17~45% yields [445]. The electroreductive conversion of aromatic esters 399 and amides 400 to their corresponding aldehydes 401 has been performed in an MeCN-Pr$_4$NClO$_4$(0.02M)-(Hg) system in the presence of chlorotrimethylsilane in moderate to good yields (eq. (2.167)) [446].

$$2 \ Ar{-}CO_2R \quad \xrightarrow[\substack{SmCl_3 \\ 60\text{\textasciitilde}90\% \ yields}]{DMF/NMP\text{-}Bu_4NBr(0.2mM)\text{-}(Mg/Ni)} \quad \underset{\textbf{398}}{Ar\overset{O}{\overset{\|}{C}}\overset{O}{\overset{\|}{C}}Ar} \qquad (2.166)$$

$$\underset{\textbf{397}}{\phantom{2 \ Ar{-}CO_2R}}$$

Ar = Ph, p-CF$_3$C$_6$H$_4$, p-MeOC$_6$H$_4$, p-t-BuC$_6$H$_4$, o-FC$_6$H$_4$
R = Me, Et, Ar

$$(2.167)$$

R^1, R^2, R^3 = H, Me, i-Pr
R^4 = Me, Ph

N,N-Dimethylamide 402 can be electrochemically reduced in the presence of tert-BuOH (3.5 equiv.) to give the aldehyde 403 in 65% yield (eq. (2.168)) [434]. The electroreduction of pyridine 2- and 4-carboxylamides is performed in an aqueous sulfuric acid solution [428]. The procedure for the electroreduction of pyridine carboxamide bases has been patented [447]. The electroreductive cleavage of the amide group of pristinamycin I$_A$ 404 was carried out in an aqueous EtOH(50%)-H$_2$SO$_4$(1.5M/L)-(Pt/Hg) system to give the free amino derivative 405 in 90% yield (eq. (2.169)) [448,449]. It is interesting to note that the electroreductive amidation of esters 407 with amine 408 takes place in an MeCN-Et$_4$NClO$_4$ system in good conversion yields giving the amide 409 (eq. (2.170)) [450]. Electrocatalytic amidation reactions between esters and amines has been realized in an MeCN-Et$_4$NClO$_4$-(Pt) system [451]. The pre-treatment of Pb-cathode in an aqueous perchloric acid solution is found to be essential for electroreduction of pyroglutamic acid, leading to proline [452]. Preparative electroreduction of N-methyl(or phenyl)phthalimide 410 in an MeCN-Et$_4$NCl-(Pt/Hg) system in the presence of dimethylsulfate or trimethylchlorosilane affords either isoindolinon 411, dimers 412 or silyloxy isoindole 413, independently (eq. (2.171)) [453].   The reduction of thiobenzamides under the same conditions as below results in the formation of β-dialkylaminocinnamonitrile in 10~36% yields [445]. Polarographic half-wave potentials E1/2 of S-phenyl benzothioates have been investigated [454].

$$\text{Ph–(CH}_2\text{)}_2\text{CONMe}_2 \xrightarrow[\text{7.5 F/mol}]{\text{THF/}t\text{-BuOH-LiClO}_4\text{-(Mg)}} \text{Ph(CH}_2\text{)}_2\text{CHO} \qquad (2.168)$$

402   **403**

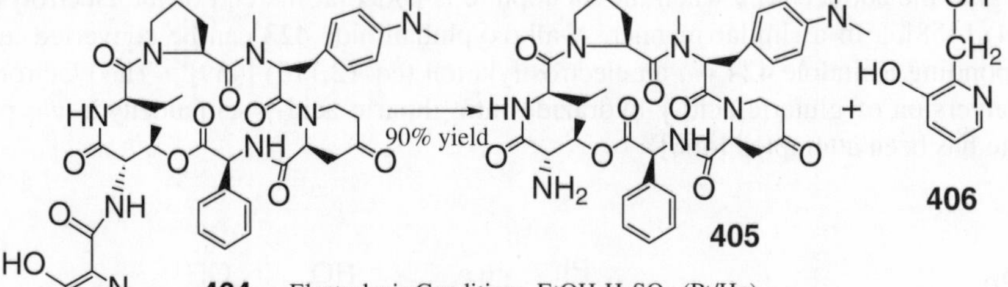

(2.169)

404   Electrolysis Condition: EtOH-H$_2$SO$_4$-(Pt/Hg)

$$\text{Ph–CO}_2\text{Et} + \text{H}_2\text{N(CH}_2\text{)}_2\text{OH} \xrightarrow[\text{75\% yield}]{\text{MeCN-Et}_4\text{NClO}_4\text{-(?)}} \text{Ph–CONH(CH}_2\text{)}_2\text{OH} \qquad (2.170)$$

407   **408**   **409**

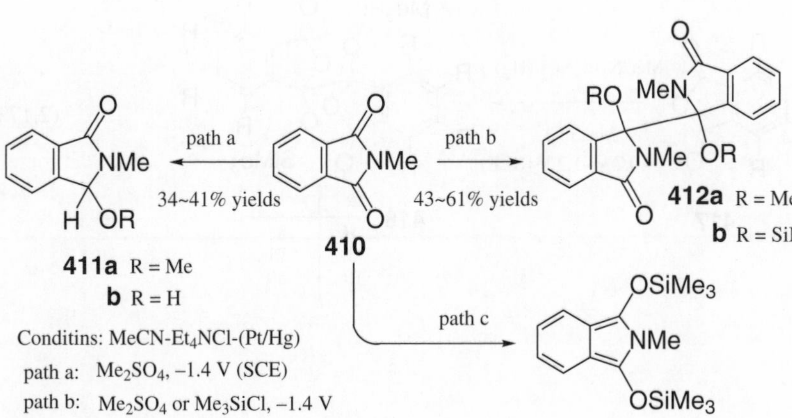

411a R = Me
   b R = H

Conditins: MeCN-Et$_4$NCl-(Pt/Hg)

path a: Me$_2$SO$_4$, –1.4 V (SCE)
path b: Me$_2$SO$_4$ or Me$_3$SiCl, –1.4 V
path c: excess Me$_3$SiCl, –1.3 V

412a R = Me
   b R = SiMe$_3$

(2.171)

413a

2.6.3 Electroreduction of Acid Anhydrides and Imides

Glutaric anhydride can be electrochemically reduced to δ-valerolactone [455]. The electroreduction of 3,6-diphenylphthalic anhydride **414** in an MeCN-Et$_4$NCl/Me$_3$SiCl-(Hg) system affords a mixture of **415** and **416** (eq. (2.172)) [456]. The relative electron-acceptor properties of aromatic anhydrides have been investigated in a DMF-Et$_4$NClO$_4$(0.1M)-(C) system [457]. The electroreduction of phthalic acid anhydride **417** in an MeCN-Et$_4$NCl-(Hg) system in the presence of chlorotrimethylsilane at $-0.7 \sim -1.2$ V (SCE) gives rise to dimerization to form diphthalyl acid lactone ether **418** (eq. (2.173)) [458]. The electroreduction of phthalic imides **419** in an MeCN-Et$_4$NCl-(Hg) system in the presence of an excess amount of chlorotrimethylsilane may provide 1,3-bis-trimethylsiloxyisoindoles **420** as intermediates which can be trapped with N-phenyl-maleinimide **421** via a cycloaddition sequence to give the adducts **422** when the dienophile is added at the end of the electrolysis (eq. (2.174) [458]. In a similar manner, N-alkoxyphthalimide **423** can be converted into the corresponding isoindole **424** via an electrosilylation (eq. (2.175) [459]. The electroreductive conversion of glutaric acid-γ-hydrazide into glutaric acid-γ-semialdehyde via pyroglutamate has been attempted [460].

$$(2.172)$$

$$(2.173)$$

R	yield,%
H	81
Cl	71
Br	59

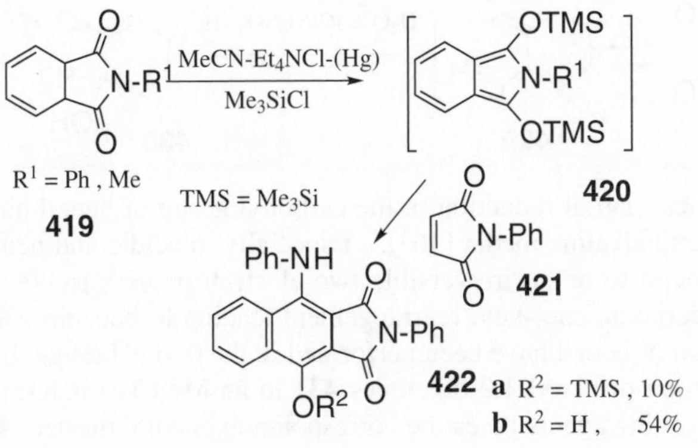

$$R^1 = Ph, Me$$

TMS = Me$_3$Si

419

420   (2.174)

421

Ph-NH

422 a R^2 = TMS, 10%

b R^2 = H,   54%

R = Me, Bz

423   **424**   (2.175)

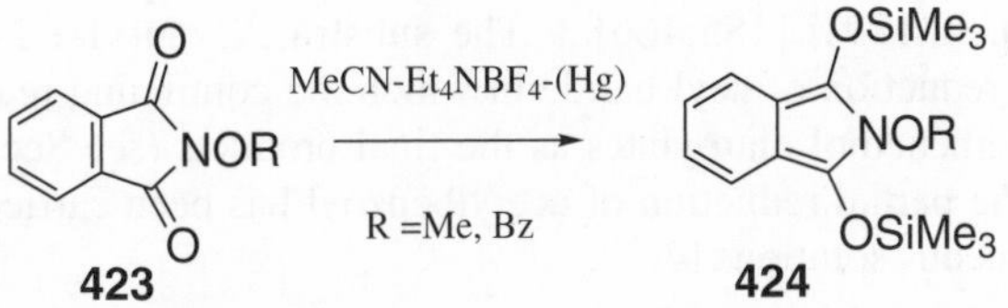

2.7 Miscellaneous

2.7.1 Electroreduction of 1,2-Dicarbonyls

The electroreduction transformation of 2,3-butanedione **425**, leading to butane **427**, has been investigated in acid media at a platinized platinum electrode [190b]. The major end-product from **425** in an $H_2O-H_2SO_4$(1M)-(Pt) system is butane via acetoine **426** (eq. (2.176)). 1,2-Cyclohexadione **428** as a cyclic 1,2-diketone exists 40% in a monoenol form **429** in aqueous media. The protonated form of **429** is reduced in a wide range of pH values and the primary product is an enediol which is a precursor of the ketol **430** (eq. (2.177)) [461]. Polarographic reduction of the symmetrical diprotonated form of 1-phenyl-1,2-propanedione occurs in two waves at pH < 8 in a reversible two-electron process. The product of this reduction is a 1-phenyl-1,2-dihydroxy-1-propene enediol, which is converted into 1-phenyl-1-hydroxy-2-propanone (α-ketol) and 1-phenyl-2-hydroxy-1-propanone (β-ketol) [462].

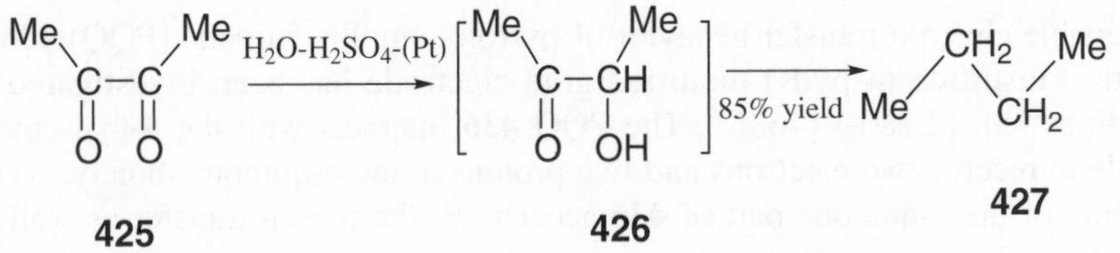

$H_2O-H_2SO_4$-(Pt)

85% yield

425   **426**   **427**   (2.176)

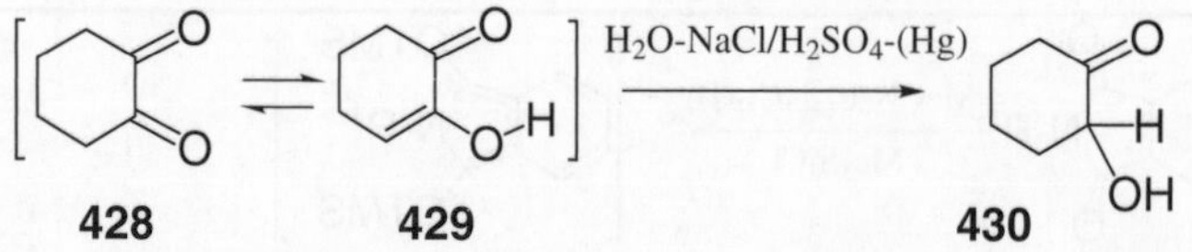
H2O-NaCl/H2SO4-(Hg)
428
429
430

(2.177)

The electrochemical partial reduction of the carbonyl group of benzil has been investigated in strong acid and alkaline media [463]. Especially in acidic and neutral media, the electroreduction is found to be an irreversible two-electron process to *cis*- and *trans*-stilbenediols, which undergo an enol-keto rearrangement leading to benzoin. Kinetic studies on the electroreduction of benzil have been performed in the 0~6 pH range [464]

The electroreduction of diaryl-1,2-diketones **431** in an Me_2CO-$LiClO_4$-(Pt/Hg) system in the presence of acylating agents gives the corresponding enediol diesters **432** in 73~98% yields (eq. (2.178)) [465]. The diaryl 1,2-diketones **433** can lead to the corresponding *N*-substituted iminocarbonates **435** *via* electroreduction of **433** in the presence of carbonimidoyl chlorides **434** (eq. (2.179)) [185,466]. The substrates, diaryl-1,2-diketones, are formed by the cathodic reduction of acyl halide in which the continuing reaction may produce the 1,2-diaryl-1,2-ethenediol diaroylates as the final products (see Section 2.7.3). A mechanistic survey of the partial reduction of acetylbenzoyl has been carried out in a wide pH range in buffered aqueous solutions [467].

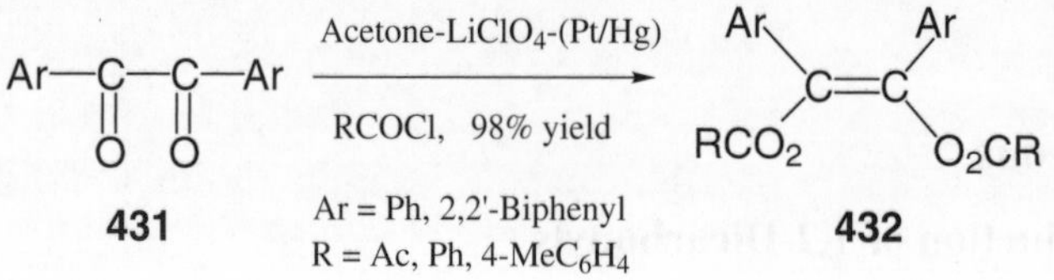
Acetone-LiClO4-(Pt/Hg)
RCOCl, 98% yield
Ar = Ph, 2,2'-Biphenyl
R = Ac, Ph, 4-MeC6H4
431
432

(2.178)

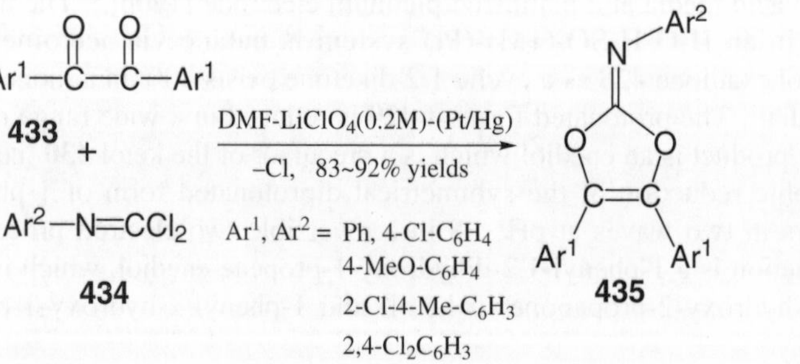
DMF-LiClO4(0.2M)-(Pt/Hg)
–Cl, 83~92% yields
Ar1, Ar2 = Ph, 4-Cl-C6H4
4-MeO-C6H4
2-Cl-4-Me-C6H3
2,4-Cl2C6H3
433
434
435

(2.179)

The reversible electron transfer behavior of pyrroloquinolinequinone (PQQ) **436** using a di-(4-pyridyl) disulfide (4-pyds) modified gold electrode has been investigated under acidic conditions (eq. (2.180)) [468]. The PQQ **436** interacts with the 4-pyds modified gold electrode to receive two electrons and two protons at the o-quinone moiety. The hydrogen bonding of the o-quinone part of **436** accelerates the proton transfer as well as the electron transfer.

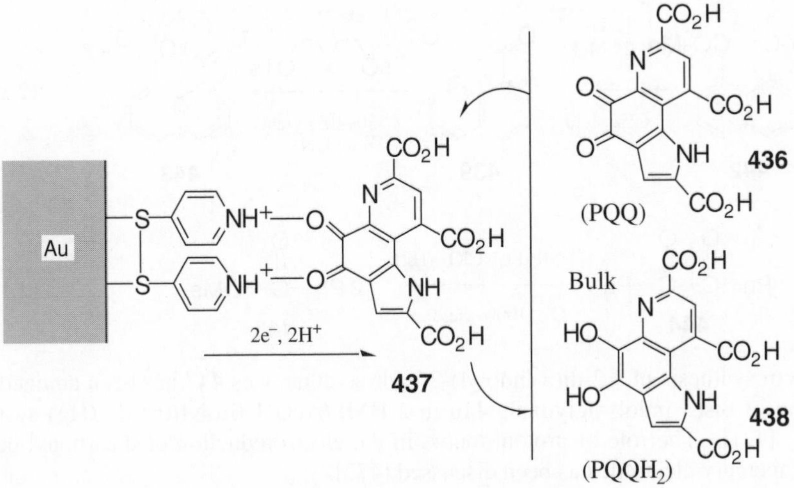

(2.180)

Electroreductive acylation of 1,2-acenaphthenedione **439** in the presence of non-electroactive acylating reagents provides two different types of products in good yields (eq. (2.181)) [247]. The acylation of the radical anion intermediate is fast with benzoyl chloride, and the resulting neutral intermediate may undergo a second electron transfer, leading to an anion which reacts with a further molecule of benzoyl chloride to afford the product **440**. The relatively slower reaction of the radical anion intermediate with acetic anhydride prevents a second electron transfer, and the acylation gives the radical intermediate, whose coupling gives the homocoupling product **441**.

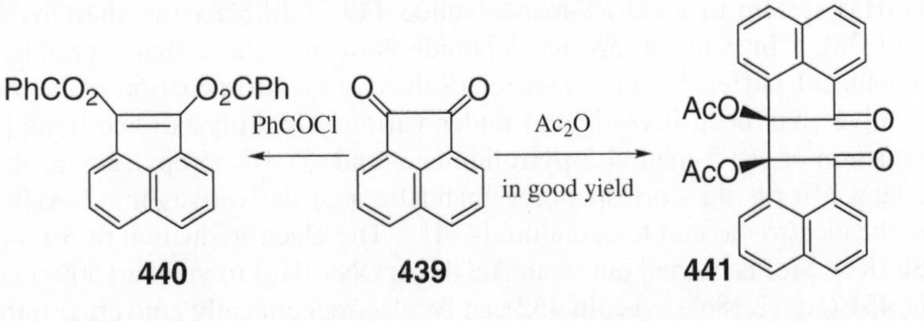

(2.181)

Aromatic α-diketones, e.g., benzil, acenaphtenequinone, and phenanthrenequinone, undergo oxidative cleavage of the medium carbon-carbon bond by electrogenerated superoxide ion $O_2^{-\cdot}$ in dry DMF [469]. For instance, the cleavage of acenaphtenequinone **439** can lead either the dimethyl ester **442** or macro-ring ester **443** in a DMF-Bu$_4$NBF$_4$(0.1M)-(Pt/Hg) system at -0.55 V vs. Ag/AgI/I$^-$ in the presence of appropriate electrophile, e.g. methyl iodide, poly(ethelene glycol) ditosylates (eq. (2.182)). The electroreduction of α-diketones in an air-saturated solution in dipolar aprotic solvents undergoes cleavage of the CO-CO bond to afford the cleavaged products [470]. The typical example (**444** → **445**) is shown in eq. (2.183). In the case of N-substituted indoles, a very stable anion radical is produced. The cyclic voltammograms demonstrate self-protonation reactions after one-electron transfer, when the nitrogen atom in the indole groups is not substituted.

$$
\begin{array}{ccc}
\underset{\textbf{442}}{\text{MeO}_2\text{C}\quad\text{CO}_2\text{Me}} & \underset{\substack{\text{1) O}_2^{-\bullet}\\ \text{2) MeI}\\ \xleftarrow{\hspace{2cm}}\\ 62\%\ \text{yield}}}{\text{(naphthalene)}} & \underset{\textbf{439}}{\overset{\text{O}\quad\text{O}}{\text{(acenaphthenequinone)}}} & \underset{\substack{\text{1) O}_2^{-\bullet}\\ \text{2)}\\ \text{TsO}\quad x\quad\text{OTs}\\ \xrightarrow{\hspace{2cm}}\\ 30\sim40\%\ \text{yields}\\ X = \text{O, N}}}{} & \underset{\textbf{443}}{}
\end{array}
\tag{2.182}
$$

$$
\underset{\textbf{444}}{\text{Ph}-\overset{\overset{\text{O}}{\|}}{\text{C}}-\overset{\overset{\text{O}}{\|}}{\text{C}}-\text{Ph}} \xrightarrow[\text{O}_2,\ 100\%\ \text{yield}]{\text{DMF-Et}_4\text{NClO}_4\text{-(Hg)}} \underset{\textbf{445}}{2\ \text{Ph}-\overset{\overset{\text{O}}{\|}}{\text{C}}-\text{OMe}}
\tag{2.183}
$$

The electrosynthesis of 1,2-di(3-indolyl)-2-hydroxyethanones **447** has been attained by the reduction of bis(3-indolyl)glyoxals **446** in a DMF/AcOH-Bu$_4$NI(0.1M)-(Hg) system (eq. (2.184)) [471]. The role of proton donors in the electroreduction of dicarbonyl compounds on a mercury electrode has been discussed [472].

$$
\textbf{446} \xrightarrow[\substack{-1.16\ \text{V (Ag/AgI/I}^-)\\ 98\%\ \text{yield}\\ R = \text{H, Me}}]{\text{DMF/AcOH-Bu}_4\text{NI(0.1M)-(Hg)}} \textbf{447}
\tag{2.184}
$$

Asymmetric induction by electroreduction of optically active S(-)-*N*-(α-methylbenzyl)benzoylformamide **448** has been attempted in an H$_2$O/EtOH(1/1)-H$_2$SO$_4$(0.5M)-(Hg) system to yield *RS*-mandelamide **449** in 12.5% ee of diastereo-excess (eq. (2.185)) [473a]. In contrast, *SS* mandelamide is obtained as a major product under acetic or ammoniacal buffer media. Diastereoselective electroreduction of benzoylformamide derivatives has been investigated under various electrolysis conditions [473b]. The electroreduction of (*R*)-5-methyl-2-pyrrolidinone- and (*S*)-4-isopropyl-2-oxazolidinone phenylglyoxylates affords the corresponding mandelic acid derivatives in 42~81% yields depending on the electrolyte and temperature [474]. The electroreduction of S-(-)-proline derivative **450** (R = Me) is carried out in an AcOH-AcONa-(Hg) to yield an 50% excess of the *SS* epimer **451** (eq. (2.186)). Isatin **452** can be electrochemically converted into either isatyde (pinacol) or dioxyindole **453** [475]. Under fresh acidic and neutral solutions, isatyde is formed through a one-electron process in quantitative yield, whereas under aged neutral and alkaline solutions, dioxyindole **453** is produced via a two-electron reduction (eq. (2.187)) [475].

$$
\underset{\textbf{448}}{\text{Ph}-\text{CO}-\text{CONH}-\overset{\overset{\text{Ph}}{|}}{\underset{\underset{\text{H}}{|}}{\overset{*}{\text{C}}}}-\text{Me}} \xrightarrow[p\text{H}<3.5,\ \text{optical yield}\ 12.5\%]{\text{EtOH-H}_2\text{SO}_4\text{(0.5M)-(Hg)}} \underset{\textbf{449}}{\text{Ph}-\overset{*}{\text{CH}}-\text{CONH}-\overset{\overset{\text{Ph}}{|}}{\underset{\underset{\text{H}}{|}}{\overset{*}{\text{C}}}}-\text{Me}}
\tag{2.185}
$$

$$Ph-CO-CO-N \xrightarrow[\substack{-1.1\ \text{V (SCE)} \\ 62\%\ \text{yield} \\ \rho(\%)\ 27\sim49}]{2e^- + 2H^+} Ph-\overset{*}{C}H-CO-N \tag{2.186}$$

450
(S)   R = H, Me, Bz   **451**   (RS + SS)

$$\xrightarrow[\substack{-1.5\ \text{V (SCE)},\ 85\%\ \text{yield}}]{\text{aq-NaOH/Na}_2\text{HPO}_4\text{-(Hg)}} \tag{2.187}$$

452   **453**

2.7.2 Ketoacids, Esters and Amides

Some carbonyl compounds (formaldehyde, α-keto acid, etc.) are appreciably hydrated owing to the strong acceptor effect on the carbonyl carbon. The hydration products (*gem*-diols) are not able to be directly reduced at the electrode, while the non-hydrated form of the carbonyl group is reduced [476]. Selective reduction of α-keto esters into the corresponding alcohols has been achieved in a Britton-Robinson buffer solution. Electroreduction of triethyl phosphonopyruvate **454** in a 20% ethanol solution of the Britton-Robinson buffer (pH 4.5) at -1.6 V (SCE) affords the alcohol **455** in 80% yield (eq. (2.188)) [477]. Electrochemical pinacolization of **454** proceeds smoothly in an acidic medium. Electrolysis of **454** in an EtOH/AcOH-AcOLi-(Hg) system undergoes radical coupling to afford the pinacol **456** in 80% yield (eq. (2.189)) [477]. Electroreductive-amination of **454** has been attained in an H_2O/NH_4OH-NH_4Cl-(Hg) system to give amino acid derivative **457** (eq. (2.190)) [477]. The electroreduction of α-ketoglutaric acid and its derivatives is discussed using polarographic data [478].

$$\xrightarrow[\substack{-1.6\ \text{V (SCE)},\ 80\%\ \text{yield}}]{\text{aq.EtOH(20\%)-Buffer*-(Hg)},\ p\text{H4.5}} \begin{array}{c} \text{OH} \\ | \\ \text{CH}_2-\text{CH}-\text{CO}_2\text{Et} \\ | \\ \text{PO(OEt)}_2 \quad \textbf{455} \end{array} \tag{2.188}$$

$$\begin{array}{c} \text{CH}_2-\text{CO}-\text{CO}_2\text{Et} \\ | \\ \text{PO(OEt)}_2 \end{array} \quad \textbf{454}$$

$$\xrightarrow[\substack{-1.4\ \text{V (SCE)},\ 80\%\ \text{yield}}]{\text{EtOH/AcOH-AcOLi-(Hg)}} \begin{array}{c} \text{(EtO)}_2\text{PO} \quad \text{CO}_2\text{Et} \\ | \qquad\quad | \\ \text{CH}_2-\text{COH} \\ | \\ \text{CH}_2-\text{COH} \\ | \qquad\quad | \\ \text{(EtO)}_2\text{PO} \quad \text{CO}_2\text{Et} \quad \textbf{456} \end{array} \tag{2.189}$$

$$\xrightarrow[\substack{-1.5\ \text{V (SCE)},\ 56\%\ \text{yield}}]{\text{aq.NH}_4\text{OH-NH}_4\text{Cl-(Hg)}} \begin{array}{c} \text{NH}_2 \\ | \\ \text{CH}_2-\text{CH}-\text{CO}_2\text{H} \\ | \\ \text{PO(OEt)}_2 \quad \textbf{457} \end{array} \tag{2.190}$$

Electroreductive behavior of phenylglyoxylic acid **458** has been investigated on mercury cathode (eq. (2.191)). Two reduction waves due to the reduction of the carbonyl group are observed around neutral pH, and mandelic acid **459** can be obtained in an aqueous KCl(1M)–(Pt/Ag) system in the controlled-potential electrolysis. More details on electrochemical parameters have been clarified (eq. (2.191)) [479]. Electroreduction of **458** in a magnetic field does not lead to the formation of optically active mandelic acid [480], in contrast to an earlier report on enantio-excess preparation of the mandelic acid [133]. Asymmetric electroreduction of phenylglyoxylic acid **458** and its ethyl ester has been attempted [481]. The electrolysis is carried out using a poly-L-valine-coated graphite electrode, and the reduced products are obtained in 30~48% yields (0.2 ~ 6.7% of ee %).

$$\text{Ph--}\underset{\underset{\text{O}}{\|}}{\text{C}}\text{--CO}_2\text{H} \quad \xrightarrow{\text{aq. KCl(1M)–(Pt/Ag)}} \quad \text{Ph--}\underset{\underset{\text{OH}}{|}}{\text{CH}}\text{--CO}_2\text{H} \qquad (2.191)$$

458   **459**

More recently, the role of Mg^{2+} ions in the pathway of the one-step electrosynthesis of 1,2-aryldiketones has been pointed out [482]. The electrolysis of aromatic ester in a DMF-Bu$_4$NBF$_4$-(Mg/Ni) system in an undivided cell affords the 1,2-diketone in over 95% yield.

Asymmetric electrohydrogenation of methyl acetoacetate **460** gives methyl 3-hydroxy-butyrate **461** in 28.5% of optical yield and in 90% chemical yield using modified Raney Ni (eq. (2.192)) [483]. A similar asymmetric reduction of ethyl acetoacetate using a uniform layer of nickel-black modified with *R,R*-tartaric acid on a platinum foil has been attempted; *R*-ethyl β-hydroxybutyrate is obtained in 8~12% optical yields [484]. Electroreduction of benzoylacetanilide **462** in an MeCN-LiClO$_4$(0.1M)-(Pt) system at –1.2 V (Ag/AgCl) affords 3-anilino-3-hydroxy-1-phenylpropan-1-one **463** in 80% yield (eq. (2.193)) [485].

$$\text{Me--}\underset{\underset{\text{O}}{\|}}{\text{C}}\text{--CH}_2\text{CO}_2\text{Me} \quad \xrightarrow[\substack{\text{chem. yield   90\%}\\\text{optical yield   28.5\%}}]{\text{Modified   Raney–Ni   Electrode}} \quad \text{Me--}\underset{\underset{\text{OH}}{|}}{\text{CH}}\text{--CH}_2\text{CO}_2\text{Me} \qquad (2.192)$$

460   **461**

$$\text{PhCOCH}_2\text{--}\underset{\underset{\text{N}}{|}}{\overset{\overset{\text{O}}{\|}}{\text{C}}}\text{--H--Ph} \quad \xrightarrow[\substack{\text{–1.2 V (Ag/AgCl)}\\\text{80\% yield}}]{\text{MeCN-LiClO}_4\text{(0.1M)-(Pt)}} \quad \text{PhCOCH}_2\text{--}\underset{\underset{\text{H}}{|}}{\overset{\overset{\text{OH}}{|}}{\text{CH}}}\text{--N--Ph} \qquad (2.193)$$

462   **463**

2.7.3   Others

The cathodic cleavage of the N-C bonds of *N*-acetylpiperidin-4-ones **464**, leading to **465** has been attained in an MeOH/PhH(7/3v/v)-NaOH-(C) system in moderate yields (eq. (2.194)) [486]. The successful electroreductive cleavage of the N-S bonds proceeds in *p*-toluenesulfonamide of polyamines and arenesulfonamides of secondary amines, giving the corresponding amines [487, 488].

$$(2.194)$$

$R^1 = R^2 = H$, Alkyl
$Ar = Ph, p\text{-}ClC_6H_4$

MeOH/PhH(7/3)-NaOH-(C)

48 ~ 69% yields

464   **465**

A selective electrocatalytic cleavage of the N-C bond of the *N*-benzyloxycarbonyl moiety of **466** as protected amino acids and peptides have been successfully carried out in an MeOH/AcOH-NaClO$_4$-(Pd/C) system to give **467** under very mild conditions (eq. (2.195)) [489].

$$\text{(2.195)}$$

MeOH/AcOH-NaClO$_4$-(Pd/C)

90~99% yields

$R^1 = R^2$ = peptide residues

466   **467** $+ CO_2 + PhCH_3$

As a bond-making reaction, a novel glycosylation of proteins **469**, leading to **471**, has been attained in an H$_2$O-NaN$_3$-(Hg) system in the presence of the carbohydrate **468** by a potential-controlled electroreduction method instead of the usual borohydride procedure (eq. (2.196)) [490]. The method can be applied to whole casein and ordinary carbohydrates.

$$\text{(2.196)}$$

Prot. = Proteins

468   **469**   **470**   **471**

The electroreduction of benzoyl chloride **472** (Ar = Ph) in an acetone-LiClO$_4$-(Hg) system occurs by one-electron cleavage of the carbon-chlorine bond to form the benzoyl radical, which dimerizes to form benzil **473**. Further reduction of benzil **473** gives *cis*- and *trans*-stilbenediol dibenzoates **474** and **475** in quantitative yield (eq. (2.197 (Ar = Ph)) [491]. Similar coupling products **473** are obtained from other aroyl chlorides [492~494]. Further reduction of the enediol diesters **474** and **475** in a DMF-LiClO$_4$-(Hg) system affords 1,2-diarylacetylenes in good yields [495]. In contrast, aroyl and arylacetyl chlorides **476** can lead to the corresponding symmetrical ketones **477** in an MeCN-Bu$_4$NBF$_4$-(SUS/Ni) system by using an undivided cell, when the mercury pool cathode is replaced with other electrode materials, *i.e.*, nickel, stainless, carbon fiber (eq. (2.198)) [496]. The electroreduction of heptanoyl chloride in an MeCN-Et$_4$NClO$_4$-(Hg) system, however, affords heptanal in 48~59% yields together with heptanoic anhydride (11~46%) [497,498]. Electroreduction of glutaryl dichloride in an MeCN-Et$_4$NClO$_4$(0.1M)-(Hg) system affords a mixture of 5-chlorovalerolactone and valerolactone in 24~32% yields together with a polymer [499]. The electroreduction of benzoyl cyamides **478** in dry acetonitrile preferentially affords mandelonitrile **479** in 53-76% yields (eq. (2.199)) [500].

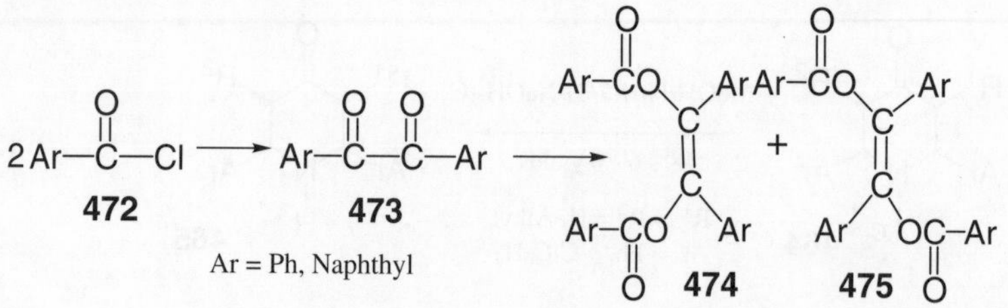

$$2\,Ar\overset{O}{\underset{}{-C}}-Cl \longrightarrow Ar\overset{O}{\underset{}{-C}}\overset{O}{\underset{}{-C}}-Ar \longrightarrow \quad + \quad \tag{2.197}$$

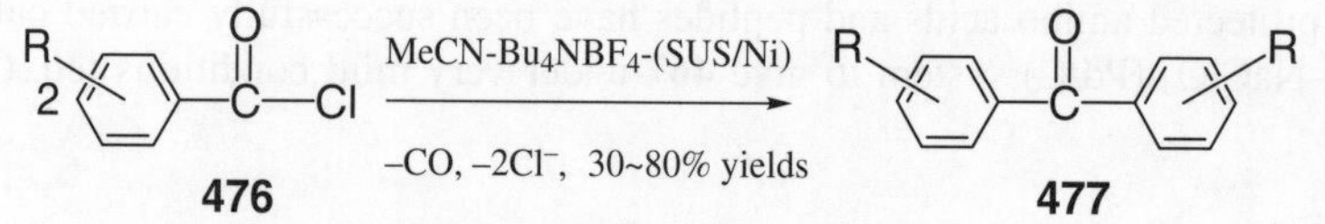

472   **473**   **474**   **475**

Ar = Ph, Naphthyl

$$2\;R\;\text{—}\;\overset{O}{\underset{}{C}}\;\text{—Cl} \quad\xrightarrow[\;-CO,\;-2Cl^-,\;30\text{~}80\%\;\text{yields}\;]{MeCN\text{-}Bu_4NBF_4\text{-}(SUS/Ni)}\quad R\;\text{—}\;\overset{O}{\underset{}{C}}\;\text{—}\;R \tag{2.198}$$

476   **477**

R = H, 4-Me, 3-Me, 4-F, 3-F, 4-Br

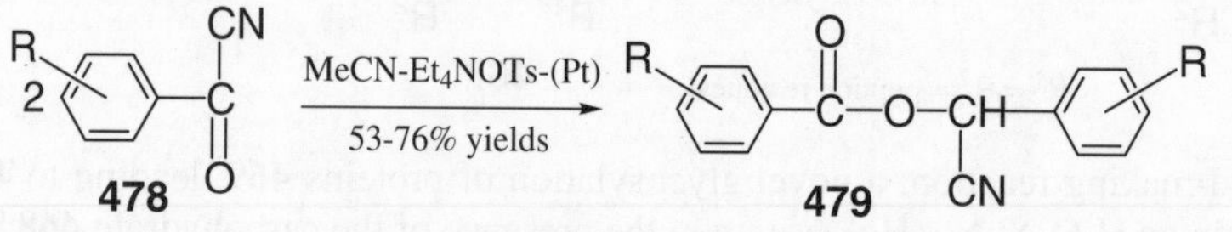

$$2\;R\;\text{—}\;\overset{CN}{\underset{O}{C}}\quad\xrightarrow[\;53\text{-}76\%\;\text{yields}\;]{MeCN\text{-}Et_4NOTs\text{-}(Pt)}\quad R\;\text{—}\;\overset{O}{\underset{}{C}}\text{—O—CH}\;\text{—}\;R \tag{2.199}$$

478   **479** CN

R = H, o-Me, p-Me, o-MeO, p-MeO, o-Cl, p-Cl, p-t-Bu,

The electroreductive cross-coupling of ketones with aliphatic O-methyl oximes has been found to occur in an i-PrOH-Et$_4$NOTs-(Sn) system in excellent yields [501].   The choice of electrode materials is important.   Lead and cadmium can also be used for the cross-coupling but others, e.g., Ag, Cu, Zn, C, etc., bring inferior results.   The electrolysis of cyclohexanone **480** (1 equiv.) and the oxime **481** (2.5 equiv.) is carried out under the above conditions to give the product **482** in 90% yield (eq. (2.200)).   The reaction proceeds through the coupling reaction of a ketyl anion radical with an azomethine group.

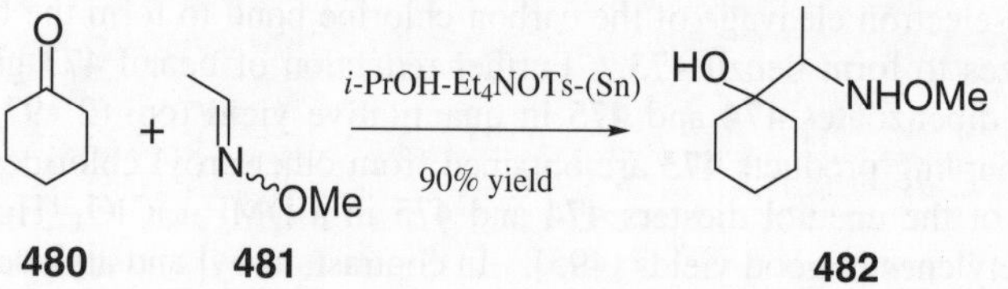

$$\text{cyclohexanone} \quad + \quad \underset{\underset{OMe}{N}}{} \quad\xrightarrow[\;90\%\;\text{yield}\;]{i\text{-PrOH-Et}_4\text{NOTs-(Sn)}}\quad \underset{}{HO}\text{—}\underset{}{}\text{NHOMe} \tag{2.200}$$

480   **481**   **482**

The activation of CO_2 by EG superoxide in aprotic solvents has been investigated [502a]. The superoxide ion-assisted activated CO_2 provides a carboxylating reagent, which is able to convert NH-protic acetamides or propanamides bearing a leaving group at the _ position into oxazole or 1,3-oxazine derivatives [502b].

The formation of carboxylic acids from elemental carbon and water by arc-discharge experiments has been attempted [503]. The growth of carbon chains of these acids proceeds by a progressive addition of one carbon species.

References

[1] F. D. Popp, H. P. Schultz, *Chem. Rev.*, **62**, 19 (1962).

[2] (a) U. Kaiser, E. Heitz, *Ber. Bunsenges. Phys. Chem.*, **77**, 818 (1973); (b) P. G. Russel, N. Kovac, S. Srinivasan, M. Steinberg, *J. Electrochem. Soc.*, **124**, 1329 (1977); (c) F. Wolf, J. Rollin, *Z. Chem.*, **17**, 337 (1977); (d) Y. Shimizu, X.-Q. Wu, M. Egashira, *Denki Kagaku*, **57**, 914 (1989); (e) Y. Hori, *Denki Kagaku*, **58**, 996 (1990); (f) K. Tanaka, *Denki Kagaku*, **58**, 989 (1990); (g) Y. Hori, *Kagaku to Kogyo*, **43**, 2016 (1990); (h) K. Ito, *Kagaku*, **46**, 74 (1991); (i) Y. Hori, *Sen-i Gakkaishi*, **48**, 38 (1992).

[3] C. Amatore, J.-M. Savéant, *J. Am. Chem. Soc.*, **103**, 5021 (1981).

[4] C. J. Stalder, S. Chao, M. S. Wrighton, *J. Am. Chem. Soc.*, **106**, 3673 (1984).

[5] (a) D. Canfield, K. W. Frese, Jr., *J. Electrochem. Soc.*, **130**, 1772 (1983); (b) S. Kapusta, N. Heckerman, *J. Electrochem. Soc.*, **130**, 607 (1983); (c) J. Ryu, T. N. Anderson, H. Eyring, *J. Phys. Chem.*, **76**, 3278 (1972).

[6] W. M. Sears, S. Roy Morrison, *J. Phys. Chem.*, **89**, 3295 (1985).

[7] (a) F. B. Leitz, H. I. Viklund, *Final Report - Phase I*, Contract No. N00014-66-C0139, AD-654146, National Technical Information Service, Springfield, VA, 1967; (b) F. H. Meller, *Final Report - Phase II*, Contract No. N00014-66-C0139, AD-678427, National Technical Information Service, Springfield, VA, 1968.

[8] M. H. Miles, A. N. Fletcher, G. E. McManis, *J. Electroanal. Chem., Interfacial Electrochem.*, **190**, 157 (1985).

[9] K. Ito, *Denki Kagaku*, **58**, 984 (1990).

[10] (a) M. Azuma, K. Hashimoto, M. Hiramoto, M. Watanabe, T. Sakata, *J. Electrochem. Soc.*, **137**, 1772 (1990); (b) K. Hara, A. Kudo, T. Sakata, *J. Electrochem. Soc.*, **142**, L57 (1995).

[11] (a) K. Ogura, M. Takagi, *J. Electroanal. Chem., Interfacial Electrochem.*, **201**, 359 (1986); (b) *Idem, ibid.*, **206**, 209 (1986); (c) K. Ogura, K. Mine, J. Yano, H. Sugihara, *J. Chem. Soc, Chem. Commun.*, **1993**, 20; (d) Idem, *Denki Kagaku*, **61**, 810 (1993); (e) K. Ogura, K. Mine, J. Yano, H. Sugihara, *Proc.- Electrochem. Soc.*, **18**, 66 (1993); (f) K. Ogura, M. Higasa, J. Yano, N. Endo, *J. Electroanal. Chem.*, **379**, 373 (1994); (g) K. Hara, A. Kudo, T. Sakata, *J. Electroanal. Chem.*, **386**, 257 (1995); (h) K. Ogura, N. Endo, M. Nakayama, H. Ootsuka, *J. Electrochem. Soc.*, **142**, 4026 (1996).

[12] D. P. Summers, S. Leach, K. W. Frese, Jr., *J. Electroanal. Chem., Interfacial Electrochem.*, **205**, 219 (1986).

[13] (a) B. P. Sullivan, C. M. Bolinger, D. Conrad, W. J. Vining, T. J. Meyer, *J. Chem. Soc., Chem. Commun.*, **1985**, 1414 ; (b) T. R. O'Toole, L. D. Margerum, T. D. Westmoreland, W. J. Vining, R. W. Murray, T. J. Meyer, *ibid.*, **1985**, 1416; (c) M. R. M. Bruce, E. Megehee, B. P. Sullivan, H. H. Thorp, T. R. O'Toole, A. Downward, F. R. Pugh, T. J. Meyer, *Inorg. Chem.*, **31**(23), 4864 (1992); (d) M. R. M. Bruce, E. Megehee, B. P. Sullivan, H. Thorp, T. R. O'Tool, A. Downard, T. J. Meyer, *Organometallics*, **7**, 238 (1988);(e) C. M. Bolinger, B. P. Sullivan, D. Conrad, J. A. Gilbert, N. Story, T. J. Meyer, *J. Chem. Soc., Chem. Commun.*, **1985**, 796; (f) C. M. Bolinger, B. P. Sullivan, D. Conrad, J. A. Gilbert, N. Story, T. J. Meyer, *Inorg. Chem.*, **27**, 4582 (1988).

[14] T. Tomohiro, K. Uoto, H. Okuno, *J. Chem. Soc., Chem. Commun.*, **1990**, 194.

[15] (a) S. Nakagawa, A. Kudo, M. Azuma, T. Sakata, *J. Electroanal. Chem. Int. Electrochem*, **308**, 339 (1991); (b) M. Todoroki, K. Hara, A. Kudo, T. Sakata, *J. Electroanal. Chem.*, **394**, 199 (1995).

[16] C. L. Bailey, R. D. Bereman, D. P. Rillema, R. Nowak, *Inorg. Chim. Acta*, **116**, L45 (1986).

[17] Y. Kiso, K. Saeki, Japanese Patent Kokai, 52-36617 (1977). [*Chem. Abstr.* **1977**, 87,84562s]

[18] (a) K. Tanaka, *Denki Kagaku*, **58**, 989 (1990);(b) H. Ishida, H. Tanaka, K. Tanaka, T. Tanaka, *J. Chem. Soc., Chem. Commun.*, **1987**, 131; (c) Y. Kushi, H. Nagao, T. Nishioka, K. Isobe, K. Tanaka, *Chem. Lett.*, **1994**, 2175; (d) H. Nagao, T. Mizukawa, K. Tanaka, *Inorg. Chem.*, **33**, 3415 (1994); (e) Y. Kushi, H. Nagao, T. Nishioka, K. Isobe, K. Tanaka, *J. Chem. Soc., Chem. Commun.*, **1995**, 1223.

[19] (a) H. C. Hurrell, A.-L. Mogstad, D. A. Usifer, K. T. Potts, H. D. Abruna, *Inorg. Chem.*, **28**, 1080 (1989); (b) J. A. R. Senda, C. R. Arana, L. Hernández, K. T. Potts, M. Keshevarz-K, H. D. Abruna, *Inorg. Chem.*, **34**, 3339 (1995).

[20] P. G. Jessop, Y. Hsiao, T. Ikariya, R. Noyori, *J. Am. Chem. Soc.*, **116**, 8851 (1994).

[21] H. Ishida, K. Fujiki, T. Ohba, K. Ohkubo, *J. Chem. Soc., Dalton Trans.*, **1990**, 2155.

[22] P. A. Christensen, S. J. Higgins, *J. Electronanal. Chem.*, **387**, 127 (1995)

[23] (a) A. Bocarsly, G. Seshadri, L. Chao, *Proc.- Electrochem. Soc.*, **18**, 30 (1993); (b) G. Seshadri, C. Lin, A. B. Bocarsly, *J. Electronanal. Chem.*, **372**, 145 (1994).

[24] D. L. Dubois, A. Miedaner, C. Haltiwanger, *J. Am. Chem. Soc.*, **113**, 8753 (1991).

[25] (a) M. Watanabe, M. Shibata, A. Katoh, M. Azuma, T. Sakata, *Denki Kagaku*, **59**, 508 (1991); (b) M. Watanabe, M. Shibata, A. Katoh, T. Sakata, M. Azuma, *J. Electronanal. Chem. Int. Electrochem.*, **305**, 319 (1991); (c)M. Watanabe, M. Shibata, A. Katoh, H. Uchida, *Proc.- Electrochem. Soc.*, **18**, 55 (1993); (d) A. Katoh, H. Uchida, M. Shibata, M. Watanabe, *J. Electrochem. Soc.*, **141**, 2054 (1994).

[26] A. Naitoh, K. Ohta, T. Mizuno, *Chem. Express*, **8**, 145 (1995).

[27] (a) A. Bandi, H.-M. KüPhne, *J. Electrochem. Soc.*, **139**, 1605 (1992); (b) A. Bandi, *J. Electrochem. Soc.*,

137, 2157 (1992)

[28] B. I. Podlovchenko, E A. Kolyadko, S. Lu, *J. Electronanal. Chem.*, **373**, 185 (1994).
[29] (a) Y. Nakato, S. Yano, T. Yamaguchi, H. Tsubomura, *Denki Kagaku*, **59**, 491 (1991); (b) Y. Nakato, T. Mori, *Proc.- Electrochem. Soc.*, **18**, 95 (1993); (c) R. Hinogami, T. Mori, S. Yae, Y. Nakato, *Chem. Lett.*, **1994**, 1725.
[30] H. Yoshitake, K. Takahashi, K. Ota, *J. Chem. Soc., Faraday Trans. I*, **90**, 155 (1994).
[31] G. Nogami, H. Itagaki, R. Shiratsuchi, *J. Electrochem. Soc.*, **141** 1138 (1994).
[32] (a) Y. Hori, K. Kikuchi, S. Suzuki, *Chem. Lett.*, **1985**, 1695; (b) Y. Hori, A. Murata, K. Kikuchi, S. Suzuki, *J. Chem. Soc., Chem. Commun.*, **1987**, 728; (c) A. Murata, Y. Hori, *Bull. Chem. Soc. Jpn.*, **64**, 123 (1991); (d) Y. Hori, O. Koga, A. Aramata M. Enyo, *Bull. Chem. Soc. Jpn.*, **65**, 3008 (1992); (e) O. Koga, Y. Hori, *Denki Kagaku*, **61**, 812 (1993).
[33] (a) K. Kudo, H. Phala, N. Sugita, Y. Takezaki, *Chem. Lett.*, **1977**, 1495; (b) H. Phala, K. Kudo, N. Sugita, *Bull. Inst. Chem. Res., Kyoto Univ.*, **59**, 88 (1981).
[34] (a) K. Hashimoto, K. Ohkawa, A. Fujishima, *Proc.- Electrochem. Soc.*, **18**, 22 (1993); (b) K. Ohkawa, Y. Noguchi, S. Nakayama, K. Hashimoto, A. Fujishima, *J. Electroanal. Chem.*, **367**, 165 (1994); (c) K. Ohkawa, Y. Noguchi, S. Nakayama, K. Hashimoto, A. Fujishima, *J. Electroanal. Chem.*, **369**, 247 (1994); (d) T. Saeki, K. Hashimoto, A. Fujishima, N. Kimura, K. Omata, K. Ohkawa, Y. Noguchi, S. Nakayama, K. Hashimoto, A. Fujishima, *J. Phys Chem.*, **99**, 8440 (1995).
[35] (a) R. L. Cook, R. C. MacDuff, A. F. Sammells, *J. Electrochem. Soc.*, **134**, 2375 (1987); (b) M. Schwartz, M. E. Vercauteren, A. F Sammells, *J. Electrochem. Soc.*, **141**, 3119 (1994).
[36] A. Miedaner, C. J. Curtis, R. M. Barkley, D. L. DuBois, *Inorg. Chem.*, **33**, 5482 (1994).
[37] R. Kostecki, J Augustynski, *Ber. Bunsenges. Phys. Chem.*, **98**, 81510 (1994).
[38] M. Fujihira, T. Noguchi, *Chem. Lett.*, **1992**, 2043.
[39] (a) K. Ito, S. Ikeda, T. Iida, H. Niwa, *Denki Kagaku*, **49**, 106; (b) K. Ito, S. Ikeda, T. Iida, A. Nomura, *Denki Kagaku*, **50**, 463 (1982); (c) K. Ito, S. Ikeda, N. Yamauchi, T. Iida, T. Takagi, *Bull. Chem. Soc. Jpn.*, **58**, 3027 (1985); (d) H. Noda, S. Ikeda, Y. Oda, K. Ito, *Chem Lett.*, **1989**, 289; (e) S. Ikeda, Y. Tomita, A. Hattori, H. Noda, M. Sakai, K. Ito, *Proc.- Electrochem. Soc.*, **18**, 78 (1993); (f) S. Ikeda, T. Ito, K. Azuma, K. Ito, H. Noda, *Denki Kagaku*, **63**, 303 (1995); (g) S. Ikeda, T. Ito, K. Azuma, N. Nishi, K. Ito, H. Noda, *Denki Kagaku*, **64**, 69 (1996).
[40] M. Chandrasekaran, T. Raju, V. Krishnan, *Bull. Electrochem.*, **8**, 124 (1992).
[41] (a) B. R. Eggins, E. M. Brown, E. A. McNeill, J. Grimshaw, *Tetrahedron Lett.*, **29**, 945 (1988); (b) B. R. Eggins, C. Ennis, J. T. S. Irvine, *J. Appl. Electrochem.*, **24**, 271 (1994).
[42] D. A. Tyssee, J. H. Wagenknecht, M. M. Baizer, J. L. Chruma, *Tetrahedron Lett.*, **13**, 4809 (1972).
[43] (a) M. Tezuka, T. Yajima, A. Tsuchiya, *J. Am. Chem. Soc.*, **104**, 6834 (1982); (b) M. Tezuka, M. Iwasaki, *Chem. Lett.*, **1993**, 427.
[44] (a) T. Fukaya, M. Kodama, M. Sugiura, *Kagaku Gijutu Kenkyusho Hokoku*, **81**, 129 (1986); (b) H. Fujiwara, A. Konno, T. Nonaka, *Chem. Lett.*, **1991**, 1843.
[45] (a) T. Yoshida, K. Tsutsumida, S. Teratani, K. Yasufuku, M. Kaneko, *J. Chem. Soc., Chem. Commun.*, **1993**, 631; (b) T. Yoshida, T. Iida, T. Shirasagi, R.-J. Lin, M. Kaneko, *J. Electroanal. Chem.*, **344**, 355 (1993).
[46] G. Kyriacou, A. Anagnostopoulos, *J. Electroanal. Chem.*, **328**, 233 (1992).
[47] M.-C. Massebieau, E. Dunach, M. Troupel, J. Périchon, *New J. Chem.*, **14**, 259 (1990)
[48] T. Atoguchi, A. Aramata, A. Kazusaka, M. Enyo, *Denki Kagaku*, **59**, 526 (1991).
[49] (a) T. E. Teeter, P. Van Rysselberghe, *J. Chem. Phys.*, **22**, 759 (1954); (b) M. Y. P. Y. Hong, *Univ. Microfilms, Ann. Arbor, Mich.*, No. **70**-2026 (1970).
[50] (a) A. Bewick, G. P. Greener, *Tetrahedron Lett.*, **10**, 4623 (1969); (b) *Idem, ibid.*, **1970**, 391; (c) K. Ito, S. Ikeda, T. Iida, H. Niwa, *Denki Kagaku*, **49**, 106 (1981).
[51] (a) S. Ikeda, T. Takagi, K. Ito, *Bull. Chem. Soc. Jpn.*, **60**, 2517 (1987); (b) H. Noda, S. Ikeda, A. Yamamoto, H. Einaga, K. Ito, *Bull. Chem. Soc. Jpn.*, **68**, 1889 (1995).
[52] S. Kapusta, N. Hackerman, *J. Electrochem. Soc.*, **131**, 1511 (1984).
[53] (a) K. Ito, S. Ikeda, M. Yoshida, S. Ohta, T. Iida, *Bull. Chem. Soc. Jpn.*, **57**, 583 (1984); (b) T. Fukaya, M. Kodama, M. Sugiura, *Kagaku Gijutu Kenkyusho Hokoku*, **81**, 255 (1986); (c) S. Ikeda, A. Yamamoto, H. Noda, M. Maeda, K. Ito, *Bull Chem Scc. Jpn.*, **66**, 2473 (1993).
[54] I. Bhugun, D. Lexa, J.-M. Savéant, *J. Am. Chem. Soc.*, **118**, 1769 (1996).
[55] (a) L. V. Haynes, D. T. Sawyer, *Anal. Chem.*, **39**, 332 (1967); (b) J. C. Gressin, D. Michelet, L. Nadjo, J. M. SavéPant, *Nouv. J. Chim.*, **3**, 545 (1979);
[56] P. G. P. Ang, A. F. Sammells, A. Morduchowitz, *U. S. Pat.*, **4**, 595, 465 (1986).
[57] J. Hawecker, J.-M. Lehn, R. Ziessel, *Helv. Chim. Acta*, **69**, 1990 (1986).
[58] (a) K. Ogura, *Chemistry*, **46**, 307 (1991); (b) K. Ogura, *J. Appl. Electrochem.*, **16**, 732 (1986); (c) K. Ogura, M. Fujita, *J. Mol. Catal.*, **41**, 303 (1987).
[59] K. Ogura, H. Sugihara, J. Yano, M. Higasa, *J. Electrochem. Soc.*, **141**, 419 (1994).

[60] K. Ogura, C. T. Migita, T. Nagaoka, *J. Mol. Catal.*, **56**, 276 (1989).
[61] (a) K. Ogura, I. Yoshida, *J. Mol. Catal.*, **34**, 67 (1986); (b) K. Ogura, K. Takamagari, *J. Chem. Soc., Dalton Trans.*, **1986**, 1519.
[62] (a) S. Kuwabata, R. Tsuda, K. Nishida, H. Yoneyama, *Chem. Lett.*, **1993**, 1631; (b) S. Kuwabata, R. Tsuda, H. Yoneyama, *J. Am. Chem. Soc.*, **116**, 5437 (1994).
[63] (a) M. Schwarz, R. L. Cook, V. M. Kehoe, R. C. Macduff, J. Patel, A. F. Sammells, *J. Electrochem. Soc*, **140**, 614 (1993); (b) S. Kuwabata, K. Nishida, R. Tsuda, H. Inoue, H. Yoneyama, *ibid.*, **141**, 1498 (1994).
[64] K. Sugimura, S. Kuwabata, H. Yoneyama, *Bioelectrochem. Bioenerg.*, **24**, 241 (1990).
[65] Yu. B. Vassiliev, V. S. Bagotzky, O. A. Khazova, N. A. Mayorova, *J. Electroanal. Chem., Interfacial Electrochem.*, **189**, 271 (1985).
[66] Yu. B. Vassiliev, V. S. Bagotzky, N. V. Osetrova, A. A. Mikhailova, *J. Electroanal. Chem., Interfacial Electrochem.*, **189**, 311 (1985).
[67] (a) Yu. B. Vassiliev, V. S. Bagotzky, O. A. Khazova, N. A. Mayorova, *J. Electroanal. Chem., Interfacial Electrochem.*, **189**, 295 (1985); (b) N. A. Maiorova, O. A. Khazova, Yu. B. Vasil'ev, *Elektrokhimiya*, **22**, 1196 (1986).
[68] M. Koudelka, A. Monnier, J. Augustynski, *J. Electrochem. Soc.*, **131**, 745 (1984).
[69] K. W. Frese, Jr., S. Leach, *J. Electrochem. Soc.*, **132**, 259 (1985).
[70] (a) H. Nagao, T. Mizukawa, K. Tanaka, *Chem. Lett.*, **1993**, 955; (b) H. Nagao, T. Mizukawa, K. Tanaka, *Inorg. Chem.*, **33**, 3415(1994).
[71] M. Hammouche, D. Lexa, M. Momenteau, J.-M. Savéant, *J. Am. Chem. Soc.*, **113**, 8455 (1991).
[72] A. Murata, Y. Hori, *Chem. Lett.*, **1991**, 181.
[73] Y. Hori, *Denki Kagaku*, **58**, 996 (1990).
[74] Y. Hori, A. Murata, R. Takahashi, *J. Chem. Soc., Faraday Trans. 1*, **85**, 2309 (1989).
[75] D. W. Dewulf, A. J. Bard, *Catalysis Lett.*, **1**, 73 (1988).
[76] (a) M. Halmann, *Nature*, **275**, 115 (1978); (b) A. Monnier, J. Augustynski, C. Stalder, *J. Electroanal. Chem., Interfacial Electrochem.*, **112**, 383 (1980); (c) Y. Taniguchi, H. Yoneyama, H. Tamura, *Bull. Chem. Soc. Jpn.*, **55**, 2034 (1982); (d) D. Canfield, K. W. Frese, Jr., *J. Electrochem. Soc.*, **130**, 1772 (1983); (e) K. W. Frese, Jr., D. Canfield, *ibid.*, **131**, 2518 (1984); (f) I. Taniguchi, B. Aurian-Blajeni, J. O'M. Bockris, *Electrochim. Acta*, **29**, 923(1984); (g) T. Saeki, K. Hashimoto, Y. Noguchi, K. Omata, A. Fujishima, *J. Electrochem. Soc.*, **141**, L130 (1994).
[77] (a) B. Fisher, R. Eisenberg, *J. Am. Chem. Soc.*, **102**, 7361 (1980); (b) C. M. Lieber, N. S. Lewis, *ibid.*, **106**, 5033 (1984).
[78] (a) J. Hawecker, J. M. Lehn, R. Ziessel, *J. Chem. Soc., Chem. Commun.*, **1984**, 328; (b) M.-N. Collomb-Dunand-Sauthier, A. Deronzier, R. Ziessel, *ibid.*, **1994**, 189.
[79] H. Ishida, K. Tanaka, T. Tanaka, *Chem. Lett.*, **1985**, 405.
[80] (a) S. Meshitsuka, M. Ichikawa, K. Tamaru, *J. C. S., Chem. Commun.*, **1974**, 158; (b) K. Hiratsuka, K. Takahashi, H. Sasaki, S. Toshima, *Chem. Lett.*, **1977**, 1137; (c) K. Takahashi, K. Hiratsuka, H. Sasaki, S. Toshima, *ibid.*, **1979**, 305; (d) D. Wang, H. Nishino, H. Segawa, K. Honda, T. Shimidzu, *Chem. Express*, **7**, 517 (1992).
[81] (a) A. P. Tomilov, *Elektrokhimiya*, **30**, 725 (1994); (b) G. Silvestri, S. Gambino, G. Filardo, *NATO ASI SER., SER. C.*, **314**, 101 (1990).
[82] (a) W. J. M. Van Tilborg, C. J. Smit, *Recl. Trav. Chim. Pays Bas*, **100**, 437 (1981); (b) D. Pletcher, J. T. Girault, *Inst. Chem. Eng. Symp. Ser.*, **98**, 13 (1986); (c) S. Dérien, E. Duñach, J. Périchon, *J. Organomet. Chem.*, **385**, C43 (1990).
[83] (a) S. Dérien, E. Duñach, J. Périchon, *J. Am. Chem. Soc.*, **113**, 8447 (1991); (b)S. Dérien, J.-C. Clinet, E. Duñach, J. Périchon, *J. Org. Chem.*, **58**, 2578 (1993); (c) E. Duñach, J. Périchon, *Synlett*, **1990**, 143; (d) S. Dérien, J.-C. Clinet, E. Duñach, J. Périchon, *J. Chem. Soc., Chem. Commun.*, **1991**, 549.
[84] J. Chaussard, M. Troupel, Y. Robin, G. Jacob, J. P. Juhasz, *J. Appl. Electrochem.*, **19**, 345 (1989).
[85] S. Dérien, J.-C. Clinet, E. Duñach, J. Périchon, *J. Organomet. Chem.*, **424**, 213 (1992).
[86] (a) S. Dérien, J.-C. Clinet, E. Duñach, J. Périchon, *Synlett*, **1990**, 361; (b) S. Dérien, J.-C. Clinet, E. Duñach, J. Périchon, *Tetrahedron*, **48**, 5235 (1992).
[87] P. Tascedda, E. Duñach, *J. Chem. Soc., Chem. Commun.*, **1995**, 43.
[88] H. Nagao, H. Miyamoto, K. Tanaka, *Chem. Lett.*, **1991**, 323.
[89] N. Komeda, H. Nagao, T. Matsui, G. Adachi, K. Tanaka, *J. Am. Chem. Soc.*, **114**, 3625 (1992).
[90] V. P. Gul'tyai, L. M. Korotaeva, T. Ya. Rubinskaya, *Dokl. Akad. Nauk SSSR*, **267**, 662 (1982).
[91] M. A. Casadei, A. Inesi, F. M. Moracci, L. Rossi, *J. Chem. Soc., Chem. Commun.*, **1996**, 2575.
[92] H. Nakajima, T. Mizukawa, H. Nagao, K. Tanaka, *Chem. Lett.*, **1995**, 251.
[93] G. S. Cano, V. Montiel, A. Aldaz, *Bull. Electrochem.*, **6**, 931 (1990).
[94] (a) C. Amatore, A. Jutand, *J. Am. Chem. Soc.*, **113**, 2819 (1991); (b) C. Amatore, A. Jutand, F. Khalil, M. F. Neilsen, *J. Am. Chem. Soc.*, **114**, 7076 (1992).
[95] T. Nagaoka, N. Nishii, K. Fujii, K. Ogura, *J. Electroanal. Chem., Interfacial Electrochem.*, **322**, 383 (1992).

[96] T. C. Simpson, R. R. Durand, Jr., *Electrochim. Acta*, **35**, 1399 (1990).
[97] H. Nakajima, T. Mizukawa, H. Nagao, K. Tanaka, *Chem. Lett.*, **1995**, 251.
[98] G. Pierre, A. Ziade, *Nouv. J. Chim.*, **10**, 233 (1986).
[99] A. Murata, Y. Hori, *Denki Kagaku*, **59**, 499 (1991).
[100] (a) Y. Hori, A. Murata, R. Takahashi, S. Suzuki, *Tetrahedron Lett.*, **1987**, 1665; (b) Y. Hori, A. Murata, R. Takahashi, S. Suzuki, *J. Am. Chem. Soc.*, **109**, 5022 (1987).
[101] T. M. Gur, R. A. Huggins, *J. Catal.*, **102**, 443 (1986).
[102] K. Ogura, M. Takagi, *Solar Energy*, **37**, 41 (1986).
[103] (a) V. D. Neff, *J. Electrochem. Soc.*, **125**, 886 (1978); (b) D. Ellis, M. Eckhoff, V. D. Neff, *J. Phys. Chem.*, **85**, 1225 (1981).
[104] K. Itaya, H. Akahoshi, S. Toshima, *J. Electrochem. Soc.*, **129**, 1498 (1982).
[105] (a) K. Ogura, M. Takagi, *J. Electroanal. Chem., Interfacial Electrochem.*, **195**, 357 (1985); (b) K. Ogura, S. Yamasaki, *J. Chem. Soc., Dalton Trans.*, **1985**, 2499; (c) K. Ogura, S. Yamasaki, *J. Mol. Catal.*, **30**, 411 (1985); (d) K. Ogura, M. Kaneko, *J. Mol. Catal.*, **31**, 49 (1985); (e) K. Ogura, H. Watanabe, *J. Chem. Soc., Faraday Trans. 1*, **81**, 1569 (1985); (f) K. Ogura, *Nippon Kagaku Kaishi*, **1986**, 478.
[106] K. Ogura, S. Yamasaki, *J. Chem. Soc., Faraday Trans. I*, **81**, 267 (1985).
[107] H. Yoneyama, K. Wakamoto, N. Hatanaka, H. Tamura, *Chem. Lett.*, **1985**, 539.
[108] S. Yamamura, H. Kojima, W. Kawai, *J. Electroanal. Chem., Interfacial Electrochem.*, **186**, 309 (1985).
[109] (a) L. Eberson, H. Schäfer, Topics in Current Chemistry, Vol. 21, Springer-Verlag, New York (1971); (b) A. J. Fry, *ibid*, Vol, 34 (1972); (c) D. H. Evans, *Encyclopedia of Electrochemistry of the Elements* (ed. A. J. Bard), Vol. 12, Chapter 1, Marcel Dekker, New York (1978); (d) L. G. Feoktistov, H. Lund, *Organic Electrochemistry* (ed. M. M. Baizer H. Lund), 2nd ed., Chapter 9, Marcel Dekker, New York (1983); (e) L. Horner, *ibid.*, Chapter 29; (f) M. Fujihira, *Topics in Organic Electrochemistry* (ed. A. J. Fry, W. E. Britton), Chapter 6, Plenum Press, New York (1986) ; (g) A. J. Fry, *ibid.*, Vol. 34 (1972).
[110] (a) S. Kuwabata, K. Nishida, H. Yoneyama, *Chem. Lett.*, **1994**, 407; (b) J. Pardillos-Guindet, S. Vidal, J. Court, P. Fouilloux, *J. Catal.*, **155**, 12 (1995).
[111] M. Ishikawa, Y. Takahashi, M. Morita, Y. Matsuda, *Denki Kagaku*, **62**, 1227 (1994).
[112] (a) F. Goodridge, K. Lister, R. E. Plimley, K. Scott, *J. Appl. Electrochem.*, **10**, 55 (1980); (b) I. Gimenez, S. Maximovitch, M. J. Barbier, Y. Christidis, G. Mattioda, *New J. Chem.*, **11**, 733 (1987); (c) D. J. Picket, K. S. Yap, *J. Appl. Electrochem.*, **4**, 17 (1974).
[113] K. Scott, *Electrochim. Acta*, **36**, 1447 (1991).
[114] E. Brillas, J. J. Novoa, M. T. Pujol J. Virgili, *Anales De Quimica*, **81**, 352 (1985).
[115] (a) E. Kariv-Miller, C. Nanjundiah, J. Eaton, K. E. Swenson, *J. Electroanal. Chem., Interfacial Electrochem.*, **167**, 141 (1984); (b) E. Kariv-Miller, T. J. Mahachi, *J. Org. Chem.*, **51**, 1041 (1986).
[116] H. Mansour, M. A. Ghandour, G. A. Noubi, M. M. Mostafa, *Electrochem. Soc. India*, **36**, 15 (1987).
[117] T. Shono, I. Nishiguchi, H. Ohmizu, M. Mitani, *J. Am. Chem. Soc.*, **100**, 545 (1978).
[118] J. E. Swartz, T. J. Mahachi, E. Kariv-Miller, *J. Am. Chem. Soc.*, **110**, 3622 (1988).
[119] P. B. Lawin, A. C. Hutson, E. Kariv-Miller, *J. Org. Chem.*, **54**, 526 (1989).
[120] N. Kise, T. Suzumoto, T. Shono, *J. Org. Chem.*, **59**, 1407 (1994).
[121] G. Horanyi, V. N. Andreev, *Electrokhimiya*, **21**, 1104 (1985).
[122] T. Chiba, M. Okimoto, H. Nagai, Y. Takata, *Bull. Chem. Soc. Jpn.*, **56**, 719 (1983).
[123] (a) T. Nonaka, M. Takahashi, T. Fuchigami, *Bull. Chem. Soc. Jpn.*, **56**, 2584 (1983); (b) idem, *Denki Kagaku*, **51**, 129 (1983).
[124] Y. Senda, M. Tateoka, H. Itoh, J. Ishiyama, *Bull. Chem. Soc. Jpn.*, **64**, 3302 (1991).
[125] A. Curulli, A. Inesi, *Electrochim. Acta*, **32**, 1117 (1987).
[126] M. A. Zón, J. M. Rodríguez Mellado, *J. Electroanal. Chem.*, **338**, 229 (1992).
[127] K. R. Bharucha, C. E. Slemon, *U. S. Pat.*, **4**, 547,271 (1985).
[128] J.-P. Pulicani, J.-D. Bourzat, H. Bouchard, A. Commerçon, *Tetrahedron Lett.*, **35**, 4999 (1994).
[129] E. Leszczynska, A. Cisak, *Polish J. Chem.*, **4**, 657 (1989)
[130] S. Bienz, A. Guggisberg, R. Walchli, M. Hesse, *Helvetica Chim. Acta*, **71**, 1708 (1988).
[131] G. M. Abou-Elenien, M. A. Aboutabl, A. O. Sherin, H. M. Fahmy, *J. Chem. Soc., Perkin Trans.* II, **1991**, 377
[132] M. Fujihira, A. Yokozawa, H. Kinoshita, T. Osa, *Chem. Lett.*, **1982**, 1089.
[133] F. Takahashi, K. Tomii, H. Takahashi, *Electrochim. Acta*, **31**, 127 (1986).
[134] W. A. Bonner, *Origins of Life and Evolution of the Biosphere*, **20**, 1 (1990).
[135] M. Fedoronko, *Chem. Papers*, **41**, 767 (1987).
[136] (a) K. Park, P. N. Pintauro, M. M. Baizer, K. Nobe, *J. Appl. Electrochem.*, **16**, 941 (1986); (b) V. Anantharaman, P. N. Pintauro, *J. Electrochem. Soc.*, **141**, 2729 (1994); (c) V. Anantharaman P. N. Pintauro, *J. Electrochem. Soc.*, **141**, 2742 (1994).
[137] A. Jokic, N. Ristic, M. M. Jaksic, M. Spasojevic, N. Krstajic, *J. Appl. Electrochem.*, **21**, 321 (1991).
[138] P. Zuman, Collect. Czecho. *Chem. Commun.*, **33**, 2548 (1968).

[139] J.-P. Seguin, J.-P. Doucet, R. Uzan, *C. R. Acad. Sci. Paris, Sér. C*, **278**, 129 (1974).

[140] J. Wiley, *Techniques of Chemistry* (ed. N. L. Weinberg) Vol. V, Part II, John Wiley & Sons, Inc., New York (1975).

[141] D. Pletcher, M. Razag, *Electrochim. Acta*, **26**, 819 (1981).

[142] (a) Y. Ikeda, E. Manda, *Chem. Lett.*, **1989**, 839; (b) Y. Ikeda, *Nippon Kagaku Kaishi*, **1990**, 1263.

[143] Y. Ikeda, *Denki Kagaku*, **59**, 117 (1991).

[144] Y. Ikeda, *Chem. Lett.*, **1990**, 1719

[145] C. Mousty, G. Mousset, *New. J. Chem.*, **16**, 1063 (1992).

[146] M. J. Lain, D. Pletcher, *Electrochim. Acta*, **32**, 109 (1987).

[147] Y. Kunugi, T. Fuchigami, T. Nonaka, *Chem. Lett.*, **1989**, 1467.

[148] M. Kimura, H. Yamagishi, Y. Sawaki, *Denki Kagaku*, **62**, 1119 (1994).

[149] A. M. Martre, G. Mousset, P. Pouillen, R. Prime, *Electrochim. Acta*, **36**, 1911 (1991).

[150] (a) V. G. García, V. Montiel, J. M. Feliu, A. Aldaz, M. Feliz, *J. Chem. Res.*(S), **1990**, 144; (b) V. G. García, V. Montiel, A. Aldaz, *J. Chem. Res.*(S), **1992**, 58.

[151] J. Simonet, H. Dupuy, *J. Electroanal. Chem.*, **327**, 201 (1992).

[152] J. M. Tanko, R. E. Drumright, *J. Am. Chem. Soc.*, **112**, 5362 (1990).

[153] (a) H. Mansour, J . *Electrochem. Soc. India*, **34**, 42 (1985); (b) H. Mansour, M. Kamal, G. Noubi, *Ibid*, **34**, 50 (1985).

[154] J. Berthelot, M. Jubault, J. Simonet, *Electrochim. Acta*, **28**, 1719 (1983).

[155] G. Capobiance, G. Farnia, A. Gambaro, M. G. Severin, *J. Chem. Soc., Perkin Trans. II*, **1980**, 1277.

[156] M. D. Birkett, A. T. Kuhn, *Electrochim. Acta*, **25**, 273 (1980) and references cited therein.

[157] (a) Y.-L. Chen, T.-C. Chou, *Ind. Eng. Chem. Res.*, **33**, 676 (1994); (b) Y.-L. Chen, T.-C. Chou, *J. Appl. Electrochem.*, **24**, 434 (1994); (c) A. M. Polcaro, S. Palmas, S. Dernini, *Electrochim. Acta*, **38**, 199 (1993).

[158] (a) H. Mansour, M. A. Ghandour, M. H. M. Abu-El-Wafa, M. M. Moustafa, *J. Indian Chem. Soc.*, **64**, 334 (1987); (b) M. Ganesan, M. Kottaisamy, S. Thangavelu, P. Manisankar, *Bull. Electrochem.*, **8**, 424 (1992).

[159] C. P. Andrieux, M. Grzeszczuk, J.-M. Savéant, *J. Am. Chem. Soc.*, **113**, 8811 (1991).

[160] T. Matsue, C. Tasaki, M. Fujihira, T. Osa, *Bull. Chem. Soc. Jpn.*, **56**, 1305 (1983).

[161] (a) Y. Kunugi, T. Fuchigami, H.-J. Tien, T. Nonaka, *Chem. Lett.*, **1989**, 757; (b) K. Matsuda, M. Atobe, T. Nonaka, *Chem. Lett.*, 1994, 1619; (c) M. Atobe, T. Nonaka, *Chem. Lett.*, **1995**, 669.

[162] P. C. Cheng, T. Nonaka, *Bull. Chem. Soc. Jpn.*, **64**, 3500 (1991).

[163] J. J. Jow, T. C. Chou, *Electrochim. Acta*, **32**, 311 (1987).

[164] M. Chandrasekaran, M. Noel, V. Krishnan, *J. Appl. Electrochem.*, **22**, 1072 (1992).

[165] M. Chandrasekaran, M. Noel, V. Krishnan, *Bull. Electrochem.*, **6**, 524 (1990).

[166] M. M. Ellaithy, M. M. Amer, *Analytical Lett.*, **18**, 135 (1985).

[167] M. Svaan, V. D. Parker, *Acta Chem. Scand.* **B39**, 401 (1985).

[168] A. Susaimanickam, M. Chandrasekaran, V. Krishinan, *J. Appl. Electrochem.*, **20**, 335 (1990).

[169] T. Nagaoka, S. Okazaki, T. Fujinaga, *J. Electroanal. Chem., Interfacial Electrochem.*, **133**, 89 (1982).

[170] B. Paduszek-Kwiatek, M. K. Kalinwski, *Electrochim. Acta*, **29**, 1439 (1984).

[171] A. Demortier, A. J. Bard, *J. Am. Chem. Soc.*, **95**, 3495 (1973).

[172] S. Wawzonek, A. Gundersen, *J. Electrochem. Soc.*, **107**, 537 (1960)

[173] (a) N. Egashira, Y. Takita, H. Hori, *Bull. Chem. Soc. Jpn.*, **55**, 3331 (1982); (b) Idem, *Denki Kagaku*, **49**, 217 (1981).

[174] (a) R. Pasternak, *Helv. Chim. Acta*, **31**, 753 (1948); (b) P. J. Elving, J. T. Leone, *J. Am. Chem. Soc.*, **80**, 1021 (1958).

[175] J. S. Jaworski, M. Malik, K. Kalinowski, *J. Phys. Org. Chem.*, **5**, 590 (1992).

[176] R. Saraswathi, R. Narayan, *Proc. Indian Acad. Sci.*, **97**, 403 (1986).

[177] F. Fournier, J. Berthelot, Y.-L. Pascal, *Can. J. Chem.*, **61**, 2121 (1983).

[178] (a) V. K. Sharma, D. K. Sharma, C. M. Gupta, *Stud. Org. Chem.*, **30**, 235 (1987); (b) M. Chandrasekaran, M. Noel, V. Krishnan, *J. Appl. Electrochem.*, **24**, 460 (1994).

[179] B. S. Chauhan, *J. Inst. Chemists*, **58**, 5 (1986).

[180] N. Takano, N. Takeno, M. Morita, *Denki Kagaku*, **53**, 139 (1985).

[181] J. M. A. Empis, B. J. Herold, *J. Chem. Soc., Perkin Trans. II*, **1986**, 425.

[182] R. Naef, *Helv. Chim. Acta*, **65**, 1734 (1982).

[183] S. L. Solodar, E. Yu, Khmel'nitskaya, I. D. Razgonyaeva, Y. Y. Urman, S. G. Alekseeva, N. T. Ioffe, *Zh. Obshch. Khim.*, **55**, 2108 (1985).

[184] N. Takano, N. Takeno, M. Morita, Y. Otsuji, *Denki Kagaku*, **53**, 807 (1985).

[185] A. Guirado, A. Zapata, *Tetrahedron*, **51**, 3641 (1995).

[186] J. Delaunay, A. Orliac-Le Moing, J. Simonet, *Electrochim. Acta*, **30**, 1109 (1985).

[187] (a) X. de Hemptinne, K. Schunck, *Ann. Soc. Sci. Brux.*, **80**, 289 (1966); (b) *Idem, Trans. Faraday Soc.*, **65**, 591 (1969).

[188] P. Florequin, H. Larmuseau, X. de Hemptinne, *Ann. Soc. Sci. Brux.*, **88**, 241 (1974).

[189] (a) G. Horányi, S. Szabó, J. Solt, F. Nagy, *Acta Chim. Acad. Sci. Hung.*, **68**, 205 (1971): (b) *Idem, ibid*, 71, 239 (1972); (c) S. Szabó, G. Horányi, *Acta Chim. Acad. Sci. Hung.*, **96**, 1 (1978).

[190] (a) G. Horányi, G. Inzelt, *J. Electroanal. Chem., Interfacial Electrochem.*, **91**, 287 (1978); (b) G. Horányi, G. Inzelt, K. Torkos, *ibid*, 106, 305 (1980); (c) G. Horányi, G. Inzelt, E. Szetey, *Acta Chim. Acad. Sci. Hung.*, **97**, 313 (1978).

[191] G. Horányi, G. Inzelt, K. Torkos, *J. Electroanal. Chem., Interfacial Electrochem.*, **106**, 319 (1980).

[192] N. L. Weinberg, M. Lipsztajn, D. J. Mazur, M. Reicher, H. R. Weinberg, E. P. Weinberg, *Stud. Org. Chem.*, **30**, 441 (1987).

[193] I. Rádoi, C. Daminescu, G. Musca, N. Vaszilcsin, Z. Popa, *Rev. Chim.*, **40**, 202 (1989).

[194] N. A, Kobyl'nik, V. T. Novikov, A. P. Tomilov, *Elektrokhimiya*, **27**, 678 (1991).

[195] N. L. Weinber, D. J. Mazur, *J. Appl. Electrochem.*, **21**, 895 (1991).

[196] J. J. Barber, *U. S. Pat.*, 4, 517, 062 (1985).

[197] N. P. Stepnova, S. M. Makarochkina, A. P. Tomilov, *Elektrokhimiya*, **28**, 951 (1992).

[198] P. K. Robertson, B. R. Eggins, *J. Chem. Soc., Perkin Trans. 2*, **1994**, 1829.

[199] V. A. Smirnov, E. S. Kagan, S. V. Kondrashov, Yu. I. Smushkevich, *Dokl. Akad. Nauk SSSR*, **288**, 1138 (1986).

[200] (a) T. J. Curphey, C. W. Amelotti, T. P. Layloff, R. L. McCartney, J. H. Williams, *J. Am. Chem. Soc.*, **91**, 2817 (1969); (b) T. J. Curphey, R. L. McCartney, *Tetrahedron Lett.*, **1969**, 5295.

[201] E. Leonard, E. Dunach, J. Perichon, *J. Chem. Soc., Chem. Commun.*, **1989**, 276.

[202] F. Fournier, J. Berthelot, Y. L. Pascal, *Tetrahedron*, **40**, 339 (1984).

[203] F. Fournier, M. Fournier, *Can. J. Chem.*, **64**, 881 (1986).

[204] Y. Ikeda, E. Manda, *Kagaku Gijutsu Kenkyusho Hokoku*, **86**, 385 (1991).

[205] J. Douch, G. Mousset, *Can. J. Chem.*, **65**, 549 (1987).

[206] A. Honnorat, P. Martinet, *Electrochim. Acta*, **28**, 1703 (1983).

[207] F. Pragst, F. Scholz, P. Woitke, V. Kollek, G. Henrion, *J. Prakt. Chem.*, **327**, 1028 (1985).

[208] E. Raoult, J. Sarrazin, A. Tallec, *Bull. Soc. Chim. Fr.*, **1985**, 1200.

[209] K. Tsunashima, T. Nonaka, *Chem. Lett.*, **1995**, 862.

[210] G. Farnia, G. Sandona, R. Fornasier, F. Marcuzzi, *Electrochim. Acta*, **35**, 1149 (1990).

[211] M. A. Michel, G. Mousset, J. Simonet, *J. Electroanal. Chem., Interfacial Electrochem.*, **98**, 319 (1979).

[212] E. M. Abbot, A. J. Bellamy, J. B. Kerr, I. S. MacKirdy, *J. Chem. Soc., Perkin Trans. II*, **1982**, 425.

[213] A. Bewick, H. P. Cleghorn, *J. Chem. Soc., Perkin Trans. II*, **1973**, 1410.

[214] V. P. Gul'tyai, A. S. Mendkovich, T. Y. Rubinskaya, *Izv. Akad. Nauk SSSR, Ser. Khim.*, **1987**, 1576.

[215] (a) C. P. Andrieux, J. M. Savéant, *Bull. Soc. Chim. Fr.*, **1973**, 2090; (b) J.-S. Yang, K.-T. Liu, Y. O. Su, *J. Phys. Org. Chem.*, **3**, 723 (1990).

[216] M. Perrin, P. Pouillen, G. Mousset, P. Martinet, *Tetrahedron*, **36**, 221 (1980).

[217] P. Pouillen, M. Perrin, P. Martinet, *C. R. Acad. Sci. Ser. C*, **284**, 955 (1977).

[218] D. W. Sopher, J. H. P. Utley, *J. C. S., Chem. Commun.*, **1979**, 1087.

[219] F. Fournier, J. Berthelot, Y.-L. Pascal, *C. R. Acad. Sci. Paris*, **294**, 849 (1982).

[220] F. Fournier, J. Berthelot, Y.-L. Pascal, *Tetrahedron*, **40**, 339 (1984).

[221] H. G. Thomas, K. Littmann, *Synlett*, **1990**, 757.

[222] B. M. Davis, P. H. Gore, K. A. K. Lott, E. L. Short, *J. Chem. Soc., Perkin Trans. II*, **1981**, 58.

[223] C. Mousty, E. Vauche, P. Pouillen, G. Mousset, *Bull. Soc. Chim. Fr.*, **1986**, 554.

[224] M. M. Srivastava, S. N. Srivastava, *Electrochim. Acta*, **30**, 1481 (1985).

[225] B. Aalstad, V. D. Parker, *Acta Chem. Scand.*, **B36**, 187 (1982);

[226] G. Capobianco, G. Farnia, A. Gambaro, M. G. Saverin, *J. Chem. Soc., Perkin Trans. II*, **1982**, 545.

[227] L. Mandell, H. Hamilton, R. A. Day, Jr., *J Org. Chem.*, **45**, 1710 (1980).

[228] S. Wawzonek, *J. Electrochem. Soc.*, **128**, 84 (1981).

[229] (a) W. Pfleiderer, R. Gottlieb, *Heterocycles*, **14**, 1603 (1980); (b) J. H. Stocker, R. M. Jenevein, *J. Org. Chem.*, **33**, 2145 (1968); (c) J. H. Stocker, R. M. Jenevein, *J. Org. Chem.*, **34**, 2807 (1969); (d) J. H. Stocker, R. M. Jenevein, D. H. Kern, *J. Org. Chem.*, **34**, 2810 (1969).

[230] V. D. Parker, O. Lerflaten, *Acta Chem. Scand.*, **B37**, 403 (1983).

[231] (a) W. R. Fawcett, A. Lasia, *Can. J. Chem.*, **59**, 3256 (1981); (b) P. Kaur, J. S. Manait, S. S. Pahil, *Bull. Electrochem.*, **7**, 562 (1991).

[232] C. Z. Smith, J. H. P. Utley, *J. C. S., Chem. Commun.*, **1981**, 492.

[233] V. K. Sharma, D. K. Sharma, C. M. Gupta, *Bull. Electrochem.*, **5**, 359 (1989).

[234] A. Lasia, A. Rami, *Can. J. Chem.*, **65**, 744 (1987).

[235] T. Lund, H. Lund, *Acta Chem. Scand.*, **B39**, 429 (1985).

[236] F. Pragst, F. Scholz, P. Woitke, V. Kollek, G. Henrion, *J. Prakt. Chem.*, **327**, 1028 (1985).

[237] M.-T. Escot, P. Pouillen, P. Martinet, *Electrochim. Acta*, **28**, 1697 (1983).

[238] M. C. Arevalo, E. Pastor, S. Gonzalez, *Bull Electrochem.*, **5**, 786 (1989).

[239] E. Sequoia, *J. Electroanal. Chem., Interfacial Electrochem.*, **266**, 349(1989).

[240] A. Sivakumar, S. J. Reddy, *Trans. SAEST*, **20**, 27 (1985).

[241] M. Blázquez, J. M. Rodríguez-Mellado, J. J. Ruiz, *Electrochim. Acta*, **30**, 1527 (1985).

[242] L. N. Vykhodtseva, L. N. Nekrasov, E. V. Bulatkina, D. V. Ioffe, *Elektrokhimiya*, **21**, 1138 (1985).

[243] T. Sato, M. Kubo, K. Nishigaki, *Bull. Chem. Soc. Jpn.*, **53**, 815 (1980).

[244] G. T. Cheek, *Proc.-Electrochem. Soc.*, **1990**, 325.

[245] G. T. Cheek, *Proc.-Electrochem. Soc.*, **1992**, 426.

[246] M. Petit, A. Mortreux, F. Petit, *J. Chem. Soc., Chem. Commun.*, **1984**, 341.

[247] A. Guirado, F. Barba, M. B. Hursthouse, A. Arcas, *J. Org. Chem.*, **54**, 3205 (1989).

[248] G. Silvestri, S. Gambino, G. Filardo, *Tetrahedron Lett.*, **27**, 3429 (1986).

[249] S. Mcharek, M. Heintz, M. Troupel, J. Périchon, *Bull. Soc. Chim. Fr.*, **1989**, 95.

[250] Y. Ikeda, E. Manda, *Chem. Lett.*, **1984**, 453.

[251] Y. Ikeda, E. Manda, T. Shimura, *Japan Kokai Tokkyo Koho*, JP 60-103193 (1985).

[252] Y. Ikeda, E. Manda, *Bull. Chem. Soc. Jpn.*, **58**, 1723 (1985).

[253] (a) J. H. Wagenknecht, *U. S. Pat.*, 4, 582, 577 (1986); (b) A. S. C. Chen, T. T. Huang, J. H. Wagenknecht, R. E. Miller, *J. Org. Chem.*, **60**, 742 (1995).

[254] G. Filardo, G. Silvestri, S. Gambino, *Eur. Pat. App. EP* 0,189,120 (1986).

[255] M. R. Rifi, F. H. Covitz, in:Introduction to *Organic Electrochemistry*, p.167, Marcel Dekker, New York, N.Y. (1974).

[256] (a) T. Shono, M. Mitani, *J. Am. Chem. Soc.*, **93**, 5284 (1971); (b) S. Kashimura, M. Ishifune, Y. Murai, N. Moriyoshi, T. Shono, *Tetrahedron Lett.*, **36**, 5041 (1995).

[257] J. E. Swartz, E. Kariv-Miller, S. J. Harrold, *J. Am. Chem. Soc.*, **111**, 1211 (1989).

[258] E. Kariv-Miller, H. Maeda, F. Lombardo, *J. Org. Chem.*, **54**, 4022 (1989).

[259] T. Shono, N. Kise, *Tetrahedron Lett.*, **31**, 1303 (1990).

[260] F. Lombardo, R. A. Newmark, E. Kariv-Miller, *J. Org. Chem.*, **56**, 2422 (1991).

[261] C. G. Sowell, R. L. Wolin, R. D. Little, *Tetrahedron Lett.*, **31**, 485 (1990).

[262] T. Shono, I. Nishiguchi, H. Ohmizu, *Chem. Lett.*, **1976**, 1233.

[263] (a) G. Pattenden, G. M. Robertson, *Tetrahedron Lett.*, **24**, 4617 (1983); (b) G. Pattenden, G. M. Robertson, *Tetrahedron*, **41**, 4001 (1985).

[264] H. Takeshita, A. Mori, Y. Goto, T. Nagao, *Bull. Chem. Soc. Jpn.*, **60**, 1747 (1987).

[265] S. G. Mairanovskii, G. K. Bishimbaeva, N. Y. Grigor'eva, I. M. Avrutov, V. N. Odinokov, G. A. Tolstikov, *Izv. Akad. Nauk SSSR, Ser. Khim.*, **1985**, 2703.

[266] T. Shono, N. Kise, T. Suzumoto, T. Morimoto, *J. Am. Chem. Soc.*, **108**, 4676 (1986).

[267] M. A. Fox, J. R. Hurst, *J. Am. Chem. Soc.*, **106**, 7626 (1984).

[268] D. Archier-Jay, N. Besbes, A. Laurent, E. Laurent, H. Stamm, R. Tardivel, *Tetrahedron Lett.*, **30**, 2271 (1989).

[269] (a) S. Kashimura, M. Ishifune, Y. Murai, T. Shono, *Chem. Lett.*, **1996**, 309; (b) S. Kashimura, M. Ishifune, Y. Murai, T. Shono, *Tetrahedron Lett.*, **37**, 6737 (1996).

[270] H. Ishida, H. Tanaka, K. Tanaka, T. Tanaka, *Chem. Lett.*, **1987**, 597.

[271] T. Shono, N. Kise, T. Fujimoto, N. Tominaga, H. Morita, *J. Org. Chem.*, **57**, 7175 (1992).

[272] P. Zhang, W. Zhang, T. Zhang, Z. Wang, W. Zhou, *J. Chem. Soc., Chem. Commun.*, **1991**, 491

[273] J. C. Gard, B. Hanquet, Y. Mugnier, J. Lessard, *J. Electroanal. Chem.*, **365**, 299 (1994).

[274] (a) T. Shono, S. Kashimura, Y. Mori, T. Hayashi, T. Soejima, Y. Yamaguchi, *J. Org. Chem.*, **54**, 6001 (1989) (b) T. Shono, N. Kise, T. Fujimoto, *Tetrahedron Lett.*, **32**, 525 (1991).

[275] T. Nonaka, K. Sugino, *Nippon Kagaku Kaishi*, **90**, 686 (1969).

[276] T. Nonaka, S. Miyaji, K. Odo, *Denki Kagaku*, **41**, 142 (1973).

[277] T. Shono, Y. Morishima, N. Moriyoshi, M. Ishifune, S. Kashimura, *J. Org. Chem.*, **59**, 273 (1994).

[278] S. Sibille, E. D'Incan, L. Leport, J. Périchon, *Tetrahedron Lett.*, **27**, 3129 (1986).

[279] T. Shono, Y. Morishima, N. Moriyoshi, M. Ishifune, S. Kashimura, *Denki Kagaku*, **38**, 105 (1970).

[280] T. Koizumi, T. Fuchigami, Z. E.-S. Kandeel, N. Sato, T. Nonaka, *Bull. Chem. Soc. Jpn.*, **59**, 757 (1986).

[281] T. Nonaka, K. Sugino, *Denki Kagaku*, **34**, 105 (1966).

[282] K. Sugino, T. Nonaka, *Electrochim. Acta*, **13**, 613 (1968).

[283] T. Nonaka, T. Sekine, K. Odo, K. Sugino, *Electrochim. Acta*, **22**, 271 (1977).

[284] T. Nonaka, K. Odo, *Denki Kagaku*, **41**, 662 (1973).

[285] (a) T. Nonaka, K. Odo, *Denki Kagaku*, **40**, 66 (1972); (b) *Idem, ibid.*, **40**, 665 (1972); (c) T. Nonaka, A. Omura, T. Fuchigami, K. Odo, *ibid.*, **45**, 111 (1977); (d) T. Nonaka, K. Yui K. Odo, *ibid.*, **42**, 160 (1974).

[286] A. O.-L. Moing, J. Delaunay, A. Lebouc, J. Simonet, *Tetrahedron*, **41**, 4483 (1985).

[287] T. J. Curphey, L. D. Trivedi, T. Layloff, *J. Org. Chem.*, **39**, 3831 (1974).

[288] T. Lund, M. L. Pedersen, L. A. Frandsen, *Tetrahedron Lett.*, **35**, 9225 (1994).

[289] G. Belot, C. Degrand, P.-L. Compagnon, *J. Org. Chem.*, **47**, 325 (1982).

[290] J. Hoffmann, J. Voss, *Electrochim. Acta*, **36**, 465 (1991).

[291] M. Tokuda, S. Satoh, Y. Katoh, H. Suginome, *Electroorg. Synth.*, **1991**, 83.

[292] C. Degrand, F. Gasquez, P.-L. Compagnon, *J. Organomet. Chem.*, **280**, 87 (1985).
[293] (a) R. Saiganesh, K. K. Balasubramanian, C. S. Venkatachalam, *Tetrahedron Lett.*, **30**, 1711 (1989); (b) F. Barba, K. L. de la Fuente, *J. Org. Chem.*, **61**, 8662 (1996).
[294] P. Zuman, L. Spritzer, *J. Electroanal. Chem*, **69**, 433 (1976).
[295] B. Mahdavi, P. Chambrion, J. Binette, E. Martel, J. Lessard, *Can. J. Chem.*, **73**, 846 (1995)
[296] P. Boutote, G. Mousset, *Can. J. Chem.*, **70**, 2266 (1992).
[297] E. Raoult, J. Sarrazin, A. Tallec, *Bull. Soc. Chim. Fr.*, **1981**, 420.
[298] A. Abouelfida, J.-C. Roze, J.-P. Pradere, M. Jubault, *Phosphorus, Sulfur and Silicon, Relat. Elem.*, **54**, 123 (1990).
[299] A. Kunai, M. Yakushido, M. Nishihara, K. Sasaki, *Chem. Express*, **5**, 781 (1990).
[300] A. M. Abeysekere, S. Amaratunge, J. Grimshaw, N. Jayeweera, G. Senanayake, *J. Chem. Soc., Perkin Trans. I*, **1991**, 202.
[301] (a) T. Osa, T. Matsue, A. Yokozawa, T. Yamada, *Denki Kagaku*, **52**, 629 (1984); (b) Idem. *ibid.*, **54**, 484 (1986).
[302] M. M. Srivastava, I. Bala, M. Singh, S. N. Srivastava, *Indian J. Chem.*, **24**, 297 (1985).
[303] V. I. Shinkarenko, Y. V. Samusenko, V. I. Magda, G. F. Dzhurka, F. G. Yaremenko, V. D. Bezuglyi, *Zh. Obshch. Khim.*, **55**, 2378 (1985).
[304] (a) N. Takano, M. Sugawara, N. Takeno, Y. Otsuji, *Denki Kagaku*, **54**, 604 (1986); (b) N. Takano, N. Takeno, Y. Otsuji, *Stud. Org. Chem*, **30**, 211 (1987).
[305] M. Largeron, H. Dupuy M.-B. Fleury, *Tetrahedron*, **51**, 4953 (1995).
[306] S. Wawzonek, A. Gundersen, *J. Electrochem. Soc.*, **111**, 324 (1964).
[307] J. Harada, Y. Sakakibara, A. Kunai, K. Sasaki, *Bull. Chem. Soc. Jpn.*, **57**, 611 (1984).
[308] H. Matschiner, H. H. Ruttinger, S. Austen, *J. Prakt. Chem.*, **327**, 45 (1985).
[309] A. Kunugi, M. Yasuzawa, H. Matsui, *Electrochim. Acta*, **36**, 1341 (1968).
[310] I. M. Kolthoff, J. J. Lingane, *Polarography*, Vol. 2, Chapter 38, Interscience Publishers, New York (1952).
[311] L. Toppare, B. Hacioglu, U. Akbulut, *J. Macromol. Sci. Chem.*, **A27**, 317 (1990).
[312] P. Zuman, D. Barnes, A. R.-Kejharova, *Discuss. Faraday Soc.*, **45**, 202 (1968).
[313] E. Brillas, J. J. Ruiz, *J. Electroanal. Chem., Interfacial Electrochem.*, **215**, 293 (1986).
[314] (a) P. Tissot, P. Margaretha, *Helvetica Chim. Acta*, **60**, 1472 (1977); (b) P. Tissot, P. Margaretha, *Electrochim. Acta*, **23**, 1049 (1978); (c) P. Tissot P. Margaretha, *Nouv. J. Chim*, **3**, 13 (1979).
[315] E. J. Denney, B. Mooney, *J. Chem. Soc. B*, **1968**, 1410.
[316] E. Brillas, A. Ortiz, *Electrochim. Acta*, **30**, 1185 (1985);
[317] E. Touboul, F. Weisbuch, J. Wiemann, *C. R. Acad. Sci. Paris, Sér. C*, **268**, 1170 (1969).
[318] E. Brillas, A. Ortiz, *J. Chem. Soc., Faraday Trans.1*, **82**, 495 (1986).
[319] P. Tissot, J.-P. Surbech, F. O. Gulacar, *Helvetica Chim. Acta*, **64**, 1570 (1981).
[320] N. Egashira, H. Kanaeda, F. Hori, *Denki Kagaku*, **52**, 772 (1984).
[321] P. Margaretha, P. Tissot, *Helvetica Chim. Acta*, **65**, 1949, (1982).
[322] R. M. Bastida, E. Brillas, J. M. Costa, *J. Electrochem. Soc.*, **138**, 2296 (1991).
[323] T. Troll, W. Elbe, G. W. Ollmann, *Tetrahedron Lett.*, **22**, 2961 (1981).
[324] M. Bauscher W. Mäntele, *J. Phys. Chem.*, **96**, 11101 (1992).
[325] M. Gómez-Kaifer, P. A. Reddy, C. D. Gutsche, L. Echegoyen, *J. Am. Chem. Soc.*, **116**, 3580 (1994).
[326] D. Bethell, G. Dougherty, D. C. Cupertino, *J. Chem. Soc., Chem. Commun.*, **1995**, 675.
[327] A. Mori, Y. Goto, H. Takeshita, *Bull. Chem. Soc. Jpn.*, **60**, 2497 (1987).
[328] M. E, Peover, *J. Chem. Soc.*, **1960**, 5020.
[329] K. L. N. Phani, R. Narayan, *J. Electroanal. Chem., Interfacial Electrochem.*, **187**, 187 (1985).
[330] K. Suga, S. Aoyagui, *Bull. Chem. Soc. Jpn.*, **59**, 1937 (1986).
[331] K. Pekmez, M. Can, A. Yildiz, *Electrochim. Acta*, **38**, 607 (1993).
[332] O. H. Given, M. E. Peover, J. Schoen, *J. Chem. Soc.*, **1958**, 2674.
[333] J. Duda, *Pol. J. Chem.*, **65**, 1701 (1991).
[334] M. B, Mizen, M. S. Wrighton, *J. Electrochem. Soc.*, **136**, 941 (1989).
[335] T. Bechtold, E. Burtscher, A. Amann, O. Bobleter, *Angew. Chem., Int. Ed. Engl.*, **31**, 1068 (1992).
[336] N. Takano, N. Takeno, M. Morita, Y. Otsuji, *Bull. Chem. Soc. Jpn.*, **58**, 2417 (1985).
[337] N. Takano, N. Takeno, M. Morita, *Nippon Kagaku Kaishi*, **1984**, 2003.
[338] M.-A. O.-L. Moing, J. Delaunay, J. Simonet, *Nouv. J. Chim.*, **8**, 217 (1984).
[339] M. A. Fox, K. Campbell, G. Maier, L. H. Franz, *J. Org. Chem.*, **48**, 1762 (1983).
[340] N. Takano, N. Takeno, M. Morita, *Denki Kagaku*, **51**, 483 (1983).
[341] (a) N. Takano, N. Takeno, M. Morita, *Denki Kagaku*, **51**, 779 (1983); (b) Idem, *Denki Kagaku*, **52**, 529 (1984).
[342] (a) G. Mabon, G. Le Guillanton, J. Simonet, *Nouv. J. Chim.*, **7**, 305 (1983); (b) G. Mabon, G. Le Guillanton, J. Simonet, *J. Chem. Soc., Chem. Commun.*, **1982**, 571.
[343] E. Yu. Khmel'nitskaya, I. D. Razgonyaeva, S. L. Solodar, *Zh. Obshch. Khim.*, **54**, 1679 (1984).

[344] A. Mazzenga, D. Lomnitz, J. Villegas, C. J. Polowczyk, *Tetrahedron Lett.*, **10**, 1665 (1969).

[345] Y. J. Abul-Hajj, K. Tabakovic, I. Tabakovic, *J. Am. Chem. Soc.*, **117**, 6144 (1995).

[346] M. J. Gasic, D. Sladic, *Croat. Chem. Acta*, **58**, 531 (1985).

[347] K. Sugihara, H. Kamiya, M. Yamaguchi, T. Kaneda, S. Misumi, *Tetrahedron Lett.*, **22**, 1619 (1981).

[348] R. E. Wolf, Jr., S. R. Cooper, *J. Am. Chem. Soc.*, **106**, 4646 (1984).

[349] K. Maruyama, H. Sohmiya, H. Tsukube, *Tetrahedron Lett.*, **26**, 3583 (1985).

[350] R. L. Blankespoor, R. Hsung, D. L. Schutt, *J. Org. Chem.*, **53**, 3032 (1988).

[351] R. L. Blankespoor, E. L. Kosters, A. J. Post, D. P. VanMeurs, *J. Org. Chem.*, **56**, 1609 (1991).

[352] A. V. Bulatov, G. P. Voskerchyan, S. N. Dobryakov, A. T. Nikitaev, *Izv. Akad. Nauk SSSR, Ser. Khim.*, **1986**, 822

[353] T. Iyoda, T. Saika, K. Honda, T. Shimizu, *Tetrahedron Lett*, **30**, 5429 (1989).

[354] H. Lund, *The Chemistry of Hydroxyl Group*, p. 274, Interscience Publishers Ins., New York (1971).

[355] (a) Ch. Comninellis, E. Plattner, *J. Appl. Chem.*, **15**, 771 (1985); (b) J. Revenga, F. Rodriguez, J. Tijero, *J. Electrochem. Soc.*, **141**, 330 (1994).

[356] (a) C. Van der Stouwe, H. J. Schafer, *Tetrahedron Lett.*, **20**, 2643 (1979); (b) C. Van der Stouwe, H. J. Schafer, *Chem. Ber.*, **114**, 946 (1981).

[357] M. H. Khalifa, A. Rieker, *Tetrahedron Lett.*, **25**, 1027 (1984).

[358] P. A. Andrews, Su-shu Pan, N. R. Bachur, *J. Am. Chem. Soc.*, **108**, 4158 (1986).

[359] H. Kohn, N. Zein, X. Q. Lin, J.-Q. Ding, K. M. Kadish, *J. Am. Chem. Soc.*, **109**, 1833 (1987).

[360] A. A. Volod'kin, V. V. Ershov, R. D. Malysheva, *Izv. Akad. Nauk SSSR, Ser. Khim.*, **1985**, 1086.

[361] A. Anne, J. Moiroux, *Nouv. J. Chim.*, **9**, 83 (1985).

[362] R. J. Robbins, D. E. Falvey, *J. Org. Chem.*, **58**, 3616 (1993).

[363] A. Anne, J. Moiroux, *J. Chem. Soc., Perkin Trans. II*, **1989**, 2097.

[364] J. A. Richards, D. H. Evans, *J. Electroanal. Chem., Interfacial Electrochem.*, **81**, 171 (1977).

[365] L. I. Kudinova, A. A. Volod'kin, V. V. Ershov, T. I. Prokofeva, *Izv. Akad. Nauk SSSR, Ser. Khim.*, **1978**, 1503.

[366] M. O. F. Goulart, J. H. P. Utley, *J. Org. Chem.*, **53**, 2520 (1988).

[367] S. M. Mattar, D. Sutherland, *J. Phys. Chem.*, **95**, 5129 (1991).

[368] B. Uno, A. Kawabata, K. Kano, *Chem. Lett.* **1992**, 1017.

[369] K. Komaguchi, Y. Katsusegawa, A. Kitani, K. Sasaki, *Bull. Chem. Soc. Jpn.*, **64**, 2686 (1991).

[370] S. F. Rak, T. H. Jozefiak, L. L. Miller, *J. Org. Chem.*, **55**, 4794 (1990).

[371] A. R.-Kejharova, P. Zuman, *J. Electroanal. Chem. Int. Electrochem.*, **21**, 197 (1969).

[372] K. Butkiewicz, *J. Electroanal. Chem. Int. Electrochem.*, **39**, 407 (1972).

[373] M. L. Ash, F. L. O'Brien, D. W. Boykin, Jr., *J. Org. Chem.*, **37**, 106 (1972).

[374] (a) S. S. Katiyar, M. Lalithambika, G. C. Joshi, *J. Electroanal. Chem. Int. Electrochem.*, **47**, 439 (1973); (b) S. S. Katiyar, M. Lalithambika, D. N. Dhar, *J. Electroanal. Chem. Int. Electrochem.*, **53**, 449 (1974).

[375] A. Rusina, J. Volke, J. Cernak, J. Kovac, V. Kollar, *J. Electroanal. Chem. Int. Electrochem.*, **50**, 351 (1974).

[376] K. Butkiewicz, *J. Electroanal. Chem., Interfacial Electrochem.*, **89**, 379 (1978).

[377] B. A. Abo-El-Nabey, A. A. Khalaf, T. A. Amireh, A. A. Ashy, M. M. Almalki, *Bull. Electrochem.*, **6**, 689 (1990).

[378] H. D. Tabba, K. R. Barqawi, M. M. Al-Arab, *Bull. Soc. Chim. Fr.*, **1990**, 381.

[379] J. P. Petrovich, M. M. Baizer, M. R. Ort, *J. Electrochem. Soc.*, **116**, 749 (1969).

[380] V. D. Parker, *Acta Chem. Scand.*, **B35**. 279 (1981).

[381] (a) V. D. Parker, *Acta Chem. Scand.*, **B35**, 147 (1981); (b) M. Svaan, V. D. Parker, *Acta Chem. Scand.*, **B39**, 445 (1985).

[382] D. K. Johnson, R. E. W. Jansson, *J. Electrochem. Soc.*, **128**, 1885 (1981).

[383] (a) S. Takenaka, M. Uchida, C. Shimakawa, *Japan. Kokai Tokkyo Koho* , *JP* 05, 156, 478 (1993); (b) N. Tateyama S. Takenaka, C. Shimakawa, *Japan. Kokai Tokkyo Koho, Jp* 05, 25, 672 (1993).

[384] P. R. Unwin, R. G. Compton, *J. Chem. Soc. Faraday Trans.* **86**, 657 (1990).

[385] V. D. Parker, *Acta Chem. Scand.*, **B35**, 295 (1981).

[386] M. M. Baizer, J. D. Anderson, *J. Electrochem. Soc.*, **111**, 223(1964).

[387] L. N. Nekrasov, A. P. Korotkov, *Elektrokhimiya*, **21**, 1464 (1985).

[388] Y. Kunugi, T. Nonaka, *Denki Kagaku*, **58**, 182 (1990).

[389] P-C. Cheng, T. Nonaka, *Denki Kagaku*, **61**, 218 (1993).

[390] M. N. Elinson, S. K. Feducovich, A. A. Zakharenkov, B. I. Ugrak, G. I. Nikishin, S. V. Lindeman, J. T. Struchkov, *Tetrahedron*, **51**, 5035 (1995).

[391] J. Delaunay, A. Lebouc, F. Le Guillanton, I. Mavoungou Gomes, J. Simonet, *Electrochim. Acta.* **27**, 287 (1982).

[392] A. J. Fry, R. D. Little, J. Leonetti, *J. Org. Chem.*, **59**, 5017 (1994).

[393] P. G. Gassman, C.-J. Lee, *Synth. Commun.*, **24**, 1465 (1994).

[394] D. P. Fox, R. D. Little, M. M. Baizer, *J. Org. Chem.*, **50**, 2202(1985).

[395] (a) R. D. Little, D. P. Fox, L. Moens, R. Wolin, M. M. Baizer, *Stud. Org. Chem.*, **30**, 171(1987); (b) R. D. Little, D. P. Fox, L. Van Hijfte, R. Dannecker, G. Sowell, R. L. Wolin, L. Moens, M. M. Baizer, *J. Org. Chem.*, **53**, 2287(1988).

[396] P. G. Gassman, C. Lee, *J. Am. Chem. Soc*, **111**, 739(1989).

[397] J. D. Anderson, M. M. Baizer, E. J. Prill, *J. Org. Chem.*, **30**, 1645(1965).

[398] M. M. Baizer, J. P. Petrovich, D. A. Tyssee, *J. Electrochem. Soc.*, **117**, 173(1970).

[399] B. M. Prasad, R. A. Misra, *Indian J. Chem.*, **29**, 392(1990).

[400] S. DeKeczer, D. Kertesz, H. Parnes, *J. Labelled Compd. Radiopharm.*, **33**, 219 (1993).

[401] O. R. Brown, J. A. Harrison, K. S. Sastry, *J. Electroanal. Chem., Interfacial Electrochem.*, **58**, 387(1975).

[402] M. Bhatti, *J. Electrochem. Soc. India*, **34**, 35(1985).

[403] Yu. D. Smirnov, A. P. Tomilov, *Zh. Org. Khim.*, **28**, 51(1992).

[404] J. A. Harrison, D. W. Shoesmith, *J. Electroanal. Chem., Interfacial Electrochem.*, **32**, 125(1971).

[405] J. H. Wagenknecht, *J. Org. Chem.*, **37**, 1513(1972).

[406] H. Maeda, T. Maki, H. Ohmori, *Denki Kagaku*, **62**, 1109 (1994).

[407] M. J. Allen, *Organic Electrode Processes*, Chapman and Hall, London (1958).

[408] D. Deprez, R. Margraff, J. -P. Pulicani, *Tetrahedron Lett.*, **28**, 4679(1987).

[409] L. Horner, H. Hönl, *Leibigs Ann. Chem.*, **1977**, 2036.

[410] R. Shundo, Y. Matsubara, I. Nishiguchi, T. Hirashima, *Bull. Chem. Soc. Jpn.*, **65**, 530(1992).

[411] K. Scott, A. P. Colbourne, S. D. Perry, *Electrochim. Acta*, **35**, 621(1990).

[412] B. R. Eggins, E. A. O'Neill, *Anal. Chem. Symp. Ser.*, **25**, 111 (1986).

[413] A. Morduchowitz, A. F. Sammells, *U. S. Pat.*, 4, 560, 450 (1985).

[414] K. Scott, *Chem. Eng Res Des.*, **64**, 266 (1986).

[415] B. Scharbert, S. Dapperheld, P. Babusiaux, *Pct Int. Appl.* WO, 93-17151 (1993).

[416] E. Gagyi-Pálffy, E. Prépostffy, Gy. Korányi, *Period. Polytech., Chem. Eng.*, **29**, 95 (1985).

[417] J. R. Ochoa, A. De Diego, J. Santa-Olalla, *J. Appl. Electrochem.*, **23**, 905 (1993).

[418] M. Matsuoka, Y. Kokusenya, *Japan Kokai Tokkyo Koho*, 51133229 (1976).

[419] J. W. Park, M. H. Choi, K. K. Park, *Tetrahedron*, **36**, 2637 (1995).

[420] (a) T. Iwasaki, M. Sasaki, N. Sato, A. Yoshiyama, T. Fuchigami, T .Nonaka, *Denki Kagaku*, **58**, 83 (1990); (b) N. Sato, A. Yoshiyama, P.-C. Cheng, T. Nonaka, M. Sasaki, *J. Appl. Electrochem.*, **22**, 1082 (1992).

[421] A. Chaintreau, G. Adrian D. Couturier, *Synth. Commun.*, **11**, 439(1981).

[422] S. Takenaka, R. Oi, C. Shimakawa, Y. Shimokawa, *Stud. Org. Chem.*, **30**, 215(1987).

[423] R. Oi, C. Shimakawa, Y. Shimokawa, S. Takenaka, *Bull. Chem. Soc. pn*, **60**, 4193 (1987).

[424] S. Takenaka, Y. Kouno, T. Uchida, A. Takagi, *Denki Kagaku.*, **59**, 570 (1991).

[425] (a) S. Takenaka, C. Shimakawa, *Japan Kokai Tokkyo Koho*, S60-234987 (1985); (b) S. Takenaka, C. Shimakawa, *Japan. Kokai Tokkyo Koho*, S60-243293 (1985).

[426] K. Naito, *Japan. Kokai Tokkyo Koho*, H4-116188 (1992).

[427] R. Oi, K. San-Nohe, S. Takenaka, *Bull. Chem. Soc. Jpn.*, **61**, 3773 (1988).

[428] T. Nonaka, T. Kato, T. Fuchigami, T. Sekine, *Electrochim. Acta*, **26**, 887 (1981).

[429] T. Awata, M. M. Baizer, T. Nonaka, T. Fuchigami, *Chem. Lett.*, **1985**, 371.

[430] T. Fuchigami, T. Awata, T. Nonaka, M. Baizer, *Bull. Chem. Soc. Jpn.*, **59**, 2873 (1986).

[431] (a) V. G. Mairanovsky, *Angew. Chem., Int. Ed. Engl.*, **15**, 281 (1976); (b) C. J. Salomon, E. G. Mata, O. A. Mascaretti, *Tetrahedron*, **49**, 3691 (1993).

[432] R. A. Benkeser, H. Watanabe, S. J. Mels, M. A. Sabol, *J. Org. Chem.*, **35**, 1210 (1970).

[433] J. Chaussard, C. Combellas, A. Thiebault. *Tetrahedron Lett.*, **28**, 1173 (1987).

[434] (a) T. Shono, H. Masuda, H. Murase, M. Shimomura, S. Kashimura, *J. Org. Chem.*, **57**, 1061(1992); (b) T. Shono, M. Ishifune, H. Kinugasa, S. Kashimura, *J. Org. Chem.*, **57**, 5561 (1992).

[435] G. H. Coleman, H. L. Johnson, *Organic Syntheses*, **3**, 61 (1955).

[436] M. L. Vincent, D. G. Peters, *J. Electroanal. Chem.*, **327**, 121 (1992).

[437] (a) J. P. Coleman, H. G. Gilde, J. H. P. Utley, B. C. L. Weedon, *J. C. S., Chem. Commun.*, **1970**, 738; (b) D. F. Corbett A. J. Eglington, *J. C. S., Chem. Commun.*, **1980**, 1083.

[438] V. K. Sharma, D. K. Sharma, C. M. Gupta, *Bull. Electrochem.*, **5**, 605 (1989).

[439] T. Shono, N. Kise, N. Inakoshi, *Denki Kagaku*, **61**, 870 (1993).

[440] J. H. Wagenknecht, R. D. Goodin, P. J. Kinlen F. E. Woodard, *J. Electrochem. Soc.*, **131**, 1559 (1984).

[441] R. Seeber, F. Magno, G. Bontempelli, G. Mazzocchin, *J. Electroanal. Chem., Interfacial Electrochem.*, **72**, 219 (1976).

[442] H. Hébri, E. Duñach, M. Heintz, M. Troupel, J. Périchon, *Synlett*, **1991**, 901.

[443] M. Heintz, M. Devaud, H. Hébri, E. Duñach, M. Troupel, *Tetrahedron*, **49**, 2249 (1993).

[444] S. Kashimura, Y. Murai, M. Ishifune, H. Masuda, H. Murase, T. Shono, *Tetrahedron Lett.*, **36**, 4805 (1995).

[445] L. Kistenbrügger, P. Mischke, J. Voss G. Wiegand, *Liebigs Ann. Chem.*, **1980**, 461.

[446] P.-R. G.-Schatowitz, G. Struth, J. Voss, G. Wiegand, *J. Prakt. Chem.*, **335**, 230 (1993).

[447] J. E. Toomey, *Eur. Pat. Appl.*, **189**, 678 (1986).

[448] J. C. Barriere, M. C. Dubroeucq, M. Fleury, M. Largeron, J. M. Paris, *Fr. Demande Fr.* 2, 664, 894 (1992).
[449] M. Largeron, M. Vuilhorgne, I. Le Potier, N. Auzeil,l E. Bacqué, J. M. Paris, M. B. Fleury, *Tetrahedron*, **50**, 6307 (1994).
[450] S. Nakajima, T. Masumizu, K. Arai, C.-H. Shaw, K. Nozawa, K.-I. Kawai, *Stud. Org. Chem.*, **30**, 265 (1987).
[451] K. Arai, C.-H. Shaw, K. Nozawa, K.-I. Kawai, S. Nakajima, *Tetrahedron Lett.*, **28**, 441(1987).
[452] Y. Kokusenya, M. Matsuoka, *Denki Kagaku*, **55**, 257 (1987).
[453] T. Troll, G. W. Olimann, *Electrochim. Acta*, **29**, 467 (1984).
[454] L. Prangova, T. Strelow, J. Voss, *J. Chem. Research (S)*, **1985**, 118.
[455] O. I. Shirobokova, A. A. Adamov, G. N. Freidlin, N. S. Antonenko, Yu. D. Grudtsyn, *Zh. Prikl. Khim.*, **60**, 2619 (1987).
[456] T. Troll, G. W. Ollmann, H. Leffler, *Angew. Chem., Int. Ed. Engl.*, **23**, 622 (1984).
[457] C. De Luca, C. Giomini, L. Rampazzo, *Stud. Org. Chem.*, **30**, 219 (1987).
[458] T. Troll, G. W. Ollmann, *Tetrahedron Lett.*, **22**, 3497 (1981).
[459] T. Troll, K. Schmid, I. Rasch, *Z. Naturforsch., B, Anorg. Chem.*, **42**, 1027 (1987).
[460] Y. Kokusenya, M. Matsuoka, *Denki Kagaku*, **55**, 165 (1987).
[461] J. P. Segretario, N. Sleszynski, P. Zuman, *J. Electroanal. Chem., Interfacial Electrochem.*, **214**, 259 (1986).
[462] J. P. Segretario, P. Zuman, *J. Electroanal. Chem., Interfacial Electrochem.*, **214**, 237(1986).
[463] M. Dominguez, E. Roldan, J. Carbajo, J. Calvente, D. Gonzalez-Arjona, R. Andreu, *J. Eectroanal. Chem.*, **316**, 133 (1991).
[464] M. Blazquez, J. L. Avila, J. J. Ruiz, *Anales De Quimica*, **80**, 692 (1984).
[465] A. Guirado, F. Barba A. Tévar, *Synth. Commun.*, **14**, 333 (1984).
[466] A. Guirado, A. Zapata, J. Galvez, *Tetrahedron Lett.*, **35**, 2365 (1994).
[467] (a)J. M. Rodriguez-Mellado, J. L. Avila, J. J. Ruiz, *Can. J. Chem.*, **63**, 891 (1985); (b) J. M. R.-Mellado, J. L. Avila, J. J. Ruiz, *Can. J. Chem*, **63**, 891 (1984); (c) J. M. R.-Mellado, M. Blazquez, J. J. Ruiz, *J. Electroanal. Chem., Interfacial Electrochem.*, **195**, 363 (1985).
[468] N. Nakamura, T. Kohzuma, S. Suzuki, *Bull Chem. Soc. Jpn.*, **66**, 1289(1993).
[469] L. Ferian, R. Kossai, K. Boujlel, *Electrochim. Acta*, **36**, 783 (1991).
[470] K. Boujlel, J. Simonet, *Tetrahedron Lett.*, **12**, 1063(1979).
[471] A.-M. Martre, G. Mousset, *New. J. Chem.*, **17**, 207(1993).
[472] J. M. Rodriguez-Mellado, M. Blázquez, J. J. Ruiz, *Electrochim. Acta*, **31**, 1473 (1986).
[473] (a) A. Boulmedais, M. Jubault, A. Tallec, *Bull. Soc. Chim. Fr.*, **1988**, 610; (b) Idem. *ibid.*, **1989**, 185.
[474] C. Zielinski, H. J. Schäfer, *Tetrahedron Lett.*, **35**, 5621 (1994).
[475] E. Leszczynska A. Cisak., *Pol. J. Chem.*, **63**, 657 (1990).
[476] K. Kunchev, Ya. I. Tur'yan, Kh. Dinkov, *Elektrokhimiya*, **27**, 96(1991).
[477] D. Colletta, M. Devaud, *Electrochim. Acta*, **30**, 253(1985).
[478] J. Kozlowski, P. Zuman, *J. Electroanal. Chem., Interfacial Electrochem.*, **226**, 69(1987).
[479] S. Thangavelu, *Bull. Electrochem.*, **4**, 249(1988).
[480] W. A. Bonner, *Electrochim. Acta*, **35**, 683(1990).
[481] S. Abe, T. Fuchigami, T. Nonaka, *Chem. Lett.*, **1983**, 1033.
[482] M. Heintz, M. Devaud, H. Hébri, E. Duñach, M. Troupel, *Tetrahedron*, **49**, 2249(1993).
[483] T. Osa, T. Matsue, A. Yokozawa, T. Yamada, M. Fujihira, *Denki Kagaku*, **53**, 104(1985).
[484] M. D. Baturova, T. I. Kuznetsova, A. A. Vedenyapin, E. I. Klabunovskii, *Izv. Akad. Nauk SSSR, Ser. Khim.*, **12**, 2674 (1987).
[485] G. M. Abou-Elenien, B. E. El-Anadouli, R. M. Baraka, *Electroanalysis*, **6**, 515 (1994).
[486] A. Rajanarayanan, R. Jeyaraman, *Tetrahedron Lett.*, **32**, 3873(1991).
[487] R. Kossai, J. Simonet, G. Jeminet, *Tetrahedron Lett.*, **12**, 1059(1979).
[488] K. S. Quaal, S. JI, Y. M. Kim, W. D. Closson, J. A. Zubieta, *J. Org. Chem.*, **43**, 1311(1978).
[489] M. A. Casadei, D. Platcher, *Synthesis*, **1987**, 1856.
[490] G. Tainturier, L. Roullier, J-P.Martenot, D. Lorient, *J. Agric. Food Chem*, **40**, 760(1992).
[491] A. Guirado, F. Barba, C. Manzanera, M. D. Velasco, *J. Org. Chem.*, **47**, 142(1982).
[492] A. Guirado, F. Barba, J. Martin, *Synth. Commun.*, **13**, 327(1983).
[493] A. Guirado, F. Barba, J. Martin, *Electrochim. Acta*, **29**, 587(1984).
[494] G. T. Cheek, P. A. Horine, *J. Electrochem. Soc.*, **131**, 1796(1984).
[495] C. Polo, M. Quintanilla, F. Barba, *Synth. Commun.*, **24**, 907(1994).
[496] J-Claude Folest, E. P-Martins, M. Troupel, J. Périchon, *Tetrahedron Lett.*, **34**, 7571(1993).
[497] G. A. Urove, D. G. Peters, *J. Org. Chem.*, **57**, 786(1992).
[498] M. Mubarak, G. A. Urove, D. G. Peters, *J. Electroanal. Chem.*, **350**, 205(1993).
[499] G. A. Urove, D. G. Peters, *Tetrahedron Lett.*, **34**, 1271(1993).
[500] M. Okimoto, T. Itoh, T. Chiba, *J. Org. Chem.*, **61**, 4835 (1996).
[501] T. Shono, N. Kise, T. Fujimoto, A. Yamanami, R. Nomura, *J. Org. Chem.*, **59**, 1730 (1994).

[502] (a) J. T. Roberts, Jr., T. S. Calderwood, D. T. Sawyer, *J. Am. Chem. Soc.*, **106**, 4667 (1984); (b) M. A. Casadei, S. Cesa, F. M. Moracci, *J. Org. Chem.*, **61**, 380 (1996).

[503] A. Shimoyama, H. Ikeda, A. Nomoto, K. Harada, *Bull. Chem. Soc. Jpn.*, **67**, 257 (1994).

3. Electroreductive Reaction of Olefins

3.1 Electroreduction of Alkenes, Alkynes and Allenes

3.1.1 Electroreductive Hydrogenation of Alkenes

The electroreduction of olefins to their corresponding dihydro derivatives still plays a minor role as a synthetic tool.

Ethylene can be reduced at the liquid-solid interface on the platinum cathode surface by adsorbed hydrogen atoms [1]. The electrohydrogenation of ethylene on Pt(111) and Pt(100) oriented electrodes as the most active of the surfaces has been found [2,3]. The electrochemical hydrogenation of allyl alcohol has been well characterized by a double galvanostatic pulse technique at a platinum electrode [4]. The electroreduction of 3-ethyl-1-penten-3-ol **1** can convert it into the corresponding hydrocarbon **2** by electrolysis in a DMF-Bu$_4$NI(0.1M)-(Hg) system at –2.80 V (SCE) in 90% yield (eq. (3.1)) [5]. The electroreductive dimerization of ethylene in a propylene carbonate-Bu$_4$NClO$_4$(0.1M)-(Pt) system in the presence of the Ni(II)Br$_2$(PPh$_3$)$_2$ complex has been attained to give 2-butene **5** as a major product together with **4** (eq. (3.2)) [6a]. The electrolysis is performed in a

$$\underset{\underset{\textbf{1}}{OH}}{Et_2C\!-\!CH\!=\!CH_2} \xrightarrow[\text{–2.80 V (SCE), 90\% yield}]{\text{DMF-Bu}_4\text{NI(0.1M)-(Hg)}} \underset{\textbf{2}}{Et_2CH\!-\!CH_2\!-\!CH_3} \qquad (3.1)$$

$$\underset{\textbf{3}}{CH_2\!=\!CH_2} \xrightarrow[\text{Ni(II)Br}_2\text{(PPh}_3)_2]{\text{Propylene Carbonate-Bu}_4\text{NClO}_4\text{(0.1M)-(Pt)}} \begin{array}{c} CH_2\!=\!CHCH_2CH_3 \\ + \qquad \textbf{4} \\ CH_3CH\!=\!CHCH_3 \\ \textbf{5} \end{array} \qquad (3.2)$$

one-compartment cell with a suitable sacrificial anode (Zn, Cd, Al). The cobalt-based catalysts show an unusual behavior, as they are particularly selective for 1-butene production (> 90% yield) [6b]. The gas-liquid and the liquid-solid processes, both heterogeneous processes, involve mass transfer to and from the surface, adsorption, desorption, and chemical reactions at the surface. The product distribution in liquid-solid ethylene hydrogenation over deuterium-adsorbed platinum cathodes has been investigated [7].

The redox catalysis phenomenon of allocimene and 2,3-dimethylbutadiene has been discussed based on the indirect electroreduction in the presence of different catalysts such as triphenylene, naphthalene, 2,6-dimethylnaphthalene, and methylnaphthalene in a DMF-Bu$_4$NBF$_4$(0.1M)-(Hg) system [8]. The electroreductive catalytic hydrogenation of methyl linoleate on palladium catalysts in an acetone-Et$_4$NClO$_4$-(Pt) system in the presence of hydrogen leads mainly to a saturated product such as methyl stearate [9]. The potential controlled electrolysis of methyl linoleate at –1.0 V (SCE) affords the monoenic intermediate as an end product.

Perfluoroalkyl-substituted olefins undergo a one-electron reduction in an MeCN-Et$_4$NClO$_4$(0.1M)-(Hg) system to the corresponding anion radicals, which are stable enough (0 to –30 °C) to be accumulated by a macroscale electrolysis [10]. A further one-electron transfer, affording the corresponding dianions, which give rise to very fast decay processes, is observed at more negative potentials.

A novel successive system for hydrogenation of styrene leading to ethylbenzene has been attained by using Pd sheet electrode [11]. The electroreductive hydrogenation of 1,1-diphenylethylene **6** in a DMF-Bu$_4$NBF$_4$(0.1M)-(Hg) system in the presence of methanol undergoes a one-electron reduction to give the anion radical intermediate **7**, which uptakes a proton to generate methyldiphenyl-methyl radical **8**, a precursor of 1,1-diphenylethane **9** (eq. (3.3)) [12]. Tetraphenylethene derivatives can be used as two-electron transfer catalysts since they can undergo rapid two-electron reduction, and electron transfer can be initiated photochemically [13]. The reduction of 1,2-bis(4-acetylphenyl)-1,2-diphenylethylenes results in the formation of tetraarylethanes [13d].

$$Ph_2C=CH_2 + MeOH \xrightarrow[-2.1\ V,\ -7\sim39°C]{DMF\text{-}Bu_4NBF_4(0.1M)\text{-}(Hg)} [Ph_2C=CH_2]^{-•} + MeOH$$

6   **7**

$$Ph_2\overset{•}{C}-CH_3 + MeO^- \qquad (3.3)$$

8

$$Ph_2CH-CH_3$$

9

Protonation of the dianion of tetraphenylethylene (TPE) by alcohols and water in an MeCN-Et$_4$NClO$_4$(0.1M)-(Hg) system has been investigated [14]. The anion radical derived from TPE by electrolysis in HMPA or DMF is more stable than the dianion, whereas in acetonitrile, the dianion is directly produced at the electrode in a reversible two-electron step. The reactivity sequence of the proton donors ROH is in the following order: MeOH > EtOH > H$_2$O > i-PrOH > tert-BuOH.

Electrolysis of 2-phenyl-1,3-butadiene **10** in an aqueous DMF-Bu$_3$MeNOTs-(Hg) system results in 2-phenyl-2-butene **11** in 85% current yield (eq. (3.4)) [15]. The electrohydrogenation of 1,1,4,4-tetraphenyl-1,3-butadiene to 1,1,4,4-tetraphenyl-1-butene has been performed in a DMF-Bu$_4$MeNClO$_4$(0.1M)-(Hg) system in 98% yield [16]. Cyclobutadienopleiadiene **12** is reduced to its dihydro derivative, cyclobutenopleiadiene **13**, in a DMF-Bu$_4$NBF$_4$-(C) system (eq. (3.5)) [17]. Mechanistic investigation of cyclooctatetraene (COT) in liquid ammonia by cyclic voltammetric and controlled potential coulometric techniques reveals a two-step two-electron transfer reaction with a strong ion-pairing of the COT [18].

$$\underset{\underset{Ph}{|}}{CH_2=C}-CH=CH_2 \xrightarrow[85\%\ yield]{aq.\ DMF\text{-}Bu_3MeNOTs\text{-}(Hg)} \underset{\underset{Ph}{|}}{Me-C}=C-Me \qquad (3.4)$$

10   **11**

(3.5)

12   **13**

The interpretation of the stereochemical course of the electroreduction processes of substituted indenes has been attempted [19a]. Two different processes occur, depending on the presence of more or less acidic proton donors: a process *via* protonation of the radical anion formed in the first electron transfer, which affords preferentially or exclusively indans of *cis* configuration around the C_2–C_3 bond under a kinetic control, and a process *via* protonation of the dianion formed by disproportion action of the radical anion, which affords exclusively indans of *trans* configuration around the C_2–C_3 bond under a thermodynamic control [19a]. The cathodic hydrogenation of 1,1,3-triphenylindene **14** has been performed in a DMF-Et₄NClO₄(0.1M)-(Hg) system in the presence of phenol as a proton donor to give 1,1,3-triphenylindan **15** in a quantitative yield (eq. (3.6)) [19b].

$$\text{DMF-Bu}_4\text{NClO}_4(0.1\text{M})/\text{PhOH-(Hg)} \qquad -2.6 \text{ V (SCE)}, 100\% \text{ yield}$$

(3.6)

14   **15**

The cathodic behavior of 1,1-diaryl-substituted ethene **16** in a DMF-Et₄NF(0.1M)-(Hg) system in the presence of deuterated water affords exclusively the deuterated product **17** in 90% yield (eq. (3.7)). However, a similar electrolysis without adding any deuterated material proceeds to give the corresponding dimer **18** in a quantitative yield (eq. (3.8)) [20]. The electroreduction of *N*-vinyl derivatives of pyrazole, imidazole, triazoles, and tetrazole in acetonitrile has been investigated [21].

$$\text{DMF-Et}_4\text{NF}(0.1\text{M})\text{-(Hg)} \qquad \text{D}_2\text{O}, 90\% \text{ yield}$$

(3.7)

17

16

$$\text{DMF-Et}_4\text{NF}(0.1\text{M})\text{-(Hg)} \qquad 100\% \text{ yield}$$

(3.8)

18

The electroreduction of 5*H*-dibenz[b,f]azepine derivative **20** proceeds in a DMF/H₂O(20%)-Et₄NClO₄(0.1M)-(Hg) system in the presence of ammonium chloride to give the 10,11-dihydrocarbamazepine **21** in 85% yield (eq. (3.9)) [22]. The indirect cathodic hydrogenation proceeds *via* ammonium amalgam **19** (eq. (3.9)).

$$2 \text{ NH}_4^+ + 2 e^- + 2 n\text{Hg} \longrightarrow 2 \text{ NH}_4(\text{Hg})_n \qquad \text{- - - - (a)}$$

19

(3.9)

$$2 \text{ NH}_4(\text{Hg})_n \qquad 85\% \text{ yield} \qquad + 2 \text{ NH}_3 + 2 n\text{Hg} \quad \text{- - - - (b)}$$

20 CONH₂   **21** CONH₂

$$2 \text{ NH}_4(\text{Hg})_n \longrightarrow 2 \text{ NH}_3 + \text{H}_2 + 2 n\text{Hg} \qquad \text{- - - - - (c)}$$

The electrohydrogenation of *N*-methylindole **22** in a THF/H$_2$O(2%)-Bu$_4$NBF$_4$(0.25M)-(Hg) system preferentially gives *N*-methylindoline **23** in 85% yield together with 15% of the corresponding Birch reduction product (eq. (3.10)) [23].   In a similar manner, the

$$\text{(3.10)}$$

electroreduction of benzofuran affords the corresponding dihydrobenzofuran in 83% yield. (*E*)-5-(2-Bromovinyl)-2'-deoxyuridine (BVDU) **24** is electrochemically reduced in an MeOH-Et$_4$NBr-(Hg) system at pH 5.85 at an applied potential of −1.8 V (SCE) (eq. (3.11)) [24].   The reduction of **24** involves a four-electron transfer, which may be attributed to the fission of the C–Br bond followed by hydrogenation of the vinyl double bond, giving **25**.

$$\text{(3.11)}$$

3.1.2   Electroreduction of Alkynes and Allenes

A selective hydrogenation of acetylene has been performed in which acetylene is hydrogenated to ethylene with minimized formation of ethane.   The hydrogen pumped to a Cu-cathode in a membrane reactor fitted with a phosphoric acid containing silica-wool disk hydrogenates acetylene to ethylene very selectively with only slight production of ethane [25]. The electroreduction of acetylenic compounds has been shown to produce *cis*- [26b] and *trans*-olefins [26b] or saturated hydrocarbons [26] under various conditions.   Conjugated aromatic acetylenes are readily reduced to the corresponding alkyl benzenes [26b-d].

Electroreductive hydrogenation of the terminal acetylenic bond of **26** on a Cu-disk electrode coated with Raney nickel in alkaline ethanol at 40 °C under a constant current of 2000 A/m^2 yields the corresponding vinyl derivative **27** as the sole product (eq. (3.12)) [27].   The electroreductive hydrogenation of ethyl 4-hydroxy-2-alkynoates **28** in a DMF-Et$_4$NClO$_4$-(0.1M)/PhCO$_2$H-(Pb) system affords ethyl (*E*)-4-hydroxy-2-alkenoates **29** in 50~75% yields (eq. (3.13)) [28].

$$\text{(3.12)}$$

$$\underset{\textbf{28}}{\overset{R^2}{\underset{HO}{R^1}}\overline{\quad}\equiv-CO_2Et} \xrightarrow[\substack{PhCO_2H,\ 50\text{~}75\%\ \text{yields} \\ R^1, R^2 = \text{alkyl, aryl}}]{DMF\text{-}Et_4NClO_4(0.1M)\text{-}(Pb)} \underset{\textbf{29}}{\overset{R^1\quad R^2}{HO}}CO_2Et \qquad (3.13)$$

The electroreduction of phenylpropadiene **30** undergoes electrocatalytic hydrogenation of the allene moiety to the corresponding propene **31**. The polarograms and cyclic voltammograms demonstrate two waves for the reduction of **30** in a DMF-Bu$_4$NClO$_4$-(0.1M)-(Hg) system: the first wave signals reduction of phenylpropadiene **30** to 1-phenyl-1-propene **31** (eq. (3.14)) and the second wave is attributable to reduction of **31** to 1-phenyl-1-propane **32** (eq. (3.15)). However, the first wave is abnormally small because phenylpropadiene **30** undergoes substantial rearrangement to 1-phenyl-1-propyne, which is reducible to 1-phenylpropane **32** at nearly the same potential as 1-phenyl-1-propene **31** (eqs. (3.14),(3.15)) [29]. 1-Phenyl-1,2-butadiene undergoes electroreductive hydrogenation in a DMF-Bu$_4$NClO$_4$-(0.1M)-(Hg) system to form *trans*-1-phenyl-1-butene in 83% yield [30]. A survey of the electrochemical reaction of allenic compounds at the cathode has been reviewed recently [31]. Allenic functional groups are inherently electroinactive. Most of them take up an electron at the potential less than –2.0 V (SCE) to lead to the corresponding anion radicals **34** and **35** (eq. (3.16)) [31].

$$\underset{\textbf{30}}{Ph-CH=C=CH_2} \xrightarrow[-1.73\ V\ (SCE),\ 92\%\ \text{yield}]{DMF\text{-}Bu_4NClO_4\text{-}(Hg)} \underset{\textbf{31}}{Ph-CH=CH-CH_3} \qquad (3.14)$$

$$\underset{\textbf{31}}{Ph-CH=CH-CH_3} \xrightarrow[-2.00\ V\ (SCE),\ 95\%\ \text{yield}]{DMF\text{-}Bu_4NClO_4\text{-}(Hg)} \underset{\textbf{32}}{Ph-CH_2CH_2CH_3} \qquad (3.15)$$

$$\underset{\textbf{33}}{\diagdown C=C=C\diagup} \xrightarrow{e^-} \underset{\textbf{34}}{\diagdown C=\overset{\bullet}{C}-\overset{\ominus}{C}\diagup} \longleftrightarrow \underset{\textbf{35}}{\diagdown C=\overset{\ominus}{C}-\overset{\bullet}{C}\diagup} \qquad (3.16)$$

The anion radical derived from tetraphenylallene **36** is rapidly protonated and further reduced at the same potential (–2.1 V (SCE)) to give the corresponding carbanion [32].

Tetraphynylallene **36** shows a single 2e^- polarographic wave in the DMF-Bu$_4$NClO$_4$-(Hg) system at –2.11 V (SCE), upon further protonation, to yield 1,1,3,3-tetraphenyl-1-propene **37** in good yield (eq. (3.17)) [32~34].

$$\underset{\textbf{36}}{Ph_2C=C=CPh_2} \xrightarrow[-2.22\ V\ (SCE),\ 86\%\ \text{yield}]{DMF\text{-}Bu_4NClO_4\text{-}(Hg)} \underset{\textbf{37}}{Ph_2CH-CH=CPh_2} \qquad (3.17)$$

The controlled potential electrolysis of allenyl *p*-tolyl sulfone **38** proceeds in an MeOH-Me$_4$NBr-(Hg) system in a potassium dihydrogen phosphate-potassium hydroxide buffer (pH = 6.55) at –1.40 V (SCE) for 7 days to give allyl p-tolylsulfone **39** in a quantitative yield (eq. (3.18)) [35].

$$\text{TsSO}_2\text{–CH=C=CH}_2 \quad \xrightarrow[\text{–1.40 V (SCE), 84\% yield}]{\text{MeOH-Me}_4\text{NBr/KOH/KH}_2\text{PO}_4\text{(pH6.55)-(Hg)}} \quad \text{TsSO}_2\text{–CH}_2\text{–CH=CH}_2 \tag{3.18}$$

38   **39**

The two-electron reduction of tetraphenylbutatriene **40** in a DMF-Bu$_4$NI(0.2M)-(Hg) system affords 1,1,4,4-tetraphenyl-1,2-butadiene **43**, and subsequent two-electron reduction gives 1,1,4,4-tetraphenyl-1-butene **42** in 56.2% yield [36].   Recently, the more precise behavior of the reduction of 1,1,4,4-tetraphenylbutatriene **40** has been found to depend on the identity of the solvent and on the availability and strength of the added proton donors [14,36a,37a,b].   The electroreduction of **40** at –1.10 V (SCE) in a DMF-Bu$_4$NClO$_4$ (0.1M)-(Pt) system yields 1,1,4,4-tetraphenyl-1,3-butadiene **41** exclusively (eq. (3.19)).   At a slightly more negative potential (–1.25~–1.50 V), further reduction of **41** proceeds to give 1,1,4,4-tetraphenyl-1-butene **42** in 96% yield (eq. (3.20)).   Under similar electrolysis conditions, the presence of activated neutral alumina together with a trace amount of water produces the allene derivative **43** in 90% yield along with **41** (10%) (eq. (3.21)) [37].

The electrochemically induced hydrogenation of 1,1,4,4-tetraphenyl-1,2,3-butatriene **40** to 1,1,4,4-tetraphenyl-1,3-butadiene **41** occurs *via* an indirect self-protonation mechanism which reveals that hydroxide ion derived from residual water in the system acts as a base toward the starting material to initiate the isomerization [38].

$$
\begin{array}{c}
\text{DMF-Bu}_4\text{NClO}_4\text{(0.1M)-(Pt)} \\
\xrightarrow{\text{–1.10 V (SCE), ~100\% yield}} \quad \text{Ph}_2\text{C=CH–CH=CPh}_2 \quad (3.19) \\
\textbf{41}
\end{array}
$$

$$\text{Ph}_2\text{C=C=C=CPh}_2$$

40

$$
\begin{array}{c}
\text{DMF-Bu}_4\text{NClO}_4\text{(0.1M)-(Pt)} \\
\xrightarrow{\text{–1.25 V~–1.50 V, 96\% yield}} \quad \text{Ph}_2\text{C=CH–CH}_2\text{–CHPh}_2 \quad (3.20) \\
\textbf{42}
\end{array}
$$

$$
\begin{array}{c}
\text{DMF-Bu}_4\text{NClO}_4\text{(0.1M)-(Pt)} \\
\xrightarrow{\text{–1.10 V, 90\% yield}} \quad \text{Ph}_2\text{C=C=CH–CHPh}_2 \quad (3.21) \\
\textbf{43}
\end{array}
$$

Phenylacetylene **44** can be reduced electrochemically in the region of high cathode potentials with cleavage of the C-H bond, leading to the formation of a phenylacetylide, Ph≡C$^-$ [39].   The alkylation of phenylacetylide has been performed in an HMPA-Bu$_4$NI-(Pt) system in the presence of alkyl halides in an undivided cell to give the alkylated products **45** in excellent yields (eq. (3.22)) [40].

$$\text{Ph–C≡CH} \quad \xrightarrow[\text{RX, 79~100\% yields}]{\text{HMPA-Bu}_4\text{NI}_4\text{-(Pt)}} \quad \text{Ph–C≡C–R} \tag{3.22}$$

44   **45**

R = Me, Et, Bu

The electroreductive silylation of phenylacetylene **44** has been performed in a PN-Bu$_4$NBPh$_4$-(Pt) system in the presence of chlorotrimethylsilane **46** to give the silylated product **47** in 91% yield (eq. (3.23)) [41].   Phenylethynyl carbanion is proposed as the reactive intermediate.

$$Ph-C\equiv CH + Me_3SiCl \xrightarrow[\substack{91\% \text{ yield}}]{PN-Bu_4NBPh_4-(Pt)} Ph-C\equiv C-SiMe_3 \qquad (3.23)$$

44   **46**   **47**

PN = pivalonitrile

3.2 Electroreduction of Enones and Enoates

3.2.1 Electroreductive Hydrogenation of Enones and Enoates

The electroreductive hydrogenation of maleic and fumaric acids in aqueous buffered solutions has been intensively studied over the last half-century. The reduction product of both acids in all cases has been proven to be succinic acid [42]. The chlorinated acids give the corresponding chlorosuccinic or succinic acid, depending on the pH value [43]. Dihydroxyfumaric acid can be reduced only in strong acid solutions, leading to tartaric acid [44]. The electroreduction of maleic and fumaric acids and their dimethyl esters in methanol has also been investigated [45].

Methyl 4-*tert*-butylcyclo-1-hexenecarboxylate **48** is hydrogenated in a DMF-Bu_4NBF_4(0.5M)-(Hg) system in the presence of methanol and *p*-toluenesulfonic acid as proton donors to yield methyl 4-*tert*-butylcyclohexanecarboxylate **49** in 97% yield (*cis/trans* = 45/55) (eq. (3.24)). The *cis/trans* ratio becomes 62/28 when hydroquinone is used as the proton donor [46a]. The stereoselective hydrodimerization of the methyl ester proceeds in a DMF-Et_4NBr(0.1M)-(Pt/Hg) system to give the corresponding dimer in 36% yield [46b,c].

$$tert\text{-Bu}-\langle\ \rangle-CO_2Me \xrightarrow[\substack{97\% \text{ yield}}]{DMF-Bu_4NBF_4(0.5M)-(Hg)} tert\text{-Bu}-\langle\ \rangle-CO_2Me$$

48   *cis/trans* = 45/55 **49**   (3.24)

Both the α,β-unsaturated monoester **50a** and mononitrile **50b** fail to cyclize under electroreduction in an MeCN-Bu_4NBr-(Hg) system, giving the hydrogenated compounds **51** in good yields (eq. (3.25)) [47]. In contrast, more activated systems undergo the intramolecular cyclization (cf. Section 3.2.2).

$$\xrightarrow[\substack{-2.3 \text{ V (SCE)}}]{\substack{MeCN-n-Bu_4NBr-(Hg)\\ CH_2(CO_2CH_3)_2}} \qquad (3.25)$$

50   **50a:** Y = CO_2Me, 84%   **51**
          b: CN,   95%

Electrochemically activated Raney nickel has been used for the hydrogenation of maleic acid in aqueous potassium hydroxide solution at 50 °C [48].

The electrohydrogenation of benzalacetones has been attempted by activated nickel electrodes [49]. The electrocatalytic hydrogenation of 4-substituted benzalacetones in an aqueous EtOH-(Ni-black) system affords the corresponding hydrogenated products in

60~70% yields [50].   The Ni-black cathode is prepared by an electrochemical deposition method.

The electrochemical hydrogenation of 2,3-diphenylindenone **52** has been carried out in a DMF/AcOH-Bu$_4$NBF$_4$-(Hg) system at −1.2 V (SCE) to yield the indanone **53** in quantitative yield (eq. (3.26)) [51].

$$\text{(structure 52)} \xrightarrow[\text{−1.2 V (SCE), 98\% yield}]{\text{DMF/AcOH-Bu}_4\text{NBF}_4\text{-(Hg)}} \text{(structure 53)} \qquad (3.26)$$

52   **53**

The portonation of methyl cinnamate anion radical derived from the one-electron reduction of methyl cinnamate **54** in a DMF-Bu$_4$NBF$_4$(0.1M)-(Hg) system in the presence of phenol as a proton donor has been investigated [52].   The overall reaction is observed to involve $2e^-$ by molecule of **54** consumed, producing the hydrogenated product **55** (eq. (3.27)).

$$\text{PhCH=CHCO}_2\text{Me} \xrightarrow{\text{DMF-Bu}_4\text{NBF}_4\text{(0.1M)-(Hg)}} \text{PhCH}_2\text{CH}_2\text{CO}_2\text{Me} \qquad (3.27)$$

54   **55**

The asymmetric reduction of 4-methylcoumarin **56** has been attempted in an EtOH-NaH$_2$PO$_4$(0.2M)-(C) system whose graphite electrode is coated by poly-L-valine (eq. (3.28)) [53].   The maximum optical yield of 3,4-dihydro-4-methylcoumarin **57** was 43%

$$\text{(structure 56)} \xrightarrow[\text{−1.6 V (SCE), pH 2~3}]{\text{EtOH-NaH}_2\text{PO}_4\text{(0.2M)-(C)}} \text{(structure 57)} \qquad (3.28)$$

56   **57**   18% (58%ee)

(8% conv. yield) in a phosphate-buffer solution (pH = 6).   The reduction of citraconic acid **58** affords methylsuccinic acid **59** in 53% enantiomeric excess (2.4 conv.) (eq. (3.29)) [54]. The hydrogenation of the double bond of demethyl *N,N*-diisopropylaminomaleates proceeds in an MeCN-Et$_4$NBr(0.1M)-(Hg) system to yield dimethyl *N,N*-diisopropylaminobutanedioic acid in 75% yield [55].

$$\underset{\textbf{58}}{\overset{\text{HO}_2\text{C}\diagdown\diagup\text{CO}_2\text{H}}{\underset{\text{Me}\text{H}}{\text{C=C}}}} \xrightarrow[\substack{\text{2.4\% conv. yield}\\\text{53\% asymm. yield}}]{\text{aq. EtOH, Buffer}} \underset{\textbf{59}}{\overset{\text{Me—CHCO}_2\text{H}}{\underset{\text{CH}_2\text{CO}_2\text{H}}{|}}} \qquad (3.29)$$

The selective electrohydrogenation of the conjugated double bond of 4*H*-1,3-thiazine-4-ones **60** has been attained [56a].   The electrolysis of 2-ethyl-6-methoxycarbonyl-4*H*-1,3-

thiazine-4-one **60** can lead to the corresponding 5,6-dihydro derivative **61** in 92% yield (eq. (3.30)).   The controlled potential electrolysis of 6*H*-1,3-thiazines **62** in an EtOH/AcOH(1/1)-AcONa(0.5M)-(Hg) system at -0.8 V (SCE) affords the corresponding hydrogenated products **63** in good yields (eq. (3.31)) [56b].

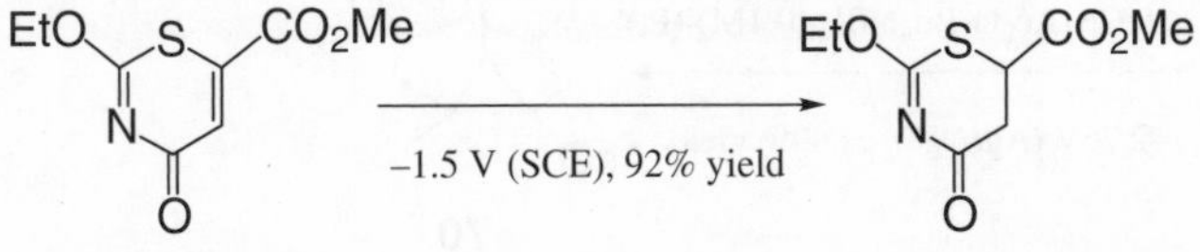

$$(3.30)$$

-1.5 V (SCE), 92% yield

60   **61**

EtOH/AcOH(1/1)-AcONa(0.5M)-(Hg)

-0.8 V (SCE), 75~80% yields

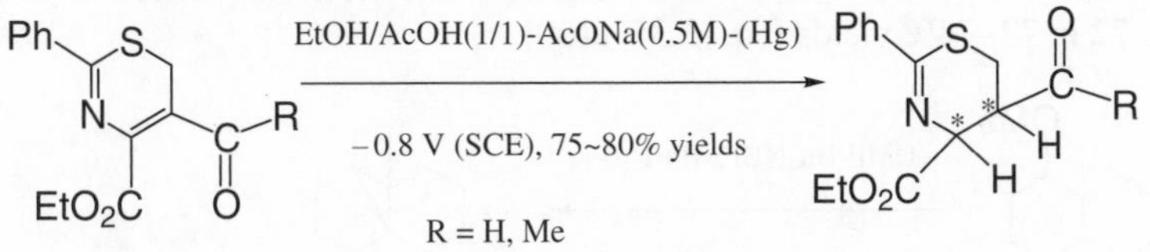

R = H, Me

$$(3.31)$$

62   **63**

The hydrogenation of the enamino double bond of **64** conjugated with the catechol moiety has been performed in an MeOH-Me$_4$NCl-(Hg) system at -0.8 V to give the hydrogenated product **65** in 78.3% yield (eq.(3.32)) [57].

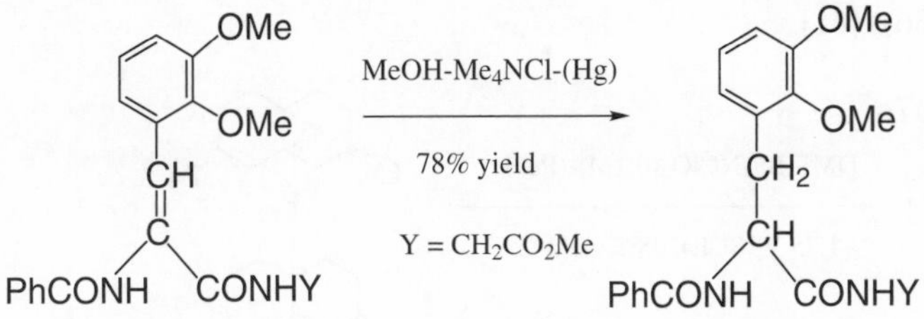

MeOH-Me$_4$NCl-(Hg)

78% yield

Y = CH$_2$CO$_2$Me

$$(3.32)$$

64   **65**

3.2.2   Electrophilic Substitution of Conjugated Olefins

The reaction of 2,3-dimethylindenone **66** with terminal halides proceeds with a low consumption of electricity [58].   The formation of the adducts suggests a chain process induced by the cathodic reduction.   The electroreduction of **66** in a DMF-Bu$_4$NI(0.1M)-(Hg) system in the presence of ethyl bromide yields the 2-ethylated product **67** in 62% yield together with saturated product **68** (6%) (eq. (3.33)).

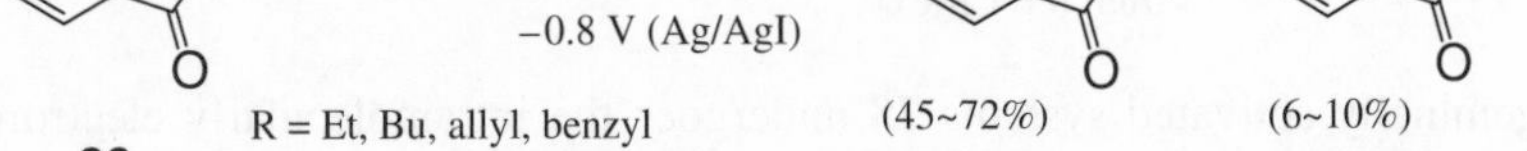

DMF-Bu$_4$NI(0.1M)-(Hg)

-0.8 V (Ag/AgI)

R = Et, Bu, allyl, benzyl

(45~72%)   (6~10%)

$$(3.33)$$

66   **67**   **68**

The intramolecular Michael addition of the conjugated olefin **69** has been attained in an $MeCN/Ac_2O$-Bu_4NBF_4(0.1M)-(Pt) system at –2.75 V (Ag/Ag^+) to yield the keto acetate **70** in *ca.* 40% yield (eq. (3.34)) [59].

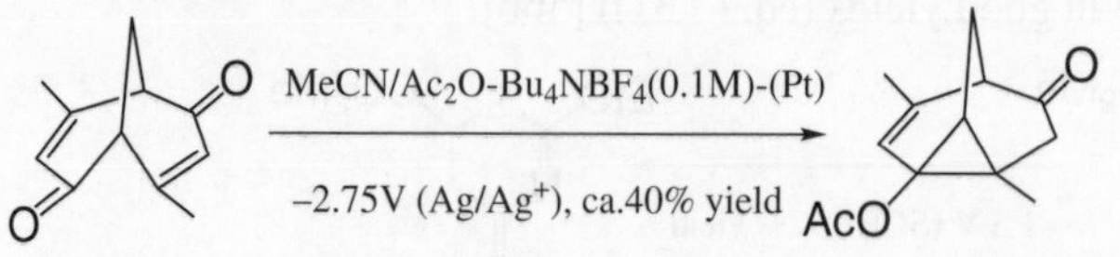

69 **70** (3.34)

The electroreductive intramolecular displacement of the mesyloxy group of *d,l-cis*-5-mesyloxy-10-methyl-1(9)-octal-2-one **71** undergoes a one-electron transfer to give the tricyclic ketone **72** in 72~74% yields (eq. (3.35)) [60].

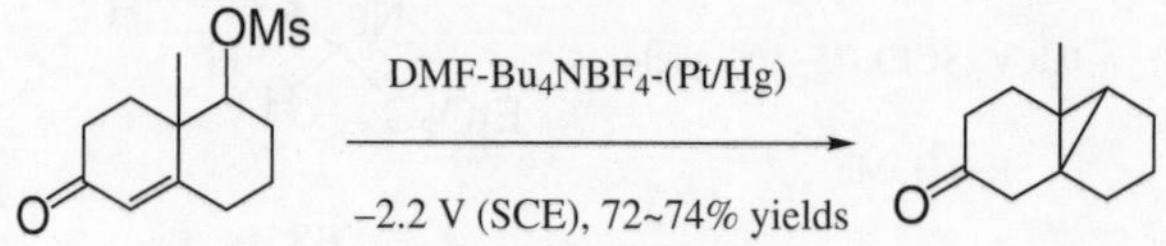

71 **72** (3.35)

The electroreductive intramolecular cyclization of the conjugated cyclohexenone sulfonate ester **73** has been attained to produce the bicyclo[4.1.0]-heptan-3-ones **74** in a DMF-Bu_4NClO_4(0.1M)(or Bu_4NBF_4)-(Pt/Hg) system gives the [4.4.1]tricyclo compounds in excellent yield (eq. (3.36)) [61].

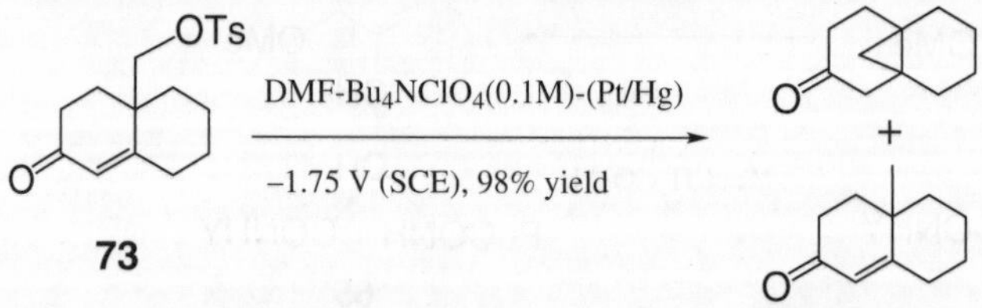

73 **74** **75** (74/75 = 99/1) (3.36)

The electroreduction of steroidal alkenylnitro compound **76** in a DMF-Bu_4NClO_4 (0.1M)-(Pt) system yields a cyclosteroid **77** in 36~43% yields (eq. (3.37)) [62]. The cyclopropane formation reaction is found to be highly dependent on the leaving group, R (Ts> CF_3CO).

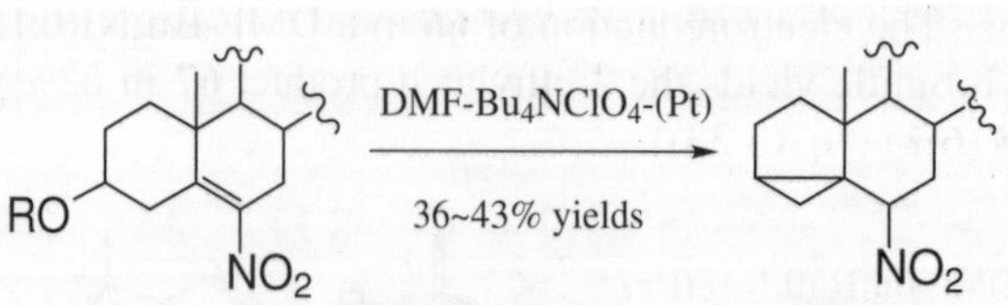

76 **77** (3.37)

76a: R = Ts
76b: R = CF_3CO

The geminally activated systems **78** undergoes the intramolecularly electroreductive cyclization to yield tricyclic systems **79**, **80** in good yields (eq. (3.38)) [47].

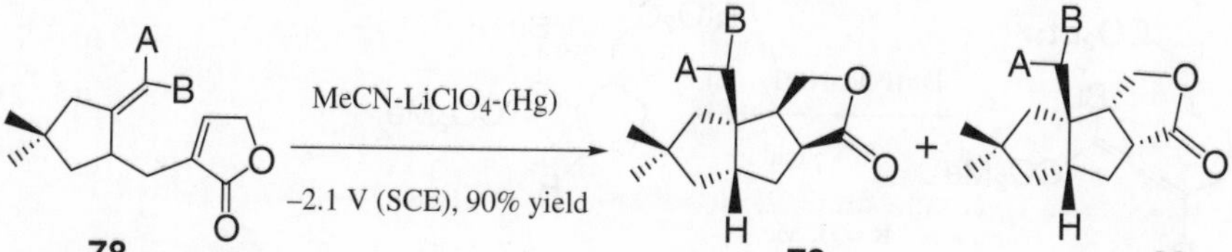

(3.38)

78

78a, A = B = CO$_2$CH$_3$
   b, A = B = CN

79

79a, A = B = CO$_2$C$_2$H$_5$
   b, A = B = CN

80

80a, A = B = CO$_2$C$_2$H$_5$
   b, A = B = CN

A tandem cyclization of β-ketoester enol phosphate **81** has been attained by electroreduction in a DMF-Bu$_4$NClO$_4$-(Hg) system at –2.3 V (SCE) to give the bicyclo product **82** in 77% yield (eq. (3.39)) [63]. The cyclization probably proceeds through intramolecular addition of an electroreductively produced vinyl radical to the double bond of **81**.

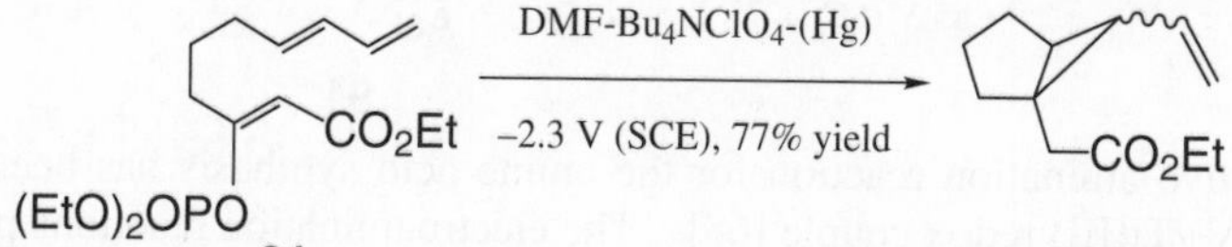

(3.39)

81

82

The electroreductive conjugated addition of allyl halides to α,β-unsaturated esters has been attained in a DMF-Et$_4$NOTs(0.2M)-(Pt) system [64]. The allylation of diethyl fumarate **83** proceeds to give the adduct **85** in 15~70% yields, depending on the choice of halide (eq. (3.40)).

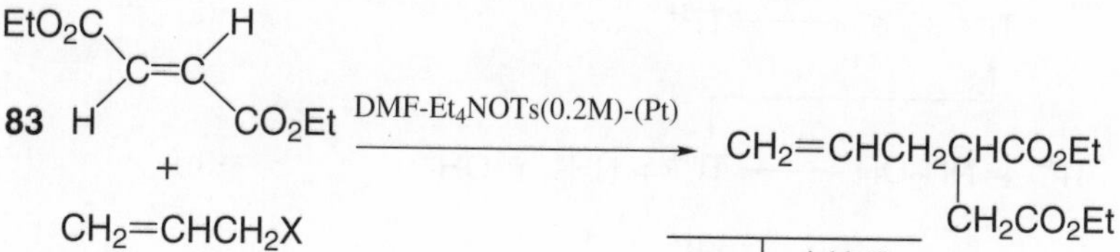

(3.40)

83

84

85

X	yield, %
Cl	15
Br	70
I	39

The intramolecular reductive coupling of the bisactivated olefins **86**, in an aqueous MeCN(50%)-Et$_4$NOTs-(Hg) system results in β- to β-coupling (when n = 1,2,3, and 4) to give high yields of the cyclized compounds **87** (eq. (3.41)) [65].

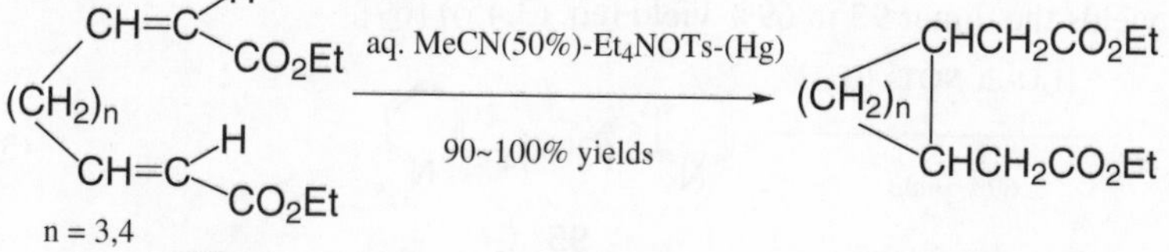

(3.41)

n = 3,4

86

87

The electroreductive intramolecular cyclization of allenic carboxylates **88** has been performed in a DMF-Et$_4$NOTs-(C) system in the presence of phenol as a proton donor to give the cyclopentene derivative **89** in 93% yield (eq. (3.42)) [66].

$$\underset{\textbf{88}}{\text{(structure)}} \xrightarrow[\substack{\text{96\% yield} \\ \\ R = H, Me}]{\text{DMF-Et}_4\text{NOTs-(C)}} \underset{\textbf{89}}{\text{(structure)}} \qquad (3.42)$$

The electroreductive acetylation of ethyl cinnamate **90** has been attained in a DMF-Ac$_2$O-(Hg) system in the presence of aluminum oxide, Al$_2$O$_3$, at −1.85 V (SCE) to yield the corresponding acetylated product **91** in 75% yield (eq. (3.43)) [67].

$$\underset{\textbf{90}}{\text{Ph—HC=CHCO}_2\text{Et}} \xrightarrow[\text{−1.85 V (SCE), 75\% yield}]{\text{DMF-Ac}_2\text{O-(Hg)}} \underset{\textbf{91}}{\underset{\overset{|}{\text{Ac}}}{\text{Ph—CH—CH}_2\text{CO}_2\text{Et}}} \qquad (3.43)$$

The electroreductive amination reaction for the amino acid synthesis has been developed by use of a Ti(IV)/Ti(III) redox couple [68].   The electroamination reactions proceeds at a mercury electrode in aqueous sulfuric acid (2M) solution containing hydroxylamine sulfate (0.5M) and a catalytic amount of Ti^{4+} (0.005 M).   The one-electron transfer from Ti(III) species to hydroxyl amine probably causes the production of an amino radical, NH$_2$·, together with Ti(IV) species (eq. (3.44b)).   The produced NH$_2$· species would be trapped by cinnamic acid **92** to form *dl*-phenylalanine **93** in 9.4% faradic efficiency.

$$\text{Ti}^{4+} + e^- \longrightarrow \text{Ti}^{3+} \qquad\qquad \text{---(a)}$$

$$\text{Ti}^{3+} + \text{NH}_2\text{OH} \longrightarrow \text{Ti}^{4+} + \text{NH}_2^{\bullet} + \text{OH}^- \qquad \text{---(b)} \qquad (3.44)$$

$$\underset{\textbf{92}}{\text{Ph—CH=CHCO}_2\text{H}} + \text{NH}_2^{\bullet} \xrightarrow[\text{Ti}^{3+}]{\text{H}^+} \underset{\textbf{93}}{\underset{\overset{|}{\text{NH}_2}}{\text{Ph—CH}_2\text{CHCO}_2\text{H}}} \quad \text{---(c)}$$

3.3   Electroreductive Coupling

3.3.1   Electroreductive Dimerization

Electroreductive hydrodimerization of 2-vinylpyridine **94** in an aqueous Et$_4$NOTs-(Hg) system at 25-30 °C yields the dimer **95** in 69% yield (eq. (3.45)) [69].

$$\underset{\textbf{94}}{\text{(structure)}} \xrightarrow[\text{69\% yield}]{\text{H}_2\text{O-Et}_4\text{NOTs-(Hg)}} \underset{\textbf{95}}{\text{(structure)}—(\text{CH}_2)_4—(\text{structure})} \qquad (3.45)$$

The electroreduction of 2,2'-distyrylbiphenyl **96** in a DMA-Bu$_4$NBr-(Pt) system at −2.4 V (Ag/AgCl) at −30 °C undergoes intramolecular cyclization to yield dihydrophenanthrene derivative **97** in 85% yield (eq. (3.46)) [70].

$$\text{(3.46)}$$

96   **97**

Hydrodimerization of methyl vinyl ketone **98** has been reported (eq. (3.47)) [71]. The electrolysis of **98** is carried out in a DMF-Bu₄NI-(Hg) system to give the dimer **99** in 80% yield. The electrohydrodimerization of 2-cyclopenten-1-one **100** in the pH range 8.0~13.5 proceeds to form the hydrodimer of (1,1'-bicyclopentyl)-3,3'-dione **101** in 45~50% yield (eq. (3.48)) [72a]. The process follows two different radical-radical coupling pathways,

$$\text{H}_2\text{C=CH–COMe} \xrightarrow[\text{pH} < 4, \ 80\% \text{ yield}]{\text{DMF-Bu}_4\text{NI-(Hg)}} \begin{array}{c} \text{CH}_2\text{–CH}_2\text{COMe} \\ | \\ \text{CH}_2\text{–CH}_2\text{COMe} \end{array} \qquad \text{(3.47)}$$

98   **99**

$$\text{(3.48)}$$

100   45~50% yields   **101**

depending on the pH of the solution. In the pH range 8.0~10.5, the final hydrodimer **101** is formed *via* protonation of the radical anion to generate the neutral radical, followed by coupling of the radical anion. At pH > 10.5, the same final hydrodimer is obtained *via* dimerization of the radical anion and subsequent protonation of the dimer dianion formed [72a]. A similar dimerization over the pH range 3.5~12.5 has been also investigated [72b]. The hydrodimerization of isophorone **102** leading to the hydrodimer **103** has been achieved (eq. (3.49)) [71,73].

$$\text{(3.49)}$$

102   **103**

Electoreduction of benzalacetone and benzalacetophenone in an aqueous EtOH(50%)-H₂SO₄(1M)-(C) system has been carried out in the pH range 0.50~10.5. At low pH less than 7, the protonated form is reduced to give the corresponding neutral radical, which subsequently couples to give a diketone. At neutral and higher pH more than 7, the electroreduction leads to the formation of saturated ketone [74].

Surfactant effect on the electroreduction of benzalacetone **104** in an AcOH/AcONa buffer system has been studied [75]. Greater changes in the yield of **105** are obtained when hexadecyltrimethylammonium bromide (HTABr) is added to electrolysis media (eq. (3.50)). The efficient hydrodimerization of 1-phenyl-4,4-dimethyl-1-penten-3-one **107**

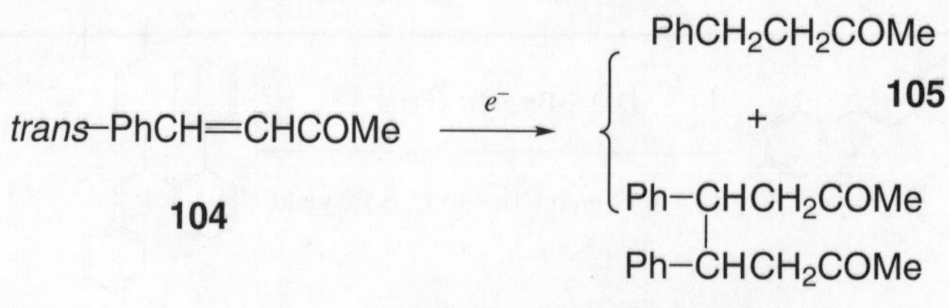

$$trans\text{-}PhCH\text{=}CHCOMe \xrightarrow{e^-} \begin{cases} PhCH_2CH_2COMe \quad \textbf{105} \\ + \\ Ph\text{-}\underset{|}{\overset{}{C}}HCH_2COMe \\ Ph\text{-}\overset{}{C}HCH_2COMe \end{cases}$$

(3.50)

104

106

EtOH/H₂O(1/1)	Yield, %	
	105	**106**
—	3	24
Me₄NBr/HTABr	95	1
HTABr	94	1

leading to the dimer **108** can be achieved in a DMF-LiClO₄-(Hg) system in the presence of chloride salts, MCln (M: Li⁺, Cr²⁺, Fe²⁺, Mn²⁺, Co²⁺, Zn²⁺, Ni²⁺) in good yields (eq. (3.51)) [76].   An acid-base interaction between ketone and metallic cation as a Lewis acid

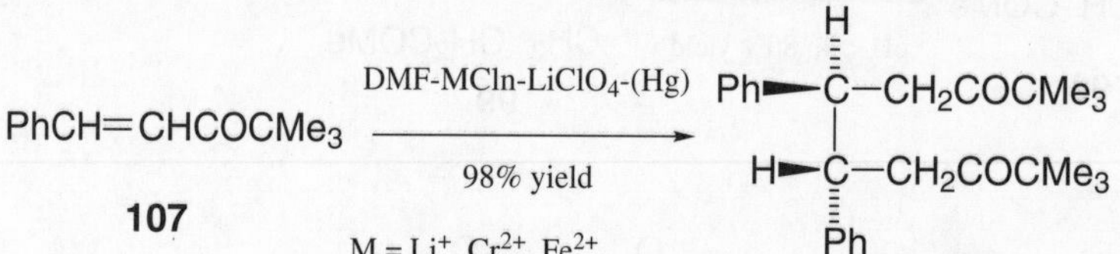

$$PhCH\text{=}CHCOCMe_3 \xrightarrow[\substack{98\% \text{ yield} \\ M = Li^+, Cr^{2+}, Fe^{2+}, \\ Mn^{2+}, Co^{2+}, Zn^{2+}, Ni^{2+}}]{DMF\text{-}MCln\text{-}LiClO_4\text{-}(Hg)} \begin{array}{c} H \\ Ph\text{-}\overset{|}{C}\text{-}CH_2COCMe_3 \\ H\text{-}\overset{|}{C}\text{-}CH_2COCMe_3 \\ Ph \end{array}$$

(3.51)

107

108

is suggested.   The electroreductive dimerization of 1-phenyl-1-penten-3-one **109** proceeds in a DMF-LiClO₄(0.1M)-(Hg) system to afford the cyclohexanone derivative **110** in quantitative yield (eq. (3.52)) [77].   Lithium ion assisted hydrodimerization of coumarin **111** has

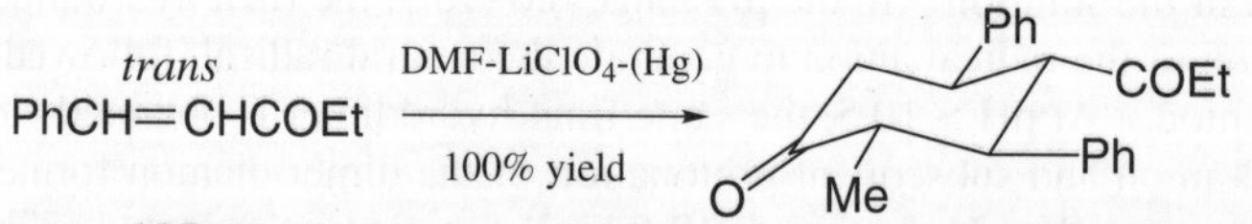

$$\substack{trans \\ PhCH\text{=}CHCOEt} \xrightarrow[100\% \text{ yield}]{DMF\text{-}LiClO_4\text{-}(Hg)}$$

(3.52)

109

110

been developed.   It is assumed that the lithium ions are coordinated with anion radicals derived from coumarin **111** by a one-electron reduction.   The electrolysis of coumarin **111** in an MeOH/H₂O(3/1)-LiOAc-(Pt/Hg) system gives coumarin hydrodimer **112** in 89% yield along with dihydrocoumarin **113** (1.5%) (eq. (3.53)) [78].   Alkaloid induced chiral

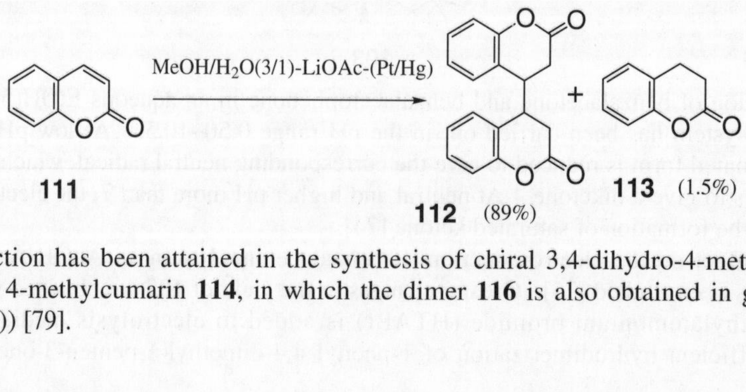

$$\xrightarrow{MeOH/H_2O(3/1)\text{-}LiOAc\text{-}(Pt/Hg)}$$

(3.53)

111

112 (89%)

113 (1.5%)

induction has been attained in the synthesis of chiral 3,4-dihydro-4-methylcumarin **115** from 4-methylcumarin **114**, in which the dimer **116** is also obtained in good yields (eq. (3.54)) [79].

$$(3.54)$$

	115	116
Yohimbin	18% (58 %ee)	42%
Spartein	4% (17 %ee)	89%

Adiponitrile has been manufactured by electrohydrodimerization of acrylonitrile which was developed by M. M. Baizer at Monsanto Co. (USA).   Asahi Chem. Ind. Co. developed an emulsion electrolysis process and entered commercial operation in 1971.   Advantages of the undivided cell process have been found to be a further developed technique [80]. Some benzylidenemalononitriles have been hydrodimerized in a DMF-Bu$_4$BF$_4$(0.1M)/ AcOH(1M)-(Hg) system to give *cis*- and *trans*-2-amino-4,5-diaryl-2-cyclopentene-1,1,3-tricarbonitrile [81].

The controlled potential electrolysis of β-methoxy-α,β-cyclohexenones **117** in an aqueous EtOH(50%)-Me$_4$NCl-(Hg) system at –2.2 V (SCE) undergoes β-coupling to form bisenone systems **118** in 33~55% yields (eq. (3.55)) [82].

$$(3.55)$$

The controlled potential electroreduction of 2,6-diphenyl-4-pyrone **119** in an aqueous MeCN-Et$_4$NClO$_4$(0.2M)-(Hg) system leads to the formation of the dimers **120** and **121** in 85% yields (eq. (3.56)) [83].   Hydrodimerization of dimethyl maleate in an aqueous MeOH(*ca*.85%)-R$_4$NOTs-(Pt/Pb) system has been patented [84a,b].   Hydrophobic zinc and lead electrodes are used for the hydrodimerization of dimethyl maleate [84c].   A pilot scale process of electrohydrodimerization of dimethyl maleate has been developed in an undivided bipolar cell using an MeOH-AcONa-(C) system as a several hundred kilogram scale [85].

$$(3.56)$$

The constant-potential electroreduction of *N*-phenylmaleimide **122** in an aqueous EtOH(50%)-HCl-(Hg) system gives the corresponding dimer **123** in 60% yield (eq. (3.57)) [86].

$$(3.57)$$

The electroreduction of 5-arylmethylene-3-pyrrolin-3-ones **128** in a DMF-LiClO$_4$-(Hg) system at −1.8 V (SCE) yields a mixture of dimer **125** and **126** in quantitative yield (eq. (3.58)) [87].

$$(3.58)$$

The cross-coupling of α,β-enone with a variety of Michael acceptors is reported. For methyl vinyl ketone it has been proposed that hydrodimerization occurs *via* a radical intermediate [88]. A successful electroreductive cross-coupling of methyl vinyl ketone **127** (−1.4 V (SCE)) with diethyl fumarate **128** (−1.2 V (SCE)) has, however, been found to proceed in an MeCN/H$_2$O(5%)-Et$_4$NOTs-(Hg) system to give the coupling products **129** in 85% yield (eq. (3.59)) [89].

$$(3.59)$$

The electroreduction of α,β-enoates in an aqueous medium leads to the corresponding hydrodimer, *i.e.*, adipic acid derivatives, as a major product [90,91]. Electrolytic hydrodimerization of α,β-unsaturated acids in quaternary ammonium electroltes is listed in Table 3.1 [91]. Ethyl acrylate can be hydrodimerized in a DMF-Et$_3$MeNOTs-(Hg) system to give the corresponding hydrodimer in good yield [92].

For the electrohydrodimerization of activated olefins, the mechanisms of the dimerization of anion radicals [93], the radical-substrate coupling [94], and the anion radical-proton donor complex coupling [95] have been discussed taking into consideration the effect of water and ion species in solvent-electrolyte systems [96].

143

Table 3.1 Electrolytic hydrodimerization of a,b-unsaturated acids in quaternary ammonium electrolytes

W	X	Y	Z	Electrolysis Conditions	− V (SCE)	Hydrodimer[a], %	Ref.
H	H	H	CN	MeCN–Et$_4$NOTs–(Hg)	1.81~1.91	75~100	[91]
H	H	Me	CN	MeCN/DMF–Et$_4$NOTs–(Hg)	2.01~2.05	75.3[c,*]	[91]
Me	Me	H	CN	MeCN/DMF–Et$_4$NOTs–(Hg)	2.08~2.11	87–93[d]	[91]
H	H	H	CO$_2$Et	MeCN/DMF–Et$_4$NOTs–(Hg)	1.85	74–87	[91]
Me	Me	H	CO$_2$Et	MeCN–Et$_4$NOTs–(Hg)	2.10~2.18	66.2[e]	[91]
H	CO$_2$Et	H	CO$_2$Et	MeCN–Et$_4$NOTs–(Hg)	1.32~1.40	61.5[f]	[91]
Ph	H	H	CO$_2$Et	aq. MeCN–MeBu$_3$NOTs–(Hg)	1.57~1.61	28[f]	[91]
Me	H	H	CONEt$_2$	aq. MeCN–Et$_4$NOTs–(Hg)	2.03~2.12	61.4[g,*]	[91]
H	H	H	CONEt$_2$	aq. MeCN–Et$_4$NOTs–(Hg)	1.91~1.95	73.3[h]	[91]
H	H	H	CONH$_2$	aq. MeCN–Et$_4$NOTs–(Hg)	1.82~2.00	39.6	[91]
H	–(CH$_2$)$_4$–		CN	MeCN/DMF–Et$_4$NOTs–(Hg)	2.15~2.20	66.4	[91]
H	–(CH$_2$)$_3$–		CN	MeCN/DMF–Et$_4$NOTs–(Hg)	2.13	29	[91]
EtH	H	H	CN	MeCN/DMF–Et$_4$NOTs–(Hg)	1.97~2.20	58	[91]
EtO	H	H	CO$_2$Et	MeCN/DMF–Et$_4$NOTs–(Hg)	2.22	57	[91]

In aprotic solvent such as acetonitrile or dimethylformamide, methyl cinnamate **130** undergoes hydrodimerization and subsequent intramolecular cyclization to give methyl oxocyclopentane carboxylate **131** in 52~78% yields (eq. (3.60)) [97].

$$(3.60)$$

130   R = H, Me, Et, Ph
Ar = Ph, Py, p-MeC$_6$H$_4$,
p-MeOC$_6$H$_4$, p-ClC$_6$H$_4$,   **131**

A cyclic hydrodimer, methyl d,l-3,4-diphenylcyclopentanone-2-carboxylate **132** (R = Me) in which the ring substituents are all *trans*, has been first isolated under relatively aprotic conditions [76a].   Using chiral auxiliaries, dimethyl (3R,4R)-diphenyladipate **133** is synthesized enantioselectively by electroreductive intermolecular hydrocoupling of chiral *N-trans*-cinnamoyl-2-oxazolidones **134**, giving cyclized product **135** and subsequent methanolysis (eq. (3.61)) [98,99].   The stereoselectively and mechanism in the electrohydrodimerization of cinnamates have been clarified by rate constants and reaction orders [100].

130 (R = H; Ar = Ph)   **132**

$$(3.61)$$

134

Y = $-$N   R1 = Alkyl, Ph
R2 = H, Ph   **133**   **135**

The substituted cyclobutanetetracarboxylates have been electrosynthesized from alkylidenemalonates **136** in an MeOH-NaI-(C) system in an undivided cell (eq. (3.62) [101,102]). The reaction proceeds *via* the reductive coupling of two substrate molecules at the cathode and the cyclization of a hydrodimer dianion by its interaction with an active form of a mediator, an anode-generated halogen, to give **137** and **138**.

$$(3.62)$$

136   R = Me, Et, Pr,
Ph, C$_7$H$_{15}$ **137**   **138**

The cathodic reduction of cinnamate esters formed with chiral alcohols proceeds smoothly *via* coupling and intramolecular condensation, with high stereoselectivity, to yield diastereoisomeric mixtures of the esters of the all-*trans dl*-2-carboxy-3,4-diphenylcyclopentanone [100].

The electroreductive cyclization of 5-(1-cyano-2-arylvinyl)-1,3,4-thiadiazoles **139** has been attained in an aqueous DMF/EtOH (45/15 v/v)-HCl(0.03M)-(Hg) system to afford cyclopentene carboxylic acid derivatives **140** in good yields (eq. (3.63)) [103].

$$2 \left(Ar-CH=C \begin{smallmatrix} X \\ CN \end{smallmatrix} \right) \xrightarrow[\text{43% yield}]{\text{aq. DMF/EtOH(3/1)-HCl(0.03M)-(Hg)}} \text{140}$$

(3.63)

139

$$X = \begin{smallmatrix} N=C-NHCOPh \\ | \quad S \\ N=C \end{smallmatrix}$$

Ar = Ph, *p*-ClC$_6$H$_4$
p-MeOC$_6$H$_4$, *p*-O$_2$NC$_6$H$_4$

140

The cathodic hydrodimerization of alkynoates has been studied. Products are formed by competing hydrodimerization, hydrogenation, and nucleophilic addition to the triple bond and their distribution depends strongly on the composition of the electrolyte.

The dimerization occurs mostly at the β,β-positions and the initially formed 1,3-dienes are reduced further to monoolefins. The electrolysis of ethyl propyonates **141** in a DMF/H$_2$O(96/4)-Bu$_4$NBF$_4$-(Pt/Hg) system at -1.8 V (SCE) results in the formation of a mixture of hydrodimers **142** and **143** together with the water adduct **144** (eq. (3.64)) [104].

$$HC\equiv C-CO_2R \xrightarrow[\substack{\text{50% yield} \\ \text{R = Me, Et} \\ \text{Y = CH}_2\text{CO}_2\text{R}}]{\text{DMF/H}_2\text{O(96/4)-Bu}_4\text{NBF}_4\text{-(Hg)}} \quad \text{142} \quad + \quad \text{143} \quad + \quad \text{144}$$

(3.64)

141

142/143/144 = 1/0.7/0.4

Electroreduction of acetylene derivatives **145** using a conductive sulfur-carbon cathode yields mainly thiophene derivatives **146** together with other sulfur containing products (eq. (3.65)) [105]. The reaction is influenced by the nature of the substituents at R^1 and R^2.

$$\begin{smallmatrix} R^1 \\ ||| \\ R^2 \end{smallmatrix} \xrightarrow[\substack{-0.9 \text{ V (SCE), } \sim 85\% \text{ yield}}]{\text{DMF-Et}_4\text{-ClO}_4\text{-(S-C)}} \text{146}$$

(3.65)

145

R^1 = H, Ph, CO$_2$Me
R^2 = CHO, CN, Ph, CO$_2$Et, CO$_2$Me

146

The electroreductive coupling of vinylic compounds **147** can lead to substituted butadienes **148** by a nucleophilic addition of the anion radical intermediate to the substrate followed by loss of X$^-$ [106]. The electrolysis is performed in a column type flow cell which consists of a carbon felt anode, a porous glass microfiber insulator, and a porous disc of amalgamated copper in a DMF-Bu$_4$NBF$_4$-(Hg/Cu) system (eq. (3.66)) [106].

$$
\underset{\mathbf{147}}{\underset{E}{\overset{Ph}{\diagdown}}C=C\underset{H}{\overset{X}{\diagup}}} \xrightarrow[25\sim34\% \text{ yields}]{DMF\text{-}Bu_4NBF_4\text{-}(Hg/Cu)} \underset{\mathbf{148}}{\underset{E}{\overset{Ph}{\diagdown}}C=CH-CH=C\underset{E}{\overset{Ph}{\diagup}}} \tag{3.66}
$$

E = CN, X = CO$_2$Me

Polyvinyl acetate **150** in the range of 1,500,000 Mw is electroreductively synthesized from the monomer **149** in an aqueous Et$_4$NOTs-(Al) system in 60% total yield (eq. (3.67)) [107].

$$
\underset{\mathbf{149}}{n\ CH_2{=}CH\underset{|}{\underset{OAc}{}}} \xrightarrow[{-1.8\ V\ (SCE),\ 60\%\ yield}]{aq.\ Et_4NOTs\text{-}(Al)} \underset{\mathbf{150}}{\left[CH_2{-}CH\underset{|}{\underset{OAc}{}}\right]_n} \tag{3.67}
$$

3.3.2   Electroreductive Coupling with Carbonyls

The electroreductive carboxylation of activated olefins which are reduced at less negative potentials than carbon dioxide has been investigated in solvents of low proton availability [108].   The β-carboxylated enoates are obtained as a major product in good yields.   The hydrogenation of the double bond is a competing reaction path whose importance increases with the water content of the solvent.   The electrocarboxylation proceeds as a competing reaction between two major routs as follows: 1) carbon dioxide is reduced by the olefin anion radical which then couples with $CO_2{}^-$ while, 2) carbon dioxide acts as an electrophile adding to the olefin anion radical.   The latter mechanism is shown to be predominant for activated olefins based on kinetic evidence [108].

The carboxylation of 1,3-butadiene proceeds in an MeCN-Bu$_4$NBr (0.1M)-(Sn) system at –20 °C to give 3-hexenedioic acid [109].   Electroreductive carboxylation of styrene **151** has been performed in a DMF-Bu$_4$NBr-(Al/C) system by continuously bubbling CO$_2$ between the electrodes to give phenylsuccinic acid **152** as a major product together with **153** (eq. (3.68)) [110].   The electrochemical incorporation of carbon dioxide (CO$_2$) into alkenes using electrogenerated Ni(0) complexes has been attained [111].   The electrolysis of styrene **151** is carried out in a DMF-Bu$_4$NBF$_4$(0.3M)-(Mg/C) system in the presence of NiBr$_2$ and PMDTA (pentamethyldiethylene triamine) under CO$_2$ atmosphere to give the corresponding dicarboxylic acid **152** in 85% yield.

$$
\underset{\mathbf{151}}{Ph\diagup\!\!\!\diagdown} \xrightarrow{DMF\text{-}Bu_4NBr\text{-}(Al/C)} \underset{\mathbf{152}}{Ph\overset{CO_2H}{\diagdown}\!\!\diagup\!\!\diagdown CO_2H} + \underset{\mathbf{153}}{Ph\diagdown\!\!\diagup\!\!\diagdown CO_2H} \tag{3.68}
$$

The electrocarboxylation of acenaphthylene **154** to *trans*-acenaphthene-1,2-dicarboxylic acid **155** (R = CO$_2$H) and acenaphthene-1-carboxylic acid **155** (R = H) has been attained in a preparative scale [112].   The electrolysis is performed in a DMF-Bu$_4$NBr-(Zn) systems in a gas-lift type cell to yield the dicarboxylic acid **155** (R = CO$_2$H) with current yields greater than 80% (current density: 40 mA/cm^2) (eq. (3.69)).

$$\text{154} \quad \xrightarrow[\substack{CO_2,\ 69\sim81\%\ yields \\ R = H\ or\ CO_2H}]{DMF\text{-}Bu_4NBr\text{-}(Zn)} \quad \text{155} \tag{3.69}$$

The electroreductive carboxylation of acenaphthylene **154** is carried out in a DMF-Et_4NI-(Hg) system in the presence of carbon dioxide and methyl chloride to give 1,2-dimethyoxycarbonyl-1,2-dihydroacenaphthene **156** in 59.4% yield (eq. (3.70)) [113].   In a similar manner, the coupling of **154** with acid anhydride occurs to form the corresponding acylated products.

$$\text{154} \quad \xrightarrow[\substack{-1.7\ V\ (SCE),\ 59.4\%\ yield}]{\substack{DMF\text{-}Et_4NI\text{-}(Hg) \\ CO_2,\ MeCl,}} \quad \text{156} \tag{3.70}$$

The electroreductive carboxylation of allenes **157** has been performed in the presence of Ni(II) complexes [114].   The electrolysis of the terminal allene **157** in a DMF-Bu_4NBF_4-(Mg/C) system in the presence of nickel dibromide and N,N,N',N'',N''-pentamethyldiethylene triamine (PMDTA) together with carbon dioxide yields a mixture of carboxylic acids **158**, **159**, and **160** in 80% yields (eq. (3.71)).

$$\text{157}\ (R = C_8H_{17}) \quad \xrightarrow[\substack{Ni(II)Br_2\text{-}DMF/PMDTA,\ 80\%\ yield}]{CO_2,\ DMF\text{-}Bu_4NBF_4\text{-}(Mg/C)} \quad \text{Carboxylic Acids}$$

$$\text{158}\ (45\%) \quad + \quad \text{159}\ (30\%) \quad + \quad \text{160}\ (5\,\%) \tag{3.71}$$

The carboxylation of the anionic intermediate derived from the electroreduction of tetraphenylallene **161** has been performed in a DMF-Bu_4NClO_4-(Hg) system in the presence of carbon dioxide to yield 2,2,4,4-tetraphenyl-3-butanoic acid **162** in 54% yield (eq.(3.72)) [33].

$$Ph_2C=C=CPh_2 \quad \xrightarrow[\substack{-2.17\ V\ (SCE),\ 54\%\ yield}]{CO_2,\ DMF\text{-}Bu_4NClO_4\text{-}(Hg)} \quad Ph_2C-CH=CPh_2 \atop \underset{CO_2H}{|} \tag{3.72}$$

161   **162**

The nickel-catalyzed carboxylation of 1,3-enynes **163** undergoes hydrocarboxylation of the triple bond through a stereoselective *cis*-addition [115)ER92116].   The electrocarboxylation of 163 in a DMF-Bu_4NBF_4-NiBr$_2$DME/PMDTA-(Mg/C) system in the presence of carbon dioxide provides **164** and **165** in good yields (eq. (3.73)).

$$163 \xrightarrow[\text{NiBr}_2\text{-DME/PMDTA, 80\% yield}]{\text{CO}_2, \text{DMF-Bu}_4\text{NBF}_4\text{-(Mg/C)}} 164 \ (72\%) + 165 \ (8\%)$$

(3.73)

The hydroformylation of olefins **166** (R = C_4H_9) with activated homogeneous catalysts, involving electrogenerated rhodium and platinum species, has been attained [116]. The role of $SnCl_2$ co-catalyst in both activity and selectivity has been discussed. The electrore-duced catalysts derived from $PtCl_2L_2$-$SnCl_2$ systems have been applied to 1-hexene hydro-formylation to give **167** and **168**. For a substrate-to-catalyst ratio of 500, a 98% selectivity into 1-heptanal **167** is obtained with the Pt-Sn system (eq. (3.74)) [116].

$$\text{RCH}{=}\text{CH}_2 + \text{CO} + \text{H}_2 \xrightarrow{\text{PtL}_2\text{Cl}_2\text{-(Sn)}} \begin{array}{c} \text{RCH}_2\text{CH}_2\text{CHO} \\ \text{n} \quad \textbf{167} \\ + \\ \text{RCH(CHO)CH}_3 \\ \text{b} \quad \textbf{168} \end{array}$$

(3.74)

166
R = C_4H_9
L = PPh_3, $(CH_2)_nPPh_2$, n = 1~4

The electroreductive coupling of substituted 1,3-dienes **169** with methyl esters **170** to give cyclopent-3-en-1-ols **171** has been attained by use of magnesium electrodes (eq. (3.75)) [117]. The electrolysis is carried out in a THF-LiClO$_4$-(Mg) system in an undivid-ed cell, and a constant current electricity (50 mA/ 35 cm^2) is passed during the electrolysis.

$$\begin{array}{c} R^1 \\ R^2 \end{array} + R^3CO_2Me \xrightarrow[\text{56~88\% yields}]{\text{THF-LiClO}_4\text{-(Mg)}} \begin{array}{c} R^1 \quad R^3 \\ R^2 \quad \text{OH} \end{array}$$

(3.75)

169   **170**   **171**

R^1 = Me, $Me_2C{=}CH(CH_2)_2$
R^2 = H, Me; R^3 = Et, Pr, Bu, Ph, $Ph(CH_2)_2$

3.3.3 Electroreductive Coupling with Electrophiles

The vicinal methylation of styrene **151** has been achieved in an HMPA-LiCl-(Pt) system in the presence of methyl iodide to give 2-phenylbutane **172** in 43% yield (eq. (3.76)) [118a]. Arylalkenes are alkylated with alkyl bromides or alkane-α, ω-dinyl dibromide using a sac-rificial aluminium anode in a DMF-Bu$_4$NI system affords a mono-alkylated or cyclic prod-ucts, respectively [118b].

$$\text{Ph} \diagup\!\!\!\diagdown \xrightarrow[\text{96\% yield}]{\text{HMPA-LiCl-(Pt)}} \begin{array}{c} \text{Me} \\ \text{Me} \end{array}$$

(3.76)

151   **172**

The electroreductive alkylation of cycloheptatriene systems has been developed [119]. The electrolysis of 1-methoxy-1,3,5-cycloheptatriene **173** in a DMF-Et₄NOTs-(Pt) system in the presence of isopropyl chloride yields a mixture of the alkylated products **174** and **175** in 45% yield (eq. (3.77)) [119].

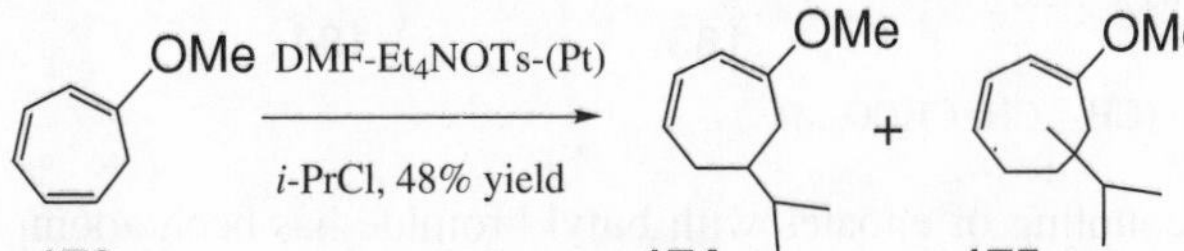

$$\text{(3.77)}$$

The electroreductive coupling of the enoates **177** such as dimethyl maleate, methyl cinnamate, 4-phenyl-3-buten-2-one, or methyl acrylate with the substituted alkyl dihalides **176** such as dibromomethane, 1,3-dibromopropane, 1,4-dibromobutane gives the corresponding cyclic products **178** in satisfactory yields (eq. (3.78) [120]. For example, dimethyl 1,2-cyclopentanedicarboxylate **178** (n = 3) is electrosynthesized in a NMP-Bu₄NBF₄/Bu₄NI-(Al/SUS) system in the presence of 1,3-dibromopropane and diethylmaleate with a current density of 2 A/dm² (eq. (3.78)).

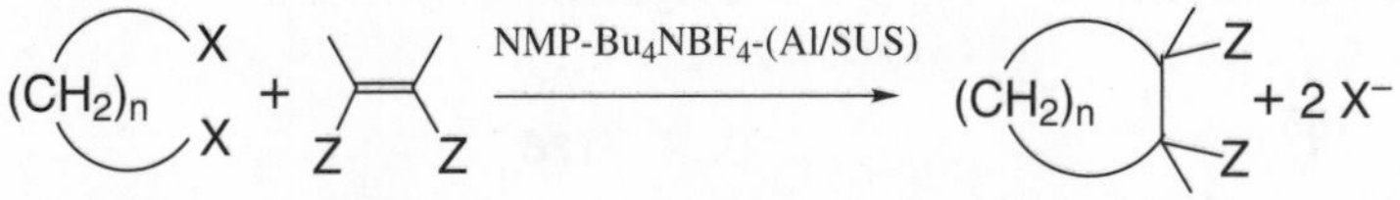

$$\text{(3.78)}$$

X = Br, Cl,   n = 1~4
Z = electon withdrawing group
NMP: *N*-methylpyrrolidone

The electroreduction of ω-bromo-butylidenemalonate **179** in a DMF-Bu₄NBr-(Pt/Hg) system at a potential of –1.85 V (SCE) at room temperature gives methyl cyclobutylylmalonate **180** in 65~80% yields (eq. (3.79)) [121a]. The bicyclo[3,2,1]carbon framework is synthesized by the electroreductive cyclization [121b].

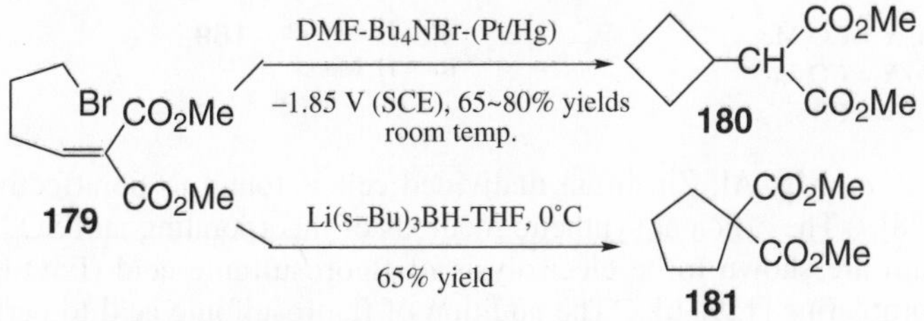

$$\text{(3.79)}$$

Electroreductive cyclization of β-substituted butenolides bearing either an α,β-unsaturated ester, an α,β,γ,δ-unsaturated ester, allylic bromide, mesylate or aldehyde functionality on the appending side chain has been investigated [122]. For example, the controlled potential electrolysis of the butenolide **182** is carried out in an MeCN-Bu₄NBr-(Pt/Hg) system at –2.22 V (SCE) to give the spirocyclic lactones **183** and **184** (1.0/1.1) in 35~41% yields (eq. (3.80)).

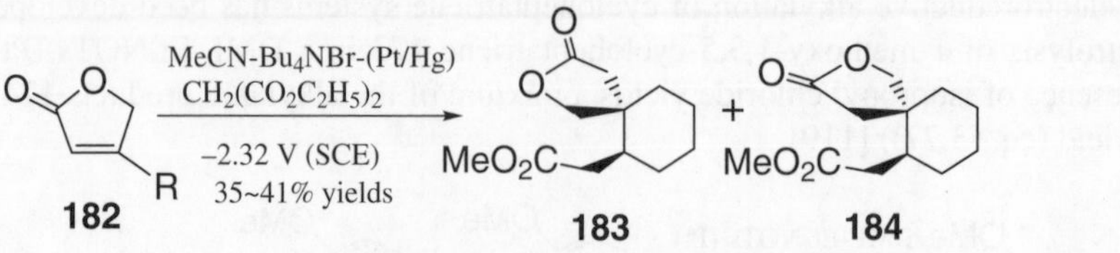

$$\text{182} \xrightarrow[\substack{-2.32 \text{ V (SCE)} \\ 35\sim41\% \text{ yields}}]{\substack{\text{MeCN-Bu}_4\text{NBr-(Pt/Hg)} \\ \text{CH}_2(\text{CO}_2\text{C}_2\text{H}_5)_2}} \text{183} + \text{184} \tag{3.80}$$

R = (E)-(CH$_2$)$_4$CH=CHCO$_2$Me

The electroreductive coupling of enoates with butyl bromide has been attempted in an HMPA-LiClO$_4$-(Pb) system to yield the corresponding butylated products in 7~45% yields [123].

The electroreductive acylation of aromatic hydrocarbon [124], activated olefins [67,125a], activated halides [125b], and ketones [126] provides a promising method for the preparation of functionalized carbonyl compounds. The acetylation of arylated olefin and/or enoates has been carried out in a DMF-Bu$_4$NI(0.1M)-(?) system in the presence of acetic anhydride [127]. The electrolytic acetylation of ethyl 3-acetoxycinnamate **185** affords ethyl 3-acetoxy-3-phenyl-4-oxopentanoate **186** in 68% yield (eq. (3.81)) [127]. The

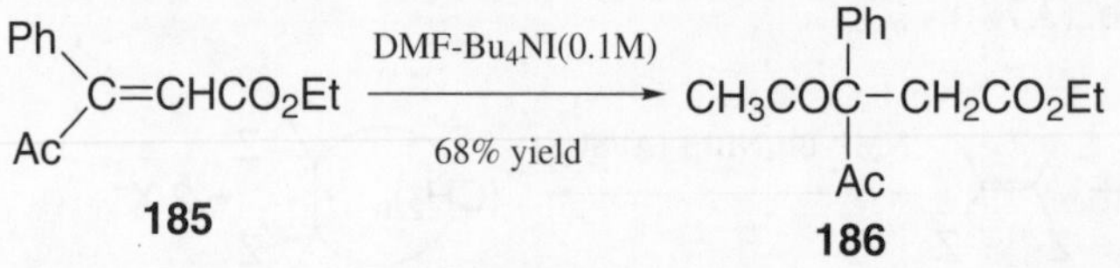

$$\underset{\text{Ac}}{\overset{\text{Ph}}{\diagdown}}\text{C=CHCO}_2\text{Et} \xrightarrow[\text{68\% yield}]{\text{DMF-Bu}_4\text{NI(0.1M)}} \text{CH}_3\text{CO}\underset{\underset{\text{Ac}}{|}}{\overset{\overset{\text{Ph}}{|}}{\text{C}}}\text{-CH}_2\text{CO}_2\text{Et} \tag{3.81}$$

185   **186**

electroreductive acylation of activated olefins has been performed by the electrolysis of the olefins **187** in an MeCN-Et$_4$NOTs-(C) system in the presence of anhydride **188** to give the corresponding γ-keto-esters **189** in 50~82% yields (eq. (3.82)) [125a]. The sacrificial type

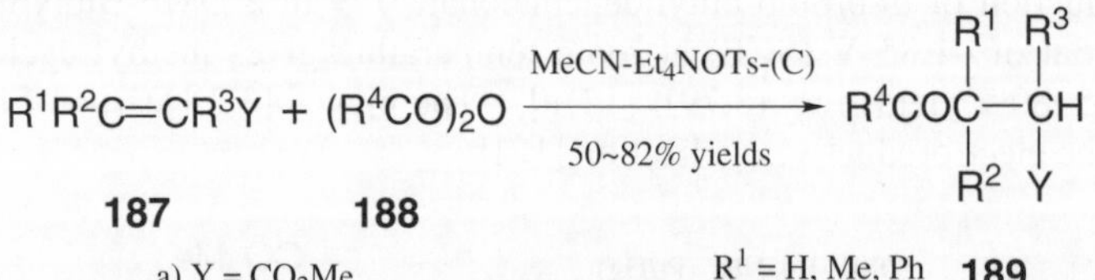

$$\text{R}^1\text{R}^2\text{C=CR}^3\text{Y} + (\text{R}^4\text{CO})_2\text{O} \xrightarrow[\text{50\textasciitilde82\% yields}]{\text{MeCN-Et}_4\text{NOTs-(C)}} \text{R}^4\text{CO}\underset{\underset{\text{R}^2}{|}}{\overset{\overset{\text{R}^1}{|}}{\text{C}}}\text{-}\underset{\underset{\text{Y}}{|}}{\overset{\overset{\text{R}^3}{|}}{\text{CH}}} \tag{3.82}$$

187   **188**

a) Y = CO$_2$Me   R^1 = H, Me, Ph **189**
b) Y = CO$_2$Et   R^2 = H, Me
c) Y = CN

of anode materials, i.e., Mg, Al, Zn, in an undivided cell is found to be effective in the above reactions [128]. The efficient synthetic methods of incorporating an FSO$_3$ group in a fluorocarbon chain are shown to be electrolysis of fluorosulfonic acid (FSO$_3$H) in the presence of perfluoroolefins [129a,b]. The addition of fluorosulfonic acid to perfluoro-2-methyl-3-isopropyl-2-pentene **190** is performed by electrolysis in an FSO$_3$H solution in an undivided cell to give a mixture of products **191** and **192** in 60% yield (based on GLC) (eq. (3.83)) [129c].

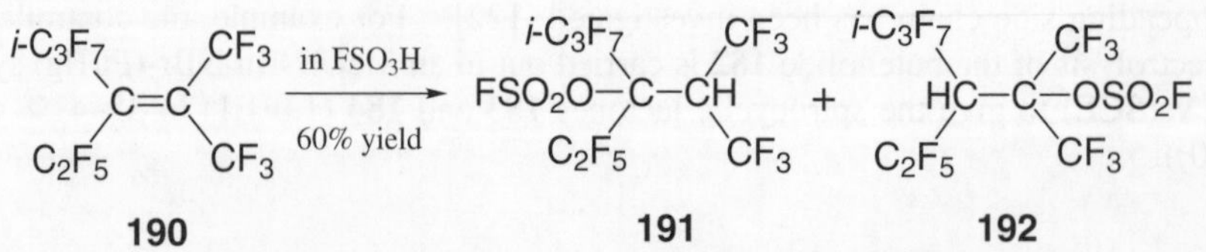

$$\underset{\text{C}_2\text{F}_5}{\overset{i\text{-C}_3\text{F}_7}{\diagdown}}\text{C=C}\underset{\text{CF}_3}{\overset{\text{CF}_3}{\diagup}} \xrightarrow[\text{60\% yield}]{\text{in FSO}_3\text{H}} \text{FSO}_2\text{O}-\underset{\underset{\text{C}_2\text{F}_5}{|}}{\text{C}}-\underset{\underset{\text{CF}_3}{|}}{\overset{\overset{i\text{-C}_3\text{F}_7}{|}}{\text{CH}}}\overset{\text{CF}_3}{} + \underset{\text{C}_2\text{F}_5}{\text{HC}}-\underset{\underset{\text{CF}_3}{|}}{\text{C}}-\text{OSO}_2\text{F} \tag{3.83}$$

190   **191**   **192**

A silylating method introducing a silyl group to the β-position of activated olefins has been developed.   Electroreductive silylation at the β position of α,β-unsaturated esters, nitriles, and ketones **193** is performed in a DMF-Bu$_4$NBr-(Mg) system in the presence of chlorotrimethylsilane **194**, yielding the trimethylsilylated products **195** in 34~79% yields (eq. (3.84)) [130].   The cyclopentadienyl anions, C$_5$H$_5^-$, derived from cyclopentadiene by

$$\begin{array}{c} R^1 \quad R^2 \\ \diagup\!\!\!\diagdown \\ H \quad X \end{array} + \text{Me}_3\text{SiCl} \xrightarrow[\text{34~79% yields}]{\text{DMF-Bu}_4\text{NBr-(Mg)}} \begin{array}{c} R^1 \quad R^2 \\ \diagdown\!\!\diagup \\ \text{Me}_3\text{Si} \quad X \end{array} \qquad (3.84)$$

194

193   X = CO$_2$Et, CN, COCH$_3$ **195**
                    R^1 = Ph, p-MeOC$_6$H$_4$, p-ClC$_6$H$_4$,
                         p-MeC$_6$H$_4$, 1-naphthyl, etc.
             R^2 = H, CO$_2$Et

electroreduction, can be trapped by silylation.   The electrolysis of **196** is carried out in an MeCN-Et$_4$NBr-(SUS) system in the presence of chlorotrimethylsilane **194** to give trimethylsilylcyclopentadiene **197** in 96% yield (eq. (3.85)) [131].

$$\bigcirc\!\!\!\!\diagdown + \text{Me}_3\text{SiCl} \xrightarrow[\text{96% yield}]{\text{MeCN-Et}_4\text{NBr-(SUS)}} \qquad (3.85)$$

196   **194**   **197**   SiMe$_3$
                 SUS = Stainless Steel

A short-lived anion radical derived from activated olefin under electroreduction in DMSO has been investigated [132].

3.4   Electroreduction of Other Activated Olefins Other Than Enones and Enoates

3.4.1   Reaction of Cyanoolefins

The electrochemical hydrodimerization of acrylonitrile **198** to adiponitrile **199** has been widely investigated (eq. (3.86)) [133~135].   The electroreduction taking place in an aqueous medium using sodium phosphate in the presence of a small concentration of an ammonium salt (R$_4$N$^+$) derived from hexamethylenediamine is a well-known industrial process established by Baizer at Monsanto [91,92,134].   The exclusive hydrodimerization of acrylonitrile is performed in an aqueous Et$_4$NOTs-(Pb) system to yield adiponitrile in virtually quantititative yields and at current efficiencies close to 100% [92,134a,136].   An interpretation of the kinetics in the electrohydrodimerization of acrylonitrile to adiponitrile has been attempted [137].   Detection of acrylonitrile anion radical intermediate performed by means of a scanning electrochemical microscopy [138].   The tail-to-tail hydrodimerization of conjugated nitriles, MeCH=CHCN, EtCH=CHCN etc., preferentially proceeds in protic media, whereas the head-to-tail dimerization mainly occurs in acetonitrile [139].   The use of a gas depolarized anode, which comprises a mixture of binder (a fluorocarbon polymer) and 5% platinum catalyst supported on a conductive electrode substrate (carbon paper, stainless steel, nickel etc.), has been patented [140].   A theoretical and flow-cell experimental study for adiponitrile electrosynthesis revealed that the most favored route is *via* an anionic intermediate of acrylonitrile [141].   Practical procedures for the hydrodimerization of acrylonitrile, methacrylonitrile, and 2-pentenenitrile have been patented [142].   The trimerization of methacrylonitrile has been attained in a DMSO-Et$_4$NOSO$_2$Et-(Pb) system [143].

$$CH_2=CHCN \xrightarrow[\substack{-1.80\sim-1.91\ V\ (SCE)}]{aq.\ Et_4NOTs-(Hg)} NC-(CH_2)_4-CN \qquad (3.86)$$

198   75~100% yields   **199**

The electroreduction of α,β-unsaturated nitriles in acetonitrile undergoes either cyanomethylation, hydrogenation, or hydrodimerization, depending on the structure of the nitrile. For example, the electrolysis of cinnamonitrile **200** in an MeCN-Et$_4$NBF$_4$(0.1M)-(Hg) system yields the cyanomethylated product **201** in 80% yield (eq. (3.87)) [144]. However, the electroreduction of 3-methylcinnamonitrile in a similar medium affords the corresponding cyanomethylated product in 30% yield together with the hydrogenated product (23%) [144].

$$PhCH=CHCN \xrightarrow[\substack{80\%\ yield}]{MeCN-Et_4NBF_4(0.1M)-(Hg)} PhCH\begin{smallmatrix}CH_2CN\\ \\CH_2CN\end{smallmatrix} \qquad (3.87)$$

200   **201**

The electroreductive coupling of α,β-unsaturated nitriles with carbonyl compounds takes place in a DMF-Et$_4$NOTs-(Pb) system in the presence of chlorotrimethylsilane to afford the corresponding coupling products in good yield [145]. The coupling of acrylonitrile **198** and butanal **202** gives the adduct **203** in 81% yield (eq. (3.88)).

$$CH_2=CHCN + Pr-CHO \xrightarrow[\substack{Me_3SiCl,\ 81\%\ yield}]{DMF-Et_4NOTs-(Pb)} Pr\overset{OH}{\diagup\diagdown}CN \qquad (3.88)$$

198   **202**   **203**

The electroreductive hydrogenation of acrylonitrile, leading to propionitrile, has been controlled by the nature and concentration of the cationic surfactant and the control of pH value at neutral [146].

The electroreduction of 2-methoxy-3-phenylpropenonitrile **204** in an aqueous MeCN(40%)-LiCl(0.15M)-(Hg) system yields 2-methoxy-3-phenylpropanonitrile **205** as a major product together with **206** (eq. (3.89)) [147].

$$Ph-CH=C\begin{smallmatrix}CN\\ \\OMe\end{smallmatrix} \xrightarrow[\substack{-1.90\ V\ (SCE)}]{aq.\ MeCN(40\%)-LiCl-(Hg)} \begin{array}{c} Ph-CH_2-CH\begin{smallmatrix}CN\\ \\OMe\end{smallmatrix}\ (70\%) \\ + \\ Ph-CH=CH-CN\ (30\%) \end{array} \qquad (3.89)$$

204   **205**

206

The hydrodimerization of 1,2-diactivated olefins, XCH=CHY (X, Y = electron-withdrawing groups), preferentially proceeds with tetraethylammonium ion rather than alkali metal cations [148]. The presence of alkali metal cations increases the quantity of propionitrile as a by-product. 1-Cyano-1,3-butadiene undergoes hydrodimeriation to give coupling products in 65~70% yield [134b].

The cyclohydrodimerization of activated carbon-carbon double bonds of benzylidene malononitrile homologs has been attained [149,150].   The electroreduction of β-phenyl-benzylidene malononitrile **207** (R = Ph) in a DMF-Bu$_4$NI-(Hg) system affords the cyclized products **208** and **209** in over 90% yields (eq. (3.90)) [149a].   The mechanism of the

207   R = Ph,   **208**   **209**   (3.90)

cyclohydrodimerization of the p-methylbenzylidene malononitrile anion radical has been discussed on the bases of kinetic and activation parameters [150].   The electroreduction of 2-substituted propionitriles **210** has been attempted in different pH media.   Under strong acidic conditions, the electrolysis of **210** in an aqueous EtOH(50%)-HCl(0.01M)-(Hg) system undergoes hydrodimerization followed by intramolecular cyclization to give the cyclopentadiene derivative **211** in 65% yield (eq. (3.91)) [151].   More recently, the controlled potential electrolysis of ethyl benzylidenecyanoacetate **210** in an aqueous EtOH(50%)-Britton-Robinson buffer-(Hg) system at −1.2 V (SCE) affords 1-amino-2,5-dicarbethoxy-5-cyano-3,4-diphenylcyclopent-1-ene in ~70% yield [152].   Electrohydrogenation of the double bond of **210** occurs at pH 8.6 to give the compound **212** in around 10% yield (eq. (3.92)).

(3.91)

210   **211**

(3.92)

212

α-Methyleneglutaronitrile has been electrohydrodimerized to yield 1,3,6,8-tetracyanooctane in 93% current yield [153].

Diethyl fumarate couples with two molecules of acrylonitrile to yield diethyl α,α'-bis(2-cyanoethyl)succinate in good yield [154].   For diethyl fumarate and cinnamonitrile, evidence for the formation of an anion radical which undergoes dimerization and polymerization reactions and of the dianion formation at more negative potentials has been presented [155].

An attempt has been made to couple acrylonitrile electroreductively with a more easily reduced partner, *e.g.*, acetone, styrene, 1,1-diphenylethylene, benzophenone, benzaldehyde, at the cathode voltage [156].   Relatively low-melting polyacrylonitriles (average mol. wt. 600~1300) are formed under very low water concentrations [157].   Electro-copolymerization of acrylonitrile and methyl acrylate onto graphite fibers has been performed in an aqueous sulfuric acid solution [158].

3.4.2   Reaction of Nitroolefins

Nitro alkenes can be converted into the corresponding ketones or oximes by the combination of electroreduction followed by chemical reaction [159].   For example, the electroreduction of nitro alkenes **213** is carried out in an aqueous CH_2Cl_2/dioxane (4/1)-$HClO_4$(20%)-(Pb) system and then the electrolyte is treated with 37% formaldehyde under stirring for 20 min without passing current to give the ketones **214** in 63~95% yields (eq. (3.93)).   On the other hand, after electrolysis (3.5 F/mol of electricity passed), the organic phase is treated with a mixture of $NH_2OH \cdot HCl$ and sodium acetate to give the corresponding oximes **215** in good yields (eq. (3.94)) [159].

The electroreduction of nitro olefins derived from asomatic aldehydes can be converted into the corresponding oximes in an MeOH-H_2SO_4(20%)-(Pt) system in the presence of *p*-toluenesulfonic acid [160].

3.4.3   Reaction of Miscellaneous Activated Olefins

The electroreductive dimerization of phenyl vinyl sulfone **216** has been attempted in a DMF-Et_4NBF_4(0.1M)-(Hg) system in the presence of phenol as a proton donor, giving two-major products **217** and **218** (eq. (3.95)) [161].

The manner of the electroreductive coupling of α,β-unsaturated sulfones depends on the nature of the substituents at the β position.   The electrocatalytic cyclodimerization has been observed with vinyl sulfones [162a].   The electrolysis of aryl vinyl sulfones **219** in a DMF-Et_4NClO_4(0.1M)-(Hg) system at -1.35 V (Ag/AgI/I$^-$) affords the cyclodimerized product **220** in 55~95% yields (eq. (3.96)) [162a].

$$2 \; ArSO_2CH=CH_2 \xrightarrow[\text{55~95\% yields}]{\text{cathode}} \quad \begin{array}{c} SO_2Ar \\ \square \\ SO_2Ar \end{array} \qquad (3.96)$$

219   **220**

Dimerization and hydrodimerization processes are found in the electroreduction of β-substituted α,β-unsaturated sulfones. For example, the electrolysis of β-ethylvinyl sulfone **221** in a DMF-Et$_4$NClO$_4$(0.1M)-(Hg) system yields the corresponding hydrodimer **222** in 75% yield (eq. (3.97)) [162b,c].

$$2 \; ArSO_2CH=CHEt \xrightarrow[\text{75\% yield}]{\text{DMF-Et}_4\text{NClO}_4\text{(0.1M)-(Hg)}} \begin{array}{c} ArSOCH_2-CHEt \\ | \\ ArSOCH_2-CHEt \end{array} \qquad (3.97)$$

221   **222**

The electroreduction of 1-fluoro-2-arylvinyl phenyl sulfones **223** in an MeCN-Bu$_4$NBF$_4$(0.1M)-(Pt) system in the presence of a proton donor (e.g., PhOH, AcOH, etc.) involves the cleavage of carbon-sulfur and/or carbon-fluorine bonds, resulting in the formation of 1-fluoro-2-arylethylenes **224** and stylenes **225** (eq. (3.98)) [163].

$$Ar-CH=C\begin{array}{c} F \\ \backslash \\ SO_2Ph \end{array} \xrightarrow{\text{MeCN-Bu}_4\text{NBF}_4\text{(0.1M)-(Pt)}} \begin{array}{c} Ar-CH=C\begin{array}{c} F \\ \backslash \\ H \end{array} \quad (50\%) \quad \textbf{224} \\ + \\ Ar-CH=CH_4 \quad (46\%) \\ \textbf{225} \end{array} \qquad (3.98)$$

223   Ar = Ph, p-MeC$_6$H$_4$, p-MeOC$_6$H$_4$,
               p-ClC$_6$H$_4$, p-CNC$_6$H$_4$,

The cathodic activation of 2-substituted vinyl phenyl sulfoxides **226** in deuteriated solvents (e.g., DC3CN, DMSOd6) allows exclusive conversion into monodeuteriated analogs **227** when electrolyses are stopped after a catalytic amount of electricity has passed through the electrolysis cell (eq. (3.99)) [164].

$$PhSOCH=C\begin{array}{c} R^1 \\ \backslash \\ R^2 \end{array} \xrightarrow[\text{45\% yield}]{\text{CD}_3\text{CN(or DMSOd)}_6\text{-Me}_4\text{NBF}_4\text{(0.1M)-(Hg)}} PhSOCD=C\begin{array}{c} R^1 \\ \backslash \\ R^2 \end{array} \qquad (3.99)$$

226   **227**

R^1, R^2 = H, Ph

The preparative electrolysis of the vinyl triflate **228** in a DMF-Bu$_4$NBF$_4$(0.2M)-(Mg/C) system in the presence of a catalytic amount of PdCl$_2$(PPh$_3$)$_2$ together with a stoichiometric amount of benzoic acid affords selectively the steroidal alkene **229** in 80% yield (eq. (3.100)) [165]. The reaction probably proceeds via a catalytic cycle initiated by the one-step two-electron reduction of the Pd complex followed by oxidative addition of the vinyl triflates. The palladium-catalyzed reaction of allyl acetates with electrophiles has been reported [166,167].

$$\text{228} \xrightarrow[\substack{\text{PdCl}_2\text{L}_2,\ 80\%\ \text{yield} \\ \text{L = PPh}_3}]{\text{DMF-Bu}_4\text{NBF}_4(0.2\text{M})\text{-(Mg/C)}} \text{229} \tag{3.100}$$

3.5   Miscellaneous Reactions

The *cis/trans* isomerization of *cis*-cinnamyl alcohol has been attained by electrolysis in an aqueous MeOH(70%)-LiClO$_4$(0.1M)-(Hg) system in a divided cell to give the *trans*-isomer in 95% ratio [168].

The addition of alcohol to double bonds has been realized by use of paired electrochemical reaction with a Raney-nickel electrode [169].   The electrolysis of cyclohexene **230** in an MeOH-Et$_4$NBr-(Raney Ni) system in an undivided cell affords methoxycyclohexane **233** in 83% yield *via* bromomethoxide **232** (eq. (3.101)) [169].

$$2\text{Br}^- \longrightarrow \text{Br}_2 + 2\ e^- \quad \text{----- (a)}$$

$$\text{Br}_2 + \underset{\textbf{230}}{\bigcirc} \longrightarrow \underset{\textbf{231}}{\bigcirc}^{\text{Br}^+} + \text{Br}^- \quad \text{----- (b)}$$

$$\textbf{231} + \text{MeOH} \longrightarrow \underset{\textbf{232}}{\bigcirc}^{\text{Br}\ \text{OMe}} \quad \text{----- (c)}$$ (3.101)

$$\underset{\textbf{232}}{\bigcirc}^{\text{Br}\ \text{OMe}} + 2\ e^- + \text{H}^+ \longrightarrow \underset{\textbf{233}}{\bigcirc}^{\text{OMe}} + \text{Br}^- \quad \text{----- (d)}$$

The electrosynthesis of propylene oxide by operating the electrode potential as a sine wave with 60 Hz transmission power has been developed [170].

The preparation of keto-alcohols from α-alkylcycloalkenes has been attained by the combination of ozonization and electroreduction [171].   The electrolysis of the ozonide derived from p-menthene **234** is performed in an aqueous AcONa-(Pb) system to give the keto-alcohol **235** in 70% yield (eq. (3.102)).

$$\underset{\textbf{234}}{} \xrightarrow[\text{AcOH}]{\text{O}_3} \text{Ozonide} \xrightarrow[\text{70\% yield}]{\text{aq. AcONa-(Pb)}} \underset{\textbf{235}}{} \tag{3.102}$$

References

[1] A. Wieckowski, S. D. Rosasco, G. N. Salaita, A. Hubbard, B. E. Bent, F. Zaera, D. Godbey, G. A. Somorjai, *J. Am. Chem. Soc.*, **107**, 5910 (1985).

[2] M. Hourani, A. Wieckowski, *Langmuir*, **6**, 379 (1990).

[3] P. S. Fedkiw, J. M. Potente, W.-H. Her, *J. Electrochem. Soc.*, **137**, 1451 (1990).

[4] I. Willner, M. Rosen, Y. Eichen, *J. Electrochem. Soc.*, **138**, 434 (1991).

[5] H. Lund, H. Doupeux, M. A. Michel, G. Mousset, J. Simonet, *Electrochim. Acta*, **19**, 629 (1974).

[6] (a) S. Sibille, J. Coulombeix, J. Perichon, J.-M. Fuchs, A. Mortreux, F. Petit, *J. Mol. Catal.*, **32**, 239 (1985); (b) H. Masotti, J. C. Wallet, G. Peiffer, F. Petit, A. Mortreux, G. Buono, *J. Organomet. Chem.*, **308**, 241 (1986).

[7] S. H. Langer, I. Feiz, C. P. Quinn, *J. Am. Chem. Soc.*, **93**, 1092 (1971).

[8] R. Carlier, K. Boujlel, J. Simonet, *J. Electroanal. Chem. Interfacial Electrochem.*, **221**, 275 (1987).

[9] A. Froling, F. Hornung, R. O. De Jongh, *Recl. Trav. Chim. Pays-bas*, **105**, 123 (1986).

[10] C. Corvaja, G. Farnia, G. Formenton, W. Navarrini, G. Sandona, V. Tortelli, *J. Phys. Chem.*, **98**, 2307 (1994).

[11] C. Iwakura, T. Abe, H. Inoue, *J. Electrochem. Soc.*, **143**, L71 (1996).

[12] O. Lerflaten, V. D. Parker, *Acta Chem. Scand.*, **B36**, 193 (1982).

[13] (a) M. A. Fox, D. Shultz, *J. Org. Chem.*, 53, 4386 (1988); (b) D. A. Shultz, M. A. Fox, *J. Org. Chem.*, 55, 1047 (1990); (c) J. L. Muzyka, M. A. Fox, *J. Org. Chem.*, **56**, 4549 (1991); (d) M. O. Wolf, H. H. Fox, M. A. Fox, *J. Org. Chem.*, **61**, 287 (1996).

[14] G. Farnia, F. Maran, G. Sandona, *J. Chem. Soc., Faraday Trans. I*, **82**, 1885 (1986).

[15] M. M. Baizer, J. D. Anderson, *J. Org. Chem.*, **30**, 1348 (1965).

[16] T. R. Chen, M. R. Anderson, D. G. Peters, *J. Electroanal. Chem. Interfacial Electrochem.*, **197**, 341 (1986).

[17] R. L. Blankespoor, *J. Org. Chem.*, **50**, 3010 (1985).

[18] W. H. Smith, A. J. Bard, *J. Electroanal. Chem. Interfacial Electrochem.*, **76**, 19 (1977).

[19] (a) G. Farnia, F. Marcuzzi, G. Melloni, G. Sandona, *J. Am. Chem. Soc.*, **106**, 6503 (1984); (b) G. Farnia, F. Marcuzzi, G. Melloni, G. Sandona, M. V. Zucca, *J. Am. Chem. Soc.*, **111**, 918 (1989).

[20] M. Fruianu, M. Marchetti, G. Melloni, G. Sanna, R. Seeber, *J. Chem. Soc., Perkin Trans. II*, **1994**, 2039.

[21] V. A. Lopyrev, T. G. Ermakova, T. N. Kashik, L. E. Protasova, T. I. Vakul'skaya, *Khim. Geterotsikl. Soedin.*, **1986**, 315.

[22] (a) U. Duennbier, W. Jugelt, *Pharmazie*, **46**, 512 (1991); (b) W. Jugelt, U. Dünnbier, *Z. Chem.*, **30**, 173 (1990).

[23] E. Kariv-Miller, D. F. Dedolph, C. M. Ryan, T. J. Mahachi, *J. Heterocyclic Chem.*, **22**, 1389 (1985).

[24] L. G. Chatten, L. Amankwa, *Analyst*, **110**, 1369 (1985).

[25] K. Otsuka, T. Yagi, *J. Catal.*, **145**, 289 (1994).

[26] (a) K. N. Campbell, E. E. Young, *J. Am. Chem. Soc.*, **65**, 965 (1943); (b) R. A. Benkeser, C. A. Tincher, *J. Org. Chem.*, **33**, 2727 (1968); (c) S. Wawzonek, D. Wearing, *J. Am. Chem. Soc.*, **81**, 2067 (1959); (d) W. M. Moore, D. G. Peters, *J. Am. Chem. Soc.*, **97**, 139 (1975).

[27] V. I. Filimonova, I. V. Kirilyus, *Deposited Doc.*, **1983**, 93.

[28] A. Arcadi, S. Cacchi, I. Carelli, A. Curulli, A. Inesi, F. Marinelli, *Synlett.*, **1990**, 408.

[29] T. R. Chen, M. R. Anderson, S. Grossman, D. G. Peters, *J. Org. Chem.*, **52**, 1231 (1987).

[30] J. Z. Stemple, D. G. Peters, *J. Org. Chem.*, **54**, 5318, (1989).

[31] J. Y. Becker, *Isr. J. Chem.*, **26**, 196 (1985).

[32] A. E. J. Forno, *Chem. Ind.*, **1968**, 1728.

[33] G. Schlegel, H. J. Schafer, *Chem. Ber.*, **116**, 960 (1983).

[34] (a) R. Dietz, M. E. Peover, R. Wilson, *J. Chem. Soc., B*, **1968**, 75; (b) R. Dietz, M. E. Peover, *J. Chem. Soc., B*, **1970**, 1369; (c) A. Zweig and A. K. Hoffmann, *J. Am. Chem. Soc.*, **84**, 3278 (1962).

[35] R. W. Howsam, C. J. M. Stirling, *J. Chem. Soc., Perkin Trans. II*, **1972**, 847.

[36] (a) W. Kemula, J. Kornacki, *Rocz. Chem.*, **36**, 1835 (1962); (b) W. Kemura, J. Kornacki, *Rocz. Chem.*, **36**, 1849 (1962).

[37] (a) U. Hayat, P. N. Bartlett, G. H. Dodd, J. Barker, *J. Electroanal. Chem. Interfacial Electrochem.*, **220**, 287 (1987); (b) M. L. Vincent, D. G. Peters, *J. Electroanal. Chem.*, **324**, 93, (1992); (c) D. G. Peters, *Stud. Org. Chem.*, **30**, 261 (1987).

[38] M. L. Vincent, D. G. Peters, *J. Electroanal. Chem.*, **328**, 63 (1992).

[39] P. C. Wang, J. M. Renga, *U. S. PAT.* 4638064 (1987).

[40] D. Serve, *Electrochim. Acta*, **21**, 1171 (1976).

[41] A. Kunai, O. Ohnishi, T. Sakurai, M. Ishikawa, *Chem. Lett.*, **1995**, 1051.

[42] F. Mizutani, N. Sato, T. Sekine, *Denki Kagaku Oyobi Kogyo Butsuri Kagaku*, **46**, 274 (1978).

[43] R. Annino, R. J. Boczkowski, D. J. Bolton, W. E. Geiger Jr., D. T. Jackson Jr., J. Mahler, *J. Electroanal. Chem. Interfacial Electrochem.*, **38**, 403 (1972).

[44] D. Sazou, P. Karabinas, D. Jannakoudakis, *J. Electroanal. Chem. Interfacial Electrochem.*, **176**, 225 (1984).
[45] D. Sazou, P. Karabinas, *Collect. Czech., Chem. Commun.*, **52**, 2132 (1987).
[46] (a) C. I. De Matteis, J. H. P. Utley, *J. Chem. Soc., Perkin Trans. II*, **1992**, 879; (b) J. H. P. Utley, M. Gullu, C. I. De Matteis, M. Motevalli, M. F. Nielsen, *Tetrahedron*, **51**, 11873 (1995); (c) C. I. De Matteis, J. H. P. Utley, *J. Chem. Soc., Perkin Trans. II*, **1992**, 879.
[47] H. E. Bode, C. G. Sowell, R. D. Little, *Tetrahedron Lett.*, **31**, 2525 (1990).
[48] E. N. Pervii, A. N. Sofronkov, N. M. Fedyshina, *Zh. Prinkl. Khim.*, **58**, 1905 (1985).
[49] S. Ishiwata, T. Nozaki, *J. Pharm. Soc. Japan*, **71**, 1257 (1951).
[50] A. Bryan, J. Grimshaw, *Electrochim. Acta*, **1996**, in contribution.
[51] N. Takano, N. Takeno, M. Morita, *Nippon Kagaku Kaishi* (Japanese), **1983**, 1753.
[52] V. D. Parker, *Acta Chem. Scand.*, **B35**, 295 (1981).
[53] (a) S. Abe, T. Nonaka, T. Fuchigami, *J. Am. Chem. Soc.*, **105**, 3630 (1983); (b) N. Schoo, H.-J. Schäfer, *Dechema Monogr.*, **125**, 815 (1992); (c) N. Schoo, H.-J. Schäfer, *Liebigs Ann. Chem.*, 1993, 601; (d) U. Höweler, N. Schoo, H.-J. Schäfer, *Liebigs Ann. Chem.*, **1993**, 609; (e) R. N. Gourley, J. Grimshaw, P. G. Millar, *J. C. S., Chem. Commun.*, **1967**, 1278.
[54] (a) S. Abe, T. Nonaka, *Chem. Lett.*, 1983, 1541; (b) T. Nonaka, S. Abe, T. Fuchigami, *Bull. Chem. Soc. Jpn.*, **56**, 2778 (1983).
[55] K. Hu, M. E. Niyazymbetov, D. H. Evans, *J. Electroanal. Chem.*, **396**, 457 (1995).
[56] (a) A. Aboulfida, J. C. Roze, J. P. Pradere, M. Jubault, *Phosphorus, Sulfur, Silicon Relat. Elem.*, **54**, 123 (1990); (b) M. Jubault, A. Tallec, B. Bujoli, J.-C. Rozé, J.-P. Pradére, *Tetrahedron Lett.*, **26**, 745 (1985).
[57] C. D'Silva, K. T. Douglas, *J. Org. Chem.*, **48**, 263 (1983).
[58] J. Delaunay, A. O. Moing, J. Simonet, *J. Chem. Soc., Chem. Commun.*, **1983**, 820.
[59] J. M. Mellor, B. S. Pons, J. H. A. Stibbard, *J. Chem. Soc., Perkin Trans. I*, **1981**, 3092.
[60] R. A. J. Smith, D. J. Hannah, *Tetrahedron Lett.*, **21**, 1081 (1980).
[61] P. G. Gassman, O. M. Rasmy, T. O. Murdock, K. Saito, *J. Org. Chem.*, **46**, 5455 (1981).
[62] T. Sato, Y. Komeichi, S. Kobayashi, A. Omura, *Bull. Chem. Soc. Jpn.*, **55**, 520 (1982).
[63] (a) P. G. Gassman, C-J. Lee, *Synth. Commun.*, **24**, 1465 (1994); (b) P. G. Gassman, C-J. Lee, *Synth. Commun.*, **24**, 1457 (1994); (c) P. G. Gassman, C-J. Lee, *J. Am. Chem. Soc.*, **111**, 739 (1989).
[64] S. Satoh, H. Suginome, M. Tokuda, *Bull. Chem. Soc. Jpn.*, **54**, 3456 (1981).
[65] (a) J. P. Petrovich, J. D. Anderson, M. M. Baizer, *J. Org. Chem.*, **31**, 3897 (1966); (b) J. D. Anderson, M. M. Baizer, J. P. Petrovich, *J. Org. Chem.*, **31**, 3890 (1966).
[66] S. Torii, H. Okumoto, Md. A. Rashid, M. Mohri, *Synlett*, **1992**, 721.
[67] H. Lund, C. Degrand, *Tetrahedron Lett.*, **1977**, 3593.
[68] R. L. Cook, A. F. Sammells, *J. Electrochem. Soc.*, **136**, 1845 (1989).
[69] J. D. Anderson, M. M. Baizer, E. J. Prill, *J. Org. Chem.*, **30**, 1645 (1965).
[70] A. Böhm, K. Meerholz, J. Heinze, K. Müellen, *J. Am. Chem. Soc.*, **114**, 688 (1992).
[71] J. Simonet, *C. R. Acad. Sci., Ser. C*, **267**, 1548 (1968).
[72] (a) R. M. Bastida, E. Brillas, J. M. Costa, *J. Electroanal. Chem. Interfacial Electrochem.*, **227**, 55 (1987); (b) R. M. Bastida, E. Brillas, J. M. Costa, *J. Electrochem. Soc.*, **138**, 2289 (1991).
[73] J. Grimshaw, R. J. Haslett, *J. Chem. Soc., Perkin Trans., I*, **1979**, 395.
[74] P. Manisankar, A. Gomathi, T. Vasudevan, D. Velayutham, R. K. Srinivasan, S. Chidambaram, *Trans. SAEST*, **28**, 152 (1993).
[75] D. A. Jaeger, D. Bolikal, B. Nath, *J. Org. Chem.*, **52**, 276 (1987).
[76] (a) L. H. Klemm, D. R. Olson, *J. Org. Chem.*, **38**, 3390 (1973); (b) J. Berthelot, *Electrochim. Acta*, **32**, 179 (1987).
[77] F. Fournier, D. Davoust, J.-J. Basselier, *Tetrahedron*, **23**, 5677 (1985).
[78] L. Horner, C. Franz, *Z. Naturforsch., B: Anorg. Chem. Org. Chem.*, **40B**, 822 (1985).
[79] N. Schoo, H. -J. Schäfer, *Dechema Monogr.*, **125**, 815 (1992).
[80] (a) T. Isoya, K. Nakagawa, A. Aoshima, *Stud. Org. Chem.*, **30**, 415 (1987); (b) V. A. Klimov, A. P. Tomilov, *Russ. J. Appl. Chem.*, **66**, 898 (1993).
[81] S. U. Pedersen, R. G. Hazell, H. Lund, *Acta Chem. Scand.*, **B41**, 336 (1987).
[82] L. Mandell, F. J. Heldrich, R. A. Day, JR., *Synth. Commun.*, **11**, 55 (1981).
[83] G. Mason, G. Le Guillanton, J. Simonet, *J. Chem. Soc., Chem. Commun.*, **1982**, 571.
[84] (a) S. Takenaka, C. Shimakawa, M. Uchida, *Japanese Patent Kokai* JP 05-25672 (1993); (b) S. Takenaka, M. Uchida, C. Shimakawa, H. Kamaike, *Japanese Patent Kokai* JP 05-156478 (1993); (c) Y. Kunugi, T. Nonaka, *J. Electroanal. Chem.*, **356**, 163 (1993).
[85] E. A. Casanova, M. C. Dutton, D. J. Kalota, J. H. Wagenknecht, *J. Appl. Electrochem.*, **25**, 479 (1995).
[86] H. Lund, *J. Electroanal. Chem. Interfacial Electrochem.*, **202**, 299 (1986).
[87] J. Claret, J. M. Feliu, C. Muller, J. M. Ribo, X. Serra, *Tetrahedron*, **41**, 1713 (1985).
[88] (a) J. Wiemann, M. Monot, J. Gardan, *C. R. Acad. Sci.*, **245**, 172 (1957); (b) J. Wiemann, *Bull. Soc. Chim. Fr.*, **1964**, 2545.

[89] M. M. Baizer, J. D. Anderson, *J. Org. Chem.*, **30**, 3138 (1965).
[90] N. L. Weinberg, *Technique of Electroorganic Synthesis, Part II*. p. 87, John Wiley & Sons, Inc., New York (1974), and references cited therein.
[91] M. M. Baizer, J. D. Anderson, *J. Electrochem. Soc.*, **111**, 223 (1964).
[92] M. M. Baizer, *J. Electrochem. Soc.*, **111**, 215 (1964).
[93] (a) E. Lamy, L. Nadjo, J. M. Saveant, *J. Electroanal. Chem. Interfacial Electrochem.*, **42**, 189 (1973); (b) E. Lamy, L. Nadjo, J. M. Saveant, *J. Electroanal. Chem.*, Interfacial Electrochem., **50**, 141 (1974).
[94] V. D. Parker, *Acta Chem. Scand.*, **B35**, 149 (1981).
[95] M. D. Koppang, G. A. Ross, N. F. Woolsey, D. E. Bartak, *J. Am. Chem. Soc.*, **108**, 1441 (1986).
[96] (a) V. D. Parker, *Acta Chem. Scand.*, B35, 147 (1981); (b) V. D. Parker, *Acta Chem. Scand.*, **B37**, 393 (1983).
[97] I. Nishiguchi, T. Hirashima, *Angew. Chem., Int. Ed. Engl.*, **22**, 52 (1983).
[98] (a) N. Kise, M. Echigo, T. Shono, *Tetrahedron Lett.*, **35**, 1897 (1994).
[99] J. H. P. Utley, M. Güllü, M. Motevalli, *J. Chem. Soc., Perkin Trans. 1*, **1995**, 1961.
[100] I. Fussing, M. Güllü, O. Hammerich, A. Hussain, M. F. Nielsen, J. H. P. Utley, *J. Chem. Soc., Perkin Trans. II*, **1996**, 649.
[101] M. N. Elinson, S. K. Feducovich, A. A. Zakharenkov, B. I. Ugrak, G. I. Nikishin, S. V. Lindeman, J. T. Struchkov, *Tetrahedron*, **51**, 5035 (1995).
[102] (a) G. I. Nikishin, M. N. Elinson, S. K. Feducovich, *Tetrahedron Lett.*, **32**, 799 (1991); (b) G. I. Nikishin, M. N. Elinson, S. K. Feducovich, B. I. Ugrak, *Tetrahedron Lett.*, **33**, 3223 (1992).
[103] H. M. Fahmy, N. F. A. Fattah, M. R. H. Elmoghayar, M. A. Azzem, *J. Chem. Soc., Perkin Trans. II*, **1988**, 1.
[104] J. M. Kern, H. J. Sch_fer, *Electrochim. Acta*, **30**, 81 (1985).
[105] G. Le Guillanton, Q. T. Do, J. Simonet, *Tetrahedron Lett.*, **27**, 2261 (1986).
[106] G. Mabon, C. Moinet, J. Simonet, *J. C. S., Chem. Commun.*, **1981**, 1040.
[107] M. C. S. Pitchumani, V. Krishnan, *Bull. Electrochem.*, **2**, 395 (1986).
[108] E. Lamy, L. Nadjo, J. M. Savent, *Nouv. J. Chim.*, **3**, 21 (1979).
[109] H. T. Chen, H. J. Tien, T. T. Lai, T. C. Chou, *J. Chin. Inst. Chem. Eng.*, **16**, 25 (1985).
[110] (a) G. Filardo, S. Gambino, G. Silvestri, A. Gennaro, E. Vianello, *J. Electroanal. Chem. Interfacial Electrochem.*, **177**, 303 (1984); (b) S. Gambino, A. Gennaro, G. Filardo, G. Silvestri, E. Vianello, *J. Electrochem. Soc.*, **134**, 2172 (1987).
[111] H. L. S. Maia, L. S. Monteiro, L. Grehn, U. Ragnarsson, *Pept., Proc. Eur. Pept. Symp., 22nd*, **1992**, 211.
[112] S. Gambino, G. Filardo, G. Silvestri, *J. Appl. Electrochem.*, **12**, 549 (1982).
[113] C. Degrand, R. Mora, H. Lund, *Acta Chem. Scand.*, **B37**, 429 (1983).
[114] S. Derien, J. C. Clinet, E. Duñach, J. Périchon, *Synlett.*, **1990**, 361.
[115] S. Derien, J. C. Clinet, E. Duñach, J. Périchon, *J. Organomet. Chem.*, **424**, 213 (1992).
[116] A. Mortreux, F. Petit, *New Science in Transition Metal Catalyzed Reactions, Adv. Chem. Series*, **230**, 261 (1992).
[117] T. Shono, M. Ishifune, H. Kinugasa, S. Kashimura, *J. Org. Chem.*, **57**, 5561 (1992).
[118] (a) S. Satoh, T. Taguchi, M. Itoh, M. Tokuda, *Bull. Chem. Soc. Jpn.*, **52**, 951 (1979); (b) E. Leonel, J. P. Paugam, J.-Y. Nedelec, J. Perichon, *J. Chem. Research (S)*, **1995**, 278.
[119] T. Shono, T. Nozoe, Y. Yamaguchi, M. Ishifune, M. Sakaguchi, H. Masuda, S. Kashimura, *Tetrahedron Lett.*, **32**, 1051 (1991).
[120] Y.-W. Lu, J.-Y. Nedelec, J.-C. Folest, J. Perichon, *J. Org. Chem.*, **55**, 2503 (1990).
[121] (a) S. T. Nugent, M. M. Baizer, R. D. Little, *Tetrahedron Lett.*, **23**, 1339 (1982); (b) R. D. Little, R. Wolin, G. Sowell, *Denki Kagaku*, **62**, 1105 (1994).
[122] M. A. Amputch, R. D. Little, *Tetrahedron*, **47**, 383 (1991).
[123] T. Shono, M. Mitani, *Nippon Kagaku Kaishi* (Japanese), **1972**, 2370.
[124] H. Lund, *Acta Chem. Scand.*, **B31**, 424 (1977).
[125] (a) T. Shono, I. Nishiguchi, H. Ohmizu *J. Am. Chem. Soc.*, **99**, 7396 (1977); (b) T. Shono, I. Nishiguchi, H. Ohmizu, *Chem. Lett.*, **1977**, 1021.
[126] T. J. Curphey, L. D. Trivadi, T. Layloff, *J. Org. Chem.*, **39**, 3831 (1974).
[127] H. Lund, C. Degrand, *Acta Chem. Scand.*, **B33**, 57 (1979).
[128] T. Ohno, H. Aramaki, H. Nakahiro, I. Nishiguchi, *Tetrahedron*, **52**, 1943 (1996).
[129] (a) H. Schwertfeger, G. Siegemund, H. Millauer, *J. Fluor. Chem.*, **29**, 104 (1985); (b) V. M. Rogovik, V. F. Cherstkov, V. A. Grinberg, Yu. B. Vasil'ev, M. G. Petrleitner, S. R. Sterlin, L. S. German, *Izv. Akad. Nauk SSSR, Ser. Khim.*, **1991**, 2362; (c) E. A. Avetisyan, A. F. Aerov, V. F. Cherstkov, B. L. Tumanskii, S. R. Sterlin, L. S. German, *Izv. Akad. Nauk, Ser. Khim.*, **1992**, 1208.
[130] T. Ohno, H. Nakahiro, K. Sanemitsu, T. Hirashima, I. Nishiguchi, *Tetrahedron Lett.*, **33**, 5515 (1992).
[131] H. G. Thomas, S. Kessel, E. Muller, *Chem. Ber.*, **119**, 2173 (1986).
[132] S. Margel, M. Levy, *J. Electroanal. Chem. Interfacial Electrochem.*, **56**, 259 (1974).
[133] P. Margaretha, V. D. Parker, *Acta Chem. Scand.*, **B36**, 260 (1982).

[134] (a) M. M. Baizer, *Tetrahedron Lett.*, **1963**, 973; (b) M. M. Baizer, J. D. Anderson, *J. Electrochem. Soc.*, **111**, 226 (1964); (c) M. M. Baizer (ed.), Organic Electrochemistry, Marcel Dekker, New York, (1973), Chapter XIX for a summary of the preparative aspects of the reaction.

[135] (a) L. R. Yeh, A. J. J. Bard, *J. Electrochem. Soc.*, **124**, 189 (1977); (b) L. R. Yeh and A. J. J. Bard, *J. Electrochem. Soc.*, **124**, 355 (1977).

[136] M. M. Baizer, *Chemtech.*, **1980**, 161.

[137] A. N. Haines, I. F. McConvey, K. Scott, *Electrochim. Acta*, **30**, 291 (1985).

[138] F. Zhou, A. J. Bard, *J. Am. Chem. Soc.*, **116**, 393 (1994).

[139] J. Y. Becker, T. A. Koch, *Electrochim. Acta*, **39**, 2067 (1994).

[140] J. C. Trocciola, *U. S. Pat.*, 4,566,957 (1972).

[141] K. Scott, B. Hayati, A. N. Haines, I. F. McConvey, *Chem. Eng. Technol.*, **13**, 376 (1990).

[142] (a) H. M. Fox, F. N. Ruehlen, *U. S. Pat.*, 3,689,382 (1972); (b) G. T. Miller, N. Y. Lewiston, *U. S. Pat.*, 3,492,209 (1970); (c) N. Kitaguchi, T. Shimizu, *Japanese Patent Kokai*, 60-50190 (1985); (d) N. Kitaguchi, T. Shimizu, *Japanese Patent Kokai*, 60-52587 (1985).

[143] (a) S. Ando, N. Kitaguchi, *Japanese Patent Kokai*, 60-94945 (1985); (b) S. Ando, N. Kitaguchi, *Japanese Patent Kokai*, 60-94954 (1985).

[144] A. J. Bellamy, J. B. Kerr, C. J. Mcgregor, I. S. Mackirdy, *J. Chem. Soc.*, *Perkin Trans. II*, **1982**, 161.

[145] T. Shono, H. Ohmizu, S. Kawakami, H. Sugiyama, *Tetrahedron Lett.*, **21**, 5029 (1980).

[146] L. Oniciu, D. A. Lowy, M. Jitaru, I. A. Silberg, B. C. Toma, *Rev. Roum. Chim.*, **32**, 701 (1987).

[147] M. Cariou, G. Mabon, G. Guillanton, *Tetrahedron*, **39**, 1551 (1983).

[148] J. P. Petrovich, M. M. Baizer, *J. Electrochem. Soc.*, **118**, 447 (1971).

[149] (a) L. A. Avaca, J. H. P. Utley, *J. Chem. Soc.*, *Perkin Trans. I*, **1975**, 971; (b) L. A. Avaca, J. H. P. Utley, *J. Chem. Soc.*, *Perkin Trans. I*, **1975**, 169.

[150] O. Lerflaten, V. D. Parker, *Acta Chem. Scand.*, **B36**, 225 (1982).

[151] M. A. Azzem, M. M. M. Ramiz, E. A. Ghali, H. M. Fahmy, M. R. H. Elmoghayar, *Monatsh. Chem.*, **118**, 229 (1987).

[152] H. M. Fahmy, M. A. Aboutabl, *Ind. J. Chem.*, **33**, 657 (1994).

[153] M. M. Baizer, J. D. Anderson, *J. Org. Chem.*, **30**, 1357 (1965).

[154] (a) M. M. Baizer, *J. Org. Chem.*, **29**, 1670 (1964); (b) J. P. Petrovich, M. M. Baizer, M. R. Ort, *J. Electrochem. Soc.*, **116**, 743 (1969).

[155] I. Vartieres, W. H. Smith, A. J. Bard, *J. Electrochem. Soc.*, **122**, 894 (1975).

[156] M. M. Baizer, J. L Ghruma, *J. Electrochem. Soc.*, **118**, 450 (1971).

[157] M. M. Baizer, J. D. Anderson, *J. Org. Chem.*, **30**, 1351 (1965).

[158] J. Chang, J. P. Bell, S. Shkolnik, *J. Appl. Pol. Sci.*, **34**, 2105 (1987).

[159] S. Torii, H. Tanaka, T. Kato, *Chem. Lett.*, **1983**, 607.

[160] T. Shono, H. Hamaguchi, H. Mikami, H. Nogusa, S. Kashimura, *J. Org. Chem.*, **48**, 2103 (1983).

[161] N. Djeghidjegh, J. Simonet, *J. Chem. Soc., Chem. Commun.*, **1988**, 1317.

[162] (a) J. Delaunay, G. Mabon, A. Orliac, J. Simonet, *Tetrahedron Lett.*, **31**, 667 (1990); (b) J. Delaunay, A. O.-L. Moing, J. Simonet, *New J. Chem.*, **17**, 393 (1993); (c) J. Delaunay, A. Orliac, J. Simonet, *Tetrahedron Lett.*, **36**, 2083 (1995).

[163] (a) A. Kunigi, K. Yamane, M. Yasuzawa, H. Matsui, H. Uno, K. Sakamoto, *Electrochim. Acta*, **38**, 1037 (1993); (b) A. Kunugi, S. Mori, S. Komatsu, H. Matsui, H. Uno, K. Sakamoto, *Electrochim. Acta*, **40**, 829 (1995).

[164] S. Diederichs, J. Delaunay, G. Mabon, J. Simonet, *Tetrahedron Lett.*, **36**, 8423 (1995).

[165] I. Chiarotto, I. Carelli, S. Cacchi, P. Pace, *J. Electroanal. Chem.*, **385**, 235 (1995).

[166] S. Torii, H. Tanaka, T. Katoh, K. Morisaki, *Tetrahedron Lett.*, **25**, 3207 (1984).

[167] P. Zhang, W. Zhang, T. Zhang, Z. Wang, W. Zhou, *J. Chem. Soc., Chem. Commun.*, **1991**, 491.

[168] K. Shyamsunder, K. K. Balasubramanian, C. S. Venkatachalam, *Stud. Org. Chem.*, **30**, 223 (1987).

[169] T. Yamada, T. Osa, T. Matsue, *Chem. Lett.*, **1987**, 995.

[170] J. D. Lisius, P. W. Hart, *J. Electrochem. Soc.*, **138**, 3678 (1991).

[171] J. Gora, K. Smigielski, J. Kura, *Synthesis*, **1982**, 310.

4. Electroreductive Reaction of Aromatic Compounds

4.1 Electroreduction of Aromatic Nuclei

4.1.1 Electroreductive Hydrogenation of Aromatic Nuclei

The electrocatalytic hydrogenation of unsaturated organic molecules in aqueous or mixed aqueous-organic media involves first the formation of chemisorbed hydrogen at a cathode with a low hydrogen over voltage (transition metal) by reduction of water or hydronium ions. Hydrogenation then proceeds as in catalytic hydrogenation by the reaction of the adsorbed substrate with chemisorbed hydrogen. The hydrogen adsorbed on the platinum electrode surface prior to the evolution of hydrogen can be used for the hydrogenation as an effective reagent [1].

The electrocatalytic hydrogenation of fused polycyclic aromatic compounds at Raney nickel electrodes has been investigated in details [2]. The electrohydrogenation of phenanthrene, anthracene, and naphthalene can stop at derivatives with a single aromatic ring, namely octahydrophenanthrene, octahydroanthracene, and tetralin. The best conditions are ethylene glycol or propylene glycol as the co-solvent containing between 1.5 to 5% water, a neutral or slightly acidic medium containing boric acid (0.1 M) as the buffer, sodium chloride or tetrabutylammonium chloride as the supporting electrolyte, at 80°C, and a current density of 42~48 mA/cm^2 [2]. The cathodic reduction of phenanthrene **1** in an aqueous Bu$_4$NOH-(Hg) system under passage of 10 F/mol of electricities affords the corresponding octahydrophenanthrene **2** in 74% yield together with decahydrophenanthrene (17%) (eq. (4.1)) [3]. The electrohydrogenation of phenanthrene in an aqueous (CH$_2$OH)$_2$(96%)-NaCl(0.1M)/H$_2$BO$_3$-(Raney-Ni) system at 80°C yields 9,10-dihydro-, tetrahydro-, and octahydrophenanthrenes in 57% current efficiency [4].

$$\text{(4.1)}$$

1   **2**

Platinum and rhodium on carbon, which are known to hydrogenate aromatic nuclei at atmosphere pressure, are also effective under the applied electrochemical conditions. The hydrogenation of benzene has been achieved preferentially in the presence of tetraethylammonium p-toluenesulfonate. Aniline is electrocatalytically hydrogenated to cyclohexylamine in the presence of a quaternary ammonium ion supporting electrolyte (containing either Br$^-$ or TsO$^-$ anions) with ~40% of product current efficiencies. The nitro group reduction of nitrobenzene proceeds with a sodium p-toluenesulphonate supporting electrolyte [5]. Phenols, anisole, aniline, benzoic acid, cumene, and tert-butylbenzene are reduced to the corresponding cyclohexyl compounds in an aqueous H$_2$SO$_4$(0.2M)-(C) system in the presence of RhCl$_3$(2%) in good yields [6].

Phenol has been efficiently converted to cyclohexanol in an aqueous HClO$_4$-(Pt) system in the presence of quaternary ammonium bromide (69.5% current efficiency) [7a]. An improved procedure for the electrohydrogenation of aromatic rings in an EtOH/HMPA-LiCl-(C/Al) system has been developed [7b]. Phenol, anisole, and toluene are reduced electrochemically into cyclohexanol, methoxycyclohexane, and methylcyclohexane in

45~52% yields.   The role of solvated electrons has been suggested.   The substituted phenols **3** are electrochemically hydrogenated to form the corresponding cyclohexanols **4** with a 100% current efficiency on platinum electrodes in aqueous acid solutions (eq. (4.2)) [8].

$$
\begin{array}{cc}
\underset{\underset{\textbf{3}}{X}}{\overset{OH}{\bigcirc}} & \xrightarrow[\text{100\% current eff.}]{H_2O\text{-}H_2SO_4\text{-}(Pt)} \qquad \underset{\underset{\textbf{4}}{X}}{\overset{OH}{\bigcirc}}
\end{array}
\qquad (4.2)
$$

X = OMe, OH, Ac, Me
Et, CO$_2$H, *tert*-Bu, CN
NO$_2$

The reaction mechanism has been concluded to be the surface reaction between adsorbed phenols and hydrogen.   The hydrogenation is most favored for unsubstituted phenol, suggesting that the polarity of adsorbed hydrogen atoms are essentially neutral.   In order to obtain reasonable current efficiencies, the use of low current densities is favored, which in turn implies the use of high amounts of platinum.   The electrocatalytic hydrogenation of phenol to cyclohexanol on highly dispersed platinum particles on graphite has been developed [9].   Current efficiencies as high as 85% are obtained on the electrode with platinum loading as low as 2%.   The high performance of the electrodes is due to their high surface area, which allows the application of very low current densities without decreasing the total current.   The electroreduction of 1-naphthol **5** has been performed in an HMPA/EtOH(67/33v/v)-(Al) system to give *cis*-1-decalol **6** (47%) and 5,6,7,8-tetrahydro-1-naphthol **7** (35% yield) (eq. (4.3)) [10].   The role of the solvated electrons generated at the cathode has been pointed out.

$$
\underset{\textbf{5}}{\overset{OH}{\bigcirc\bigcirc}} \xrightarrow{\text{HMPA/EtOH(67/33)-(Al)}} \underset{\textbf{6}\ \ 47\%}{\overset{OH}{\bigcirc\bigcirc}} + \underset{\textbf{7}\ \ 35\%}{\overset{OH}{\bigcirc\bigcirc}} \qquad (4.3)
$$

Electrohydrogenation of pyridine proceeds on the surface of skeletal nickel and cobalt catalysts in an aqueous alkaline medium to yield piperidine in 80~92% material yields [11]. An increase in hydroxyl ion concentration tends to inhibit the adsorption of pyridine in the working surface of the catalyst.   Flow-cell electrolysis of pyridine using a Raney-Ni electrode in an aqueous NaOH-(Ni) system in a divided cell affords piperidine in 100% yield (70% current efficiency) [12].   A comparison of the Raney-Ni electrode with the Devarda-Cu electrode has been made in a basic aqueous medium (NaOH, 0.15M) [13], indicating that the lifetime of the Raney-Ni electrode is 60 days and that of the Devarda-Cu electrode 53 days (nitrobenzene reduction).

The electrocatalytic reduction of nicotinic acid **8** leading to piperidine-3-carboxylic acid **9** at a platinized platinum electrode has been surveyed by a radio tracer study (eq. (4.4)) [14].   The character of the adsorption of nicotinic acid is very similar to that of benzoic acid.   The aromatic ring plays a predominant role in the adsorption behavior.   The reductive saturation and the subsequent desorption of adsorbed species are characteristic features occurring at low potentials ($E < 100$ mV).   The rate determining process in the reduction reaction is a step involving adsorbed organic species and adsorbed hydrogen atoms. The saturation of the pyridine ring, *i.e.* the formation of a piperidine ring, leads to a significant decrease in the adsorbability.   In the case of piperidine-3-carboxylic acid, the adsorption occurs *via* the CO_2H group and the ring plays only a secondary role.

$$\text{(4.4)}$$

The electroreduction of bianthryl, bianthrylalkanes, and the corresponding trisanthrylene in a THF-NaBPh$_4$(0.1M)-(Pt) system demonstrated that, in all cases, tetraanions are generated and coulombic interactions between excess charges localized in the anthracene moieties depend on the number of excess charges, ion-pairing effects, solvent influences, and structural peculiarities [15]. Electrochemical formation of dianions from unsaturated cyclophanes [16], pyrene [17], 15,16-dihydropyrenes [18], 2,5-bis(dicyanovinyl)furans [19], and dimethoxyacenazulenes [20], as well as tetraanions from acepleiadylene and pyrene [21] has been reported.

Poly(anion radicals) and poly(dianions) of linear oligoimides have been electrochemically prepared [22]. Stable two- and four-electron reduction species of solvable aromatic oligoimides have been obtained in a DMF-Bu$_4$NBF$_4$-(C) system.

4.1.2 Electroreductive Birch-type Reduction

The electrochemical Birch-type reduction of aromatic compounds is of synthetic importance as an alternative in large scale procedures. Initially, liquid ammonia [23], ethylenediamine [24a,b], and HMPA [24c] have been used as amine-type solvents. The conversion of benzene to 1,4-dihydrobenzene proceeds preferentially on mercury cathode with a combination of alkylammonium salts [25,26]. The use of tetramethyl or tetraethyl ammonium tetrafluoroborate results in extreme lowering of yields, whereas tetrapropyl- and tetrabutylammonium tetrafluoroborates are significantly effective in an aqueous HMPA solution [27]. The anion species are found to be inert in most cases. The electrolysis of benzene in aqueous HMPA(1/1v/v)-Pr$_4$NBF$_4$(0.1M)-(Hg) system at a potential of –3.2 V (SCE) gives 1,4-dihydrobenzene in 60% yield. The role of solvated electrons as well as the formation of quaternary ammonium amalgams (or anionic mercury salts) is discussed as reducing species. The electrochemical Birch reduction of alkylbenzenes has been also carried out in a THF-LiClO$_4$-(Mg or Al) system [28]. A variety of Birch-type reactions are summarized in Table 4.1 [3,23,26,28~39].

An electrochemical alternative to the Birch reduction using tetrabutylammonium hydroxide in water has been developed [26,29]. The electroreduction of methoxybenzene, 1,2,3,4-tetrahydro-6-methoxynaphthalene, and steroids in an aqueous Bu$_4$NOH-(Hg) system affords the corresponding Birch-reduction products in 80~95% yields. Temperature and electrolyte concentrations are found to have a strong effect on the reaction rate [26].

Substituted benzylamines and phenethylamine are electrochemically reduced to the corresponding 1,4-dihydro derivatives. The differences in current efficiency are explained by differences in the stabilization of the anion radicals and also by differences in protonation rate [34]. The electroreduction of phenethylamine **10** is carried out in an MeNH$_2$-LiCl-(Pt) system in an undivided cell to yield the 1,4-dihydro product **11** in 85% yield (eq. (4.5)) [34].

$$\text{(4.5)}$$

Table 4.1 Birch Reduction

Substrate	Electrolysis Conditions	Products	Yield, % Conv. (Current)	Ref.
(benzene)	aq. HMPA-Pr$_4$NBF$_4$(0.1M)-(Hg)	(1,4-cyclohexadiene)	60	[23]
OMe (anisole)	H$_2$O-Bu$_4$NOH-(Hg)	OMe	80	[29]
OMe (anisole)	THF/HMPA(4/1)-Bu$_4$NPF$_6$(1mM)-(Al/SUS) Me$_3$SiCl	SiMe$_3$ / OMe / SiMe$_3$	70	[30]
CH$_2$OMe	THF/H$_2$O-Bu$_4$NBF$_4$-(Hg)	CH$_3$		[31]
R^1 / R^2 R^1 = H, Me, (CH$_2$)CO$_2$H R^2 = H, Me	H$_2$O-Bu$_4$NOH-(Hg)	R^1 / R^2	90~33	[25]
NH$_2$ / CH / CO$_2$H	liq. NH$_3$-LiOAc/t-BuOH-(Mg/Al)	NH$_2$ / CH / CO$_2$H	72	[32] [33]
X / Y CH$_2$NR$_2$ R = H, Me X = Y = H, MeO	MeNH$_2$-LiCl-(Pt)	X / Y CH$_2$NR$_2$	19~70	[34]
(p-cymene)	THF-LiClO$_4$-(Mg or Al)		80	[28]
(1,4-diisopropylbenzene)	THF-LiClO$_4$-(Mg or Al)		83	[28]
OH (1-naphthol)	liq. NH$_3$-NaCl/NaI(60/4)-(Mg/Al)	OH	70	[32] [33]
(naphthalene/tetralin)	THF-LiClO$_4$-(Mg or Al)		94	[28]
MeO (6-methoxytetralin)	H$_2$O-Bu$_4$NOH-(Hg)	MeO	80 / 85	[29] [26]
MeO / OMe / OMe	MeOH/MeCN(20/3)-Et$_4$NCl-(Pt/Pb)	MeO / OMe / OMe	80	[35]
		MeO / OMe / OMe	13	
(phenanthrene)	H$_2$O-Bu$_4$NOH-(Hg) 3 F/mol, 80 °C		59	[3]
(anthracene)	DMF/ROH-Bu$_4$NBF$_4$-(Hg)		100	[36]
	H$_2$O-Bu$_4$NOH-(Pb or Zn, C)		>90	[37]
	H$_2$O-Bu$_4$NOH-(Hg)		75	[3]

Table 4.1 Birch Reduction (continued)

Substrate	Electrolysis Conditions	Products	Yield, % Conv. (Current)	Ref.
(anthracene-Me)	DMF-Et₄NClO₄-(Hg)	(dihydroanthracene-Me)		[38]
(Me-anthracene)	DMF-Et₄NClO₄-(Hg)	(Me-dihydroanthracene)		[38]
(acridone)	MeOH-Et₄NOH-(Hg)	(dihydroacridine)		[39]
(triisopropyl benzene)	THF-LiClO₄-(Mg or Al)	(triisopropyl cyclohexadiene)	84	[28]
(steroid OH)	H₂O-Bu₄NOH-(Hg)	(steroid OH product)	95	[29]
(steroid C≡CH)	THF/H₂O-Bu₄NBF₄-(Hg)	(steroid C≡CH product)	90	[31]

The electrochemical Birch-type reduction of anthracene **12** has been achieved in aqueous solutions [3].   The electrolysis of **12** is carried out in an aqueous Bu₄NOH-(Hg) system in a divided cell to give the Birch-type product **14** in 75% yield together with 15% of the starting material (eq. (4.6)).   Anthracene forms initially 9,10-dihydroanthracene **13**, which can be reduced with additional charge to 1,4,5,8,9,10-hexahydroanthracene **14** in high yield.

$$\text{(anthracene } \mathbf{12}\text{)} \xrightarrow{\text{aq. Bu}_4\text{NOH-(Hg)}} \text{(}\mathbf{13}\text{)} + \text{(}\mathbf{14}\text{)} + \text{others} \qquad (4.6)$$

12   **13** 3%   **14** 75%   6%

The kinetics and mechanism of the protonation of anthracene [40] as well as 9-phenyl- and 9,10-diphenylanthracene anion radicals by phenol have been discussed [40].

The electroreduction of pyrene, coronene, fluorathene, and picene in a DMF-Et₄NClO₄-(Hg) system undergoes the ECE mechanism to the final products [41].   The first step is a reversible one-electron transfer to yield an anion radical.   This is protonated and reduced irreversibly to the carbanion and finally protonated rapidly to the corresponding dihydro products.

The electroreduction of 1-methylindole **15** is performed in a THF/H$_2$O(4%)-Bu$_4$NBF$_4$ (0.25M)-(Hg) system to give 1-methylindoline **16** (89%) as a major product together with a small amount of Birch reduction product **17** (6%) (eq. (4.7)) [42].   Cyclic voltammogram for the electroreduction of aromatic hydrocarbons in liquid dimethylamine as a solvent has been carried out without problems due to side reactions [43].

$$(4.7)$$

The electroreduction of 2-methyl-, 2-phenyl-, 2-phenyl-3-methyl-, 2,3-diphenyl-, and 2,3-dimethylquinoxalines in alkaline or neutral medium leads to 1,4-, 1,2-, and 3,4-dihydro derivatives or to 1,2,3,4-tetrahydro derivatives.   The electrolysis of 2,3-diphenylquinoxa-line **18** in an aqueous MeOH(50%)-NaOH(10%)-(Hg) system in pH 7 at a potential of −0.80 V (SCE) yields 1,4-dihydro derivative **19** in 50% yield (eq. (4.8)), whereas at −1.50 V (pH 13) it gives 1,2,3,4-tetrahydro derivatives **20** in 50% yield (eq. (4.9)) [44].

$$(4.8)$$

$$(4.9)$$

Electrosynthesis of N-benzyl-4-carbamoyldihydropyridine **22** has been carried out by electroreduction of N-benzyl-4-carbamoylpyridinium **21** in an aqueous PhH-NH$_3$(0.1M)/ Et$_4$NCl(0.5M)-(Hg) system to give the dihydropyridine **22** in 90% yield (eq. (4.10)) [45].

$$(4.10)$$

4.1.3   Electroreductive Syntheses of Activated Intermediates

Anion radicals are reactive intermediates in a variety of chemical, electrolytic, and photolytic reactions. An easy method for the electrochemical generation of a stable anion radical has been reported as follows: the electroreduction of 1,4-benzoquinone **23** in a dry MeCN-LiClO$_4$-(Hg) system at −0.8 V (SCE) yields stable anion radical species characterized as 1,4-dihydroxybenzene dibenzoate lithium anion radicals **24** (eq. (4.11)) [46].

$$\mathbf{23} \quad + \quad PhCOCl \xrightarrow[-0.8 \text{ V (SCE)}]{*} Li^+ \left[\mathbf{24a} \rightleftharpoons \mathbf{24b} \rightleftharpoons \mathbf{24c} \rightleftharpoons \right] \qquad (4.11)$$

23 * MeCN-LiClO$_4$-(Pt/Hg)

Electrosynthesis of a novel molecular semiconductor, the lithium naphthalocyanin radical, has been developed [47]. This new material with semiconducting properties is discussed in terms of an intermolecular charge transfer transition.

Substituted 1,2-dihydrophthalic acids **26** as a precursor of the Diels-Alder addition are electrosynthesized by the reduction of *o*-phthalic acids **25** in an aqueous dioxane-H$_2$SO$_4$(5%)-(Pb) system at 25 °C in 83~90% yields (eq. (4.12)) [48,49].

$$\mathbf{25} \xrightarrow[83\sim90\% \text{ yields}]{aq. \text{ Dioxane-H}_2SO_4\text{-(Pb)}} \mathbf{26} \qquad (4.12)$$

25 R^1, R^2, R^3, R^4 = H, F, Cl   **26**
R^1, R^3, R^4 = H; R^2 = Me, CF$_3$, *tert*-Bu
R^1, R^4= H; R^2, R^3 = Me

4.2 Electroreductive Coupling of Aromatic Nuclei

4.2.1 Electroreductive Homo-Coupling of Aromatic Nuclei

The anion radicals of aromatic halides are of interest due to their role in the mechanism of reductive dehalogenation through an S$_{RN}$1 nucleophilic substitution [50,51]. The reduction potentials of 4-substituted nitrobenzenes bearing fluorine atoms at both *ortho* positions to the nitro group decrease for the potentials corresponding to the formation of anion radicals during the reduction of the respective nitrobenzenes [52].

The dimerization of pentafluoronitrobenzene anion radical has been investigated [53]. The unstable anion radical derived from pentafluoronitrobenzene **27** by one-electron reduction in a DMF-Bu$_4$NClO$_4$-(Hg) system is readily dimerized to give a mixture such as 3-hydroxy- and 3,3'-dihydroxydiphenyl derivatives **28** in almost quantitative yield (eq. (4.13)) [53]. The electroreduction of 1,3-benzenedicarbonitrile in a DMF-Bu$_4$NClO$_4$(0.1M)-(Hg) system provides an anion radical formed by *quasi*-reversible one-electron transfer which undergoes dimerization through radical-radical coupling [54].

$$\mathbf{27} \xrightarrow[\text{quantitative yield}]{DMF\text{-Bu}_4NClO_4\text{-(Hg)}} \mathbf{28} \qquad (4.13)$$

27 X = F, OH   **28**

The electroreductive conversion of 9,9'-bianthryl to the corresponding mono-, di-, and trianions has been reported [55a,b].   The dianion of the bianthryl is found to be stable enough in aprotic solvents (MeCN, DMF, etc.) with tetraalkylammonium salts, while the trianion (a blue-violet solution) of the bianthryl is either protonated by trace water or attacked the tetraalkylammonium ion *via* Hofmann elimination.   These complications can be avoided by the use of NaBPh$_4$ in super dry THF.   Under such super dry conditions, the generation of 9,9'-bianthryl tetraanion (a green solution) has been attained [55c].

The electroreduction of aryl trifluoromethyl sulfones by preparative controlled-potential electrolysis in an MeCN-Bu$_4$NBr-(Pt) system undergoes one-electron reduction to yield dimeric products [56].

The electrochemical study of the reduction of substituted anthracene, bearing a nitro, cyano, or formyl group at the **9** position, reveals that the initially formed anion radicals dimerize rapidly to give dimeric dianions as an intermediate of the dimer [57b].   A reversible dimerization of anthracene anion radicals substituted with electron-withdrawing substituents has been discussed in terms of a mechanistic survey [58].   The preparative electrolysis of 9-nitroanthracene **29** is carried out in DMF with tetrabutylammonium salts, and after exhaustive electrolysis, the catholite is treated with a 1% acetic acid solution to yield the dimer **30** in 86~90% yields as a precipitate (eq. (4.14)) [57].   9-Cyano-10-haloanthracene anion radical undergoes unimolecular cleavage of the carbon-halogen bond in

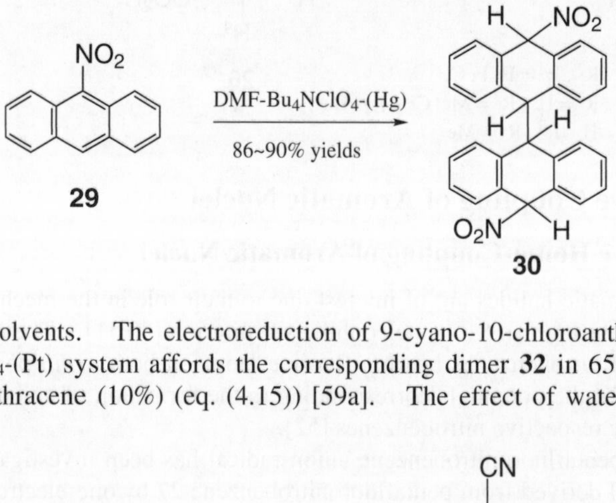

(4.14)

aprotic solvents.   The electroreduction of 9-cyano-10-chloroanthracene **31** in an MeCN-Bu$_4$NBF$_4$-(Pt) system affords the corresponding dimer **32** in 65% yield together with 9-cyanoanthracene (10%) (eq. (4.15)) [59a].   The effect of water for the dimerization of

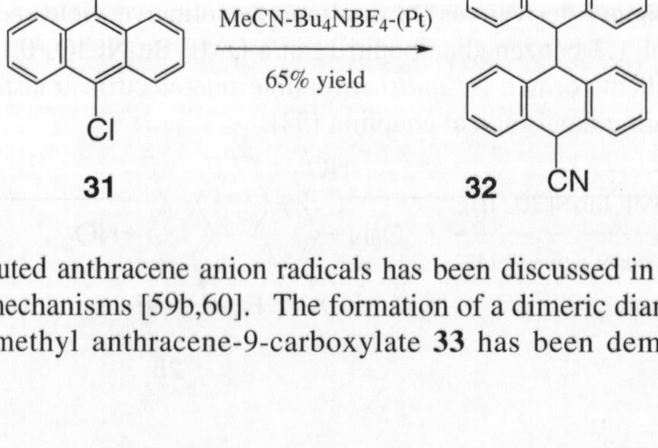

(4.15)

9-substituted anthracene anion radicals has been discussed in terms of kinetics and dimerization mechanisms [59b,60].   The formation of a dimeric dianion **34** in the cathodic reduction of methyl anthracene-9-carboxylate **33** has been demonstrated (eq. (4.16)) [61].

$$2 \quad \text{(structure 33, with } CO_2Me\text{)} \xrightarrow[\substack{-1.28 \text{ V } vs.\text{ Ag/AgI} \\ 0.8\text{F/mol}}]{\text{DMSO-Bu}_4\text{NBF}_4(0.1\text{M})\text{-(Hg)}} \quad \text{(structure 34, with } CO_2Me\text{)} \qquad (4.16)$$

The evidence for the dimeric dianion has been obtained by taking ^{1}H-NMR, in which significant shifts to upfield of protons (H-10, CH$_3$) are consistent with the shielding effect of the delocalized negative charges. The species **37** formed at the cathode is re-oxidized by air to give **35** (eq. (4.17)). Actually, the reduction peak appeared at about half the original peak height, when the electrolyte from exhaustive electrolysis in DMSO was allowed to stand for 1 h [61].

$$2 \text{ ArCO}_2\text{Me} \underset{\substack{\uparrow \\ \text{Air Oxidation } \mathbf{36}}}{\overset{2e^-}{\rightleftharpoons}} 2 \text{ [ArCO}_2\text{Me]}^{-\bullet} \longrightarrow (\text{ArCO}_2\text{Me})_2^{2-} \qquad (4.17)$$

$$\mathbf{35} \qquad\qquad\qquad\qquad\qquad \mathbf{37}$$

The electroreductive intramolecular cyclization of aryl halides **38** occurs with a mild steel cathode in an MeCN-Et$_4$NBF$_4$(0.1M) system [62]. The electrolysis of **38** is carried out in an H-type divided cell to give the compounds **39** in 81~91% yields together with **40** and **41** (0~19% yields) (eq. (4.18)).

$$\text{(structure 38, X = Cl, Br, or I)} \xrightarrow[\substack{81\sim91\% \text{ yields} \\ 64\sim72\% \text{ current yields}}]{e^-,\ -X^-} \text{(structure 39)} + \text{(structure 40)} + \text{(structure 41)} \qquad (4.18)$$

The first electroreduction step of 6,6-diphenylfulvene and 6,6-dimethylfulvene is the formation of the corresponding anion radicals which differ in their stabilities. The anion radical of 6,6-diphenylfulvene is stable and is converted to dianion after addition of a second electron. The dimerization reaction is predominant for the anion radical of 6,6-dimethylfulvene [63].

Both 2,4,6-trimethyl- and 2,4,6-tri-*tert*-butylpyrylium tetrafluoroborates have been shown to undergo a reversible one-electron reduction followed by dimerization of the pyranyl radical. In the case of **42** (R = Me), the dimerization is irreversible whereas the tri-*tert*-butyl radical **43** is in equilibrium with its dimer **44**. The electrolysis of **42** (R = Me) is carried out in an MeCN-Bu$_4$NClO$_4$-(Hg) system to yield the dimer **44** (R = Me) in 62% yield (eq. (4.19)) [64,65,66]. The electrosynthesis of 2,2',6,6'-tetraaryl-4,4'-bipyranylidenes **46** has been performed by reduction of 2,6-diarylpyrylium tetrafluoroborates **45** (eq. (4.20)) [67]. The electroreduction of **45** is conducted in a CH_2Cl_2-Bu_4NBF_4 (0.1M)-(Au) system in a divided cell to give the coupling products **46** in 45~68% yields.

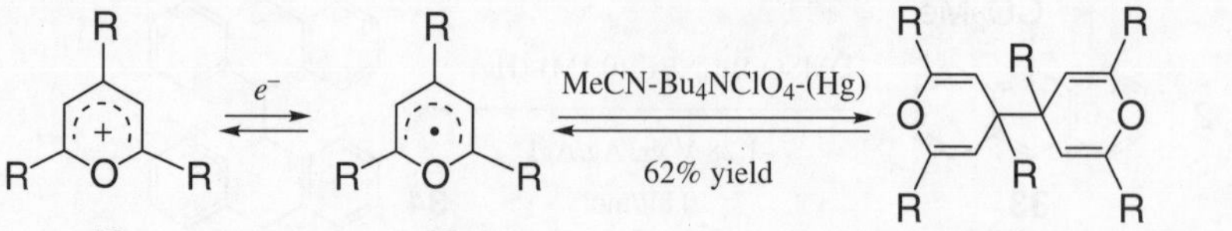

42 R = Me, *tert*-Bu **43** **44** (4.19)

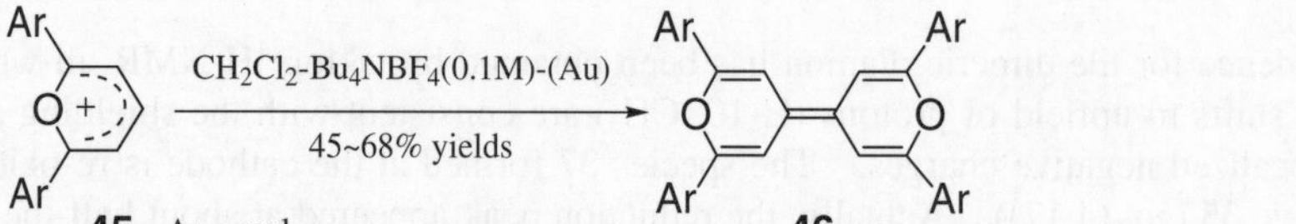

45 Ar = aryl **46** (4.20)

4.2.2 Electroreductive Hetero-Coupling of Aromatic Nuclei

Cathodic alkylation has proven a useful method for preparing alkylated aromatic and hydroaromatic polycyclic hydrocarbons. The electrosynthesis of 1-*tert*-butylpyrene [68b] and isopropylpyrene [69] provides an alternative way in place of the Friedel-Crafts alkylation reaction. The electrochemical *tert*-butylation of pyrene **47** is carried out in a DMF-Bu_4NI-(Hg) system in the presence of isopropyl chloride to give 1-*tert*-butylpyrene **48** in 52% yield (eq. (4.21)) [68]. The electroreductive methylation of anthracene proceeds in a

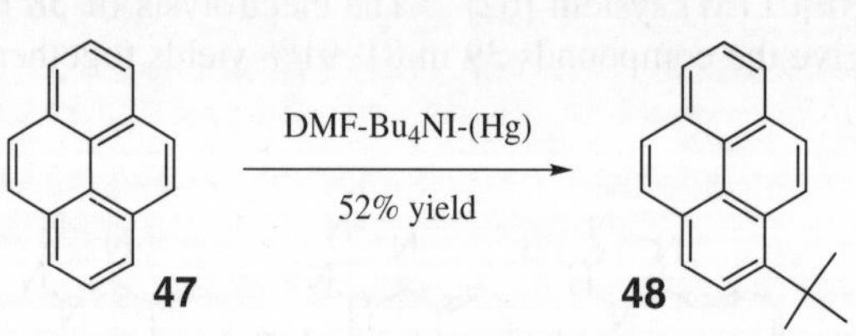

47 **48** (4.21)

DMF-Bu_4NI-(Hg) system in the presence of methyl chloride at −2.0 V (SCE) to give mono- and dimethylated 9,10-dihydroanthracenes (85%) together with 1- or 2-methyldihydroanthracene (15%) [70]. The electroreductive alkylation of anthracene **49** by optically active 2-octyl halides (I, Br, Cl, F) and methanesulfonate yields 9-(2-octyl)-9,10-dihydroanthracene with partial inversion of configuration (eq. (4.22)) [71]. The alkylation is carried out in a DMF-Et_4NClO_4(0.1M)-(Hg) system in the presence of octyl bromide to yield **50** in

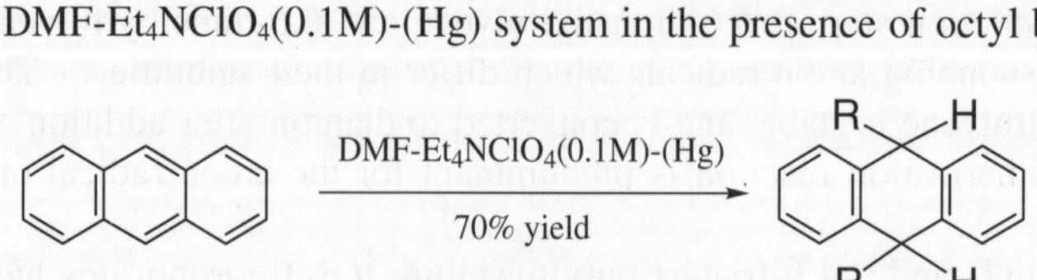

RX, R = octyl
X = Cl, Br, I

49 **50** (4.22)

70% yield. Electroreductive acetylation of anthracene **51** has been carried out in a DMF-Bu_4NI-(Hg) system in the presence of acetic anhydride to give the enol acetate of 9-acetyl-9,10-dihydroanthracene **52** in 66~75% yields (eq. (4.23)) [72].

$$\text{51} \xrightarrow[\text{Ac}_2\text{O, 66~75\% yields}]{\text{DMF-Bu}_4\text{NI-(Hg)}} \text{52} \tag{4.23}$$

The electrosynthesis of unsymmetrical biaryls **59** by an aryl-phenol cross-coupling *via* an electrochemically induced $S_{RN}1$ reaction has been accomplished in liquid ammonia [73]. The preparative scale electro-coupling of 2,6-di-*tert*-butyl phenoxide **53** with chlorobenzonitrile **57** or chloropyridines proceeds in a liquid NH_3-KBr/*tert*-BuOK-(Pt) system to give 4-arylated 2,6-di-*tert*-butyl phenols **59** in 50~95% yields. The mechanism of the reaction is summarized in eq. (4.24). The reaction loop is induced by the reduction of aromatic

$$\tag{4.24}$$

X = Cl;   Y = H, CO_2Me, Cl, CN, $PhSO_2$
$MeSO_2$, CF_3

halides **57** by the reduced form as a redox mediator **55**. The reactions involve (1) the formation of a radical species Ar· **56** by the reductive cleavage of [ArX]⁻· **55**, (2) which combines with the phenoxide **53** to give [ArNu]⁻· **54**, (3) which in turn reduces the starting aromatic halide **57**. The substitution of 1,4-dichlorobenzene **55** (Y = Cl) by ArO⁻ **53** is carried out by means of an electro-induced $S_{RN}1$ reaction in liquid ammonia to give the biphenyl derivative **59** in 67% yield [74]. The electrosynthesis of 4-(3,5-dipentyl-4-hydroxyphenyl)pyridine **62** has been attained in a liquid NH_3/THF-KBr(0.1M)-(Mg) system (eq. (4.25)) [75]. The hetero-coupling of 4-chloropyridine **60** and 2,6-dipentylphenoxide **61** undergoes an electrochemically induced $S_{RN}1$ reaction in which 2,4'-bipyridine acts as a mediator. A $S_{RN}1$ reaction of indolyl anion with 4-chloropyridine **60** has been attained in a

$$\text{60} + \text{61} \xrightarrow[\text{-Cl}^-\text{, 30\% yield}]{\substack{\text{liq. NH}_3\text{/THF-} \\ \text{KBr(0.1M)-(Mg)}}} \text{62} \tag{4.25}$$

liquid NH_3-KBr/KOBu-*tert*-(Mg/Pt) system in the presence of 2,4'-bipyridyl as a mediator at $-40°C$ in an undivided cell to give 3-(4-pyridyl)indole **64** in 60% yield (eq. (4.26)) [76]. The electrochemically induced $S_{RN}1$ reaction of aryl halides **65** with uracil anion **66** takes place in a DMSO-Et_4NBF_4-(Pt) system to give **67** in 30~55% yields (eq. (4.27)) [77].   The coupling of aryl halides with 1-naphthoxide is also recorded [78].

Ar—Cl +   (indole **63**) $\xrightarrow[\text{60\% yield}]{\text{liq. } NH_3\text{-KBr/KOBu-}t\text{-(Mg/Pt)}}$ **64**   (4.26)

60   **63**

Ar = 4-pyridyl

ArX + **65** (uracil **66**) $\xrightarrow[\text{30~55\% yields}]{\text{DMSO-}Et_4NBF_4\text{-(Pt)}}$ **67** + X⁻   (4.27)

66 Ar = aryl
X = Cl, Br, I

The electroreductive nucleophilic aromatic substitution reactions of aromatic halides are reviewed in Chapter 7.4.2.

4.3   Reduction of Quinones and Quinonemethides

4.3.1   Electroreduction of Benzoquinones, Naphthoquinones and Anthraquinones

p-Benzoquinones (BQ) have been investigated intensively due to their importance as an electron acceptor in many chemical reactions.   The electroreduction of BQ undergoes an electrochemical-chemical-electrochemical (ECE) reaction when reduced in nonaqueous media [79] (eq. (4.28)).   The mechanism of quinone and hydroquinone conversions has been shown to depend on pH value of media [80].

$$BQ + e^- \longrightarrow [BQ]^{-\bullet} \quad \text{------ (a)}$$
$$[BQ]^{-\bullet} + H^+ \longrightarrow [BQH]^{\bullet} \quad \text{------ (b)} \qquad (4.28)$$
$$[BQH]^{\bullet} + e^- \longrightarrow [BQH]^- \quad \text{------ (c)}$$
$$[BQH]^- + H^+ \longrightarrow BQH_2 \quad \text{------ (d)}$$

Similar intermediates are observed in aqueous media based on spectroelectrochemical study [81a].   It was found that the electrogenerated $[BQ]^{-\bullet}$ is reasonably stable even in neutral water [81b].

The potential-pH diagram for benzoquinone indicating BQ–BQH_2 equilibria and associated acid-base equilibria has been investigated [82].

The rate constants of quinoid compounds are found to be independent of the electrode material in nonaqueous media [83a,b].   The correlation between steady state potentials and transfer coefficients related to benzoquinone reduction has been elucidated [83c].

The correlation of half-wave potentials of the quinone-quinone anion radical system with substituent constants for mono- and poly-substituted *p*-benzoquinones has been investigated in a DMF solution [84].

The controlled potential reduction of the bridged benzoquinone **68** in an CH_2Cl_2-Bu_4NBF_4-(Hg)-system and subsequent acetylation affords the cyclized product **69** in 14% yield (eq. (4.29)) [85].

$$
\text{68} \xrightarrow[\substack{2) Ac_2O \\ -0.6\ V\ (SCE),\ 14\%\ yield}]{1)\ CH_2Cl_2\text{-}Bu_4NBF_4\text{-}(Hg)} \text{69}
\tag{4.29}
$$

The electroreduction of intramolecular charge-transfer complexes derived from 1,4-naphthoquinone derivatives has been investigated in DMF and DMF/H_2O solutions. In aprotic media, the electroreduction proceeds in two successive steps: the first process leads to formation of a stable anion radical, the second step leads to the formation of an unstable primary product [86]. The electroreductive alkylthiolation of vitamin K **70**, leading to sulfur containing analogues **72**, has been attained as described by a scheme comprising cathodic generation of the thiolate anion. The electrolysis of **70** is carried out in an MeCN-Et_4NBr-(Mg/Pt) system to give **72** *via* **71** in 85~91% yields (eq. (4.30)) [87]. Nucleophilic substitution reactions of 1,4-dimethoxybenzene and 1,3,5-trimethoxybenzene by the anions of 1*H*-tetrazoles have been carried out efficiently by paired electrosynthesis (See Chapter 5) [88].

$$
\tag{4.30}
$$

$$
R = C_8H_{17},\ C_{14}H_{29},\ C_{16}H_{33},\ CH_2CH_2OH
$$

The electroreduction of anthraquinone (AQ) **73** in an H_2O/EtOH(1/1)-H_2SO_4(50vol%)-(Pt or Ag) system proceeds as a two-step process. The initial two-electron reduction provides the conversion of AQ to anthrahydroquinone. The second step of the two-electron of protonated anthrahydroquinone to anthranol (anthrone) **74** proceeds at -0.4 V (SCE) (eq. (4.31)) [89a]. The preparative scale electrolysis has been investigated in concentrated sulfuric acid using different cathodic materials [89b]. Platinum and silver are the best choice under the employed conditions. The coupling reaction of electrogenerated anion radicals of anthracene, 9,10-diphenylanthracene, benzophenone, and quinoxaniline and hexenyl and 2,2-dimethylhexenyl radicals is found to proceed in a radical-radical coupling fashion [90].

$$
\text{73} \xrightarrow[\substack{65\text{~}74\%\ current\ eff.}]{H_2O/EtOH(1/1)\text{-}H_2SO_4(50\%)\text{-}(Pt\ or\ Ag)} \text{74}
\tag{4.31}
$$

The electroreductive cleavage at the 2-methyl position of the carbon-oxygen bond of 2-acetoxymethyl-1-methoxy-9,10-anthraquinone **75** proceeds in a DMF-LiClO$_4$-(C) system to yield 2-methyl-1-methoxy-9,10-anthraquinone **76** in good yield (eq. (4.32)) [91b,c]. The cyclic voltammograms of the acetate **75** exhibits two reduction waves, the first resulting from the formation of Li$^+$ ion pairs of their anion radicals and the second from Li$^+$ ion pairs of their dianions, which undergo air oxidation to give the cleavage product **76**. The present procedure can be used as an alternative method for deprotection of substituted 2-methylanthraquinone of carbamates as the protecting group of primary amines [91a].

$$\text{(4.32)}$$

Pulping with anthraquinone additives appeared in 1977. Further studies reveal more details for the chemistry of wood digestion, which involves a dianion and anion radical assisted acceleration of lignin degradation in an AQ/AQH$_2$ redox catalytic system [92].

Anthraquinone derivatives bearing photochemically isomerisable moieties have also been the focus of interest regarding their electrochromism properties for molecular switching. The electrochemical interconversion of anthraquinone derivatives, containing 4'-substituted stilbene at the β position, has been investigated [93]. The cation binding enhancement by electrochemical switching in a lariat ether has been found under one- or two-electron reduction of an anthraquinone moiety [94].

The behavior of anion radical dianion and dianion radical intermediates derived from the electroreduction of alizarin in an aprotic solvent has been discussed [95]. Ubiquinone-10 shows a well-defined, pH-dependent cathodic reduction wave in various organic solvents and the electroreduction undergoes two-electron and a proton transfer around the neutral pH region [96].

The two-electron electroreduction of daunamycin **77** undergoes glycoxide elimination in a DMF-Bu$_4$NOH-(Hg) system to form a quinone methide form which has been characterized by means of spectroelectrochemistry (eq. (4.33)) [97]. The rate of glycoside elimination of **77**, giving **78** under electroreduction in an aprotic medium has been compared with known anthracyclines. The reversible addition of the first electron which occurs at *ca.* –6.30 mV (SCE) yields a stable semiquinone anion radical. The second electron addition produces a dianionic species which undergoes a rather fast first-order reaction of glycoside elimination [98].

$$\text{(4.33)}$$

The kinetic information on the redox reaction of adriamycin adsorbed on a mercury electrode surface has been reported [99]. The effect of the pH and the temperature on the redox properties of the quinone moiety in adriamycin has been discussed [100].

The redox properties of unsymmetrically substituted 2,5-bis(4-oxo-2,5-cyclohexadien-1-ylidene)-2,5-dihydrothiophenes **79** has been investigated (eq. (4.34)) [101]. Compounds **79** are the first examples of terphenoquinone analogues undergoing enphoteric four-stage single-electron redox systems (**79** ⇄ **80** ⇄ **81**).

$$R = H, Cl, Br, SMe \tag{4.34}$$

The effect of magnetic field on the electroreduction of 1,4-benzoquinone has been investigated. The rate-determining step is found to shift from the diffusion step into the chemical reaction step of $[O_2]^{-\bullet}$ and diethyl malonate with increasing magnetic fields [102].

4.3.2   Electroreduction of Quinonemetides

The chemical reactions of quinonemethides mostly involve the electrophilic character of the intermediates. The cathodic reduction brings about *umpolung*; the quinonemethides are converted into nucleophiles and bases. It has been found that a large-scale electroreduction of quinonemethides preferentially undergoes hydrodimerization at the terminal carbon [103a,b]. The electroreduction of more hindered quinonemethides **82** undergoes one- and two-electron reduction to give **83** *via* relatively stable anion radical and dianion intermediates (eq. (4.35)) [103c]. Less hindered analogues efficiently and rapidly dimerize at carbon, with concomitant protonation or *O*-methylation depending on the electrophile employed.

$$\tag{4.35}$$

The formation of relatively stable protonated anion radicals has been observed in the cathodic hydrogenation of quinonemethides [104].

The finding of electrical conductivity in certain anion radical salts of TCNQ **84** paved the way for the development of organic conductors [105].   Charge-transfer salts of TCNQ exhibit a wide array of electrical properties.   The acceptor abilities of TCNQ homologues are discussed in terms of both steric and electronic effects [106].   Cyclic voltammetric data of new TCNQ homologues have been taken and discussions have been made with the data in comparison with TCNQ as a reference [105,107,108].

4.4   Electroreductive Cleavage Reactions

4.4.1   Electroreductive Cleavage of Carbon-Carbon Bonds

The electroreductive cleavage of the C-CN bond of 1,2- and 1,4-benzenedicarbonitriles **85** undergoes a *quasi*-reversible one-electron transfer accompanying a preliminary proton transfer in a DMF-Bu$_4$NClO$_4$-(Hg) system to yield the corresponding monocyanides 86 *via* dianion intermediates (eq. (4.36)) [109].

$$\text{85} \xrightarrow[\text{70\% yield}]{\text{DMF-Bu}_4\text{NClO}_4\text{-(Hg)}} \text{86} \tag{4.36}$$

4.4.2   Electroreductive Cleavage of Carbon-Heteroatom Bonds

The C-S bond cleavage of nitrogen-containing hetero aromatic sulfones has been investigated in an MeCN-Bu$_4$NBF$_4$(0.1M)-(Hg) system [110].

Electroreductive activation is found to be feasible for a "polar" nucleophile substitution of *p*-nitrobenzonitrile **87** with phenolate, leading to the formation of *p*-phenoxybenzonitrile **89** in 74% yield (eq. (4.37)) [111].   The electroreduction of *o*-bis(alkylsulfonyl)benzenes **90** undergoes decomposition of anion radical intermediates to give alkylated products **91** as major products (eq. (4.38)) [112].

$$\text{O}_2\text{N}-\!\!\!\!\bigcirc\!\!\!\!-\text{CN} \xrightarrow[\text{74\% yield}]{\text{88}, \text{ DMF-Bu}_4\text{NBF}_4\text{-(C)}} \text{PhO}-\!\!\!\!\bigcirc\!\!\!\!-\text{CN} \tag{4.37}$$

87   **89**

$$\text{90} \xrightarrow[\text{30\textasciitilde33\% yields}]{\text{DMF-Bu}_4\text{NBF}_4(0.1\text{M})\text{-(Hg)}} \text{91} + \text{92} \tag{4.38}$$

90 R = Et, Bu, Oct   **91** 30~33%   **92** 11~20%

Phenylselenobenzonitriles **94** have been electrosynthesized in terms of electrochemically induced aromatic nucleophilic substitution ($97 + 96 \rightarrow 98$) in acetonitrile.   The electrolysis of 2-, 3-, and 4-chlorobenzonitriles **93** has been performed in an MeCN-Bu$_4$NPF$_6$(0.1M)-(C) system with sonication in the presence of an equivalent amount of benzeneselenate **97** initially prepared by electroreduction of diphenyl diselenide to afford 2-,

3-, and 4-(phenylseleno)benzonitriles **94** in 36~59% yields *via* anion radical intermediates **98**. The yields can be improved by electrolysis of chlorobenzonitrile **93**. A yield of 70% has been obtained in the case of 4-(phenylseleno)benzonitirle (eq. (4.39)) [113].

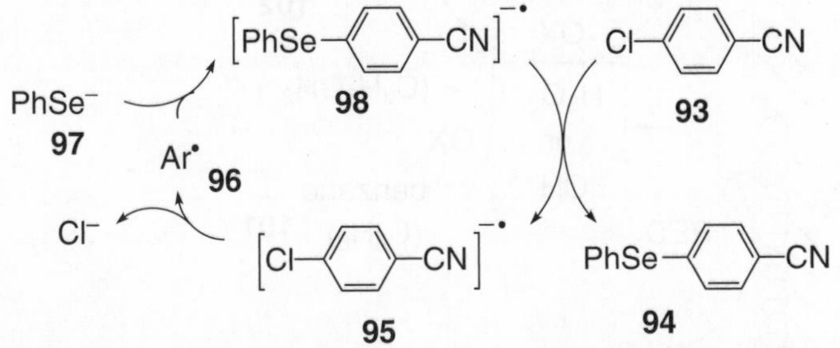

(4.39)

Electroreductive deamination of primary arylamines has been achieved in an aqueous HNO_3-(Pb/Pt) system to give the corresponding aromatic compounds in 55~77% yields [114]. The direct removal of a nitro group on an aromatic ring is also feasible in the same medium (eq. (4.40)). The electrolysis of **99** in an HNO_3(0.5mL)-(Pb/Pt) system yields **100** in 68% yield.

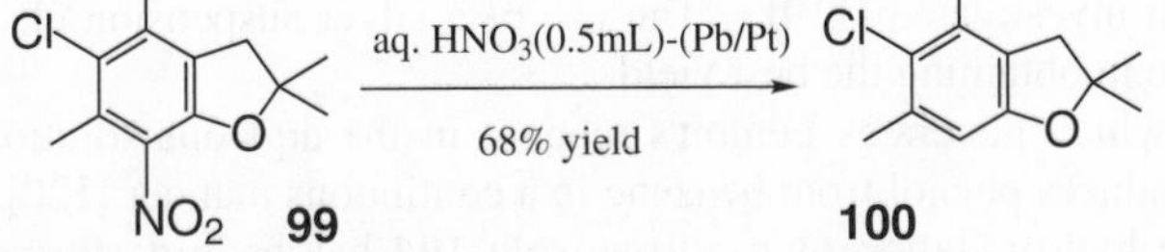

(4.40)

The electroreductive C-N bond cleavage of the nitro group of nitroarenes has been found when 1,2,4,5-tetrafluoro-3,6-dinitrobenzene is allowed to reduce in a DMF-Et_4NClO_4-(Pt) system [115].

4.5 Reductive Hydrooxylation of Aromatic Nuclei

One-step oxidation of benzene to phenol is an alternative to the cumene process. Recently, the oxidation of aromatic compounds using Fenton's reagent has attracted much attention. The electrolytic recycle use of Fe^{2+} ions has been attempted under several conditions [116]. In the presence of Cu^{2+} ions with Fenton's reagent at 0.1 V (SCE) in an aqueous MeCN-H_2SO_4(0.1N) system, benzene **101** can be oxidized into phenol **102** (46%) together with hydroquinone **103** (18%) (eq. (4.41)) [117].

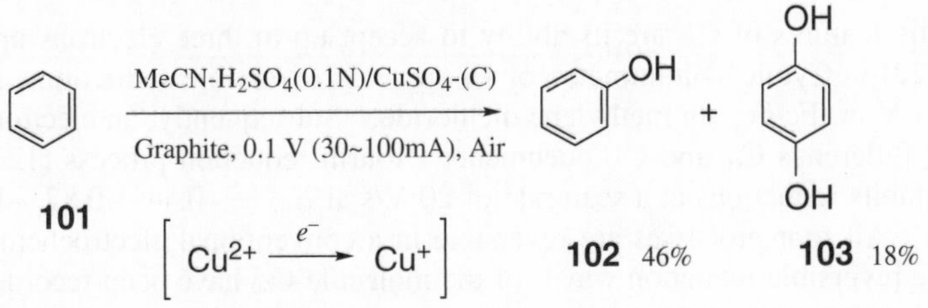

(4.41)

Direct phenol **102** synthesis from benzene **101** has been achieved in 22~41% current efficiency on graphite cathode with hydrogen peroxide electrochemically produced from oxygen reduction in a highly acidic solution, CF_3SO_3H, at room temperature [118]. The proposed reaction mechanism is illustrated in Scheme 4.1 [118].

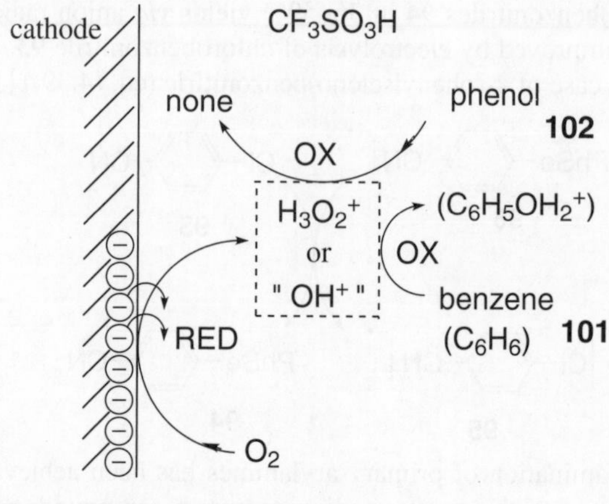

Scheme 4.1 Electroreductive Phenol Formation from Benzene.
[Reproduced with permission from R. Ohnishi, A. Aramata, *Stud. Org. Chem.*, **30**, 337 (1987)]

The hydroxylation of benzene and phenol during oxygen electroredution on mercury, lead, copper, and silver electrode in an aqueous H_2SO_4(0.1M) system in the presence of Fenton's reagent has been investigated [119]. The use of a silver suspension electrode is found to be most effective in obtaining the best yield.

A fuel cell system, which generates Fenton's reagent in the aqueous solution of the cathode compartment, produces phenol from benzene in a continuous manner [120].

The electroreductive hydroxylation of *p*-nitroanisole **104** has been performed in an aqueous MeCN-HClO$_4$-(Pt/SUS) system in the presence of Fenton's reagent to give 5-nitroguaiacol **105** in 26% yield (eq. (4.42)) [121].

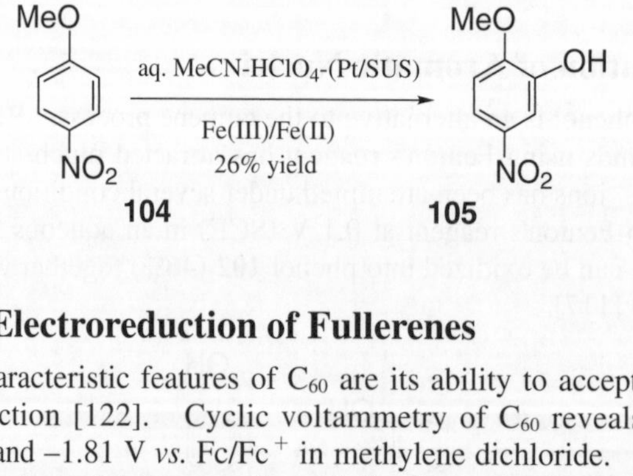

$$(4.42)$$

4.6 Electroreduction of Fullerenes

The characteristic features of C_{60} are its ability to accept up to three electrons upon electroreduction [122]. Cyclic voltammetry of C_{60} reveals reversible reductions at -0.92, -1.32, and -1.81 V *vs.* Fc/Fc$^+$ in methylene dichloride. Subsequently, an electrochemical investigation of fullerenes C_{60} and C_{70} documents a fourth reduction process [123]. The fullerene C_{60} exhibits reductions at a scan rate of 20 V/s at $E_{1/2} = -0.44$, -0.82, -1.25, and -1.72 V (SCE). All four processes are reversible in a conventional electrochemical cell. Futhermore, five reversible reduction waves of the molecule C_{60} have been recorded [124]. Surprisingly, low temperature cyclic voltammetry of C_{60} in a water-free DMF/ Toluene(2/3)-Et$_4$NPF$_6$(0.1M) system shows a complete set of six one-electron reversible waves, providing evidence for the existence of $C_{60}{}^{6-}$ [125]. It is shown that the anions resulting from the first and second reduction steps, namely $C_{60}{}^-$ and $C_{60}{}^{2-}$, are associated with two or more tetraalkylammonium cations, the strength of the resulting association increasing with increasing cation size [126].   Both C_{60} and C_{70} exhibit identical cyclic voltammetry.   Singly, doubly, and triply reduced fullerenes (C_{60}^-, C_{60}^{2-} and C_{60}^{3-}) are prepared by electroreduction in a Py-Bu_4NClO_4(0.1M)-(Pt) system in an H-type divided cell under oxygen-free nitrogen atomosphere.   The generated anions are stable in the inert atmosphere for at least several hours (and very often several days) [127].   The electroreduction of fullerene C_{60} in an $MeCN/MeC_6H_5$(2/3)-Bu_4NBr(0.05M)-(Pt) system affords a series of three EG bases of successively increasing basicity [128].   The base-catalyzed reaction of ethyl malonate 106 with acrylonitrile 107 proceeds in a similar electrolysis conditions in the presence of fullerene to give the addition product 108 in an excellent yield (eq. (4.43)) [128].

$$(4.43)$$

There is keen interest in the functionalization and conversion of fullerenes, which involves cycloaddition, radical addition, addition across double bonds, metal complex formation, nucleophilic substitution (alkylation, nitration, fluorination etc.), hydrogenation, oxidation, and polymerization as chemical reactions.   Among the nucleophilic substitutions, methylation involving chemically generated C_{60}^{n-} anions affords a mixture of polymethylated $(CH_3)_mC_{60}$ products (m < 24).   Difficulty stems from no control over the number of addends on the fullerene cage by the chemical method.   The large difference in formal redox potentials between consecutive $C_{60}^{n-/(n+1)-}$ redox couples is promising for electrochemical generation of C_{60}^{n-} anions, which can be employed as nucleophiles in substitution reactions.   The selective electrosynthesis of $(CH_3)_2C_{60}$ has been achieved by control of the charge n, on C_{60}^{n-}, generated prior to methylation.

The electroreductive methylation of the fullerene C_{60} has been performed in a PhCN-Bu_4NClO_4-(Pt) system at −1.1 V (SCE) by the addition of a 100-fold excess of methyl iodide after electrolysis was stopped to yield a mixture of 1,2-$(CH_3)_2C_{60}$ 109 (25%), 1,4-$(CH_3)_2C_{60}$ 110 (9%), and 1,11-$(CH_3)_2C_{60}$(9%) (eq. (4.44)) [129].

$$(4.44)$$

109 1,2-$(CH_3)_2C_{60}$ 110 1,4-$(CH_3)_2C_{60}$

Electroreduction of fullerenes in the presence of dioxygen and water results in polyoxygenation and fragmentation of the C_{60} framework [130].   It has been found that (1) the electrochemically produced monoanions of C_{60} and C_{70} fullerenes can react with water; (2)

the dianion reacts with dioxygen to yield the corresponding $C_{60}O$, $C_{60}O_2$, $C_{60}O_3$, and $C_{60}O_4$; (3) prolonged electroreduction in the presence of dioxygen and water leads to the loss of a C_{10} fragments for C_{50} products that have cyclic voltammetry similar to that of C_{60} and C_{70} [130].

It is known that the reaction of substituted diazo compounds with C_{60} generates a mixture of monoaddition products across the 5,6(fulleroids) and the 6,6(methanofullerenes) ring fusions. The electrochemical addition of a third electron to a fulleroid isomerizes it to the methanofullerene [131]. The higher fullerene, chiral C_{76}, is expected to exhibit organic electron donor properties [132].

References

[1] (a) H. Kita, T. Nakamura, H. Itoh, H. Kano, *Electrochim. Acta*, **23**, 405 (1978); (b) K. Shimazu, H. Kita, *Electrochim. Acta*, **24**, 1085 (1979).
[2] D. Robin, M. Comtois, A. Martel, R. Lemieux, A. K. Cheong, G. Belot, J. Lessard, *Can. J. Chem.*, **68**, 1218 (1990).
[3] E. Kariv-Miller, R. I. Pacut, *Tetrahedron*, **42**, 2185 (1986).
[4] B. Mahdavi, J.-M. Chapuzet, L. Brossard, A. Martel, P. L. G. Capuano, J. Lessard, *Proc. -Electrochem. Soc.*, **94-21**, 220 (1994).
[5] P. N. Pintauro, J. R. Bontha, *J. Appl. Electrochem.*, **21**, 799 (1991).
[6] L. L. Miller, L. Christensen, *J. Org. Chem.*, **43**, 2059 (1978).
[7] (a) R. A. Misra, B. L. Sharma, *Electrochim. Acta*, **24**, 727 (1978); (b) R. A. Misra, A. K. Yadav, *Bull. Chem. Soc. Jpn.*, **55**, 347 (1982).
[8] K. Sasaki, A. Kunai, J. Harada, S. Nakabori, *Electrochim. Acta*, **28**, 671 (1983).
[9] K. Amouzegar, O. Savadogo, *Electrochim. Acta*, **39**, 557 (1994).
[10] R. A. Misra, A. Jain, *Trans. SAEST*, **19**, 189 (1984).
[11] G. K. Murzatova, I. V. Kirilyus, D. V. Sokol'skii, R. G. Baisheva, *Elektrokhimiya*, **21**, 791 (1985).
[12] J. Lessard, G. Belot, Y. Couture, S. Desjardins, C. Roy, Int. J. *Hydrogen Energy*, **18**, 681 (1993).
[13] G. Belot, S. Desjardins, J. Lessard, *Tetrahedron Lett.*, **25**, 5347 (1984).
[14] G. Horanyi, *J. Electroanal. Chem., Interfacial Electrochem.*, **284**, 481 (1990).
[15] J. Mortensen, J. Heinze, H. Herbst, K. Müllen, *J. Electroanal. Chem.*, **324**, 201 (1992).
[16] K. Ankner, B. Lamm, B. Thulin, O. Wennerström, *J. Chem. Soc., Perkin Trans. II*, **1980**, 1301.
[17] A. M. Waller, R. G. Compton, *Electrochim. Acta*, **33**, 1335 (1988).
[18] A. J. Fry, J. Simon, M. Tashirp, T. Yamato, R. H. Mitchell, T. W. Dingle, R. V. Williams, R. Mahedevan, *Acta Chem. Scand.*, **B37**, 445 (1983).
[19] J. Daub, J. Salbeck, T. Knöchel, C. Fischer, H. Kunkely, K. M. Rapp, *Angew. Chem., Int. Ed. Engl.*, **28**, 1494 (1989).
[20] J. Salbeck, J. Daub, *Chem. Ber.*, **122**, 1681 (1989).
[21] J. Mortensen, J. Heinze, *Tetrahedron Lett.*, **26**, 415 (1987).
[22] T. M. Dietz, B. J. Stallman, W. Sum, V. Kwan, J. F. Penneau, L. L. Miller, *J. Chem. Soc., Chem. Commun.*, **1990**, 367.
[23] A. J. Birch, *Nature*, **158**, 60 (1946).
[24] (a) H. W. Sternberg, R. Markby, I. Wender, *J. Electrochem. Soc.*, **110**, 425 (1963); (b) R. A. Benkeser, J. Mels, *J. Org. Chem.*, **34**, 3970 (1969); (c) H. W. Sternberg, R. Markby, I. Wender, D. M. Mohilner, *J. Am. Chem. Soc.*, **91**, 4191 (1969).
[25] J. P. Coleman, J. H. Wagenknecht, *J. Electrochem. Soc.*, **128**, 322 (1981).
[26] E. Kariv-Miller, K. E. Swenson, D. Demach, *J. Org. Chem.*, **48**, 4210 (1983).
[27] M. Tezuka, T. Yajima, A. Tsuchiya, *Stud. Org. Chem.*, **30**, 295 (1987).
[28] T. Shono, S. Kashimura, *Japan Kokai Tokkyo Koho* JP 05,255,878 (1993).
[29] R. W. Binkley, M. G. Ambrose, *J. Org. Chem.*, **48**, 1777 (1983).
[30] C. Biran, M. Bordeau, F. Seren-Spirau, M.-P. Léger-Lambert, J. Dunogués, *Synth. Commun.*, **23**, 1727 (1993).
[31] E. Kariv-Miller, K. E. Swenson, G. K. Lehman, R. Andruzzi, *J. Org. Chem.*, **50**, 556 (1985).
[32] J. Chaussard, C. Combellas, A. Thiebault, *Tetrahedron Lett.*, **28**, 1173 (1987).
[33] C. Combellas, H. Marzouk, A. Thiebault, *J. Appl. Electrochem.*, **21**, 267 (1991).
[34] P. J. M. Van Andel-Scheffer, A. H. Wonders, E. Barendrecht, *J. Electroanal. Chem.*, **366**, 135 (1994).
[35] A. P. Tomilov, I. N. Chernykh, S. E. Zabusova, V. L. Sigachevam, N. M. Alpatova, *Elektrokhimiya*, **23**,

1330 (1987).
[36] M. F. Nielsen, O. Hammerich, V. D. Parker, *Acta Chem. Scand.*, **B38**, 809 (1984).
[37] E. G. Palffy, P. Starzewski, A. Labani, A. Fontana, *J. Appl. Electrochem.*, **24**, 337 (1994).
[38] R. Abdel-Hamid, *J. Electrochem. Soc. India*, **35**, 47 (1986).
[39] L. I. Svyatkina, L. L. Dmitrieva, G. N. Kurov, *Zh. Obshch. Khim.*, **58**, 2357 (1988).
[40] (a) V. D. Parker, *Acta Chem. Scand.*, **B35**, 583 (1981); (b) M. F. Nielsen, O. Hammerich, V. D. Parker, *Acta Chem. Scand.*, **B40**, 101 (1986).
[41] R. Abdel-Hamid, *J. Electrochem. Soc. India*, **34**, 88 (1985).
[42] E. Kariv-Miller, D. F. Dedolph, C. M. Ryan, T. J. Mahachi, *J. Heterocycl. Chem.*, **22**, 1389 (1985).
[43] K. Meerholz, J. Heinze, *J. Am. Chem. Soc.*, **111**, 2325 (1989).
[44] J. Pinson and J. Armand, *Collect. Czech. Chem. Commun.*, **36**, 585 (1971).
[45] (a) F. M. Moracci, S. Tortorella, *Synth. Commun.*, **13**, 1225 (1983); (b) F. M. Moracci, S. Tortorella, B. Di Rienzo, I. Carelli, *Synth. Commun.*, **11**, 329 (1981); (c) F. M. Moracci, S. Tortorella, B. Di Rienzo, F. Liberatore, I. Carelli, *Ann. Chim.*, **71**, 499 (1981).
[46] A. Guirado, F. Barba, J. M. Cuadrado, *Electrochim. Acta*, **28, 761 (1983).**
[47] M. A. Petit, M. Bouvet, D. Nakache, *J. Chem. Soc., Chem. Commun.*, **1991**, 442.
[48] T. Ohno, M. Ozaki, A. Inagaki, T. Hirashima, I. Nishiguchi, *Tetrahedron Lett.*, **34**, 2629 (1993).
[49] Y. Kunugi, T. Nonaka, *J. Electroanal. Chem.*, **356**, 163 (1993).
[50] J,-M. Savéant, *Acc. Chem. Res.*, **13**, 323 (1980).
[51] R. A. Rossi, *Acc. Chem. Res.*, **15**, 164 (1982).
[52] G. A. Selivanova, V. F. Starichenko, V. A. Ryabinin, V. D. Shteingarts, *Zh. Org. Khim.*, **28**, 1445 (1992).
[53] G. A. Selivanova, V. F. Starichenko, V. D. Shteingarts, *Izv. Akad. Nauk SSSR, Ser. Khim.*, **1988**, 1155.
[54] A. Gennaro, A. M. Romanin, M. G. Severin, E. Vianello, *J. Electroanal. Chem., Interfacial Electrochem.*, **169**, 279 (1984).
[55] (a) K. Itaya, A. J. Bard, M. Szwarc, *Z. Phys. Chem. N. F.*, **112**, 1 (1978); (b) O. Hammerich, J.-M. Savéant, *J. C. S., Chem. Commun.*, **1979**, 938; (c) J. Mortensen, J. Heinze, *J. Electroanal. Chem., Interfacial Electrochem.*, **175**, 333 (1984).
[56] N. V. Ignat'ev, L. A. Nechitailo, G. M. Shchupak, L. M. Yagupol'skii, *Elektrokhimiya*, **26**, 1112 (1990).
[57] (a) L. V. Michalchenko, A. S. Mendkovich, V. P. Gultyai, *Izv. Akad. Nauk SSSR, Ser. Khim.*, **1985**, 2158; (b) A. S. Mendkovich, L. V. Michalchenko, V. P. Gultyai, *J. Electroanal. Chem., Interfacial Electrochem.*, **224**, 273 (1987).
[58] O. Hammerich, V. D. Parker, *Acta Chem. Scand.*, **B35**, 341 (1981).
[59] (a) O. Hammerich, V. D. Parker, *Acta Chem. Scand.*, **B37**, 851 (1983); (b) *Idem. ibid.*, **B37**, 379 (1983).
[60] J.-M. Savéant, *Acta Chem. Scand.*, **B37**, 365 (1983).
[61] C. Z. Smith, J. H. P. Utley, *J. Chem. Res. (S)*, **1982**, 18.
[62] S. Connelly, J. Grimshaw, J. Trocha-Grimshaw, *Electrochim. Acta*, **41**, 489 (1996).
[63] M. V. Yarosh, T. P. Konovalova, V. L. Shirokii, A. N. Ryabtsev, N. A. Maier, *Elektrokhimiya*, **26**, 1167 (1990).
[64] W. Neil, C. Garrard, F. G. Thomas, *Aust. J. Chem.*, **36**, 1983 (1983).
[65] (a) F. D. Saeva, G. R. Olin, *J. Am. Chem. Soc.*, **102**, 299 (1980); (b) V. Wintgens, J. Pouliquen, J. Kossanyi, M. Heintz, *New J. Chem.*, **10**, 345 (1986).
[66] M. I. Ismail, *Tetrahedron*, **47**, 1957 (1991).
[67] C. Amatore, A. Jutand, F. Pfüger, C. Jallabert, H Strzelecka, M. Veber, *Tetrahedron Lett.*, **30**, 1383 (1989).
[68] (a) P. E. Iversen, *J. Chem. Educ.*, **48**, 136 (1971); (b) P. E. Hansen, A. Berg, H. Lund, *Acta Chem. Scand.*, **B30**, 267 (1976).
[69] A. Berg, J. Lam, P. E. Hansen, *Acta Chem. Scand.*, **B40**, 665 (1986).
[70] (a) H. Lund, J. Simonet, *Bull. Soc. Chim. Fr.*, **1973**, 1843; (b) J. Simonet, M.-A. Michel, H. Lund, *Acta Chem. Scand.*, **B29**, 489 (1975).
[71] E. Hebert,- J.-P. Mazaleyrat, Z. Welvart, *New J. Chem.*, **9**, 75 (1985).
[72] (a) H. Lund, J. Simonet, *C. R. Acad. Sci. Ser. C*, **277**, 1387 (1973); (b) H. Lund, *Acta Chem. Scand.*, **B31**, 424 (1977).
[73] P. Boy, C. Combellas, C. Suba, A. Thiébault, *J. Org. Chem.*, **59**, 4482 (1994).
[74] (a) C. Combellas, H. Marzouk, C. Suba, A. Thiébault, *Synthesis*, **1993**, 788; (b) C. Amatore, C. Combellas, N.-E. Lebbar, A. Thiébalut, J.-N. Verpeaux, *J. Org. Chem.*, **60**, 18 (1995).
[75] (a) C. Amatore, C. Combellas, J. Pinson, J.-M. Savéant, A. Thiébault, *J. Chem. Soc., Chem. Commun.*, **1988**, 7; (b) N. Alam, C. Amatore, C. Combellas, J. Pinson, J.-M. Savéant, A. Thiébault, J.-N. Verpeaux, *J. Org. Chem.*, **53**, 1496 (1988); (c) N. Alam, C. Amatore, C. Combellas, A. Thiébault, J.-N. Verpeaux, *J. Org. Chem.*, **55**, 6347 (1990); (d) C. Combellas, C. Suba, A. Thiébault, *Tetrahedron Lett.*, **33**, 4923 (1992).
[76] (a) M. Chahma, C. Combellas, H. Marzouk, A. Thiébault, **32**, 6121 (1991); (b) M. Chahma, C. Combellas, A. Thiébault, *Synthesis*, **1994**, 366.
[77] M. Médebielle, M. A. Oturan, J. Pinson, J.-M. Savéant, *Tetrahedron Lett.*, **34**, 3409 (1993).
[78] C. Combellas, C. Suba, A. Thiébault, *Tetrahedron Lett.*, **35**, 5217 (1994).

[79] V. J. Koshy, V. Swayambunathan, N. Periasamy, *J. Electrochem. Soc.*, **127**, 2761 (1980).

[80] G. S. Reddy, S. J. Reddy, *J. Electrochem. Soc. India*, **41-3**, 160 (1992).

[81] (a) C.-H. Pyun, S.-M. Park, *J. Electrochem. Soc.*, **132**, 426 (1985); (b) S.-I. Fukuzumi, Y. Ono, T. Keii, *Bull. Chem. Soc. Jpn.*, **46**, 3353 (1973).

[82] S. I. Bailey, I. M. Ritchie, F. R. Hewgill, *J. Chem. Soc., Perkin Trans. II*, **1983**, 645.

[83] (a) A. Capon, R. Parsons, *J. Electroanal. Chem., Interfacial Electrochem.*, **46**, 215 (1973); (b) T. W. Rosanske, D. H. Evans, *J. Electroanal. Chem., Interfacial Electrochem.*, **72**, 277 (1976); (c) I. Biryol, M. Kabasakaloglu, S. Uneri, *Bull. Electrochem.*, **6**, 793 (1990).

[84] B. D. Sviridov, T. D. Nikolaeva, V. F. Venderina, S. I. Zhdanov, *Zh. Obshch. Khim.*, **55**, 821 (1985).

[85] L. Mandell, S. M. Cooper, B. Rubin. C. F. Campana, R. A. Day, *J. Org. Chem.*, **48**, 3132 (1983).

[86] V. Glezer, J. Stradins, J. Freimanis, L. Baider, *Electrochim. Acta*, **28**, 87 (1983).

[87] M. E. Niyazymbetov, I. V. Aref'eva, E. I. Zakharova, L. D. Konyushkin, S. M. Alekseev, V. P. Litvinov, R. P. Evstigneeva, *Izv. Akad. Nauk Ser. Khim.*, **1992**, 2605.

[88] K. Hu, M. E. Niyazymbetov, D. H. Evans, *Tetrahedron Lett.*, **36**, 7027 (1995).

[89] (a) H. Lund, *The Chemistry of Hydroxyl Group*, pp 274, Interscience Publishers Inc., New York (1971); (b) Ch. Comninellis, E. Plattner, *J. Appl. Electrochem.*, **15**, 771 (1985).

[90] S. U. Pedersen, T. Lund, *Acta Chem. Scand.*, **45**, 397 (1991).

[91] (a) R. L. Blankespoor, A. N. K. Lau, L. L. Miller, *J. Org. Chem.*, **49**, 4441 (1984); (b) R. L. Blankespoor, D. L. Schutt, M. B. Tubergen, R. L. de Jong, *J. Org. Chem.*, **52**, 2059 (1987); (c) R. L. Blankespoor, R. Hsung, D. L. Schutt, *J. Org. Chem.*, **53**, 3032 (1988).

[92] J. Haggin, *C&EN*, **62**, 20 (1984).

[93] A. K. Newell, J. H. P. Utley, *J. Chem. Soc., Chem. Commun.*, **1992**, 800.

[94] L. Echegoyen, D. A. Gustowski, V. J. Gatto, G. W. Gokel, *J. Chem. Soc., Chem. Commun.*, **1986**, 220.

[95] A. V. Bulatov, I. M. Sosonkin, A. T. Nikitaev, G. A. Kalb, M. L. Khidekel, *Izv. Akad. Nauk SSSR, Ser. Khim.*, **1985**, 983.

[96] T. Erabi, M. Tanaka, *Bull. Chem. Soc. Jpn.*, **56**, 15 (1983).

[97] A. Anne, J. Moiroux, *New J. Chem.*, **9**, 83 (1985).

[98] A. Anne, F. Bennani, J.-C. Florent, J. Moiroux, C. Monneret, *Tetrahedron Lett.*, **26**, 2641 (1985).

[99] K. Kano, T. Konse, N. Nishimura, T. Kubota, *Bull. Chem. Soc. Jpn.*, **57**, 2383 (1984).

[100] K. Kano, T. Konse, T. Kubota, *Bull. Chem. Soc. Jpn.*, **58**, 424 (1985).

[101] (a) K. Takahashi, T. Sakai, *Chem. Lett.*, **1993**, 157; (b) K. Takahashi, T. Suzuki, K. Akiyama, Y. Ikegami, Y. Fukazawa, *J. Am. Chem. Soc.*, **113**, 4576 (1991).

[102] I. Mogi, Y. Nakagawa, *Nippon Kagaku Kaishi* (Japanese), **11**, 1218 (1990).

[103] (a) J. A. Richards, D. H. Evans, *J. Electroanal. Chem., Interfacial Electrochem.*, **81**, 171 (1977); (b) L. I. Kudinova, A. A. Volod'kin, V. V. Ershov, T. I. Prokofeva, *Izv. Akad. Nauk SSSR, Ser. Khim.*, **1978**, 1503; (c) M. F. O. Goulart, J. H. P. Utley, *J. Org. Chem.*, **53**, 2520 (1988).

[104] M. F. Nielsen, S. Spriggs, J. H. P. Utley, Y. Gao, *J. Chem. Soc., Chem. Commun.*, **1994**, 1395.

[105] A. M. Kini, D. O. Cowan, F. Gerson, R. Möckel, *J. Am. Chem. Soc.*, **107**, 556 (1969).

[106] N. Martin, J. A. Navarro, C. Seoane, A. Albert, F. H. Cano, J. Y. Becker, V. Khodorkovsky, E. Harlev, M. Hanack, *J. Org. Chem.*, **57**, 5726 (1992).

[107] N. Martin, R. Behnisch, M. Hanack, *J. Org. Chem.*, **54**, 2563 (1989).

[108] P. de la Cruz, N. Martin, F. Miguel, C. Seoane, A. Albert, F. H. Cano, A. Leverenz, M. Hanack, *Synth. Metals*, **48**, 59 (1992).

[109] A. Gennaro, F. Maran, A. Maye, E. Vianello, *J. Electroanal. Chem., Interfacial Electrochem.*, **185**, 353 (1985).

[110] J. Simonet, M. C. E. Badre, M. Cariou, *J. Electroanal. Chem.*, **334**, 169 (1992).

[111] M. Mir, M. Espin, J. Marquet, I. Gallardo, C. Tomasi, *Tetrahedron Lett.*, **35**, 9055 (1994).

[112] A. Belkasmioui, J. Simonet, *Tetrahedron Lett.*, **32**, 2481 (1991).

[113] C. Degrand, *J. Org. Chem.*, **52**, 1421 (1987).

[114] S. Torii, H. Okumoto, H. Satoh, T. Minoshima, S. Kurozumi, *Synlett*, **1995**, 439.

[115] V. A. Ryabinin, V. F. Starichenko, V. D. Shteingarts, *Mendeleev Commun.*, **1992**, 37.

[116] (a) C. Walling, R. A. Johnson, *J. Am. Chem. Soc.*, **97**, 363 (1975); (b) J. Wellmann, E. Steckhan, *Chem. Ber.*, **110**, 3561 (1977).

[117] (a) T. Kinoshita, J. Harada, S. Ito, K. Sasaki, *Angew. Chem., Int. Ed. Engl.*, **22**, 502 (1983); (b) K. Sasaki, S. Ito, Y. Saheki, T. Kinoshita, T. Yamasaki, J. Harada, *Chem. Lett.*, **1983**, 37.

[118] R. Ohnishi, A. Aramata, *Stud. Org. Chem.*, **30**, 337 (1987).

[119] B. Fleszar, A. Sobkowiak, *Electrochim. Acta*, **28**, 1315 (1983).

[120] K. Otsuka, I. Yamanaka, *Chem. Lett.*, **1990**, 509.

[121] Y. Matsuda, K. Nishi, T. Azuma, *Denki Kagaku*, **52**, 635 (1984).

[122] P.-M. Allemand, A. Koch, F. Wudl, Y. Rubin, F. Diederich, M. M. Alvarez, S. J. Anz, L. Whetten, *J. Am. Chem. Soc.*, **113**, 1050 (1991).

[123] D. Dubois, K. M. Kadish, *J. Am. Chem. Soc.*, **113**, 4364 (1991).
[124] D. Dubois, K. M. Kadish, S. Flanagan, R. E. Haufler, *J. Am. Chem. Soc.*, **113**, 7773 (1991).
[125] (a) Y. Ohsawa, T. Saji, *J. Chem. Soc., Chem. Commun.*, **1992**, 781; (b) Q. Xie, E. Pérez-Cordero, L. Echegoyen, *J. Am. Chem. Soc.*, **114**, 3978 (1992).
[126] W. R. Fawcett, M. Opallo, M. Fedurco, J. W. Lee, *J. Am. Chem. Soc.*, **115**, 196 (1993).
[127] D. Dubois, M. T. Jones, K. M. Kadish, *J. Am. Chem. Soc.*, **114**, 6446 (1992).
[128] M. E. Niyazymbetov, D. H. Evans, *J. Electrochem. Soc.*, **1995**, 2655.
[129] C. Caronm, R. Subramanian, F. D'Souza, J. Kim, W. Kutner, M. T. Jones, K. M. Kadish, *J. Am. Chem. Soc.*, **115**, 8505 (1993).
[130] W. A. Kalsbeck, H. H. Thorp, *J. Electroanal. Chem., Interfacial Electrochem.*, **314**, 363 (1991).
[131] M. Eiermann, F. Wudl, *J. Am. Chem. Soc.*, **116**, 8364 (1994).
[132] Q. Li, F. Wudl, C. Thilgen, R. L. Whetten, F. Diederich, *J. Am. Chem. Soc.*, **114**, 3994 (1992).

5.   Electroreduction of Nitrogen Compounds

5.1   Reduction of Nitro and Nitroso Compounds

5.1.1   Electroreduction of Nitro Compounds

5.1.1.1   Aliphatic Nitro Compounds

Primary and secondary aliphatic nitro compounds **1** have been electrochemically reduced into the nitroso intermediates **2**, which can undergo a transformation into the oxime **5** [1]. Under an aqueous acidic conditions, the generation of nitroso derivatives has been confirmed on electroreduction of tertiary aliphatic nitro compounds [2].   The continuing two-electron reduction of the nitroso compounds **2** can lead to the formation of the corresponding hydroxylamine **3** as a precursor of the amine **4** [3].   The electroreduction of primary and secondary nitroalkanes takes place according to eq. (5.1).   The choice of adequate potential, electrode material, and pH value of electrolytes is found to be influential factors for the product selectivity [4].   The behavior of nitro and nitroso compounds in various solvent systems, *i.e.*, MeCN, DMF, H_2O, liq. NH_3, AcOH, conc. H_2SO_4 etc., has been well documented in monograph [5].   The electroreduction of tertiary nitroalkanes in aqueous buffer solutions may provide either hydroxylamines (pH 1.8~4.2, 1.15~1.30 V *vs.* SCE) or amines (pH 1.8, 1.65 V), depending upon the reduction potential [6].

$$\underset{\textbf{1}}{RCH_2-NO_2} \xrightarrow[-H_2O]{2e^-,\ 2H^+} \underset{\textbf{2}}{RCH_2-NO} \xrightarrow[-0.8\ V\ vs.\ SCE]{2e^-,\ 2H^+} \underset{\textbf{3}}{RCH_2-NHOH} \tag{5.1}$$

$$\underset{\textbf{5}}{R-CH=N-OH} \qquad\qquad \underset{\textbf{4}}{RCH_2-NH_2} \quad \Big| \; 2e^-,\ 2H^+ \;\; -1.6\ V\ vs.\ SCE$$

The electrosynthesis of *N*-methylhydroxylamine **7** by electroreduction of nitromethane **6** can be carried out in an aqueous H_2SO_4(10%)-(Pb) system in a divided cell in higher than 95% both Faradaic and conversion yields (eq. (5.2)) [7].   Large-scale electrosynthesis of

$$\underset{\textbf{6}}{Me-NO_2} \xrightarrow[-0.42\ V\ vs.\ Ag/AgCl,\ 95\%\ yield]{aq.\ H_2SO_4(10\%,\ w/w)-(Pb)} \underset{\textbf{7}}{Me-NHOH} \tag{5.2}$$

amino alcohols from nitro alcohols has been reported [8,9].   Amino alcohols such as tris(hydroxymethyl)aminomethane, 2-amino-2-methyl-1,3-propanediol, 2-amino-1,3-propanediol, and 2-amino-1-butanol are electrosynthesized in an aqueous sulfuric acid solution.   The influence of operating conditions such as initial reagent concentration, programming current density, ratio of the electrode surface with reagent quantity, and electrolyte flow is discussed.   Electroreductive acetylation of nitro and nitroso compounds **8** has been attained by trapping anionic intermediates with acetic anhydride, leading to *N,O*-diacetyl-*N*-substituted hydroxylamines **9** (eq. (5.3)) [10].

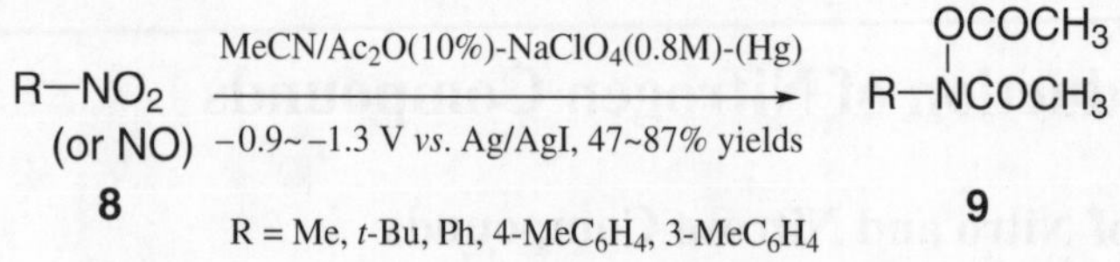

$$R-NO_2 \xrightarrow[\substack{-0.9\sim-1.3\ \text{V } vs.\ \text{Ag/AgI, 47}\sim\text{87\% yields}}]{\text{MeCN/Ac}_2\text{O(10\%)-NaClO}_4\text{(0.8M)-(Hg)}} \begin{matrix} \text{OCOCH}_3 \\ | \\ R-\text{NCOCH}_3 \end{matrix} \qquad (5.3)$$

$$\text{(or NO)}$$

8   **9**

$$R = \text{Me, } t\text{-Bu, Ph, 4-MeC}_6\text{H}_4\text{, 3-MeC}_6\text{H}_4$$

The electroreduction of γ-nitro esters **10** can be first converted into the corresponding hydroxylamine **11**, which tends to undergo a sufficient first ring closure by loss of methanol to give N-hydroxypyrrolidinones **12** in a quantitative yield (eq. (5.4)) [11].   Controlled

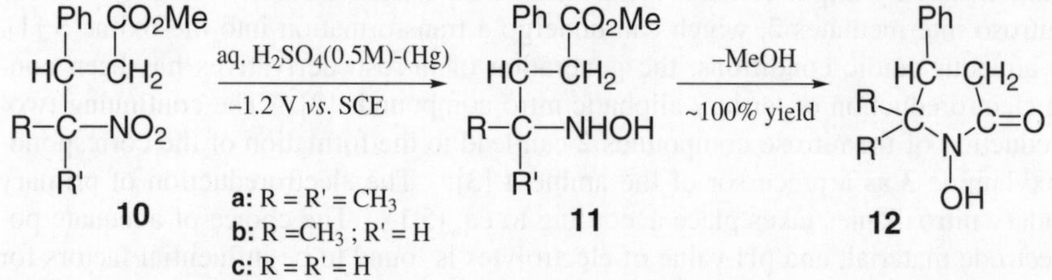

$$(5.4)$$

10   a: R = R' = CH$_3$ **11**   **12**
          b: R =CH$_3$; R' = H
          c: R = R' = H

potential electrolysis of γ-nitroalkanones **13** affords the corresponding 1-pyrroline-1-oxides **14**, pyrrolines **15**, and pyrrolidines **16** in quantitative yields [12].   The electrochemical conversion of **13** into **14** has been performed in an aqueous NH$_3$/Acetone-NH$_4$Cl-(Pt/Hg) system at –1.20 V $vs.$ SCE (eq. (5.5)).   The electrosynthesis of the pyrrolines **15** can be performed in an aqueous EtOH-H$_2$SO$_4$(0.5M)-(Pt/Hg) system at –1.01 V $vs.$ SCE (eq. (5.6)).   More drastic electroreduction conditions in an AcOH/EtOH-AcONa-(Pt/Hg) system at –1.50 V provides the pyrrolidines **16**, exclusively (eq. (5.7)) [12,13].

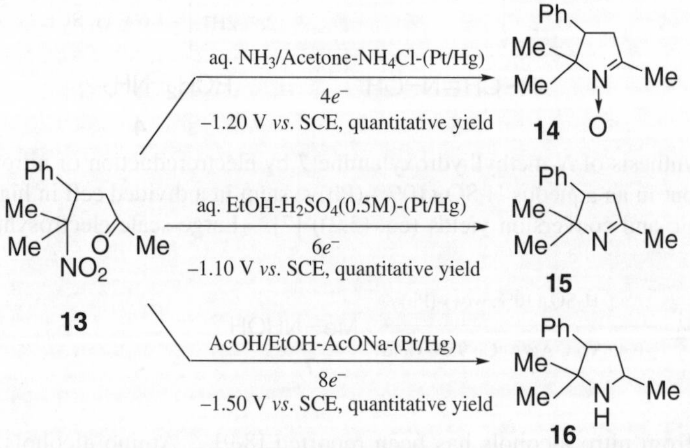

------ (5.5)

------ (5.6)

------ (5.7)

It has been found that the rate constant for the cleavage reaction increases about three orders of magnitude on going from a mono-nitroalkane to a vic-dinitroalkane.   For example, one-electron reduction of 1,1-dinitrocyclohexane **17** is followed by rapid cleavage of a C-N bond, giving nitrite and 1-nitrocyclohexyl radical as a precursor of the dimer **18** (eq. (5.8)) [14].

$$\underset{\textbf{17}}{\text{(cyclohexane)}\begin{array}{c}NO_2\\NO_2\end{array}} \xrightarrow[\text{-1.5 V vs. Ag/Ag$^+$, 71\% yield}]{\text{DMF-Bu$_4$NPF$_6$(0.1M)-(Pt)}} \underset{\textbf{18}}{\begin{array}{c}NO_2\\ \\NO_2\end{array}} \qquad (5.8)$$

Platinum electrodes deposited by heavy metal monolayers are used in the electroreduction of phenyldinitromethane **19**. The Tl, Pb, and Bi adsorbates markedly catalyze the reduction of the substrate **19**, leading to **21** via **20** (eq. (5.9)) in an aqueous HClO$_4$(0.5M) system (eq. (5.10)) [15]. Electrogenerated base-promoted Michael addition of ethyl

$$\underset{\textbf{19}}{\text{PhCH}\begin{array}{c}NO_2\\NO_2\end{array}} + 2\,e^- + 2\,H^+ \xrightarrow{\text{aq. HClO}_4} \underset{\textbf{20}}{PhCH_2NO_2} + HNO_2 \qquad \cdots (5.9)$$

$$\textbf{19} + 2\,e^- + 2H^+ \xrightarrow{\text{aq. HClO}_4} \underset{\textbf{21}}{PhCH=NOH} + H_2O \qquad \cdots (5.10)$$

	Yield, %	
	Pt/Tl	Pt/Pb
20	50~52	10~12
21	34~35	49~52

nitroacetate to ethyl acrylate, acrylonitrile, and methyl vinyl ketone, leading to either **23** and **25** followed by the electroreductive removal of the nitro group has been attained in an MeCN-Bu$_4$NBr(0.2M)-(Mg/Pt) system, giving **24** and **26** in good yields (eqs. (5.11, 5.12)) [16]. The anion radical of azobenzene can be used as an electrogenerated base to catalyze

$$\cdots (5.11)$$
$$\cdots (5.12)$$

the Michael addition of primary nitro alkanes **27** with a variety of acceptors **28**, giving the adduct **29** in good yield (eq. 5.13)) [17]. Under oxygen atmosphere, the nitroalkane **30** can

$$\underset{\textbf{27}}{\sim NO_2} + \underset{\textbf{28}}{\begin{array}{c}R^2\\ \\R^1 \quad CO_2Me\end{array}} \xrightarrow[\substack{\text{PhN = NPh/O}_2,\ 46\text{~}60\%\ \text{yields}\\ R^1 = R^2 = H,\ \text{alkyl}}]{\text{MeCN-Bu}_4\text{NBr-(Hg)}} \underset{\textbf{29}}{\begin{array}{c}O\quad R^2\\ \\R^1\quad CO_2Me\end{array}} \qquad (5.13)$$

be converted into the corresponding carbonyl group **31** (eq. (5.14)) [17].   1-Nitroalkenes **32** are electroreduced in high yields in an PrOH/H$_2$O(3/2 v/v)-H$_2$SO$_4$(0.1M)-(Hg) system to afford the corresponding oximes **33** (eq. (5.15)) [18].   The direct conversion of the

$$\underset{\mathbf{30}}{\text{(structure with NO}_2\text{ and O)}} \xrightarrow[\text{82~86\% yields}]{\text{MeCN-Bu}_4\text{NBr-O}_2\text{-(Hg)}} \underset{\mathbf{31}}{\text{(structure with two O)}} \qquad (5.14)$$

$$\underset{\mathbf{32}}{\underset{\text{R, R}^1 =\text{Ar, alkyl, alkenyl}}{\text{(structure with R, R}^1\text{, NO}_2\text{)}}} \xrightarrow[\substack{-0.25~-0.5 \text{ V vs. SCE,} \\ 85~93\% \text{ yields}}]{\substack{i\text{-PrOH/H}_2\text{O(3/2v/v)-} \\ \text{H}_2\text{SO}_4\text{(0.1M)-(Hg)}}} \underset{\mathbf{33}}{\text{(structure with NOH)}} \xrightarrow[\substack{-1.1~-1.3 \text{ V vs. SCE} \\ 60~69\% \text{ yields}}]{} \underset{\mathbf{34}}{\text{(structure with NH}_2\text{)}} \qquad (5.15)$$

nitroalkenes into the corresponding amines has been performed by controlling the potentials at $-1.1~-1.3$ V *vs.* SCE [19].   The conversion of 1-nitroalkenes **35** into nitriles **36** has been performed by electroreduction in a DMF-Et$_4$NOTs-(C) in the presence of titanium(IV) chloride (eq. (5.16)) [20].   Electroreductive conversion of nitroalkenes **37** in an aqueous

$$\underset{\mathbf{35}}{\text{R-CH=CH-NO}_2} \xrightarrow[\substack{64~95\% \text{ yields} \\ \text{R = alkyl, aryl}}]{\text{DMF-Et}_4\text{NOTs-(C), TiCL}_4} \underset{\mathbf{36}}{\text{R-CH}_2\text{CN}} \qquad (5.16)$$

CH$_2$Cl$_2$/Dioxane-HClO$_4$-(Pb) system as a two-phase affords the corresponding oximes **38** in 55~91% yields when the organic phase is treated with a mixture of NH$_2$OH•HCl and sodium acetate after passage of 3.5 F/mol of electricity (eq. (5.17)).   On the other hand, upon treatment with aqueous 37% formaldehyde, the corresponding ketones **39** are obtained in high yields (eq. (5.18)) [21].   The direct conversion of nitroalkenes to oximes has been attempted.   The electroreduction of β-substituted nitro alkanes in an aqueous MeOH(80mL)-H$_2$SO$_4$(20%,10mL)-(Pt) system affords the corresponding oximes in 17~67% yields [22].

$$\underset{\mathbf{37}}{\text{(structure with R}^1\text{, R}^2\text{, NO}_2\text{)}} \begin{cases} \xrightarrow[\substack{\text{2) Treatment with aq. NH}_2\text{OH} \\ \text{after electrolysis, 70~95\% yields}}]{\text{1) aq. CH}_2\text{Cl}_2\text{/Dioxane-HClO}_4\text{-(Pb)}} \underset{\mathbf{38}}{\text{(structure with N-OH)}} & \cdots (5.17) \\ \\ \xrightarrow[\substack{\text{3) Treatment with aq. HCHO} \\ \text{after electrolysis, 55~91\% yields}}]{\text{1) aq. CH}_2\text{Cl}_2\text{/Dioxane-HClO}_4\text{-(Pb)}} \underset{\mathbf{39}}{\text{(structure with O)}} & \cdots (5.18) \end{cases}$$

$$R^1 = \text{Ar, } R^2 = \text{alkyl}$$

The electroreduction of 1-phenyl-1-cyanonitroethylene **40** can lead to the formation of a cyclic compound, 5-amino-4-phenyl-isoxazole **42**, *via* the ene-hydroxylamine **41** as a transient (eq. (5.19)) [23].   The reaction path has been proved to be a function of pH value

$$Ph \underset{NC}{\overset{H}{C}} = C \underset{NO_2}{\overset{H}{}} \quad \xrightarrow[\substack{pH\ 0.85 \\ aq.\ MeCN(25\%) \\ 100\%\ yield}]{+4e^-,\ 4H^+} \quad Ph-\overset{H}{\underset{NC}{C}}-\overset{H}{\underset{\overset{N}{\underset{OH}{}}}{C}} \quad \xrightarrow{H^+} \quad Ph \underset{H_2N}{\overset{H}{}} \underset{O}{\overset{N}{}} \quad (5.19)$$

$$\mathbf{40} \qquad\qquad\qquad \mathbf{41} \qquad\qquad \mathbf{42}$$

in the medium.   6-Nitro-3-tosyloxycholestene **43** can be converted into the cyclopropane derivative **44** under electroreduction conditions without affecting the nitro group (eq. (5.20)) [24].   The electroreductive conversion of secondary nitroalkanes **45** into the corresponding alkanones **46** has been developed by electrolysis in an MeOH-HCO$_2$Na-(Pt) system (eq. (5.21)) [25].   By this procedure, a variety of γ-keto esters and 1,4-dikenones can be obtained in good yield.

$$\xrightarrow[\text{-1.49 V \textit{vs.} SCE, 61$\sim$70\% yields}]{\text{DMF-Bu}_4\text{NClO}_4(0.1\text{M})\text{-(Pt)}} \qquad (5.20)$$

$$\mathbf{43} \qquad X = Cl,\ Br \qquad\qquad\qquad\qquad \mathbf{44}$$

$$\underset{R^1}{\overset{R^2}{CH}}-NO_2 \quad \xrightarrow[\text{40$\sim$90\% yields}]{\text{MeOH-HCO}_2\text{Na-(Pt)}} \quad \underset{R^1}{\overset{R^2}{C}}=O \qquad (5.21)$$

$$\mathbf{45} \qquad\qquad\qquad\qquad \mathbf{46}$$

$$R^1 = Et,\ Bu,\ pentyl,\ hexyl,\ (CH_2)_2Ac$$

$$R^2 = Me,\ CO_2Bu,\ CMe_2CH_2Ac,\ (CH_2)_2CO_2Et,$$

$$(CH_2)_2COMe,\ (CH_2)_2CN$$

The N-N bond of nitroamines can be electrochemically cleaved to give disubstituted amines under the reduction conditions [26].   The electroreduction of 3-(N-nitro-N-phenyl-hydrazono)pentane-2,4-dione **47** affords N-hydroxyamino compound **48** as the sole product in a DMF-Et$_4$NBr-(C) system at -0.5 V *vs.* Ag/AgCl, pH 2.5~4.0 (eq. (5.22)) [27].

$$\underset{\mathbf{47}}{\left[\text{N}\right]-\overset{NO_2}{\underset{}{N}}-N=C\overset{COMe}{\underset{COMe}{}}} \quad \xrightarrow[\substack{-0.5\ V\ \textit{vs.}\ Ag/AgCl,\ pH\ 2.5\sim4.0 \\ quantitative\ yield}]{\text{DMF-Et}_4\text{NBr-(C)}} \quad \underset{\mathbf{48}}{Ph-\overset{R}{\underset{}{N}}-NHOH} \qquad (5.22)$$

$$R = C(COMe)_2$$

The cleavage of the N-O bond of benzyl nitrate **49** proceeds in a DMF-Bu$_4$NBF$_4$-(Hg) system at -1.0 V *vs.* Ag/Ag$^+$ to give benzyl alcohol (50~55%) and benzaldehyde (35~40%) (eq. (5.23)) [28].

$$PhCH_2ONO_2 \quad \xrightarrow[\text{-1.0 V \textit{vs.} Ag/ Ag$^+$}]{\text{DMF-Bu}_4\text{NBF}_4\text{-(Hg)}} \quad PhCH_2OH + PhCHO + NO_2^- \qquad (5.23)$$

$$\underset{\mathbf{49}}{} \qquad\qquad\qquad\qquad \underset{\text{50$\sim$55\% yields}}{} \quad \underset{\text{35$\sim$40\% yields}}{}$$

The electroreduction of the *N*-nitroso compounds may provide either hydrazines or amines, depending on the solvent and the amount of proton donor. For example, *N*-nitroso morpholine **50** can be converted into the hydrazine **51** in an MeCN-AcOH-(Hg) system, whereas it is converted into morphorine **52** in a DMF-AcOH-(Hg) system (eq. (5.24)) [29].

$$(5.24)$$

5.1.1.2 Aromatic Nitro Compounds

There are many investigations concerning the electroreduction of nitrobenzene under various conditions [30]. The reaction feature depends on the electrolyte conditions, *e.g.* solvent, pH, etc., and on the electrode surface conditions, *e.g.* electrode materials, pre-treatment, etc.. The reduction of nitrobenzene proceeds as a strictly four-electron process [31a]. The first reduction wave of nitro- and dinitrobenzenes corresponds to reversible electron transfer to the molecule to give a primary anion radical [31b,c]. Nitrobenzene is hydrogenated into aniline with a sodium *p*-toluenesulfonate supporting electrolyte, whereas the tetraethylammonium *p*-toluenesulfonate supporting electrolyte causes to produce cyclohexylamine [31d].

Electroreduction of aromatic nitro compounds in acid solution generally yields hydroxylamines, which may be reduced to amines at a more negative potential or may undergo rearrangements [3].

A high-current-density electrosynthesis of amines from nitro compounds has been attained using metal powders as mediators [32]. In particular, copper, iron, tin, and zinc powders can each be formed in very good yield by electrolysis at 0.5 A/cm^2 current density of a solution of their chloride or sulfate salts (0.4~1.0 M) in an appropriate acid solution. Tin and zinc are the best metals for this reaction; the conversion and current yields can exceed 90%. Use of Devarde copper electrodes for the electrohydrogenation of nitrobenzene, leading to aniline is also effective [33]. Electroreduction of nitrobenzene on Ti/TiO$_2$ cathodes as a heterogeneous redox catalysis proceeds efficiently [34].

In an aqueous medium, the electroreduction of nitrobenzene yields phenylhydroxylamine, azobenzene, azoxybenzene or aniline, depending on the experimental conditions and the electrode material. The intermediate nitrosobenzene, however, cannot be isolated in the above aqueous media. In an aprotic medium (THF) in the presence of an added proton donor (PhCO$_2$H), nitrosobenzene becomes the only the product under the electrolysis conditions [35]. The optimized conditions for electrosynthesis of nitrosobenzene bearing electron-withdrawing groups (CN, CO$_2{}^-$, CO$_2$Me, CHO, COMe) in meta or para position are found to be the following electrolysis system: aq. DMF(20~40%)-AcONa buffer (pH 5.7) using a porous carbon electrode [36].

The major products of the electroreduction of nitrobenzene **53** in an aqueous KOH(0.1M)-(Ag) system are found to be phenylhydroxylamine **54** and aniline **55** (eq. (5.25)), and in the course of the reduction, strong Raney scattering signals of azobenzene **56** are observed on the electrode surface (eq. (5.26)) [37,38].

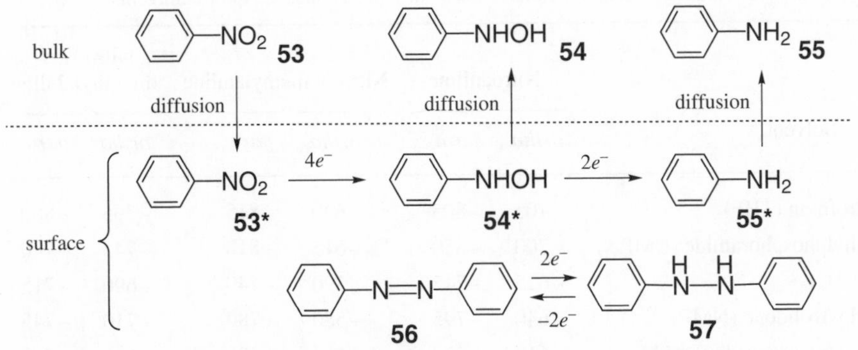

bulk

[structure 53: benzene ring—NO$_2$] **53**   [structure 54: benzene ring—NHOH] **54**   [structure 55: benzene ring—NH$_2$] **55**

(5.25)

diffusion   diffusion   diffusion

surface

[structure 53*: benzene ring—NO$_2$] **53*** $\xrightarrow{4e^-}$ [structure 54*: benzene ring—NHOH] **54*** $\xrightarrow{2e^-}$ [structure 55*: benzene ring—NH$_2$] **55***

[structure 56: benzene—N=N—benzene] **56** $\underset{-2e^-}{\overset{2e^-}{\rightleftarrows}}$ [structure 57: benzene—N(H)—N(H)—benzene] **57**

(5.26)

The use of aprotic solvents undergoes a one-electron transfer, leading to the formation of a stable anion radical. In the presence of an efficient proton donor, the corresponding hydroxylamine becomes the product. The presence of alkyl halides or acetic anhydride as an electrophile causes the course of the overall reaction to change in a different orientation.

The influence of ionic strength on the current-potential responses for the electroreduction of nitrobenzenes, *e.g.* nitrobenzene, nitrotoluene, bromonitrobenzene, in aprotic solvents has been investigated using microelectrodes. The interaction between anion radicals and tetraalkylammonium cations is found always to be weak but the cations interact strongly with more reduced intermediates [39].

The entropy of formation of the anion radicals is observed to be influenced by both steric and electronic effects of the alkyl substituents [40].

The activity of Pd-, Rh-, and Pd-Raney catalysts in the electroreduction and liquid-phase hydrogenation of nitrobenzene increases slightly on addition of Pd or Rh, passes through a maximum (at 10 atom% Pd) and its lowest value for Raney palladium. In the hydrogenation of nitrobenzene, the 10 atom% Pd alloys show the greatest catalytic activity [41].

p-Nitroaniline can be electrochemically reduced to *p*-phenylene diamine at stationary nickel and copper amalgamated electrodes in an aqueous sulfuric acid solution. The effect of current density, temperature, depolarizer ratio, and nature of the cathode material has been examined in order to optimize the electrolysis conditions to obtain over 90% yield [42].

Use of two consecutive porous electrodes and one or two counter-electrodes satisfactorily lead to the conversion of *m*-nitrobenzoic acid into *m*-nitrosobenzoic acid [43]. The reduction-peak potential values of nitroanilines in aprotic solvents are shown in Table 5.1. [44]. The results from the electroreduction of substituted nitrobenzenes are summarized in Table 5.2. [32~36, 45~78].

The electroreduction of 2- and 4-nitrodiphenylamines in the presence of a proton donor (*e.g.* phenol) in a DMF-Et$_4$NI system causes the second wave to shift in polarography (2-nitro: wave I 1.028~1.059; wave II 1.446~1.571; 4-nitro: wave I 1.201~1.245; wave II 1.538~1.629) to the more positive side, indicating that the reduction becomes facile in the presence of the proton donor [79].

The electroreduction of nitrobenzene, leading to *p*-aminophenol, is a keen research subject in the practical sense. The most typical electrolysis procedure involves the reduction of nitrobenzene in the aqueous H$_2$SO$_4$(20%)-(amalgamated Cu) system, giving 56~71% of *p*-aminophenol in the presence of bismuth chloride as additive [80]. The use of a TiO$_2$/Ti

Table 5.1 Reduction-peak Potential Values* of Nitroanilines in Aprotic Solvents

Solvent	Nitroaniline		Nitro-N-methylaniline		Nitro-N,N-dimethylaniline	
	ortho	*para*	*ortho*	*para*	*ortho*	*para*
Tetrahydrofuran (THF)	−705	−805	−640	−815	−765	−800
Hexamethylphosphoramide (HMPA)	−700	−850	−615	−810	−745	−780
Acetone (AC)	−625	−745	−590	−740	−690	−715
N-Methylpyrrolidone (NMP)	−640	−795	−580	−780	−710	−745
N,N-Dimethylacetamide (DMA)	−610	−780	−565	−775	−680	−710
Pyridine (PY)	−615	−725	−590	−735	−685	−710
Benzonitrile (BN)	−600	−720	−580	−710	−655	−700
Dimethylformamide (DMF)	−585	−710	−530	−720	−645	−680
Trimethyl phosphate (TMP)	−600	−720	−555	−740	−660	−715
1,2-Dichloroethane (DCE)	−595	−710	−575	−720	−685	−725
Acetonitrile (AN)	−530	−610	−490	−640	−595	−625
Dimethyl sulfoxide (DMSO)	−500	−625	−470	−625	−570	−610

*Ep^{red}/mV *vs.* SCE

[Reproduced with permission from M. A. Santa Ana et al., *J. Chem. Soc. Perkin Trans. II*, **1985**, 1755]

electrode is found to be effective for improving current efficiency [48].   The preparative conversion of nitrobenzene to *p*-aminophenol in an electrolyte recirculated from a holding tank to a parallel-plate reactor has been studied in light of determining the effects of period-ic cell-voltage control on the selectivity for the *p*-aminophenol [81,82]

The rearrangement of the hydroxyamine intermediates derived from the electroreduc-tion of 2,3-dimethylnitrobenzene **58** has been attained in an EtOH(2%H$_2$O)-H$_2$SO$_4$(2%)-(C) system to give 4-ethoxy(or hydroxy)-2,3-dimethylaniline **60** (or **59**) in good yield (eq. (5.27)) [83].

(5.27)

p-Nitrophenol is reduced in the first step, after the transfer of 4e^- and 4H$^+$ ions, to *p*-hydroxyphenylhydroxylamine.   This hydroxylamine can undergo dehydration followed by the transfer of 2e^- and 2H$^+$ ions, leading to *p*-aminophenol [84].   The cathodic reduction of nitrobenzoyl chloride **62** in an Me$_2$CO-LiClO$_4$(0.3M)-(Hg) system at −0.60 V *vs.* SCE brings about the formation of 4-nitrobenzoic 4-(4-nitrobenzoylamino)benzoic anhydride **63** in quantitative yield (eq. (5.28)) [85].   Electroreductive cyclization of 2,β-dinitrostyrene **64**

Table 5.2a Results from Electroreduction of Substituted Nitrobenzenes

R^1	R^2	R^3	R^4	R^5	Electrolysis Conditions	Products,	Yield, %		Ref.
							Conv.	Current	
H	H	H	H	H	$H_2O/CH_2ClCHCl_2(7/3)$-$ZnCl_2(4M)$-(Al)	C	98	90	[45]
H	H	H	H	H	aq. $H_2SO_4(30-40\%)/Ti_2(SO_4)_2(3\%)$-(Cu)	C	95	95	[46a]
H	H	H	H	H	aq. $HCl(1M)/SnCl_2(0.4M)$-(Sn)	C	89	96	[32]
H	H	H	H	H	aq. $HCl(1M)/ZnCl_2(0.4M)$-(Zn)	C	93	89	[32]
H	H	H	H	H	aq. MeOH(50%)-$NaClO_4$/NaOH −2.0 V $vs.$ SCE	B	100	–	[47]
H	H	H	H	H	MeOH(1.5% H_2O)-KOH(0.28M)-(Ra-Ni)	C	92	91	[33]
H	H	H	H	H	aq. HCl(20%)-(Ti/TiO_2)	C	65	70	[34]
H	H	H	H	H	aq. $H_2SO_4(1M)$-Ti/TiO_2	C	90	94.5	[48]
H	H	H	H	H	MeOH/H_2O(93/7)-AcOH(0.54M)/AcONa(0.35M)-(Devarda Cu)	C	100	95	[49]
H	H	H	H	H	aq. MeOH-$CuSO_4$-(0.5M)-(Cu,Pt-Nafion)	C	–	80	[50]
H	H	H	H	H	THF-Bu_4NClO_4(0.2M)/$PhCO_2H$-(Pt) −1.0 V $vs.$ SCE	A	90	–	[35]
H	H	H	H	H	aq. H_2SO_4(1.88M)-(Ti/TiO_2)	C	99.5	91.8	[51]
H	H	H	H	H	aq. H_2SO_4(10%v/v)-(Ti/TiO_2)	C	95.5	93.6	[52]
H	H	H	H	H	aq.DMF(20%)-AcONa-(Hg)	B	–	99.3	[36]
H	H	H	H	H	aq.DMF(20%)-AcONa(pH5.7)-(C)	A	–	97.9	[36]

194

Table 5.2b Results from Electroreduction of Substituted Nitrobenzenes

R¹	R²	R³	R⁴	R⁵	Electrolysis Conditions	Products,	Conv.	Yield, % Current	Ref.
H	H	Ac	H	H	MeOH(1.5% H₂O)-KOH(0.28M)-(Ra-Ni)	C	79	85	[33]
H	H	EtO	H	H	aq. H₂SO₄(30-40%)/Ti₂(SO₄)₂(3%)-(Cu)	C	90.5	81	[46a]
H	H	EtO	H	H	aq. NaOH(1N)-(Cu)	C	90	60	[53]
H	H	i-Pr	H	H	aq. H₂SO₄(25%)-(Hg)	C	good	–	[54]
CN	H	H	H	H	aq.DMF(20%)-AcONa-(Hg)	B	–	90	[36]
H	CN	H	H	H	aq. HCl(1M)/SnCl₂(0.4M)-(Sn)	C	88	75	[32]
H	CN	H	H	H	aq.DMF(20%)-AcONa-(Hg)	B	–	72	[36]
H	CN	H	H	H	aq.DMF(20%)-AcONa(pH5.7)-(C)	A	–	51.8	[36]
H	H	CN	H	H	aq. HCl(1M)/SnCl₂(0.4M)-(Sn)	C	80	30	[32]
H	H	CN	H	H	aq.DMF(20%)-AcONa-(Hg)	B	–	89	[36]
H	H	CN	H	H	aq.DMF(20%)-AcONa(pH5.7)-(C)	A	–	45.3	[36]
CONHAr	H	H	H	H	EtOH/AcOH(4/1,v/v)-AcONa(2.5M)-(Hg)	A	~93	–	[55]
F	H	H	H	H	H₂O/CH₂ClCHCl₂(7/3)-ZnCl₂(4M)-(Al)	C	98	90	[45]
F	H	H	H	H	aq. HCl(1M)/SnCl₂(0.4M)-(Sn)	C	89	96	[32]
F	H	H	H	H	aq. HCl(1M)/ZnCl₂(0.4M)-(Zn)	C	92	90	[32]
Cl	H	H	H	H	aq. HCl(1M)/SnCl₂(0.4M)-(Sn)	C	–	96	[32]
Cl	H	H	H	H	aq. NaOH(1M)-(Cu)	C	95	49	[53]
Cl	H	H	H	H	aq. H₂SO₄(50%)-(Ti/TiO₂)	C	89.1	97.2	[46b]
Cl	H	H	H	H	aq. EtOH(30%)-H₂SO₄(7%)-(Ni)	C	95	–	[56]
H	Cl	H	H	H	aq. H₂SO₄(50%)-(Ti/TiO₂)	C	82.9	83.0	[46b]
H	H	Cl	H	H	aq. H₂SO₄(30-40%)/Ti₂(SO₄)₂(3%)-(Cu)	C	82.6	98.2	[46a]
H	H	Cl	H	H	aq. NaOH(1M)-(Cu)	C	91	30	[53]
H	H	Cl	H	H	aq. H₂SO₄(50%)-(Ti/TiO₂)	C	85.9	95.4	[46b]

Table 5.2c Results from Electroreduction of Substituted Nitrobenzenes

R^1	R^2	R^3	R^4	R^5	Electrolysis Conditions	Products,	Yield, %		Ref.
							Conv.	Current	
I	H	H	H	H	aq. KOH(pH>13)-(Raney Ni)	C	98	–	[57]
CHO	H	H	H	H	aq. EtOH-H$_2$SO$_4$(0.1M)-(Pb)	C	61.4	30.7	[58]
H	CHO	H	H	H	aq.DMF(20%)-AcONa(pH5.7)-(C)	A	–	66.3	[36]
H	H	CHO	H	H	aq.DMF(20%)-AcONa(pH5.7)-(C)	A	–	44.9	[36]
CO$_2$H	H	H	H	H	aq. NaH$_2$SO$_4$(0.25M)-NaH$_2$SO$_4$(0.25M)-(C)	A	93	–	[59]
CO$_2$H	H	H	H	H	aq. NaOH(1M)-(Cu)	C	72	24	[53]
H	CO$_2$H	H	H	H	aq. NaOH(1M)-(Cu)	C	100	0	[53]
H	CO$_2$H	H	H	H	aq. AcOH-AcONa(0.5M)-(porous C)	A	94	–	[43]
H	CO$_2$H	H	H	H	aq.DMF(20%)-AcONa(pH5.7)-(C)	A	–	91.7	[36]
H	H	CO$_2$H	H	H	aq. H$_2$SO$_4$(30-40%)/Ti$_2$(SO$_4$)$_2$(3%)-(Cu)	C	85	93	[46a]
H	H	CO$_2$H	H	H	aq. NaOH(1M)-(Cu)	C	100	23	[53]
H	H	CO$_2$H	H	H	aq.DMF(20%)-AcONa(pH5.7)-(C)	A	–	71.8	[36]
H	CO$_2$H	OH	H	H	aq. NaOH(1M)-(Cu)	C	89	83	[53]
Me	H	H	H	H	aq.H$_2$SO$_4$(5%)-(Ti/TiO$_2$)	C	84.8	83.5	[60]
Me	H	H	H	H	aq.H$_2$SO$_4$(5%)-(Ti/TiO$_2$)	C	84.8	83.5	[61]
H	Me	H	H	H	aq. H$_2$SO$_4$(30-40%)/Ti$_2$(SO$_4$)$_2$(3%)-(Cu)	C	70	80	[46a]
H	Me	H	H	H	aq. H$_2$SO$_4$(30-40%)/Ti$_2$(SO$_4$)$_2$(3%)-(Cu)	C	69.8	74.9	[46a]
H	Me	H	H	H	aq.H$_2$SO$_4$(10%)-(Ti/TiO$_2$)	C	88.5	86.2	[60]
H	H	Me	H	H	aq. H$_2$SO$_4$(30-40%)/Ti$_2$(SO$_4$)$_2$(3%)-(Cu)	C	80	90	[46a]
H	H	Me	H	H	aq. MeOH-H$_2$SO$_4$(1M)-(Cu, Pt-Nafion)	C	ca. 80	–	[62]
H	H	Me	H	H	aq. NaOH(1M)-(Cu)	C	–	–	[53]
H	H	Me	H	H	aq.H$_2$SO$_4$(20%)-(Ti/TiO$_2$)	C	93.1	93.0	[60]
MeO	H	H	H	H	aq. NaOH(1M)-(Cu)	C	85	41	[53]
H	MeO	H	H	H	aq. NaOH(1M)-(Cu)	C	100	18	[53]

Table 5.2d Results from Electroreduction of Substituted Nitrobenzenes

R¹	R²	R³	R⁴	R⁵	Electrolysis Conditions	Products,	Yield, %		Ref.
							Conv.	Current	
H	H	MeO	H	H	aq. NaOH(1M)-(Cu)	C	88	63	[53]
MeO	H	H	H	H	aq. H_2SO_4(30-40%)/$Ti_2(SO_4)_2$(3%)-(Cu)	C	70	67	[46a]
MeO	H	H	H	H	aq. HCl(1M)/$SnCl_2$(0.4M)-(Sn)	C	82	99	[32]
NH_2	H	H	H	H	aq. EtOH(50%)-HCl(1.3M)-(Cu)	C	51.6	62	[63]
NH_2	H	H	H	H	aq. H_2SO_4(0.94M)-(Ti/TiO_2)	C	92	91.9	[64]
H	NH_2	H	H	H	aq. H_2SO_4(2M)/TiO_2(1%)-(Cu)	C	90	82	[65]
H	NH_2	H	H	H	aq. H_2SO_4(0.94M)-(Ti/TiO_2)	C	93.7	74.7	[64]
H	H	NH_2	H	H	aq. H_2SO_4(10%)-(Pb/Cu or Ni)	C	95	–	[66]
H	H	NMe_2	H	H	aq. H_2SO_4(10%)-(Cu)	C	75	75	[67]
H	NO_2	H	H	H	aq. H_2SO_4(30-40%)/$Ti_2(SO_4)_2$(3%)-(Cu)	C	85	97	[46a]
H	NO_2	H	H	H	aq. H_2SO_4(3.8M)-(Ti/TiO_2)	C	92.2	88.9	[51]
H	NO_2	H	H	H	aq. H_2SO_4(20%v/v)-(Ti/TiO_2)	C	92.2	91.1	[52]
H	H	NO_2	H	H	aq. H_2SO_4(1M)-(C)	B	28	–	[68]
H	NO_2	Me	H	H	aq. H_2SO_4(20%)-(Ti/TiO_2)	C	94.6	92.5	[60]
H	NO_2	Cl	H	H	aq. H_2SO_4(20%)-(Ti/TiO_2)	C	92.7	88.3	[69]
OH	H	H	H	H	aq. H_2SO_4(30-40%)/$Ti_2(SO_4)_2$(3%)-(Cu)	C	97	97	[46a]
OH	H	H	H	H	aq. H_2SO_4(10%)-(Cu)	C	96	93.3	[70]
OH	H	H	H	H	aq. NaOH(1M)-(Cu)	C	86	81	[53]
OH	H	H	H	H	aq. H_2SO_4(2M)-(Ti/TiO_2)	C	95	93.2	[71]
H	OH	H	H	H	aq. NaOH(1M)-(Cu)	C	81	44	[53]
H	H	OH	H	H	aq. HCl(1M)/$SnCl_2$(0.4M)-(Sn)	C	80	89	[32]
H	H	OH	H	H	aq. NaOH(1M)-(Cu)	C	95	93	[53]
H	H	OH	H	H	aq. H_2SO_4(10%)-(TiO_2/Ti)	C	97	98	[72]
H	H	OH	H	H	aq. H_2SO_4(10%)-(TiO_2/Ti)	C	97	92	[73]

Table 5.2e Results from Electroreduction of Substituted Nitrobenzenes

R¹	R²	R³	R⁴	R⁵	Electrolysis Conditions	Products,	Yield, %		Ref.
							Conv.	Current	
SO₃H	H	H	H	H	aq. H₂SO₄(10M)-(Ti/Ti₂)	C	95.2	93.4	[61]
H	SO₃H	H	H	H	aq. H₂SO₄(30–40%)/Ti₂(SO₄)₂(3%)-(Cu)	C	76.3	87.5	[46a]
H	SO₃H	H	H	H	aq. H₂SO₄(0.5M)-(Pt-SPE)	C	–	80	[74]
H	H	SO₃H	H	H	aq. NaOH(1M)-(Cu)	C	99	14	[53]
Cl	H	H	Cl	H	aq. HCl(1M)/SnCl₂(0.4M)-(Sn)	C	90	31	[32]
CO₂H	CO₂H	H	H	H	aq. NaH₂PO₄(0.25M)-NaH₂PO₄(0.25M)-(C)	A	95	–	[59]
CO₂H	H	H	CO₂H	H	aq. NaH₂PO₄(0.25M)-NaH₂PO₄(0.25M)-(C)	A	87	–	[59]
CO₂H	H	Cl	H	H	aq. NaH₂PO₄(0.25M)-NaH₂PO₄(0.25M)-(C)	A	97	–	[59]
CO₂H	H	H	Cl	H	aq. NaH₂PO₄(0.25M)-NaH₂PO₄(0.25M)-(C)	A	94	–	[59]
CO₂H	H	Me	H	H	aq. NaH₂PO₄(0.25M)-NaH₂PO₄(0.25M)-(C)	A	93	–	[59]
OH	H	H	H	H	aq. NaOH(1M)-(Cu)	C	92	88	[53]
OH	H	CO₂H	H	H	aq. NaOH(1M)-(Cu)	C	83	85	[53]
OH	H	H	CO₂H	H	aq. EtOH(50%)-HCl(0.1M)-(Cu)	C	98	90	[75]
OH	H	Me	H	H	aq. NaOH(1M)-(Cu)	C	95	103	[53]
OH	H	H	Me	H	aq. NaOH(1M)-(Cu)	C	88	73	[53]
OH	H	H	H	Me	aq. NaOH(1M)-(Cu)	C	89	85	[53]
Me	OH	H	H	Me	aq. NaOH(1M)-(Cu)	C	98	36	[53]
Me	H	OH	H	Me	aq. NaOH(1M)-(Cu)	C	94	93	[53]
Me	Me	H	H	H	aq. H₂SO₄(30–40%)/Ti₂(SO₄)₂(3%)-(Cu)	C	73	79	[46a]
Me	H	H	NO₂	H	MeOH(7% H₂O)-NaOH(0.14M)-(Cu or Ni) –1.4 V vs. SCE	C	87-92	100	[76]
Br	Me	NO₂	Me	Me	aq. AcOH-H₂SO₄(15%)-(Hg)	C	85	–	[77]
F	F	NO₂	F	F	DMF-Et₄NClO₄(0.1M)-(Pt)	C	25	–	[78]

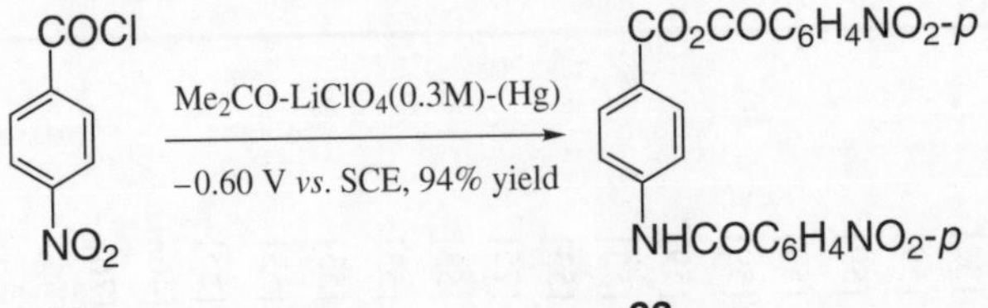

$$(5.28)$$

62   **63**

has been performed in an aqueous MeCN-H$_2$SO$_4$(10%)-(Cu) system to yield indole **65** in 56% yield (eq. (5.29)) [86].   Electroreductive cyclization of methyl 2-(2-nitro-phenylthio)acetate **66** is performed in an aqueous EtOH-H$_2$SO$_4$(0.5M)-(Hg) system at

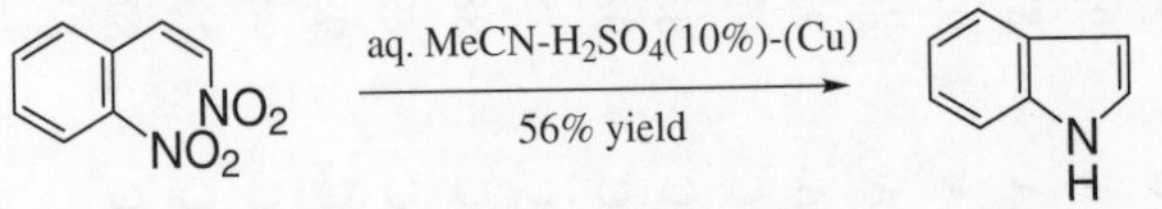

$$\xrightarrow{\substack{\text{aq. MeCN-H}_2\text{SO}_4(10\%)\text{-(Cu)} \\ 56\% \text{ yield}}} \quad (5.29)$$

64   **65**

-0.4 V (SCE) to give **67** in 89% yield (eq. (5.30)) [87].   The electrosynthesis of 3,3'-dichloroazobenzene **69** directly from 3-chloronitrobenzene **68** is accomplished in an aqueous EtOH-H$_2$SO$_4$(7%w/v)-(Hg-Cu) system (eq. (5.31)) [88].

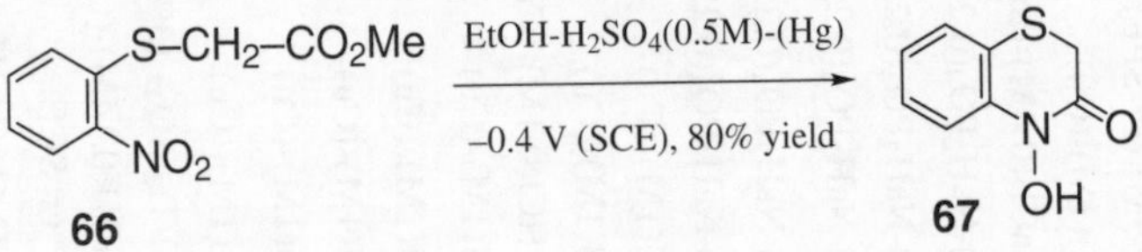

$$\xrightarrow{\substack{\text{EtOH-H}_2\text{SO}_4(0.5\text{M})\text{-(Hg)} \\ -0.4 \text{ V (SCE), } 80\% \text{ yield}}} \quad (5.30)$$

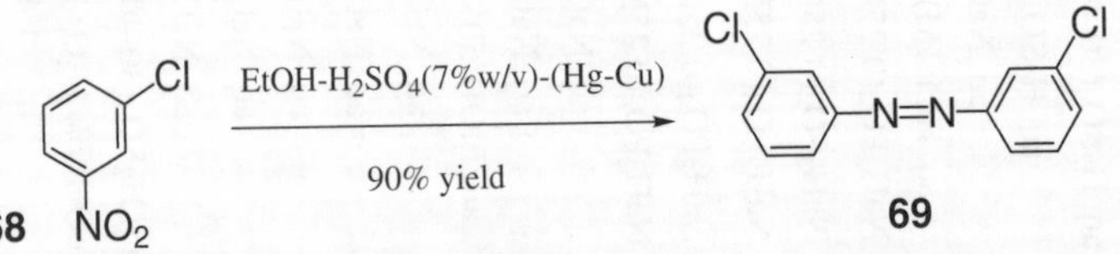

66   **67**

$$\xrightarrow{\substack{\text{EtOH-H}_2\text{SO}_4(7\%\text{w/v})\text{-(Hg-Cu)} \\ 90\% \text{ yield}}} \quad (5.31)$$

68   **69**

Effect of various parameters such as current density, temperature, medium concentration, depolarizer ratio, and nature of cathode material on yield percentage of **70** are demonstrated.   The conversion of 4,4'-dinitrostilbene-2,2'-disulfonic acid **71** has been attained by electrolysis in an aqueous H$_2$SO$_4$(1.25M)-(Hg) system to give 4,4'-diaminostilbene-2,2'-disulfonic acid **71** in 90% yield (eq. (5.32)) [89].   The electroreduction of 2-nitrohydra-zobenzene **72** in a DMF-Bu$_4$NClO$_4$(0.04M)-(Hg) system affords 1,3-dihydroxy-2-phenyl-benztriazole **73**, whereas the product of the reduction becomes 2-phenylbenztriazole-N-oxide **74** in the presence of alkali metal salts, $e.g.$, LiClO$_4$ and NaClO$_4$ (eq. (5.33)) [90].   Further dehydration reaction proceeds under the latter conditions.   The electroreduction of 2-nitroazobenzene proceeds in a DMF-Bu$_4$NClO$_4$(0.04M)-(Hg) system to give 1,3-dihy-droxyl-2-phenylbenztriazole [91].

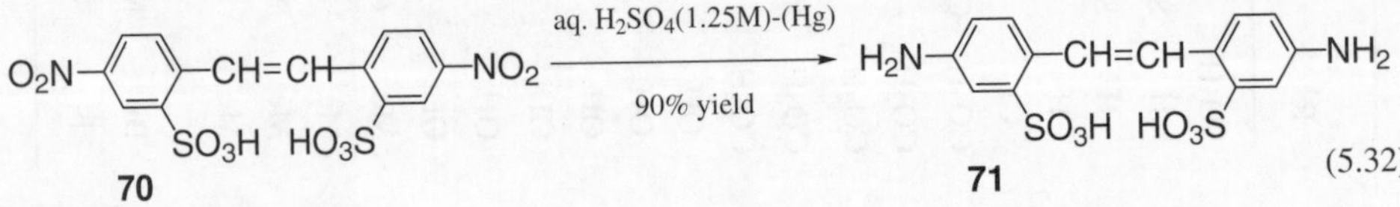

$$\xrightarrow{\substack{\text{aq. H}_2\text{SO}_4(1.25\text{M})\text{-(Hg)} \\ 90\% \text{ yield}}}$$

70   **71**

$$(5.32)$$

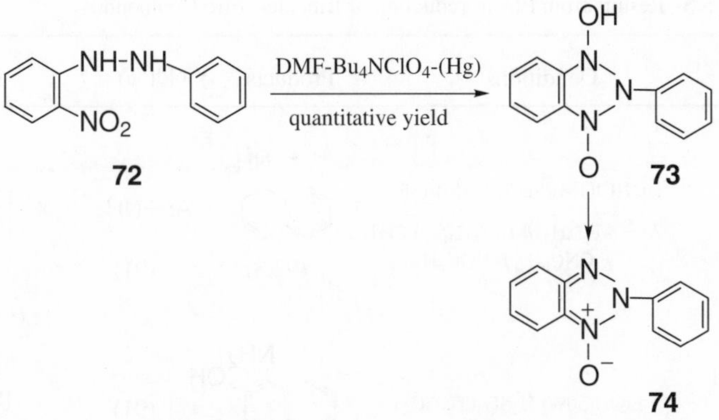

$$
\text{(5.33)}
$$

Electroreduction of aromatic nitro compounds having more than twelve carbon atoms are listed in Table 5.3. [32, 92~99]. In a practical sense, the combination of TiO_2 coated over titanium and sulfuric acid solution is one choice for the large-scale production of aromatic amines [92]. The use of the Ti(III)/Ti(IV) redox system is also valuable [94]. Aprotic solvents, *e.g.* DMF, are effective for the preparation of aromatic hydroxylamines and nitroso derivatives [96,98]. The rate constants of the $PhNO_2^-\cdot/PhNHOH$ and $PhNHOH/PhNO$ processes decrease with increasing HMDA concentration [100]. Table 5.4. [101~114] and 5.5. [115~122] demonstrates the results from the electroreductive cyclization of *o*-nitrobenzene derivatives. The CV data of two nitroamphetamine derivatives (2-nitro-4,5-dimethoxyamphetamine and 2-nitro-4,5-methlenedioxyamphetamine) demonstrate that a reversible one-electron reduction takes place to form a stable nitro anion radical. At more negative potential values, a further three-electron reduction occurs irreversibly to give the hydroxyamine derivative [123].

The electroreduction of α- and β-*meso*-nitrodeuterophoephyrius in an MeOH-H_2SO_4-(C) system affords the corresponding amines (64%) and hydroxyamine (24%), respectively [124]. 1,2-Dialkyl-4-nitroimidazoles are reduced between pH 0~13 in a four-electron wave. The resulting arylhydroxylamine is stable and in acidic media, reduced further to the amine [125].

The electroreduction of 2-nitroimidazole in a strong acid solution (aq. $HClO_4(0.5M)$-(Pt)) undergoes two-electron transfer to yield the corresponding *N,N*-dihydroxylamine derivative (the hydrated form of the corresponding nitroso compound) [126,127]. Electroreduction of 5-methyl-1-nitroaminotetrazole 75 in an aqueous $H_2SO_4(0.5M)$-(Hg) system affords 5-methyl-1-aminotetrazole 76 in 79.5% yield (eq. (5.34)) [128].

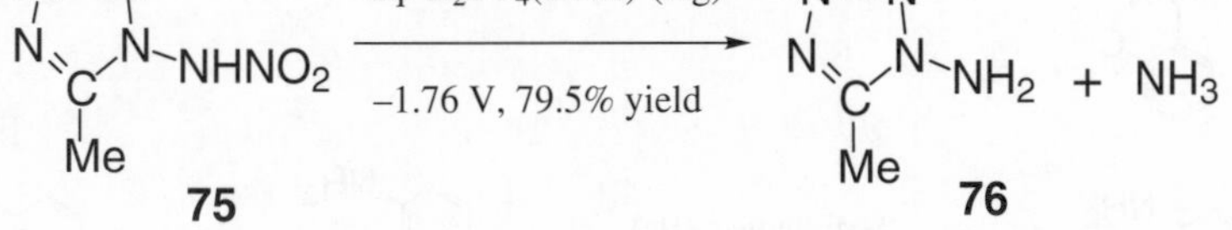

$$
\text{(5.34)}
$$

Table 5.3 Results from Electroreduction of Binuclei Nitro Compounds

Substrate	Conditions	Products	Yield(%)	Ref.
NO$_2$ (naphthalene)	aq. HCl(1M)/SnCl$_2$(0.4M)-(Sn) Ar = o-FC$_6$H$_4$, o-BrC$_6$H$_4$, o-FC$_6$H$_4$ p-CNC$_6$H$_4$, p-HOC$_6$H$_4$	NH$_2$ (naphthalene) (92.3) Ar—NH$_2$ (91)		[32] [92]
NO$_2$ OH (naphthalene)	EtOH(50%)-H$_2$SO$_4$(Ti/TiO$_2$)	NH$_2$ OH (naphthalene) (91)		[93]
NO$_2$ NO$_2$ (naphthalene)	aq. H$_2$SO$_4$/TiOSO$_4$-(C) preparative scale	NH$_2$ NH$_2$ (naphthalene)		[94]
R (carbazole) NO$_2$	aq. MeCN-H$_2$SO$_4$(15%)-(C) R = Me, Et, Pr, Bu	R (carbazole) NH$_2$ (83)		[95]
H$_2$N—⟨⟩—⟨⟩—NH$_2$ NO$_2$	1) DMF-Bu$_4$NI-(Pt) 2) aq. MeCH(20%)-KCl(0.1M)-(Hg) pH 4.5	H$_2$N—⟨⟩—⟨⟩—NH$_2$ NHOH (2) NH$_2$)		[96] [97]
H$_2$N—⟨⟩—⟨⟩—NH$_2$ NO$_2$ (two NO$_2$)	DMF-Bu$_4$NI(0.1M)-(Hg)	H$_2$N—⟨⟩—⟨⟩—NH$_2$ NO (two NO)		[98]
(diazepam structure)	DMF-Buffer*-(Hg) pH 4.8 *Britton-Robinson Buffer	O$_2$N— NH$_2$ C=O Cl		[99]
O$_2$N— NH$_2$ C=O Cl	DMF-Buffer*-(Hg) pH 4.8 *Britton-Robinson Buffer	H$_2$N— NH$_2$ HC—OH Cl HOHN— NH$_2$ HC—OH Cl		[99]

Table 5.4a Results from Electroreductive Cyclization of *o*-Nitrobenzene Derivatives

Substrate	Conditions	Product (Yield, %)	Ref.
NHNHCOR / NO₂ R = H, Me, Ph	aq. EtOH(50%)-NH₄OH/Buffer-(Hg)	(75)	[101]
CONHCOMe / NO₂	aq. EtOH(50%)-H₂SO₄(0.5M)-(Hg) −0.26 V *vs.* SCE	(~100)	[102]
OH / N / Me / O	aq. EtOH(50%)-H₂SO₄(0.5M)-(Hg) −1.10 V *vs.* SCE		[102]
OCHCO₂Me₂ / NO₂	aq. EtOH(50%)-H₂SO₄(0.5M)-(Hg) −0.4 V *vs.* SCE	(~100)	[103]
OCH₂CO₂H / NO₂	aq. EtOH(50%)-H₂SO₄(0.5M)-(Hg) −0.9 V *vs.* SCE	(~100)	[103]
OCH₂CN / NO₂	aq. EtOH(50%)-NH₄OH/Buffer-(Hg) −0.9 V *vs.* SCE	(60)	[103]
OCH₂COPh / NO₂	aq. EtOH(50%)-AcOH/Buffer-(Hg) −0.8 V *vs.* SCE	(~100)	[103]
COPh / COCH / COPh / NO₂	aq. EtOH(50%)-H₂SO₄(0.5M)-(Hg) −0.4 V *vs.* SCE	(70)	[104]
O / C₂H / COMe / COMe / NO₂	aq. NH₄OH/Buffer-(Hg)	(80)	[104]

Table 5.4b Results from Electroreductive Cyclization of *o*-Nitrobenzene Derivatives

Substrate	Conditions	Product (Yield, %)	Ref.
(*o*-nitrobenzene with C₂H bearing CO₂Me, COMe, and CHO)	aq. EtOH(50%)-H₂SO₄(0.5M)-(Hg) −0.25 V *vs.* SCE	(quinoxaline N-oxide with OH, CO₂Me, Me) (65); (benzo ring with OH, COMe, N–OH) (35)	[104]; [104]
(*o*-nitrobenzene with C₂H bearing CN, COPh)	aq. EtOH(50%)-H₂SO₄(0.5M)-(Hg) −0.25 V *vs.* SCE	(quinoxaline N-oxide with OH, CN, Ph) (90)	[104]
(*o*-nitrobenzene with C₂H bearing CN, CO₂Et)	aq. EtOH(50%)-H₂SO₄(0.5M)-(Hg) −1.1 V *vs.* SCE	(OH, CN, N–OH, O) (70); (OH, CO₂Et, NH₂, O) (30)	[104]; [104]
(*o*-nitrobenzene with CH=C bearing COPh, CN)	aq. EtOH/AcOH(1/1)-H₂SO₄(0.5M)-(Hg)	(quinoline N-oxide with COPh, NH₂) (10); (quinoline N-oxide with CN, Ph) (50)	[105]; [105]
(*o*-nitrobenzene with CH=C bearing COMe, COMe)	aq. EtOH(40%)-HCl(1M)-(Hg) −0.35 V *vs.* SCE	(quinoline N-oxide with Me, COMe) (80~86)	[106]

Table 5.4c Results from Electroreductive Cyclization of *o*-Nitrobenzene Derivatives

Substrate	Conditions	Product (Yield, %)	Ref.
	aq. EtOH(70%)-HCl(1M)-(Hg) −0.2 V *vs.* SCE	(77)	[106]
	aq. EtOH(50%)-H₂SO₄(0.5M)-(Hg) −1.10 V *vs.* SCE Nu = OEt,OH, NH₂		[102]
	aq. EtOH(50%)-H₂SO₄(0.5M)-(Hg) −1.10 V *vs.* SCE	(85)	[102]
	aq. EtOH(50%)-NH₄OH/Buffer-(Hg) −1.7 V *vs.* SCE	(100)	[102]
 X = H, Me, Cl, CO₂H R = NHCH₂CO₂Et	MeOH/H₂O(5%)-H₂SO₄(20%)-(Cu)	(69~98)	[107]
	aq. EtOH(50%)-H₂SO₄(0.5M)-(Hg) −0.4 V *vs.* SCE	(80)	[108]
 X = H, Cl, CF₃, CO₂Me	MeOH/H₂O-H₂SO₄-(C) −0.81~−1.28 V *vs.* SCE	(71~99)	[107]
	MeOH/H₂O(5%)-H₂SO₄(20%)-(Zn) X = H, Cl, Me	(98, 99)	[107]
	MeOH/H₂O(5%)-H₂SO₄(20%)-(Zn) X = CF₃, CO₂Me	(94, 96)	[107]

Table 5.4d Results from Electroreductive Cyclization of *o*-Nitrobenzene Derivatives

Substrate	Conditions	Product (Yield, %)	Ref.
R⎯NO₂, N-CN, H; R = H	MeOH-H₂SO₄-(Pt/Hg)	R⎯ benzimidazole OH, NH₂ (69)	[109]
NO₂, SCN	aq. EtOH(65%)-HCl(0.4M)-(Hg) 0.4 V *vs.* SCE	benzothiazole N→O, NH₂ (75)	[106]
NO₂, NCS	aq. EtOH(40~75%)-HCl(0.5M)-(Hg) −0.6 V *vs.* SCE	benzimidazole O, SH (58)	[106]
N-C(O)Me, H; NO₂	aq. AcOH(90%)-(Pt/Hg)	benzimidazole, Me (94)	[110]
NO₂, R³; R¹⎯CON-CHCO₂Et, R²; R¹ = H, Cl, Me; R² = H, Me, Ph; R³ = H, Me	AcOH-Zn powder-(Zn)	R¹⎯ benzodiazepinedione, R³, R² (80~98)	[107]
NO₂, CO₂Me, Me; H; MeO₂C, N-Me, Me	aq. EtOH(50%)-H₂SO₄-(Hg) −0.9 V (SCE)	quinoline N→O, Me, CO₂Me (80)	[111]
NO₂, CN, Me; H; NC, N-Me, Me	aq. EtOH(50%)-H₂SO₄-(Hg) −0.9 V (SCE)	N→O, NH₂, Me, N-Me, NC, Me (75)	[112]
O, NO₂; R⎯X triazole-N-N, Y; X = N, CH; Y = H, Me, AcO(CH₂)₄, Na; R = H, Me, SMe, SEt	DMF-Bu₄NClO₄-(Pt)	O, NH₂; R⎯X triazole-N-N, Y (45~72)	[113]

Table 5.4e Results from Electroreductive Cyclization of *o*-Nitrobenzene Derivatives

Substrate	Conditions	Product (Yield, %)	Ref.
R^1, R^2 = H, Me	aq. MeCN-Et$_4$NBF$_4$(0.1M)-(Hg)	(5)	[114]
		(10)	[114]
		(10)	[114]

Table 5.5 Results from Electroreduction of Miscellaneous Nitro Compounds

Substrate	Conditions	Products Yield(%)	Ref.
	aq. EtOH-HCl-(Hg)	(74)	[115]
	Ti$_2$(SO$_4$)$_3$-H$_2$SO$_4$-(Cu/Pb)	(83~89)	[116]
	aq. H$_2$SO$_4$(1M)-(Hg)	(59)	[117]
	aq. DMF(50%)-H$_2$SO$_4$(0.05M)-(Pb)	(91)	[118]
	aq. MeCN/Ac$_2$O-NaClO$_4$-(Pt/Hg)	(74~80)	[119]
	aq. MeCN/Ac$_2$O-NaClO$_4$-(Pt/Hg)	(83)	[119]

Table 5.5   Results from Electroreduction of Miscellaneous Nitro Compounds (continued)

5.1.2   Electroreduction of Nitroso Compounds

The electroreduction of *N*-nitrosoamines in an aqueous organic media has been shown to give the corresponding hydrazines and amines [129a].   The influence of the medium (solvent, nature and concentration of the protonating agent) and the structure of substrates have been investigated [129b].   For example, the electrolysis of 4-nitrosomorphorine **77** in an MeCN-AcOH(0.83M)-(Hg) affords the corresponding hydrazine **78** in good yield (eq. (5.35)) [129b].   The electroreduction of nitrosobenzene to phenylhydroxylamine has been

attained in an aqueous medium between pH 0.4~13 [130].   The nitrobenzene anion radical formed by the electroreduction of nitrobenzene **79** reacts rapidly with alkyl halides to give the *N,O*-dialkylhydroxylamines **80** *via* a nitroso intermediate (eq. (5.36)) [131].   Anion

radicals of nitrosobenzene derived from nitrobenzene **79** react with alkyl halide to afford, in both cases, *N,O*-dialkylphenylhydroxylamines **80** [131,132].   The electroreduction of *p*-nitrosophenol to aminophenol is carried out by using copper, lead, and Ti/TiO$_2$ electrodes [133].   Reduction of nitrobenzene **79** as a precursor of nitrosobenzene in the presence of 1,3-dibromopropane gives rise to the isoxazolidine derivative **81** in 48% yield (eq. (5.37)), but in the presence of 1,4-dibromobutane, affords two major products **82** and **83** together with a small amount of azobenzene (eq. (5.38)).   On the other hand, intermolecular reactions are faster than intramolecular cyclization when 4-bromobutyryl chloride is used. In DMF solution, the bromine-chloride exchange reaction takes place during the electrolysis, giving the diacylated product **84** in 72% yield (eq. (5.39)).

$$\begin{array}{c} \text{DMF-Bu}_4\text{NI(or Bu}_4\text{NPF}_6)\text{-(Pt/Hg)} \\ \hline \text{Br(CH}_2)_3\text{Br} \quad -1.14 \text{ V } vs. \text{ SCE} \end{array}$$

81

(48%)

(5.37)

79

$$\begin{array}{c} \text{DMF-Bu}_4\text{NI(or Bu}_4\text{NPF}_6)\text{-(Pt/Hg)} \\ \hline \text{Br(CH}_2)_3\text{Br} \quad -1.3 \text{ V } vs. \text{ SCE} \end{array}$$

82   +   **83**

(42%)   (28%)

(5.38)

$$\begin{array}{c} \text{DMF-Bu}_4\text{NI(or Bu}_4\text{NPF}_6)\text{-(Pt/Hg)} \\ \hline \text{ClCO(CH}_2)_3\text{Cl} \quad -0.8 \text{ V } vs. \text{ SCE} \end{array}$$

N–O$_2$C(CH$_2$)$_3$Cl
|
CO(CH$_2$)$_3$Cl

84

(72%)

(5.39)

A preparative way for obtaining unsymmetrical hydrazines is the electroreduction of the corresponding N,N-disubstituted N-nitrosamines. Unsymmetrical dialkylhydrazines have been electrosynthesized in 65% yield in acidic media using a lead cathode [134].

An efficient electoreduction method for the preparation of N,N-dimethyl(or diphenyl) hydrazines **86** from the corresponding N,N-disubstituted nitrosamine **85** in an aqueous MeOH(or EtOH)-HCl(10%)-(Cu) system has been proposed (eq. (5.40)) [135]. Large scale electroreduction of N-nitrosopiperazine in an aqueous H$_2$SO$_4$(20%,w/w)-(Pb) system affords the corresponding hydrazine in 95% yield [136].

$$\begin{array}{c} \text{R} \\ \text{R} \end{array}\!\!\!N\!-\!NO \xrightarrow[\text{61~62\% yields, 30°C}]{\text{MeOH(or EtOH)-HCl(10\%)-(Cu)}} \begin{array}{c} \text{R} \\ \text{R} \end{array}\!\!\!N\!-\!NH_2 \qquad (5.40)$$

85   **86**

R = i-Pr, Bu, i-Bu, Ph

The electroreduction of N-ethyl-N-nitrosourea (ENU) **87** in a DMF-Et$_4$NFeCl$_4$-(C) system has been shown to afford a stable dimeric Fe^{2+} complex [Fe(ENU)$_2$]$^{2+}$ **88** (eq. (5.41)) [137]. The anion radicals of nitrosoureas formed at the first electrochemical step give rise

$$\begin{array}{c} \text{Et} \\ \text{ON} \end{array}\!\!\!N\!-\!CO\!-\!NH_2 \xrightarrow{\text{DMF-Et}_4\text{NFeCl}_4\text{-(C)}} \left[\begin{array}{c} \text{NH}_2\ \text{NH}_2 \\ |\quad\ | \\ \text{CO}\ \ \text{CO} \\ |\quad\ | \\ \text{Et–N}\!-\!\!-\!\!\text{N–Et} \\ \text{O}\diagdown\ \diagup\text{O} \\ \text{Fe} \end{array} \right]^{2+} \qquad (5.41)$$

87   **88**

to a self-protonation process, and the deprotonation form of the substrate readily cleaves into several products [138]. The electroreduction of 4-nitrosodiphenylamine **89** in a DMF-HCl(0.1M)-(C) system in a Britton-Robinson buffer medium can lead to 4-aminodiphenylamine **91** *via* the hydroxyamine **90** (eq. (5.42)) [139]. The electroreduction of 4-nitrosodiphenylamine in cationic micellar media at solid electrodes provides a useful preparative method for the synthesis of 4-aminodiphenylamine [140]. The controlled potential electrolysis of 2-nitroso-1-naphthol **92** undergoes four-electron reduction, resulting in the formation of 2-amino-1-naphthol **93** (eq. (5.43)) [141].

$$(5.42)$$

$$(5.43)$$

Electroreductive cyclization of *N*-benzyl-*N*-nitrosoanthranilic acid **94** takes place at a mercury cathode to give 1-benzyl-1,2-dihydro-3H-indazol-3-one **96** in good yield after elimination of water (eq. (5.44)) [142].   The preparative electroreduction of 1-nitroso-2-methylindoline **97** is performed in an EtOH/AcOH(1/1)-AcONa(0.5M)-(Hg/Cu) system to give 1-amino-2-methyl-indoline **100** in 92% yield *via* **98** and **99** (eq. (5.45)) [143].   For the electroreduction, a continuous process through porous electrodes has been developed [144]. The electroreduction of 4-nitrosoantipyrine **101** in an aqueous HCl(1M)-(Pb) system yields 4-aminoantipyrine **103** in quantitative yield *via* the corresponding hydroxyamine **102** (eq. (5.46)) [145].

$$(5.44)$$

*EtOH/AcOH(1/1)-AcONa(0.5M)-(Cu), −1.2 V *vs*. SCE.

$$(5.45)$$

$$(5.46)$$

5.2   Electroreduction of Nitrogen Atom Containing Double Bonds

5.2.1   Electroreduction of N=N Bonds

5.2.1.1   Electroreductive Cleavage of N=N Bonds Leading to Amines

Azobenzene **104** is reduced in two one-electron steps in a MeCN-Et4NClO4 system.   The polarogram of **104** in an MeCN-Bu$_4$NClO$_4$-(Hg) system shows two well-defined waves as follows: first wave −1.36 V *vs.* SCE (eq. (5.47)) and second wave −2.03 V *vs.* SCE [146]. The overall reaction paths for the electroreduction are assumed to be as shown in eqs. (5.48) and (5.49) [147a].   A stable anion radical **105** forms by the first electron transfer from **104** and the second electron transfer to produce the dianion **106** is followed by a chemical reaction producing a protonated species **107** or a combination with electrophile [146].   The half-wave potentials first and second waves of the reversible steps for the reduction of the substituted azobenzenes in acetonitrile are directly related to the Hammett substituent constants [147b].   The stepwise electroreduction of azobenzene has been achived by use of copper electrodes [148].   The reduction of azobenzene on copper electrode affords dihydroazobenzene in quantitative yield, whereas use of Devarda copper gives aniline selectively in 91% yield.

$$\text{Ph—N=N—Ph} \underset{\longleftarrow}{\overset{e^-}{\longrightarrow}} \left[\text{Ph—N=N—Ph}\right]^{-\bullet} \quad \text{(first wave)} \qquad (5.47)$$
$$\textbf{104} \hspace{5cm} \textbf{105}$$

$$\textbf{105} \underset{\longleftarrow}{\overset{e^-}{\longrightarrow}} \left[\text{Ph—N—N—Ph}\right]^{2-} \qquad \cdots\cdots (5.48) \left.\begin{array}{c} \\ \\ \\ \\ \\ \end{array}\right\} \text{(second wave)}$$
$$\textbf{106}$$

$$\textbf{106} \overset{CH_3CN}{\longrightarrow} \left[\text{Ph—NH—N—Ph}\right]^- + CH_2CN^- \cdots\cdots (5.49)$$
$$\textbf{107}$$

4-Monosubstituted azobenzenes **108** such as azobenzene **104** can be electrochemically reduced in a MeCN-Et$_4$NClO$_4$-(Hg) system through two one-electron reduction steps, and the first and second reduction steps (eqs. (5.50) and (5.51)) are reversible while the third step (eq. (5.52)) is irreversible, except for 4-nitroazobenzene [147b,149,150].   In the presence of strong proton donors, diffusion controlled waves of 4-substituted azobenzene are observed at potentials up to −1.7 V (SCE) more positive than the normal azobenzene reduction waves [147c].   A variety of azo compounds has been electrochemically reduced and the results are summarized in Table 5.6. [151~163].

$$\text{Ar—N=N—Ph} \underset{\longleftarrow}{\overset{e^-}{\longrightarrow}} \left[\text{Ar—N=N—Ph}\right]^{-\bullet} \qquad (5.50)$$
$$\textbf{108} \hspace{4cm} \textbf{109}$$

$$\textbf{109} \underset{\longleftarrow}{\overset{e^-}{\longrightarrow}} \left[\text{Ar—N=N—Ph}\right]^{2-} \qquad (5.51)$$
$$\textbf{110}$$

$$\textbf{110} + HS \longrightarrow \left[\text{Ar—NH—N—Ph}\right]^- + S^- \qquad (5.52)$$
$$\textbf{111}$$
$$\text{Ar = Y–C}_6\text{H}_4 \text{ (Y = Cl, Me}_2\text{N, SO}_3^-, \text{NO}_2, \text{MeO)}$$

The reductive cleavage of the N=N bond of ethyl 4-arylazo-5-oxo-4,5-dihydropyrazol-3-ylacetate **112** in an aqueous DMF/EtOH(20/30%)-HCl(50%)-(Hg) system affords the cleavage products **113** and **114** (eq. (5.53)) [164].

Table 5.6a Results from Electroreduction of Azo Compounds

Diazo Compound	Conditions	Product	Ref.
CH₃O–C₆H₄–N=N–(Meldrum's acid)	aq. MeOH(60%)-KCl(0.2M)-(Hg), Buffer* pH 4.30 *Britton-Robinson	CH₃O–C₆H₄–NH₂, H₂N–(Meldrum's acid)	[151]
(2,4-dihydroxyphenyl)azo, OH / HO	aq. NaNO₃-Buffer*-(Hg) *Britton-Robinson	OH/NH₂, HO/NH₂ aminophenols	[152]
benzothiazolone–N=N–C₆H₄–Y	aq. EtOH(50%)-HCl(0.001M)/Buffer*-(Hg) *Britton-Robinson	benzothiazolone–S–NH₂ + Y–C₆H₄–NH₂; Y = H, 4-Me, 3-Me, 4-MeO, 4-Cl, 3-NO₂, 4-CO₂H	[153]
4-hydroxy-3-(arylazo)coumarin, OH–N=N–C₆H₄X	aq. EtOH(50%)-HCl(0.01M)-(Hg)	3-amino-4-hydroxycoumarin (OH/NH₂), H₂NC₆H₄–X; X = H, 4-Cl, 4-Me, 4-NO₂	[154]
isoquinolinedione–N=N–C₆H₄X, OH, SC₂H₅	aq. EtOH(50%)-HCl(0.01M)/Buffer*-(Hg) *Britton-Robinson	isoquinolinedione–NH₂/NH, H₂NC₆H₄–X; X = H, 4-Cl, 2-Me, 4-Me, 4-MeO, 4-NO₂	[155]
R–C₆H₄–NH–N=C–C(CH₃)=oxazolone	aq. DMF(50%)-(Hg) pH 4.1, Buffer* *Britton-Robinson	R–C₆H₄ with NH₂, HC–C–C=CH₂ oxazolone; R = Me, MeO, EtO, Cl	[156]

Table 5.6b Results from Electroreduction of Azo Compounds

Diazo Compound	Conditions	Product	Ref.
XC$_6$H$_4$-N=N- (thiazole with C$_6$H$_5$, NH$_2$)	aq. EtOH-Buffer*-(Hg) *Britton-Robinson	H$_2$N- (thiazole with C$_6$H$_5$, NH$_2$) , H$_2$N-C$_6$H$_4$X X = H, 4-Me, 4-MeO 4-Cl, 3-Cl	[157]
Ar-N=N-CH-C=S, Ph-N-C=O NH	aq. EtOH(50%)-HCl/Buffer*-(Hg) *Britton-Robinson Ar = Y- (phenyl) Y = H, 3-Me, 4-Me 4-Cl, 4-Br, 4-MeO 4-NO$_2$, 4-CO$_2$H	H$_2$N-CH-C=O, Ph-N-C=O NH , Ar-NH$_2$	[158]
R- N=N-CH-C-NH$_2$, N-C=O (phenyl)	aq. DMF(30ml)-Bu$_4$NClO$_4$(0.1M)-(Hg) −1.10 V vs.SCE	NH$_2$ NH$_2$, HC-C=N, C-N (O=)-phenyl ; R- -NH$_2$ R = H, Cl, Me, MeO EtO, 4-Br	[159]
R- -N=N-CH-C-CH$_3$, N-C=O, C=O Ar	aq. DMF(50%)-BuNClO$_4$(0.1M)-(Hg)	NH$_2$ CH$_3$, C-C=N, HO-C-N, C=O Ar ; R- -NH$_2$ R = H, Me, MeO, Cl, EtO Ar = 2-Py Ar = Ph	[160]
OH, naphthalene-N=N-naphthalene, $^-$O$_3$S, SO$_3^-$, Y	aq. H$_2$SO$_4$-(Hg)	OH, H$_2$N, naphthalene, $^-$O$_3$S, NH$_2$, SO$_3^-$, Y Y = SO$_3^-$ Y = H	[161]

Table 5.6c Results from Electroreduction of Azo Compounds

Diazo Compound	Conditions	Product	Ref.
(structure)	aq. Buffer*-(Hg) −1.2 V *vs.*SCE *pH 3.0~10.4, Phosphate	(structure)	[162]
(structure)	aq. Buffer*-(Hg) pH 3.0~10.3 *Phosphate Buffer	(structure)	[163]

$$ (5.53) $$

5.2.1.2   Electroreductive Hydrogenation of N=N Bonds Leading to N-N Bonds

The electrosynthesis of phenylhydrazine **117** from aniline has been established by employing an aqueous THF/MeOH-NaOH-(C) system as an electrolyte in a practical procedure [165a]. The total scheme of the production of **117** is outlined in eq. (5.54). The electroreduction of diazoaminobenzene **116** derived from the ammonium sulfate **115** by diazotization undergoes a two-electron transfer process [165b]. The electroreduction of p-aminoazobenzene is kinetically controlled under acidic conditions (pH 3.0), indicating that slow protonation of the reactant before electron transfer controls the overall process, and the unprotonated p-aminoazobenzene undergoes electron transfer under alkaline conditions [166]. The preparative electrolysis of p-aminoazobenzene in an aqueous methanolic alkaline condition undergoes two-electron reduction to afford p-aminohydrazobenzene, exclusively [166]. Δ^3-1,3,4-Thiadiazoline-2-spiro-2'-adamantane **118** can be reduced to the corresponding thiadiazolidine **119** on mercury electrode in an aqueous EtOH(45%)-Britton-Robinson Buffer system (eq. (5.55)) [167].

$$ (5.54) $$

$$ (5.55) $$

A preparative scale of the reduction of 4-benzyloxyazobenzene **120** has been carried out in an aqueous 20% NaOH solution at a spongy lead-coated electrode (eq. (5.56)) to give the corresponding 4-benzyloxyhydrazobenzene **121** in 92% yield [168].

$$ (5.56) $$

syn- and *anti*-2,2'-Azopyridines **122** are reduced in phosphate buffers at pyrolyte graphite electrode through a *quasi*-reversible step to give hydroazopyridine **123**, which is further reduced under controlled potential conditions to 2-aminopyridine **124** (eq. (5.57)) [169].   Azopyridines and phenylazopyridines are converted into the corresponding dianions which cause a proton to be abstracted from acetonitrile to give [CH₂CN]- as an EG base [170].   Such anions can be rapidly trapped with the carbonyl group of acetophenone **125a** and benzophenone **125b**, leading to the cyanomethylation products **128** and **129** (eq. (5.58)) [171].   A variety of azo dyes are electrochemically reduced to the corresponding hydrazo derivatives under appropriate buffer solutions [172].   Electroreduction of 1-dinitrophenyl arylazopyrazoles is also attempted [173].

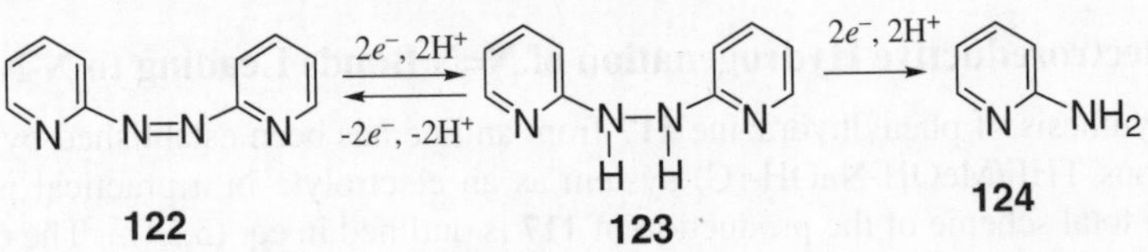

(5.57)

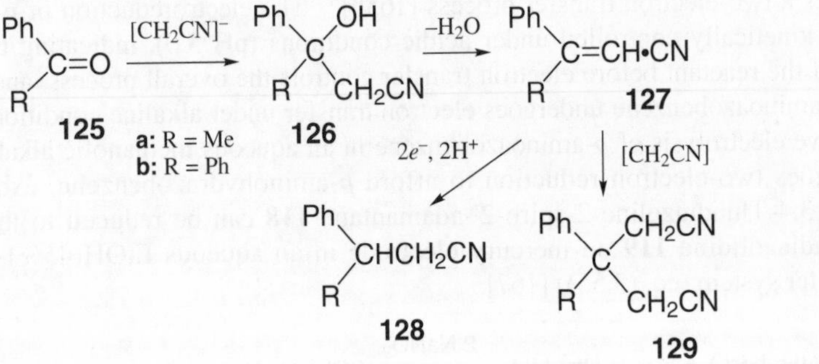

(5.58)

5.2.1.3   Trapping Intermediates with Electrophiles

Azobenzene **130** is reduced in two discrete one-electron steps in DMF or MeCN in the presence of tetraalkylammonium salts.   The first reduction leads to a stable anion radical and the second reduction is ascribable to a dianion.

The electrogenerated nitrogen anion can be trapped by methylation in a DMF-Me₄NBF₄-(Pt/Hg) system in the presence of methyl iodide, giving **131** and **132** in good yields (eq. (5.59)) [174].   The *N*-methylation of tetraazo-4,5,9,10-pyrene **133** proceeds in a

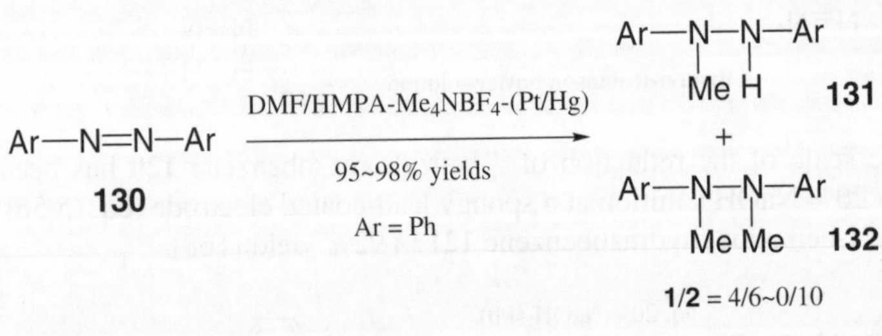

(5.59)

1/2 = 4/6~0/10

THF-Bu₄NBF₄-(Pt) system in the presence of methyl iodide to give **134** in 45% yield (eq. (5.60)) [175].   The ring compounds **136** containing two vicinal nitrogen atoms are synthesized through the alkylation or acyloxylation of the nitrogen anion produced by the electroreduction of azobenzene **130** in the presence of 1,ω-dibromo alkanes or succinoyl chloride (eq. (5.61)) [176].

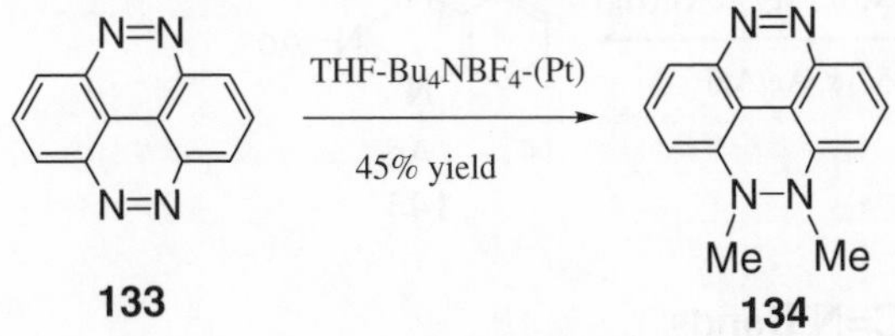

$$\text{(5.60)}$$

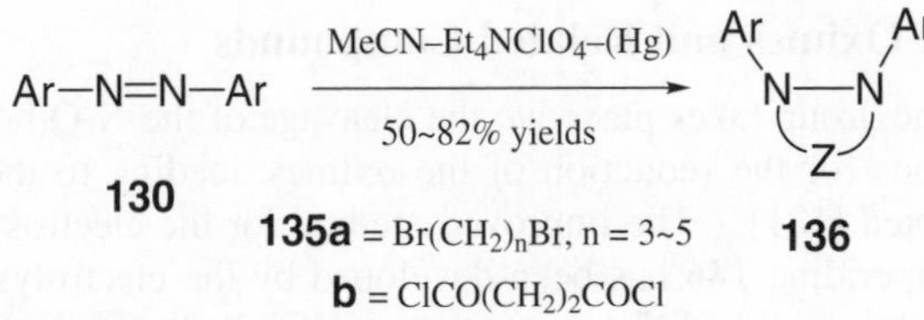

Ar—N=N—Ar $\xrightarrow[\text{50~82\% yields}]{\text{MeCN–Et}_4\text{NClO}_4\text{–(Hg)}}$

130

135a = Br(CH$_2$)$_n$Br, n = 3~5   **136**

b = ClCO(CH$_2$)$_2$COCl

$$\text{(5.61)}$$

The electrochemical carboxylation to a nitrogen atom is an important tool in organic synthesis. The electroreductive carboxylation of the heterocycles **137** can lead to the formation of *N*-methoxycarbonyl compounds **138** saturated with CO$_2$ in the presence of ethyl chloride (eq. (5.62)) [177]. The ethoxycarbonylation of azobenzene occurs under electroreduction in a DMF-Et$_4$NCl-(Hg) system in the presence of ethyl phenylacetate at –1.9 V *vs.* SCE to give diethyl *N,N'*-hydrazobenzenedicarboxylate in 64% yield [178].

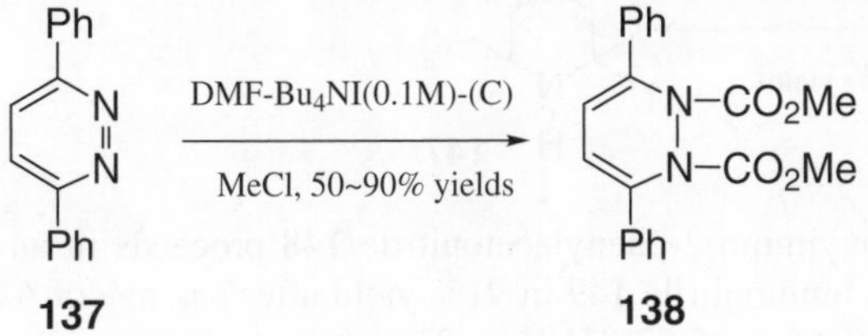

$$\text{(5.62)}$$

137   **138**

Electrosynthesis of the thiocarboxylated *N,N*-diphenylhydrazine **141** and **143** has been performed in a MeCN-Et$_4$NBr-(Hg) system in the presence of carbon disulfide and methyl iodide (eqs. (5.63, 5.64)) [179]. The electroreduction of benzotriazole **144** in the presence of acetic anhydride can lead to the corresponding triacetylated derivative **145** at –1.3 V *vs.* Ag/AgI (eq. (5.65)) [180].

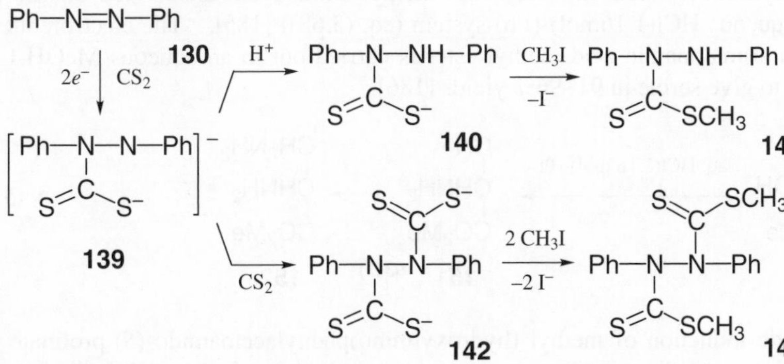

$$\text{(5.63)}$$

$$\text{(5.64)}$$

$$
\text{144} \quad \xrightarrow[\text{–1.3V } vs.\ \text{Ag/AgI}]{\text{DMF-Bu}_4\text{NI(0.1M)/Ac}_2\text{O-(Hg)}} \quad \text{145} \tag{5.65}
$$

5.2.2   Electroreduction of C=N Bonds

5.2.2.1   Electroreduction of Oximes and Related Compounds

The electroreduction of the oxime group takes place *via* the cleavage of the N-O bond, and the consumption of four electrons for the reduction of the oximes, leading to the corresponding amines has been accepted [181].   The improved method for the electrosynthesis of 4-amino-2,2,6,6-tetramethylpiperidine **146** has been developed by the electrolysis of 4-hydroxyimino-2,2,6,6-tetramethylpiperidine **147** in an aqueous HCl(20%)-(Pb/Hg) system in 80~90% yields (eq. (5.66)) [182].   The yield of **147** strongly depends on the acid concentration.   The imine is likely the intermediate in the oxime reduction to the amines. Better yields of amines are obtained at pH 8 and at lower current density with nickel black and iron black cathodes rather than with cobalt black cathode [183].

$$
\text{146} \quad \xrightarrow[\text{80~90% yields}]{\text{aq. HCl(2~20%)-(Pb/Hg)}} \quad \text{147} \tag{5.66}
$$

Electroreduction of 2-hydroxyimino-2-phenylacetonitrile **148** proceeds in an aqueous HCl(1M)-(C) system to give the aminonitrile **149** in 91% yield after passage of 6 F/mol of electricity (61% current efficiency) (eq. (5.67)) [184].   The electroreductive conversion of

$$
\text{Ph-C(=NOH)CN} \xrightarrow[\text{91% yield}]{\text{aq. HCl(1M)-(C)}} \text{Ph-CH(NH}_2\text{)CN} \tag{5.67}
$$
$$
\text{148} \qquad\qquad\qquad\qquad \text{149}
$$

methyl (hydroxyimino)cyanoacetate **150** into methyl aminocyanoacetate **151** can be performed in an aqueous HCl(1.16mol)-(Pb) system (eq. (5.68)) [185].   The electroreduction of (hydroxyimino)malonamic acid methyl ester is carried out in an aqueous MeOH-LiCl-(Pt/Pb) system to give serine in 91~96% yields [186].

$$
\begin{array}{c}
\text{CN} \\
| \\
\text{C=NOH} \\
| \\
\text{CO}_2\text{Me}
\end{array}
\xrightarrow{\text{aq. HCl(1.16 mol)-(Pb)}}
\begin{array}{c}
\text{CN} \\
| \\
\text{CHNH}_2 \\
| \\
\text{CO}_2\text{Me}
\end{array}
\longrightarrow
\begin{array}{c}
\text{CH}_2\text{NH}_2 \\
| \\
\text{CHNH}_2 \\
| \\
\text{CO}_2\text{Me}
\end{array}
\tag{5.68}
$$
$$
\text{150} \qquad\qquad\qquad \text{151} \ ^{(93\%)} \quad \text{152}
$$

Assymmetric induction of methyl (hydroxyimino)phenylacetoamido-(*S*)-prolinate **153** leading to a mixture of the two diastereomers of pyrrolo[1,2a]-3-phenyl-1,4-diketopoperazine **154** has been attempted in an aqueous EtOH/AcOH-H$_2$SO$_4$/AcCl-(Hg) system, and the preferential formation of the SS isomer is observed in a 34% yield (eq. (5.69)) [187].

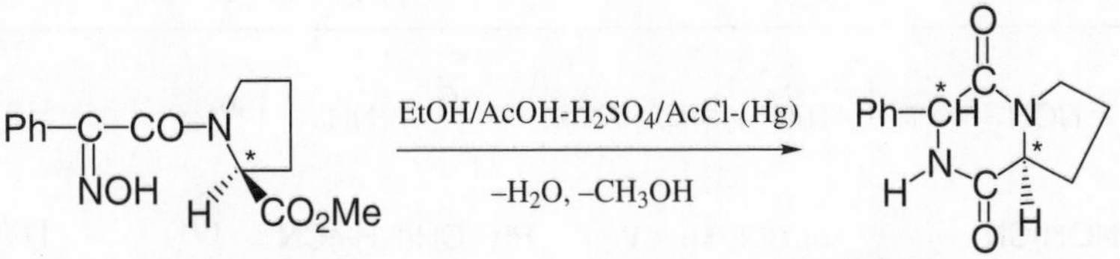

$$\text{(5.69)}$$

153   **154**

The conversion of the hydroxyimino group of 2,4-diamino-5-hydroxyimino-6-oxypyrimidine **155** into the amino derivative **156** has been attained in good yield (eq. (5.70)) [188]. Electroreduction of other oximes are shown in Table 5.7. [183, 189~196].

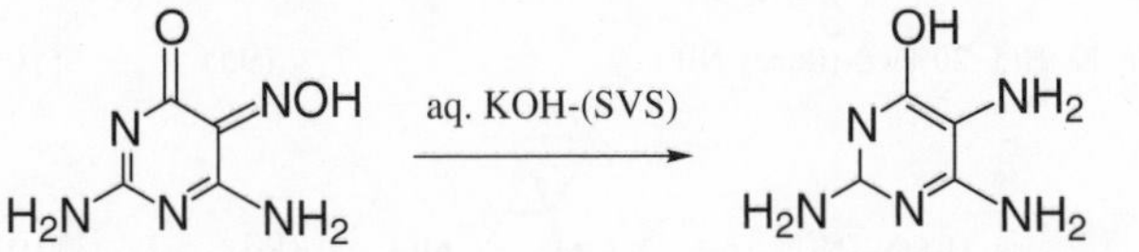

$$\text{(5.70)}$$

155   **156**   (96.6%)

Electrochemical conversion of aldoximes **157** into the corresponding nitriles **158** involves a pair of oxidation and reduction process as shown in Fig. 5.1. (eq. (5.71)) [197]. Initially, the aldoximes **157** are oxidized to nitrile oxides **159** followed by the cathode reduction to the nitriles **158**.

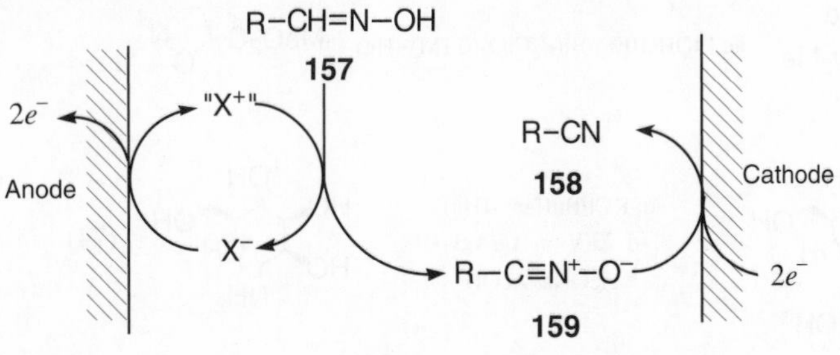

R–CH=N–OH

157

$2e^-$ "X$^+$"

Anode   R–CN

158   Cathode

X$^-$

R–C≡N$^+$–O$^-$ $2e^-$

159

Fig. 5.1 Paired reaction of aldoximes

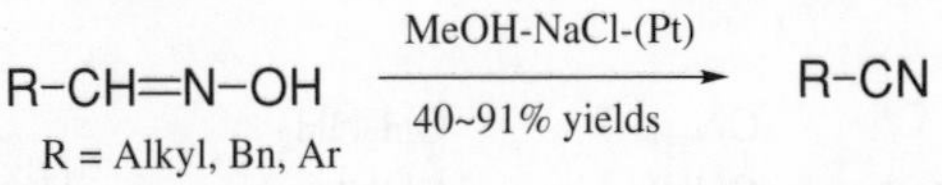

R–CH=N–OH $\xrightarrow[\text{40~91\% yields}]{\text{MeOH-NaCl-(Pt)}}$ R–CN

R = Alkyl, Bn, Ar

157   **158**

$$\text{(5.71)}$$

The electroreduction of the vicinal oxime groups proceeds *via* a similar mechanism as shown in the mono oximes. Some are recorded in Table 5.8. [198~201].

Table 5.7 Results from Electroreduction of Oximes

Oximes	Electrolysis Conditions	Products Yield, (%)	Ref.
(cyclohexanone)=NOH	aq. H_2SO_4(5%)/$(NH_4)_2SO_4$-(Ni)	(cyclohexyl)-NH_2 (59)	[183]
Ph-C(=NOH)CN	aq. HCl(1M)-(C)	Ph-CH(NH_2)CN (91)	[184]
$\underset{Me}{\overset{PhCH_2}{C}}$=NOH	aq. EtOH(50%)-$(NH_4)_2SO_4$(5%)-(Ni)	$\underset{Me}{\overset{PhCH_2}{C}}HNH_2$ (30)	[189]
	aq. Na_3PO_4/$ZnSO_4$-(Pb/Zn)	(80)	[190]
	aq. KOH(5~20%wt)-(Raney Ni)	(95)	[191]
HN (ring) =NOH	aq. H_2SO_4-(Ni)	HN (ring)-NH_2 (90)	[192]
$\underset{}{\overset{NOH}{Me-C-CO_2H}}$	aq. $HClO_4$(1M)-(Hg), pH 4	$\underset{}{\overset{NH_2}{Me-CH-CO_2H}}$	[193]
HO-CHCO$_2$Me CH$_2$-CCO$_2$Me ‖ NOH	aq.EtOH(10%)-Bu_4NClO_4(0.1M)-(Hg)	MeO$_2$C-(isoxazoline)-CO$_2$Me	[194]
(sugar chain with OH, HO, =N-OH)	aq. KCl/Buffer* -(Hg) –1.73 V *vs.* Ag/AgI * AcONa/ AcOH	(sugar chain with OH, HO, NH_2) (74)	[195]
$\underset{Py}{\overset{Ph}{C}}$=NOH	aq. EtOH(10%)-KCl(1M)/Buffer*-( ) * Buffer-Robinsin	$\underset{Py}{\overset{Ph}{C}}HNH_2$	[196]
CN C=NOH CO$_2$Me	aq. HCl(1.16 mol)-(Pb)	CN CHNH$_2$ CO$_2$Me , CH$_2$NH$_2$ CHNH$_2$ CO$_2$Me (93)	[185] [186]

Table 5.8 Results from Electroreduction of Vicinal Oximes

Oximes	Conditions	Product	Ref.
H–C=N–OH H–C=N–OH	aq. NaClO$_4$/Buffer*-(Hg)	H$_2$N–CH$_2$–CH$_2$–NH$_2$	[198]
(benzene ring) NOH, NOH	aq. HClO$_4$(0.5 M)/Bi-(Pt)	(benzene ring) NH$_2$, NH$_2$	[199]
(ring) NOH, NOH, NOH, NOH	aq. HClO$_4$-(Pt or Au)* * modified electrode	(ring) NH$_2$, NH$_2$, NH$_2$, NH$_2$	[200]
(ring) NOH, NH, NOH	aq. Acid*-(DME) * pH 1.45~12.36	(ring) H NH$_2$, NH, H NH$_2$	[201]

5.2.2.2 Electroreduction of Hydrazones and Related Compounds

The electroreduction of the hydrazone group takes places *via* the cleavege of the N-N bond, and the consumption of four electrons for the reduction of the hydrazones has been accepted as an initial reduction in acidic as well as in alkaline media [202].   The electroreduction of anisylidene benzohydrazide **160** has been performed in an aqueous EtOH(40%)/AcOH-HCl-(?) system to afford the corresponding *p*-methoxybenzylamine **161** in 61% yield (eq. (5.72)) [203].   Ethyl 2-phenylhydrazono-3-semicarbazono-2,3-dioxobutyrate **163** undergoes electroreductive cleavage of the N-N bond of the hydrazine moiety to give the amino-semicarbazone derivative **164** (eq. (5.73)) [204].   The electroreduction of *N,N*-disubstituted

$$\text{MeO–}\!\!\left\langle\right\rangle\!\!\text{–CH=NNH} \xrightarrow[\text{pH 4.25, }-1.3\text{ V }vs.\text{ SCE}]{\text{EtOH(40\%)-AcOH/HCl-( )}} \quad \text{CH}_2\text{NH}_2 \text{ (MeO, 61\%)} \;+\; \text{H}_2\text{NCOPh}$$

(5.72)

160 COPh   **161**   **162** (88%)

$$\text{Ph–NH–N=C}\begin{smallmatrix}\text{CO}_2\text{Et}\\\text{C–CH}_3\end{smallmatrix} \xrightarrow[\text{*Phosphate Buffer}\;-\;\text{Ph–NH}_2]{\text{DMF/H}_2\text{O(0.1\%)-Buffer*-(Pt/C)}} \begin{smallmatrix}\text{H CO}_2\text{Et}\\\text{C}\\\text{H}_2\text{N C–CH}_3\end{smallmatrix}$$

(5.73)

163 N–NHCONH$_2$ **164** N–NHCONH$_2$

hydrazones **165** of nitrobenzaldehyde and nitrofurfural in DMF may provide the corresponding nitriles **166** (eq. (5.74)) [205].   The electroreductive hydrogenation of arylhydrazones of α-ketocyano ketones **167** (R = COPh) and α-cyano esters **167** (R = CO$_2$Et) in a

$$Ar-CH=N-NR \xrightarrow[-NHR_2]{DMF-(Hg)} Ar-C\equiv N \qquad (5.74)$$

165   hydrazone   **166**

Ar = 4-nitrophenyl   R = Me, Ph
4-nitrobutyl

PhCN-Bu$_4$NClO$_4$-(Pt) system, giving **168**, proceeds under controlled potentials in which a single two-electron or two one-electron waves are observed (eq. (5.75)) [206].   Further reduction of the amino compound produced by the four-electron reduction undergoes two-electron reduction to afford ammonia and carbonyl compounds.   Recently, the reduction has been preformed at solid electrodes [202,204,207].   The typical examples of the electroreduction of the hydrazones are listed in Table 5.9. [208~216].

$$(5.75)$$

167   **168**

X = H, NO$_2$, X, Me, MeO, NH$_2$ etc.
R = CN, CO$_2$Et, COPh

Quanylhydrazone is converted into the corresponding quanylhydrazine by electroreduction in an aqueous AcONa/AcOH(or NH$_4$Cl/NH$_4$OH etc.)-Buffer-(Hg) system [217].

The electrochemical reduction of fluorenone triphenylphosphazine **169** in a DMF (0.1M)-Bu$_4$NClO$_4$-(Pt') system affords initially the corresponding anion radical [**169**]$^{-\bullet}$, which undergoes nitrogen-phosphorus bond cleavage to give 9-diazofluorene anion radical **172** together with triphenylphosphine.   The rapid coupling of the former species gives a stable dimeric dianion **173**, which is oxidized in successive two-electron steps to (FlN$_2$)$_2$ **174** (eq. (5.77)).   The tetraazatriene **174** slowly loses nitrogen gas to give fluorenone azine **170** as a major product (eq. (5.76)) [218].   Fluorenone tosylhydrazone (Fl = N-NHTs, **175**) undergoes one-electron reductive dehydrogenation in a DMF-Bu$_4$NClO$_4$(0.1M)-(Pt) system to give hydrogen and Fl=NNTs as products.   The presence of methanol as a proton donor gives FlHNH$_2$ **176** and TsNH$_2$ **177** as the principal products (eq. (5.78)) [219].

$$Fl=N-N=PPh_3 \xrightarrow{DMF-Bu_4NClO_4-(Pt)} \qquad =N-N= \quad + \quad =NNH_2 \qquad (5.76)$$

169

Fl =

170 (66%)   **171** (12%)

$$[169]^{-\bullet} \xrightarrow[-PPh_3]{} Fl=N_2^{-\bullet} \longrightarrow FlN_2=(FlN_2)^{2-} \xrightarrow{2e^-} Fl=N-N=N-N=Fl \qquad (5.77)$$

172   **173**   **174**

$$Fl=N-NHTs \xrightarrow[45~50\% \text{ yields}]{DMF/MeOH-Bu_4NClO_4(0.1M)-(Pt)} FlHNH_2 + TsNH_2 \qquad (5.78)$$

175   (70%)   (75%)

                                176   **177**

Table 5.9 Electroreduction of Hydrazones and Their Related Compounds

Hydrazone	Electrolysis Conditions	Products Yield, (%)	Ref.
Ph–C(Py)=N–NH₂	aq. EtOH(10%)-KCl(1.0M)-(Hg)	Ph–CH₂(Py) + 2NH₃	[208]
ArCO–NH–N=CHAr	DMF-Et₄NClO₄(0.08M)-(Hg) −1.70~−1.75 V, −2.3~2.5 V *vs.* Ag⁺/AgCl	ArCONH₂ + H₂NCH₂Ar	[209]
HN=C(tetramethylpiperidine)=N–NH₂	aqueous solution-(Pb)	HN–CH–NH₂ (100)	[210] [192]
X–C₆H₄–N(H)–N=(Meldrum's acid)	MeOH/H₂O(6/4~8/2)-KCl-(Hg) −0.44~−0.94 V *vs.* SCE, pH 1.24~6.91	X–C₆H₄–NH₂ + H₂N–(Meldrum's acid)	[211]
quinoxalinone-C=N–NH–C(O)–Ar, (CHOH)₂, CH₂OH	MeOH-Buffer*-(Hg) *Britton-Robinson −0.40~−0.65 V *vs.* SCE	quinoxalinone-CHNH₂, (CHOH)₂, CH₂OH + H₂N–C(O)–Ar	[212]
2-C₅H₄N·HC=N–NH–(pyridinyl)	EtOH(50%)-Buffer*-(Hg) *Britton-Robinson −1.50~−1.90 V *vs.* SCE	2-OHC-C₅H₄N 2H₂N-C₅H₄N NH₃	[213] [214]
R–C₆H₄NH–N=C(CONHPh)(COMe) R = H, Me, MeO, Cl	aq. DMF(40%)-Et₄NBr(1M)-(C) −1.15~−1.20 V *vs.* Ag/AgCl, pH 8.5	R–C₆H₄NH₂, H₂N–CH(CONHPh)(COMe)	[202]
(pyridinyl)CH=N–NH–(quinolinyl)	aq. MeOH-Buffer*-(C) * Phosphate, pH 3.0, −0.90 V	H₂N–(quinolinyl), (pyridinyl)CH₂NH₂	[207b]
(benzothiazolyl)-C(-CN)=N–NH–(C₆H₄Y) Y = H, 2-MeO, 2-Me, 2-Cl, 2-NO₂	aq. DMF/EtOH(1/4)-HCl(10N,10%)-(Hg) pH 1.6, −0.7 V *vs.* SCE	[(benzothiazoline)-CH(CO₂H)]₂ (50)	[215]
R–C₆H₄CH=C–CO, S–C–NH, ‖ NNHPh R = H, *p*-Me, *p*-MeO, *p*-OH	aq. DMF/EtOH(2/3,50%)-NaOH-(?)	ArCH₂–CH–CO, S–C–NH, PhNH₂ NH₂	[216]

5.2.2.3   Electroreductive Hydrogenation of C=N Bonds

The electroreductive hydrogenation procedure can be used for the amination of ketones in aqueous media.   (Methylamino)cyclohexane **180** has been synthesized from a mixture of cyclohexanone (0.35 M) and methylamine (0.42M) *via* imine **179** by electroreduction in an aqueous KOH/K$_3$PO$_4$(1N)/H$_3$PO$_4$(25%)-(Pb) system at pH 12 (eqs. (5.79, 5.80)) [220]. The electroreduction of benzophenone imine in a DMF-Me$_4$NBF$_4$(0.1M)-(C) is shown to undergo two successive one-electron transfers initially to provide the corresponding anion radical as a strong EG base in the absence of a proton donor [221].   The electroreduction of benzylidene-4'-nitroaniline in an MeOH-LiCl(0.1M)-(Pt/Hg) system undergoes first reduction of the nitro group by taking four electrons, prior to two electrons reduction of the azomethine bond [222].

$$\text{(5.79)}$$

$$\text{(5.80)}$$

*K$_3$PO$_4$(1N)/H$_3$PO$_4$(25%) Buffer

The choice of electrolyte (acid) for the electroreduction of the C=N bond of **181** is important.   The reduction becomes more and more irreversible with increase in the electrolyte concentration.   The case of reduction of cimetidine **181** is found to give **182**.   The yield of **182** increases in the following order HClO$_4$ > HCl > H$_2$SO$_4$ (eq. (5.81)) [223].   The reaction trends for alcoholic solvents in electroreduction of Schiff bases are observed: methanol > ethanol/water (1/1) > ethanol [224].

$$\text{(5.81)}$$

Preparative electrosynthesis of *N*-(4-chlorobenzyl)-2-hydroxylmethyl-cyclohexylamine **184** is carried out by electrolysis of the 1,3-oxazin derivatives **183** in an aqueous EtOH (40%)-H$_2$SO$_4$(6M)-(Hg) system (eq. (5.82)) [225].

$$\text{(5.82)}$$

Electrohydrogenation of the C=N bond of thiazoline-azetidinone derivatives **185** has been attained in an aqueous CH$_2$Cl$_2$-HClO$_4$-(Pb) biphase system, yielding the thiazolidine-azetidinones **186** in 77~100% yields (eq. (5.83)) [226].

$$(5.83)$$

The preparative electroreduction of 2-phenyl-4,4-dimethyl-2-oxazole **187** in a THF/H_2O(4%)-Bu_4NBF_4(0.25M)-(Hg) system affords N-benzyl-2-amino-2-methylpropanol **188** in 91% yield (eq. (5.84)) [227].

$$(5.84)$$

In an acidic medium, the electroreduction of pyridazin-3-ones **189** takes place at the 4,5-double bond to give the dihydro compounds **190** in 80% yield (eq. (5.85)) [228]. In a basic ammonical buffer, the electrolysis of **189** can lead to the tetrahydro derivatives **191** in good yields (eq. (5.86)).

$$(5.85)$$

$$(5.86)$$

a : $R^1 = R^2 = R^3 = H$; $R^4 = Ph$
b : $R^1 = R^2 = R^4 = H$; $R^3 = Ph$
c : $R^1 = R^2 = Me$, $R^3 = H$; $R^4 = Ph$

The electroreduction of 2-phenyl-5-acetyl (or formyl)-6H-1,3-thiazine **192** undergoes either hydrodimerization or hydrogenation, depending on the reduction potential [229]. The lower reduction potential at $-0.65\sim -0.8$ V vs. SCE provides the dimer **193** in 70~80% yields (eq. (5.87)), whereas the higher reduction potential at $-1.15\sim -1.4$ V affords the hydrogenation product **195** via **194** in good yield (eq. (5.88)). The electrohydrogenation of the C=N bond together with the C=C bond of 2-substituted 6-methoxycarbonyl-4H-1,3-thiazine-4-ones **196** proceeds in an AcOH/EtOH(1/1v/v)-AcONa(0.5M)-(Hg) system to give the corresponding 2,3,5,6-tetrahydro-4H-1,3-thiazine-4-ones **197** in 40~56% yields (eq. (5.89)) [230].

$$\text{(5.87)}$$

$$\text{(5.88)}$$

$$\text{(5.89)}$$

The electroreduction of 5H-2,3-benzodiazepine **198** can lead to different compounds **199** and **200** by controlling the reduction potential as well as the electrolyte [231]. The di-hydroisoquinoline **199** is obtained in an aqueous KCl(0.5M)/HCl(0.2M) system at –1.2 V vs. SCE (eq. (5.90)). In contrast, in DMF, the electrolysis of **198** affords the benzodi-azepine derivative **200**, exclusively (eq. (5.91)). The electrocatalyzed cyanomethylation of azomethines **201** with acetonitrile anions proceeds in an MeCN-Bu$_4$NBr(0.1M)-(Hg) to give cyanomethylation products **202** in 45~85% yields (eq. (5.92)) [232].

$$\text{(5.90)}$$

$$\text{(5.91)}$$

$$\text{(5.92)}$$

Ar^1 = Ph, p-FC$_6$H$_4$, o-MeOC$_6$H$_4$, Chinolin

Ar^2 = Me, Et, CF$_3$, Ph, o-MeOC$_6$H$_4$

Upon electroreduction phenyl carbodimides **203** can be electrochemically reduced to the corresponding diphenylurea **204** (eq. (5.93)) [233]. A successful substitution of 1,3,5-

trimethoxybenzene **208** with 1H-tetrazole **206** has been achived by a paired electrosynthesis (eq. (5.94)) [234].   The electrolysis is carried out in an MeCN-(Pt) system in the presence of tetrabutylammonium salt of tetrazole **207** in an undivided cell.   The salt is prepared by the reaction of tetrazole and tetrabutylammonium bromide in the presence of sodium methoxide.

$$(5.93)$$

$$(5.94)$$

An unexpected stable anion-radical has been obtained when 2-(arylidineamino)furans are electroreduced in a DMSO-LiClO$_4$-(Hg) system [235].   The results from electroreduction of other azomethine compounds are summarized in Table 5.10. [236~249].

Table 5.10a Results from Electroreduction of Azomethine Compounds

	Conditions	Products	Yields(%)	Ref.
	aq. EtOH(40%)-(Hg) –PhNH$_2$			[236]
	aq. Et OH(60%)-H$_2$SO$_4$(0.25M)-(Hg) –0.1V *vs.* SCE, quant. yield			[237]
	aq. EtOH(40%)-H$_2$SO$_4$-(Hg) pH 1.0		(30~60)	[238]

226

Table 5.10b Results from Electroreduction of Azomethine Compounds

Conditions	Products	Yields(%)	Ref.
aq. EtOH(40%)-H$_2$SO$_4$-(Hg) pH 1.0		(16)	[238]
aq. EtOH(40%)-H$_2$SO$_4$/Buffer-(Hg) −1.15 V *vs.* SCE, pH 2 Buffer : Britton-Robinson		(68)	[239]
MeCN-LiClO$_4$(0.5M)/CF$_3$CO$_2$H(1.14M)-(?)			[240]
aq. MeCN-H$_3$PO$_4$/NaCl-(Hg) −1.15 V, pH 1.4		(81)	[241]
aq. MeOH(50%)-Na$_2$HPO$_4$(0.5M)-(?)		(85)	[242]
aq. EtOH(40%)-H$_2$SO$_4$-(Hg)			[243]
aq. DMF(30%)-Buffer-(Hg) pH 1.6, −1.25 V *vs.* SCE		(38)	[244]
aq. MeOH(20%)-Buffer-(Hg) pH 1.81, −0.9 V *vs.* SCE		(88)	[245]
aq. EtOH-Buffer-(Hg) −1.10 V *vs.* SCE, pH 4.6		(35)	[246]

Table 5.10c Results from Electroreduction of Azomethine Compounds

	Conditions	Products	Yields(%)	Ref.
	aq. EtOH(20%)-Buffer*-(Hg) *Britton-Robinson Buffer		(50)	[247]
	aq. EtOH-LiClO$_4$/Buffer-(Hg) −1.14 V *vs.* SCE			[248]
R = CH(CH$_3$)CO$_2$CH$_2$CH$_3$	aq. EtOH-NH$_4$Cl/NH$_3$-(Hg) pH 9.25, E$_{1/2}$ −0.95 V *vs.* SCE			[249]

5.2.2.4   Inter- and Intramolecular Coupling Reactions

The intermolecular hydrodimerization of Schiff bases is thought to be a short-cut way to prepare 1,2-vicinal diamines.   The optimal conditions for such hydrodimerization have been investigated with *N*-benzylidine-p-toludine **210**, leading to 1,2-diphenyl-*N*,*N*-di(*p*-tolyl)ethylene diamine **211** by systematic variation of the electrolysis parameters (eq. (5.95)) [250].   The product composition depends upon 1) the proton availability in the electrical double layer (increased yield of hydrodimer in non aqueous base electrolyte systems and hydrophobic supporting electrolyte anions and cations); 2) the electrode material (on mercury, the hydrodimer is the major product, while on copper, lead and glassy carbon, the reduction to the secondary amine is favored); 3) the number of coulombs passed (the current yield in relation to the hydrodimer is constant until about 60% of the theoretical conversion); 4) the concentration of depolarizer and the catholyte temperature.

$$\text{(5.95)}$$

The electroreductive carboxylation of Schiff bases may provide a method for the synthesis of isotopically labeled α-amino acids.   For instance, the electroreduction of *N*-(benzylidene)aniline **212** in an MeCN-Et$_4$NClO$_4$-(Hg) system under CO$_2$ atmosphere gives the *C*-carboxylation product **213** in 84% yield (eq. (5.96)) [251,252].   The scale-up of the electrocarboxylation of benzalaniline to the α-amino acids has been examined using a filter-press electrolytic cell with semi-continuous renewal of sacrificial electrodes [253].

$$
\underset{\substack{\text{H}\\\textbf{212}}}{\overset{\text{Ph}}{}}\text{C=N-Ph} \xrightarrow[\substack{\text{CO}_2,\ \text{H}^+\\84\%\ \text{yield}}]{2e^-} \underset{\substack{\text{H}\quad\textbf{213}}}{\overset{\text{CO}_2^-\ \text{H}}{\text{Ph-C-----N-Ph}}}
\tag{5.96}
$$

An attempt to trap the two-electron reduction intermediates of *N*-benzalaniline by alkylation with 1,3-dibromopropane has been made in a DMF-Bu$_4$NI-(Hg) system (eqs. (5.97, 5.98)) [254].   The characteristic behaviors of the anion radical **215** and anion **220** which are derived from benzophenone anil **214** by electroreduction, have been investigated.   The

$$
\underset{\textbf{214}}{\text{ArCH=NPh}} \underset{}{\overset{e^-}{\rightleftarrows}} \underset{\textbf{215}}{\text{Ar}\overset{\bullet}{\text{C}}\text{H-}\overset{-}{\text{N}}\text{Ph}} \overset{\text{H}^+}{\longrightarrow} \underset{\textbf{216}}{\text{Ar}\overset{\bullet}{\text{C}}\text{HNHPh}} \overset{\text{Dim., Dispr.}}{\longrightarrow} \underset{\substack{\text{PhNH}\quad\text{NHPh}\\\textbf{217}}}{\overset{\substack{\text{Ar}\quad\text{Ar}}}{\text{CH-CH}}}
$$

(Ar = Ph)
(Ar = 2-Py)

$$+$$

$$\underset{\textbf{218}}{\text{ArCH}_2\text{NHPh}}$$

$$\tag{5.97}$$

Br(CH$_2$)$_x$Br

$$
\underset{\substack{\textbf{219}\\(\text{CH}_2)_3\text{Br}}}{\text{Ar}\overset{\bullet}{\text{C}}\text{H-NPh}} \overset{e^-}{\underset{}{\rightleftarrows}} \underset{\substack{\textbf{220}\\(\text{CH}_2)_3\text{Br}}}{\text{Ar}\overset{-}{\text{C}}\text{H-NPh}} \overset{\text{cyclis.}}{\longrightarrow} \underset{\substack{\textbf{221}\\(\text{CH}_2)_n}}{\text{ArCH-NPh}} \quad (59\%)
\tag{5.98}
$$

(Ar = Ph, n = 4)
(Ar = 2-Py, n = 3)

electroreduction of **222** in a DMF-LiClO$_4$(0.1M)-(Hg) system in the presence of methyl chloride affords a mixture of **223**, **224**, **225**, and **226** in quantitative total yield (eq. (5.99)) [255].   The courses of electroreduction of **227** are influenced by the choice of electrolytes to give **228** or **229** (eq. (5.100)) [256].

$$
\underset{\textbf{222}}{\overset{\text{Ph}}{\underset{\text{Ph}}{}}}\text{C=N-Ph} \xrightarrow{\text{DMF-LiClO}_4(0.1\text{M}),\ \text{MeCl}} \underset{\substack{\text{MeMe}\\\textbf{223}\ (7\%)}}{\overset{\text{Ph}}{\underset{\text{Ph}}{}}\text{C-N-Ph}} + \underset{\substack{\text{H}\ \text{Me}\\\textbf{224}\ (33\%)}}{\overset{\text{Ph}}{\underset{\text{Ph}}{}}\text{C-N-Ph}} +
$$

$$
\underset{\substack{\text{Me}\ \text{H}\\\textbf{225}\ (29\%)}}{\overset{\text{Ph}}{\underset{\text{Ph}}{}}\text{C-N-Ph}} + \underset{\substack{\text{H}\ \text{H}\\\textbf{226}\ (31\%)}}{\overset{\text{Ph}}{\underset{\text{Ph}}{}}\text{C-C-Ph}}
$$

$$\tag{5.99}$$

$$
\underset{\substack{\textbf{228}\ \ \text{N-Me}\\\text{Ph}}}{} \xleftarrow[\text{MeCl}]{\text{Bu}_4\text{NI}(0.15\text{M})} \underset{\substack{\textbf{227}\ \ \text{N}\\\text{Ph}}}{} \xrightarrow[\text{MeCl}]{\text{LiClO}_4(0.25\text{M})} \underset{\substack{\textbf{229}\ \ \text{NH}\\\text{Ph}}}{}
\tag{5.100}
$$

The electroreductive intermolecular coupling of aromatic imine **230** with the aldehyde **231** is found to be effectively promoted in the presence of chlorotrimethylsilane, giving **232** in 81% yield (eq. (5.101)) [257].   The electrolysis is carried out in a DMF-Et$_4$NOTs-(Pb) system in the presence of equimolar amounts of chlorotrimethylsilane and trimethylamine as additives.

$$(5.101)$$

In connection with electrosynthetic carboxylation and bridgehead alkylation experiments, the lowest free orbitals and the free unpaired electron densities are correlated with the electrochemical data of thirty-five substituted Schiffbases [258].

The intramolecular and transannular cyclizations of a series of di-Schiff bases provide a useful synthetic pathway for the preparation of piperazines, indoloindoles, diazepines and diazocines.   The electroreductive criss-cross addition of aldazones **233** to N-aryl-maleimides **234** has been attained in good yields [259].   The electrolysis is performed in an aqueous EtOH(50%)-HCl-(Hg) system to yield tetracyclo derivatives **235** (eq. (5.102)).

$$(5.102)$$

Preparative scale electrosynthesis of **237** is carried out in a Glyme-Bu$_4$NClO$_4$-(Hg) system in second wave potential at –2.90 V vs. Ag/Ag$^+$ of **236** to give a red-yellow anion intermediate, which is immediately destroyed by the addition of 4 equiv. of methyl iodide, yielding the 1,4-piperazine (eq. (5.103)) [260].   The intramolecular coupling reaction of **238** readily occurs to give the bridged product **239** (eq. (5.104)).

$$(5.103)$$

$$(5.104)$$

The cathodic acylation of the bridged 1,5-benzodiazepines **240** derived from the condensation of *o*-phenylenediamines with 4,6-dimethylbicyclo- [3.3.1]nona-3,6-diene-2,8-dione in an MeCN-Bu$_4$NBF$_4$-(Pt) system in the presence of acetic anhydride gives rise to the substituted barbaralane **241** (eq. (5.105)) [261].   The electroreduction proceeds at –2.35 V in a single two-electron wave ascribable to the reduction potential of Schiff bases.

$$
\begin{array}{c}
\text{240}
\end{array}
\xrightarrow[\text{–2.35~ –2.16 V }vs.,\text{ Ag/Ag}^+]{\text{MeCN-Bu}_4\text{NBF}_4/\text{Ac}_2\text{O-(Pt)}}
\begin{array}{c}
\text{241}
\end{array}
\tag{5.105}
$$

R^1	R^2	Yield, %
H	H	39
H	Cl	45
Me	Me	95

Electroreductive carboxylation of nitrogen-containing heteroaromatic compounds has been realized in a DMF-Et$_4$NBr(0.2M)-(C/SUS) system in the presence of carbon dioxide and cyclohexene [262].   The electrolysis of quinoline **242** affords the carboxylated products **243** and **244** in total 60% yields after treating with ethyl bromide (eq. (5.106)).

$$
\begin{array}{c}
\text{242}
\end{array}
\xrightarrow[\text{2) EtBr, 60\% yield}]{\substack{\text{1) CO}_2,\text{ c-C}_6\text{H}_{10}, \\ \text{DMF-Et}_4\text{NBr(0.2M)-(C/SUS)}}}
\begin{array}{c}
\text{(32\%)} \\
\text{243}
\end{array}
+
\begin{array}{c}
\text{(28\%)} \\
\text{244}
\end{array}
\tag{5.106}
$$

Zinc powder has been found to be effective for the intramolecular cyclization of aromatic diimines, yielding 1,4-diaza crown ethers.   The ring closure of the diimines **245** is performed in a DMF/THF(1/1)-MsOH-(Zn) system at –20 °C to form macrocyclic products **246** in good yields (eq. (5.107)) [263].

$$
\begin{array}{c}
\text{245}
\end{array}
\xrightarrow[\substack{\text{–20 °C, 52~86\% yields} \\ n = 0~4 \\ \text{Ar} = p\text{-MeOC}_6\text{H}_4, p\text{-ClC}_6\text{H}_4, \text{1-Furyl}}]{\text{DMF/THF(1/1)-MsOH-(Zn)}}
\begin{array}{c}
\text{246}
\end{array}
\tag{5.107}
$$

The intermolecular electroreduction coupling of 2-phenyl-6*H*-1,3-thiazines **247** has been attained at –0.65 V *vs.* SCE, 0 °C in an aqueous EtOH(50%)/AcOH-AcONa(0.5M)-(Hg) system to give the dimer **248** in 70~80% yields (eq. (5.108)) [264].   The two-electron electroreduction of the 1,3-thiazines products at –1.15~1.40 V *vs.* SCE to afford the 2,3-dihydro compounds **249** in 70~75% yields (eq. (5.109)).

$$\text{(5.108)}$$

$$\text{(5.109)}$$

5.3 Electroreductive Cleavage of Nitrogen Atom Containing a Single Bond

5.3.1 Electroreductive Cleavage of N-N Bond

The electroreductive cleavage of the N-N bond of 1-amino-4,6-diphenyl-2-pyridone **250** in an aqueous EtOH(20%)-Britton-Robinson Buffer-(Hg) system proceeds to give 4,6-diphenyl-2-pyridone **251** smoothly (eq. (5.110)) [265].   The electroreduction cleavage of

$$\text{(5.110)}$$

the N-N bond of the hydrazone **252** has been investigated in an aqueous EtOH(50%)-Britton-Robinson Buffer-(Hg) system in an acidic medium to give 2,2'-dipyridylketone **254** *via* the ketimine **253** (eq. (5.111)) [214].

$$\text{(5.111)}$$

A mixture of the *p*-tolyl derivative **255** is electroreduced at −1.6 V *vs.* SCE in a DMF/EtOH(1/5)-NaOH(0.2M)-(Hg) system at pH 12.9 to give *p*-toluidine **256** and α-thiocarboxanidoglycine **257** (eq. (5.112)) [266].   *p*-Hydroxybenzaldehyde isonicotinyl hydrazone **258** undergoes reductive cleavage of the N-N linkage, leading to the corresponding isonicotinamide **259** and 4-hydroxybenzaldehyde imine **260** (eq. (5.113)) [267].

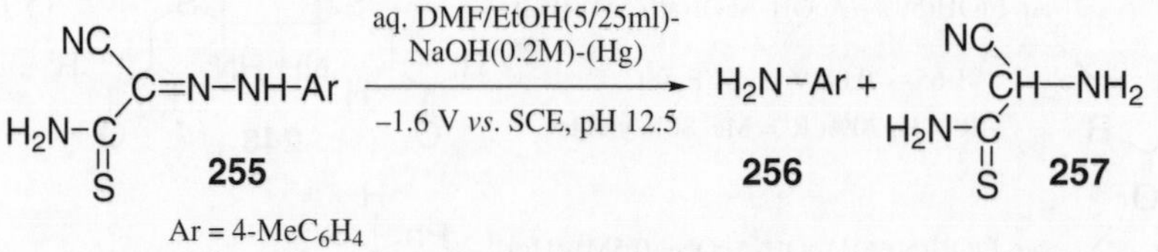

$$\text{(5.112)}$$

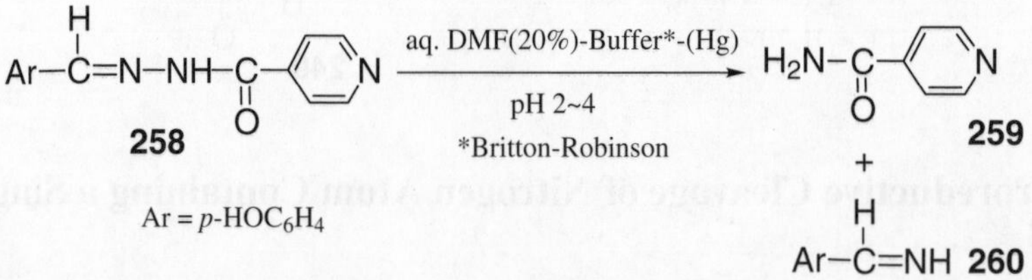

Ar = 4-MeC$_6$H$_4$

$$\text{(5.113)}$$

Ar = p-HOC$_6$H$_4$

The electroreduction of 4-phenylazo-3-amino-2-pyrazolin-5-one **261** undergoes two two-electron irreversible diffusion-controlled waves.   The first wave is attributed to the cleavage of the single bond of the C=N-NHPh group, yielding 4-iminopyrazolone **262** and aniline **263** (eq. (5.114)) [268].   The former compound **262** undergoes further reduction at

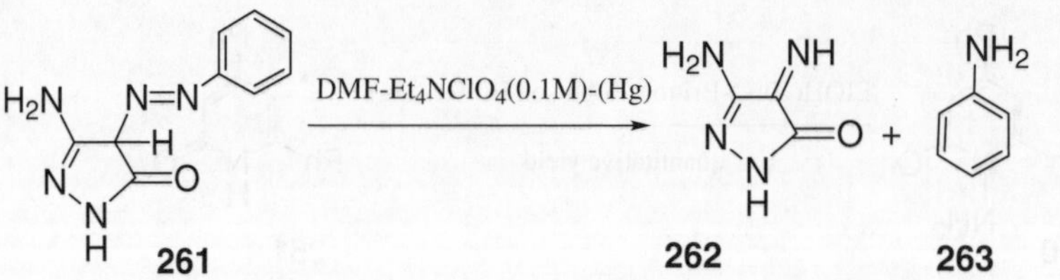

$$\text{(5.114)}$$

a more negative potential (the second wave) to give 4-aminopyrazolone.   5-(*N*-2-Benzothiazolylhydrazono)-1,3-dimethylbarbituric acid **264** can be electrochemically reduced to a mixture of 2-aminobenzothiazole **265** and 5-amino-1,3-dimethylbarbituric acid **266** (eq. (5.115)) [207c].   2-(*N*-Phenylhydrazo)-1,3-indandione **267** can be electroreduced

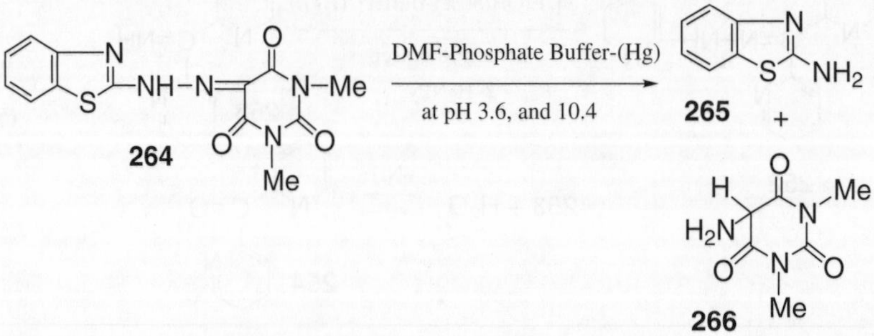

$$\text{(5.115)}$$

into aniline and 2-amino-1,3-indandione **270** *via* the ketimine **268**.   The result clearly indicates that the molecule **267** has been cleaved at the N-N bond followed by hydrogenation of the azomethine moiety of **268**, giving **270** (eq. (5.116)) [269].   The electroreduction of α-

and β-naphthylhydrazonodimedones **271** in a DMSO-Et$_4$NI(0.1M)-(Hg) system yields the corresponding naphthylamines **272** in 45~50% yields (eq. (5.117)) [270].   The reduction of

$$268 \xrightarrow{2e^- + 2H^+} \text{(structure) } \mathbf{270} \tag{5.116}$$

phenylhydrazonodimedone gives aniline [271].   The controlled potential reduction of 1,3,4-thiadiazole ring compounds **273** in an aqueous NaOH(0.2M)-(Hg) system at –1.60 V (SCE) affords 2-benzoylamino-5-(α-aminocarboxymethyl)-1,3,4-thiadiazole **274** in 60% yield (eq. (5.118)) [272].   Benzalazine **275** undergoes electroreductive N-N bond fission

and subsequent hydrolysis below pH 3.0 to give benzaldehyde *via* **276**, which is then reduced in a two-electron step to give benzyl alcohol.   In contrast, the six-electron reduction occurs at pH 7.0 and 10.4, causing the formation of benzylamine **277** as the final product (eq. (5.119)) [273].   2-Benzoylpyridine azine can be electrochemically reduced to the corresponding amine in a two-step manner involving a four- and two-electron reduction, successively [208].

$$\text{Ph–CH=N–N=CH–Ph} \xrightarrow[\substack{\text{Phosphate Buffer} \\ \text{pH 7.0, } E_{1/2} = -1.04 \text{ V } vs. \text{ SCE} \\ \text{pH 10.4, } E_{1/2} = -1.19 \text{ V}}]{\text{MeOH/H}_2\text{O(4/1)-(Pt/Hg)}} \begin{array}{c} \text{2-Ph–CH=NH} \quad \mathbf{276} \downarrow \\ \text{Ph–CH}_2\text{NH}_2 \\ \mathbf{277} \end{array} \tag{5.119}$$

275

5.3.2 Electroreductive Cleavage of C-N Single Bond

Electroreductive removal of protecting groups is an important subject in electroorganic synthesis [274]. N-benzyloxycarbonyl function of **278** can be removed by electroreduction in DMF at a high negative potential (eq. (5.120)) [275].

$$PhCH_2\text{--}O_2C\text{--}NHR \xrightarrow[\substack{70\sim80\% \text{ yields}}]{2H^+, -CO_2} Ph\text{--}Me + R\text{--}NH_2 \tag{5.120}$$

278

$$R = alkyl$$

The electroreduction of N-benzoylaziridine **279** in a THF-Bu$_4$NClO$_4$(0.5M)-(Hg) undergoes the cleavage of the C-N bond by a ketyl radical induction under electroreduction conditions, leading to a mixture of the oxazoline **280**, the ethenamide **281**, and the saturated amide **282** (eq. (5.121)) [276]. The electroreductive cleavage of the C-N bond of nitromycin C **283** has been proved by the isolation of the corresponding reduced products **284** in 90% yield (eq. (5.122)) [277]. The prolonged reduction of **283** yields the fully reduced product **285** as the major product (> 90%).

$$(5.121)$$

*THF-Bu$_4$NClO$_4$(0.5M)-(Hg)

$$(5.122)$$

Electrochemical cleavage of protecting groups, either used temporarily in the synthesis of complex multifunctional target molecules or retained until a late or final step, prior to liberation of the end products, is a promising alternative method to the chemical ones already available [278]. Regioselective deprotection of N-substituted N-(t-butoxy carbonyl)amides **287**, whose substituent is either a benzyloxycarbonyl, 2-nitrophenylsulfenyl, diphenylphosphinyl, toluene-4-sulfonyl, or benzyl group, has been performed in a DMF(orMeCN)-Et$_4$NCl-(C) system in the presence of MeOH (or Et$_3$NHCl) as the proton donor, leading to the amine **288** (eq. (5.123)) [279,280]. In all cases, the aromatic protecting group attached to the primary amine group can be cleaned selectively with

$$R–NH–Ph \xrightarrow[\text{DMAP}]{\text{Boc}_2O} R–N(Boc)–Ph \xrightarrow[\text{60~95\% yields}]{2e^-,\ -R} Boc–NH–Ph \qquad (5.123)$$

286   **287**   **288**

R = PhCH$_2$O$_2$C, 2-NO$_2$C$_6$H$_4$S, Ph$_2$P,
4-MeC$_6$H$_4$SO$_2$, PhCH$_2$

respect to the group at the secondary amine function. The electroreductive elimination of
N-oxazolone protecting group of *N*-tert-butoxycarboxyl-4-[*N*-(tert-butoxycarbonyl)indol-3-
yl]-oxazol-2-one **289** in a DMF-Bu$_4$NBF$_4$(0.1M)-(Hg) system yields the deprotected product **290** in 75% yield (eq. (5.124)) [281]. The selective cathodic cleavage of one of the

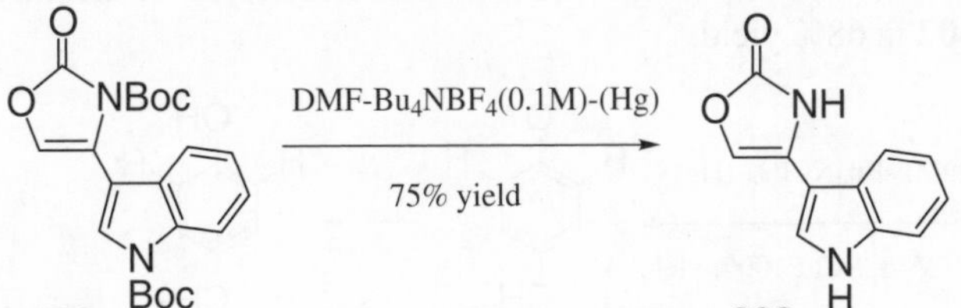

(5.124)

289   **290**

alkoxycarbonyl or acyl groups from various imidodicarbonates, acylamides, and diacylamides has been investigated in a DMF-Et$_4$NCl-(C) system in the presence of MeOH or
Et$_3$NHCl as a proton donor (eq. (5.125)) [282]. All substrates **291** except for those having
trichloroethoxycarbonyl groups can be selectively cleaved, giving the iminds **292** in
89~100% yields.

$$Y-N(Boc)-COPh \xrightarrow[\text{90~100\% yields}]{\text{DMF-Et}_4\text{NCl/Et}_3\text{NHCl-(Pt/C)}} Y-NH-COPh \qquad (5.125)$$

291   **292**

Y = PhCO, Ts, PhCH$_2$OCO,
p-NO$_2$-C$_6$H$_4$CH$_2$OCO, CCl$_3$CH$_2$OCO,

Electroreductive cleavage of the C-N bond of **293** which is attached to the cyclic and
conjugated 6*H*-1,3-thiazine moiety can be induced in acidic medium, giving **294** and **295**
(eq. (5.126)) [283].

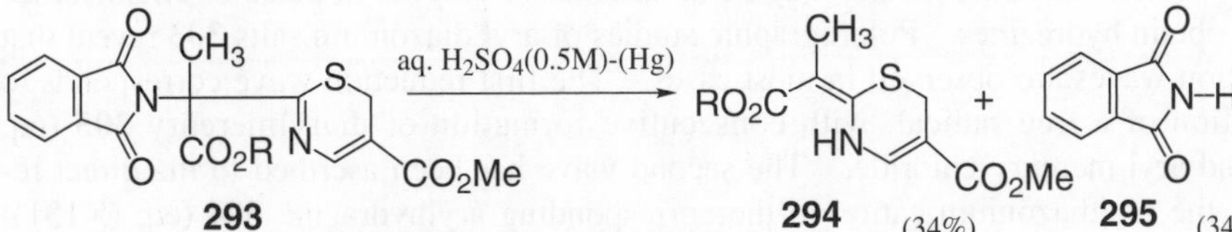

(5.126)

293   **294** (34%)   **295** (34%)

R = CH(CH$_3$)CO$_2$Et

Based on voltammogram results, the removal of the trityl group from 1-tritylpyrrole
296 to lead to **297** is viable on a preparative scale, but a solvent and electrolyte capable of
withstanding the rather large negative potentials is required (eq. (5.127)) [284].

$$
\textbf{296} \quad \xrightarrow[\text{–2.24 V}(E_{1/2}) \; vs. \; \text{Ag/Ag}]{\text{MeCN-Bu}_4\text{NBF}_4\text{-(Au)}} \quad \textbf{297} \qquad (5.127)
$$

R = H, CO₂H, COCF₃ → R = H, CO$_2$H, COCF$_3$

The electroreductive cleavage of the C-N bond of the quinames **298** occurs in a DMF-Bu$_4$NClO$_4$-(Hg) system at –2.0 V *vs.* SCE to give β-aryl propionic acid **265** in quantitative yields *via* the intermediate **299** (eq. (5.128)) [285].   The electroreductive removal of amino and nitro groups attached to the aromatic ring of **301** has been attained to afford the corresponding aromatic compounds **302** (eq. (5.129)) [286].   The electrolysis of the nitro derivative **301** may also give **302** in 68% yield.

$$
\textbf{298} \quad \xrightarrow[\substack{\text{–2.0 V } vs. \text{ SCE, 100\% yield} \\ \text{–RHN}^- \\ \text{R = H, Me}}]{\text{DMF-Bu}_4\text{NClO}_4\text{-(Hg)}} \quad \textbf{299} \quad \xrightarrow{\text{H}^+} \quad \textbf{300} \qquad (5.128)
$$

$$
\textbf{301} \quad \xrightarrow[\text{84\% yield}]{\text{aq. MeOH-HNO}_3(60\%)\text{-(Pt)}} \quad \textbf{302} \qquad (5.129)
$$

5.4   Electroreduction of Nitrogen Atom Containing Triple Bonds

5.4.1   Electroreduction of Diazo and Azide Compounds

5.4.1.1   Electroreduction of Diazo Compounds

Electroreduction of diazonium salts may be an alternative process in place of chemical reduction to obtain hydrazines.   Polarographic studies of aryl diazonium salts **303** reveal that two reduction waves are observed in most cases.   The first reduction wave corresponds to the formation of a free radical, with consecutive formation of diarylmercury **305** (eq. (5.132)) and aryl mercury chloride.   The second wave has been ascribed to the direct reduction of the aryldiazonium cation to the corresponding arylhydrazine **304** (eq. (5.131)) [287].   Arylhydrazines have been electrosynthesized using a mercury-pool cathode [288]. The electroreduction of 4-methyl- and 4-methoxybenzenediazonium tetrafluoroborate **303** in an aqueous NaClO$_4$(0.1M)-(Hg) system at 4 °C with a moderate stirring of the mercury-pool in a buffer of pH 2 yields hydrazines **304** (4-methyl-, 57%; 4-methoxy-, 80%, respectively) together with diarylmercury (eq. (5.132)) and the dimer **306** (eq. (5.133)) as minor products (eq. (5.134)) [289].

$$Ar-H \qquad\qquad (5.130)$$

aq. NaClO$_4$(0.1M)-(Hg)
$$\xrightarrow{-0.90\ V,\ pH\ 2,\ \sim 80\%\ yield} Ar-NHNH_2 \qquad (5.131)$$
304

$$Ar-N_2^+\ Y^-$$
303

aq. NaClO$_4$(0.1M)-(Hg)
$$\xrightarrow{0.05\ V,\ \sim 80\%\ yield,\ Met\ =\ Hg} Ar-Met-Ar \qquad (5.132)$$
305

$$Ar-Ar \qquad\qquad (5.133)$$
306

$$\left[R-\!\!\langle\ \rangle\!\!-N\!\equiv\!N \right]^+ BF_4^- \longrightarrow R-\!\!\langle\ \rangle\!\!-NHNH_2 \qquad (5.134)$$
303 \qquad R = Me or MeO \qquad **304**

The polarographic reduction of arenyldiazonium ions reveals that product elucidations implicate the corresponding radicals as intermediates [290]. Deactivation of the metallic surfaces at the cathode by covering the radicals has been encountered in every case of voltammetric measurement at platinum, gold, and mercury electrodes (eq. (5.130)) [291]. The electroreduction of the diazonium cation in aprotic solvents results in one-electron transfer to yield the aryl radical and nitrogen as the only products [292]. The electrore-duciton of 1-naphthalenediazonium tetrafluorobotate in an MeCN/CCl$_4$-Bu$_4$NI(1mmol)-(Hg) system affords a mixture of naphthalene, 1-chloro-naphthalene, and 1-iodonaphtha-lene, accounting for 94% yield of the consumed diazonium salt.

The electroreduction of benzenediazonium chloride **307** in the presence of unsaturated compounds **308** at Cu, Fe, and Ti anodes has been investigated. The reaction undergoes chloro-arylation, giving **309** (eq. (5.135)) along with additive dimerization, giving **310** (eq. (5.136)) catalyzed by an anodically dissolved metal species [293].

$$\begin{array}{c} \\ \text{Me} \\ | \\ \xrightarrow{\phantom{M^{n+}}} PhCH_2C\!=\!CHCH_2Cl \qquad (5.135) \\ \textbf{309} \quad (45\%) \end{array}$$

Me
$$|$$
$$PhN_2^+Cl^- +\ CH_2\!=\!C\!-\!CH\!=\!CH_2 \xrightarrow{M^{n+}}$$
307 \qquad\qquad **308**

$$\begin{array}{c} \text{Me} \\ | \\ \xrightarrow{} [PhCH_2C\!=\!CHCH_2]_2 \\ \textbf{310} \quad (5\%) \end{array} \qquad (5.136)$$

$$M^{n+} = Cu^+,\ Fe^{2+},\ Ti^{3+}$$

The modification of carbon surfaces is performed by electroreduction of diazonium salts, which leads to a very solid and noncorrosive covalent attachment of aryl groups onto the carbon surface [294].

The electroreduction of diazodiphenylmethane **311** principally gives three kind of products **312**, **313**, and **314** (eq. (5.137)) [295a,b]. Thus, the diazodiphenylmethane **311**

$$\text{Ph}_2\text{CN}_2 \quad \begin{array}{c} \xrightarrow[\text{-1.00V (Cd/Hg) 81\% yield}]{\text{DMF-Bu}_4\text{NClO}_4\text{-(Pt)}} \quad \text{Ph}_2\text{C}=\text{N}-\text{N}=\text{CPh}_2 \\ \textbf{312} \\[2mm] \xrightarrow[\text{-1.00V, 88\% yield}]{\text{DMF-Bu}_4\text{NClO}_4\text{-TEF-(Pt)}} \quad \text{Ph}_2\text{CH}_2 \\ \textbf{313} \\[2mm] \xrightarrow[\text{-1.1V, 86\% yield}]{\text{DMF-Bu}_4\text{NClO}_4\text{-PhCO}_2\text{H-(Pt)}} \quad \text{Ph}_2\text{C}=\text{NNH}_2 \\ \textbf{314} \end{array} \qquad (5.137)$$

311

TEF = trifluoroethanol

undergoes successive one-electron reductions in a DMF-Bu$_4$NClO$_4$(or Me$_4$NBF$_4$(0.1M))-(Pt or C) system to afford a relatively stable anion radical **315**, [Ph$_2$C=N$_2$]$^{-\bullet}$ and an unstable dianion **316** [Ph$_2$C=N$_2$]$^{2-}$ (eq. (5.138a)) [295]. In the absence of proton donors, benzophenone azine **318** is the principal product as a result of a chain reaction (eq. (5.138)). That is,

$$\text{Ph}_2\text{C}=\text{N}_2 \underset{-e^-}{\overset{e^-}{\rightleftarrows}} [\text{Ph}_2\text{C}=\text{N}_2]^{-\bullet} \underset{-e^-}{\overset{e^-}{\rightleftarrows}} [\text{Ph}_2\text{C}=\text{N}_2]^{2-} \quad \cdots\cdots\text{(a)}$$
$$\textbf{311} \qquad\qquad \textbf{315} \qquad\qquad \textbf{316}$$

$$\textbf{316} + \text{HA} \xrightarrow{\text{fast}} \text{N}_2 + \text{A}^- + \text{Ph}_2\text{CH}^- \quad \cdots\cdots\text{(b)} \qquad (5.138)$$
$$\textbf{317}$$

$$\textbf{317} + \textbf{311} \xrightarrow{k} [\text{Ph}_2\text{CHN-N}=\text{CPh}_2]^- \quad \cdots\cdots\text{(c)}$$
$$\textbf{318}$$

under the absence of proton donors, the dianion **316** undergoes rapid reaction accompanying to give an anion species **317** that reacts with **311** to give **318**, and then the final product, benzophenone azine. On the other hand, the anion radical **315** decomposes relatively rapidly to give N$_2$ and [Ph$_2$C]$^{-\bullet}$, then Ph$_2$CH$_2$ **322** and PhC=NNH$_2$ **320** probably arise from a protonation and one-electron reduction sequence of [Ph$_2$C]$^{-\bullet}$ and the unreacted **315**, respectively (eq. (5.139)). The electrochemical dimerization of diazofluorene **323** undergoes

$$[\text{Ph}_2\text{C}=\text{N}_2]^{-\bullet} \begin{array}{c} \xrightarrow{\text{H}^+} [\text{Ph}_2\text{C}=\text{NNH}]^\bullet \xrightarrow{e^-, \text{H}^+} \text{Ph}_2\text{C}=\text{NNH}_2 \quad \cdots\cdots\text{(a)} \\ \qquad\qquad \textbf{319} \qquad\qquad\qquad \textbf{320} \\[3mm] \xrightarrow{\text{H}^+} [\text{Ph}_2\text{CHN}_2]^\bullet \xrightarrow{-\text{N}_2, e^-, \text{H}^+} \text{Ph}_2\text{CH}_2 \quad \cdots\cdots\text{(b)} \\ \qquad\qquad \textbf{321} \qquad\qquad\qquad \textbf{322} \end{array} \qquad (5.139)$$

315

one-electron reduction in an MeCN-Bu$_4$NBF$_4$-(Pt) system to form fluorenone azine **324** via a long-lived anion radical (eq. (5.140)) [296]. Azibenzil **325** can be converted into the corresponding azine **326** and deoxybenzoin **327** by electroreduction in an MeCN-Et$_4$NBF$_4$-(Pt) system via carbon anion radical intermediates (eq. (5.141)) [297].

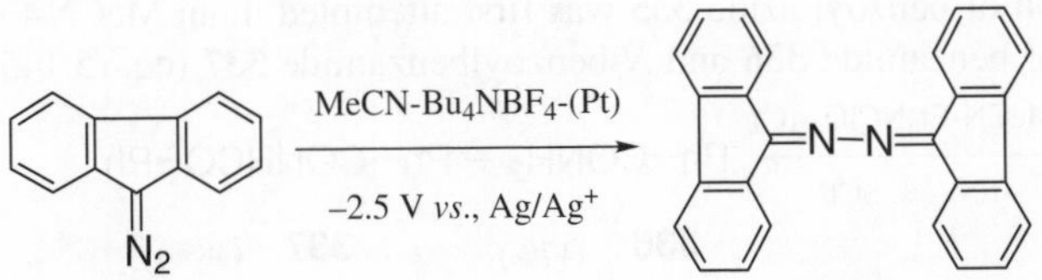

$$\text{(5.140)}$$

323   **324**

PhCOC(N$_2$)Ph $\xrightarrow{\text{MeCN-Et}_4\text{NBF}_4\text{-(Pt)}}$

$$\underset{\textbf{326}}{\text{PhCOC=N–N=CCOPh}} \quad 40\% \text{ yield}$$

325

$+$

PhCOCH$_2$Ph

327   28% yield

$$\text{(5.141)}$$

5.4.1.2   Electroreduction of Azide Compounds

Alkylazides **328** (R = Me) can be reduced into alkylamines **329** (R = Me) at either a (Bu$_4$N)$_3$[Mo-Fe]-(Hg) or (Bu$_4$N)$_3$[Mo$_2$Fe$_6$S$_8$(SPh)$_9$]-modified glassy carbon electrode system in an MeCN/THF or water solution (eq. (5.142)) [298].   Noble metal (Pt and Pd)

$$\text{R–N}_3 \xrightarrow[\substack{(n\text{-Bu}_4\text{N})_3[\text{Mo-Fe}] \\ \text{quantitative yield}}]{\text{MeOH/THF-LiCl-(Hg)}} \text{R–NH}_2 + \text{N}_2 \qquad \text{(5.142)}$$

328   **329**

R = Me, C$_2$H$_4$OH

microparticles dispersed in poly(pyrrole-alkylammonium) films electrodeposited on carbon felt electrodes have proved to be effective for the electrocatalytic hydrogenation of organic azides in aqueous media at weakly negative potentials (−0.5 V vs. SCE) [299].   Electroreduction of azido nucleosides results in the formation of the corresponding amine smoothly. The electrolysis of the azide **330** in an aqueous i-PrOH-NH$_4$OH-(Hg) system at −1.05 V (SCE) affortds the amine **331** in 95% yield (eq. (5.143)) [300].

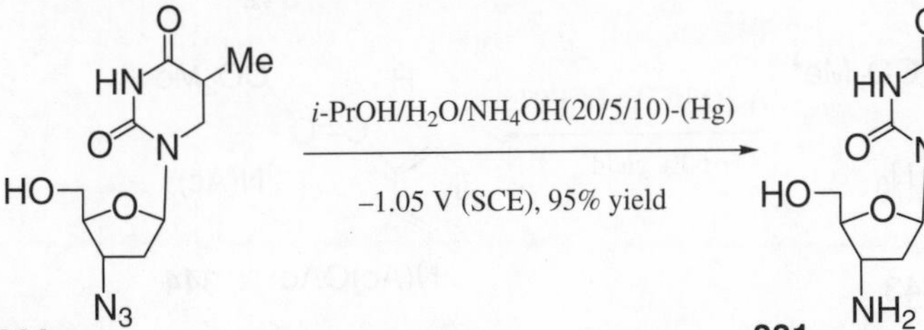

$$\text{(5.143)}$$

330   **331**

The cathode reduction of the acylazide **332** in an aprotic solvent in the presence of anhydride is a convenient way to form triacylamine derivatives **333** (eq. (5.144)) [301].

$$\text{Ar–CON}_3 \xrightarrow[\substack{85\% \text{ yield} \\ \text{Ar = 4-Cl-C}_6\text{H}_4}]{\substack{\text{MeCN/Ac}_2\text{O-LiClO}_4(1\text{M})\text{-(Hg)} \\ e^-, -\text{N}_2}} \text{Ar–CON(Ac)}_2 + \text{Ar–CONHAc} \qquad \text{(5.144)}$$

332   **333**   **334**

The electroreduction of benzoyl azide **335** was first attempted in an MeCN-Et$_4$NClO$_4$-(C) to give a mixture of benzamide **336** and N-benzoylbenzamide **337** (eq. (5.145)) [302].

$$\text{Ph—CON}_3 \xrightarrow[\text{–1.5V } vs. \text{ SCE}]{\text{MeCN-Et}_4\text{NClO}_4\text{-(C)}} \text{Ph—CONH}_2 + \text{Ph—CONHCO—Ph} \qquad (5.145)$$

335   **336** (17%)   **337** (48%)

The conversion yield of benzoyl azide into benzamide has been improved by a controlled potential electrolysis in an MeCN/Ac$_2$O(4/1)-Et$_4$NBr-(Hg) system at –1.15 V vs. SCE (eq. (5.145)). α-Azido-4-chloroacetophenone **338** affords 4-chloroacetophenone **339** in 70% yield under similar electrolysis conditions (eq. (5.146)) [303a].

$$(5.146)$$

338   **339** (70%)   **340** (9%)

Several methyl-α-azidocinnamates **341** and β-heterocyclic α-azidocinnamates can be electrochemically reduced under aprotic and protic media to give stable N,N-diacetylated enamino esters **342** (eq. (5.147)) [303b]. Nitro and azide groups of **343** behave as independent entities, leading to hydroxyamino-enamine compound **344** (eq. (5.148) [303c]. Electroreduction of α-azido-chalcones under slightly protic conditions proves to be an excellent method for a selective conversion of the azido function into the amino group without affecting other reductive parts of the molecules [303d].

$$(5.147)$$

341   **342**

$$(5.148)$$

343   **344**

The cathodic synthesis of the enanimo ketone **346** from the enazido ketone **345** proceeds through reductive elimination of dinitrogen and subsequent reductive acetylation of the amino group (eq. (5.149)) [303e].

$$(5.149)$$

345   **346**

The electrochemical reduction of 4-nitrophenyl azide undergoes initially a one-electron transfer to afford the corresponding anion radical as a transient intermediate [304]. When 4-nitrophenyl azide **347** is reduced by controlled potential electrolysis at –1.5 V *vs*. SCE in the presence of the electro-inactive proton donor as $(CF_3)_2CHOH$, 4-nitroaniline **348** is formed quantitatively in an overall two-electron process (eq. (5.150)). The development of an electrochemical method for the determination of 3'-azido-3'-deoxythymidine in human whole blood has been attempted [305].

$$p\text{-}O_2NC_6H_4\text{-}N_3 \xrightarrow[\substack{-1.5 \text{ V } vs. \text{ SCE} \\ \sim 100\% \text{ yield}}]{\text{MeCN-Bu}_4\text{NClO}_4(0.1\text{M})\text{-(Pt)}} p\text{-}O_2NC_6H_4NH_2 \qquad (5.150)$$

$$\quad\;\textbf{347} \qquad\qquad\qquad\qquad\qquad\qquad\qquad \textbf{348}$$

5.4.2 Electroreduction of C≡N Bonds

A survey of the influence of the cathode material for the electroreduction of nitriles and the effects of current density on the yield of primary amine has been reviewed [306]. Choice of metal for the cathode material is important. The reduction efficiency of both aliphatic and aromatic nitriles is lower at a lead rather than at a platinum-coated cathode with spongy palladium. At a mercury cathode, the reduction does not take place. The preparation of primary amines from organic nitriles using a Raney nickel cathode has been attempted [307]. The electroreduction of benzyl cyanide to β-phenethyl amine proceeds in 86% current efficiency. Adiponitrile **349** can be electrochemically hydrogenated to the corresponding 6-aminocapronitrile **350** using Raney nickel (R-Ni) powders as cathode materials [308]. When adiponitrile is hydrogeneted in a $ROH/H_2O\text{-AcONH}_4$-(C/Raney-Ni) system at 35~45 °C, 6-aminocapronitrile **350** is produced in a 79-97% selectivity (eq. (5.151)). 9-Aminononanonitirle is obtained in 80-93% selectivity from azelanitrile.

$$NC(CH_2)_4CN \xrightarrow[\substack{79\text{~}97\% \text{ selectivity}}]{\text{aq. AcONH}_4\text{-(Raney-Ni)}} NC(CH_2)_5NH_2 + H_2N(CH_2)_6NH_2 \qquad (5.151)$$

$$\quad\;\textbf{349} \qquad\qquad\qquad\qquad\qquad\qquad \textbf{350} \qquad\qquad \textbf{351}$$

The electrocatalytic hydrogenation of nitriles proceeds effectively in an $H_2O/THF(1/9)$-$NaClO_4$-(Ni/C) system [309]. The R-Ni cathode brings selective hydrogenation of most nitriles to the corresponding primary amines except 4-nitro-benzonitrile. The Pd-C cathode gives slow hydrogenation of 4-nitro-benzonitrile **352**, but the selective hydrogenation of both the nitro group and nitrile group of the benzonitrile **352** leading to 4-aminobenzylamine **353** can be attained in 77% yield (eq. (5.152)). Electroreductive conversion of

$$O_2N\text{-}\langle\rangle\text{-}CN \xrightarrow[\substack{77\% \text{ yield}}]{\substack{H_2O/THF(1/9)\text{-NaClO}_4\text{-(Ni/C)} \\ Me_3CCO_2H}} H_2N\text{-}\langle\rangle\text{-}CH_2NH_2 \qquad (5.152)$$

$$\qquad\qquad \textbf{352} \qquad\qquad\qquad\qquad\qquad\qquad\qquad \textbf{353}$$

methyl 4-cyanobenzoate **354** into methyl 4-aminomethylbenzoate **355** has been attained in strong acidic media (eq. (5.153)) [310]. In contrast, under neutral and alkaline conditions, the electroreduction occurs by losing the nitrile group to give methylbenzoate products.

$$MeCO_2\text{-}\langle\rangle\text{-}NC \xrightarrow[\substack{97\% \text{ yield}}]{\text{aq. HCl}(0.1\text{M})\text{-(C)}} MeCO_2\text{-}\langle\rangle\text{-}CH_2NH_2 \qquad (5.153)$$

$$\qquad\quad \textbf{354} \qquad\qquad\qquad\qquad\qquad\qquad \textbf{355}$$

The pilot scale production of 4-aminomethylpyridine **357** has been developed using a well-designed cell [311]. The cathodic reduction of 2- and 4-cyanopyridines **356** at high surface area electrodes is carried out in an aqueous MeOH-H$_2$SO$_4$(50%wt)-(Pt) system, leading to the corresponding amines **357** over 89% yield (*c.e.* 83%) (eq. (5.154)) [311]. The electroreduction of 4-amino-5-cyano-2-methylpyrimidine **358** can be preferentially converted

$$
\underset{\textbf{356}}{\text{CN-pyridine}} \quad \xrightarrow[\text{92\% yield}]{\text{aq. MeOH-H}_2\text{SO}_4\text{(50\% wt)-(Pt)}} \quad \underset{\textbf{357}}{\text{CH}_2\text{NH}_2\text{-pyridine}} \tag{5.154}
$$

into the corresponding 5-aminomethyl derivative **360** under the following conditions: cathode, Ni; additives PdCl$_2$ concentration, 0.01%; HCl concentration, 10%; current density, 2 A/dm^2; reaction temp., 20 °C (eq. (5.155)) [312]. In contrast, 4-amino-5-hydroxymethyl-2-propylpyrimidine **361** has been synthesized under slightly different electrolysis conditions as follows: cathode, Ni sintered; HCl concentration, 10%; current density, 0.25 A/dm^2; reaction temp., 60 °C (eq. (5.156)). The electrosynthesis of phthalocyanines by the

$$
\underset{\textbf{358}}{\text{Me-pyrimidine-C≡N}} \xrightarrow{\text{2H ads}} \left[\underset{\textbf{359}}{\text{Me-pyrimidine-CH=NH}} \right] \xrightarrow[\text{78\% yield}]{\text{2H ads}} \underset{\textbf{360}}{\text{Me-pyrimidine-CH}_2\text{NH}_2} \tag{5.155}
$$

$$
\xrightarrow[\text{77\% yield}]{\text{aq. HCl/PdCl}_2\text{-(C/Ni)}} \underset{\textbf{361}}{\text{Me-pyrimidine-CH}_2\text{OH}} \tag{5.156}
$$

reduction of phthalonitrile has been performed in hot absolute ethanol in 60~70% yields [313]. The controlled potential electrolysis of 2,4,6-trimethyl-benzonitrile *N*-oxide **362** has been investigated in a wide range of pH values, giving the nitrile **363** in good yield (eq. (5.157)) [314].

$$
\underset{\textbf{362}}{\text{Me-(Me)(Me)-C≡N→O}} \xrightarrow[\text{~100\% yield}]{\text{DMF(or H}_2\text{O})\text{-Pr}_4\text{NClO}_4\text{-(0.1M)-(Hg)}} \underset{\textbf{363}}{\text{Me-(Me)(Me)-C≡N}} \tag{5.157}
$$

Electroreductive removal of the nitrile group of **364** can be performed in an DMF-Et$_4$NOTs-(Zn) system to give **365** in 80% yield (eq. (5.158)) [315]. The use of zinc metal as the cathode material is important.

$$
\underset{\textbf{364}}{\text{Et}_2\text{C(C}_{10}\text{H}_{21}\text{)CN}} \xrightarrow[\text{80\% yield}]{\text{DMF-Et}_4\text{NOTs-(Zn)}} \underset{\textbf{365}}{\text{Et}_2\text{C(C}_{10}\text{H}_{21}\text{)H}} \tag{5.158}
$$

Electroreductive synthesis of 5-amino-4-cyanopyrazole **368** has been attained by the coupling of 1,1-dicyanoalkenes **366** and dinitrogen *via* a tungsten complex **367** (eq. (5.159)) [316]. The electroreduction of the dinitrogen-derived dicyanovinyl hydrazido complex [WF(dppe)$_2$NNH(CH=CCN$_2$)$^+$[BF$_4$]$^-$ in a THF-Bu$_4$NBF$_4$-(0.1M)-(Hg) system yields the cyanopyrazole **368** in 50% yield.

$$
\textbf{366} + \textbf{367} \xrightarrow[\text{50\% yield}]{\text{THF-Bu}_4\text{NBF}_4(0.1\text{M})\text{-(Hg)}} \textbf{368} \tag{5.159}
$$

5.5 Electroreduction of Nitrogen-containing Aromatic Nuclei

5.5.1 Electroreduction of Pyridine, Pyrrole and Their Derivatives

Pyridine **369** is reducible in an MeCN-Et$_4$NCl(0.1M)-(Hg) system along an irreversible two-electrons at –3.1 V *vs.* Ag/Ag$^+$ first to give an anion radical **370** (eq. (5.160a) [317]). The electroreduction in the presence of ethyl chloroformate may afford the two products **374** and **375** *via* the reaction paths (eq. (5.160b,c)). A large scale process of the electrocatalytic reduction of pyridine on an electrode activated by Raney nickel has been devised [318]. The electroreduction of pyridine **369** in alkaline solution has been reported.

$$
\underset{\textbf{369}}{\text{C}_5\text{H}_5\text{N}} + e^- \longrightarrow \underset{\textbf{370}}{\text{C}_5\text{H}_5\text{N}^{\bullet-}} \xrightarrow{\text{MeCN}} \underset{\textbf{371}}{\text{C}_5\text{H}_6\text{N}^\bullet} + \underset{\textbf{372}}{{}^-\text{CH}_2\text{CN}} \quad \cdots\cdots \text{(a)}
$$

$$
\textbf{371} + e^- \longrightarrow \underset{\textbf{373}}{\text{C}_5\text{H}_6\text{N}^-} \xrightarrow{\text{MeCN}} \textbf{369} + \textbf{372} \quad \cdots\cdots \text{(b)} \tag{5.160}
$$

$$
\textbf{369} \xrightarrow[\text{3.1 V } vs. \text{ Ag/Ag}^+]{\substack{\text{MeCN-Et}_4\text{NCl(0.1M)-(Hg)} \\ \text{ClCO}_2\text{Et}}} \underset{\substack{\textbf{374} \\ (42\%)}}{} + \underset{\substack{\textbf{375} \\ (58\%)}}{} \quad \cdots\cdots \text{(c)}
$$

The electrolysis is carried out in an aqueous PhH/Py-NaOH-(Cd) system to give a mixture of dimers **376**, **377**, and **378** in 98.6% yield (*c.e.*) (eq. (5.161)) [319]. Pyridine **369** can be

$$
\textbf{369} \xrightarrow[\text{98.6\% yield (}c.e.\text{)}]{\text{aq. PhH/Py(45/35 v/v)-NaOH-(Cd)}}
\begin{cases}
\textbf{376} \ (75.2\%) \\
+ \\
\textbf{377} \ (4.6\%) \\
+ \\
\textbf{378} \ (20\%)
\end{cases} \tag{5.161}
$$

reduced cathodically in liquid ammonia to the corresponding radical and dimerized quickly yielding a dimeric dianion, which can be oxidized to 4,4'-bipyridine by O_2 bubbling [320]. The electroreduction of pyridine in aqueous medium and the influence of proton concentration have been investigated [321].   The MeCN-Et$_4$NBr(0.1M)-(C/Pb) system is a typical reduction condition for the conversion of pyridine into 4,4'-bipyridine as a major product [322].   The adsorption of pyridine and nicotinic acid on Au and Pt electrodes has been the subject of investigations.   The reductive elimination of the chemisorbed species is encountered in the overall process of the electrocatalytic reduction of nicotinic acid [323].

Electroreduction of quinoline **379** in anhydrous liquid ammonia in the presence of an alkylating agent, *i.e.*, ethyl bromide and *n*-butyl bromide, affords the alkylated products consisting of approximately equal quantities of 1,2-diethyl-1,2-dihydroquinoline **380** and 1,4-diethyl-1,4-dihydroquinoline **381** (eq. (5.162)) [324].

$$(5.162)$$

The electroreduction of *N*-oxide of nicotinamide and pyridine monocarboxylic acids, *i.e.*, picolinic, nicotinic, and isonicotinic acids, has been investigated [325].   The *N*-oxide group of **382** is reduced in acidic and neutral media to yield the corresponding pyridine derivatives **384** *via* **383**.   The electroreduction of nicotinic acid *N*-oxide **382** in basic solutions has also been attempted in an aqueous NaOH/H$_3$PO$_4$/H$_3$BO$_3$/NaCl-(Hg) system (eq. (5.163)) [326].   Picolinic and dipicolinic acids are electrochemically reduced into their corresponding aldehydes [327].   Detail behaviors of picolinic acid have been studied in various solvent systems [328].

$$(5.163)$$

The electroreduction of niflumic aldehyde **385a** (Y = CHO) and niflumic ester **385b** (Y = CO$_2$Et) in THF(or DMF)-Et$_4$NClO$_4$-(Hg) system undergoes one-electron transfer reaction followed by proton abstraction from the amine to yield the reduction products **386** in a quantitative manner (eq. (5.164)) [329].   The trianion radical of 4,7-phenanthroline, forming by electroreduction has been characterized [330].   As a high-potential NADH/NAD$^+$ analogue, the 9-phenyl-10-methyl-9,10-dihydroacridine/acridinium redox system has been found to be cycled electrochemically between the oxidized (AcPh$^+$) and reduced state (AcPhH) without any apparent side reaction [331].

$$(5.164)$$

Electroreduction can convert nicotinic and isonicotinic hydrazines into the corresponding amides at negative potentials [332].   The electroreduction of 8-, 5-, and 2-hydroxyquinolines (8-, 5-, and 2-QOH, respectively) undergoes a fast proton transfer from the parent compounds to the electrogenerated basic intermediates (self-protonation mechanism), with the formation of the conjugate base of the former, QO-, together with the two-electron reduction products QH$_2$OH [333].

5.5.2   Electroreduction of Two- or Three-Nitrogen Atom-containing Compounds

Pyrazine **387** can be converted by cathodic reduction in a DMF-Et$_4$NBr-(Hg) system in the presence of acylating agents into 1,4-diacyl-1,4-dihydropyrazine **388** (eq. (5.165)) [334].

$$\tag{5.165}$$

387   **388**

The electrochemical behavior of 2-hydroxy-3-phenyl-6-methylpyrazine **389** derived from β-lactam antibiotics as a metabolite has been studied in order to clarify the mechanism related to penicillin-induced allergies.   In the pH < 0.7 range, a two-electron reduction proceeds to give the protonated dihydropyrazine derivative **390** (eq. (5.166)) [335].   1,4-Diaza-1,4-dihydronaphthalene is electrosynthesized in 92% yield from 1,4-diazanaphthalene in a DMF-Et$_4$NI system [336].

$$\tag{5.166}$$

389   *aq. EtOH(10%)-HClO$_4$-(Hg)   **390**

The electroreduction of 2-methyl, 2-phenyl, 2-phenyl-3-methyl, 2,3-diphenyl, and 2,3-dimethylquinoxalins **391** in an alkaline or a neutral medium can lead to 1,4-, 1,2- or 3,4-dihydro derivatives **392** and **393** or to 1,2,3,4-tetrahydro derivatives **394** (eqs. (5.167~5.169)) [337].   In the 7~13 pH range, the primary reduction process corresponds to the eq. (5.167),

$$\tag{5.167}$$

391   **392**

$$\tag{5.168}$$

392   ⟶   **393**

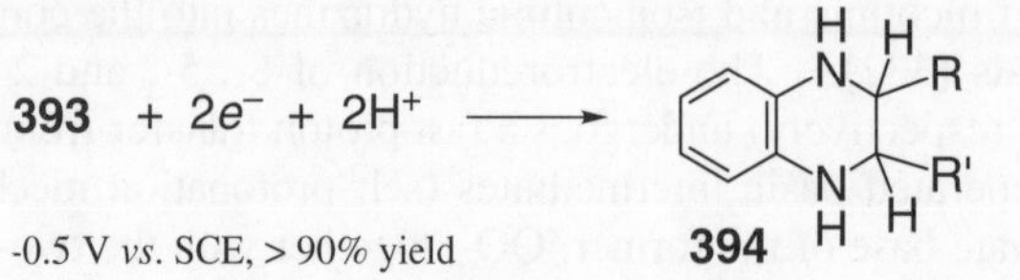

$$393 \ + 2e^- + 2H^+ \ \longrightarrow \tag{5.169}$$

-0.5 V *vs*. SCE, > 90% yield   **394**

R = R' = H, Me or Ph

that is, to the formation of a 1,4-dihydro derivatives **392**.   All the 1,4-dihydro derivatives can undergo a rearrangement into 1,2- or 3,4-dihydro derivatives **393** (eq. (5.168)) with the exception of 1,4-dihydroquinoxaline.   The reaction examples are shown in Table 5.11. [334, 337~344].   A linear inverse relationship between basicity constants and E1/2 has been found for the pyrazine derivatives [345].

The reduction of phenazine **395** proceeds in organic chloroaluminate melts through a two-electron process coupled with two protonations to form dihydrophenazine **396** (eq. (5.170) [346].

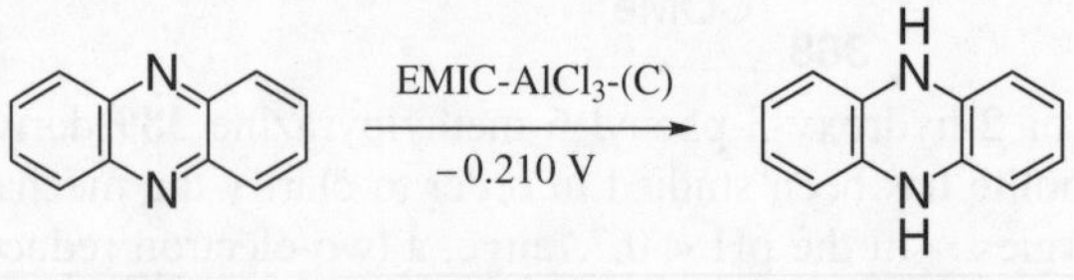

$$\xrightarrow[\ -0.210 \text{ V}\ ]{\text{EMIC-AlCl}_3\text{-(C)}} \tag{5.170}$$

395   **396**

EMIC = 1-ethyl-3-methyl-1*H*-imidazolium chloride

A preparative electrolysis of quinoxalino[2,3-b]quinoxaline **397** is carried out in aqueous DMF(50%)-(C) at pH 6.75 and E = −0.5 V *vs*. SCE to afford 5,12-dihydroquinoxalino[2,3-b]quinoxaline **398** in almost quantitative yield (eq. (5.171)) [344].   The electroreduction of the quinoxaline **398** in an acidic medium leads to 5,5a,6,11,11a,12-hexahydroquinoxalino[2,3-b]quinoxaline **399** (eq. (5.171)).   When **397** is heated in acetic anhydride, it furnishes triacetyl derivatives **400**.

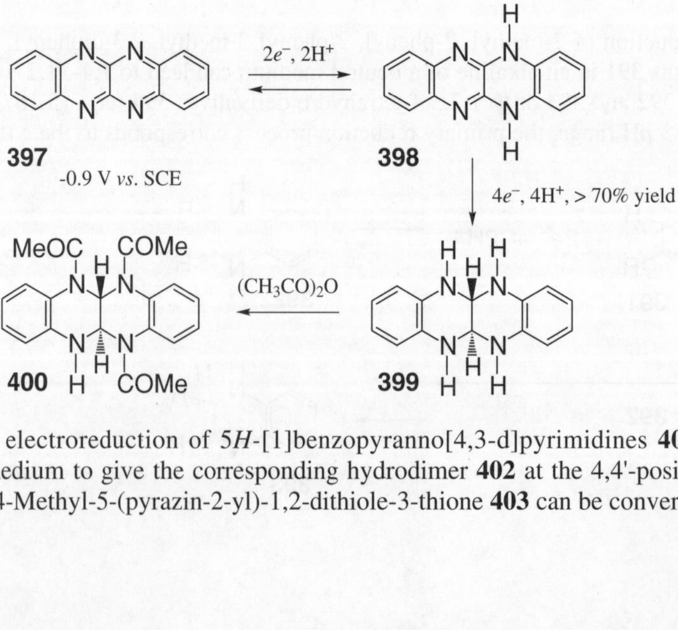

$$\xrightleftharpoons{2e^-,\ 2H^+} \tag{5.171}$$

397   **398**

-0.9 V *vs*. SCE

$4e^-, 4H^+, > 70\%$ yield

400 $\xleftarrow{(CH_3CO)_2O}$ **399**

The electroreduction of 5*H*-[1]benzopyranno[4,3-d]pyrimidines **401** takes place in acidic medium to give the corresponding hydrodimer **402** at the 4,4'-position (eq. (5.172)) [347].   4-Methyl-5-(pyrazin-2-yl)-1,2-dithiole-3-thione **403** can be converted into the novel

Table 5.11a Electroreduction of Pyrazine and It's Related Compounds

Starting compound		Products	Yield, (%)	Ref.
	DMF-Et$_4$NBr-(Hg), Ac$_2$O 44% yield			[334]
	aq. MeOH(50%)-NaOH(10%)-(Hg) -1.30V *vs.* SCE, pH 13.5		(20)	[337]
	aq. MeOH(80%)-NaOH(0.5M)-(Hg) -1.15V *vs.* SCE, pH 13.7	 	(15) (5)	[337]
	aq. MeOH(80%)-NaOH(0.5M)-(Hg) -1.50V *vs.* SCE, pH 13.7	 	(20) (65) (45)	[337]
	aq. MeOH(50%) pH 7.20, -0.6 V *vs.* SCE		(65)	[338]
	aq. MeOH(80%)-NaOH(0.1M)-(Hg)	 R^1 = R^2 = Ph R^1 = H, R^2 = Ph R^1 = R^2 = CH$_3$		[339]
 R =	aq. Buffer*-(Hg) *Britton-Robinson, pH 4			[340]

Table 5.11b Electroreduction of Pyrazine and It's Related Compounds

Starting compound		Products	Yield, (%)	Ref.
	aq. EtOH(40%)-H₂CH₂CH₂NH₂/AcOH-(Hg) -1.60 V, pH 7.5		(75) (50)	[337]
			(13)	
	aq. MeOH(80%)-NaOH( )-(Hg) -1.35V *vs*. SCE, pH 13.97		(20)	[337]
			(60)	
	aq. MeOH(80%)-NaOH(0.5M)-(Hg) -1.50V *vs*. SCE, pH 13.7		(50)	[337]
			(50)	
	aq. HCl(0.1M)-(Hg)			[341]
	aq. MeOH(50%)-NaOH(0.1M)-(Hg) -1.55V *vs*. SCE		(52.5)	[342]
	X = O; R¹ = C₆H₅; R² = H X = S; R¹ = C₆H₅; R² = H		(36)	[342]

Table 5.11c Electroreduction of Pyrazine and It's Related Compounds

Starting compound		Products	Yield, (%)	Ref.
	aq. MeOH(50%)-NaOH(0.1M)-(Hg) −0.8V *vs.* SCE		(81.6)	[342]
	R^1 = Ph, R^2 = H; R^1 = R^2 = Me			
	aq. MeOH(50%)-NaOH(0.1M)-(Hg)		(65)	[342]
	X = O, S R^1 = R^2 = H, Me			
R = H, Ac	aq. MeCN-H$_2$SO$_4$(0.1M)-(C) −0.7 V *vs.* SCE			[343]
	−0.5 V *vs.* SCE 2e^-, 2H$^+$,		(>90)	[344]
	−0.9 V *vs.* SCE 4e^-, 4H$^+$,		(>70)	
	(CH$_3$CO)$_2$O			

401 R^1, R^2 = H, Me, Ph	aq. MeOH-H$_2$SO$_4$-(Hg) pH 1.43, 72~76% yields	**402**		(5.172)

tri-iodomercurate (II) complex **404** in good yield on methylation with methyl iodide in the course of an electrolytic reaction (eq. (5.173)) [348].   The electrolysis is performed in a DMF-Bu$_4$NBF$_4$(0.1M)-(C) system in the presence of methyl iodide.

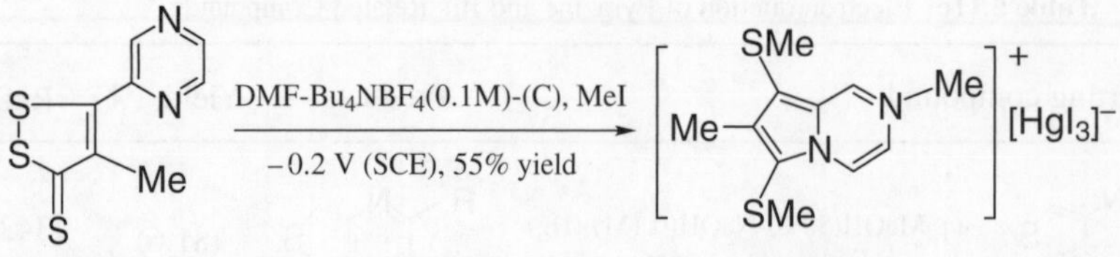

$$(5.173)$$

403   **404**

The electroreduction of the triazine **405** has been performed in a DMF-Bu$_4$NClO$_4$ (0.1M)-(Hg) system in the presence of ethyl bromide, forming the *N*- and *C*-alkylated products **406** and **407** in 80% total yield (eq. (5.174)) [349].   In the presence of a protic acid, the electroreduction undergoes a two-electron transfer from the triazine ring to give

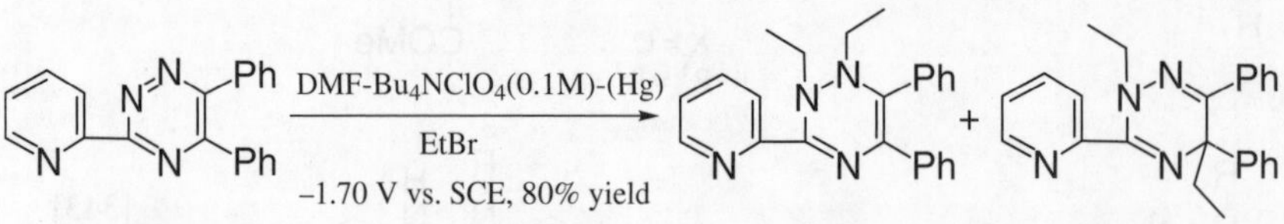

$$(5.174)$$

405   **406**   **407**

the dihydrotriazine derivatives.   The electrochemical reduction of pyrido-*as*-triazines **408** and **410** in aqueous methanol or in acetonitrile in the presence of phenol leads to the corresponding 1,4-dihydro derivatives **409** and **411**, respectively, in quantitative yield.   The electrolysis of **410** in a MeCN-Ac$_2$O system affords a 1:3 mixture of 1,4-diacetyl-1,4-dihydro and 1,2-diacetyl-1,2-dihydro compounds in 29% yield (eq. (5.175)) [350].   The 1,4-dihydro derivatives obtained through the electrochemical reduction of quinoxanilines or of

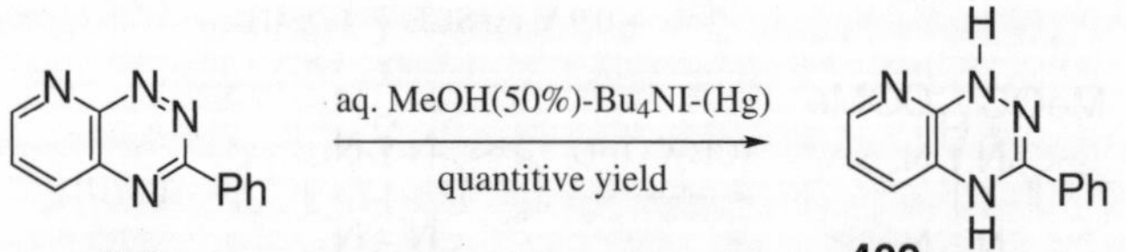

408   **409**

$$(5.175)$$

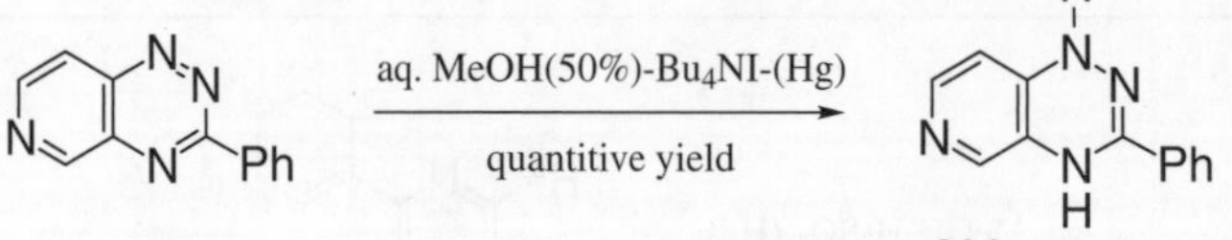

410   **411**

pyridopyrazines easily isomerize into 1,2- or 3,4-dihydro compounds [339].   The electroreductive characteristics of novel tetradentate ligands, which are built with one 2,2'-bipyridine bonded to one pyridyl triazine, have been investigated [351].   The substituted pyrazino[2,3-*g*]quinoxalines **412** undergo electroreduction in a stepwise manner (eq. (5.176)) [352].   The compounds **412** are first reduced to the 1,4-dihydro derivative **413**, which is responsible for the first wave on the voltammogram.   The 1,4,6,9-tetrahydro derivative **414** corresponds to the second peak but rearranges to the isomeric products.

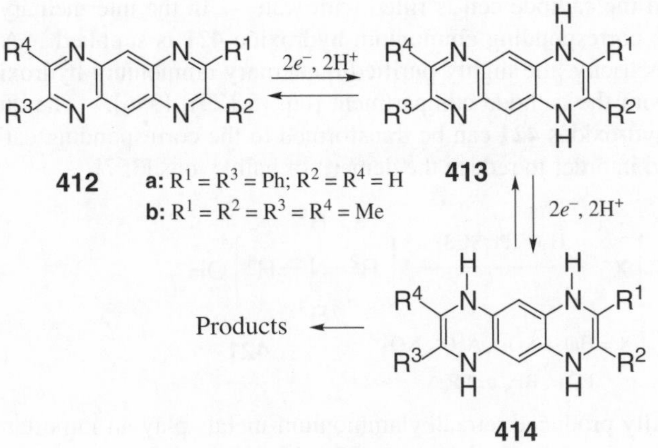

$$
\begin{array}{ccc}
412 & \xrightarrow{\;2e^-,\,2H^+\;} & 413
\end{array}
$$

(5.176)

412 a: $R^1 = R^3 = Ph$; $R^2 = R^4 = H$ **413**

b: $R^1 = R^2 = R^3 = R^4 = Me$

$$
413 \xrightarrow{\;2e^-,\,2H^+\;} 414
$$

Products ⟵

414

5.6 Electroreduction of Quaternary Nitrogen Compounds

5.6.1 Electroreduction of Ammonium and Pyridinium Salts

Electrosynthesis of benzyl alkyl ketones **416** has been performed by electroreduction of arylmethyl quaternary ammonium salts **415** in a DMF-Bu$_4$NI-(Mg/Ni) system in the presence of acetic anhydride in good yield (eq. (5.177)) [353]. The cathodic decomposition of

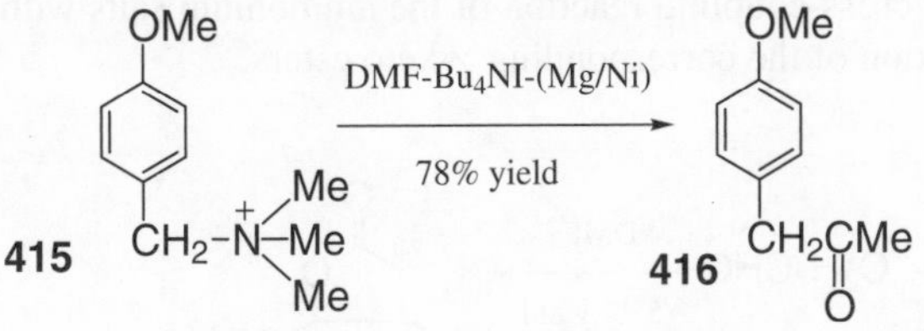

$$
\textbf{415} \xrightarrow[\text{78\% yield}]{\text{DMF-Bu}_4\text{NI-(Mg/Ni)}} \textbf{416}
$$

(5.177)

quaternary ammonium nitrates **417** containing benzyl, fluorenyl, cinnamyl, p-methoxy benzyl, etc. in DMF affords tertiary amines **419** together with the coupling products **418** of these radicals (eq. (5.178)) [354]. The reaction involves the generation and dimerization of radical species, forming a one-electron transfer at the cathode.

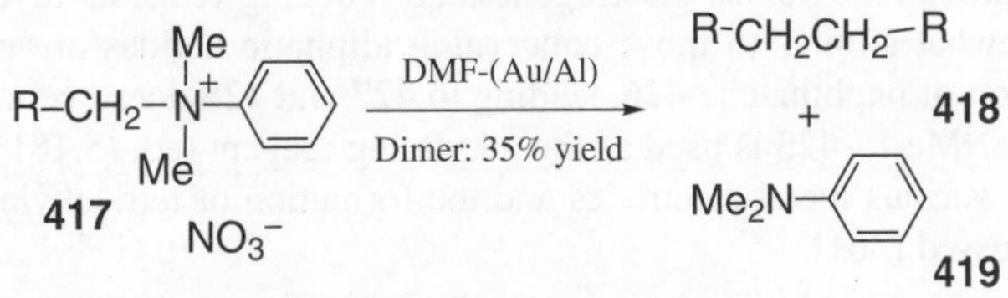

$$
\textbf{417} \xrightarrow[\text{Dimer: 35\% yield}]{\text{DMF-(Au/Al)}} \textbf{418} + \textbf{419}
$$

(5.178)

R = benzyl, fluorenyl, cinnamyl, p-methoxy benzyl, etc.

An efficient manufacturing process for the scale-up electroreduction of bis-quaternary ammonium salt, $(Me_3N^+CH_2)_2Ar$, leading to Me_2Ar and Me_3N in 85% conversion yields, has been proposed [355].

Highly purified quaternary tetraalkyl ammonium hydroxides **421** have been prepared using an electrolysis rectification technique. The electrolysis cell consists of more than three compartments divided by ion exchange membranes. The alkyl ammonium salts **420**, involving either halide ions, HSO_3^-, AcO^-, NO_3^-, PO_3^-, BF_4^- or PF_6^-, is charged in the anode compartment and the cathode cell is filled with water.   In the intermediary cell, an aqueous solution of the corresponding ammonium hydroxide **421** is supplied.   After passage of the required electricity, the highly purified quaternary ammonium hydroxide solution can be obtained from the cathode compartment (eq. (5.179)) [356].   The quaternary tetraalkyl ammonium hydroxides **421** can be transformed to the corresponding carboxylate salts, which are patented in order to reduce the demerit of halide ions [357].

$$
\left[R^2-\overset{\overset{\displaystyle R^1}{|}}{\underset{\underset{\displaystyle R^3}{|}}{N}}-R^4 \right]^+ X^- \xrightarrow{\text{H}_2\text{O-(Pb/SUS)}} \left[R^2-\overset{\overset{\displaystyle R^1}{|}}{\underset{\underset{\displaystyle R^3}{|}}{N}}-R^4 \right]^+ OH^-
$$

(5.179)

420   X = Hal, HSO$_3^-$, AcO$^-$, NO$_3^-$,   **421**

PO$_3^-$, BF$_4^-$ or PF$_6^-$

The electrochemically produced tetraalkylammonium-metals play an important role as a catalytic reducing agent in electrolysis media [358a,b].   The cathodic reduction of dimethylpyrrolidinium (DMP$^+$) in a DMF-Bu$_4$NBF$_4$(0.1M)-(Hg) system undergoes a reversible one-electron transfer forming an insoluble material as an amalgam [358c].   The DMP$^+$ amaglam is reactive and transfer electrons to suitable substrate with generation of DMP$^+$ [358d].   Both Birch type and carbonyl reductions proceed preferentially with the electrogenerated reducing agent to give the corresponding hydrogenated products [358e].

Electroreductive cross-coupling of the quaternary ammonium salts of methyl 3-aminoalkanoate **422** with butanal **423** affords the corresponding γ-lactones **424** (eq. (5.180)) [359].   Similarly, the cross-coupling reaction of the ammonium salts with acid anhydrides can lead to the formation of the corresponding γ-keto esters.

$$
\underset{\textbf{422}}{\overset{\overset{\displaystyle R^1 \quad R^2}{\text{Me}-\overset{+}{N}}}{\underset{\displaystyle \text{Me}}{\quad}} \text{CO}_2\text{Me} \quad I^-} + \underset{\textbf{423}}{\text{C}_3\text{H}_7\text{CHO}} \xrightarrow[\text{55\% yield}]{\text{DMF}} \underset{\textbf{424}}{\text{Me} \quad \text{C}_3\text{H}_7}
$$

(5.180)

a: R^1 = Et, R^2 = H   R^1 = H (36%)
b: R^1, R^2 = –(CH$_2$)$_4$–   Me (55%)

The behavior of onium cations versus electrogenerated reducing reagents reveals that alkylation reactions somewhat similar to those concerning aliphatic halides are possible. For example, *tert*-butylation of naphthalene **426**, leading to **427** and **428** does occur prior to methylation when [*tert*-BuNMe$_3$]$^+$ **425** is used as the alkylating reagent (eq. (5.181)) [360]. The cathodic corrosion of various metal electrodes and the formation of tetralakylammonium metals have been discussed [361].

$$
\textit{tert}\text{-Bu}\overset{+}{N}\text{Me}_3 + \underset{\textbf{426}}{\text{[naphthalene]}} \xrightarrow[\text{83\% yield}]{\text{DMF-Bu}_4\text{NI(0.1M)-(Hg)}} \underset{\textbf{427}}{\text{[tert-Bu]}} + \underset{\textbf{428}}{\text{[Bu-tert]}}
$$

(5.181)

425   **426**   **427/428** = 50/50   **427**   **428**

Electrogenerated pyridinyl radicals have attracted much attention in considering biological redox reactions as well as the mechanism by which important herbicides act on plants.   The mechanism for the electroreduction of *N*-alkylpyridinium salts involves the addition of an electron to the pyridinium cation and the formation of a rapidly dimering neutral radicals [362].   A similar mechanism has been extended to the unsubstituted and ring-substituted pyridinium salts.   Half-wave potentials of substituted pyridinium salts reported in the literature are listed in Table 5.12. [363~370].   The electroreductive coupling reactions of pyridinium cations are shown in Table 5.13. [371~373].

Table 5.12a Half-Wave Potentials of Substituted Pyridinium Cations

R^1	R^2	R^3	R^4	X$^-$	Reduction Properties (SCE) Wave 1 E$_{1/2}$, V	Wave 2 E$_{1/2}$, V	Ref.
Me	H	H	H	Cl	—	—	[363]
Me	Me	Me	Me	Cl	—	−1.57	[364]
Me	Me	Ph	Me	Cl	−1.18	−1.82	[364]
Me	Et	Ph	Et	Cl	−1.25	−1.95	[364]
Me	CN	CN	H	?	−0.21	−1.10	[365]
Me	CN	H	H	ClO$_4$	−0.66*	−1.29*	[366]
Me	CN	Me	H	ClO$_4$	−0.78*	−1.36*	[366]
Me	CN	t-Bu	H	ClO$_4$	−0.78*	−1.39*	[366]
Me	CO$_2$Et	H	H	ClO$_4$	−0.84*	−1.60*	[366]
Me	Ac	H	H	ClO$_4$	−0.78*	−1.45*	[366]
Bu	H	H	H	Cl	—	—	[367]
Me	H	Me$_2$N	H	BF$_4$	—	−1.76	[368]
Me	Ph	Ph	Me	ClO$_4$	−1.15	−1.70	[364]
Me	Ph	Ph	Ph	ClO$_4$	−0.98	−1.43	[364]
CONMe$_2$	H	Me$_2$N	H	BF$_4$	—	−1.36	[368]
COCMe$_2$	H	Me$_2$N	H	BF$_4$	—	−1.13	[368]
CO$_2$Et	H	Me$_2$N	H	BF$_4$	—	−1.16	[368]
CN	H	Me$_2$N	H	BF$_4$	—	−1.03	[368]
Pr	Me	Ph	Me	BF$_4$	−1.27	—	[369]
Pr	Ph	Me	Ph	BF$_4$	−1.19	—	[369]
Pr	Ph	Ph	Me	ClO$_4$	−1.11	−1.74	[364]
Pr	Me	Ph	Me	ClO$_4$	−1.22	−1.89	[364]
Pr	Et	Ph	Et	Cl	−1.26	−2.02	[364]
Pr	Et	Ph	Et	BF$_4$	−1.27	—	[369]
Pr	Ph	Ph	Ph	ClO$_4$	−0.94	−1.48	[364]
Pr	Ph	Ph	Ph	BF$_4$	−0.96	−1.52	[370]
Pr	Ph	Ph	Ph	BF$_4$	−0.99	—	[369]

Table 5.12b Half-Wave Potentials of Substituted Pyridinium Cations

$$R^4 \underset{R^1}{\overset{R^3}{\underset{N^+}{\bigcirc}}} R^2 \quad X^-$$

| | | | | | Reduction Properties (SCE) | | |
| | | | | | Wave 1 | Wave 2 | |
R^1	R^2	R^3	R^4	X^-	$E_{1/2}$, V	$E_{1/2}$, V	Ref.
Allyl	Ph	Ph	Ph	ClO_4	−0.91	−1.44	[364]
Allyl	Ph	Ph	Ph	BF_4	−0.90	—	[370]
Bu	Ph	Ph	Ph	ClO_4	−0.91	−1.45	[364]
Bu	Ph	Ph	Ph	ClO_4	−0.91	−1.51	[364]
C_5H_{11}	Ph	Ph	Ph	ClO_4	−0.89	−1.02	[364]
Ph	Ph	Ph	Ph	ClO_4	−0.93	−1.09	[364]
Ph	Ph	Ph	Ph	BF_4	−0.97	−1.15	[370]
Bz	Ph	Ph	Ph	ClO_4	−0.80	−1.41	[364]
Bz	Ph	Ph	Ph	BF_4	−0.92	−1.25	[370]

Table 5.13 Electroreductive Coupling of Pyridinium Cations

	Conditions	Products	Yield, (%)	Ref.
Bu–N⁺(pyridinium)	MeCN, $AlCl_3$, (W)-(C) 40°C	Bu–N⟨⟩–⟨⟩N–Bu	(45)	[371]
Me–N⁺(CN-pyridinium) I⁻	aq. NH_4(0.1M)/ NH_4Cl(0.1M)-(Hg)	Me–N⟨CN⟩–⟨Me,CN⟩	(good)	[372]
N⁺(CO_2Me), Me (pyridinium)	MeCN-$LiClO_4$-(Hg) −0.48 V, −1.166 V (Ag/Ag⁺)	Me–N⟨CO_2Me⟩–⟨CO_2Me⟩N–Me	(good)	[373]

4,4'-Dipyridinium salts (viologens) are known as naturally occurring redox mediators, some of which show very powerful activity as herbicides, marketed as paraquat (1,1'-dimethyl-4,4'-bipyridylium dichloride, MV^{2+} $2Cl^-$) and diquat (dibromide). Methyl viologens (MV^{2+}) undergo two one-electron reduction processes as shown below: The two-electron reduced species (a neutral quinoid form, MV) is less soluble in aqueous solution than the one-electron reduced species, $MV^{+\cdot}$. One characteristic of MV is its instability in the presence of reactive materials. The presence of β-cyclodextrin in an MV solution brings about the stabilization of the electrogenerated active species $MV^{+\cdot}$ and MV by an inclusion effect [374].

The electrosynthesis of 1,1'-dimethyl-4,4'-bipyridinium dichloride **430** has been performed by electroreduction of methyl pyridinium chloride **429** in an MeCN-MPCl(0.5M)-(C) system in an undivided cell at 40 °C (eq. (5.182)) [375]. The redox charactor of MV^{2+} $2Cl^-$ in the presence of $SnCl_4$ has been recorded [376]. Redox behavior of viologen-pendent polypyrrole on Pt electrode has been investigated [377a]. The electrons are transferred from the modified electrode to O_2 through the viologen moiety. The modified electrode works as reactive electrodes for the selective electrocatalytic reduction of O_2 to H_2O_2 [377b]. The viologen-based redox polymers immobilized on a single-crystal silicone electrode surface are found to be effective for the catalytic reduction of bicarbonate to formate [378]. The electroreduction of N-butylpyridine cation **431** in an MeCN-AlCl$_3$-(C) system at 40 °C affrods 4,4'-tetrahydrobipyridine **432** in about 45% yield (eq. (5.183)) [371].

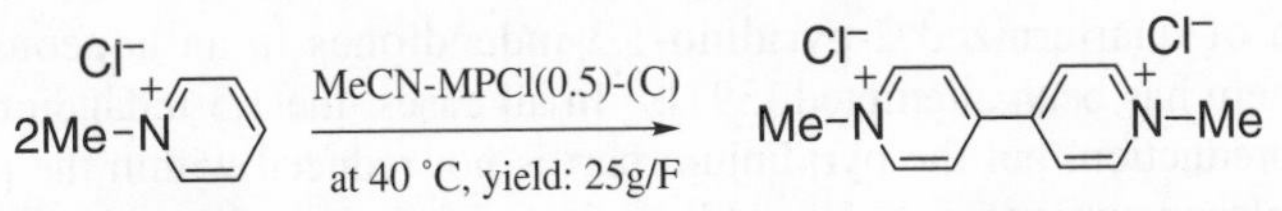

$$\text{2Me-N}^+ \text{Cl}^- \xrightarrow[\text{at 40 °C, yield: 25g/F}]{\text{MeCN-MPCl(0.5)-(C)}} \text{Me-N}^+ \text{Cl}^- \quad \text{Cl}^- \text{N-Me} \qquad (5.182)$$

$$\textbf{429} \qquad\qquad\qquad\qquad \textbf{430}$$

$$\text{Bu-N}^+ \xrightarrow[\text{40° C}]{\text{MeCN, AlCl}_3\text{, (W)-(C)}} \text{Bu-N} \quad \text{N-Bu} \qquad (5.183)$$

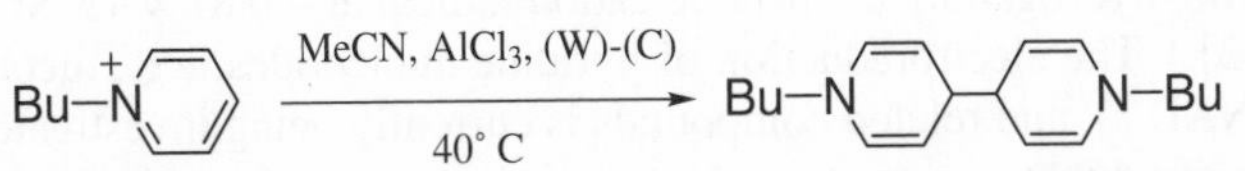

$$\textbf{431} \qquad\qquad\qquad \textbf{432} \quad (45\%)$$

The methylviologen-assisted indirect electroreduction of benzalmalononitrile **436** in a DMF-AcOH(1M) system selectively affords *trans*-2-amino-4,5-diphenyl-2-cyclopentene-1,3,3-tricarbonitrile **438** (eq. (5.184)) [379]. The one-electron reduced species of N-propyl-4,4'-bipyridinium cation is found to be able to react with CO_2 to form a stable CO_2 adduct [380]. N-Monosubstituted 4,4'-bipyridinium ions ($RBPY^+$: R = Me, PhCH$_2$, ethyl, dodecyl) is protonated at acidic conditions and the protonated species are reduced by two consecutive one-electron processes [381]. The two-electron reduced species undergoes a chemical reaction with H^+. At high pH, the electrode reaction of $RBPY^+$ is one-step two-electron transfer process with concomitant addition of H^+.

$$MV^{2+} \underset{}{\overset{e^-}{\rightleftarrows}} MV^{+\cdot} \underset{}{\overset{e^-}{\rightleftarrows}} MV \qquad \cdots\cdots (a)$$

$$\mathbf{433} \qquad\qquad \mathbf{434} \qquad\qquad \mathbf{435}$$

(5.184)

$$2\ \mathbf{435} + 2PhCH{=}C(CN)_2 \longrightarrow 2\ MV^+ + \qquad \cdots\cdots (b)$$

$$\mathbf{436} \qquad\qquad\qquad \mathbf{437}$$

438

Apparently, viologen homologues are the most important redox compounds used widely not only for basic researches on electrochemical and photoelectro-chemical processes but also for electrochromic materials [382,383], for electron transfer mediators, and for functional elements of molecular devices [384] due to their reversible single electron accepting ability, producing stable and colored cation radicals.   Intensive efforts for the preparation of various types of viologen homologue have been made.   So far some vinylogous viologens [385], heterocyclic and benzologue three-ring assembling viologens [386], alkyl chain bridge dimers [387], o-xylene bridge dimer [388], inclusion complexation with β-cyclodextrin [389] and their related redox systems [390] have been synthesized with the prospect of developing a novel and superior function of viologens.

Electroreduction of quarternized 2-pyridino-1,3-indendiones in an aqueous DMF-$H_2SO_4(1M)$-(Hg) system has been attempted [391].   In all cases, the 1,3-indandionyl moiety undergoes electroreduction, but the pyridinium part is not reduced within the potential range accessible to polarographic study.   Unstable radical cations have been shown to form when bis(tetrafluoroborate) salts of 2,3-dipyridylnorbornadienes are electrochemically reduced [392].   1,1'-Dimethyl-3,3'-oxybispyridinium diiodide can be reduced by a one-electron transfer not involving hydrogen to an unstable cation radical at -0.81 V vs. SCE in the pH range 6.3~12.0 [393].   The electroreduction of pyridine nucleotides, e.g., nicotinamide adenine dinucleotide (NAD$^+$), and related compounds is currently being investigated intensively.   It has been clarified that one- or two-electron reduction products are formed, i.e., tetrahydrobipyridine derivatives and dihydropyridines, respectively.   The electroreduction of 1-benzyl-3-carbamoylpyridinium chloride has been characterized by two reduction waves, the first one (wave A) pH independent and the second (wave B) showing only alkaline pH values.   The first step implies a reversible one-electron transfer to the pyridinium cation to give a radical intermediate which irreversibly dimerizes [394a].   In the course of neutralization of the positive charge of the pyridinium cation adsorbed at the mercury/water interface, the perpendicularly adsorbed radical is found to be electrochemically inactive [395].   The electrochemical regeneration procedure of the N-phenyl-1,4-dihydronicotinamide **440** from the pyridinium cation **444** via **439** has been developed in a two-phase medium (H_2O/CH_2Cl_2) (eq. (5.185)) for the reduction of **442** to **443** [396].   This procedure allows the catalytic use of the electrochemically produced reducing agent.

$$(5.185)$$

Electrochemical reduction of 1-benzyl-3-carbamoylpyridinium chloride **445** in an aqueous Me$_4$NOH/Me$_4$NOH-(Hg) system at pH 10.3 affords a mixture of dihydropyridines **446** and **447**, and tetrahydrobipyridines **448** and **449**, as major product (eq. (5.186)) [397]. The product distribution highly depends on the molar concentration of the substrate **445**, as shown in Table 5.14. [394].

$$(5.186)$$

Table 5.14 Electroreduction of N-Benzyl-3-carbamoylpyridinium chloride

445	Product, % yield				
M	446	447	448	449	Others
0.1	45	20	13	10	12
0.001	69	21	4	4	2

The enolate ion **452** derived from 1-methyl-4-(methoxycarbonyl)pyridinium iodide **450** by two successive one-electron reductions *via* **451** is found to be an excellent electron donor and reacts with *tert*-butyl bromide through one-electron transfer in a DMF-Bu$_4$NI-(Pt) system to give 4-*tert*-butyl-1,4-dihydro-4-(methoxycarbonyl)-1-methylpyridine **453** in 85% yield (eq. (5.187)) [398].   1-Benzyl-4-carbamoyl-1,4-dihydropyridine **455** and **456**

$$(5.187)$$

* DMF-Bu₄NI-(Pt)

has been synthesized by electrolysis of the pyridinium chloride **454** in good yield in an aqueous $NH_3(0.1M)/NH_4Cl(0.1M)/Et_4NCl(0.5M)$-(Hg) system (eq. (5.188a)) [399].   The chemical reduction of **454** with sodium borohydride, however, gives the corresponding 1,6-dihydropyridine **456** (eq. (5.188b)).   Electrochemical properties of alkaloid jatrorubinium ion **457** have been investigated.   The alkaloid cation is reduced irreversibly in an aqueous $NaCl(0.1M)/HCl(pH2)$-(Hg) system to tetrahydrojatrorubine **458** (eq. (5.189)) [400].

$$(5.188)$$

$$(5.189)$$

The voltammetric study of the cefsulodin as a cephem derivative bearing an isonicotinamide pendant yields two two-electron signals which may be accounted for by an ECE mechanism [401].   The electrochemical behavior of methylene blue at a carbon fiber microcylinder electrode has been studied by cyclic voltammetry.   The electrode reaction mechanism for methylene blue dye at various pH ranges is shown in Fig. 5.2. [402].   The compound **463** is a mediator in the acid-base valance between **460** and **462**.   The adenine moiety of the coenzyme NAD^+ has been found to undergo electroreduction in acidic solutions [403].   The reduction pattern for NAD^+, however, is different in alkaline solutions. The first one-electron reduction of NAD^+ corresponding to wave **1** of the coenzyme occurs in the absorbed state in the 2~11 pH range.   The difference between the two mechanisms in acid and alkaline media refers only to the second one-electron and proton transfer to the NAD^+ molecule.   In acid medium, the second electron transfer to the nicotinamide moiety occurs indirectly *via* reduction of the adenine moiety [403].   Quinone-mediated bioelectrochemical reduction of $NAD(P)^+$ is found to be catalyzed by flavoproteins [404].

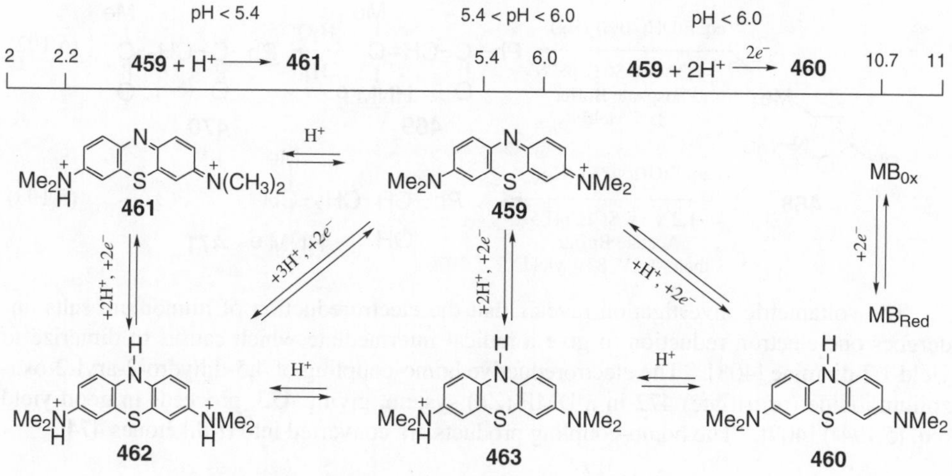

2   2.2   **459** + H$^+$ ⟶ **461**   5.4   6.0   **459** + 2H$^+$ $\xrightarrow{2e^-}$ **460**   10.7   11

Fig. 5.2 The electroreaction of methylene blue at various pH ranges.

5.6.2   Electroreduction of Imminium and Immonium Cations

The electroreduction of *N*-cyclohexylideneaniline **464** in an MeCN-Et$_4$NClO$_4$-(Pt/Hg) in the presence of acrylonitrile gives the adduct **465** in good yields (eq. (5.190)) [405].

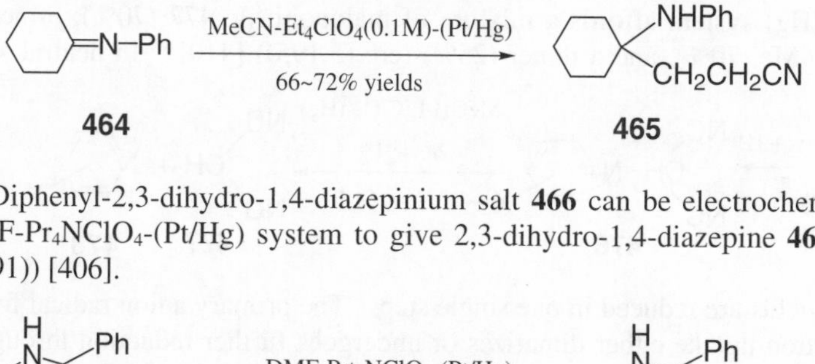

$$\text{(5.190)}$$

464   **465**

5,7-Diphenyl-2,3-dihydro-1,4-diazepinium salt **466** can be electrochemically reduced in a DMF-Pr$_4$NClO$_4$-(Pt/Hg) system to give 2,3-dihydro-1,4-diazepine **467** in 47% yield (eq. (5.191)) [406].

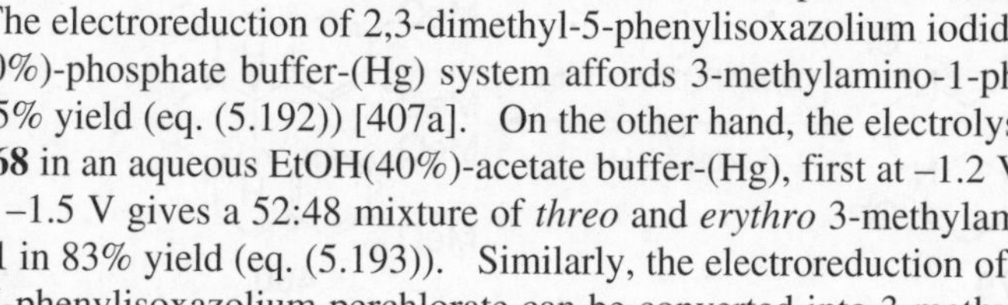

$$\text{(5.191)}$$

466   **467**

Preparative reduction of isoxazolium iodide has been performed under different conditions. The electroreduction of 2,3-dimethyl-5-phenylisoxazolium iodide **468** in an aqueous EtOH(40%)-phosphate buffer-(Hg) system affords 3-methylamino-1-phenyl-2-buten-1-one **469** in 85% yield (eq. (5.192)) [407a]. On the other hand, the electrolysis of the oxazolium iodide **468** in an aqueous EtOH(40%)-acetate buffer-(Hg), first at −1.2 V *vs.* SCE and afterwards at −1.5 V gives a 52:48 mixture of *threo* and *erythro* 3-methylamino-1-phenyl-1-butanol **471** in 83% yield (eq. (5.193)). Similarly, the electroreduction of 4,5-dihydro-2,3-dimethyl-5-phenylisoxazolium perchlorate can be converted into 3-methylamino-1-phenyl-1-butanol in 72% yield [407].

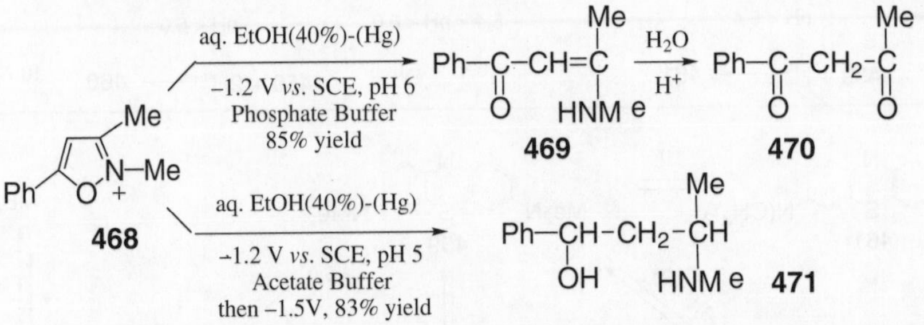

$$Ph-C-CH=C \quad \xrightarrow[H^+]{H_2O} \quad Ph-C-CH_2-C \qquad (5.192)$$

469   **470**

$$Ph-CH-CH_2-CH \qquad (5.193)$$

471

The voltametric investigation reveals that the electroreduction of immonium salts undergoes one-electron reduction to give a radical intermediate which causes to dimerize to yield 1,2-diamine [408]. The electroreductive homo-coupling of 4,5-dihydro-2-aryl-2-oxazolium iodide (or triflate) **472** in a DMF-(Zn) system, giving **473**, proceeds in good yield (eq. (5.194)) [409]. The homo-coupling products are converted into α-diketones **474**.

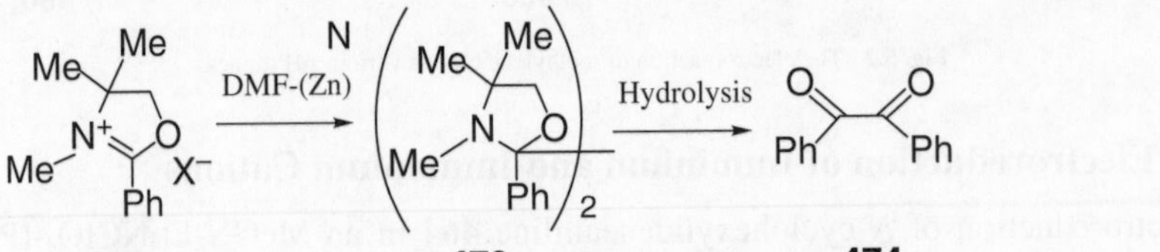

$$(5.194)$$

472   **473**   **474**

The controlled potential electroreduction of pyridinium dicyanomethylide **475** in an MeOH-LiClO$_4$-(Hg) system affords a mixture of malononitrile **477** (70%), p-methylpyridine **478** (X = p-Me, 70%), and a dimer (25%) (eq. (5.195)) [410]. In neutral solutions,

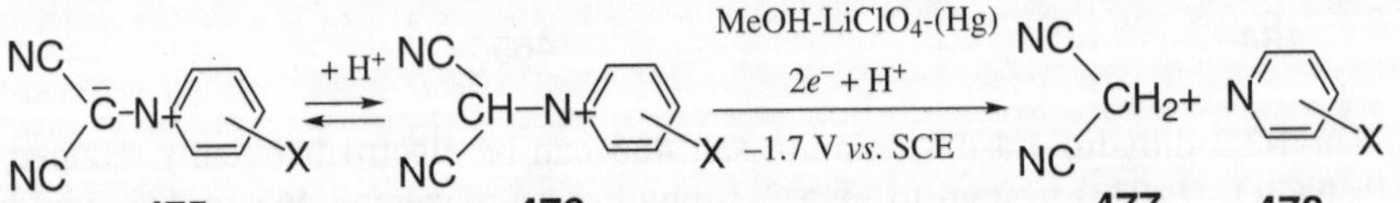

$$(5.195)$$

475   **476**   **477** **478**

the pyridinium yields are reduced in one single step. The primary anion radical formed after the first electron uptake either dimerizes or undergoes further reduction through protonation and cleavage of the C-N bond after the second electron uptake. In acidic solutions, protonation in the bulk solution is followed by electroreduction exclusively through cleavage of the C-N bond.

The electroreduction of diisoquinolinium chloride **479** undergoes intramolecular coupling reaction to afford the cyclized product **480** in 80% yield (eq. (5.196)) [411].

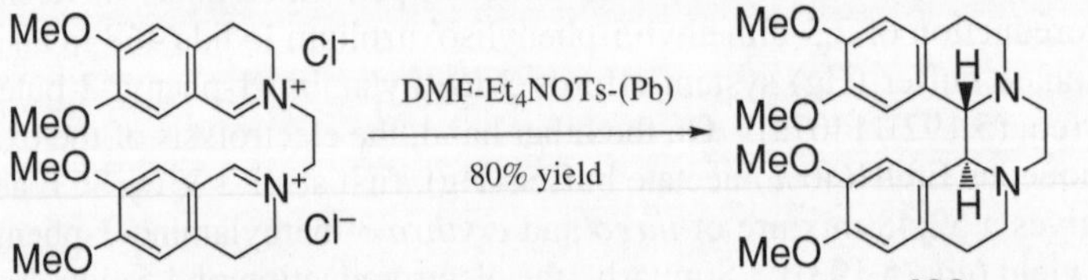

$$(5.196)$$

479   **480**

The electroreductive cross-coupling of α-hydroxycarbamates **481** with activated olefins has been attained in a DMF-Et$_4$NOTs-(0.83M)-(Pb) system in the presence of chlorotrimethylsilane, giving the corresponding cross-coupling products **483** in 39~81% yields (eq. (5.197)) [412]. The coupling reaction undergoes the formation of an acyliminium intermediate which provides a radical species after one-electron reduction. The preparative electroreduction of N,N-dimethyl-N'-phenylformamidinium perchlorate **484** proceeds in an

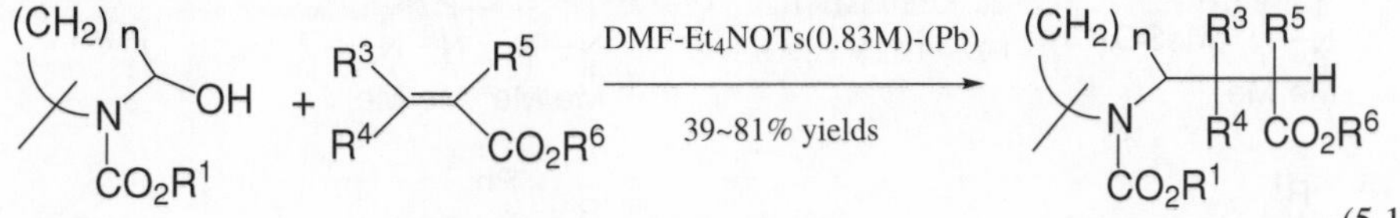

$$
\begin{array}{ccc}
\textbf{481} & \textbf{482} & \textbf{483}
\end{array}
$$

(5.197)

MeCN-LiClO$_4$(0.1M)-(Pt) system smoothly to give the corresponding amidine **485** in 95~100% yields (eq. (5.198)) [413]. Succinimide **486** in an MeCN-Bu$_4$NBF$_4$(0.2M)-(Pt)

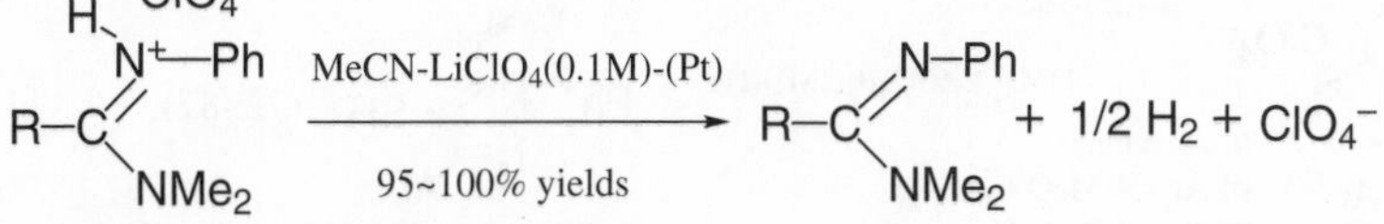

$$
\textbf{484} \qquad R = Me, Pr, Ph, PhCH_2 \qquad \textbf{485}
$$

(5.198)

system shows an irreversible reduction wave by cyclic voltammetry [414]. The product of the electrolytic reaction is succinimide anion **487** which is stable in the electrolysis media, but reacts with tetrabutylammonium ion during vpc analysis or under refluxing in DMF to form N-butylsuccinimide **488** (eq. (5.199)) [414].

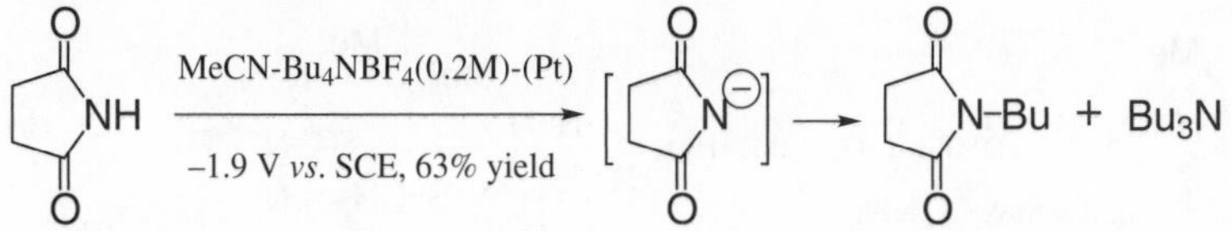

$$
\textbf{486} \qquad\qquad \textbf{487} \qquad \textbf{488}
$$

(5.199)

Chiral tri- and tetrasubstituted piperazines **492** are electrosynthesized effectively from chiral 1,2-diamines **490**. The potential-controlled electroreduction method has been used as a reducing agent instead of the usual boronhydride in N-alkylation of proteins (eq. (5.200)) [415]. The electroassisted modification of bouine casien is performed in an aqueous NaN$_3$(0.2%)-Britton Robinson buffer (pH 9.15)-(Hg) system for a period of 1-5 days. The results and conditions on electroreduction of imminium and immonium cations are summarized in Table 5.15. [344, 345, 416~420].

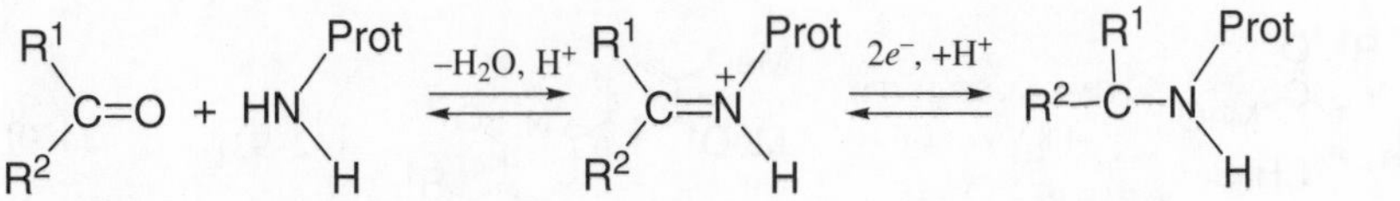

$$
\textbf{489} \qquad \textbf{490} \qquad \textbf{491} \qquad \textbf{492}
$$

(5.200)

Prot = proteins
R^1, R^2 = carbohydrate residue

Table 5.15a Results and Conditions in Electroreduction of Imminium and Immonium Salts

	Conditions	Products	Yields(%)	Ref.
Ph, Ph with $+N-N$, Me Me, MeSO$_4^-$	aq. KCl(0.5M)-(Hg) −1.27~−1.52 V vs. Ag/AgCl	Ph, Ph, Ph, Ph structure		[416]
R^1, R^2, Ph, ClO$_4^-$	DMF-Pr$_4$NClO$_4$-(Hg) $R^1 = R^2 = $ H or Me	R^2, Ph, Ph, R^2 structure		[417]
Ph, Ph, ClO$_4^-$, S, N, Ar **a:** Ar = Ph **b:** Ar = 4-MeO-C$_6$H$_4$ **c:** Ar = 4-Br-C$_6$H$_4$	DMF-LiClO$_4$(0.2M)-(Hg)	Ph, Ph, SH, Ar structure	(72~87)	[418]
Ph, S, ClO$_4^-$, N, S, Ar **a:** Ar = Ph **b:** Ar = 4-Br-C$_6$H$_4$	DMF-LiClO$_4$(0.2M)-(Hg)	Ph, S, SH, Ar structure	(77~86)	[418]
Me, N–N, ClO$_4^-$, R, S, Ar **a:** R = MeS Ar = Ph **b:** R = MeS Ar = 4-MeO-C$_6$H$_4$ **c:** R = H Ar = Ph **d:** R = H Ar = 4-MeO-C$_6$H$_4$	DMF-LiClO$_4$(0.2M)-(Hg)	Me, N–N, S, R, S–CH$_2$, N, Ar structure	(63~78)	[418]
N, S, CN, Ar, H **a:** Ph **b:** 4-ClC$_6$H$_4$ **c:** 4-MeOC$_6$H$_4$ **d:** 4-NO$_2$C$_6$H$_4$	aq. EtOH(60%)-HCl- −1.0 V vs. SCE	[N, S, CN, H, Ar]$_2$ structure	(70)	[419]
R^2, R^1, O, C–MeO, R^3, CH$_2$Br **a:** $R^1 = R^2 = R^3 = $ H **b:** $R^1 = R^2 = $ MeO, $R^3 = $ H **c:** R^1, $R^2 = $ OCH$_2$O, $R^3 = $ H **d:** $R^1 = $ H, $R^2 = R^3 = $ MeO	DMF-MeSO$_2$H-(Pt) −1.8 V vs. SCE	MeO, MeO, N, O, R^1, R^2, R^3 structure	(62~93)	[420]

Table 5.15b Results and Conditions in Electroreduction of Imminium and Immonium Salts

	Conditions	Products	Yields(%)	Ref.

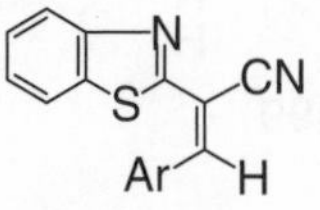

aq. EtOH(60%)-HCl-
−1.0 V *vs.* SCE

(70)   [344]

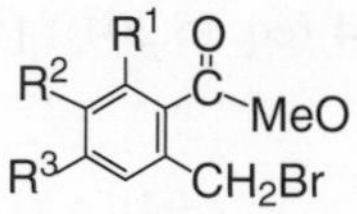

a: Ph
b: 4-ClC$_6$H$_4$
c: 4-MeOC$_6$H$_4$
d: 4-NO$_2$C$_6$H$_4$

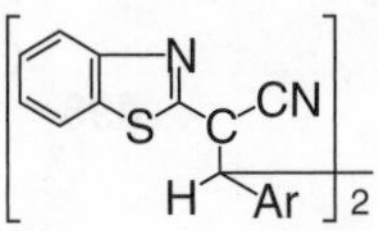

DMF-MeSO$_2$H-(Pt)
−1.8 V *vs.* SCE

(62~93)   [345]

a: R^1 = R^2 = R^3 = H
b: R^1 = R^2 = MeO, R^3 = H
c: R^1 , R^2 = OCH$_2$O, R^3 = H
d: R^1 = H, R^2 = R^3 = MeO

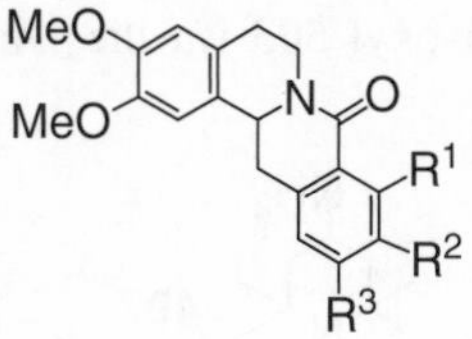

5.6.3 Electroreduction of *N*-Oxides

The electroreduction of 2,2,6,6-tetramethyl-4-oxopiperidine-1-oxyl radical **493** to the dihydroxypiperidine **495** *via* the keto alcohols intermediate **494** has been attained in aqueous NaOH(0.1M)-(Hg or amalg. Cu) system in quantitative yield (eq. (5.201)) [421a]. Under acidic conditions, the electroreduction undergoes the cleavage of the N-O bond at −1.1 V to give the 4-oxopyperidine derivative **496** in quantitative yield [421b].   The 1-oxyl-4-aminopiperidine (4-amino-TEMPO) **497** can be converted into the 1-hydroxy-4-aminopiperidine derivative **498** in 97% yield, whereas the electroreduction of **497** in an acidic medium affords the corresponding 4-aminopiperidine **499** in good yield (eq. (5.202)) [422].   The 4-amino-TEMPO **497** has been used for immobilization of poly(acrylic acid)

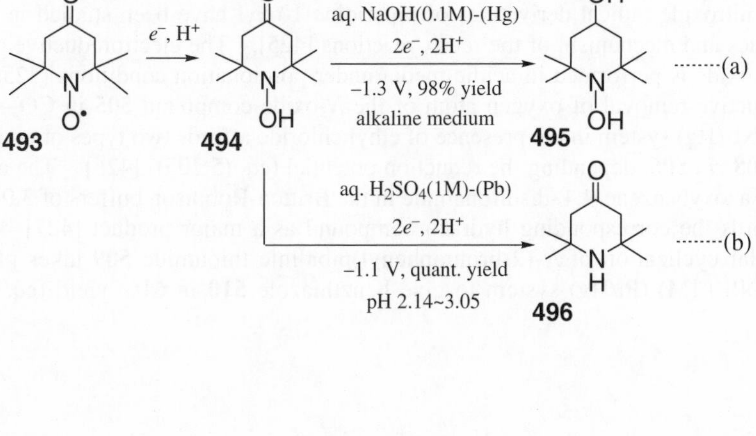

(5.201)

$$(5.202)$$

497   **498**   **499**

500 in order to prepare a redox polymer **502** [423].   The polymer is prepared by the reaction of 4-amino-TEMPO by an amide-linkage with the coupling reagent, *N,N*-dicyclohexyl-carbodiimide (DCC) **501**, in DMF (eq. (5.203)).   1-Hydroxy-4-amino piperidine, the precursor of the 1-oxyl-4-aminopiperidine **498**, can be obtained by the electroreduction of 4-hydroxyiminopiperidine-1-oxyl **503** *via* the 1-hydroxy derivative **504** (eq. (5.204)) [424].

$$(5.203)$$

500   **501**   **502**

$$(5.204)$$

503   **504**   **498**

The eight nitroxide radical derivatives of piperidine-1-oxyl have been studied in terms of their kinetics and mechanism of the redox reactions [425].   The electroreductive of nicotinamide *N*-oxide is performed in acidic media under porotonation conditions [325a].   The electroreductive removal of oxygen atom of the *N*-oxide compound **505** in CO_2-saturated DMF-Bu₄NI-(Hg) system in the presence of ethyl chloride affords two types of compounds, **507** and **508** *via* **506**, depending the reduction potential (eq. (5.205)) [426].   The electroreduction of azoxybenzene-4,4'-disulfonamide in the Britton-Robinson buffers of 3.0-10.6 pH range affords the corresponding hydrazo compound as a major product [427].   The intramolecular cyclization of *N*-(2-bromophenyl)pibalinic thioamide **509** takes place in a DMF-Bu₄NI(0.1M)-(Pt/Hg) system to give benzthiazole **510** in 61% yield (eq. (5.206)) [428].

$$(5.205)$$

$$(5.206)$$

5.7 Miscellaneous

The electrochemical *N*-alkylation has been performed by the reaction of arylamines with carbonyl compounds followed by the electroreduction of the Schiff base. For instance, the electroreduction of the Schiff base **512**, formed by the reaction of 4-aminodiphenylamine **511** with acetone, in an aqueous $Me_2CO(70\%)$-HCl(1M)-(C) system at pH 2 gives 4-iso-propylaminodiphenylamine **513** (eq. (5.207)) [429].   The electroreductive alkylation of a primary amine with aliphatic aldehydes affords the corresponding secondary amines in 60~75% yield [430].   The electroreduction of the Schiff bases occurs successfully at electrodes with high and medium values of hydrogen overvoltages such as at lead and copper electrodes at pH *ca.* 12.

$$(5.207)$$

Electroreductive amidation of the esters **514** with the amines **515** has been attained in an $MeCN$-Et_4NClO_4-(Pt) or CH_2Cl_2-Bu_4NI-(Pt) system at room temperature at –1.7 V *vs.* SCE (eq. (5.208)) [431].   Recently, the method has been extended to the preparation of ureides **518** from substituted malonates **517** (eq. (5.209)).   The electrolysis of **517** is carried out in a DMF-Et$_4$NClO$_4$-(Pt) system by passage of 5 F/mol of electricity to give barbituric acids **518** in 14~71% yields [432].

$$R^1-CO_2R^2 + R^3-NH_2 \xrightarrow[\substack{1.7\ V\ vs.\ SCE,\ 42\text{~}93\%\ yields}]{\substack{MeCN-Et_4NClO_4\text{-}(Pt) \\ or\ CH_2Cl_2\text{-}Bu_4NI\text{-}(Pt)}} R^2-CONH-R^3 \qquad (5.208)$$

514   **515**   **516**

R^1 = Et, Ph, PhCH$_2$; R^2 = Me, Et, Ph, tolyl

R^3 = (CH$_2$)$_2$OH, (CH$_2$)$_3$OH, PhCH$_2$

$$(5.209)$$

517   **518**

R^1 = Et, allyl; R^2 = Et, allyl, Bz; R^3 = H, Ph

The catalytic electroreduction of tertiary *p*-toluenesulfonamides and *gem-N*-di-p-toluenesulfonamides has been performed in a DMF-LiClO$_4$-(Hg) system to give the corresponding amines.   The tertiary tosylamides are not directly reducible by electrochemical means. Therefore the electrogenerated pyren anion radical may assist the cleavage of the S-N bond in a catalytic manner [433].

The electroreductive cleavage of the Mo-N bond of the Mo complex **519** occurs in the presence of acetic acid to release amino acid esters **520** and form η^2-acetatomolybdenum hydrides **521** (eq. (5.210)) [434].   An electroreductive nitrogen-fixing cycle has been attained by reducing *trans*-[MoBr(N$_2$R$_2$)L$_2$]$^+$ complex under dinitrogen to yield a dialkylhydrazine and *trans*-[Mo(0)(N$_2$)$_2$L$_2$] complex which may lead to the *trans*-[MoBr(N$_2$R$_2$)L$_2$]$^+$ complex by the reaction with alkyl bromides (L = (Ph$_2$PCH$_2$)$_2$) [435].

$$(5.210)$$

519   **520**   **521**

R = H, Me

The electrosynthesis of magnesium diacetamide, Mg(HNCOMe)$_2$ has been performed in an MeCN-Et$_4$NBr-(Mg/Pt) system at 50 °C in high current efficiency [436].

Pyridines are electrochemically synthesized by the reaction of an alkyne and a cyanide in the presence of cobalt complex catalysts which are activated by electroreduction [437]. The S-N bond cleavage of sulfapyridine has been examined in a DMF-Et$_4$NClO$_4$-(Hg) system to give the corresponding sulfinate and aminopyridine [438].

References

[1]   F. M.-Roch, A. Tallec, R. Tardivel, *Electrochim. Acta*, **40**, 1877 (1995).
[2]   (a) P. E. Iversen, H. Lund, *Tetrahedron Lett.*, **41**, 4027 (1967); (b) P. Martigny, J. Simonet, G. Mousset, *J. Electroanal. Chem. Interfacial Electrochem.*, **148**, 51 (1983).
[3]   H. Lund, M. M. Baizer (ed.), *Organic Electrochemistry*, p. 316, Marcel Dekker, New York (1973).
[4]   (a) M. Masui, H. Sayo, K. Kishi, *Chem. Pharm. Bull.*, **12**, 1397 (1964); (b) M. Masui, H. Sayo, K. Kishi, *Tetrahedron*, **21**, 2831 (1965).
[5]   C. K. Mann, K. K. Barnes, *Electrochemical Reactions in Nonaqueous Systems*, Chapt. 11, Marcel Dekker, New York (1970).
[6]   H. Ohmori, S. Furusako, M. Kashu, C. Ueda, M. Masui, *Chem. Pharm. Bull.*, **32**, 3345 (1984).
[7]   J. R. O. Gomez, *J. Appl. Electrochem.*, **21**, 331 (1991).
[8]   (a)*French Patent* 2577242; (b) *French Patent* 2614044.
[9]   A. Savall, J. Quesado, M. Rignon, J. Malafosse, *J. Appl. Electrochem.*, **21**, 805 (1991).
[10] (a) L. H. Klemm, P. E. Iversen, H. Lund, *Acta Chem. Scand.*, **B28**, 593 (1974); (b) L. Christensen, P. E. Iversen, *Acta Chem. Scand.*, **B33**, 352 (1979).
[11] (a) M. Cariou, R. Hazard, M. Jubault, A. Tallec, *J. Electroanal. Chem.*, **182**, 345 (1985); (b) M. Cariou, R. Hazard, M. Jubault, A. Tallec, *J. Chem. Research*, **1986**, 184.
[12] M. Cariou, R. Hazard, M. Jubault, A. Tallec, *Tetrahedron Lett.*, **22**, 3961 (1981).
[13] M. Cariou, R. Hazard, M. Jubault, A. Tallec, *Can. J. Chem.*, **61**, 2359 (1983).
[14] (a) W. J. Bowyer, D. H. Evans, *J. Org. Chem.*, **53**, 5234 (1988); (b) J. C. Rühl, D. H. Evans, P. Hapiot, P. Neta, *J. Am. Chem. Soc.*, **113**, 5188 (1991).
[15] G. Kokkinidis, E. Coutouli-Argyropoulou, *Electrochim. Acta*, **30**, 493 (1985).
[16] M. E. Niyazymbetov, D. H. Evans, *Denki Kagaku*, **62**, 1139 (1994).
[17] W. T. Monte, M. M. Baizer, R. D. Little, *J. Org. Chem.*, **48**, 803 (1983).
[18] M. Wessling, H. J. Schäfer, *Chem. Ber.*, **124**, 2303 (1991).
[19] M. Wessling, H. J. Schäfer, *Dechema Monogr.*, **125**, 807 (1992).
[20] A. Sera, H. Tani, I. Nishiguchi, T. Hirashima, *Synthesis*, **1987**, 631.
[21] S. Torii, H. Tanaka, T. Katoh, *Chem. Lett.*, **1983**, 607.
[22] T. Shono, H. Hamaguchi, H. Mikami, H. Nogusa, S. Kashimura, *J. Org. Chem.*, **48**, 2103 (1983).
[23] (a) C. Bellec, R. Colan, S. Deswarte, J. C. Dore, C. Viel, *C. R. Acad. Sc. Paris*, **281**, 885 (1975); (b) S. Deswarte, C. Bellec, J. Pucheault, C. Ferradini, L. Gilles, *J. Heterocyclic Chem.*, **17**, 891 (1980).
[24] (a) T. Sato, Y. Komeichi, T. Wada, *Nippon Kagaku kaishi*, **1984**, 1815 (in Japanese); (b) T. Sato, T. Wada, Y. Komeichi, M. Kainosho, *Bull. Chem. Soc. Jpn.*, **58**, 1452 (1985).
[25] J. Nokami, T. Sonoda, S. Wakabayashi, *Synth. Commun.*, **13**, 763 (1983).
[26] (a) S. S. Khripko, *Zh. Org. Khim.*, **22**, 50 (1986); (b) N. I. Semakhina, T. A. Podkovyrina, L. O. Toktaulova, Yu. M. Kargin, *Zh. Obshch. Khim.*, **56**, 2764 (1986).
[27] R. Jain, D. D. Agarwal, R. K. Shrivastava, *J. Chem. Soc., Perkin Trans. II*, **1990**, 1353.
[28] H. Balslev, H. Lund, *Acta Chem. Scand.*, **45**, 436 (1991).
[29] Yu. M. Kargin, V. Z. Latypova, A. V. Supyrev, V. V. Zhuikov, *Zh. Obshch. Khim.*, **54**, 1695 (1984).
[30] (a) W. Kemula, T. M. Krygowski, A. J. Bard, H. Lund (eds.), *Encyclopedia of Electrochemistry of the Elements*, Marcel Dekker, New York and Basel, **13**, 78 (1979); (b) W. H. Smith, A. J. Bard, *J. Am. Chem. Soc.*, **97**, 5203 (1975); (c) M. Maggini, C. Paradisi, G. Scorrano, S. Daniele, F. Magno, *J. Chem. Soc., Perkin Trans. II*, **1986**, 267; (d) V. P. Gul'tyai, V. N. Leibzon, *Elektrokhimiya*, **32**, 65 (1996).
[31] (a) P. Zuman, *Collect. Czech. Chem. Commun.*, **58**, 41 (1993); (b) N. E. Minina, A. Falakh, V. M. Kazakova, *Zh. Obshch. Khim.*, **62**, 2582 (1992); (c) A. V. Il'yasov, M. Yu, Kitaeva, A. A. Vafina, R. M. Zaripova, Yu. P. Kitaev, *Izv. Akad. Nauk, Ser. Khim.*, **6**, 1067 (1993); (d) P. N. Pintauro, J. R. Bontha, *J. Appl. Electrochem.*, **21**, 799 (1991).
[32] N. E. Gunawardena, D. Pletcher, *Acta Chem. Scand.*, **B37**, 549 (1982).
[33] G. Belot, S. Desjardins, J. Lessard, *Tetrahedron Lett.*, **25**, 5347 (1984).
[34] F. Beck, W. Gabriel, *Angew. Chem. Int. Ed. Engl.*, **24**, 771 (1985).
[35] J.-C. Gard, J. Lessard, Y. Mugnier, *Electrochim. Acta*, **38**, 677 (1993).
[36] C. Karakus, P. Zuman, *J. Electrochem. Soc.*, **142**, 4018 (1995).
[37] C. Nishihara, H. Shindo, *J. Electroanal. Chem. Interfacial Electrochem.*, **202**, 231 (1986).
[38] H. Shindo, C. Nishihara, *Surf. Sci.*, **158**, 393 (1985).
[39] M. F. Bento, M. J. Medeiros, M. I. Montenegro, *J. Electroanal. Chem.*, **345**, 273 (1993).
[40] M. Svaan, V. D. Parker, *Acta Chem. Scand.*, **B36**, 357 (1982).
[41] T. M. Grishina, L. I. Lazareva, *Russ. J. Phys. Chem.*, **59**, 576 (1985).
[42] V. K. Sharma, D. K. Sharma, C. M. Gupta, *J. Electrochem. Soc. India*, **34**, 103 (1985).
[43] (a) C. Lamoureux, C. Moinet, A. Tallec, *Electrochim. Acta*, **31**, 1 (1986); (b) C. Lamoureux, C. Moinet, *Bull. Soc. Chim. Fr.*, **128**, 599 (1991).

[44] M. A. Santa Ana, I. Chadwick, G. Gonzalez, *J. Chem. Soc. Perkin Trans. II*, **1985**, 1755.

[45] D. Pletcher, M. Razaq, G. D. Smilgin, *J. Appl. Electrochem.*, **11**, 601 (1981).

[46] (a) M. Noel, P. N. Anantharaman, H. V. K. Udupa, *J. Appl. Electrochem.*, **12**, 291 (1982); (b) C. Ravichandran, D. Vasudevan, S. Thangavelu, P. N. Anantharman, *J. Appl. Electrochem.*, **22**, 1087 (1992).

[47] K. Kobayakawa, T. Yamabe, A. Fujishima, *Nippon Kagaku Kaishi*, **1983**, 351.

[48] C. Ravichandran, C. Chellammal, P. N. Anantharaman, *J. Appl. Electrochem.*, **19**, 465 (1989).

[49] A. Cyr, P. Huot, G. Belot, J. Lessard, *Electrochim. Acta*, **35**, 147 (1990).

[50] M. Inaba, Z. Ogumi, Z. Takehara, *J. Electrochem. Soc.*, **140**, 19 (1993).

[51] C. Ravichandran, D. Vasudevan, P. N. Anantharaman, *Electrochem. Soc. India.*, **42-3**, 193 (1993).

[52] M. Noel, C. Ravichandran, P. N. Anantharaman, *J. Appl. Electrochem.*, **24**, 1256 (1994).

[53] K. J. Stutts, C. L. Scortishini, C. M. Repucci, *J. Org. Chem.*, **54**, 3740 (1989).

[54] A. Petsom and H. Lund, *Acta Chem. Scand.*, **B34**, 693 (1980).

[55] A. G.-Criqui, C. Moinet, *Bull. Soc. Chim. Fr.*, **130**, 101 (1993).

[56] V. K. Sharma, M. S. Madhu, *Bull. Electrochem.*, **8**, 323 (1992).

[57] B. J. Cote, D. Despres, R. Labrecque, J. Lamothe, J.-M. Chapuzet, J. Lessard, *J. Electroanal. Chem.*, **355**, 219 (1993).

[58] A. Susaimanickam, M. Chandrasekaran, V. Krishnan, *J. Appl. Electrochem.*, **20**, 335 (1990).

[59] C. Gault, C. Moinet, *Tetrahedron*, **45**, 3429 (1989).

[60] C. Ravichandran, M. Noel, P. N. Anantharaman, *J. Appl. Electrochem.*, **24**, 965 (1994).

[61] M. Noel, C. Ravichandran, P. N. Anantharaman, *J. Appl. Electrochem.*, **25**, 690 (1995).

[62] Z. Takehara, Z. Ogumi, M. Inaba, *Stud. Org. Chem.*, **30**, 409 (1987).

[63] (a) Y. Matsuda, H. Kimura, Y. Okuhama, *Denki Kagaku*, **52**, 796 (1984); (b) Y. Nakamura, J.-Y. Cheng, I. Tabata, S. Suye, M. Senda, *Denki Kagaku*, **62**, 1235 (1994).

[64] C. Ravichandran, D. Vasudevan, P. N. Anantharaman, *J. Appl. Electrochem.*, **22**, 1192 (1992).

[65] M. D. Ravi, V. N. S. Pillai, P. N. Anantharaman, *Bull. Electrochem.*, **4**, 241 (1988).

[66] S. Bencheikh-Sayarah, B. Cheminat, G. Mousset, P. Pouillen, *Electrochim. Acta*, **29**, 1225 (1984).

[67] E. Manda, T. Shimura, *J. Jpn. Chem. Soc.*, **1981**, 1337.

[68] R. H. Dahm, R. J. Latham, S. E. Mosley, *J. Appl. Electrochem.*, **16**, 213 (1986).

[69] C. Ravichandran, M. Noel, P. N. Anantharaman, *Bull. Electrochem.*, **10**, 283 (1994).

[70] V. L. Kornienko, A. P. Tomilov, G. P. Vakar, N. V. Kalinichenko, *Elektrokhimiya*, **22**, 666 (1986).

[71] C. Ravichandran, C. J. Kennady, S. Chellammal, S. Thangavelu, P. N. Anantharaman, *J. Appl. Electrochem.*, **21**, 60 (1991).

[72] S. Muralidharan, C. Ravichandran, S. Chellammal, P. N. Anantharaman, *J. Electrochem. Soc., India*, **38**, 217 (1985).

[73] P. N. Anantharaman, M. Noel, *Electrochemicals, Bull.*, **1-4**, 125 (1983).

[74] Z. Ogumi, H. Yamashita, K. Nishio, Z. Takahara, S. Yoshizawa, *Denki Kagaku*, **52**, 180 (1984).

[75] K. J. Stutts, C. L. Scortichini, T. D. Gregory, S. J. Babinic, C. M. Repucci, R. F. Phillips, *J. Appl. Electrochem.*, **19**, 349 (1989).

[76] A. Vélin-Prikodanovics, J. Lessard, *J. Appl. Electrochem.*, **20**, 527 (1990).

[77] M. Suprina, M. Lacan, *Glas. Hem. Drus. Beograd*, **49**, 491 (1984).

[78] V. A. Ryabinin, V. F. Starichenko, V. D. Shteingarts, *Zh. Org. Khim.*, **29**, 1379 (1993).

[79] V. J. Koshy, C. S. Venkatacharam, C. Kalidas, *Indian J. Chem.*, **24**, 134 (1985).

[80] (a) K. Jayaraman, K. S. Udupa, H. V. K. Udupa, *Trans. SAEST*, **12**, 143 (1977); (b) J. Marquez P., D. Pletcher, *Acta Cient. Venez.*, **37**, 391 (1986).

[81] (a) J. C. Smeltzer, P. S. Fedkiw, *J. Electrochem. Soc.*, **139**, 1366 (1992); (b) J. C. Smeltzer, P. S. Fedkiw, *J. Electrochem. Soc.*, **139**, 1358 (1992).

[82] Y.-P. Sun, W.-L. Xu, K. Scott, *Electrochim. Acta*, **38**, 1753 (1993).

[83] Y. Matsuda, T. Sakoda, K. Yano, K. Nakagawa, *Japanese Patent Kokai*, 60-13025 (1985).

[84] A. M. Heras, E. Munoz, J. L. Avila, L. Camacho, J. L. Cruz, *J. Electroanal. Chem. Interfacial Electrochem.*, **243**, 293 (1988).

[85] F. Barba, A. Guirado, J. I. Lozano, A. Zapata, J. Escudero, *J. Chem. Research*, **1991**, 290.

[86] S. C. Mishra, R. A. Mishra, *J. Electrochem. Soc. India*, **39**, 51 (1990).

[87] D. Sicker, H. Hartenstein, R. Hazard, A. Tallec, *J. Heterocyclic Chem.*, **31**, 809 (1994).

[88] V. K. Sharma, Madhu, D. K. Sharma, *Bull. Electrochem.*, **9**, 117 (1993).

[89] A.-M. Martre, G. Mousset, V. Cosoveanu, V. Danciu, *New J. Chem.*, **18**, 1221 (1994).

[90] (a) M. Studnickova, V. N. Flerov, O. Fischer, *J. Electroanal. Chem. Interfacial Electrochem.*, **187**, 307 (1985); (b) M. Studnickova, V. N. Flerov, O. Fischer, M. Potacek, *J. Electroanal. Chem. Interfacial Electrochem.*, **187**, 297 (1985); (c) S. A. Shcherbakov, A. B. Kilimnik, L. G. Feoktistov, B. N. Gorbunov, *Elektrokhimiya*, **23**, 1448 (1987).

[91] M. Studnickova, V. N. Flerov, O. Fischer, *J. Electroanal. Chem. Interfacial Electrochem.*, **187**, 297 (1985).

[92] V. Vijayakumaran, S. Muralldharan, C. Ravlchandran, S. Chellammal, P. N. Anantharman, *Bull.*

Electrochem., **6**, 522 (1990).

[93] D. Vasudevan, P. N. Anantharaman, *J. Appl. Electrochem.*, **24**, 559 (1994).

[94] J. Chaussard, R. Rouget, M. Tassin, *J. Appl. Electrochem.*, **16**, 803 (1986).

[95] K. Kitahara, H. Nishi, *Nippon Kagaku Kaishi* (in Japanese), **1983**, 1684.

[96] S. Aravamuthan, C. Kalidas, C. S. Venkatachalam, *J. Electroanal. Chem. Interfacial Electrochem.*, **171**, 293 (1984).

[97] S. Aravamuthan, M. S. Sashidar, C. Kalidas, C. S. Venkatachalam, *Indian Acad. Sci.*, **97**, 395 (1986).

[98] S. Aravamuthan, C. Kalidas, C. S. Venkatachalam, *Ber. Bunsenges. Phys. Chem.*, **89**, 880 (1985).

[99] H. Oelshläger, J. Volke, M. Berthold, W. Schmidt, *Arch. Pharm.*, **324**, 417 (1991).

[100] A. Kalandyk, J. Stroka, *J. Electroanal. Chem.*, **346**, 323 (1993).

[101] A. Chibani, R. Hazard, A. Tallec, *Bull. Soc. Chim. Fr.*, **129**, 343 (1992).

[102] A. Chibani, R. Hazard, A. Tallec, *Bull. Soc. Chim. Fr.*, **128**, 814 (1991).

[103] C. Mouats, R. Hazard, E. Raoult, A. Tallec, *Bull. Soc. Chim. Fr.*, **131**, 71 (1994).

[104] R. Hazard, M. Jubault, C. Mouats, A. Tallec, *Electrochim. Acta*, **33**, 1335 (1988).

[105] A. Chibani, R. Hazard, M. Jubault, A. Tallec, *Bull. Soc. Chim. Fr.*, **1987**, 795.

[106] H. Lund, L. G. Feoktistov, *Acta Chem. Scand.*, **23**, 3482 (1969).

[107] I. Nishiguchi, T. Hirashima, *Stud. Org. Chem.*, **30**, 303 (1987).

[108] R. Hazard, M. Jubault, C. Mouats, A. Tallec, *Electrochim. Acta*, **31**, 489 (1986).

[109] H. Schilling, K. Trautner, P. Gallien, H. Matschiner, *German Patent, (East)* DD 149,520 (1981).

[110] A. Alberti, P. Carloni, L. Greci, P. Stipa, R. Andruzzi, G. Marrosu, A. Trazza, *J. Chem. Soc., Perkin Trans. II*, **1991**, 1019.

[111] (a) R. Hazard, J. P. Hurvois, C. Moinet, A. Tallec, J. L. Burgot, G. Eon-Burgot, *Electrochim. Acta*, **36**, 1135 (1991); (b) J. Y. David, J. P. Hurvois, A. Tallec, *Tetrahedron*, **51**, 3181 (1995).

[112] J. P. Hurvois, A. Tallec, L. Toupet, *Bull. Soc. Chim. Fr.*, **129**, 406 (1992).

[113] E. N. Ulomskii, E. V. Tsoi, V. L. Rusinov, O. N. Chupakhin, G. L. Kalb, I. M. Sosonkin, *Khim. Geterotsikl. Soedin.*, **1992**, 674.

[114] P. Pazdera, M. Studnickova, I. Rackova, O. Fischer, *J. Electroanal. Chem. Interfacial Electrochem.*, **207**, 189 (1986).

[115] H. Matschiner, H. Biering, H. Schilling, H. Tanneberg, K. Trautner, C.-P. Maschmeier, *German Patent*, DD 220345 (1985).

[116] M. Noel, P. N. Anantharaman, H. V. K. Udupa, *Indian J. Technol.*, **19**, 100 (1981).

[117] (a) J. Hlavaty, J. Volke, O. Manousek, *Electrochim. Acta*, **23**, 589 (1978); (b) J. Mugnier, E. Laviron, *Electrochim. Acta*, **25**, 1329 (1980); (c) J. Hlavaty, J. Volke, *Electrochim. Acta*, **29**, 1399 (1984).

[118] Yu. B. Khokhryakov, I. A. Avrutskaya, *Elektrokhimiya*, **25**, 574 (1989).

[119] L. H. Klemm, Q. N. Porter, *J. Org. Chem.*, **46**, 2184 (1981).

[120] A. A. Konarev, V. Kh. Katunin, L. S. Pomogaeva, I. A. Avrutskaya, *Elektrokhimiya*, **23**, 991 (1987).

[121] M. Largeron, M.-B. Fleury, *Tetrahedron Lett.*, **32**, 631 (1991).

[122] C. Moinet, G. Simonneaux, M. Autret, F. Hindre, M. Le Plouzennec, *Electrochim. Acta*, **38**, 325 (1993).

[123] L. J. Nunez-Vergara, C. Matus, A. F. Alvarez-Lueje, B. K. Cassels, J. A. Squella, *Electroanalysis*, **6**, 509 (1994)

[124] C. Moinet, M. Autret, D. Floner, M. Le Plouzennec, G. Simonneaux, *Electrochim. Acta*, **39**, 673 (1994).

[125] D. Dumanovic, J. Jovanovic, D. Suznjevic, M. Erceg, P. Zuman, *Electroanalysis*, **4**, 889 (1992).

[126] C. Hasiotis, G. Kokkinidis, *Electrochim. Acta*, **37**, 1231 (1992).

[127] R. Panicucci, R. A. McClelland, *Can. J. Chem.*, **67**, 2128 (1989).

[128] S. S. Gordeichuk, V. N. Leibzon, A. G. Mayants, *Khim. Geterotsikl. Soedin.*, **1992**, 944.

[129] (a) Yu. M. Kargin, V. Z. Latypova, A. V. Supyrev, G. A. Bogoveeva, N. S. Sagitova, *Zh. Obshch. Khim.*, **54**, 836 (1984); (b) Yu. M. Kargin, V. Z. Latypova, A. V. Supyrev, V. V. Zhuikov, *Zh. Obshch. Khim.*, **54**, 1695 (1984).

[130] E. Laviron, A. Vallat, R. Meunier-Prest, *J. Electroanal. Chem.*, **379**, 427 (1994).

[131] J. H. Wagenknecht, *J. Org. Chem.*, **42**, 1836 (1977).

[132] C. Degrand, P.-L. Compagnon, G. Belot, D. Jacquin, *J. Org. Chem.*, **45**, 1189 (1980).

[133] S. Muralidharan, S. Chellammal, P. N. Anantharaman, *Bull. Electrochem.*, **7**, 222 (1991).

[134] (a) R. Pachori, S. C. Mishra, R. A. Mishra, *J. Electrochem. Soc. India*, **34**, 99 (1985). The references are cited there in; (b) V. N. Nikulin, V. N. Klochkova, *Sov. Electrochem.*, **8**, 481 (1972).

[135] (a) P. A. K. Nedungadi, A. Gupta, S. K. Mukherji, K. Zutshi, *J. Electrochem. Soc., India*, **35**, 203 (1986); (b) A. Gupta, K. Zutshi, S. N. Swami, *J. Electrochem. Soc., India*, **39**, 119 (1990); (c) Yu. M. Kargin, V. Z. Latypova, A. V. Supyrev, *Zh. Obshch. Khim.*, **52**, 2623 (1982).

[136] M. Jitaru, B. C. Toma, M. Toma, D. A. Lowy, *Bull. Electrochem.*, **11**, 573 (1995).

[137] M.-T. Escot, A.-M. Marte, P. Poullen, P. Martinet, *Bull. Soc. Chim. Fr.*, **1986**, 684.

[138] M.-T. Escot, A.-M. Marte, P. Poullen, P. Martinet, *Bull. Soc. Chim. Fr.*, **1986**, 548.

[139] S. S. Kucherov, I. A. Avrutskaya, *Elektrokhimiya*, **23**, 1141 (1987).

[140] A. Davidovic, D. Davidovic, I. Tabakovic, *J. Serb. Chem. Soc.*, **56**, 677 (1991).
[141] S. Antoniadou, A. D. Jannakoudakis, P. Karabinas, E. Theodoridou, *J. Electroanal. Chem. Interfacial Electrochem.*, **207**, 203 (1986).
[142] G. Marrosu, R. Petrucci, A. Trazza, *Electrochim. Acta*, **40**, 923 (1995).
[143] G. Jacob, C. Moinet, A. Tallec, *Electrochim. Acta*, **27**, 1417 (1982).
[144] G. Jacob, C. Moinet, A. Tallec, *Electrochim. Acta*, **28**, 635 (1983).
[145] (a) T. A. Arkhipova, I. A. Avrutskaya, M. Ya. Fioshin, *Elektrokhimiya*, **21**, 1367 (1985); (b) T. A. Arkhipova, I. A. Avrutskaya, *Elektrokhimiya*, **25**, 787 (1985); (c) T. A. Arkhipova, I. A. Avrutskaya, N. A. Nankov, *Elektrokhimiya*, **28**, 1038 (1992).
[146] J. L. Sadler, A. J. Bard, *J. Am. Chem. Soc.*, **90**, 1979 (1968).
[147] (a) K. G. Boto, F. G. Thomas, *Aust. J. Chem.*, **26**, 1251 (1973); (b) K. G. Boto, F. G. Thomas, *Aust. J. Chem.*, **24**, 975 (1971); (c) K. G. Boto, F. G. Thomas, *Aust. J. Chem.*, **26**, 1669 (1973).
[148] B. Mahdavi, J.-M. Chapuzet, L. Brossard, A. Martel, P. Los, G. Capuano, J. Lessard, *Proc.-Electrochem. Soc.*, **94-21**, 220 (1994).
[149] G. H. Aylward, J. L. Garnett, J. H. Sharp, *Anal. Chem.*, **39**, 457 (1967).
[150] C. Nishihara, h. Shindo, J. Hiraishi, *J. Electroanal. Chem. Interfacial Electrochem.*, **191**, 425 (1985).
[151] V. K. Rao, S. R. Ramadas, C. S. Venkatachalam, C. Kalidas, *Trans. SAEST*, **19**, 287 (1984).
[152] P. N. Gupta, A. Raina, *J. Indian Chem. Soc.*, **62**, 363 (1985).
[153] (a) H. M. Fahmy, H. A. Daboun, K. Azziz, *J. Chem. Soc., Perkin Trans. II*, **1983**, 425; (b) H. M. Fahmy, H. A. Daboun, *Electrochim. Acta*, **28**, 605 (1983).
[154] M. H. Elnagdi, H. M. Fahmy, M. A. Morsi, S. K. El-ees, *Indian J. Chem.*, **16B**, 295 (1978).
[155] H. M. Fahmy, H. A. Daboun, M. Abdel Azzem, *Can. J. Chem.*, **62**, 2904 (1984).
[156] (a) P. Venkata Ramana, D. N. Satyanarayana, L. K. Ravindranath, *J. Electrochem. Soc.*, **141**, 1114 (1994); (b) P. V. Ramana, D. Vasudevan, L. K. Ravindranath, *J. Indian Chem. Soc.*, **71**, 123 (1994); (c) P. V. Ramana, B. S. Suryanarayana, L. K. Ravindranath, D. Vasudevan, S. B. Rao, *Bull. Electrochem.*, **8**, 490 (1992).
[157] B. E. Elanadouli, A. O. Abdelhamid, A. S. Shawali, *J. Heterocyclic Chem.*, **21**, 1087 (1984).
[158] S. Darwish, H. M. Fahmy, M. A. Abdel Aziz, A. A. El Maghraby, *J. Chem. Soc., Perkin Trans. II*, **1981**, 344.
[159] (a) P. Venkata Ramana, B. Sathya Suryanarayana, L. K. Ravidranath, V. Seshagiri, S. Brahmaji Rao, *Proc. Indian natn. Sci. Acad.*, **58**, 375 (1992); (b) P. Venkata Ramana, B. Sathya Suryanarayana, L. K. Ravidranath, S. Brahnaji Rao, S. R. Ramadas, *Indian J. Chem.*, **29A**, 864 (1990).
[160] (a) B. Sathya Suryanarayana, P. Venkata Ramana, L. K. Ravidranath, V. Seshagiri, S. Brahnaji Rao, *Indian J. Chem.*, **29A**, 895 (1990); (b) P. Venkata Ramana, B. Sathya Suryanaraya, L. K. Ravidranath, V. Seshagiri, S. Brahnaji Rao, *J. Indian Chem. Soc.*, **67**, 730 (1990).
[161] (a) Y. Castrillejo, D. González, E. Barrado, R. Pardo, P. S. Batanero, *Bull. Soc. Chim. Fr.*, **127**, 609 (1990); (b) A. K. A. El Kader, M. G. A. El Wahed, G. El Sayed, *Acta Chim. Hung.*, **119**, 285 (1985).
[162] R. N. Goyal, A. Minocha, *J. Electroanal. Chem.*, **193**, 231 (1985).
[163] R. N. Goyal, A. Kumar, *Bull. Soc. Chim. Fr.*, **1987**, 577.
[164] H. M. Fahmy, M. E. Sharaf, M. A. Aboutabl, M. M. M. Ramiz, M. Abdel El-Azzem, H. A. El-Rahman, *J. Chem. Soc., Perkin Trans. II*, **1990**, 1607.
[165] (a) H. Alt, J. Cramer, *Chem. Ing. Tech.*, **52**, 58 (1980); (b) M. Kottaisamy, S. Thangavelu, P. Subbiah, G. Chandramohan, *Bull. Electrochem.*, **6**, 722 (1990).
[166] M. Kottaisamy, S. Thangavelu, R. V. Raju, G. Chandramohan, *Bull. Electrochem.*, **7**, 24 (1991).
[167] M. Turowska, G. Mloston, J. Raczak, *Pol. J. Chem.*, **67**, 1105 (1993).
[168] B. Venkataraman, P. Thirunavukkarasu, K. S. Udupa, S. J. Arulraj, *Trans. SAEST*, **21**, 33 (1986)
[169] R. N. Goyal, *Indian J. Chem.*, **27**, 858 (1988).
[170] A. J. Bellamy, I. S. MacKirdy, C. E. Niven, *J. Chem. Soc., Perkin Trans. II*, **1983**, 183.
[171] A. J. Bellamy, *J. C. S., Chem. Commun.*, **1975**, 944.
[172] (a) R. N. Goyal, S. K. Srivastava, R. Agarwal, *Bull. Soc. Chim. Fr.*, **1985**, 656; (b) R. Jain, N. Dua, *Bull. Electrochem.*, **7**, 224 (1991); (c) R. N. Goyal, S. K. Srivastava, R. Agarwal, *Trans. SAEST*, **20**, 51 (1985).
[173] R. Jain, U. Jain, *Bull. Electrochem.*, **11**, 326 (1995).
[174] T. Troll, M. M. Baizer, *Electrochim. Acta*, **20**, 33 (1975).
[175] Y. Mugnier, E. Laviron, *New J. Chem.*, **13**, 579 (1989).
[176] B. Roloff, W. jugelt, W. Duczek, *Z. Chem.*, **26**, 438 (1986).
[177] P. Fuchs, U, Hess, H. H. Holst, H. Lund, *Acta Chem. Scand.*, **35**, 185 (1981).
[178] R. C. Hallche, M. M. Baizer, *Liebigs Ann. Chem.*, **1977**, 737.
[179] P. Jeroschewski, *Z. Chem.*, **269**, 452 (1989).
[180] S. U. Pedersen, H. Lund, *Acta Chem. Scand.*, **B42**, 319 (1988).
[181] H. Lund, *The Chemistry of Carbon-Nitrogen Bond*, S. Patai (ed.), -Interscience, Chapt. 11, London (1970).
[182] V. T. Novikov, I. A. Avrutskaya, I. I. Surov, L. G. Petrova, E. S. Orekhova, *Elektrokhimiya*, **28**, 527 (1992).

[183] N. Ayyaswami, *Trans. SAEST*, **21**, 13 (1986).

[184] A. Kunai, M. Ikemoto, K. Sasaki, *Denki Kagaku*, **58**, 957 (1990).

[185] Y. Kokusenya, S. Nakajima, M. Matsuoka, *Denki Kagaku*, **55**, 853 (1987).

[186] Y. Kokusenya, S. Nakajima, M. Matsuoka, *Denki Kagaku*, **55**, 235 (1987).

[187] A. Boulmedais, M. Jubault, A. Tallec, *Tetrahedron*, **45**, 5510 (1989).

[188] (a) K. Yoshida, M. Sueoka, *Japanese Patent Kokai,* 61-09587 (1986); (b) K. Yoshida, M. Sueoka, *Japanese Patent Kokai,* 61-09586 (1986).

[189] A. Muthukumaran, V. Krishnan, *Bull. Electrochem.*, **8**, 276 (1992).

[190] Yu. D. Smirnov, A. P. Tomilov, *Russ. J. Electrochem.*, **32**, 106 (1996).

[191] G. K. Tusupbekova, *Russ. J. Electrochem.*, **32**, 139 (1996).

[192] M. Ya. Fioshin, I. A. Avrutskaya, I. I. Surov, V. T. Novikov, *Collect. Czech. Chem. Commun.*, **52**, 182 (1987).

[193] C. S. Reddy, S. J. Reddy, *J. Indian Chem. Soc.*, **68**, 610 (1991).

[194] M. V. Yarosh, N. I. Shipitsina, V. N. Leibzon, *Elektrokhimiya*, **25**, 1509 (1989).

[195] G. Ryan, J. H. P. Utlay, *Tetrahedron Lett.*, **29**, 3699 (1988).

[196] M. A. Gomez-Nieto, M. D. Luque de Castro, M. Vacarcel, *Electrochim. Acta*, **28**, 325 (1983).

[197] T. Shono, Y. Matsumura, K. Tsubata, T. Kamada, K. Kishi, *J. Org. Chem.*, **54**, 2249 (1989).

[198] F. Mánok, L. Balás, C. Várhelyi, *Acta Chim. Hung.*, **127**, 629 (1990).

[199] G. Kokkinidis, G. Papanastasiou, C. Hasiotis, N. Papadopoulos, *J. Electroanal. Chem. Interfacial Electrochem.*, **309**, 263 (1991).

[200] C. Hasiotis, G. Kokkinidis, *Electrochim. Acta*, **39**, 639 (1994).

[201] R. Forreza, V. Cerda, *Bull. Soc. Chim. Fr.*, **1984**, I-97.

[202] R. Jain, U. Jain, *J. Electroanal. Chem. Interfacial Electrochem.*, **313**, 259 (1991), the related literature cited therein.

[203] H. Lund, *Electrochim. Acta*, **28**, 395 (1983).

[204] (a) R. Jain, H. K. Pardasani, *J. Electrochem. Soc. India*, **41-3** 175 (1992); (b) R. N. Goyal, A. Minocha, *J. Indian Chem. Soc.*, **62**, 202 (1985).

[205] Yu. M. Kargin, V. Z. Latypova, M. Yu. Kitaeva, A. A. Vafina, R. M. Zaripova, A. V. Il'yasov, *Izv. Akad. Nauk SSSR, Ser. Khim.*, **1984**, 2410, the related references cited therein.

[206] G. M. Abou-Elenien. N. A. Ismail, T. S. Hafez, *Bull. Chem. Soc. Jpn.*, **64**, 651 (1991).

[207] (a) R. N. Goyal, A. Minocha, *J. Electroanal. Chem. Interfacial Electrochem.*, **172**, 373 (1984); (b) N. C. Mathur, R. N. Goyal, W. U. Malik, *Indian J. Chem.*, **29A**, 765 (1990); (c) W. U. Malik, R. N. Goyal, M. Rajeshwari, *Bull. Soc. Chim. Fr.*, **1987**, 78.

[208] M. A. Gomez-Nieto, M. D. Luque de Castro, M. Vacarcel, *Electrochim. Acta*, **28**, 1725 (1983).

[209] Yu. P. Kitaev, T. V. Troepol'skaya, L. V. Ermolaeva, E. N. Munin, *Izv. Akad. Nauk SSSR, Ser. Khim.*, **1985**, 1736.

[210] I. I. Surov, I. A. Avrutskaya, E. Yu. Kodintseva, M. Ya. Fioshin, *Elektrokhimiya*, **20**, 1276 (1984).

[211] V. Kameswara Rao, C. S. Venkatachalam, C. Kalidas, *Bull. Chem. Soc. Jpn.*, **61**, 612 (1988).

[212] B. A. Abd-El-Nabey, M. A. M-Nassr, S. Hamidona, *Bull. Electrohem.*, **2**, 71 (1986).

[213] A. Z. Abu Zuhri, J. S. Shalabi, *J. Chem. Soc., Perkin Trans. II*, **1985**, 499.

[214] A. Z. Abu Zuhri, J. S. Shalabi, *Monatsh. Chem.*, **118**, 1335 (1987).

[215] H. M. Fahmy, G. E. H. Elgemeie, M. A. Aboutabl, B. N. Barsoum, Z. M. El-Massry, *Indian J. Chem.*, **33B**, 859 (1994).

[216] (a) H. M. Fahmy, M. A. W. Negeid, *Gazz. Chim. Ital.*, **116**, 63 (1986); (b) H. M. Fahmy, H. A.-R. Ead, M. A.-Wahab, *J. Chem. Soc., Perkin Trans. II*, **1985**, 45.

[217] F. Salinas, A. Sanchez Misiego, J. J. Berzas Nevado, A. Espinosa, P. Valiente, *Ann. Chim.*, **75**, 163 (1985).

[218] D. E. Herbranson, F. J. Theisen, M. D. Hawley, R. N. McDonald, *J. Am. Chem. Soc.*, **105**, 2544 (1983).

[219] D. A. Van Galen, J. H. Barnes, M. D. Hawley, *J. Org. Chem.*, **51**, 2544 (1986).

[220] Yu. D. Smirnov, A. P. Tomilov, *Zh. Org. Khim.*, **28**, 51 (1992).

[221] S. Zhan, M. D. Hawley, *J. Electroanal. Chem. Interfacial Electrochem.*, **319**, 275 (1991).

[222] C. J. Patil, A. S. Madhav, G. Ramachandriah, D. N. Vyas, *Indian J. Chem.*, **33A**, 1037 (1994).

[223] C. Sridevi, S. J. Reddy, *Proc. Natl. Acad. Sci., Ind., Sect. A*, **59**, 507 (1989).

[224] C. J. Patil, A. S. Madhava, D. N. Vyas, *Bull. Electrochem.*, **9**, 95 (1993).

[225] H. P. Richter, P. Pflegel, F. Fülöp, G. Bernáth, *Pharmazie*, **41**, 432 (1986).

[226] S. Torii, H. Tanaka, M. Satsuki, T. Siroi, N. Saito, M. Sasaoka, J. Nokami, *Chem. Lett.*, **1981**, 1575.

[227] C. M. Ryan, E. K-Miller, *Tetrahedron*, **44**, 6807 (1988).

[228] R. Hazard, A. Tallec, R. Tardivel, *Electrochim. Acta*, **35**, 1907 (1990).

[229] (a) M. Jubault, A. Tallec, B. Bujoli, J.-C. Rozé, J.-P. Pradére, *Tetrahedron Lett.*, **26**, 745 (1985); (b) B. Bujoli, M. Chehna, M. Jubault, A. Tallec, *J. Electroanal. Chem. Interfacial Electrochem.*, **199**, 461 (1986).

[230] A. Abouelfida, J.-C. Roze, J.-P. Pradere, M. Jubault, *Phosphorus, Sulfur, Silicon Relat. Elem.*, **54**, 123 (1990).

[231] R. Fuhlendorff, H. Lund, *Acta Chem. Scand.*, **B42**, 52 (1988).

[232] U. Hess, A. K. Raasch, M. Schulze, *J. Prakt. Chem./Chem-Ztg.*, **334**, 487 (1992).

[233] R. C. Duty, M. Garrosian, *J. Electrochem. Soc.*, **130**, 1848 (1983).

[234] K. Hu, M. E. Niyazymbetov, D. H. Evans, *Tetrahedron Lett.*, **36**, 7027 (1995).

[235] F. Barba, J. R. Diaz, *J. Org. Chem.*, **57**, 4287 (1992).

[236] M. A. Morsi, A. M. A. Helmy, H. M. Fahmy, *J. Electroanal. Chem.*, **148**, 133 (1983).

[237] M. Jubault, A. Lebouc, A. Tallec, *Electrochim. Acta*, **27**, 1339 (1982).

[238] A. Pricken, F. Fülöp, P. Pflegel, G. Bernáth, *Pharmazie*, **44**, 454 (1989).

[239] A. Pricken, G. Stájer, A. E. Szabó, F. Fülöp, P. Pflegel, G. Bernáth, *Pharmazie*, **45**, 568 (1990).

[240] M. V. Mirifico, J. A. Caram, E. J. Vasini, *Electrochim. Acta*, **36**, 167 (1991).

[241] C. Bellec, N. Vinot, P. Maitte, *J. Heterocyclic Chem.*, **23**, 491 (1986).

[242] N. Vinot, C. Bellec, P. Maitte, *J. Heterocyclic, Chem.*, **27**, 1917 (1990).

[243] A. Pricken, P. Franke, I. Huber, F. Fulop, P. Pflegel, G. Bernath, *Pharmazie*, **45**, 740 (1990).

[244] J. Armand, Y. Armand, L. Boulares, *J. Heterocyclic Chem.*, **22**, 1519 (1985).

[245] L. Thomas, J. L. Vilchez, G. Crovetto, J. Thomas, *Proc. Indian Acad. Sci.*, **98**, 221 (1987).

[246] P. Pelegel, C. Kühmstedt, C. Stolpe, P. Richter, O. Morgenstern, *Pharmazie*, **44**, 825 (1989).

[247] P. Pflegel, C. Kühmstedt, M. Mann, P. Richter, *Pharmazie*, **40**, 710 (1985).

[248] C. Triballet, P. Boucly, M. Guernet, *Bull. Soc. Chim. Fr.*, **1981**, I-113.

[249] P. Hudhomme, V. Poisson, E. Raoult, J.-P. Pradère, A. Tallec, M. Jubault, G. Duguay, *Bull. Soc. Chim. Fr.*, **131**, 816 (1994).

[250] L. Horner and D. H. Skaletz, *Liebigs Ann. Chem.*, **1975**, 1210.

[251] D. K. Root and W. H. Smith, *J. Electrochem. Soc.*, **129**, 1231 (1982).

[252] R. Engels, C. J. Smit, W. J. M. van Tilborg, *Angew. Chem. Int. Ed. Engl.*, **22**, 492 (1983).

[253] G. Silvestri, S. Gambino, G. Filardo, F. Tedeschi, *J. Appl. Electrochem.*, **19**, 946 (1989).

[254] C. Degrand, C. Grosdemouge, P.-L. Compagnon, *Tetrahedron Lett.*, **1978**, 3023.

[255] H. Lund, J. Simonet, *Bull. Soc. Chim. Fr.*, **1973**, 1843.

[256] H. Lund, J. Simonet, *C. R. Acad. Sci. Paris*, **t. 275**, 837 (1972).

[257] T. Shono, N. Kise, N. Kunimi, R. Nomura, *Chem. Lett.*, **1991**, 2191.

[258] J. Komenda, R. Fiala, U. Hess, *Z. Phys. Chem.*, **268**, 48 (1987).

[259] H. M. Fahmy, B. Nashed, F. A. Khalifa, A. A. Moneim, *J. Electroanal. Chem.*, **184**, 147 (1985).

[260] R. W. Koch, R. E. Dessy, *J. Org. Chem.*, **47**, 4452 (1982).

[261] J. M. Mellor, B. S. Pons, J. H. A. Stibbard, *J. Chem. Soc., Perkin Trans. I*, **1981**, 3097.

[262] D. Michelet, *Fr. Demande* FR. 2444030 (1978).

[263] T. Shono, N. Kise, E. Okazaki, *Tetrahedron Lett.*, **33**, 3347 (1992).

[264] B. Bujolo, M. Jubault, J.-C. Roze, A. Tallec, *Tetrahedron*, **43**, 2709 (1987).

[265] J. Vera, F. Martinez-Ortiz, A Ripoll, P. Molina, *Electrochim. Acta*, **31**, 1231 (1986).

[266] M. A. Abdel Aziz, H. M. Fahmy, *J. Chin. Chem. Soc.*, **37**, 411 (1990).

[267] P. Raveendra Reddy, S. Brahmaji Rao, *Bull. Electrochem.*, **7**, 173 (1991).

[268] R. Abdel-Hamid, *Bull. Soc. Chim. Fr.*, **1986**, 390.

[269] M. A. Morsi, A. M. A. Helmy, *J. Electroanal. Chem. Interfacial Electrochem.*, **148**, 123 (1983).

[270] D. Vasudevan, C. S. Venkatachalam, C. Kalidas, *Ber. Bunsenges. Phys. Chem.*, **95**, 1633 (1991).

[271] R. N. Goyal, R. Bhushan, A. Agarwal, *Indian J. Chem.*, **24A**, 435 (1985).

[272] H. M. Fahmy, M. Abdel Azzem, M. Abdul Wahab, N. Abdel Fattah, M. A. Aboutabl, *Bull. Chem. Soc. Jpn.*, **62**, 2650 (1989).

[273] R. N. Goyal, A. Minocha, U. R. Singh, *Indian J. Chem.*, **24**, 661 (1985).

[274] V. G. Mairanovsky, *Angew. Chem. Int. Ed. Engl.*, **15**, 281 (1976).

[275] V. G. Mairanovsky, S. Ya. Melik, A. Ya. veinberg, G. I. Samokhvalov, *Avt. Svid.*, N 226628 (1966), see also Ref. 274) and the references cited therein.

[276] D. Archier-Jay, N. Besbes, A. Laurent, E. Laurent, S. Lesniak, R. Tardivel, *Bull. Soc. Chim. Fr.*, **1989**, 537.

[277] I. Han, H. Kohn, *J. Org. Chem.*, **56**, 4648 (1991).

[278] M. I. Montenegro, *Electrochim. Acta*, **31**, 607 (1986).

[279] L. Grehn, K. Gunnarsson, H. L. S. Maia, M. I. Montenegro, L. Pedro, U. Ragnarsson, *J. Chem. Research*, **1988**, 399.

[280] (a) H. L. S. Maia, M. J. Medeirs, M. I. Montenegro, D. Court, D. Pletcher, *J. Electroanal. Chem. Interfacial Electrochem.*, **164**, 347 (1984); (b) H. L. S. Maia, M. J. Medeirs, M. I. Montenegro, D. Pletcher, *J. Electroanal. Chem., Interfacial Electrochem.*, **200**, 363 (1986).

[281] A. M. Martre, G. Mousset, M. Prudhomme, E. Rodrigues-Pereira, *Electrochim. Acta*, **40**, 1805 (1995).

[282] H. L. S. Maia, L. S. Monteiro, F. Degerbeck, L. Grehn, U. Ragnarsson, *J. Chem. Soc., Perkin Trans. II*, **1993**, 495.

[283] P. Hudhomme, V. Poisson, E. Raoult, J.-P. Pradère, A. Tallec, M. Jubault, G. Duguay, *Bull. Soc. Chim. Fr.*, **131**, 816 (1994).

[284] D. J. Chadwick, S. T. Hodgson, *J. Chem. Soc., Perkin Trans. I*, **1983**, 93.
[285] A. A. Volod'kin, V. V. Ershov, R. D. Malysheva, *Izv. Akad. Nauk SSSR, Ser. Khim.*, **1985**, 1187.
[286] S. Torii, H. Okumoto, H. Satoh, T. Minoshima, S. Kurozumi, *Synlett*, **1995**, 439.
[287] (a) O. Orange, C. Elfakir-Hamet, C. Caullet, *J. Electrochem. Soc.*, **128**, 1889 (1981); (b) C. Elfakir-Hamet, C. Caullet, *Bull. Soc. Chim. Fr.*, **1986**, 688.
[288] P. Rüetschi, G. Trümpler, *Helv. Chim. Acta*, **36**, 1649 (1953).
[289] E. P. Koval'chuk, N. D. Obushak, N. I. Gaunshchak, P. Yanderka, *Zh. Obshch. Khim.*, **56**, 1891 (1986).
[290] R. M. Elofson, F. F. Gadallah, K. F. Schulz, *J. Org. Chem.*, **36**, 1526 (1971).
[291] E. Ahlberg, B. Helgée, V. D. Parker, *Acta Chem. Scand.*, **B34**, 181 (1980).
[292] R. M. Elofson, F. F. Gadallah, K. F. Schulz, J. K. Laidler, *Can. J. Chem.*, **62**, 1772 (1984).
[293] N. I. Gaunshchak, N. D. Obushak, E. P. Koval'chuk, G. V. Trifonova, *Zh. Obshch. Khim.*, **54**, 2334 (1984).
[294] M. Delamar, R. Hitmi, J. Pinson, J. M. Savéant, *J. Am. Chem. Soc.*, **114**, 5883 (1992)
[295] (a) R. N. McDonald, F. M. Triebe, J. R. January, K. J. Borhani, M. D. Hawley, *J. Am. Chem. Soc.*, **102**, 7867 (1980); (b) S. Cheng, M. D. Hawley, *J. Org. Chem.*, **51**, 3799 (1986); (c) V. D. Parker, D. Bethell, *Acta Chem. Scand.*, **B35**, 72 (1981); (d) D. Bethell, V. D. Parker, *J. Chem. Soc., Perkin Trans. II*, **1982**, 841; (e) K. Ishiguro, M. Ikeda, Y. Sawaki, *Chem. Lett.*, **1991**, 511.
[296] (a) V. D. Parker, D. Bethell, *Acta Chem. Scand.*, **B34**, 617 (1980); (b) V. D. Parker, D. Bethell, *Acta Chem. Scand.*, **B35**, 691 (1981); (c) D. Bethell, V. D. Parker, *J. Am. Chem. Soc.*, **108**, 895 (1986).
[297] (a) D. Bethell, L. J. Mcdowall, V. D. Parker, *J. Chem. Soc., Chem. Commun.*, **1984**, 308; (b) D. Bethell, V. D. Parker, *J. Am. Chem. Soc.*, **108**, 7194 (1986).
[298] S. Kuwabata, Y. Hozumi, K. Tanaka, T. Tanaka, *Chem. Lett.*, **1985**, 401.
[299] J.-C. Moutet, A. Ourari, A. Zouaoui, *Electrochim. Acta*, **37**, 1261 (1992).
[300] W. Kawzynski, B. Czochralska, D. Shugar, *Electrochim. Acta*, **40**, 213 (1993).
[301] D. Knittel, *Monatsh. Chem.*, **119**, 379 (1988).
[302] T. Arai, T. Shingaki, M. Inagaki, *Chem. Lett.*, **1981**, 765.
[303] (a) D. Knittel, *Monatsh. Chem.*, **117**, 679 (1986); (b) D. Knittel, *Monatsh. Chem.*, **116**, 1133 (1985); (c) D. Knittel, *Monatsh. Chem.*, **117**, 491 (1986); (d) D. Knittel, V. Suryanarayana Rao, *Monatsh. Chem.*, **117**, 1185 (1986); (e) D. Knittel, V. Suryanarayana Rao, *Monatsh. Chem.* **119**, 223 (1988).
[304] D. E. Herbranson, M. D. Hawley, *J. Org. Chem.*, **55**, 4297 (1990).
[305] G. C. Barone III, H. B. Halsall, W. R. Heineman, *Anal. Chim. Acta*, **248**, 399 (1991).
[306] T. L. Antonova, L. N. Ivanovskaya, M. Ya. Foishin, N. N. Savushkina, *Elektrokhimiya*, **23**, 606 (1987).
[307] A. Muthukumaran, V. Krishnan, *Bull. Electrochem.*, **7**, 410 (1991).
[308] (a) Y. Song, P. N. Pintauro, *J. Appl. Electrochem.*, **21**, 21 (1991); (b) T. Chiba, M. Okimoto, H. Nagai, Y. Tanaka, *Bull. Chem. Soc. Jpn.*, **56**, 719 (1983); (c) A. P. Tomilov, S. L. Varshavskii, M. T. Kulikov, Yu. D. Smirnov, *Khim. Prom.*, **41**, 329 (1965); (d) A. P. Tomilov, I. V. Kirilyus, I. P. Andriyanova, *Sov. Electrochem.*, **8** 1050 (1972).
[309] T. Yamada, N. Fujimoto, T. Matsue, T. Osa, *Denki Kagaku*, **56**, 175 (1988) (in Japanese).
[310] N. A. Vishnyakova, I. A. Avrutskaya, S. S. Kucherov, Yu. M. Chunaev, *Elektrokhimiya*, **28**, 287 (1992).
[311] J. E. Toomey, Jr., G. A. Chaney, M. Wilcox, *Stud. Org. Chem.*, **30**, 245 (1987).
[312] (a) T. I. Antonova, L. N. Ivanovskaya, I. A. Avrutskaya, M. Ya. Fioshin, L. I. Grobacheve, *Elektrokhimiya*, **22**, 546 (1986); (b) T. I. Antonova, N. N. Savushkina, L. N. Ivanovskaya, I. A. Avrutskaya, O. V. Maksimova, M. Ya. Fioshin, *Elektrokhimiya*, **23**, 294 (1987).
[313] M. A. Petit, V. Plichon, H. Belkacemi, *New J. Chem.*, **13**, 459 (1989).
[314] T. Kubota, S. Hiramatsu, K. Kano, B. Uno, H. Miyasaki, *Chem. Pharm. Bull.*, **32**, 3830 (1984).
[315] T. Shono, J. Terauchi, K. Kitayama, Y. Takeshima, Y. Matsumura, *Tetrahedron*, **48**, 8253 (1992).
[316] H. M. Colquhoun, A. E. Crease, S. A. Taylor, *J. Chem. Soc., Chem. Commun.*, **1983**, 1158.
[317] J. Nadra, H. Givadinovitch, M. Devaud, *J. Chem. Research*, **1983**, 192.
[318] I. V. Kirilyus, V. I. Filimonova, A. N. Negoda, *Russ. J. Electrochem.*, **32**, 136 (1996).
[319] Yu. D. Smirnov, É. N. Shitova, G. F. Shaidulina, A. P. Tomilov, V. F. Pavlichenko, *Elektrokhimiya*, **28**, 1073 (1992).
[320] O. R. Brown, R. J. Butterfield, J. P. Millington, *Electrochim. Acta*, **27**, 1655 (1982).
[321] J. O'Reilly, P. Elving, *J. Am. Chem. Soc.*, **94**, 7941 (1972).
[322] G. S. Cano, V. Montiel, A. Aldaz, *Bull. Electrochem.*, **6**, 931 (1990).
[323] G. Horanyi, *J. Electroanal. Chem. Interfacial Electrochem.*, **284**, 481 (1990).
[324] W. H. Smith, A. J. Bard, *J. Am. Chem. Soc.*, **97**, 6491 (1975).
[325] (a) M. R. Montoya, J. M. R. Mellado, R. M. Galvin, *J. Electroanal. Chem. Interfacial Electrochem.*, **293**, 185 (1990); (b) M. A. Zon, M. Angulo, J. M. Rodriguez Mellado, *J. Electroanal. Chem.*, **365**, 213 (1994); (c) W. Z.-Hao, H. Z.-Bin, *Electrochim. Acta*, **30**, 779 (1985).
[326] M. Angulo, J. M. Rodriguez Mellado, *Denki Kagaku*, **62**, 1148 (1994).
[327] O. R. Brown, J. A. Harrison, K. S. Sastry, *J. Electroanal. Chem.*, **58**, 387 (1975).
[328] G. Veerabhadram, K. S. Sastry, *J. Electrochem. Soc. India*, **34**, 30 (1985).

[329] E. Viriot, P. Boucly, E. Samuel, *New J. Chem.*, **10**, 265 (1986).
[330] H. Fujita, H. O.-Nishiguchi, *J. Chem. Soc., Chem. Commun.*, **1989**, 1091.
[331] N. W. Koper, S. A. Jonker, J. W. Verhoeven, *Recl. Trav. Chim. Pays-Bas*, **104**, 296 (1985).
[332] J. M. R. Mellado, R. M. Galvin, *Bull. Electrochem.*, **7**, 142 (1991).
[333] S. Roffia, V. Concalini, C. Paradisi, *J. Electroanal. Chem. Interfacial Electrochem.*, **302**, 115 (1991).
[334] R. Gottlieb, W. Pfleiderer, *Liebigs Ann. Chem.*, **1981**, 1451.
[335] L. J. Nunez-Vergara, J. A. Squella, M. Dominguez, M. Blazquez, *J. Electroanal. Chem. Interfacial Electrochem.*, **243**, 133 (1988).
[336] H. Lund, J. Simonet, *C. R. Acad. Sc. Paris*, **t. 277**, 1387 (1973).
[337] J. Pinson, J. Armand, *Collect. Czech. Chem. Commun.*, **36**, 585 (1971).
[338] J. Armand, L. Boulares, *Can. J. Chem.*, **66**, 1500 (1988).
[339] J. Armand, K. Chekir, J. Pinson, *Can. J. Chem.*, **56**, 1804 (1978).
[340] (a) R. C. Gurira, L. D. Bowers, *J. Electroanal. Chem. Interfacial Electrochem.*, **146**, 109 (1983); (b) R. C. Gurira, C. Montgomery, R. Winston, *J. Electroanal. Chem.*, **333**, 217 (1992).
[341] T. Priyamvada Devi, C. Kalidas, C. S. Venkatachalam, *Electrochim. Acta*, **28**, 1161 (1983).
[342] J. Armand, C. Bellec, L. Boulares, P. Chaquin, D. Masure, J. Pinson, *J. Org. Chem.*, **56**, 4840 (1991).
[343] J. Armand, L. Boulares, C. Bellec, C. Bois, M. P.-Levisalles, J. Pinson, *Can. J. Chem.*, **62**, 1028 (1984).
[344] J. Armand, L. Boulares, C. Bellec, J. Pinson, *Can. J. Chem.*, **60**, 2797 (1982).
[345] S. J. Gumbley, T. W. S. Lee, R. Stewart, *J. Heterocyclic Chem.*, **22**, 1143 (1985).
[346] J. E. Coffield, G. Mamantov, S. P. Zingg, G. P. Smith, *J. Electrochem. Soc.*, **138**, 2543 (1991).
[347] C. Bellec, M.-C. Bellassoued-Fargeau, B. Graffe, M.-C. Sacquet, J. Pinson, *J. Heterocyclic Chem.*, **27**, 551 (1990).
[348] C. Vaccher, P. Berthelot, M. Debaert, A. Darchen, J. L. Burgot, G. Evrard, F. Durant, *J. Chem. Soc., Perkin Trans. II*, **1989**, 391.
[349] A. El Jammal, E. Graf, M. Gross, *Electrochim. Acta*, **31**, 1457 (1986).
[350] J. Armand, K. Chekir, N. Ple, G. Queguiner, M. P. Simonnin, *J. Org. Chem.*, **46**, 4754 (1981).
[351] A. Nurhadi, E. Graf, M. Gross, *Electrochim. Acta*, **36**, 1997 (1991).
[352] J. Armand, L. Boulares, C. Bellec, J. Pinson, *Can. J. Chem.*, **65**, 1619 (1987).
[353] T. Robin, J. Chaussard, M. Troupel, P. Guitton, *Fr. Demande*, Fr 2629474 (1989).
[354] (a) M. Finkelstein, R. C. Petersen, S. D. Ross, *J. Am. Chem. Soc.*, **81**, 2361 (1959); (b) S. D. Ross, M. Finkelstein, R. C. Petersen, *J. Am. Chem. Soc.*, **82**, 1582 (1960).
[355] A. Thompson, *Stud. Org. Chem.*, **30**, 449 (1987).
[356] (a) Y. Takahashi, T. Kurozumi, *Japanese Patent Kokai*, 60-131986 (1985); (b) Y. Takahashi, T. Kurozumi, *Japanese Patent Kokai*, 60-131985 (1985).
[357] S. Shimizu, *Japanese Patent Kokai*, 60-100690 (1985).
[358] (a) V. Svetlicic, E. Kariv-Miller, *J. Electroanal. Chem. Interfacial Electrochem.*, **209**, 91 (1986); (b) E. Kariv-Miller, V. Svetlicic, *J. Electroanal. Chem. Interfacial Electrochem.*, **205**, 319 (1986); (c) E. Kariv-Miller, C. Nanjundiah, J. Eaton, K. E. Swenson, *J. Electroanal. Chem. Interfacial Electrochem.*, **167**, 141 (1984); (d) E. Kariv-Miller, R. Andruzzi, *J. Electroanal. Chem. Interfacial Electrochem.*, **187**, 175 (1985); (e) E. Kariv-Miller, *Stud. Org. Chem.*, **30**, 283 (1987).
[359] T. Shono, Y. Matsumura, S. Kashimura, *J. Chem. Research*, **1984**, 216.
[360] P. Martigny, J. Simonet, *J. Electroanal. Chem. Interfacial Electrochem.*, **101**, 275 (1979).
[361] E. K.-Miller, P. B. Lawin, Z. Vajtner, *J. Electroanal. Chem. Interfacial Electrochem.*, **195**, 435 (1985).
[362] S. G. Mairanovskii, *Dokl. Akad. Nauk SSSR*, **110**, 593 (1956).
[363] G. F. Shaidulina, T. N. Khomchenko, A. L. Dribin, L. N. Nekrasov, A. P. Tomilov, V. D. Simonov, É. N. Shitova, Ya. B. Yasman, *Elektrokhimiya*, **22**, 1062 (1986).
[364] J. Volke, J. Urban, V. Volkeova, *Electrochim. Acta*, **39**, 2049 (1994).
[365] I. Carelli, *Electrochim. Acta*, **35**, 1185 (1990).
[366] L. Greci, A. Alberti, I. Carelli, A. Trazza, A. Casini, *J. Chem. Soc., Perkin Trans. II*, **1984**, 2013.
[367] M. Lipsztajin, R. A. Osteryoung, *Electrochim. Acta*, **29**, 1349 (1984).
[368] S. Hunig, B. Ort, H. Wenner, *Liebigs Ann. Chem.*, **1985**, 751.
[369] J. Grimshaw, S. Moore, N. Thompson, J.-T. Grimshaw, *J. Chem. Soc., Chem. Commun.*, **1983**, 783.
[370] J. Grimshaw, S. Moore, J. T.-Grimshaw, *Acta Chem. Scand.*, **37**, 485 (1983).
[371] R. J. Gale, R. A. Osteryoung, *J. Electrochem. Soc.*, **127**, 2167 (1980).
[372] (a) I. Carelli, M. E. Cardinali, *J. Org. Chem.*, **41**, 3967 (1976); (b) M. Schwarz, J. Hermolin, E. K.-Eisner, *J. Electroanal. Chem. Interfacial Electrochem.*, **220**, 139 (1987).
[373] S. K.-Kaplan, J. Hermolin, E. K.-Eisner, *J. Electrochem. Soc.*, **128**, 802 (1981).
[374] T. Matsue, T. Kato, U. Akiba, T. Osa, *Chem. Lett.*, **1985**, 1825.
[375] (a) F. R. Yu, Y. Y. Wang, C. C. Wan, *Electrochim. Acta*, **30**, 1693 (1985); (b) C. S. Lin, Y. Y. Wang, C. C. Wan, *J. Chin. Inst. Eng.*, **10**, 385 (1987).
[376] N. Burke, K. C. Molloy, T. G. Purcell, M. R. Smyth, *Inorg. Chim. Acta*, **106**, 129 (1985).

[377] (a) M. Kijima, A. Sakawaki, T. Sato, *Chem. Lett.*, **1991**, 499; (b) M. Kijima, A. Sakawaki, T. Sato, *Bull. Chem. Soc. Jpn.*, **67**, 2323 (1994).

[378] J.-F. André, M. S. Wrighton, *Inorg. Chem.*, **24**, 4288 (1985).

[379] (a) L. A. Avaca, J. H. P. Utley, *J. Chem. Soc., Perkin Trans. II*, **1975**, 161; (b) L. A. Avaca, J. H. P. Utley, *J. Chem. Soc., Perkin Trans. I*, **1975**, 971; (c) J. Delaunay, A. Lebonc, G. LeGuillanton, L. M. Gromes, J. Simonet, *Electrochim. Acta*, **27**, 287 (1982); (d) S. U. Pedersen, H. Lund, *12th Sandbjerg Meeting, Extended Abstract*, 93 (1985); (e) H. Lund, *Stud. Org. Chem.*, **30**, 179 (1987).

[380] H. Ishida, T. Ohba, T. Yamaguchi, K. Ohkubo, *Chem. Lett.*, **1994**, 905.

[381] J. W. Park, Y. Kim, C. Lee, *Bull. Korean Chem. Soc.*, **15**, 896 (1994).

[382] J. A. Barltrop, A. C. Jackson, *J. Chem. Soc., Perkin Trans. II*, **1984**, 367.

[383] H. X. Wang, T. Sagara, H. Sato, K. Niki, *J. Electroanal. Chem. Interfacial Electrochem.*, **331**, 925 (1992).

[384] X. Marguerettaz, G. Redmond, S. N. Rao, D. Fitzmaurice, *Chem. Eur. J.*, **2**, 420 (1996).

[385] S. Hunig, H. Berneth, *Top. Curr. Chem.*, **92**, 1 (1980).

[386] K. Takahashi, T. Nihira, K. Akiyama, Y. Ikegami, E. Fukuyo, *J. Chem. Soc., Chem. Commun.*, **1992**, 620.

[387] A. Deronzier, B. Galland, M. Vieira, *Electrochim. Acta*, **28**, 805 (1983).

[388] P. Crouigneau, O. Enea, C. Lamy, *New J. Chem.*, **10**, 539 (1986).

[389] C. Lee, C. Kim, J. W. Park, *J. Electroanal. Chem.*, **374**, 115 (1994).

[390] A. Schulz, W. Kaim, *Chem. Ber.*, **124**, 129 (1991).

[391] (a) J. Stradins, J. Benders, V. Kadysh, *Electrochim. Acta*, **31**, 637 (1986); (b) J. Benders, V. Kadysh, E. Lavrinovics, J. Stradins, *Electrochim. Acta*, **31**, 1369 (1986).

[392] Y. Yamashita, T. Hanaoka, Y. Takeda, T. Mukai, T. Miyashi, *Bull. Chem. Soc. Jpn.*, **61**, 2451 (1988).

[393] D. J. Barker, L. A. Summers, *J. Heterocyclic Chem.*, **20**, 1411 (1983).

[394] (a) F. M. Moracci, F. Liberatore, V. Carelli, A. Arnone, I. Carelli, M. E. Cardinali, *J. Org. Chem.*, **43**, 3420 (1978); (b) I. Carelli, *J. Electroanal. Chem. Interfacial Electrochem.*, **107**, 391 (1980).

[395] J. Moiroux, S. Deycard, *J. Electroanal. Chem. Interfacial Electrochem.*, **194**, 99 (1985).

[396] P. Chautemps, J.-L. Pierre, *New J. Chem.*, **9**, 389 (1985).

[397] (a) F. M. Moracci, S. Tortorella, B. Di Rienzo, *Synth. Commun.*, **11**, 329 (1981); (b) F. M. Moracci, S. Tortorella, B. Di Rienzo, F. Liberatore, I. Carelli, *Ann. Chim.*, **71**, 499 (1981).

[398] (a) H. Lund, L. H. Kristensen, *Acta Chem. Scand.*, **B33**, 495 (1979); (b) T. Lund, H. Lund, *Acta Chem. Scand.*, **B40**, 470 (1986); (c) H. Lund, K. Daasbjerg, T. Lund, S. U. Pedersen, *Acc. Chem. Res.*, **28**, 313 (1995).

[399] F. M. Moracci, S. Tortrorella, *Synth. Commun.*, **13**, 1225 (1983).

[400] S. Komorsky-Lovric, *J. Serb. Chem. Soc.*, **52**, 43 (1987).

[401] E. Munoz, J. L. Avila, L. Camacho, *J. Electroanal. Chem. Interfacial Electrochem.*, **284**, 445 (1990).

[402] H. Ju, J. Zhou, C. Cai, H. Chen, *Electroanalysis*, **7**, 1165 (1995).

[403] B. Czochralska, E. Bojarska, P. Valenta, H. W. Nurnberg, *Bioelectrochem. Bioenerg.*, **14**, 503 (1985).

[404] K. Kano, K. Takagi, Y. Ogino, T. Ikeda, *Chem. Lett.*, **1995**, 589.

[405] A. Kunai, J. Harada, M. Nishihara, Y. Yanagi, K. Sasaki, *Bull. Chem. Soc. Jpn.*, **56**, 2442 (1983).

[406] D. Lloyd, C. A. Vincent, D. J. Walton, *J. Chem. Soc., Perkin Trans. II*, **1980**, 668.

[407] (a) I. Surov, H. Lund, *Acta Chem. Scand.*, **B40**, 831 (1986); (b) H. Lund, J. N. Hansen, *Stud. Org. Chem.*, **30**, 249 (1987).

[408] (a) C. P. Andrieux, J.-M. Saveant, *Bull. Soc. Chim. Fr.*, **1968**, 4671; (b) P. W. Crawford, P. Kovacic, S. Rault, M. Robba, M. D. Ryan, *Bull. Soc. Chim. Fr.*, **1986**, 756.

[409] (a) T. Shono, S. Kahimura, Y. Yamaguchi, O. Ishige, H. Uyama, F. Kuwata, *Chem. Lett.*, **1987**, 1511; (b) T. Shono, N. Kise, R. Nomura, A. Yamanami, *Tetrahedron Lett.*, **34**, 3577 (1993).

[410] (a) F. Pragst, H. Köppel, E. Walkhoff, E. Boche, *J. Prakt. Chem.*, **329**, 665 (1987); (b) G. Kokkindis, E. Hatzigrigoriou, D. Sazou, A. Varvoglis, *Electrochim. Acta*, **36**, 101 (1991).

[411] (a) T. Shono, K. Yoshida, K. Ando, Y. Usui, H. Hamaguchi, *Tetrahedron Lett.*, **1978**, 4819; (b) T. Shono, T. Miyamoto, Y. Usui, H. Hamaguchi, *Heterocycles*, **14**, 180 (1981); (c) T. Shono, T. Miyamoto, M. Mizukami, H. Hamaguchi, *Tetrahedron Lett.*, **22**, 2385 (1981); (d) T. Shono, E. Shirakawa, H. Matsumoto, E. Okazaki, *J. Org. Chem.*, **56**, 3063 (1991).

[412] H. Ohmizu, M. Takahashi, O. Ohtsuki, *Stud. Org. Chem.*, **30**, 241 (1987).

[413] B. Daoust, J. Lessard, *Can. J. Chem.*, **73**, 362 (1995).

[414] W. M. Moore, M. Finkelstein, S. D. Ross, *Tetrahedron*, **36**, 727 (1980).

[415] G. Tainturierm L. Roullier, J.-P. Martenot, D. Lorient, *J. Agric. Food Chem.*, **40**, 760 (1992).

[416] L. Pospisil, M. P. Colombini, R. Fuoco, V. V. Strelets, *J. Electroanal. Chem. Interfacial Electrochem.*, **310**, 169 (1991).

[417] (a) D. Lloyd, C. Nyns, C. A. Vincent, D. J. Walton, *J. Chem. Soc., Perkin Trans. II*, **1980**, 1441; (b) D. Lloyd, C. A. Vincent, D. J. Walton, *J. Chem. Soc., Perkin Trans. II*, **1981**, 801.

[418] P. Molina, M. D. Velasco, A. Arques, *Tetrahedron*, **46**, 5797 (1990).

[419] M. A. Azzem, G. E. H. Elgemeie, M. A. Neguid, H. M. Fahmy, Z. M. Elmassry, *J. Chem. Soc., Perkin*

Trans. II, **1990**, 545.

[420] T. Shono, Y. Usui, T. Mizutani, H. Hamaguchi, *Tetrahedron Lett.*, **21**, 3073 (1980).

[421] (a) N. P. Bogdanova, E. A. Popova, I. A. Avrutskaya, M. Ya. Fioshin, *Elektrokhimiya*, **21**, 1214 (1985); (b) N. P. Bogdanova, E. A. Popova, I. A. Avrutskaya, M. Ya. Fioshin, *Elektrokhimiya*, **21**, 1095 (1985); (c) S. V. Kondrashov, V. A. Smirnov, *Elektrokhimiya*, **28**, 923 (1992).

[422] N. P. Bogdanova, A. D. Cherednichenko, I. A. Avrutskaya, M. Ya. Fioshin, *Elektrokhimiya*, **21**, 1070 (1985).

[423] T. Osa, U. Akiba, I. Segawa, J. M. Bobbitt, *Chem. Lett.*, **1988**, 1423.

[424] (a) N. P. Bogdanova, I. A. Avrutskaya, M. Ya. Fioshin, M. A. Khrizolitova, *Elektrokhimiya*, **21**, 933 (1985); (b) N. P. Bogdanova, L. G. Petrova, I. A. Avrutskaya, M. Ya. Fioshin, *Elektrokhimiya*, **21**, 1369 (1985).

[425] P. Krzyczmonik, H. Scholl, *J. Electroanal. Chem.*, **335**, 233 (1992).

[426] U. Hess, J. Komenda, A. Hollwarth, S. Dunkel, *Pharmazie*, **47**, 848 (1992).

[427] R. N. Goyal, S. Bhargawa, *Bull. Chem. Soc. Jpn.*, **62**, 2662 (1989).

[428] T. Gade, M. Streek, J. Voss, *Chem. Ber.*, **125**, 127 (1992).

[429] S. S. Kucherov, I. A. Avrutskaya, *Elektrokhimiya*, **23**, 1143 (1987).

[430] Yu. D. Smirnov, V. F. Pavlichenko, A. P. Tomilov, *Zh. Org. Khim.*, **28**, 461 (1992)

[431] K. Arai, C. Shaw, K. Nozawa, K. Kawai, S. Nakajima, *Tetrahedron Lett.*, **28**, 441 (1987).

[432] K. Arai, S. Tamura, K. Kawai, S. Nakajima, *Chem. Pharm. Bull.*, **37**, 3117 (1989).

[433] R. Kossai, *Electrochim. Acta*, **31**, 1643 (1986).

[434] D. L. Hughes, S. K. Ibrahim, C. J. Macdonald, H. M. Ali, C. J. Pickett, *J. Chem. Soc, Chem. Commun.*, **1992**, 1762.

[435] C. J. Pickett, G. J. Leigh, *J. C. S., Chem. Commun.*, **1981**, 1033.

[436] A. V. Bukhtiarov, V. V. Mikheev, A. P. Tomilov, O. V. Kuz'min, *Elektrokhimiya*, **23**, 1639 (1987).

[437] R. B. A. Pardy, *Brit. UK Pat. Appl.* GB 2153849 (1985).

[438] G. B. Gavioli, G. Grandl, L. Benedetti, R. Andreoli, *Electrochim. Acta*, **28**, 1125 (1983).

6. Electroreduction of Sulfur, Selenium and Tellurium Compounds

6.1 Electroreduction of Divalent Sulfur Compounds

In Chapter 6, attention is mainly directed to the electroreduction of divalent, tetravalent and hexavalent sulfur-containing compounds. In particular, current topics reported in specific working articles on (1) reductive cleavage of sulfur-sulfur, selenium-selenium, tellurium-tellurium, sulfur-nitrogen, carbon-sulfur and carbon-selenium bonds, (2) reductive removal of alkylsulfonyl and arenesulfonyl groups and (3) electrosynthesis of 1,3-dithiole-2-thiones, tetrathiafulvalenes, sulfone derivatives, selenides and tellurides are reviewed from the standpoint of synthetic chemistry [1a]. A mechanistic survey of organic sulfur compounds is documented in the literature [1b].

Without using any base, alkylation of thiols proceeds electrochemically in a preparative sense. Cathodically generated thiolate has been utilized for the sulfurization of halides to give unsymmetrical sulfides in good yields [2]. The electrolysis of the thiol ($R^1 = 2$-pyridyl) **1** in an MeCN-Et$_4$NBr(0.3M)-(Pt/C) system in the presence of ethyl chloroacetate ($R^2 = CH_2CO_2Et$) **3** affords 2-ethoxycarbonylmethylthiopyridine **4** in 97% yield (eq. (6.1)).

$$R^1\text{-SH} \xrightarrow{\text{MeCN-Et}_4\text{NBr(0.3M)-(Pt/C)}} R^1\text{-S}^- + 1/2\ H_2 \quad \text{-----(a)}$$

$$\mathbf{1} \qquad\qquad\qquad \mathbf{2}$$

$$R^1\text{-S}^- + R^2\text{-X} \xrightarrow[\text{71~97\% yields}]{} R^1\text{-S-R}^2 + X^- \quad \text{-----(b)}$$

$$\mathbf{2} \qquad \mathbf{3} \qquad\qquad\qquad \mathbf{4}$$

(6.1)

X = Cl, Br
R^1 = alkyl, aryl, pyridyl
R^2 = alkyl, aryl, CH_2CO_2R (R = Me, Et), CH_2CN

6.1.1 Reductive Cleavage of Sulfur-Sulfur Bonds

Cyclic voltammetric peak potentials (E_p) of aromatic disulfides depend strongly on the electrode material, as illustrated in Fig. 6.1 [3], showing cyclic voltammetric (CV) curves of bis(4-chlorophenyl) disulfide in a DMF-Et$_4$NI system at vitreous carbon, platinum and copper electrodes. The E_p value may depend on the rate of the cleavage of the disulfide anion radical and is shifted to less negative potentials with increasing cleavage rate. The dependence of the peak potential of disulfides on the electrode material correlates with differences in the heterogeneous rate constants at the different electrode materials [3].

It is known that the rate constant of the heterogeneous electron transfer from a platinum electrode to aromatic disulfides (electron acceptors, Ar-SS-Ar) is low, whereas the homogeneous electron transfer from suitable aromatic anion radicals (electron donors, [D]$^{-\bullet}$) is much faster. A mechnism of the catalytic reduction of Ar-SS-Ar is outlined in eq. (6.2).

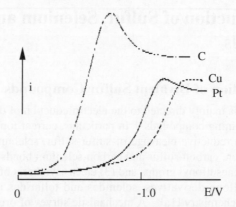

Fig. 6.1 CV curves of 4,4'-dichlorodiphenyl disulfide (0.005 M) in DMF/TBAI at vitreous carbon, platinium, and copper electrodes, $v = 10$ mV/s; reference electrode: Ag/AgI, I^- (0.2 M).
[Reproduced with permission from J. Simonet et al., *Liebigs Ann. Chem.*, 1981, 1665]

$$D + e^- \longrightarrow [D]^{\bullet} \quad \text{--------------------(a)}$$

$$[D]^{-\bullet} + Ar-SS-Ar \longrightarrow D + [Ar-SS-Ar]^{-\bullet} \quad \text{-------(b)}$$

$$[Ar-SS-Ar]^{-\bullet} \longrightarrow Ar-S^{\bullet} + Ar-S^- \quad \text{-----------(c)}$$

$$[D]^{-\bullet} + Ar-S^{\bullet} \longrightarrow D + Ar-S^- \quad \text{-------------(d)}$$ (6.2)

$$2 Ar-S^{\bullet} \longrightarrow Ar-SS-Ar \quad \text{-------------(e)}$$

Ar = aryl
D: perylene, quinoxaline, acridine, tetracene,
diacetylbenzene, azobenzene, fluorenone,
p-nitroanisole, *p*-nitrotoluene, nitrobenzene,
p-nitrocinnamaldehyde

Anion radicals [D]$^-$ of aromatic compounds as redox catalysts are electrogenerated at more negative potentials than reduction potentials of Ar-SS-Ar [3].   The complexity of the electroreduction of disulfides at a mercury cathode is, in many cases, due to the formation of mercury compounds before and/or during the electrochemical reaction [4].   For instance, dithiodipropionic acid (R-SS-R, R $= CH_2CH_2CO_2H$) in an H_2O/solvent (4:1 v/v)-electrolyte-(Hg) system can be reduced to the corresponding thiol (R-SH) according to eqs. (6.4) and (6.6), the former of which is the rate-determining step of the electrode reaction in the lower pH range (1.32-6.00).   At higher pH values (11.22-12.50), the disulfide is reversibly reduced to R-SH according to eq. (6.3).   Under a prolonged electrolysis, the R-SH formed depolarizes the mercury with the formation of RSHg, which may provide $(R-S)_2Hg$ (eq. (6.5)).   Finally, the consequent reduction of $(R-S)_2Hg$ gives the thiol according to eq. (6.6) [5].

$$R-SS-R + 2\,e^- + 2\,H^+ \rightleftharpoons 2\,R-SH \tag{6.3}$$

$$R-SS-R + e^- + H^+ \overset{Hg}{\rightleftharpoons} R-SHg + R-SH \tag{6.4}$$

$$2\,R-SHg \rightleftharpoons (R-S)_2Hg + Hg \tag{6.5}$$

$$(R-S)_2Hg + 2\,e^- + 2\,H^+ \longrightarrow 2\,R-SH + Hg \tag{6.6}$$

R: $-CH_2CH_2CO_2H$

Electrolysis System: H_2O/Solvent(4:1 v/v)-Electrolyte-(Hg)

Solvent: MeOH, EtOH, MeCN, DMSO, DMF

Electrolyte: KNO_3, KCl, NaCl, $NaClO_4$, Me_4NBr, Et_4NBr

Cathodically initiated Michael addition of thiols (RSH) to levoglucosenone **5** under passage of a small current produces novel *threo* addition product **7** in more than 90% yield (eq. (6.7)) [6]. The normal *erythro* isomer **6** identified as the kinetic product tends to be formed when a large current is used.

$$\tag{6.7}$$

6 >90% yield   * MeCN-Bu$_4$NBr-(Mg/Pt)   **7**   90% yield

R = C_8H_{17}, $C_{10}H_{21}$, p-ClC$_6$H$_4$, 2-naphthyl, etc.

Electroreduction of 2,2'-dipyridyl disulfide affords the 2-pyridylthio radical on silver and mercury cathodes which, in both cases, chemisorbs through the sulfur atom on the electrode surface, forming a thiolato framework in basic, neutral, and acidic solutions [7]. The mechanism of the electrode process for the electroreduction of pyrithioxine[3,3'-(dithiodimethylene)bis-5-hydroxy-6-methyl-4-pyridinemethanol], a cerebral metabolism activator, is surveyed by polarographic methods [8].

The electrosynthesis of L-cysteine by the electroreduction of L-cystine is one of important industrial method in obtaining the biological key-compound. The detail of electrochemistry of the conversion of L-cystine to L-cysteine has been precisely reviewed in the literatures [9a]. The industrial processes are also criticized in detail [9b]. The related analytical data have been added [9c, 9d]. The three S-S linkages of insulin are cleaved selectively by using a potentiostatic technique: the two S-S bonds between the major chains A and B can be reduced at −1.3 V (SCE) in 90% yield and the S-S bond in the A chain is capable of being cleaved at −1.8 V in 89% yield (eq. (6.8)) [10]. The electroreduction of the disulfide bonds of a vegetable protein can provide a protein of improved functionality and whiteness [11]. 1,2-Dithiole-3-thiones bearing substituents at the C(4) and C(5) positions are of great interest in pharmacology. The electroreduction of Oltipraz, 5-(2-

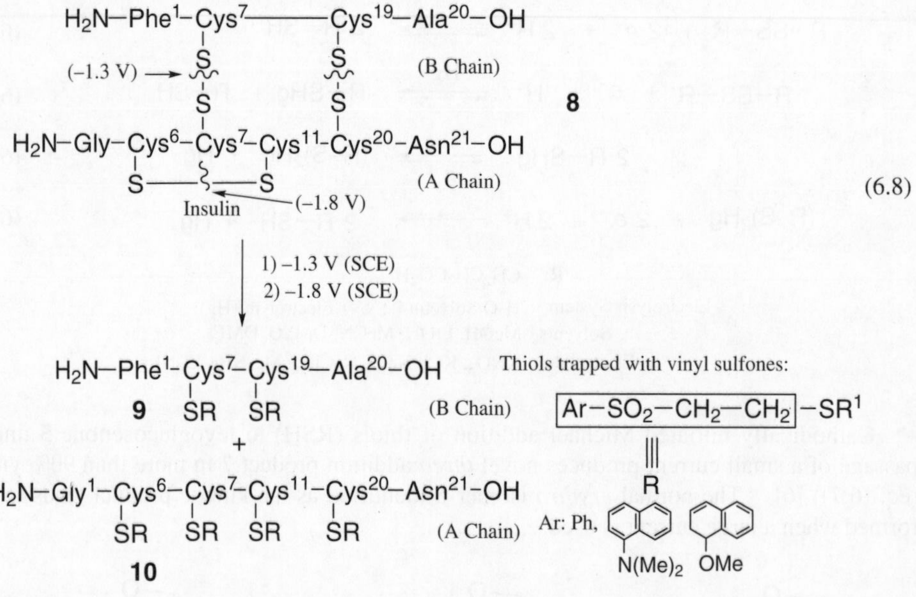

$$H_2N-Phe^1-Cys^7 \underline{\hspace{2cm}} Cys^{19}-Ala^{20}-OH$$

(–1.3 V) →

(B Chain)

8

$$H_2N-Gly-Cys^6-Cys^7-Cys^{11}-Cys^{20}-Asn^{21}-OH$$

(A Chain)

Insulin   (–1.8 V)

(6.8)

1) –1.3 V (SCE)
2) –1.8 V (SCE)

$$H_2N-Phe^1-Cys^7-Cys^{19}-Ala^{20}-OH$$

9   SR   SR   (B Chain)

+

$$H_2N-Gly^1-Cys^6-Cys^7-Cys^{11}-Cys^{20}-Asn^{21}-OH$$

| | | | |
SR   SR   SR   SR   (A Chain)

10

Thiols trapped with vinyl sulfones:

$$Ar-SO_2-CH_2-CH_2-SR^1$$

R

Ar: Ph,

N(Me)$_2$ OMe

pyrazinyl)-4-methyl-1,2-dithiole-3-thione in a DMF-Et$_4$NBr(0.1M)-(Pt) system affords preferentially the dimer **12** according to the reaction pathway shown in eq. (6.9) [12]. In contrast, the electrolysis of **11** in the presence of an excess of methyl iodide affords 7-methyl-6,8-dimethylthiopyrrolo[1,2-a]pyrazine **13** in 75% yield as the sole product (eq. (6.10)) [12]. This reduction process provides a convenient route to the pyrrolo[1,2-a]pyrazine derivatives isolated as metabolites in mouse urine [12b].

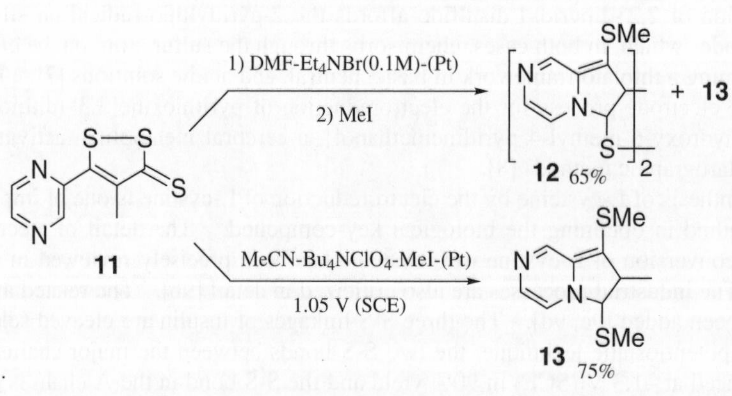

1) DMF-Et$_4$NBr(0.1M)-(Pt)

2) MeI

12 65%   + **13**

(6.9)

11

MeCN-Bu$_4$NClO$_4$-MeI-(Pt)

1.05 V (SCE)

13 75%

(6.10)

The electrochemical behavior of 1,2-dithiole-3-thiones at the mercury electrode in aqueous ethanol media has been investigated at pH 7.0 [13]. The dianionic intermediates have recently been trapped by methylation. For example, electroreduction of 5-(4-methoxyphenyl)-1,2-dithiole-3-thione **14** in a DMF-LiClO$_4$(0.2M)-(Hg) system in the presence of methyl iodide affords an *E* and *Z* mixture of methyl(3-methylthio)-3-(4-methoxyphenyl)-2-propene dithioate **15** in 62% yield (eq. (6.11)) [14].

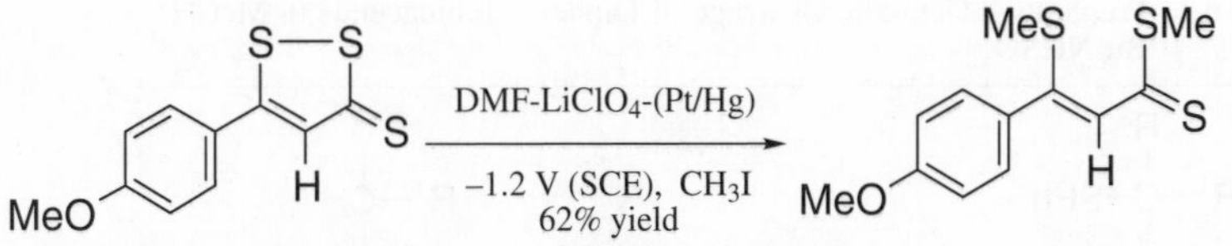

$$\text{(6.11)}$$

14 → **15**

6.1.2 Reductive Cleavage of Carbon-Sulfur Bonds

Little is known about the reductive cleavage of a simple carbon-sulfur bond. Electroreductive cleavage of aliphatic diphenyldithioacetals **16** into the corresponding phenylthioethers **17** has, however, been obtained in good isolated yields by replacing one thiophenyl group with a proton using a mercury or glassy carbon electrode [15]. For instance, the electrolysis of **16** is carried out in an MeCN-Bu$_4$NHSO$_4$-(Hg or C) system at −2.43~ −2.48 V vs. NHE under passage of 2.1~2.7 F/mol electricities to give the corresponding phenylthioethers **17** in 76~93% yields (eq. (6.12)). Some results are shown in

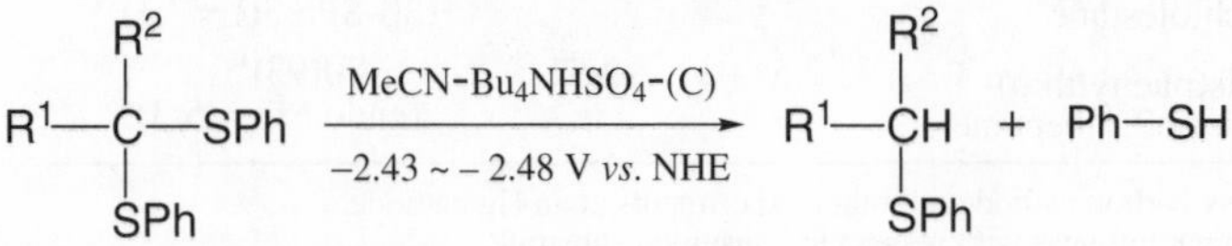

$$\text{(6.12)}$$

16 R^1 = alkyl
R^2 = H, Me

17 76~93%

Table 6.1 [15]. α-Ketodiphenyldithioacetals **18** can be reduced to the corresponding phenylthiomethylketones **19** at a less negative potential (eq. (6.13)) [15]. A survey of the electroreduction of **18** is presented in Table 6.2 [15].

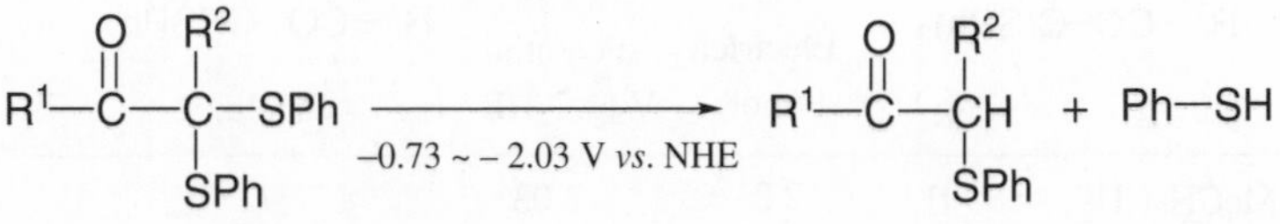

$$\text{(6.13)}$$

18 R^1 = alkyl
R^2 = H, Me

19 66~76%

The 1,2-elimination of β-hydroxysulfides **20** leading to olefins by taking an electroreductive method has been exploited. A variety of exo-methylene derivatives **21** are nicely synthesized in a DMF-Et$_4$NOTs-(Pt/Pb) system (eq. (6.14)) [16]. The cathodic elimination

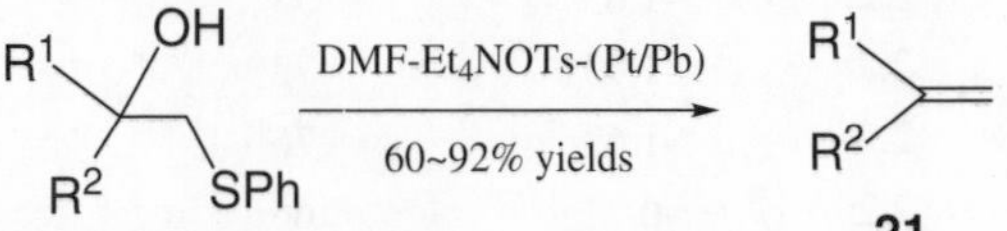

$$\text{(6.14)}$$

20 R^1, R^2 = alkyl

21

of β-acetoxysulfides **22** proceeds in a DMF-Et$_4$NClO$_4$-(Hg) system at −1.7 V vs. Ag/AgI/I$^-$ to give trans-stilbene **23**, preferentially (eq. (6.15)) [17]. The elimination

Table 6.1 Preparative Cathodic Cleavage of Diphenyldithioacetals in MeCN-Bu$_4$NHSO$_4$

R^1—C(SPh)$_2$—R^2				R^1—CH(SPh)—R^2
R^1	R^2	Electricity F/mol	Potential V vs. NHE[c]	Yield, %
Ph(CH$_2$)$_2$	Me	2.1	-2.43	76
-(CH$_2$)$_5$-		2.3	-2.43	82
-(CH$_2$)$_5$-		2.4	-2.43	73 [a]
Me$_3$C	Me	2.3	-2.48	77
Me$_2$CH	Me	2.4	-2.43	78
Me(CH$_2$)$_2$	H	2.7	-2.48	79
EtO$_2$CCH$_2$	H	2.3	-2.03	85(89)[b]
3,3-bis(phenylthio)-5α-chlolestane		2.3	-2.48	86 (3β-SPh:3α = 7:1)
2,2-bis(phenylthio)-bicyclo[2.2.1]heptane		2.5	-2.53	70(93)[b] (endo:exo = 5:3)

[a] At a glassy carbon cathode; all other experiments at an Hg cathode.
[b] Values in parenthesis with respect to consumed substrate.
[c] Calculated from values obtained with an Ag/AgNO$_3$(0.1 M) reference electrode (570 mV vs. NHE).
[Reproduced with permission from N. S.-Von Itter et al., *Tetrahedron*, **43**, 2475 (1987)]

Table 6.2 Preparative Cathodic Cleavage of α-Carbonyldiphenyl-dithioacetals in MeCN-Bu$_4$NHSO$_4$

R^1—CO—C(SPh)$_2$—R^2				R^1—CO—CHSPh—R^2
R^1	R^2	Electricity F/mol	Potential V vs. NHE[c]	Yield, %
MeCH$_2$CH$_2$	H	2.2	-2.03	72
PhCH$_2$	H	2.0	-1.73	70
1-Cyclohexen-ylmethyl	H	2.0	-1.43	72
Cyclohexyl	H	2.3	-2.03	73
Cyclohexyl	H	2.0	-1.73	74[a]
Ph	H	2.0	-0.73	69
Ph	H	2.25	-1.63	53[a,b]
MeC(OEt)$_2$	H	2.7	-1.73	72
MeC(OEt)$_2$	H	2.2	-1.63	76[a]
Me	MeCO	2.2	-0.73	66

[a] At a glassy carbon cathode; all other experiments at an Hg cathode.
[b] Accompanied by 20% acetophenone.
[c] Calculated from values obtained with an Ag/AgNO$_3$(0.1 M) reference electrode (570 mV vs. NHE).
[Reproduced with permission from N. S.-Von Itter et al., *Tetrahedron*, **43**, 2475 (1987)]

$$\underset{\underset{\textbf{22}}{\underset{PhS \quad OAc}{Ph-\overset{\overset{H}{|}}{C}-\overset{\overset{H}{|}}{C}-Ph}}}{} \xrightarrow[\text{-1.7 V } vs. \text{ Ag/AgI/I}^-]{\text{DMF-Et}_4\text{NClO}_4(0.2M)\text{-(Hg)}} Ph-CH=CH-Ph + \text{Bibenzyl} \qquad (6.15)$$

23 trans : 92%   7%
    cis : traces

method can be used for the elongation of **24** such as aldehydes, ketones and esters, leading to the next higher homologous aldehydes **27** via **25**→**26** or **28**→**29** (eq. (6.16)) [18].

$$(6.16)$$

28 R^1, R^2 = alkyl or hydrogen

Intramolecular 1,2-elimination of β-oxysulfides **30** also proceeds in an MeCN-Et$_4$NBr-(Hg) system to give 1-isopent-2-enyl-2-naphthol **31** in 53% yield (eq. (6.17)) [19].

$$\text{(structures)} \xrightarrow[\text{53\% yield}]{\text{MeCN-Et}_4\text{NBr-(Hg)}} \text{(product 31)} \qquad (6.17)$$

30   SPh   **31**

3-Methylenecephams **33** are useful precursors of 3-norcephalosporins, unnatural β-lactam antibiotics and the reductive elimination of S-substituents at the C(3') position of **32** may give straightforward access to the 3-methylenecephams. The electroreductive conversion of cephalosporinates **32** into **33** proceeds smoothly in an aqueous THF-LiClO$_4$/NH$_4$ClO$_4$-(Pb) and/or aqueous MeCN/EtOH-LiClO$_4$/NH$_4$ClO$_4$-(Pb) system, giving **33** in 78~90% yields (Table 6.3 [20, 21a]) (eq. (6.18)) [20].

$$\text{(structure 32)} \xrightarrow[e^-]{\text{78~90\% yields,}} \text{(structure 33)} \qquad (6.18)$$

32
BT = 2-benzothioazolyl   **33**

Electroreduction of sodium C(3')-S-substituted cephalosporanic acids in an H$_2$O-NaHCO$_3$/Na$_2$HPO$_4$-(Pt/Hg) system can also lead to the corresponding 3-methylenecepham derivatives **33** in 36~54% yields [21]. The preferential formation of 3-methylenecephams

Table 6.3 Electroreductive Removal of C(3')-Substituents of Cephalosporins

R^1	R^2	R^3	Electrolysis Conditions	Yield, %	Ref.
2-Thienyl	Ar	Na	AcOH-AcONa(0.1M)-(Pt/Hg)	36	[21a]
2-Thienyl	(CO-pyridyl)	Na	aq. AcOH-LiBr/AcONa-(Pt/Hg)	54	[21a]
2-Thienyl	SO₂ONa	Na	aq. HCl-Na₂HPO₄(0.1M)-(Pt/Hg)	53	[21a]
PhOCH₂	BT[a]	PhCH₂	THF/H₂O(4:1)-LiClO₄/NH₄ClO₄-(Pt/Pb)	78	[20]
PhOCH₂	BT[a]	PhCH₂	MeCN/H₂O/EtOH(4:1:0.07)-LiClO₄/NH₄ClO₄-(Pt/Pb)	90	[20]

[a] BT: 2-Benzothiazoyl.

33, not of 3-methyl-3-cephems, may be explained in terms of a kinetically controlled protonation to an anion intermediate at the C(4) position.

Divalent sulfur pendant (SY) attached to position α of electron-withdrawing groups (W = RCO, ROCO, CN, etc.) can be removed electrochemically. In most cases, the reduction of the C-S bond (34→35) occurs at less negative potentials than that of the carbonyl group (eq. (6.19)) [22]. The stereochemistry of the electroreduction of optically active

$$(6.19)$$

34 W = RCO, ROCO, CN, etc.
      R^1, R^2 = H, alkyl, aryl

ethyl 2-phenylmercaptopropionate **34** (R^1 = Me; R^2 = Y = Ph, W = CO₂Et) has been investigated. The reduction in 95% EtOH-Et₄NBr proceeds with the formation of ethyl 2-phenylpropionate **35** (90% yield) of low optical activity (2~4% stereospecificities) [23]. As part of a search for the development of an electroreductive removal of alkylthio moieties which can work on a preparative scale, the cathodic reduction of 2-(2-benzoylthio)alkanoates and 2-(2-benzoylthio)alkanones has been investigated in an MeOH-H₂SO₄-(C) and/or an MeOH-Et₄NOTs-(Pt) system as a facile handling medium. The reductive cleavage of the carbon-sulfur bond of 2-substituted 2-(2-benzothiazolylthio)alkanoates **36** proceeds smoothly in a MeOH-H₂SO₄-(C/C) system to give the corresponding desulfurization products **37** (eq. (6.20)) [24a]. Some results are indicated in Table 6.4 [23, 24a]. In contrast, electroreduction of 2-(2-pentynyl)-2-(2-benzothiazolylthio)cyclopentanone **38** in an MeOH-H₂SO₄-(C/C or Pt/Pt) system affords a tarry product. The reduction of the α-thioketone **38** in an MeOH-Et₄NOTs-(Pt) system may, however, provide the desulfurylated product **39** in 74% yield (eq. (6.21)) [24a]. Regioselective cleavage of the carbon-sulfur bond

$$R^2 - \underset{\underset{CO_2Me}{|}}{\overset{\overset{R^1}{|}}{C}} - SBT \quad \xrightarrow{MeOH-H_2SO_4-(C/C)} \quad R^2 - \underset{\underset{CO_2Me}{|}}{\overset{\overset{R^1}{|}}{C}} - H \; + \; (BTS)_2 \; + \; BTSH \qquad (6.20)$$

36 R^1 = alkyl
R^2 = alkyl, alkenyl, vinyl

37 53~87%

Table 6.4 Desulfurization of α-(2-Benzothiazolylthio)alkanoates[a]

R^1	R^2	Y	R^1, H, R^2, CO_2Me Yield, %	$(YS)_2$ Yield, %	YSH Yield, %	Ref.
Me	Ph	Ph	90	–	90	[23]
$(CH_2)_3MeO_2C$	C_5H_{11}	BT	87	31	53	[24a]
$(CH_2)_3MeO_2C$	$EtC\equiv CCH_2$	BT	82	50	22	[24a]
$(CH_2)_3MeO_2C$	$Me_2C=CHCH_2$	BT	74	96	–	[24a]
$(CH_2)_3MeO_2C$	$CH_2=CHCH_2$	BT	66	53	8	[24a]
C_6H_{13}	$CH_2=CHCH_2$	BT	72	36	25	[24a]
C_6H_{13}	$Me_2C=CHCH_2$	BT	84	63	24	[24a]
C_6H_{13}	$MeCO(CH_2)_2$	BT	80	33	25	[24a]
$(CH_2)_3CO-N\underset{}{O}$	$EtC\equiv CCH_2$	BT	53	78	–	[24a]
$(CH_2)_3CO-N\underset{}{O}$	$CH_2=CHCH_2$	BT	55	73	–	[24a]

[a] Carried out with a divided dell fitted with carbon electrode in methanol (20 mL) containing 98% H_2SO_4 (120 mg)

$$\text{38} \quad \xrightarrow{\underset{74\% \; yield}{MeOH-Et_4NOTs-(Pt/Pt)}} \quad \text{39} \qquad (6.21)$$

has been achieved in the electroreduction of 2-(2-benzothiazolyl)thioethyl alkanoates **40** (eq. (6.22)) [24b].   Under neutral conditions, the fragmentation takes place at site a to give three segments **41**, **42**, and **43** in good yields, whereas under acidic conditions, the reductive cleavage of **40** occurs at site b to produce 2-mercaptoethyl alkanoates **44** and dihydrobenzothiazol **45**.   Electroreductive cross-coupling of methyl 3-phenylthioalkanoates with aldehydes gives the corresponding γ-lactones directly.   The reduction of methyl 3-phenylthiobutyrate **46** in a DMF-Et$_4$NOTs-(Pt/Pb) system affords 3-methyl-4-propyl-γ-butyrolactone **47** in 42% yield (eq. (6.23)), while methyl 3-phenylthiopropionate gives 4-hexyl-γ-butyrolactone in only 5% yield [25].

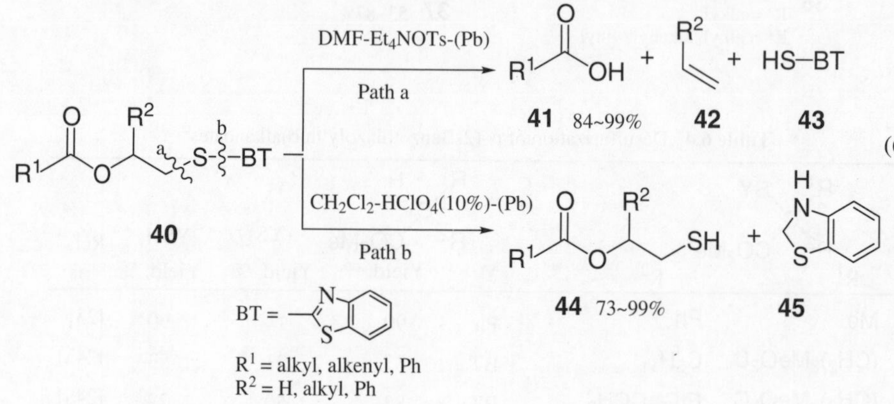

(6.22)

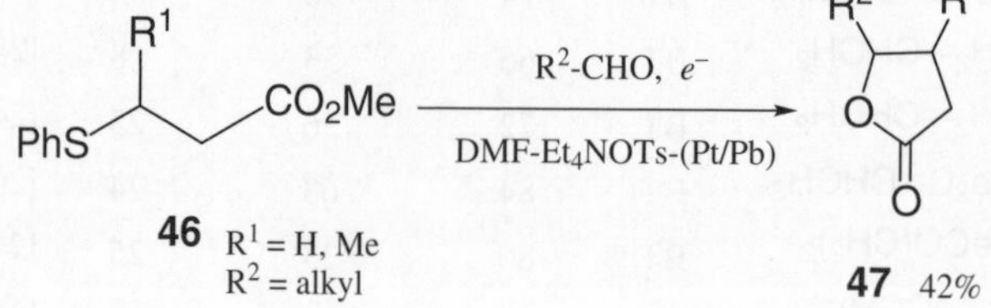

$$BT = \text{(benzothiazole structure)}$$

R^1 = alkyl, alkenyl, Ph
R^2 = H, alkyl, Ph

(6.23)

46 R^1 = H, Me
R^2 = alkyl

47 42%

Electroreductive desulfurization of 2-phenyl-1,3-thiazine-6-ones **48** takes place in an aqueous EtOH(50%)-H$_2$SO$_4$-(Hg) system to yield 5-ethoxycarbonyl-3-hydroxy-2-phenyl-1*H*-pyrrole **49** in good yields (eq. (6.24)) [26].   A similar phenomenon has been observed when substituted 4*H*-1,3-thiazines **50** are subjected to electroreduction under an acidic conditions, giving 1-pyrrole derivatives **51** (eq. (6.25)) [27].

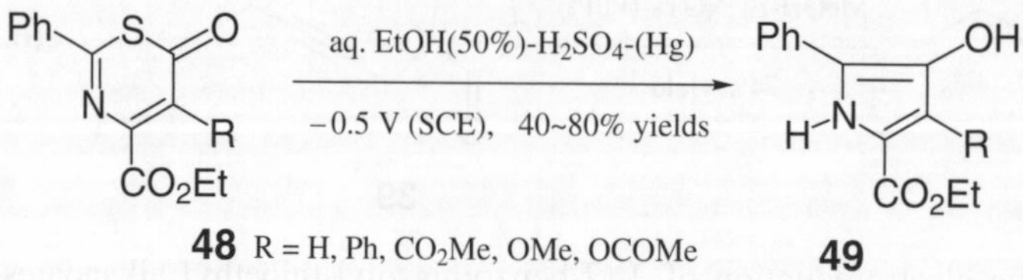

(6.24)

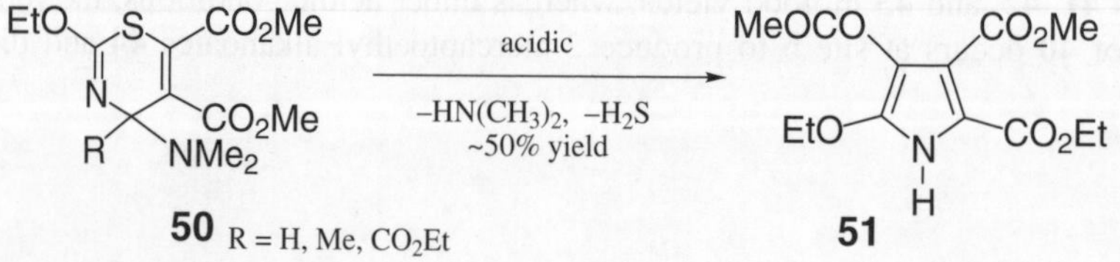

48 R = H, Ph, CO$_2$Me, OMe, OCOMe

49

(6.25)

50 R = H, Me, CO$_2$Et

51

The preparative reduction of benzo[b]thiophene **52** in aqueous or mixed organic-aqueous media was carried out in an H_2O-Bu_4NOH(2.1M)-Hg or a THF/H_2O-Bu_4NBF_4-(Hg) system. The products formed are the dihydro derivative **53** and the carbon-sulfur bond cleaved benzenethiol **54** (eq. (6.26)) [28]. Loss of thiolate [RS$^-$] from activated vinyl

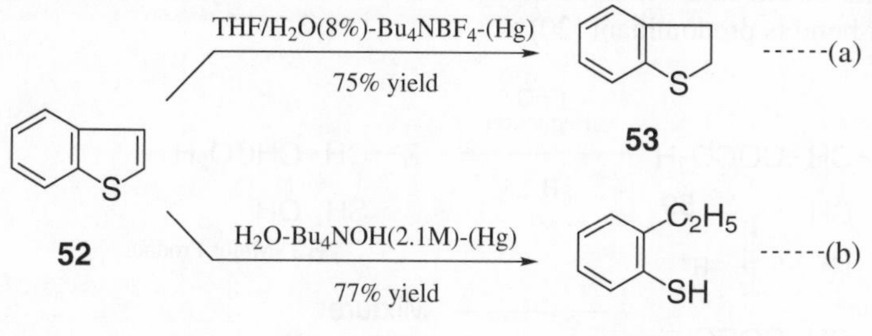

THF/H_2O(8%)-Bu_4NBF_4-(Hg)

75% yield

------(a)

53

(6.26)

52

H_2O-Bu_4NOH(2.1M)-(Hg)

77% yield

------(b)

54

thioether derivatives is also recorded [29]. Mercury cathode reduction of ethyl 2-phenyl-2-(phenylthio)propionate [23], ethyl 3-(p-tolylthio)pyruvate [7a, 30b], and 2-(thiocyanato) acetophenone [22] has been reported in an electroanalytical sense. The electroreductive cleavage of carbon-sulfur bonds of thiophene and thiazole derivatives has been performed successfully. The conversion of **55** into **56** is carried out in a DMF-$NaClO_4$-(Pt/Hg) system to give 2-mercaptothiazoles **56** in 53~70% yields (eq. (6.27)) [31]. Similarly,

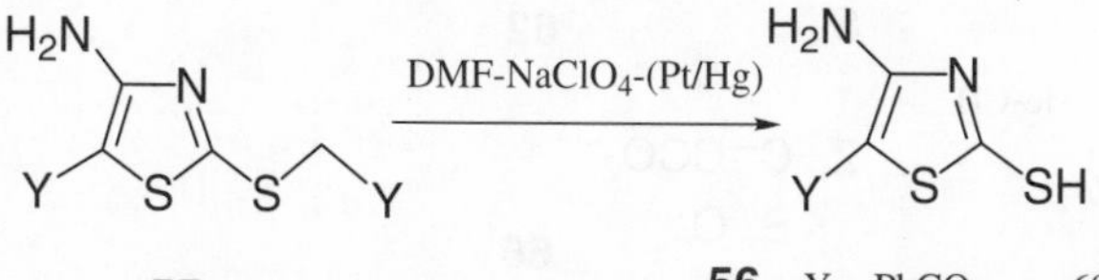

DMF-$NaClO_4$-(Pt/Hg)

(6.27)

55

56 Y = PhCO   69 %
Y = CN   53 %
Y = CO$_2$Et   70 %

thiophene derivatives **57** afford the corresponding disulfides **58** in 62~93% yields (eq. (6.28)) [31].

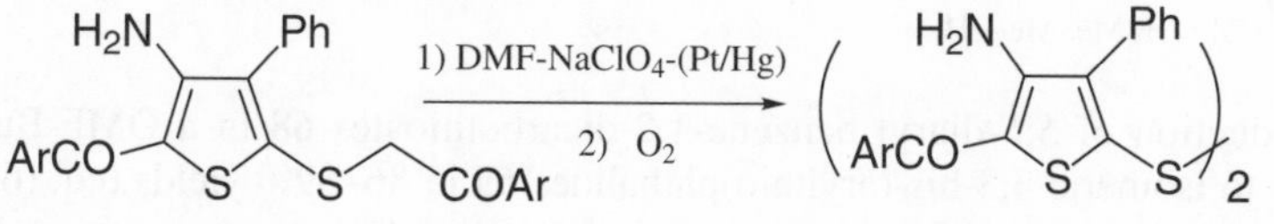

1) DMF-$NaClO_4$-(Pt/Hg)

2) O_2

(6.28)

57

58 Ar = Ph   93 %
Ar = 4-ClC$_6$H$_4$ 86 %
Ar = 4-MeC$_6$H$_4$ 62 %
Ar = Ph—N—C—   70 %

A competitive reduction of the carbonyl group and the carbon-sulfur bond in 3-thio-2-oxoalkanoic acids and their derivatives at mercury pool electrode has been investigated [32]. As shown in eq. (6.29), 3-thio-2-oxoalkanoic acids **59** undergo dissociation and form a carboxylate **60**, a thiolate-carboxylate dianion **61**, and even a trianion **62**, depending on the pH values of electrolysis media. The carbonyl group is reduced preferentially in acidic media to give the compound **63** as a major product, but under basic conditions, the carbon-sulfur linkage cleavage proceeds predominantly, giving the keto acid **64** in 75% yield. Controlled-potential electrolysis of the thioether **67** in an EtOH-H$_2$SO$_4$-(Pt/Hg) system at −1.20 V (SCE) at pH 4.5 yields the hydroxy acid **63** in a quantitative yield (eq. (6.30)) [30b, 32].   The behavior of thioethers is not uniform.   In the case of aryl thioethers, cleavage of the carbon-sulfur bond is predominant [30].

$$\text{(6.29)}$$

$$\text{(6.30)}$$

The electroreduction of *S,S*-diaryl benzene-1,2-dicarbothioates **68** in a DMF-Bu$_4$NI-(Hg) system leads to isomeric 3,3-bis-(arylthio)phthalides **69** in 86~89% yields (eq. (6.31)) [33].   The reaction is an example of an electrochemically induced rearrangement, involving a carbon-sulfur bond disconnection-reconnection sequence, for which the required amount of charge is less than 0.1 F/mol.

$$\text{(6.31)}$$

Y = H	87 %	(-1.05 V *vs.* Ag/AgI)
Y = 4-Me	89 %	(-1.15 V)
Y = 4-Cl	86 %	(-1.00 V)

Electrochemical carboxylation of ketone S,S-acetals **70** has been realized by cathodic reduction. The replacement of the alkylthio group by a nucleophile is facilitated because of the electron-withdrawing carbonyl group at a vinylogous position. The preparative scale electrolysis of the ketene S,S-acetals **70** in an MeCN-R_4NBr-(Al/Pb or Al/Pt) system gives the carboxylic acids **71** in 30~72% yields (eq. (6.32)) [34]. The cathodic reduction

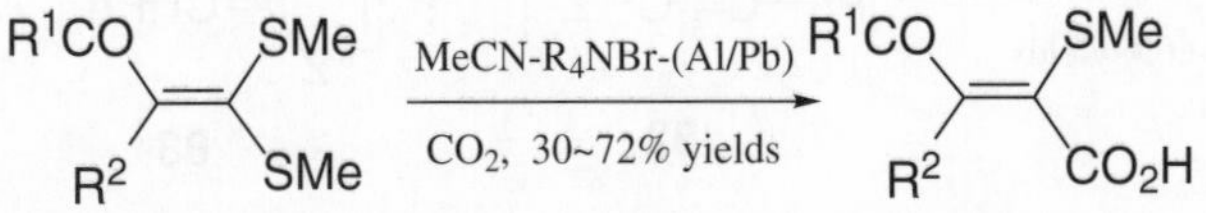

$$(6.32)$$

70 R^1 = alkyl, aryl ; R^2 = H, Me   **71**

of 2-bis(methylthio)methylene-1-tetralone **72** in a MeCN-Bu_4NHSO$_4$-(Hg) system gives the corresponding methylthioketones **73** and **74** in 94% yield, but the electroreduction of **72** in an MeCN-Bu_4NBr-(Hg) system in the presence of carbon dioxide affords the corresponding carboxylate ions which are trapped as methyl esters **75** and **76** by treatment with methyl iodide (eq. (6.33)) [15]. Further extension work has been done with bis(methylthio)methylidene-1,2,3,4-tetrahydronaphthalenones **77** and **79** [35]. It is interesting to note that the

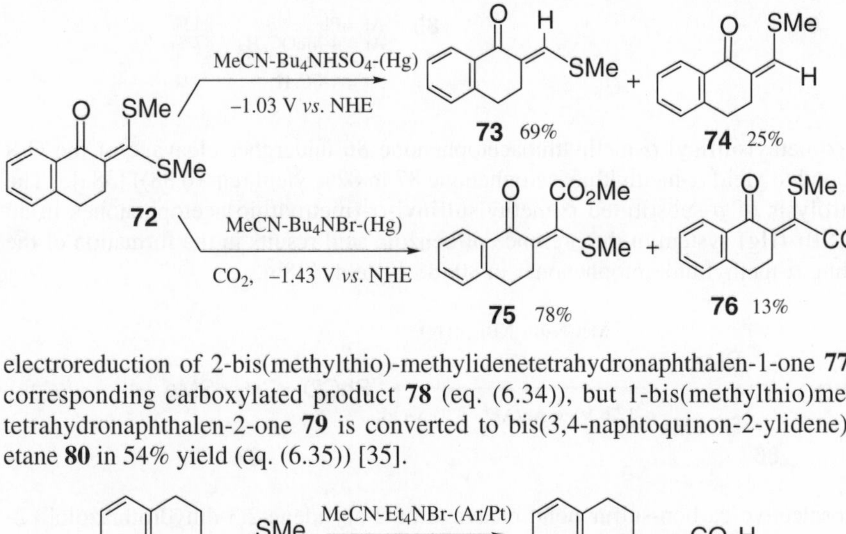

$$(6.33)$$

73 69%   **74** 25%

75 78%   **76** 13%

electroreduction of 2-bis(methylthio)-methylidenetetrahydronaphthalen-1-one **77** gives the corresponding carboxylated product **78** (eq. (6.34)), but 1-bis(methylthio)methylidene-tetrahydronaphthalen-2-one **79** is converted to bis(3,4-naphtoquinon-2-ylidene)-1,3-dithietane **80** in 54% yield (eq. (6.35)) [35].

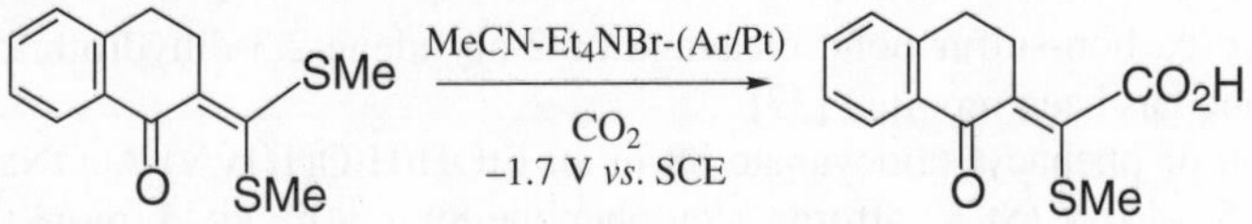

$$(6.34)$$

77   **78** 40%

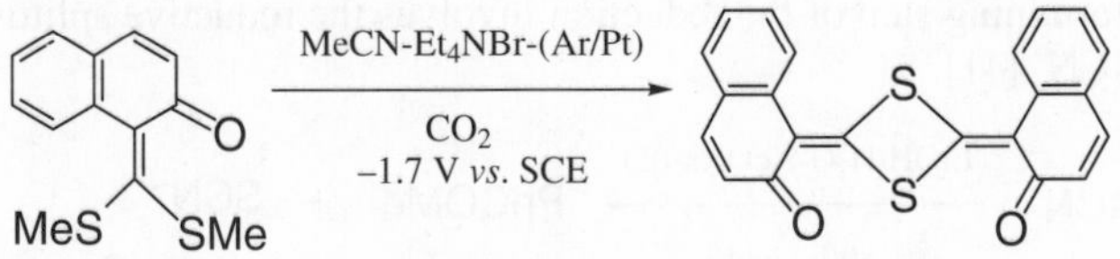

$$(6.35)$$

79   **80** 54%

The dimerization of electrochemically produced phenylethynechalcogenolate anions **82** derived from **81** preferentially proceeds to give 2-benzylidene-4-phenyl-1,3-dithiole (or diselenole) **83** in 50~60% yields (eq. (6.36)) [36].

$$2 \quad \underset{\textbf{81} \ Z = S, \ Se}{\overset{\text{Ph}}{\underset{Z}{\bigvee}}\!\!\!\!\!\!\!\!\!\!\!\!\!\overset{N}{\underset{N}{}}} \quad \xrightarrow[\text{50~60\% yields}]{\text{MeCN-Bu}_4\text{NClO}_4(0.1\text{M})\text{-(?)}} \quad \left[\text{Ph}-\text{C}\!\equiv\!\text{C}-\text{Z}^-\right] \longrightarrow \underset{\textbf{83}}{\overset{\text{Ph}}{\underset{Z}{\bigvee}}\!\!\!\!\!\!\!\!=\!\text{CHPh}} \qquad (6.36)$$

82

Electroreductive cleavage of the carbon-sulfur bond of 1-methylsulfinyl-1-methylthio-2-arylethenes **84** proceeds selectively at the carbon-sulfur (*S*-oxide) linkage, resulting in the formation of *E*-1-methylthio-2-arylethenes **85** in 73~87% yields (eq. (6.37)) [37].

$$\underset{\textbf{84}}{\overset{\text{Ar}}{\underset{\text{H}}{\bigvee}}\!\!\!\!\!\!\!\!\overset{\text{SMe}}{\underset{\text{SO-Me}}{}}} \quad \xrightarrow[\text{2 } e^-, \ \text{H}^+]{\text{MeCN-Bu}_4\text{NBF}_4\text{-PhOH-(Pt/Hg)}} \quad \underset{\textbf{85}}{\overset{\text{Ar}}{\underset{\text{H}}{\bigvee}}\!\!\!\!\!\!\!\!\overset{\text{H}}{\underset{\text{SMe}}{}}} + \ \text{MeSO}^- \qquad (6.37)$$

Ar = Ph	85 %
Ar = 4-MeOC$_6$H$_4$	87 %
Ar = 3-MeOC$_6$H$_4$	73 %
Ar = 4-FC$_6$H$_4$	79 %

Similarly, α-methylsulfinyl-α-methylthioacetophenone **86** undergoes cleavage of the C-S (divalent) bond to yield α-methylthioacetophenone **87** in 82% yield (eq. (6.38)) [38a]. The macroelectrolysis of *p*-substituted α-methylsulfinyl-α-(methylthio)acetophenones in an MeCN-Bu$_4$NBr-(Hg) system in the presence of benzoic acid results in the formation of the corresponding α-methylthioacetophenones in 80~86% yields [38b].

$$\underset{\textbf{86}}{\text{PhCO}-\text{CH}\!\!\overset{\text{SMe}}{\underset{\text{SO-Me}}{}}} \quad \xrightarrow[\text{-1.76 V } vs. \ \text{Ag/Ag}^+, \ 82\% \ \text{yield}]{\overset{\text{MeCN-Bu}_4\text{NBF}_4\text{-(Hg),}}{\text{PhCO}_2\text{H}}} \quad \underset{\textbf{87}}{\text{PhCO}-\text{CH}_2-\text{SMe}} \qquad (6.38)$$

A regioselective carbon-sulfur bond cleavage of 2-arylidene-2,3-dihydrothiazolo[3,2-a]benzimidazol-3-one has been reported [39].

Electroreduction of phenacyl thiocyanate **88** in an EtOH/H$_2$O(1/1v/v)-AcONa(buffer)-(C) system at pH 6.5, −1.0 V (SCE) affords acetophenone **89** in 80% yield (more than 95% of current efficiency) (eq. (6.39)) [40].

The cathodic decomposition of molten potassium thiocyanate at 200 °C results in a film of K$_2$S$_2$ and the rate determining step of the reduction involves the reductive splitting of the carbon-sulfur bond in SCN$^-$ [41].

$$\underset{\textbf{88}}{\text{PhCOCH}_2-\text{SCN}} \quad \xrightarrow[\text{80\% yield}]{\text{EtOH/H}_2\text{O-AcONa-(C)}} \quad \underset{\textbf{89}}{\text{PhCOMe}} \ + \ \text{SCN}^- \qquad (6.39)$$

6.1.3 Electroreduction of Carbon-Sulfur Double Bonds

6.1.3.1 Electroreductive Reactions of Carbon Disulfides

Two mechanisms are proposed for the formation of 4,5-bis(methylthio)-1,3-dithiole-2-thione **90**, dimethyl tetrathiooxalate **91**, dimethyl trithiocarbonate **92**, and carbon monosulfide by the electroreduction of carbon disulfide in DMF(or MeCN)-R$_4$NX systems, and the product distribution has been clarified after methylation of electrolysis products. One mechanism suggests the formation of an anion radical **93** by the addition of a one-electron to carbon disulfide followed by dimerization to the tetrathiooxalate ion **94** (eq. (6.40)). This intermediate **94** is considered to react further with carbon disulfide and to form a product **96** *via* **95** which, on two-electron reduction and the subsequent methylation, would form **90** [42a,b].

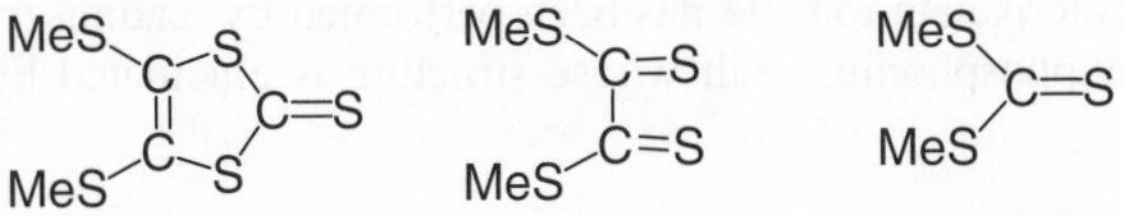

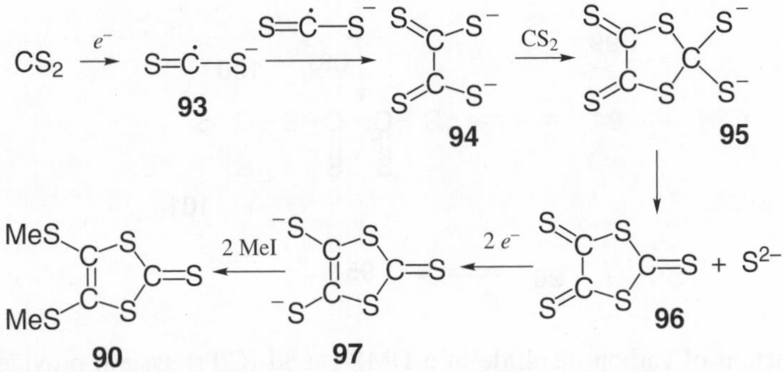

(6.40)

The other insists, on the basis of cyclic voltammetric studies of the reaction of carbon disulfide in a DMF(or MeCN)-Bu$_4$NClO$_4$-(Pt) system, coulometry and isolation of carbon monosulfide, that initial species of the electroreduction of carbon disulfide are trithiocarbonate ion **98** and carbon monosulfide, and the former species gives also **92** after methylation (eq. (6.41)) [43]. The formation of the dimer of trithiocarbonate ammonium salt is

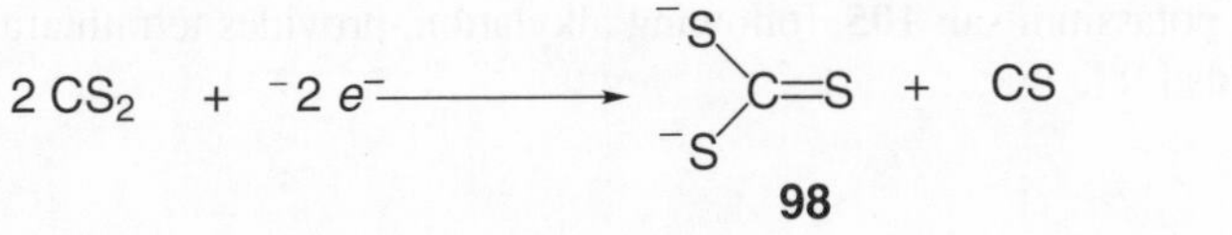

(6.41)

also recorded [44]. The voltammetric data obtained at a hanging mercury drop electrode is consistent with an initial reduction of carbon disulfide to the anion radical **93** (eq. (6.42)) [45]. The subsequent fast dimerization with carbon disulfide forms a new anion radical species **99** which undergoes further one-electron reduction to give a dianion **100**. This dianion **100** may dissociate into carbon monosulfide and trithiocarbonate **98**.   The dianion **100** can react further with carbon disulfide and form a derivative of tetrathiooxalate **101** which can dissociate to carbon disulfide and tetrathiooxalate ion **94** or cyclize to **95**.   The ring intermediate **95** can lose sulfide ion and form **96**.   The formation of carbon monosulfide as well as trithiocarbonate **98** and tetrathiooxalate ions **94** has been confirmed in a large scale electrolysis.   4,5-Dimercapto-1,3-dithiole-2-thione **96** produced in an almost quantitative yield has been identified as a major intermediate in the carbon disulfide electrolysis in a DMF-Bu$_4$NI-(Pt) system [42b].   Recently, a high-performance liquid chromatographic study revealed that the initial intermediate of the electroreduction of carbon disulfide is assigned to be tetrathiooxalate dianion **94** and not to be trithiocarbonate **98** [42b].

The latter compound **98** becomes a major product under conditions of more negative potential and longer electrolysis time, but the compound is not present in the solutions which are sampled at the initial stage of electrolysis.   The confirmation of the electrochemically generated tetrathiooxalate ions **94** has been performed by leading them to the corresponding ditetraphenylphosphonium salt whose structure is elucidated by X-ray analysis [46].

$$
CS_2 \;\rightleftharpoons\; 93 \;\xrightarrow[\text{fast}]{CS_2}\; S{=}\overset{\bullet}{C}{-}\underset{\underset{S^-}{|}}{S}{-}C{=}S \;\xrightarrow{e^-}\; S{=}C{-}\underset{\underset{S^-}{|}}{S}{-}C{=}S \;\rightleftharpoons\; 98 + CS
$$

$$
\qquad\qquad\qquad\qquad\qquad 99 \qquad\qquad\qquad\qquad\qquad\qquad 100
$$

$$
\downarrow CS_2
$$

$$
CS_2 + 94 \;\rightleftharpoons\; {}^-S{-}\underset{\underset{S}{\|}}{C}{-}\underset{\underset{S}{\|}}{C}{-}\underset{\underset{S^-}{}}{S}{-}C{=}S \qquad\qquad (6.42)
$$

$$
101
$$

$$
\downarrow
$$

$$
S^{2-} + 96 \;\rightleftharpoons\; 95
$$

The electroreduction of carbon disulfide in a DMF-Bu$_4$NI-(C/Pt) system provides a 1,3-dithiole ring intermediate **97** as shown in eq. (6.43) [1b, 42c~g].   Alkylation yields thiones **102**, while treatment of **97** with thiophosgene in ethanol gives tetrathiapentalene-2,6-dithione **104** in a good yield (eq. (6.43)).   Oxidation of the electrolysis solution of **97** and **98** by combination of the cathodic and anodic compartments, the latter of which contains a I3$^-$/I$^-$ mixture, affords 4,5-dimercapto-1,3-dithiole-2-thione **103** in a quantitative yield [47].   Substituted 1,3-dithiole-2-thiones can lead to a variety of tetrathiafulvalenes **106** which are known to be good electron donors.   The chemical conversion of the thiones **104** *via* the corresponding potassium salt **105**, following alkylation, provides tetrathiafulvalenes **106** in reasonable yields [48].

$$4 \text{ CS}_2$$

The scheme shows compound **97** (dithiolate) plus **98** reacting through several pathways:

- Via RX in DMF (DMF-Bu$_4$NI-(C/Pt) or MeCN, $4e^-$) to give **102** with R = Me (90%), CH$_2$CH$_2$SMe (70%), (CH$_2$)$_3$SCOMe (70%)
- Via **98**, I$_3^-$ in DMF to give **103** (~100%)
- Via Cl$_2$CS in EtOH to give **104**, then 1) Hg(OAc)$_2$ 2) P(OEt)$_3$ 3) KOH to give **105** (90%), then RX to give **106** with R = CH$_2$CH$_2$SMe (60%), (CH$_2$)$_3$SCOMe (60%)

$$(6.43)$$

The electrochemical conversion of **90** into **109** has been developed as a three-step process.   First, the ethylated product **107** derived from **90** by treatment with Meerwein's reagent is subjected to electroreductive coupling to give orthothiooxalate **108** and the subsequent pyrolysis in a sealed tube affords tetrathiomethoxytetrathiofulvalene **109** in 75% yield (eq. (6.44)) [49].

The scheme shows **90** (MeS-substituted dithiole-2-thione) reacting with Et$_3$O$^+$ BF$_4^-$ to give **107**, then MeCN-Et$_4$NClO$_4$-(Pt), -0.6 V (SCE), 99.5% yield to give **108**, then forming **109**.

$$(6.44)$$

Electrochemical reduction of 4,5-bis(alkylthio)-1,3-dithiole-2-thiones **110** in a DMF-Bu$_4$NI-(Pt/Hg) system, followed by alkylation, gives tetrathioethylenes **111** in good yields (eq. (6.45)).   The starting materials **110** can be smoothly electrosynthesized from 1,3,4,6-tetrathiapentalene-2,5-dione **112** under similar electrolysis conditions (eq. (6.46)) [50]. 4,5-Bis(alkylthio)-1,3-dithiole-2-thiones, however, gives the corresponding tetrathioethylenes in inferior yields under similar conditions.

$$
\text{110} \quad \xrightarrow[\substack{\text{2) Excess RI}\\ \text{93~99\% yields}}]{\substack{\text{1) DMF-Bu}_4\text{NI-(Pt/Hg)}\\ -2.25 \sim -2.3\text{V}}} \quad \text{111}
$$

(6.45)

111   R = Me   93%
         = Isopr   99%

$$
\text{112} \quad \xrightarrow[\substack{\text{2) Excess RI}}]{\substack{\text{1) DMF-Bu}_4\text{NI-(Pt/Hg)}\\ -1.7 \text{ V(Ag/AgI)}}} \quad \text{110}
$$

(6.46)

110   R = Me   93%
         = Isopr   94%

6.1.3.2   Electroreduction of Thioketones, Thioesters and Thioamides

In general, the polarization of thiocarbonyl group is highly affected by the substituent. The electronic structure of the anion radical of thioketone reflects the electronic effect of the substituent. For example, aromatic thioketone provides a carbanion-like structure of an anion radical, whereas aliphatic thioketone tends to give a thiolate-like structure of an anion radical [51]. The behavior of electrochemically generated anion radicals from thioketones has been investigated by trapping with halides and tosylates, and the electronic structure of anion radicals is discussed on the basis of the products. It is noted that alkyl aryl or dialkyl thioketones afford preferentially C-alkylation products, whereas diaryl thioketones tend to give S-alkylated products [51].

Polarographical analysis of the reduction of a 3-methylthio-2-cyclohexene-1-thione in a MeOH/H$_2$O-NaClO$_4$ system has been studied, suggesting a sequence of protonation to thione and uptake of one electron to form a radical intermediate [52].

The electroreduction of unsymmetrical thioketone 113 in methanol in the presence of tetraethylammonium bromide affords a mixture of the corresponding thioether 114 and monothioacetal 115 in 83% total yield (eq. (6.47)) [53]. The electrolysis of benzoylthioketone 116 also gives similar products 117 and 118 (eq. (6.48)). The formation of the monothioacetals may be due to disproportionation of a carbo-radical intermediate.

$$
\underset{\text{113}}{\text{Ph}-\overset{\text{S}}{\underset{\|}{\text{C}}}-\text{Bu-}tert} \quad \xrightarrow[\text{MeOH}]{e^-,\text{ MeI}} \quad \underset{\text{114}\;43\%}{\text{Ph}-\overset{\text{H}}{\underset{\text{SMe}}{\overset{|}{\text{C}}}}-\text{Bu-}tert} \;+\; \underset{\text{115}\;40\%}{\text{Ph}-\overset{\text{OMe}}{\underset{\text{SMe}}{\overset{|}{\text{C}}}}-\text{Bu-}tert}
$$

(6.47)

$$
\underset{\text{116}}{\text{PhCO}-\overset{\text{S}}{\underset{\|}{\text{C}}}-\text{Bu-}tert} \quad \xrightarrow[\text{MeOH}]{e^-,\text{ MeI}} \quad \underset{\text{117}\;43\%}{\text{PhCO}-\overset{\text{H}}{\underset{\text{SMe}}{\overset{|}{\text{C}}}}-\text{Bu-}tert} \;+\; \underset{\text{118}\;40\%}{\text{PhCO}-\overset{\text{OMe}}{\underset{\text{SMe}}{\overset{|}{\text{C}}}}-\text{Bu-}tert}
$$

(6.48)

Electroreduction of O-methyl thiobenzoate 119 can be partially converted into the corresponding O,S-acetal 120 in 25% yield together with unreacted 119 (27%) (eq. (6.49)) [54].

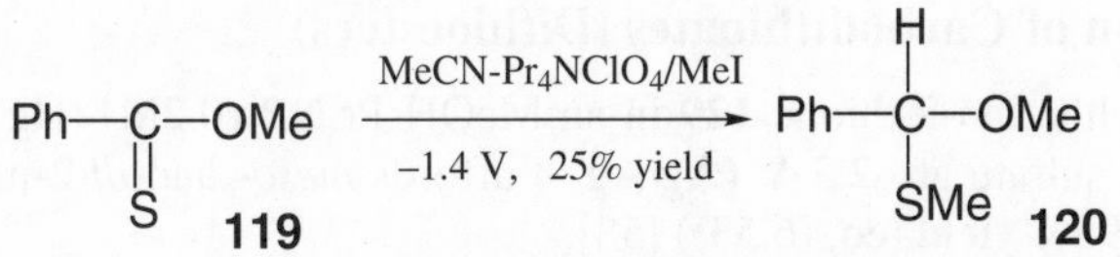

$$Ph-\underset{\underset{S}{\parallel}}{C}-OMe \xrightarrow[-1.4 \text{ V, } 25\% \text{ yield}]{MeCN-Pr_4NClO_4/MeI} Ph-\underset{\underset{SMe}{\mid}}{\overset{\overset{H}{\mid}}{C}}-OMe \qquad (6.49)$$

$$\underset{119}{} \qquad\qquad \underset{120}{}$$

Constant potential electrolysis of *S*-methyl benzoate **121** in a DMF-Bu$_4$NI-(Hg) system can be conducted to 1,2-diphenylacetylene **122** and methyl benzoate **123** in good yields (eq. (6.50)) [55].

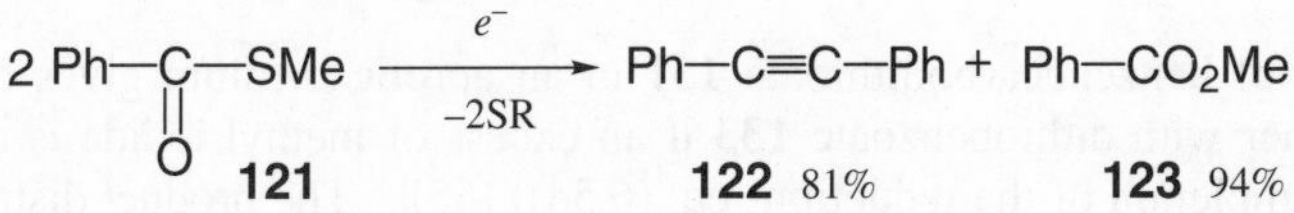

$$2 Ph-\underset{\underset{O}{\parallel}}{C}-SMe \xrightarrow[-2SR]{e^-} Ph-C\equiv C-Ph + Ph-CO_2Me \qquad (6.50)$$

$$\underset{121}{} \qquad\qquad \underset{122}{} 81\% \qquad \underset{123}{} 94\%$$

Electroreductive desulfuryzation of the thioamide **124** proceeds in aqueous sulfuric acid to give the corresponding amine **125** in 96% yield (eq. (6.51)) [56].

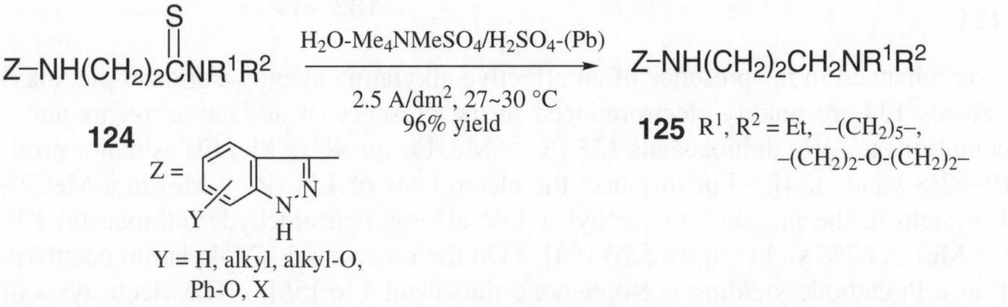

$$Z-NH(CH_2)_2\underset{\underset{124}{}}{\overset{\overset{S}{\parallel}}{C}}NR^1R^2 \xrightarrow[2.5 \text{ A/dm}^2, 27\sim30 \text{ °C}]{H_2O-Me_4NMeSO_4/H_2SO_4\text{-(Pb)}} Z-NH(CH_2)_2CH_2NR^1R^2 \qquad (6.51)$$

$$96\% \text{ yield} \qquad \underset{125}{} R^1, R^2 = Et, -(CH_2)_5-,$$
$$-(CH_2)_2-O-(CH_2)_2-$$

Z =

Y = H, alkyl, alkyl-O,
Ph-O, X

Electroreductive coupling of thiopyrones **126** has been performed in a DMF-Et$_4$NClO$_4$-(Hg) system in the presence of alkyl halides to give initial coupling products **127**. Subsequent reductive cleavage of the carbon-sulfur bonds of **127** affords bipyrannylidenes **128** as an electron-donating molecule in good yields (eq. (6.52)) [57].

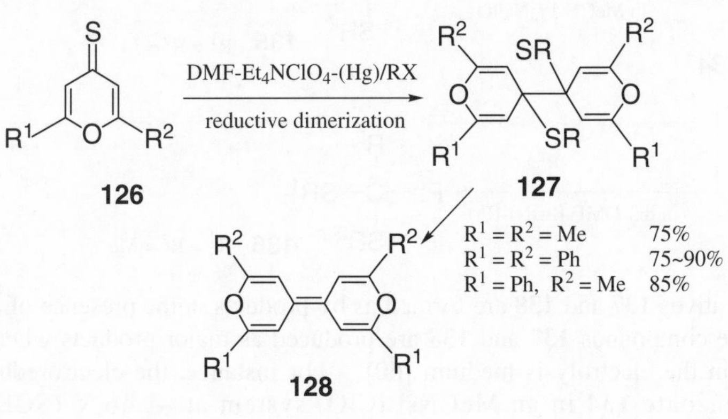

$$\xrightarrow[\text{reductive dimerization}]{DMF-Et_4NClO_4\text{-(Hg)/RX}}$$

126   **127**

$$R^1 = R^2 = Me \qquad 75\%$$
$$R^1 = R^2 = Ph \qquad 75\sim90\%$$
$$R^1 = Ph, R^2 = Me \qquad 85\%$$

128

$$(6.52)$$

6.1.3.3   Electroreduction of Carbodithioates (Dithioesters)

Electroreduction of 5-methylthiolan-2-thione **129** in an MeOH-Pr$_4$NBr(0.2M)-(Hg) system in the presence of dimethyl sulfate at –2.3 V (Ag/Ag$^+$) affords *meso-* and *dl*-2-methyl-5-(methylthio)thiolane **130** in 95% yield (eq. (6.53)) [58].

$$\text{Me}\underset{\textbf{129}}{\overset{\displaystyle \bigwedge_{S}}{}}S \xrightarrow[\text{–MeSO}_4\text{H, 95\% yield}]{\text{MeOH-Pr}_4\text{NBr(0.2M)-(Hg),}\ \text{Me}_2\text{SO}_4} \text{Me}\underset{\textbf{130}}{\overset{\displaystyle \bigwedge_{S}}{}}\text{SMe}$$

(6.53)

Electroreduction of benzenecarbodithioate **131** in an aprotic medium gives diphenylacetylene **132** together with dithiobenzoate **133** if an excess of methyl iodide is added to the catholyte after completion of the reduction (eq. (6.54)) [55].   The product distributions

$$4\ \underset{\textbf{131}}{\text{Ph}-\overset{\overset{\displaystyle \|}{S}}{\text{C}}-\text{SMe}} \xrightarrow[\text{2) MeI}]{\text{1) DMF-Bu}_4\text{NI-(Hg)}} \underset{\textbf{132}\ \ 81\%}{\text{Ph}-\text{C}\equiv\text{C}-\text{Ph}} + 2\ \underset{\textbf{133}\ \ 94\%}{\text{Ph}-\overset{\overset{\displaystyle \|}{S}}{\text{C}}-\text{S}^-} + 4\text{MeS}^-$$

(6.54)

are, however, changed in the presence of an effective alkylating agent.   For example, alkyl dithiobenzoates **134** are readily electroreduced in the presence of alkylating agents under various conditions to form dithioacetals **135** (R^1 = Me, Et, Ipr; R^2 = Me, Et) as major products in 19~67% yields [54].   For instance, the electrolysis of **134** (R^1 = Me) in a MeCN-Pr$_4$NClO$_4$ system in the presence of methyl iodide affords benzaldehyde dithioacetal **135** (R^1 = R^2 = Me) in 67% yield (eq. (6.55)) [54].   On the other hand, *C*-alkylation occurs in dry DMF at a Pt-cathode yielding acetophenone thioacetal **136** [55].   The electrolysis of **134** (R^1 = Me) in a DMF-Bu$_4$NI-(Pt) system in the presence of methyl iodide at –1.4 V *vs.* SCE gives 1,1-di(methylthio)-1-phenylethane **136** in 96% yield (eq. (6.56)) [55]. Dithioacetals are also produced as major products by the electroreduction of alkyl dithiobenzoates in an MeOH-Et$_4$NBr-(Hg) system in the presence of dimethyl sulfate [59].

$$\underset{\textbf{134}}{\text{Ph}-\overset{\overset{\displaystyle \|}{S}}{\text{C}}-\text{SR}^1} \xrightarrow[\text{in MeCN-Pr}_4\text{NClO}_4]{2\,e^-,\ \ \text{R}^2\text{X}} \underset{\underset{\textbf{135}\ \ \text{R}^1 = \text{R}^2 = \text{Et}}{67\,\%}}{\text{Ph}-\overset{\overset{\displaystyle \text{SR}^2}{|}}{\underset{|}{\text{CH}}}-\text{SR}^1}$$

(6.55)

$$\xrightarrow[\text{in dry DMF-Bu}_4\text{NI-(Pt)}]{2\,e^-,\ \ \text{R}^2\text{X}} \underset{\underset{\textbf{136}\ \ \text{R}^1 = \text{R}^2 = \text{Me}}{96\,\%}}{\text{Ph}-\overset{\overset{\displaystyle \text{R}^2}{|}}{\underset{\underset{\displaystyle \text{SR}^2}{|}}{\text{C}}}-\text{SR}^1}$$

(6.56)

Stilbene derivatives **137** and **138** are formed as by-products in the presence of alkyl iodides [54], but the compounds **137** and **138** are produced as major products when the iodides are absent in the electrolysis medium [60].   For instance, the electroreduction of methyl dithiobenzoate **134** in an MeCN-LiClO$_4$ system at –1.46 V (SCE) gives bis(alkylthio)stilbenes **137** and **138** in good yields when alkyl halides are added directly after electrolysis (eq. (6.57)) [60].

$$\underset{\textbf{134}}{\overset{S}{\underset{\|}{Ph-C-SMe}}} \quad \xrightarrow[\text{70 \% yield}]{2\,e^-,\ R^2I} \quad \underset{\textbf{137}\ 4\,\%}{\overset{Ph}{\underset{MeS}{\diagup\diagdown}}\overset{SMe}{\underset{Ph}{\diagdown\diagup}}} \quad + \quad \underset{\textbf{138}\ 66\,\%}{\overset{Ph}{\underset{MeS}{\diagup\diagdown}}\overset{Ph}{\underset{SMe}{\diagdown\diagup}}} \tag{6.57}$$

Electroreduction of methyl dithiobenzoate **131** in a MeCN-Pr$_4$NClO$_4$-(Pb) system in the presence of 1,3-dimesyloxypropane yields the cyclization product **139** in 30% yield (eq. (6.58)) [61].   Co-electrolysis of **131** with allyl chloride yields *S*-allylated product **140**

$$\underset{\textbf{131}}{\overset{S}{\underset{\|}{Ph-C-SMe}}} \quad \xrightarrow[\substack{4\,e^-,\ -2\,MeSO_3^-,\ -2\,MeS^-}]{\substack{\text{MeCN-Pr}_4\text{NClO}_4\text{-(Pb)} \\ \text{MsO-(CH}_2)_3\text{-OMs}}} \quad \underset{\textbf{139}}{\overset{Ph}{\underset{S}{\diagup}}\overset{Ph}{\underset{S}{\diagdown}}\ (CH_2)_3} \tag{6.58}$$

$$\xrightarrow[\substack{2\,e^-,\ 2\,H^+,\ -HCl}]{CH_2=CH\text{-}CH_2Cl} \quad \underset{\textbf{140}}{\underset{S-CH_2CH=CH_2}{Ph-CH-SMe}}$$

(R^1 = Me, R^2 = allyl) in 40% yield.   The electroreduction of bis-dithiobenzoates **141** in a MeOH-Et$_4$NBr-(Pb) system in the presence of methyl iodide affords the corresponding C = S bond reduced products **142** together with other complex by-products (eq. (6.59)) [61]. On the other hand, the electrolysis of **141** in dry acetonitrile in the presence of methyl iodide gives thiirane **143** in 28% yield (eq. (6.60)).   In the case of ethylene bis-dithiobenzoate **141** (n = 2), similar electrolysis affords a 1:1 mixture of seven-membered heterocyclic products **144** and **145** in 74% yield (eq. (6.61)) [61].

$$\xrightarrow[\substack{n = 3, 4}]{\text{in MeOH,\quad MeI}} \quad \underset{\textbf{142}}{\underset{SMe}{Ph-CH-S(CH_2)_nS}-\underset{SMe}{CH-Ph}} \tag{6.59}$$

$$\underset{\substack{\textbf{S}\ \textbf{141}\ \textbf{S}}}{Ph-\overset{S}{\underset{\|}{C}}-S(CH_2)_nS-\overset{S}{\underset{\|}{C}}-Ph}$$

$$\xrightarrow[\substack{n = 3, 4}]{\text{in MeCN,\quad MeI}} \quad \underset{\textbf{143}\ 28\,\%}{\overset{Ph}{\underset{MeS}{\diagup}}\overset{C-C}{\underset{S}{}}\overset{Ph}{\underset{SMe}{\diagdown}}} \tag{6.60}$$

$$\xrightarrow[\substack{74\,\%}]{\substack{\text{in MeCN,}\\ \text{MeI}\quad n = 2}} \quad \underset{\textbf{144}}{\overset{Ph}{\underset{MeS}{S}}\overset{S}{\underset{SMe}{Ph}}} \quad + \quad \underset{\textbf{145}}{\overset{SMe}{\underset{Ph}{S}}\overset{S}{\underset{SMe}{Ph}}} \tag{6.61}$$

6.1.3.4   Electroreduction of Trithiocarbonates

Trithiocarbonates are readily reduced electrochemically to the corresponding anion radicals which undergo further chemical reactions [62].   Electroreduction of ethylene trithiocarbonate **146** in a DMF-Bu$_4$NBr-(Pt) system at −1.4 to −1.6 V (SCE) affords 1,2-ethanebis(methyltrithiocarbonate) **148** as a major product after alkylation of **147** with methyl iodide (eq. (6.62)) [63].   The reduction proceeds through an anion radical of **146** which undergoes elimination of a molecule of ethylene affording bis(trithiocarbonate) dianion **147** isolated as the methyl ester **148**.

$$
2 \quad \underset{\textbf{146}}{\left[\begin{array}{c} S \\ S \end{array}\right]=S} \quad \xrightarrow[-C_2H_4]{2\,e^-} \quad \underset{\textbf{147}}{\begin{array}{c} S\text{-}CS_2^- \\ S\text{-}CS_2^- \end{array}} \quad \xrightarrow{MeI} \quad \underset{\textbf{148}}{\begin{array}{c} S\text{-}CS_2Me \\ S\text{-}CS_2Me \end{array}}
\tag{6.62}
$$

Alkyl aryl and dialkyl trithiocarbonates **149** and **150** have been subjected to electroreduction in a DMF-Bu$_4$NI-(Hg) system at –1.06 to –1.25 V (Ag/AgI) in the presence of alkylating agent to give tetraalkyltetrathioethylenes **151** in 37~47% yields (eq. (6.63)) [64].

$$
\begin{array}{c} \underset{\textbf{149}}{PhS\text{—}CS\text{—}SR} \\ or \\ \underset{\textbf{150}}{RS\text{—}CS\text{—}SR} \end{array} \quad \xrightarrow[\substack{RX, \\ 37\text{~}47\%\ yields}]{DMF\text{-}Bu_4NI\text{-}(Hg)} \quad \underset{\textbf{151}}{\begin{array}{c} RS \\ RS \end{array}\!\!\!>=<\!\!\!\begin{array}{c} SR \\ SR \end{array}} \quad + \quad Others
\tag{6.63}
$$

Electroreduction of diaryl trithiocarbonates **152** in a DMF-Bu$_4$NI-(Hg) system followed by alkylation with alkyl halides gives a mixture of Z- and E-tetrathioethylenes (Z/E = ca. 1/10) and thioanions **153a**, **153b** in reasonable yields (eq. (6.64)) [65].   The yields depend on the substituents.   Strongly electron-attracting or electron-donating substituents may provide lower yields.

$$
2\ \underset{\textbf{152}}{ArS\text{—}CS\text{—}SAr} \quad \xrightarrow[4\,RX,\ -0.9\ V\ (Ag/AgI)]{DMF\text{-}Bu_4NI\text{-}(Hg)} \quad \underset{\textbf{153a}}{\begin{array}{c} RS \\ ArS \end{array}\!\!\!>=<\!\!\!\begin{array}{c} SAr \\ SR \end{array}} \quad + \quad \underset{\textbf{153b}}{2\ ArS\text{—}R}
\tag{6.64}
$$

Ar	R	153a%	153b%
Ph	Me	41	75
Ph	Et	38	72
Ph	Isopr	38	71

Electroreduction of 4,5-bis(alkylthio)-1,3-dithiole-2-thiones **102** (R = Me) in a DMF-Bu$_4$NI-(Hg) system at –1.4 V (SCE) followed by treatment with methyl iodide after completion of the electrolysis gives tetrakis(methylthio)ethylene **151** (R = Me) in 38% yield (eq. (6.65)) [50].

$$
\underset{\textbf{102}\quad R\,=\,Me,\ isopr}{\begin{array}{c} RS \\ RS \end{array}\!\!>\!\!\underset{C-S}{\overset{C-S}{}}\!\!C=S} \quad \xrightarrow[22\text{~}43\%\ yields]{\substack{DMF\text{-}Bu_4NI\text{-}(Hg) \\ RI,\ -1.4\ V\ (SCE)}} \quad \underset{\textbf{151}}{\begin{array}{c} RS \\ RS \end{array}\!\!>\!C=C\!<\!\!\begin{array}{c} SR \\ SR \end{array}}
\tag{6.65}
$$

154   **155**

Unsymmetrical tetrathiafulvalene molecules have been electrosynthesized in an MeCN-Et$_4$NBF$_4$(0.2M)-(Pt) system at −1.0 V vs. SCE. The ethylenedithio(dimethyl) tetrathiafulvalene **154** and (ethylenedithio)benzotetrathiafulvalene **155** obtained can be conducted to their radical-cation salts by employing an electrocrystallization method [66].

A novel EG base-catalyzed rearrangement of 2,5-diaryl-1,4-dithiines **156** has been realized in an MeCN-Bu$_4$NPF$_6$-(Pt/Hg) system [67]. The rearrangement of **157** would take place under the above electrolysis conditions by deprotonation of the 1,4-dithiin with a EG base followed by ring opening to yield the thiolate intermediate, Ar − C≡C − S − C(Ar)= CH − S$^-$ as an initial step. The subsequent ring closure by intramolecular nucleophilic attack of the thiolate at the triple bond to yield the anion of 1,4-dithiafulvene **157** (eq. (6.66)).

$$(6.66)$$

6.2   Electroreduction of Tetravalent Sulfur Compounds

6.2.1   Reduction of Sulfonium Salts

Irreversible one-electron reduction of sulfonium salts **158** has been observed [29, 68]. Homolytic cleavage fragment of benzyldimethylsulfonium p-toluenesulfonate has been trapped by acrylonitrile, giving 4-phenylbutyronitrile in good yield [29]. One-electron reduction of the sulfonium salts **158** undergoes a homolytic carbon-sulfur bond cleavage to give the corresponding disulfides **159** (eq. (6.67)) [69a]. The concerted or stepwise character of the electron transfer-bond-breaking process for aryl dialkylsulfonium cations is found to be a function of molecular structure [69b].

$$(6.67)$$

158   R^1 = alkyl
            R^2 = aryl   **159**

The E$_p$ values have been shown to be extremely sensitive to the electronegativity of the fragmenting radical [R^2]$^\bullet$, suggesting that the electrochemical one-electron reductive cleavage reaction of **158** proceeds by a concerted or nearly concerted mechanism in which the bond breaking is concomitant with electron acceptance [69a]. The preparative electroreduction of (6-acetoxy-1-isopropenyl-4-methylhexyl)dimethylsulfonium perchlorate **160** leading to α-citronellyl acetate **161** as a major product has been attained in a DMF-H$_2$O(10%)-LiClO$_4$(0.3M)-(Pt/Hg) system (eq. (6.68)) [70]. Highly conductive organic charge transfer complexes are formed as complexes of the donor and acceptor, tetrathioulfulvalenes **163** and tetracyano-p-quinodimethane (TCNQ) [47].

$$
\underset{\mathbf{160}}{\text{(structure with } Me_2S^+ClO_4^-)} \xrightarrow[\text{60~80\% yields}]{\begin{array}{c}\text{DMF/H}_2\text{O(10\%)-LiClO}_4 \\ \text{(0.3M)-(Pt/Hg)}\end{array}} \underset{\mathbf{161}}{\text{(product Z)}} + \underset{\mathbf{162}}{\text{(product Z)}} \qquad (6.68)
$$

161/162 = 20/1

$$
Z = H_2C \text{(structure)} OAc , \quad H_2C \text{(structure)} OAc
$$

163

Tetrathiomethoxytetrathiofulvalene **109** is a choice for this purpose. The electrosynthesis of **109** has been carried out by the reduction of trithiosulfonium ions **106** [62] in an MeCN-Et$_4$NClO$_4$-(Pt) system to yield orthothiooxalates **108** in 99.5% yield [49] through a rapid dimerization of radical species [62, 71]. The sulfonium ions **107** can be prepared by treatment of trithiocarbonates **90** with Meerwein's reagent at room temperature in ethylene dichloride. Electroreduction of a series of perchlorate, fluorosulfonate, or tetrafluoroborate salts of cyclic trithiosulfonium ions has also been investigated [72]. Electrosynthesis of unsymmetrical tetrathiafulvalene derivatives has been performed in a similar way [66].

6.2.2 Reduction of Sulfoxides

The various methods for reducing sulfoxides to thioethers have been well reviewed in the literature [73] except for the progress of electroreduction methodology. This section will make up for some requirements of synthetic chemistry.

Electroreduciton of β-hydroxy sulfoxides **164** has been performed in an MeCN-Bu$_4$NBF$_4$-(Hg) system to give alcohols **165** and other sulfur-containing products (eq. (6.69)) [74]. The yield of alcohols can be improved by the addition of pyrene as an electron transfer mediator.

$$
\underset{\mathbf{164}}{Me\text{—}\underset{O}{\overset{}{\underset{\|}{S}}}\text{—}CH_2\text{—}\overset{R^1}{\underset{OH}{C}}\text{—}R^2} \xrightarrow{\text{method A and B}} \underset{\mathbf{165}}{Me\text{—}\overset{R^1}{\underset{OH}{C}}\text{—}R^2} + \text{Others} \qquad (6.69)
$$

Method A (Direct Electrolysis): MeCN-Bu$_4$NBF$_4$-(Hg), −2.9 V (Ag/AgNO$_3$)
Method B (Indirect Electrolysis): MeCN-Bu$_4$NBF$_4$-Pyrene-(Hg), −2.5 V

	R^1	R^2	Method A(%)	Method B(%)
165a	H	Et	5	32
165b	Me	Ph-(CH$_2$)$_2$	31	36
165c	Me	Me(CH$_2$)$_6$	41	45

Electroreductive cleavage of β-ketosulfoxides **166** has been performed in a DMF-KNO$_3$-(Pt/Hg) system in the presence of a pentane layer which serves to extract the product **167** as it is formed (eq. (6.70)) [75]. Reductive cleavage of the carbon-sulfur bond of α-

(4-tolylsulfinyl)acetophenone **168** proceeds in an EtOH-LiCl-(Pt/Hg) system at −1.15 V to give 169 and 170, but further reduction of the carbonyl group of 170 occurs at −1.55 V (eq. (6.71)) [76]. Prolonged electrolysis of **168** in 95% ethanol at pH 5.8 gives improved results (**169**, 75%; **170**, 75%) [77].

$$R-\overset{O}{\underset{}{\overset{\|}{C}}}-CH_2-\overset{O}{\underset{}{\overset{\|}{S}}}-Me \xrightarrow{\text{DMF-KNO}_3\text{-(Pt/Hg)}} R-\overset{O}{\underset{}{\overset{\|}{C}}}-Me \qquad (6.70)$$

166   **167**

R	%
C_6H_{13}	54
p-MeO-C$_6$H$_4$	74

$$Me-\!\!\!\boxed{}\!\!\!-\overset{O}{\underset{}{\overset{\|}{S}}}-CH_2-COPh \xrightarrow[\substack{\text{pH 6 (Acetate Buffer),}\\-1.15\text{ V (Ag/AgCl)}}]{\text{EtOH (95\%)-LiCl-(Pt/Hg)}} \begin{array}{c} Me-\!\!\!\boxed{}\!\!\!-SH \quad \textbf{169}\ 50\%\\ +\\ MeCOPh\\ \textbf{170}\ 50\% \end{array} \qquad (6.71)$$

168

Methyl nitrobenzenesulfenate **171** is found to be reduced electrochemically to the corresponding nitrobenzenthiolate ion **172** in a DMF-Bu$_4$NClO$_4$-(Hg) system, which can be trapped on treatment with methyl iodide to give the methyl sulfide **173** (eq. (6.72)) [78].

$$\boxed{}\substack{NO_2\\SOMe} \xrightarrow[\text{selectively}]{\text{DMF-Bu}_4\text{NClO}_4\text{-(Hg)}} \boxed{}\substack{NO_2\\S^-} \xrightarrow{MeI} \boxed{}\substack{NO_2\\SMe} \qquad (6.72)$$

171   **172**   **173**

Electroreduction of phenyl vinyl sulfoxide **174** in an aqueous acetone-H$_2$SO$_4$-(Pt/Hg) system gives 1,2-bis(phenylthio)ethane **175** in 66% yield together with 33% of the unchanged **174** (eq. (6.73)) [79]. A proposed mechanism suggests that the initially produced PhSO$^-$ would undergo Michael type addition to the conjugated thioenone system to form 1,2-bis(phenylsulfoxyl)ethane, a precursor of **175**.

$$2\ CH_2=CH-\overset{O}{\underset{}{\overset{\|}{S}}}-Ph \xrightarrow[\substack{-CH_2=CH_2,\\ 2\text{ F/mol Electricity}}]{\text{aq. Acetone-H}_2\text{SO}_4\text{-(Pt/Hg)}} Ph-S-CH_2CH_2-S-Ph \qquad (6.73)$$

174   **175** 66%

6.3   Electroreduction of Hexavalent Sulfur Compounds

6.3.1   Electroreduction of Unsymmetrical Sulfones

The reductive cleavage of cyclic and acyclic alkyl aryl sulfones at a mercury cathode has been intensively studied and different cleavage modes observed. Cleavage of the carbon-sulfur bond on unsymmetrical sulfones was first realized in an MeOH-Me$_4$NCl-(Hg) system [80]. Generally, acyclic alkyl aryl sulfones **176** undergo the electroreductive cleavage at the alkyl-sulfonyl bond (path b) (eq. (6.74)) [81, 82], and only the open-chain sulfones bearing a strongly electron-attracting substituent, *e.g.*, 4-cyano, are cleaved at the aryl-sulfonyl bond [83]. For the compounds **176** (n = 2 and 3), path a proceeds exclusively to give **177**, whereas for n = 4, an 85/15 ratio between paths a and b is observed.

In an eight-membered ring, however, compound **176** (n = 5), gives exclusively the alkyl-sulfonyl cleavage product **178** (n = 5) [84]. It is suggested that the dihedral angle between the sulfonyl group and the benzene ring determines the cleavage mode [84, 85]. A single electron transfer reduction of α-halo-, α-alkoxy-, or α-(alkylthio)alkyl phenyl sulfones producing the corresponding radical intermediates has been suggested [86]. In contrast to the behavior of open-chain alkyl aryl sulfones, the aryl-sulfonyl cleavage (path a) takes place preferentially in the compounds having a four, five, six, or, seven-membered ring with a sulfonyl group bound to the aromatic ring [81, 85].

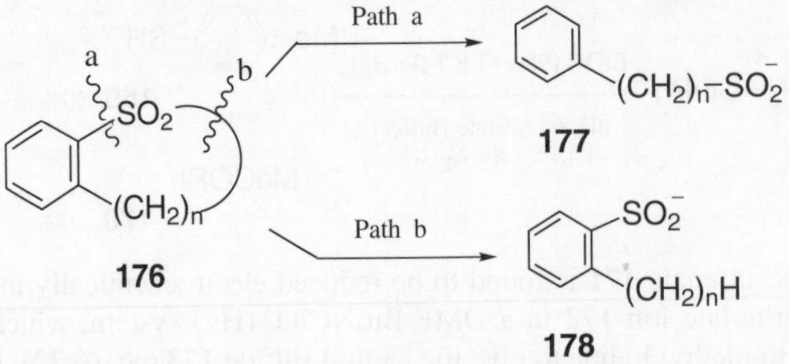

(6.74)

The preparation of 4-(2-deuterio-2-propyl)anisole **180** has been synthesized by the electroreductive cleavage of 2-(4-methoxyphenyl)-2-propyl-4-tolylsulfone **179** in a DMF-Et$_4$NClO$_4$-(Hg) system in the presence of deuterium oxide in 98.5% isotopic purity (eq. (6.75)) [87].

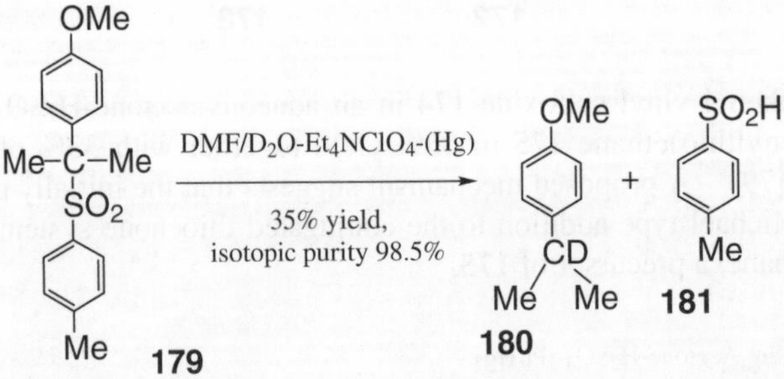

(6.75)

Intramolecular homolytic arylation of diaryl sulfones **182** has been observed. The controlled potential electrolysis of 2,3-bis(phenylsulfonyl)-1,4-dimethylbenzene **182** in a DMSO-Bu$_4$NBF$_4$/Bu$_4$NOAc-(Hg) system at -1.65 V (Ag/Ag$^+$) affords 1,4-dimethyl-2-(phenylsulfonyl)benzene **183** (23%) (eq. (6.76)) [88, 89]. Electroreductive cleavage of

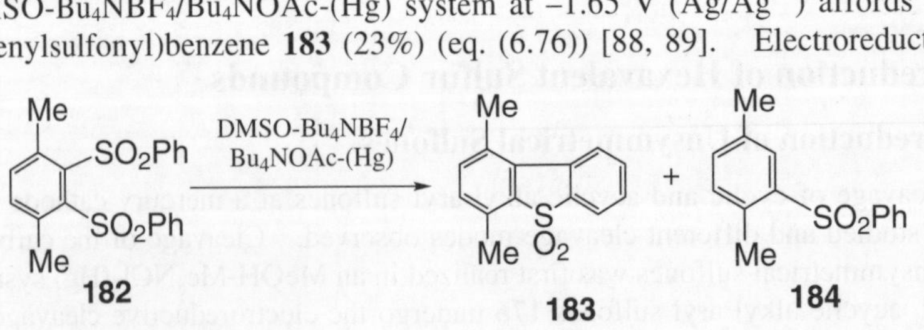

(6.76)

α-benzenesulfonylnitriles **185** proceeds exclusively at the alkyl-sulfonyl bond to give benzenesulfinate **186** and acetonitrile derivatives **187** (eq. (6.77)) [90].   A cyclic voltammetric

$$\underset{\substack{| \\ R \\ \textbf{185} \quad R = H,\ Me,\ Et,\ Ph,\ Bz}}{PhSO_2-CH-CN} \xrightarrow{\text{DMF-Et}_4\text{NClO}_4\text{-(SUS/Hg)}} \underset{\textbf{186}\ 48\sim64\%}{Ph-SO_2^-} + \underset{\textbf{187}\ 58\sim86\%}{R-CH_2-CN} \qquad (6.77)$$

study has revealed that the cleavage reaction is initiated by a one-electron reduction [91]. The regioselective cleavage of the C-S bond of p-toluenesulfonylmethylisocyanide **188** has been attained by electrochemical reduction in an MeCN-Et$_4$NBr-(C/Hg) system at $-1.36\sim-1.91$ V (Ag/Ag$^+$) to give the corresponding isocyanides **189** and **190** in 46~86% yields (eq. (6.78)) [92].

$$\underset{\textbf{188}}{\underset{\substack{| \\ R^1}}{\overset{\substack{R^2 \\ |}}{Ts-C-NC}}} \xrightarrow[\substack{46\sim86\% \text{ yields,} \\ -1.36\sim-1.91V\ vs.\ Ag/Ag^+}]{\text{MeCN-Et}_4\text{NBr-(C/Hg)}} \underset{\textbf{189}}{\underset{\substack{| \\ R^1}}{\overset{\substack{R^2 \\ |}}{H-C-NC}}} + \underset{\textbf{190}}{\underset{\substack{| \\ R^1}}{\overset{\substack{R^2 \\ |}}{Ts-C-NC}}} \qquad (6.78)$$

$$R^1 = PhCH_2,\ 4\text{-Me-C}_6\text{H}_4\text{CH}_2,\ 4\text{-Cl-C}_6\text{H}_4\text{CH}_2$$
$$R^2 = H,\ PhCH_2,\ 4\text{-Me-C}_6\text{H}_4\text{CH}_2,\ 4\text{-Cl-C}_6\text{H}_4\text{CH}_2,\ Me$$

Reductive cleavage of the C-S bond of p-toluenesulfonylmethylisocyanide

Electroreductive removal of a methylsulfonyl group from β,β-bis(methylsulfonyl) styrene **191** was carried out in a MeCN-Bu$_4$NBr-(Hg) system to afford β-methylsulfonyl styrene **192** in the presence of efficient proton donors (eq. (6.79)) [93a].   Electroreductive desulfurization of 1-methylthio-1-p-tolylsulfonyl-2-arylethens systems [93d] followed by carboxylation has been investigated [93b,c].   The electroreduction of 1,2-disubstituted compounds, bearing groups capable of leaving can lead to the corresponding unsaturated hydrocarbons.

$$\underset{\textbf{191}}{\overset{\substack{Ph \diagdown \qquad \diagup SO_2-Me \\ \diagup \qquad \diagdown}}{H \qquad SO_2-Me}} \xrightarrow[\substack{-1.7\sim-2.3\ V\ vs.\ Ag/Ag^+ \\ \text{Proton Donor : PhCO}_2\text{H}}]{\text{MeCN-Bu}_4\text{NBr-(Hg)}} \underset{\textbf{192}\ 79\%}{\overset{\substack{Ph \diagdown \qquad \diagup H \\ \diagup \qquad \diagdown}}{H \qquad SO_2-Me}} \qquad (6.79)$$

The cathodic reduction of 1,2-diphenylthiirene dioxide **193** in a DMF/AcOH-Et$_4$NClO$_4$ system affords mixed products **194** and **195**, resulting from cleavage of one or both carbon-sulfur bonds (eq. (6.80)) [94].   The cathodic behavior of β-substituted sulfones is highly

$$\underset{\textbf{193}}{\overset{\substack{Ph \qquad Ph \\ \diagdown \quad \diagup \\ \diagdown \diagup \\ S \\ O_2}}{}} \xrightarrow[\substack{2)\quad MeI}]{1)\ \text{DMF/AcOH-Et}_4\text{NClO}_4} \underset{\textbf{194}\ 40\%}{Ph-CH=CH-Ph} + \underset{\textbf{195}\ 27\%}{\overset{\substack{Ph \\ \diagup \\ \diagup \qquad \diagdown}}{Ph \qquad SO_2Me}} \qquad (6.80)$$

dependent on the medium composition and the nature of the substituent Y in the β-position to the sulfonyl group. The sulfones **196**, bearing a hydroxy, acetoxy or alkoxy group in the β-position, are electrochemically reduced with olefin formation in weakly protic media. For example, the electrolysis of **196** ($R^2 = Me$, $R^3 = Ph$, $R^4 = H$, $Y = OH$) in a DMF-Bu$_4$NI-(Hg) system at -1.6 V *vs.* Ag/AgI/I$^-$ in the presence of phenol as a proton source gives β-methylstyrene **197** ($R^1 = R^3 = H$, $R^2 = Ph$, $R^4 = Me$) in 90% yield (eq. (6.81)) [95]. Some electrolysis results in a DMF-Bu$_4$NI-(Hg) system are shown in Table 6.5 [95b]. On the other hand, the unsaturated sulfone **198** is produced in the presence of bases. For example, the electrolysis of 2-acetoxy-1,2-diphenyl-1-phenylsulfonylethane **196** in DMF-KOH system gives 1,2-diphenyl-1-phenylsulfonylethylene **198** in good yields (eq. (6.82)). On the contrary, in an aprotic medium, lack of a fast protonation causes a dramatic change in the reaction pathway, owing to the instability of the substrate toward EG bases. The action of such bases is closely related to the nature of the substituent Y in the β-position to the sulfonyl group and also of R^3 and R^4 groups. For instance, the electrolysis of β-hydroxy-sulfones undergoes a retrocondensation reaction by the action of EG base to afford the sulfones **199** and the carbonyl compounds **200** (eq. (6.83)).

Table 6.5 Reductive Elimination of β-Substituted Sulfones Leading to Olefins[†]

Substrate	Proton Donor	Potential V(Ag/AgI/I$^-$)	Product		Yield, %
OH Ph Ph–CH–CHSO$_2$Ph *erythro*	AcOH	−1.45	PhCHOH–CH$_2$Ph PhCH=CH–Ph *trans* PhCH=CH–Ph *cis* PhCH$_2$–CH$_2$Ph		30 25 traces 20
threo	AcOH	−1.45	PhCHOH–CH$_2$Ph PhCH=CH–Ph *trans* PhCH$_2$=CH$_2$Ph		80 5 15
OH Ph–CH–CH$_2$SO$_2$PhMe	PhOH	−1.75	PhCH=CH$_2$		80
OH Me Ph–CH–CHSO$_2$Ph	without	−1.65	Ph–CHOH–CHOHPh *meso + dl* PhSO$_2$CH$_2$Me		40 40
OAc Ph Ph–CH–CHSO$_2$Ph *erythro*	AcOH	−1.45	PhCH$_2$–CH$_2$Ph		95
OAc Ph Ph–CH–CHSO$_2$Ph *threo*	AcOH	−1.45	PhCH=CH–Ph *trans* PhCH=CH–Ph *cis* PhCH$_2$–CH$_2$Ph		80 15 5
OAc Ph–CH–CH$_2$SO$_2$PhMe	PhOH	−1.65	PhCH$_2$=CH$_2$		85
O O Me Ph–CH–CHSO$_2$Ph	PhOH	−1.70	PhCH=CH–Me		70

[†] Carried out in a DMF-Bu$_4$NI-(HG) system.
[Reproduced with permission from S. Gambino et al., *J. Electroanal. Chem. International Electrochem.*, **90**, 105 (1978)]

$$Ar-SO_2-\underset{\underset{R^2}{|}}{\overset{\overset{R^1}{|}}{C}}-\underset{\underset{R^4}{|}}{\overset{\overset{Y}{|}}{C}}-R^3$$

196

R^1 = H, Me, Ph
R^2 = H, Ph
Y = Oh, OMe, OAc

$\xrightarrow{\quad -ArSO_2^-,\ Y^- \quad}$

$$\underset{R_2}{\overset{R^1}{\diagdown}}C=C\underset{R^4}{\overset{R^3}{\diagup}} \qquad \textbf{197}$$

(6.81)

$\xrightarrow{\quad \text{EG Base, R}^1 = \text{H} \quad}$

$$\left[-\overset{|}{\underset{|}{C}}\overset{\frown Y}{-}\overset{|}{\underset{|}{C}}- \right]$$

$$Ar-SO_2\underset{\diagdown}{\overset{R^2}{\diagup}}C=C\underset{R^4}{\overset{R^3}{\diagdown}} \qquad \textbf{198}$$

(6.82)

$\xrightarrow{\quad \text{EG Base, Y = OH} \quad}$

$$\left[-\overset{|}{\underset{\underset{Ar-SO_2}{|}}{C}}\overset{O^-}{\underset{|}{C}}- \right]$$

199

$$\underset{\underset{Ar-SO_2}{|}}{R^2-\overset{\overset{R^1}{|}}{C^-}} \quad + \quad \underset{R^4}{\overset{R^3}{\diagup}}C=O \qquad \textbf{200}$$

(6.83)

6.3.2 Electroreductive Desulfonylation

6.3.2.1 Reductive Removal of Alkylsulfonyl Groups

The conversion of aliphatic alcohols **201** to the corresponding alkanes **203** is a routine technique in organic synthesis. The process involves tosylates, isoureas, and phosphates prepared from the alcohol as a good leaving group being exposed to several reduction procedures.

An efficient electroreductive transformation of methanesulfonates **202** of aliphatic alcohols **201** to the corresponding alkanes **203** has been reported. The electrolysis is carried out in a DMF-Et$_4$NOTs-(Pt/Pb) system at 5 °C to 10 °C and the substrate solution is carefully added into the cathodic cell in order to avoid the hydrolysis of methanesulfonates before electroreduction (eq. (6.84)) [96]. 1,3-Dimethanesulfonates **204** can lead to the corresponding cyclopropanes **205** by direct electroreduction. Electroreduction of 1,3-dimethanesulfonates **204** in a DMF-Et$_4$NOTs-(Pt/Pb) system affords cyclopropane derivatives **205** in 52~97% yields (eq. (6.85)), Table 6.6 [97]. Such 1,3-eliminative cyclopropane formation may proceed favorably in the case of 1,3-dimethanesulfonates **204**, and monomethanesulfonate derivatives **206** bearing an OTHP group at the position β bring about the formation of non-cyclized products **207** and **208** as shown in eq. (6.86) [97].

$$R-OH \longrightarrow R-OMs \xrightarrow[\text{57~87\% yields}]{\text{DMF-Et}_4\text{NOTs-(Pt/Pb)}} R-H$$

201   **202**   **203**

R = alkyl, alkenyl, aralkyl

(6.84)

$$\underset{R^2}{\overset{R^1}{\diagdown}}\overset{\displaystyle -OMs}{\underset{\displaystyle -OMs}{}} \xrightarrow[\text{52~97\% yields}]{\text{DMF-Et}_4\text{NOTs-(Pt/Pb)}} \underset{R^2}{\overset{R^1}{\diagdown}}\triangle$$

204 R^1, R^2 = alkyl, aralkyl   **205**

(6.85)

Table 6.6 Electroreductive Synthesis of Cyclopropane Derivatives[†]

R[1]	R[2]	R[3]	Electricity F/mol	Yield, %
PHCH$_2$	CH$_2$=CHCH$_2$	H	5	71
PhCH$_2$	PhCH$_2$	H	10	71
Me(CH$_2$)$_7$	H	H	6	70
Me(CH$_2$)$_{11}$	H	Me	5	76
Me(CH$_2$)$_5$	—(CH$_2$)$_3$—		4	55
H	—(CH$_2$)$_{10}$—		6	52
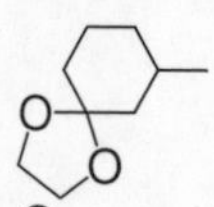 (spiro dioxolane cyclohexane)	H	H	5	97
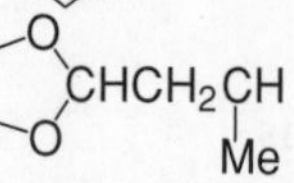 (dioxolane)CHCH$_2$CH(Me)	H	H	5	78
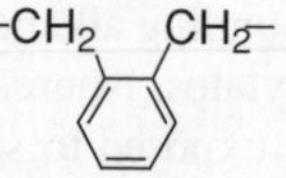 —CH$_2$—(benzene)—CH$_2$—		H	5	84

[†] Carried out in a DMF-Et$_4$NOTs-(Pt/Pb) system at 5.7 mA/cm^2, -2.5 V *vs.* SCE. [Reproduced with permission from T. Shono et al., *J. Org. Chem.*, **47**, 3090 (1982)]

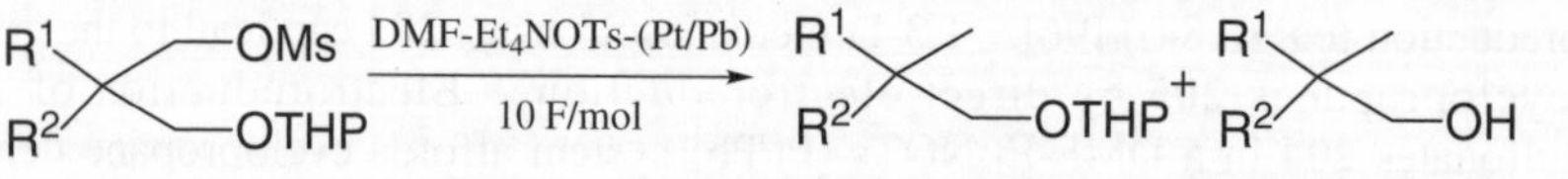

$$\underset{\textbf{206}}{\underset{R^1, R^2 = \text{alkyl, aralkyl}}{R^1\!\!\diagdown\!\!\diagup\!\!-OMs \atop R^2\!\!\diagup\!\!-OTHP}} \xrightarrow[\text{10 F/mol}]{\text{DMF-Et}_4\text{NOTs-(Pt/Pb)}} \underset{\textbf{207}\ 70\%}{R^1\!\!\diagdown\!\!\diagup\!\!-OTHP \atop R^2} + \underset{\textbf{208}\ 23\%}{R^1\!\!\diagdown\!\!\diagup\!\!-OH \atop R^2}$$

(6.86)

Electroreduction of vicinal bistosyloxy derivatives **209** in a MeCN-Bu$_4$NI-(Hg) system affords the corresponding epoxides **211** *via* **210** (eq. (6.87)) [98].   The result suggests that the cleavage occurs preferentially at the S-O bond linkage.

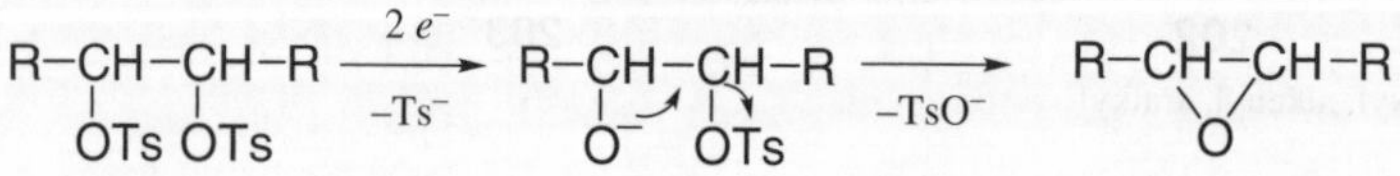

$$\underset{\textbf{209}}{\underset{\text{OTs OTs}}{R\text{-CH-CH-R}}} \xrightarrow[-\text{Ts}^-]{2\,e^-} \underset{\textbf{210}}{\underset{\text{O}\ \ \text{OTs}}{R\text{-CH-CH-R}}} \xrightarrow{-\text{TsO}^-} \underset{\textbf{211}}{\underset{\text{O}}{R\text{-CH-CH-R}}}$$

(6.87)

R	Yield, %
H	85
Me	60

Electrosynthesis of (3aS,8aR)-4-hydroxy-6-(tosyloxy)-3a,8a-dihydrofuro[2,3-b]benzo-furans **213** and **214** has been attained by electrolysis of ditosylate **212** in an MeCN-Et$_4$NBr-(Pt/Hg) system at −1.40 V (Ag/Ag$^+$) in 83% yield (eq. (6.88)) [99].   The electroreductive removal of the tosyl group proceeds in a regioselective manner (the ratio of **213/214** = 89/11).

$$(6.88)$$

212   **213** 83%   **214**

213/214 = 89/11

Electroreduction of cyclic and acyclic derivatives **215** and **216** of 2,3-butanediol ob-tained from xylose fermentation can lead to 2-butene **217**.   The cyclic sulphite and sulphate derivatives are directly reduced to **217**, but the non-cyclic derivatives can only be reduced with electrogenerated amalgam to give **217** in 39~49% yields (eq. (6.89)) [100].

$$(6.89)$$

6.3.2.2   Reductive Removal of Arenesulfonyl Groups

The controlled potential reduction in aprotic solvents has been shown to provide a selective method for the removal of reducible protecting groups from alcohols, phenol, thiol, and amine centers in polyfunctional molecules.   Many protecting groups commonly used, *e.g.*, tosyl, benzyl, and benzyloxycarbonyl, are not reduced until very negative potentials −2.0 to −3.0 V (SCE), which may lead to other difficulties, *e.g.*, reduction elsewhere in the mole-cule, generation of traces of a strong EG base.   These cause the loss of product selectivity.

The electroreduction of alkyl tosylates **218** in a MeCN-(Hg or C) system results in the formation of alcohols, ethers, and toluene together with p-toluenesulfinic acid (eq. (6.90)), and their product distributions are shown in Table 6.7 [101].   One electron reduction of alkyl tosylate (ROTs) forms a very short-lived anion radical [p-MeC$_6$H$_4$SO$_2$OR]$^{-\bullet}$ **A** (step (a)).   The anion radical **A** may decompose by scission of either an S-C bond as in step (b) or a C-O bond as in step (c).   Under the electrolysis conditions, the p-tolyl radical produced by decomposition of the anion radical is reduced to the p-tolyl carbanion [MeC$_6$H$_4$]− **B**. Protonation of the anion **B** in step (d) probably accounts for the formation of toluene as a product.

$$\text{ROTs} + e^- \longrightarrow [\, p\text{-MeC}_6\text{H}_4\text{SO}_2\text{OR} \,]^{-\bullet} \quad \cdots \text{(a)}$$
$$\underset{\textbf{218}}{} \qquad\qquad\qquad\qquad\qquad\quad \textbf{A}$$

$$\textbf{A} \longrightarrow \text{ROSO}_2^- + \text{Me}\!-\!\langle\ \rangle\!\bullet \xrightarrow{\ e^-\ } \text{Me}\!-\!\langle\ \rangle\!- \quad \cdots \text{(b)}$$
$$\qquad\qquad\qquad\qquad\qquad\qquad\qquad \textbf{B}$$

$$\textbf{A} \longrightarrow \text{RO}^- + p\text{-MeC}_6\text{H}_4\text{SO}_2{}^{\bullet} \xrightarrow{\ e^-\ } p\text{-MeC}_6\text{H}_4\text{SO}_2^- \quad \cdots \text{(c)}$$
$$\qquad\qquad\qquad\qquad\qquad\qquad\qquad\qquad\qquad \textbf{C} \qquad\qquad\qquad\qquad (6.90)$$

$$\textbf{B} \quad + \quad \text{H}^+ \longrightarrow \text{C}_6\text{H}_5\text{Me} \quad \cdots\cdots \text{(d)}$$

$$\text{RO}^- \quad + \quad \text{H}^+ \longrightarrow \text{ROH} \quad \cdots\cdots\cdots \text{(e)}$$

$$\text{RO}^- \quad + \quad \text{ROTs} \longrightarrow \text{ROR} \; + \; \text{OTs}^- \quad \cdots\cdots \text{(f)}$$

$$\text{ROSO}_2^- \; + \; \text{H}_2\text{O} \longrightarrow \text{ROH} \; + \; \text{HSO}_3^- \quad \cdots \text{(g)}$$
$$\text{R = alkyl}$$

Table 6.7 Constant Potential Reduction of Alkyl Tosylates[a]

Compound	Electrode	Electrolyte	Product, Yield, %		
			ROH	ROR	PhMe
Methyl tosylate	C	TPAP	24	36	27
	Hg	TPAP	44	–	25
	Hg	TEAB	42	35	
Ethyl tosylate	C	TPAP	32	18	30
Butyl tosylate	C	TPAP	62	–	26
	Hg	TPAP	58		28
Neopentyl tosylate	C	TPAP	68	–	34
	C	TPAI	98	–	28
	Hg	TPAI	97	–	25
Cyclohexyl tosylate	C	TPAP	75[b]	–	4
	Hg	TPAP	60[c]	–	23
	C	TEAB	96	–	trace
	Hg	TEAB	96	–	trace

[a] Carried out at -2.80 V *vs.* Ag/Ag NO₃. All yields mole % of starting tosylate. TPAP: Tetrapropylammonium perchlorate; TEBE: Tetraethylammonium bromide; TPAI; Tetrapropylammonium iodide. [b] Plus 4% cyclohexanone. [c] Plus 5% cyclohexane.
[Reproduced with permission from P. Yousefzadeh et al., *J. Org. Chem.*, **33**, 2716 (1968)]

Cleavage of an S-C bond in the decomposition of the anion radical **A** gives alkoxide ions and *p*-toluenesulfinyl radicals which are promptly reduced to sulfinate ions (step (c)). The recovery of substantial yields of alcohols reflects protonation of alkoxide during or after the electrolysis. The supporting electrolyte and the solvent, acetonitrile, probably serve at least as a source of protons for steps (d) and (e). Isolation of ethers suggests the displacement reaction shown in step (f).   Alcohol and toluene are presumably produced by steps (b), (d) and (g) (eq. (6.90)).

The introduction of an electron-withdrawing group into the protecting group can improve the selectivity.   The nitro group is an obvious choice.   The electroreduction of 2-, 3-, and 4-nitrobenzenesulfonates **219** in a DMF-Bu$_4$NBF$_4$-(Pt/Pt) system affords the corresponding detosylated alcohols in 54~94% yields (eq. (6.91)) [102].   The esters **219** are

$$NO_2\text{-}\bigcirc\text{-}SO_2\text{-}OR \xrightarrow[-1.4\sim2.1\ V\ (SCE)]{DMF\text{-}Bu_4NBF_4\text{-}(Pt/Pt)} NO_2\text{-}\bigcirc\text{-}SO_2OH + R\text{-}OH \qquad (6.91)$$

219 **a:** 2-nitro
       b: 3-nitro
       c: 4-nitro

220

R = Ph	Yield, %
a	94
b	80
c	76

reduced in two steps: the first reduction proceeds in the range of -0.70 to -0.90 V (SCE) and leads to a rather stable anion radical, and the second in the range of -1.20 to -1.70 V and leads to a stable dianion.   Cleavage of the S-O bond, affording free phenol or alcohol becomes a major and rapid process beyond the second peak at -1.20 to -1.70 V.   It is clear that introduction of the nitro group shifts the reduction potential to around -1.0 V.   The regioselective S-O cleavage of aryl tosylates has been attained by electroreduction of ditosylate **221** in an MeCN-Bu$_4$NBr-(Pt/Hg) system to yield **222** and **223** (91/9) in 85% yield (eq. (6.92)) [103].

$$\underset{\textbf{221}}{\overset{CO_2Me}{\bigcirc}} \xrightarrow[85\%\ yield]{MeCN\text{-}Bu_4NBr\text{-}(Pt/Hg)} \underset{\textbf{222}}{\overset{CO_2Me}{\bigcirc}} + \underset{\textbf{223}}{\overset{CO_2Me}{\bigcirc}} \qquad (6.92)$$

A characteristic feature of sulfones has been found in the cleavage of a carbon-sulfur bond.   Alkyl aryl sulfones are split at a mercury cathode to give arenesulfinate ions together with alkanes [80, 82a].   The electroreduction of allyl phenyl sulfone affords benzenesulfinate ion and propene [80].   Electroreductive hydrogenation of phenyl vinyl sulfone is known to give ethyl phenyl sulfone [104].   However, 2,2-diphenylvinyl phenyl sulfone can lead to 1,1-diphenylethene under electrolysis in a DMF-Bu$_4$NI-(Hg) system at -1.4 V vs. Ag/AgI [105].   Other vinyl sulfone derivatives are also discussed.

Indirect electrolysis procedure has been found to be effective for a selective cleavage of the allylic benzenesulfonyl group.   Electrolysis of the disulfone **224** was carried out in a DMF-Bu$_4$NI-(Hg) system at -2.0 V vs. SCE to give 2-methyl-5-benzenesulfonyl-2-pentene in 60% yield [106].   A facile transformation of β-hydroxysulfones **225** to the corresponding olefins **226** has been performed in a DMF-Et$_4$NOTs-(Pt/Pb) system at -2.4 to -2.6 V vs. SCE in a divided cell (eq. (6.93)) [107].   Some results are shown in Table 6.8 [107].

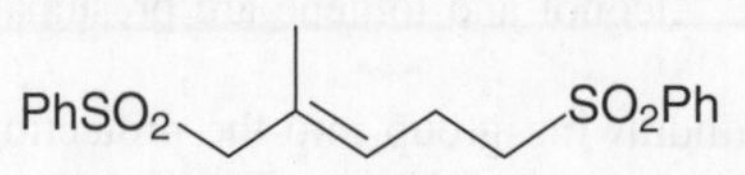

PhSO₂ ⟍ ⟋ ⟍ ⟋ ⟍ SO₂Ph

224

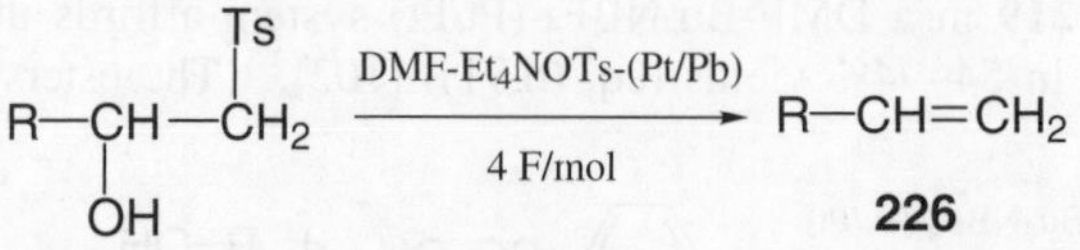

$$\underset{\underset{OH}{|}}{R-\overset{Ts}{\underset{|}{CH}}-CH_2} \xrightarrow[\text{4 F/mol}]{\text{DMF-Et}_4\text{NOTs-(Pt/Pb)}} R-CH=CH_2 \tag{6.93}$$

225   Ts = *p*-MeC₆H₄SO₂
          R = alkyl, alkenyl

226

Table 6.8 Electroreductive Preparation of Olefins from β-Hydroxysulfones[a]

$\underset{\underset{OH}{|}}{R-CH-CH_2-Ts^b}$ R	$R-CH-CH_2$ Yield, %
Me(CH₂)₁₆	82
Ph(CH₂)₂	74
CH₂(CH₂)₅CH₂	87
Me(CH₂)₇CH=CH(CH₂)₆CH₂	74
(bicyclic structure)	76
HO(CH₂)₄CH₂	73
(HO—CH(iPr)...CH₂ structure)	73
HO(CH₂)₁₀CH₂	80
Me(CH₂)₇CH₂	70
(alkenyl CH₂ structure)	71
PhCH₂(CH₂)	74

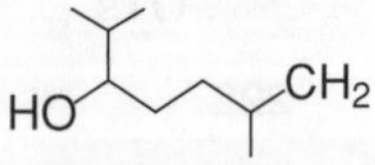

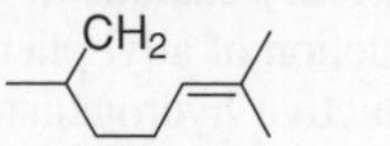

[a] Carried out in a DMF-Et₄NOTs-(Pt/Pb) system by passage of 8 F/mol electricity. [b] Ts: *p*-MeC₆H₄SO₂.
[Reproduced with permission from T. Shono et al., *Chem. Lett.*, **1978**, 69]

The electroreductive cyclization of 4-(phenylsulfonylthio)azetidin-2-one derivative **227** has been attained by electrolysis in a DMF/Py(5/1v/v)-BiCl₃-(Sn) system to afford 3-hydroxy cephem **228** in 67% yield (eq. (6.94)) [108].

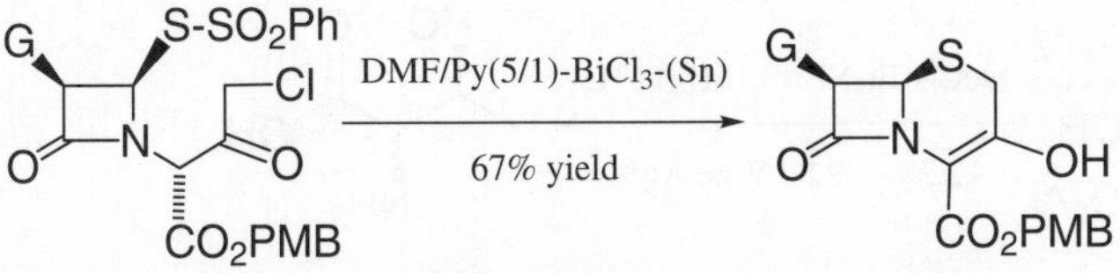

$$(6.94)$$

227   G = PhCH₂CONH
PMB = *p*-CH₃OC₆H₄CH₂

228

6.3.3 Sulfur-Nitrogen Bond Cleavage of Arenesulfonamides

Arylsulfonyl substituents are a usable group for the protection of a variety of primary amines. Indeed, *N*-arylsulfonylamino functions are stable enough for most reactive conditions. The electrochemical cleavage method of the S-N bond of arenesulfonamides has to be one of the choice because they can be cathodically hydrolyzed in alkaline as well as in acidic media [109]. The problem is the lack of a standard procedure for their removal. Some suitable methods have been devised for a specific compound and there is no proof of its versatility to other molecules.

The electroreduction of benzenesulfonamides in a MeCN-Et₄NBr-(Hg) system has been shown to be quite similar to that of alkyl and aryl halides. A two-electron process forms an intermediate [Ar-SO₂NR₂]⁻˙ which undergoes scission of the sulfur-nitrogen bond to produce sulfinate [Ar-SO₂]⁻ and amide [R₂N]⁻ ions in high yields [110]. The rate constants for the cleavage of the anion radicals lie in the range of 10^4 sec⁻¹ to > 10^8 sec⁻¹ [111]. Other mechanistic attempts at the electroreductive cleavage of arenesulfonamides in various media have been made and discussion and verification can be found in the literature [112]. For benzene- and *p*-toluenesulfonamides, one irreversible reduction step (–*Ep*, V *vs.* Ag/AgNO₃(0.1M)) in the range of –2.8 to –3.0 V is observed [110]. Polarographic and theoretical analysis of the shape of LUMO of sulfonamides reveals that the C-S-N is the electroactive reduction center and the S-N bond is, in general, the preferred cleavage site [113]. The amino radical intermediates are characterized by means of the spin marking technique [114]. The fundamental concepts and experimental directions of the method of electro-deprotection have been well documented [115].

A comparative investigation of deprotection methods has been carried out on *N*-(aryl-sulfonyl)threonines **229**. Three deprotections, namely, sodium in liquid ammonia, HBr in acetic acid, and electrolysis methods have been compared (eq. (6.95)) [116]. In the latter two methods, a (2,4-dimethylphenyl)sulfonyl group appears to be most desirable for deprotection purposes. The electroreductive deprotection of **229** takes place in a MeCN-Et₄NBr-(Hg) system in the presence of phenol as a proton source to give the corresponding α-amino acid **230** in good yields. The cleavage of the S-N bond of *tert*-butyl *N*-sulfonylcarbamates is found to take place at –1.67~ –2.64 V (SCE) in an MeCN-Et₄NCl(0.1M)-(C) system in the presence of proton donor (Et₃NHCl), giving the corresponding deprotected products (RSO₂NHCH₂Ph, R = Ph, Aryl) in 67~96% yields [117]. The tosyl protecting group (Ts) in amino acids, aromatic and aliphatic amines, and alcohols can be cleaved in a DMF-R₄NI system in 70~90% yields [115a, 118]. In general, removal of *O*-tosyl groups readily takes place, as shown above, by electroreduction without affecting *N*-tosyl groups of amino acids and aliphatic amines. The electroreductive deprotection of *N*-substituted N-tosylamides has been also investigated [119].

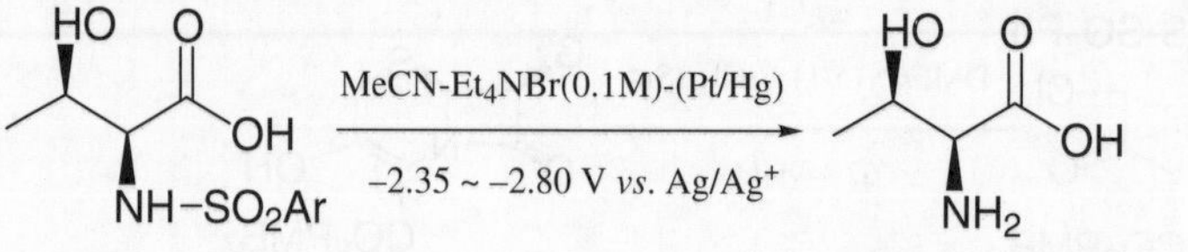

$$(6.95)$$

Ar	Yield, %
Phenyl	85
2,4-Dimethylphenyl	84

Removal of the N-tosyl-protecting group from p-toluenesulfonamides **231** was first re-alized in an MeOH-Me$_4$NCl-(Hg) system (eq. (6.96)) [102b]. Horner and Neumann have observed a cleavage of the S-N bond which shows recovery of amines **233** (55~98%) and sulfinic acids **232** (80~98%). The reaction proceeds through electrolytic formation of tetramethylammonium amalgam which can react with sulfone amides **231** [102b]. More

$$\text{Ts-}\S\text{-OAr} > \text{Ts-}\S\text{-OR} > \text{Ts-}\S\text{-NHAr} > \text{Ts-}\S\text{-NHR} > \text{Ts-}\S\text{-NHCHRCO}_2\text{H}$$

R = alkyl

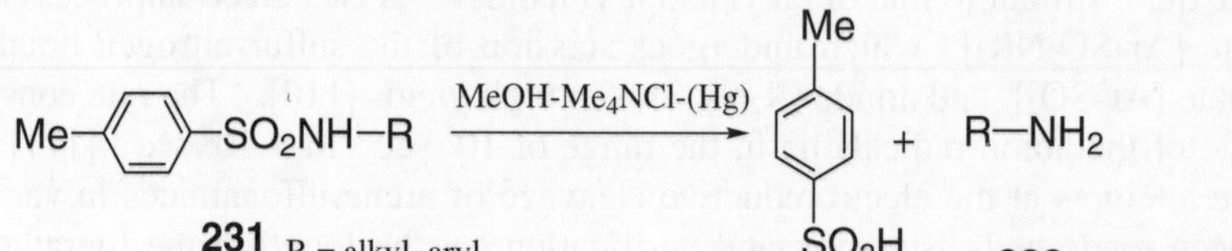

$$(6.96)$$

231 R = alkyl, aryl   **232** 80~98%   **233** 55~98%

practical electrolysis conditions other than the tetramethylammonium amalgam system [102b] have been developed (Table 6.9 [102b]). For instance, the reductive cleavage of L-tosyl serine **234** was performed in an aqueous 20% MeOH-NaCl-(C/Pb) system at pH 11 and the potential of –2.08 V to –2.0 V vs. SCE under the current density of 3.2 A/dm^2 at 25 °C to give L-serine **235** in 88% yield (eq. (6.97)) [120]. Nitrobenzenesulfonamides are

$$\text{Ts-NH-CH-COR}^2 \xrightarrow[\text{aq. 20\% MeOH-NaCl-(C/Pb)}]{2\,e^-,\,2\,H^+} \text{H}_2\text{N-CH-COR}^2 + \text{TsOH}$$

$$(6.97)$$

234   **235**

Ts: Me—⟨⟩—SO$_2^-$

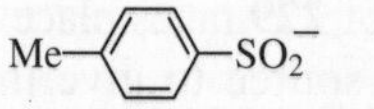

R^1: Amino acid side chains
R^2: OH or amino acid residues

reduced in two cathodic steps at –0.90 V and –1.70 V vs. SCE. The latter reduction step (–1.70 V) can lead to cleavage of the S-N bond in good yields [121]. The results obtained by the electrolytic reductive cleavage of other N-tosyl amino acids and peptides are shown in Table 6.10 [102b].

Electroreductive removal of the N-tosyl group from 2,5-bis[benzyl(p-tosylsulfonyl)amino]-4-(2,5-dioxo-1-methylcyclopentyl)toluene **236** proceeds by taking a new cooperative system of anthracene and ascorbic acids, affording 1-benzyl-8-(benzy-lamino)-6,9-dimethyl-2,5-dioxo-1,2,3,4,5,6-hexahydro-1-benzazocine **237** through a con-trolled crisscross annulation (**238→239**) (eq. (6.98)) [122].

Table 6.9 Reductive Cleavage of Sulfonamides with
Tetramethylammonium Salt

Sulfonamide	Amine %	TsOH[†] %
p-Toluenesulfonamide		
N-Hexyl	94	97
N-Butyl	55	97
N-Cyclohexyl	77	95
N-Benzyl	64	90
N-Phenyl	88	87
N,N-Diphenyl	88	86
N-Phenyl-*N*-benzyl	95	97
N-(2,6-Dimethylphenyl)	92	94
N-(2,6-Diethylphenyl)	94	94
N-Methyl-*N*-benzyl	98	96
N-(2,4,6-Trimethylphenyl)	96	95
N-*p*-Tolyl	95	98
p-Tolunenesulfonylpiperidine	68	91
Benzoylsulfonamide		
N-Benzoylsulfonamide	87	87
N-Benzylbenzoylsulfonamide	67	80
N-Methanesulfonylanilide	no reduction	

[†] TsOH = *p*-toluenesulfonic acid   [Reproduced with permission
from L. Horner et al., *Chem. Ber.*, **98**, 3462 (1965)]

(6.98)

The deprotection of sulfonamides through C-S bond cleavage has been investigated in
comparison of SmI$_2$ and electrochemical reduction methods [123].   The electrochemical
cleavage of the sulfonamide **240** proceeds in an EtOH-Bu$_4$NBr(0.1M)-(C/Hg) system at
−1.65 V (SCE) to give the deprotected product **241** in 90% yield together with the hydrogenated byproduct **242** (4%) (eq. (6.99)).

$$(6.99)$$

240 R = 2-PySO$_2$ **241** 90%   **242** 4%

Table 6.10 Reductive Cleavage of *N*-Protected Amino Acids and Peptides

Tosylated Peptide	Yield, %	Tosylated Peptide	Yield, %[†]
Tos-L(-)-Ser	88	Tos-L(-)-Tyr	96
Tos-L(-)-Thr	91	Tos-DL-Meth	87
Tos-L(-)-Phe	66	Tos-Gly-Gly	91
Tos-L(-)-Ala	84	Tos-Gly-DL-Meth	87
Tos-L(-)-Val	77	Tos-Gly-DL-Ala	89
Tos-L(-)-Leu	84	Tos-Gly-L(-)-Phe	93
Tos-L(-)-ILeu	85	Tos-DL-Ala-Gly	89
Tos-L(-)-Glu	84	Tos-*S*-Bzl-L(-)-Cys	86
Tos-L(-)-Asp	79	Tos-Gly-DL-Try	98
Tos-L(-)-Cys(Bzl)	78	Benzoyl-Gly-Gly	76
Tos-L(-)-Lys(Z)	92	Benzoyl-DL-Meth	77
Tos-L(-)-Ala-Gly	74	Benzoyl-Gly-DL-Meth	68
Tos-L(-)-Val-Gly	83	Benzoyl-Gly-DL-Phe	63
S-Bzl-glutathione	92	Z-DL-Phe	no cleavage
Tos-L(-)-β-Cl-Ala	33	Formyl-DL-Phe	no cleavage

[†] The result after isolation and chromatography of amino acids from respective dipetides.
[Reproduced with permission from L. Horner et al., *Chem. Ber.*, **98**, 3462 (1965)]

6.3.4   Electrosynthesis of Sulfone Derivatives

The electroreduction of sulfur dioxide in the presence of organic halides [RX] has been shown to result in the formation of the corresponding organic sulfones [R-SO$_2$-R] in appreciable current yield (eq. (6.100)) [124].   The sequence of the reactions has been proposed as shown in eq. (6.101) [125].   The second and fourth steps are S$_N$2 substitution reactions, while the third step is a homogeneous electron transfer reaction.   Cyclic voltammetry and other mechanistic studies suggest that the electroreduction of sulfur dioxide takes place through a primary electron transfer, giving an anion radical [SO$_2$]$^{-\bullet}$ followed by a comparatively slow nucleophilic substitution reaction [SO$_2^{-\bullet}$ + RX→R-SO$_2^\bullet$ + X$^-$] [126].

$$2\,RX + SO_2 \xrightarrow{2\,e^-} R{-}SO_2{-}R + 2\,X^-$$

$$R = \text{alkyl, alkenyl, Bz} \tag{6.100}$$

$$
\begin{aligned}
SO_2 + e^- &\longrightarrow SO_2^{-\bullet} && \text{------------ (a)}\\
SO_2^{-\bullet} + RX &\longrightarrow R{-}SO_2^{\bullet} + X^- && \text{----- (b)}\\
R{-}SO_2^{\bullet} + SO_2^{-\bullet} &\xrightarrow{S} R{-}O_2^- + SO_2 && \text{---- (c)}\\
R{-}SO_2^- + RX &\xrightarrow{S} R{-}O_2{-}R + X^- && \text{------ (d)}
\end{aligned}
\tag{6.101}
$$

$$R = \text{alkyl, alkenyl, Bz}$$

A continuous process for the electrosynthesis of sulfones has been investigated. Organic bromide, consumed in the cathode process, is regenerated at the anode cell and the bromine atom is hence recycled. For the electrosynthesis of dipropyl sulfone, sulfur dioxide and propyl alcohol are the starting materials. The alcohol, simultaneously serving as the anolyte solvent, reduces bromine to give hydrobromic acid which, in turn, regenerates propyl bromide by the reaction with further alcohol. For example, after continuous operation for 2.5 h (charge 32400 C), current yields are 53% dipropyl sulfone and 73% propyl bromide [127].

Cyclic sulfone derivatives are also obtained by the use of 1,ω-dihalides [128]. The following hetero ring compounds have been prepared: 1,2-oxathiolane-2-oxide (60~85% yield); 1,2-oxathiane-2-oxide (37~54%); tetramethylenesulfone (25~45%); thiane-1,1-dioxide (60~70%); thiepane-1,1-dioxide (25~35%).

Non-symmetric sulfones are synthesized with high selectivity by optimizing electrolysis temperature and by controlling the concentration and dosage of the two organic compounds [129].

The electrolytically prepared sulfur dioxide anion radicals are used as a reducing reagent for halide-, nitro-, and nitrosocompounds under aprotic and slightly protic conditions [130]. In the presence of excess sulfur dioxide, the initially generated anion radical $SO_2^{-\bullet}$ is in equilibrium with a dimeric ion radical $S_2O_4^{-\bullet}$, indicating a greenish-blue to deep blue color (eq. (6.102)).

$$SO_2 \underset{}{\overset{e^-}{\rightleftharpoons}} SO_2^{-\bullet} \underset{}{\overset{SO_2}{\rightleftharpoons}} S_2O_4^{-\bullet} \tag{6.102}$$

However, the monomeric anion radical is known as a real reactive species and often disables the electrode by the formation of a non-conducting layer [127]. Recently, the reducing capability of $SO_2^{-\bullet}$ species has been examined with the reduction of halides and some nitrogen compounds. For example, benzoyl chloride can be reduced to benzoic acid in an MeCN-Et$_4$NBr system in more than 90% yield (eq. (6.103)) [130]. Some results are shown in Table 6.11 [130].

$$
R{-}\overset{\displaystyle O}{\underset{\displaystyle Cl}{C}} \xrightarrow[-Cl^-]{SO_2^{-\bullet}} \left[R{-}\overset{\displaystyle O}{\underset{\displaystyle SO_2^{\bullet}}{C}} \right] \xrightarrow{SO_2^{-\bullet}} \left[R{-}\overset{\displaystyle O}{\underset{\displaystyle SO_2^{-}}{C}} \right] \xrightarrow{-[H_2SO_2]} R{-}\overset{\displaystyle O}{\underset{\displaystyle OH}{C}}
\tag{6.103}
$$

$$R = \text{aryl}$$

Table 6.11 Reduction of Organic Functional Groups with $S_2O_4^{-\cdot}$

Substrate	Product	Yield, %
4-Methyl-ω-bromoacetophenone	4-Methylacetophenone	71
4-Chloro-ω-bromoacetophenone	4-Chloroacetophenone	65
Carbon Tetrachloride	Chloroform	65
Ethyl Trichloroacetate	Ethyl Dichloroacetate	74 - 93
Benzoylchloride	Benzoic Acid	>90
Nitrobenzene	Aniline	63
4-Chloronitrobenzene	4-Chloroaniline	45 - 60
4-Bromonitrobenzene	4-Bromoaniline	35 - 42
1-Nitropropane	Propanal	37
2-Methyl-2-nitropropane	2-Amino-2-methylpropane	12
ω-Nitrostyrene	β-Phenylacetoaldehyde	15
4-Dimethylaminonitrosobenzole	N,N-Dimethylphenylenediamine	58

[Reproduced with permission from D. Knittel, *Monatsh. Chem.*, **117**, 359 (1986)]

6.4   Electroreduction of Selenium and Tellurium Compounds

Recently, organoselenium and -tellurium compounds have become highly important as chemical reagents, versatile synthetic intermediates, and promising donor molecules for conductive and photoconductive organic materials.   Their redox properties are listed in Table 6.12.   Individual polarographic data on organoselenium and -tellurium compounds are well documented in the literature [4a].

Table 6.12   Redox Properties of Organosulfur Selenium and Tellurium Compounds

	Reduction Process		Potential
Se	$\xrightarrow{e^-}$	Se_2^{2-}, Se^{2-}	-0.8V (MeCN, SCE)
Te	$\xrightarrow{e^-}$	Te_2^{2-}, Te^{2-}	-1.1V (MeCN, SCE)
PhSH	$\xrightarrow{e^-}$	PhS^- + 1/2 H_2	-1.40V (DMF, SCE)
PhSeH	$\xrightarrow{e^-}$	$PhSe^-$ + 1/2 H_2	-1.15V (DMF, SCE)
PhSSPh	$\xrightarrow{2e^-}$	$2PhS^-$	-1.75V (DMF, SCE)
PhSeSePh	$\xrightarrow{2e^-}$	$2PhSe^-$	-1.20V (DMF, SCE)
PhTeTePh	$\xrightarrow{2e^-}$	$2PhTe^-$	-1.20V (MeCN, SCE)
$PhSeH^-$	$\xrightarrow{e^-}$	1/2 $PhSeSePh$ + H^+	$+1.0$V (DMF, SCE)
PhSeSePh	$\xrightarrow{-e^-}$	$PhSeSePh^{+\cdot}$	$+1.6$V (DMF, SCE)
ArTerAr	$\xrightarrow{-e^-}$	$ArTeAr^{+\cdot}$ (Ar = $MeOC_6H_4$)	$+0.8$V (DMF, Ag/Ag(I))

6.4.1   Electrosynthesis of Selenides and Tellurides

The sacrificial Se and Te electrode techniques have been used for the electrosynthesis of di-alkyl and diaryl chalcogenides [131].   A gray selenium can be conveniently reduced to Se_2^{2-} on mercury and solid cathodes (Pt, carbon cloth) [132].

Ultrasound has been shown to increase the rate of electrolysis [133].   More recently, the use of ultrasound for the electrosynthesis of Se_2^{2-}, Se^{2-}, Te_2^{2-}, and Te^{2-} anions has been recorded.   The ultrasonic electroreduction of insoluble Se or Te powder is performed in a DMF(or MeCN, THF)-Bu$_4$NBF$_4$(or Bu$_4$NPF$_6$)-(C/Pt) system using an H-type cell under cooling in an ice bath.   After electrolysis, addition of alkyl halide as an electrophile leads to the synthesis of dialkyl diselenides, selenides, ditellurides, and tellurides according to eqs. (6.104) and (6.105) [134].

$$2\ Se \xrightarrow{2\,e^-} Se_2^{2-} \xrightarrow[-2X^-]{2RX} R-SeSe-R \quad \text{--------(a)}$$

$$Se \xrightarrow{2\,e^-} Se^{2-} \xrightarrow[-2X^-]{2RX} R-Se-R \quad \text{--------(b)}$$

$$R = Bz$$

(6.104)

$$2\ Te \xrightarrow{2\,e^-} Te_2^{2-} \xrightarrow[-2X^-]{2RX} R-TeTe-R \quad \text{-------(a)}$$

$$Te \xrightarrow{2\,e^-} Te^{2-} \xrightarrow[-2X^-]{2RX} R-Te-R \quad \text{-------(b)}$$

$$R = Bz$$

(6.105)

Trimethyltelluronium iodide **244**, a plant growth regulator, is prepared in 68% yield by electroreduction of tellurium metal **243** as a cathode in the presence of excess methyl iodide in a DMF-NaClO$_4$-(Pt/Te) system at -1.0 to -1.4 V (SCE) (eq. (6.106)) [135].

$$\text{MeI} \xrightarrow[-1.2\ \text{V (SCE),}\ 20\text{mA/cm}^2]{\text{DMF-Bu}_4\text{NClO}_4\text{-(Te)}} \text{Me}_3\text{TeI}$$

243   **244**

(6.106)

Electrolysis of bromobenzonitriles **245** in an MeCN-Bu$_4$NPF$_6$-(Pt/C) system in the presence of diphenyl diselenide gives the corresponding phenylselenobenzonitriles **246** in 36~58% yields (eq. (6.107)) [136a].   The yield of **246a** can be improved by electrolysis of chlorobenzonitrile **247** (eq. (6.108)).   In addition, electroreduction of 4-(phenylseleno)ben-zonitrile **249** followed by chemical oxidation by air can afford a mixture of 4,4'-dicyan-odiphenyl diselenide **250** (76%) and diphenyl diselenide (24%) (eq. (6.109)).   The direct synthesis of diaryl diselenides and ditellurides using tea-bag type Se and Te electrodes has been attempted [136b].

$$\text{Br}\overset{}{\longleftarrow}\text{CN} \quad \xrightarrow[\text{Sonication}]{e^-,\ \text{PhSeSePh}} \quad \text{PhSe}\overset{}{\longleftarrow}\text{CN} \tag{6.107}$$

245 **a:** *p*-CN (−1.55 V)
b: *m*-CN
c: *o*-CN

246 **a:** 58% (−2.00 V)
b: 42%
c: 36%

$$\text{Cl}\overset{}{\longleftarrow}\text{CN} \quad \xrightarrow[\substack{\text{Sonication}\\(-1.62\ \text{V})}]{\substack{e^-\\\text{PhSeSePh}}} \quad \textbf{246a} + \text{NC}\overset{}{\longleftarrow}\text{Se}\overset{}{\longleftarrow}\text{CN} \tag{6.108}$$

247   70%   **248** 8%

$$\text{PhSe}\overset{}{\longleftarrow}\text{CN} \quad \xrightarrow[2)\ O_2]{1)\ e^-,\ \text{Sonication}} \quad \text{NC}\overset{}{\longleftarrow}\text{SeSe}\overset{}{\longleftarrow}\text{CN} \tag{6.109}$$

249   +   **250** 76%

$(\text{PhSe})_2$ 24%

Electrosynthesis of novel selenoquinolone derivatives has been attained by use of a tea-bag type sacrificial Se electrode [137].   The electrolysis of 7-chloroquinolone **251** in an MeCN-Bu$_4$NPF$_6$(0.1M)-(Mg/Se) system affords the diselenide **252** in 82% yield.   The electroreductive cleavage of the diselenide **252** is carried out in an MeCN-NaClO$_4$ (0.1M)/Et$_4$NF•2H$_2$O(5M)-(Mg/C) system to yield first the selenolate **253**, upon treatment with methyl iodide, giving the selenoquinolone **254** in 83% yield (eq. (6.110)).

$$\textbf{251} \quad \xrightarrow[\text{82\% yield}]{\text{MeCN-Bu}_4\text{NPF}_6(0.1\text{M})\text{-(Mg/Se)}} \quad 1/2\left[\text{Se}\overset{}{\longleftarrow}\right]_2 \quad \text{------(a)} \qquad \textbf{252} \tag{6.110}$$

$$\textbf{252} \quad \xrightarrow[\text{83\% yield}]{\substack{\text{MeCN-NaClO}_4(0.1\text{M})/\\\text{Et}_4\text{NF}\cdot2\text{H}_2\text{O(5M)-(Mg/C)}}} \quad \textbf{253} \quad \xrightarrow[\text{Me}-\text{Se}]{\text{MeI}} \quad \textbf{254} \quad \text{------(b)}$$

Pentaselenide complexes **256** are known as suitable materials for a selenium atom transfer [138].   Ultrasound-induced electroreduction of gray Se powder in DMF or THF, followed by addition of $(\eta^5\text{-}C_5H_4R)_2TiCl_2$ **255** is shown to be effective for the formation of pentaselenide complexes such as $(\eta^5\text{-}C_5H_4R)_2TiSe_5$ **256** (eq. (6.111)) [139].   The electrochemical synthesis of **256** (R = Me) is achieved in 70% yield.   Diaryltellurium dichlorides can be reduced elecytrochemically to give diaryltelluride in a CH_2Cl_2-Bu$_4$NClO$_4$(0.1M)-(Pt) system [140].

$$(6.111)$$

R	Yield, %
H	64
Me	70
i-Pr	82

Ethenethiolates **258** and their selenium analogues derived from the corresponding *N,N*-dialkylchalcogenoacetamides **257** by the action of EG base react with acyl chloride hydrazones or nitrilimines to give 1,3,4-chalcogenadiazolines **259** in moderate yields (eq. (6.112)) [141].

$$(6.112)$$

257 Z = S, Se
* MeCN-Bu₄NClO₄(0.1M)-(Mg/Pt)

The complexes **256** are irreversibly reduced and play a role as a soluble Se reservoir. For example, the electroreduction of **256** in the presence of a large excess of benzyl chloride can lead to a mixture of dibenzyl diselenide **260** and dibenzyl selenide **261** in 65% and 13% yields, respectively (eq. (6.113)) [139]. It is known that dibenzyl selenide may arise from a further electroreduction of **260** [142].

$$\textbf{256} + PHCH_2Cl \xrightarrow[1.5\ F/mol]{e^-} (PHCH_2Se)_2 + (PHCH_2)_2Se$$

$$\textbf{260}\ 65\% \qquad \textbf{261}\ 13\%$$

$$(6.113)$$

The polarographic study of diphenyl diselenide in a buffered EtOH/H₂O system in different pH ranges reveals that the real electroactive species in the cathode is assigned to be a mercuric derivative (PhSe)₂Hg derived from a preceding chemical reaction between the mercury electrode and diphenyl diselenide [143]. Redox properties of thiolate compounds of oxomolybdenum(V) and their tungsten and selenium analogues have been investigated [144].

6.4.2 Reaction of Electroreductively Produced Selenide and Telluride Anions

The direct selenation of alkyl halides, alkenyl sulfonates, and epoxides with phenyl selenide anion derived from diphenyl diselenide by electroreduction has been performed [145]. For instance, the electrochemical selenation of 1-bromopentane **263** is carried out in an MeOH-Et₄NOTs-(Pt) system in the presence of diphenyl diselenide under a constant applied voltage of 10 V in the cathode compartment of a divided cell at 3~4 °C (eq. (6.114)) [145]. The epoxide **264** can also react with selenide anion to produce the selenol **265** (eq. (6.115)). Other results of the selenation with various electrophiles are listed in Table 6.13 [145].

$$\text{262} \xrightarrow[\text{(PhSe)}_2,\ 87\%\ \text{yield}]{\text{MeOH-Et}_4\text{NOTs-(Pt)}} \text{263} \tag{6.114}$$

$$\text{264} \xrightarrow[\text{(PhSe)}_2,\ 94\%\ \text{yield}]{\text{EtOH-NaClO}_4\text{-(Pt)}} \text{265} \tag{6.115}$$

Table 6.13 Electrolytic Phenylselenation of Halides, Tosylates and Epoxides

Electrophile	Product	Yield, %
$C_5H_{11}Br$	$C_5H_{11}SePh$	87
sec-C_4H_9Br	SePh	88
CH_2I_2	$CH_2(SePh)_2$	85
$CH_2(CH_2Br)_2$	$CH_2(CH_2SePh)_2$	84
MeO_2CCH_2Br	MeO_2CCH_2SePh	92
(allylic chloride) Cl	(allylic) SePh	94
(pinene) OTs	(pinene) SePh	98
OTs	SePh	82
epoxide C_6H_{13}	Ph...C_6H_{13} OH	98
(cyclohexene oxide)	SePh OH	84

Benzophenone derivatives **267**, bearing *o*-, *m*-, or *p*-phenyltelluro and *m*-, or *p*-phenylseleno groups, are electrosynthesized in a MeCN-Bu$_4$NPF$_6$-(Pt/C) system by the mediated cathodic reduction of bromobenzophenone **266** in the presence of electrochemically produced PhTe$^-$ or PhSe$^-$ anion under sonication [146].

Azobenzene is used as a redox catalyst and an acid such as fluorene or malononitrile is present to avoid the formation of $^-CH_2CN$ and its addition upon the bromo ketones (eq. (6.116)).

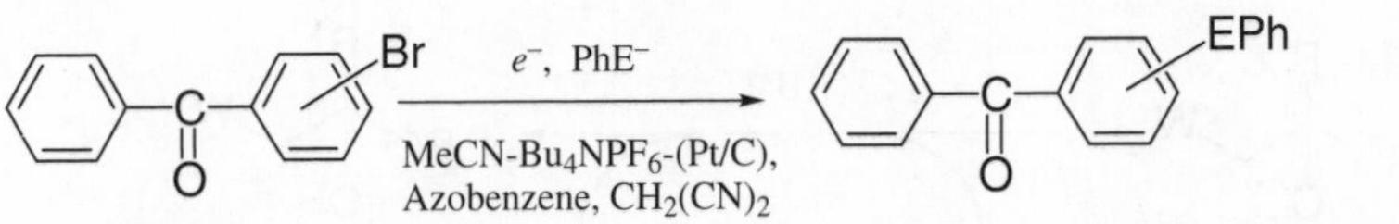

(6.116)

266   **267**

EPh	SePh	TePh
	Yield, %	
o-	–	75
m-	62	48
p-	86	45

The arylseleno group can be also introduced to enone systems. A straightforward arylselenation of **268** proceeds in a MeOH-Et$_4$NOTs-(Pt) system in the presence of diphenyl diselenide together with trimethylsilyll chloride to the corresponding arylselenides **269** in 62~92% yields (eq. (6.117), Table 6.14) [147].

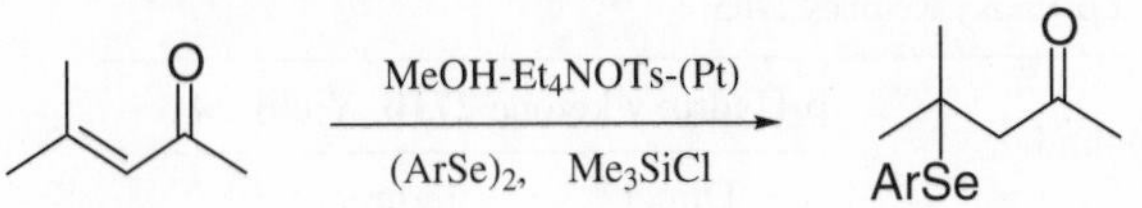

(6.117)

268   **269**

Table 6.14 Electrolytic Aryl Selenenylation of α,β-Unsaturated Carbonyls

Enone	Diselenide	mA/cm^2 (F/mol)	Yield of Adduct %
Mesityl oxide	$(C_6H_5Se)_2$	7-16 (6.6)	87
Mesityl oxide	$(p\text{-}MeOC_6H_4Se)_2$	13-14 (9.1)	92
3-Methyl-2-cyclopentenone	$(C_6H_5Se)_2$	5-7 (9.3)	47
2-Methyl-2-cyclohexenone	$(C_6H_5Se)_2$	4-10 (12.5)	87
Carvone	$(p\text{-}MeOC_6H_4Se)_2$	2-4 (3.4)	78
Pulegone	$(p\text{-}MeOC_6H_4Se)_2$	7-15 (3.6)	83
1-Methoxycarbonyl-1-cyclohexene	$(p\text{-}MeOC_6H_4Se)_2$	8-13 (16)	62

A catalytic amount of phenyl selenide or phenyl telluride anion can be used as an electron-transfer mediator for the reductive ring-opening of α,β-epoxy ketones **270a**, leading to the corresponding β-hydroxy ketones **271a** [148a]. The indirect electroreduction of α,β-epoxy ketones **270a** is carried out in a MeOH-NaClO$_4$-(Pt) system in the presence of diphenyl diselenide and dimethyl malonate. Similarly, the conversion of α,β-epoxy esters **270b** and nitriles **270c** to the corresponding β-hydroxy esters **271b** and nitriles **271c** has been accomplished. The outline of the ring-opening reaction is shown in eq. (6.118). The indirect procedure appears to be superior to the direct one in terms of their product-selectivities and yields. Some comparison data are shown in Table 6.15 [148].

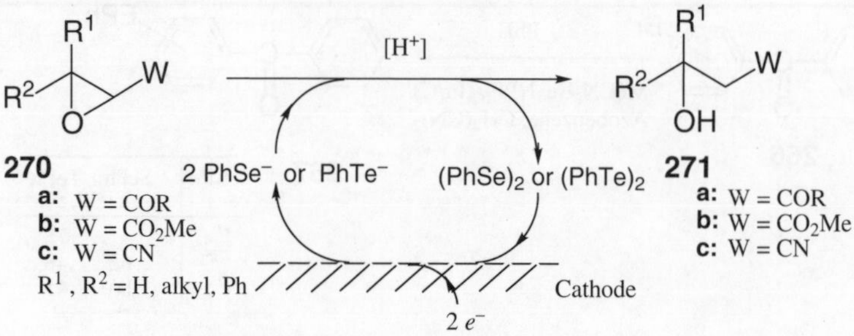

$$(6.118)$$

270
a: W = COR
b: W = CO$_2$Me
c: W = CN
R^1 , R^2 = H, alkyl, Ph

2 PhSe$^-$ or PhTe$^-$ (PhSe)$_2$ or (PhTe)$_2$

271
a: W = COR
b: W = CO$_2$Me
c: W = CN

Cathode

2 e$^-$

[H$^+$] = dimethyl malonate or acetic acid

Table 6.15 Comparison of Indirect and Direct Electroreductive Procedures with α,β-Epoxy Ketones **270a**

Compound	β-Hydroxy ketone **271b**, Yield , % [a]	
	Direct [b]	Indirect [c]
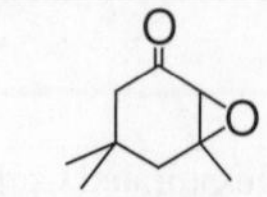	65	79
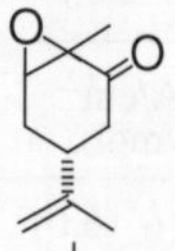	52	85
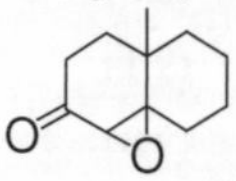	71	85
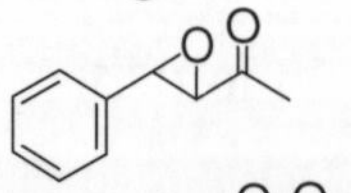	35 [d]	70
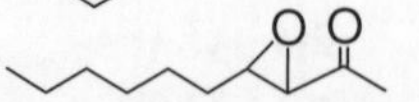	41[e]	72

[a] Based on isolated products.   [b] Direct electroreduction: substrate(1 mmol)-diethyl malonate(5 mmol) in a THF-H$_2$O(9/1 v/v)-Bu$_4$NBF$_4$(0.1 M)-(Pt/C) system at 13-16 V(12 mA/cm^2) for 10.0 F/mol of electricity.   [c] Indirect electroreduction: substrate(1 mmol)-(PhSe)$_2$(0.02 mmol)-dimethyl malonate(5 mmol) in an MeOH-NaClO$_4$(0.2 M)-(Pt/Pt) system under an applied voltage of 3 V(30-5 mA/cm^2) for 4.5 F/mol of electricity.   [d] Diol (30%) is isolated.   [e] Diol (39%) is isolated.

The indirect electroreductive ring-opening of the oxirane ring of poly-functionalized compounds **272** and **274** proceeds specifically to give the corresponding β-hydroxy ketones **273** and **275** in good yields (eqs. (6.119) and (6.120)).   Such electroreductive ring-opening reactions probably proceeds *via* an ionic mechanism involving an initial attack of electro-generated PhSe$^-$ or PhTe$^-$ anion to the α-position of electron-withdrawing groups.   The subsequent nucleophilic attack of the second PhSe$^-$ or PhTe$^-$ anion on the α-phenylseleno or -telluro moiety of the adducts results in the formation of β-hydroxy carbonyls or nitriles together with diphenyl diselenide or ditelluride, which may be recycled in the next reaction.

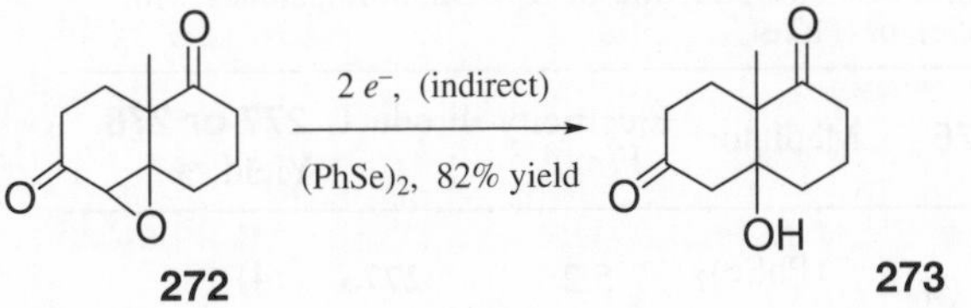

$$(6.119)$$

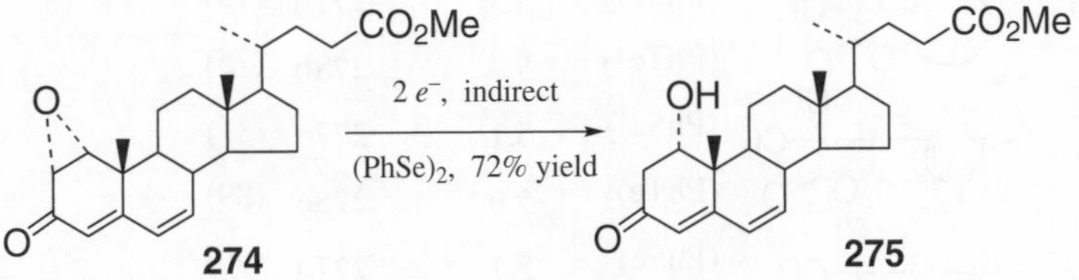

272   **273**

$$(6.120)$$

274   **275**

The indirect electrochemical procedure for dehalogenation of α,α-dichlorolactones **276** and α,α-dibromolactams **279** by using diphenyl diselenide or ditelluride as a recyclable mediator has been developed.   The electroreductive removal of chlorine atoms attached to α,α-dichlorolactones **276** is carried out in an MeOH-NaClO$_4$(0.2M)/(PhSe)$_2$(2mol%)-(Pt) system in a Nafion-separated divided cell to give the corresponding α-chlorolactones **277** in 74% yield.   Similarly, the electrolysis of **276** with diphenylditelluride results in the formation of completely dechlorinated products **278** (eq. (6.121)) [149a].   The results derived

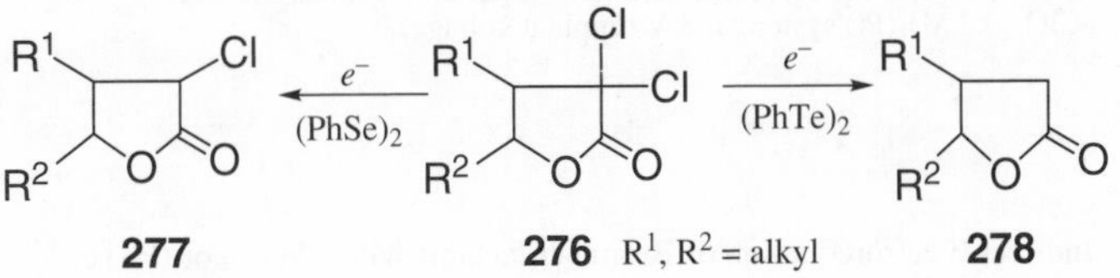

$$(6.121)$$

277   **276** R^1, R^2 = alkyl   **278**

from a variety of α,α-dichlorolactones **276** are exemplified in Table 6.16 [149a].   The same technique has been successfully applied to bromo-β-lactams **279** (eq. (6.122)) and some results are shown in Table 6.17 [149a].   The methanesulfonates of α-hydroxy esters can be converted to the corresponding deoxygenated esters in 70~88% yields by the electrolysis with diphenyl diselenide as a recyclable reducing agent in a divided cell [149b].

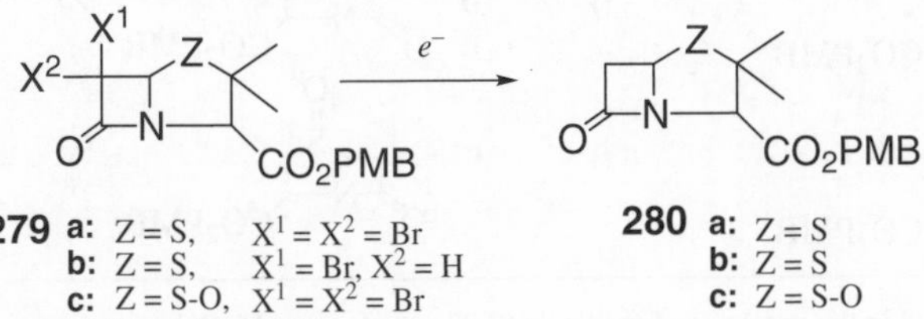

$$(6.122)$$

279 **a:** Z = S,   X^1 = X^2 = Br
    b: Z = S,   X^1 = Br, X^2 = H
    c: Z = S-O, X^1 = X^2 = Br
    PMB = p-MeOC$_6$H$_4$

280 **a:** Z = S
    b: Z = S
    c: Z = S-O

Table 6.16 Indirect Electroreduction of α,α-Dichlorolactones with $(PhSe)_2$ or $(PhTe)_2$[†]

	Substrate **276**	Mediator	Electricity F/mol	Products **277 or 278** Yield, %
a	*(cyclohexane-fused dichlorolactone)*	$(PhSe)_2$	5.2	**277a** (74)
		$(PhTe)_2$	6.0	**278a** (76)
b	*(cyclooctane-fused dichlorolactone)*	$(PhSe)_2$	5.3	**277b** (75)
		$(PhTe)_2$	5.9	**278b** (75)
c	*(benzo-fused dichlorolactone)*	$(PhSe)_2$	5.1	**277c** (81)
		$(PhTe)_2$	5.0	**278c** (89)
d	*(cyclopentene-fused dichlorolactone)*	$(PhSe)_2$	5.1	**277d** (90)
		$(PhTe)_2$	5.0	**278d** (90)
e	*(C_6H_{13} dichlorolactone)*	$(PhSe)_2$	6.1	**277e** (85)
		$(PhTe)_2$	5.5	**278e** (88)
f	*(Ph dichlorolactone)*	$(PhSe)_2$	5.1	**277f** (71)
		$(PhTe)_2$	4.9	**278f** (75)

[†] Electrolyzed in the presence of meditator (0.02 mmol) in an MEOH-$NaClO_4$(0.2 M)-(Pt) system at 3 V (applied voltage).

Table 6.17 Indirect Electroreduction of Bromo-β-lactams with $(PhSe)_2$ or $(PhTe)_2$[†]

	Substrate **279**	Electricity F/mol		Products **280**	Yield, %
a	*(Br, Br β-lactam, CO_2PMB)*	5.5	a	*(β-lactam, CO_2PMB)*	92
b	*(Br, H β-lactam, CO_2PMB)*	3.0	b	*(H β-lactam, CO_2PMB)*	95
c	*(Br, Br sulfoxide β-lactam, CO_2PMB)*	6.0	c	*(sulfoxide β-lactam, CO_2PMB)*	93

[†] Electrolyzed in the presence of mediator (0.02 mmol) in an MeOH-$NaClO_4$(0.2M)-(Pt) system at 3 V (applied voltage).

6.5 Miscellaneous Reactions

The electroreaction at the sulfur cathode allows the formation of C-S bonds. The product distribution depends both on the nature of the leaving group and that of the solvent. Methylthio dithiooxalpiperidine **282** has been electrosynthesized from **281** by a one pot reaction under mild conditions from elementary sulfur at the cathode. The electrolysis of piperidine **281** (or morpholine) is carried out in an EtOH-Mg(ClO$_4$)$_2$-(C-S) system at –0.65 to –0.70 V *vs*. SCE (eq. (6.123)) [150]. The C-S working electrode is made by melting a

$$ \text{(6.123)} $$

sulfur/graphite(4/1) mixture at 120° to 125 °C. The three extra carbon as well as the three extra sulfur atoms of **282** come from the graphite-sulfur electrode. For the preparation of sulfur containing heterocycles, a sacrificial sulfur-carbon electrode is shown to be useful [151]. For example, the electrolysis of cinnamonitrile **283** gives the bis-dithiazolidine **284** in 75% yield (eq. (6.124)) [151a].

$$ \text{(6.124)} $$

An improved electrolytic method for the preparation of bis(fluorosulfuryl) peroxide **286**, a versatile fluorosulfonating agent and an oxidizing agent from **285**, has been reported (eq. (6.125)) [152].

$$ \text{(6.125)} $$

Dihydrolipoamide(DHLAm)-Fe(II) complexes **287** formed by mixing DHLAm with ferrous ion in a slightly alkaline solution have sufficient reducing ability. Electrochemical properties of **287** have been investigated by cyclic voltammetry in an EtOH-KCl-(Au) system at pH 9.8 (buffer), revealing that the DHLAm-Fe(II) complexes are more easily oxidized than DHLAm or ferrous ion. Hydroxylamines, isoxazoles, and nitrobenzene are reduced by the DHLAm-Fe(II) complexes through their coordination transition (eq. (6.126)) [153].

$$\underset{\underset{\textbf{287}}{\underset{R^1, R^2 = H, alkyl}{Y= (CH_2)_4CONH_2}}}{\overset{Y}{\underset{S}{\overset{S}{\diagup}}Fe(II)}} \xrightarrow[-Fe^{2+}]{R-NH-OR^1} \underset{\textbf{288}}{\overset{Y}{\underset{S}{\overset{S}{\diagup}}}} + R^1\!-\!NH_2 + R^2\!-\!OH \qquad (6.126)$$

The formation of an anion radical intermediate in the electroreductive desulfurization of dimethylthiophosphinyl group has been suggested [154].   Electrogenerated polysulfide ions $S_3^{-\bullet}$ and S_8^{2-} can react with haloaromatics, ArX, activated by electron-withdrawing substituents such as a nitro group, giving arylmonosulfides and bisaryldisulfides [155].

References

[1] (a) S. Torii, *J. Synth. Org. Chem. Jpn.*, **48**, 553 (1990); (b) H. Lund, *NATO ASI Ser., Ser. A*, **197**, 93 (1990).

[2] V. A. Petrosyan, M. E. Niyazymbetov, L. D. Konyushkin, V. P. Litninov, *Synthesis*, **1990**, 841.

[3] J. Simonet, M. Carriou, H. Lund, *Liebigs Ann. Chem.*, **1981**, 1665.

[4] (a) J. Q. Chambers, *Encyclopedia of Electrochemistry of the Elements* A. J. Bard, H. Lund (eds.), Vol. 12, pp 329, Marcel Dekker, New York (1978); (b) M. Stankovich, A. J. Bard, *J. Electroanal. Chem.*, *Interfacial Electrochem.*, **75**, 487 (1977); (c) S. Yamaguchi, T. Tsukamoto, M. Senda, *Nippon Kagaku Kaishi* (in Japanese), **1981**, 372.

[5] R. S. Saxena, A. Gupta, K. S. Gupta, *Electrochim. Acta*, **28**, 1569 (1983).

[6] M. E. Niyazymbetov, A. L. Laikhter, V. V. Semenov, D. H. Evans, *Tetrahedron Lett.*, **35**, 3037 (1994).

[7] (a) M. Takahashi, M. Fujita, M. Ito, *Surface Science*, **158**, 307 (1985); (b) J. M. Antelo, F. Arce, F. Rey, M. Sastre, J. M. L. Fonseca, *Electrochim. Acta*, **30**, 927 (1985).

[8] J. M. L. Fonseca, M. L. P. Del Molino, *Electrochim. Acta*, **28**, 383 (1983).

[9] (a) T. R. Ralph, M. L. Hitchman, J. P. Millington, F. C. Walsh, *J. Electroanal. Chem.*, **375**, 1 (1994); (b) T. R. Ralph, M. L. Hitchman, J. P. Millington, F. C. Walsh, *J. Electroanal. Chem.*, **375**, 17 (1994); (c) S. Vavricka, M. Heyrovsky, *J. Electroanal. Chem.*, **375**, 371 (1994); (d) W. R. Fawcett, M. Fedurco, Z. Kovacova, Z. Borkowska, *Langmuir*, **10**, 912 (1994).

[10] L. Horner, H. Lindel, *Liebigs Ann. Chem.*, **1985**, 40.

[11] J. L. Chen, *U. S. Pat.*, 4,551,274 (1985).

[12] (a) J. Moiroux, S. Deycard, *J. Electrochem. Soc.*, **131**, 2840 (1984); (b) M. Largeron, D. Fleury, M. B. Fleury, *Tetrahedron*, **42**, 409 (1986).

[13] (a) M. Largeron, D. Fleury, M. B. Fleury, *J. Electroanal. Chem.*, *Interfacial Electrochem.*, **167**, 183 (1984); (b) J. Moiroux, S. Deycard, M. B. Fleury, *J. Electroanal. Chem.*, *Interfacial Electrochem.*, **146**, 313 (1983).

[14] A. Datchen, P. Berthelot, C. Vaccher, M. N. Viana, M. Debaert, J. L. Burgot, *J. Heterocycl. Chem.*, **23**, 1603 (1986).

[15] N. S.-Von Itter, E. Steckhan, *Tetrahedron*, **43**, 2475 (1987).

[16] T. Shono, Y. Matsumura, S. Kashimura, H. Kyutoku, *Tetrahedron Lett.*, **1978**, 2807.

[17] P. Martigny, J. Simonet, *J. Electroanal. Chem.*, *Interfacial Electrochem.*, **81**, 407 (1977).

[18] T. Shono, Y Matsumura, S. Kashimura, *Tetrahedron Lett.*, **21**, 1545 (1980).

[19] S. E. N. Mohamed, P. Thomas, D. A. Whiting, *J. Chem. Soc., Chem. Commun.*, **1983**, 738.

[20] S. Torii, H. Tanaka, T. Ohshima, M. Sasaoka, *Bull. Chem. Soc. Jpn.*, **59**, 3975 (1986).

[21] (a) M. Ochiai, O. Aki, A. Morimoto, T. Okada, K. Shinozaki, Y. Asahi, *J. Chem. Soc., Perkin Trans. I*, **1974**, 258; (b) *Idem*, *Tetrahedron Lett.*, **1972**, 2341.

[22] H. Lund, *Acta Chem. Scand.*, **14**, 1927 (1960).

[23] C. M. Fischer, R. E. Erickson, *J. Org. Chem.*, **38**, 4236 (1973).

[24] (a) S. Torii, H. Okumoto, H. Tanaka, *J. Org. Chem.*, **45**, 1330 (1980); (b) S. Torii, H. Tanaka, K. Ando, unpublished results.

[25] T. Shono, Y. Matsumura, S. Kashimura, *J. Chem. Res., (S)*, **1984**, 216; *ibid.*, *(M)*, **1984**, 1922.

[26] (a) C. Moinet, D. Peltier, *Bull. Soc. Chim. Fr.*, **1969**, 690; (b) B. Bujoli, M. Chehna, M. Jubault, A. Tallec, *J. Electroanal. Chem.*, *Interfacial Electrochem.*, **199**, 461 (1986); (c) T. Dalati, D. Rondeau, E. Raoult, A. Abouelfida, J.-P. Pradére, A. Tallec, M. Jubault, *J. Chem. Res., (S)*, **1993**, 282.

[27] A. Abouelfida, J. P. Pradére, M. Jubault, A. Tallec, *Can. J. Chem.*, **70**, 14 (1992).

[28] R. I. Pacut, E. Kariv-Miller, *J. Org. Chem.*, **51**, 3468 (1986).

[29] M. M. Baizer, *J. Org. Chem.*, **31**, 3847 (1966).

[30] (a) M. B. Fleury, J. Moiroux, *Bull. Soc. Chim. Fr.*, **1971**, 4637; (b) J. Moiroux, M. B. Fleury, *Electrochim. Acta*, **18**, 691 (1973).

[31] T. Fuchigami, Z. E.-S. Kandeel, T. Nonaka, *Bull. Chem. Soc. Jpn.*, **59**, 338 (1986).

[32] M. B. Fleury, J. Tohier, P. Zuman, *J. Electroanal. Chem., Interfacial Electrochem.*, **143**, 253 (1983).

[33] K. Praefcke, C. Weichsel, M. Falsig, H. Lund, *Acta. Chem. Scand.*, **B34**, 403 (1980).

[34] H. H. Rüttinger, W.-D. Rudorf, H. Matschiner, *Electrochim. Acta.*, **30**,155 (1985).

[35] S. Janietz, H. H. Rüttinger, H. Matschiner, *J. Prakt. Chem.*, **330**, 147 (1988).

[36] M. A. Abramov, M. E. Niyazymbetov, M. L. Petrov, *Zh. Obshch. Khim.*, **62**, 2138 (1992).

[37] (a) A. Kunugi, K. Abe, *Chem. Lett.*, **1984**, 159; (b) A. Kunugi, K. Abe, T. Hagi, T. Hirai, *Bull. Chem. Soc. Jpn.*, **59**, 2009 (1986); (c) *Idem*, *Electrochim. Acta*, **30**, 1049 (1985).

[38] (a) A. Kunugi, K. Abe, T. Hirai, *Stud. Org. Chem.*, **30**, 253 (1987); (b) A. Kunugi, N. Takahashi, K. Abe, T. Hirai, *Bull. Chem. Soc. Jpn.*, **62**, 2055 (1989).

[39] H. M. Fahmy, H. A. Daboun, M. A. Azzem, G. Pierre, *Electrochim. Acta*, **28**, 605 (1983).

[40] M. Sato, T. Karakasa, *Nippon Shika Daigaku Kiyo, Ippan Kyouiku-kei* ,**1983**, 165 (in Japanese).

[41] R. J. Potter, D. J. Schiffrin, *Electrochim. Acta*, **30**, 1285 (1985).

[42] (a) S. Wawzonek, S. M. Heilmann, *J. Org. Chem.*, **39**, 511 (1974); (b) J. C. Lodmell, W. C. Anderson, M. F. Hurley, J. Q. Chambers, *Anal. Chim. Acta*, **129**, 49 (1981); (c) G. C. Papavassiliou, *Chemica Scripta*, **25**, 167 (1985); (d) P. Jeroschewski, *Z. Chem.*, **18**, 27 (1978); (e) U. Reuter, G. Gattow, *Z. Anorg. Allg. Chem.*, **421**, 143 (1976); (f) G. Steimecke, R. Kirmse, E. Hoyer, *Z. Chem.*, **15**, 28 (1975); (g) G. C. Papavassiliou, *J. Phys., Colloq.*, **44**, 1257 (1983).

[43] G. Bontempelli, F. Magno, G.-A. Mazzocchin, R. Seeber, *J. Electroanal. Chem., Interfacial Electrochem.*, **63**, 231 (1975).

[44] P. Jeroschewski, B. Strubing, H. Berge, *Z. Chem.*, **20**, 102 (1980).

[45] S. Wawzonek, H.-F. Chang, W. Everett, M. Ryan, *J. Electrochem. Soc.*, **130**, 803 (1983).

[46] H. Lund, E. Hoyer, R. G. Hazell, *Acta Chem. Scand.*, **B36**, 207 (1982).

[47] M. F. Hurley, J. Q. Chambers, *J. Org. Chem.*, **46**, 775 (1981).

[48] R. R. Schumaker, E. M. Engler, *J. Am. Chem. Soc.*, **99**, 5521 (1977).

[49] P. R. Moses, J. Q. Chambers, *J. Am. Chem. Soc.*, **96**, 945 (1974).

[50] M. Falsig, H. Lund, *Acta Chem. Scand.*, **B34**, 591 (1980).

[51] S. Yasui, K. Nakamura, A. Ohno, S. Oka, *Bull. Chem. Soc. Jpn.*, **55**, 1981 (1982).

[52] L. E. Protasova, V. A. Usov, M. G. Voronkov, *Zh. Obshch. Khim.*, **56**, 417 (1986); *J. Gen. Chem. USSR*, **56**, 365 (1986).

[53] K. Langner, S. Tesch-Schmidtke, J. Voss, *Chem. Ber.*, **120**, 67 (1987).

[54] L. Kistenbrügger, J. Voss, *Liebigs Ann. Chem.*, **1980**, 472.

[55] M. Falsig, H. Lund, *Acta Chem. Scand.*, **B34**, 585 (1980).

[56] H. Kawakubo, K. Yamataka, F. Fukuzaki, *Japanese Patent Kokai*, 85, 172,969, 6 Sept. 1985 Appl. 84/26,111, 16 Feb. 1984.

[57] G. Mabon, J. Simonet, *Tetrahedron Lett.*, **25**, 193 (1984).

[58] A. Böge, J. Voss, *Chem. Ber.*, **123**, 1733 (1990).

[59] G. Drosten, P. Mischke, J. Voss, *Chem. Ber.*, **120**, 1757 (1987).

[60] G. Adiwidjaja, L. Kistenbrügger, J. Voss, *J. Chem. Research (S)*, **1981**, 88; *ibid.*, (*M*), **1981**, 1227.

[61] J. Voss, C. Von Bülow, T. Drews, P. Mischke, *Acta Chem. Scand.*, **B37**, 519 (1983).

[62] P. R. Moses, J. Q. Chambers, J. O. Sutherland, D. R. Williams, *J. Electrochem. Soc.*, **122**, 608 (1975).

[63] F. J. Goodman, J. Q. Chambers, *J. Org. Chem.*, **41**, 626 (1976).

[64] M. Falsig, H. Lund, *Acta Chem. Scand.*, **B34**, 545 (1980).

[65] M. Falsig, H. Lund, L. Nadjo, J. M. Savéant, *Acta Chem. Scand.*, **B34**, 685 (1980).

[66] J. P. Morand, L. Brzezinski, C. Manigand, *J. Chem. Soc., Chem. Commun.*, **1986**, 1050.

[67] M. L. Andersen, M. F. Nielsen, O. Hammerich, *Acta Chem. Scand.*, **49**, 503 (1995).

[68] (a) J. Grimshaw, *The Chemistry of Sulfonium Group* C. J. M. Stilring, S. Patai (eds.), Chapter 7, Wiley-Interscience, New York (1981); (b) J. Q. Chambers, *Encyclopedia of Electrochemistry of the Elements* (Organic Section) (eds. A. J. Bard, H. Lund), Vol. 12, pp 476, Marcel Dekker, New York (1978).

[69] (a) F. D. Saeva, B. P. Morgan, *J. Am. Chem. Soc.*, **106**, 4121 (1984); (b) C. P. Andrieux, M. Robert, F. D. Saeva, J. M. Savéant, *ibid.*, **116**, 7864 (1994).

[70] L. M. Korotaeva, T. Ya. Rubinskaya, V, P. Gultyai, *Izv. Akad. Nauk SSSR, Ser. Khim.*, **1994**, 1255.

[71] C. P. Andrieux, L. Nadjo, J. M. Savéant, *J. Electroanal. Chem., Interfacial Electrochem.*, **26**, 147 (1970).

[72] P. R. Moses, J. Q. Chambers, *J. Electroanal Chem., Interfacial Electrochem.*, **49**, 105 (1974).

[73] M. Madesclaire, *Tetrahedron*, **44**, 6537 (1988).

[74] A. Kunugi, A. Muto, T. Hirai, *Nippon Kagaku Kaishi*, **1984**, 1821 (in Japanese).

[75] B. Lamm, B. Samuelsson, *Acta Chem. Scand.*, **23**, 691 (1969).

[76]   (a) A. Kunugi, A. Muto, N. Kunieda, *Denki Kagaku*, **51**, 137 (1983); (b) A. Kunugi, N. Kunieda, *Electrochim. Acta*, **28**, 715 (1983).
[77]   A. Kunugi, T. Hirai, *New Mater. New Processes*, **3**, 303 (1985).
[78]   Z. V. Todres, *Elektrokhimiya*, **24**, 563 (1988).
[79]   T. Koizumi, T. Fuchigami, Z. E.-S. Kandeel, N. Sato, T. Nonaka, *Bull. Chem. Soc. Jpn.*, **59**, 757 (1986).
[80]   L. Horner, H. Neumann, *Chem. Ber.*, **98**, 1715 (1965).
[81]   B. Lamm, *Tetrahedron Lett.*, **1972**, 1469.
[82]   (a) J. Simonet, G. Jeminet, *Bull. Soc. Chim. Fr.*, **1971**, 2754; (b) G. Mabon, M. C. El Badre, J. Simonet, *ibid.*, **129**, 9 (1992); (c) J. Delaunay, G. Mabon, M. C. El Badre, A. Orliac, J. Simonet, *Tetrahedron Lett.*, **33**, 2149 (1992).
[83]   O. Manousek, O. Exner, P. Zuman, *Collect. Czech. Chem. Commun.*, **33**, 3988 (1968).
[84]   B. Lamm, C.-J. Aurell, *Acta Chem. Scand.*, **B36**, 561 (1982).
[85]   B. Lamm, J. Simonet, *Acta Chem. Scand.*, **B28**, 147 (1974).
[86]   C. Amatore, M. Bayachou, F. Bontejengout, J. N. Verpeaux, *Bull. Soc. Chim. Fr.*, **130**, 371 (1993).
[87]   B. Lamm, A. Nilsson, *Acta Chem. Scand.*, **B37**, 77 (1983).
[88]   (a) M. Novi, G. Garbarino, C. Dell'Erba, G. Petrillo, *J. Chem. Soc., Chem. Commun.*, **1984**, 1205; (b) M. Novi, C. Dell'Erba, G. Garbarino, G. Scarponi, G. Capodaglio, *J. Chem. Soc., Perkin Trans. II*, **1984**, 951.
[89]   (a) B. Lamm, K. Ankner, *Acta Chem. Scand.*, **B32**, 193 (1978); (b) *Idem, ibid.*, **B32**, 264 (1978).
[90]   B. Lamm, K. Ankner, *Acta Chem. Scand.*, **B32**, 31 (1978).
[91]   A. J. Bellamy, I. S. MacKirdy, *J. Chem. Soc., Perkin Trans. II*, **1981**, 1093.
[92]   U. Hess, H. Brosig, W. P. Fehlhammer, *Tetrahedron Lett.*, **132**, 5539 (1991).
[93]   (a) A. Kunugi, K. Minani, M. Yasuzawa, K. Abe, T. Hirai, *Chem Express*, **4**, 189 (1989); (b) A. Kunugi, M. Yasuzawa, *Denki Kagaku*, **58**, 264 (1990); (c) A. Kunugi, T. Myouse, M. Yasuzawa, K. Abe, *ibid.*, **59**, 64 (1991); (d) A. Kunugi, K. Yamane, M. Yasuzawa, H. Matsui, H. Uno, K. Sakamoto, *Electrochim. Acta*, **38**, 1037 (1993).
[94]   A. J. Fry, K. Ankner, V. K. Handa, *J. C. S., Chem. Commun.*, **1981**, 120.
[95]   (a) G. Jeminet, J.-G. Gourcy, J. Simonet, *Tetrahedron Lett.*, **1972**, 2975; (b) S. Gambino, P. Martigny, G. Mousset, J. Simonet, *J. Electroanal. Chem., Interfacial Electrochem.*, **90**, 105 (1978).
[96]   T. Shono, Y. Matsumura, K. Tsubata, Y. Sugihara, *Tetrahedron Lett.*, **1979**, 2157.
[97]   T. Shono, Y. Matsumura, K. Tsubata, Y. Sugihara, *J. Org. Chem.*, **47**, 3090 (1982).
[98]   R. Gerdil, *Helv. Chim. Acta*, **53**, 2097 (1970).
[99]   E. R. Civitello, H. Rapoport, *J. Org. Chem.*, **59**, 3775 (1994).
[100] (a) T. Nonaka, M. M. Baizer, *Electrochim. Acta*, **28**, 661 (1983); (b) T. Nonaka, S. Kihara, T. Fuchigami, M. M. Baizer, *Bull. Chem. Soc. Jpn.*, **57**, 3160 (1984).
[101] P. Yousefzadeh, C. K. Mann, *J. Org. Chem.*, **33**, 2716 (1968).
[102] (a) D. Pletcher, N. R. Stadiotto, *J. Electroanal. Chem., Interfacial Electrochem.*, **186**, 211 (1985); (b) L. Horner, H. Neumann, *Chem. Ber.*, **98**, 3462 (1965).
[103] E. R. Givitello, H. Rapoport, *J. Org. Chem.*, **57**, 834 (1992).
[104] A. A. Pozdeeva, S. G. Mairanovskii, L. K. Gladkova, *Elektrokhimiya*, **3**, 1127 (1967).
[105] K. Ankner, B. Lamm, J. Simonet, *Acta Chem. Scand.*, **B31**, 742 (1977).
[106] J. Simonet, H. Lund, *Acta Chem. Scand.*, **B31**, 909 (1977).
[107] T. Shono, Y. Matsumura, S. Kashimura, *Chem. Lett.*, **1978**, 69.
[108] H. Tanaka, Y. Kameyama, D. Nonen, S. Torii, *Chem. Express*, **7**, 885 (1992).
[109] J. Simonet, *Chem. Sulphonic Acids, Esters Their Deriv.*, **1991**, 553.
[110] P. T. Cottrell, C. K. Mann, *J. Am. Chem. Soc.*, **93**, 3579 (1971).
[111] H. L. S. Maia, M. J. Medeiros, M. I. Montenegro, *J. Electroanal. Chem., Interfacial Electrochem.*, **164**, 347 (1984).
[112] (a) M. R. Asirvatham, M. D. Hawley, *J. Electroanal. Chem., Interfacial Electrochem.*, **53**, 293 (1974); (b) L. Horner, R.-J. Singer, *Tetrahedron Lett.*, **1969**, 1545; (c) *Idem, Chem. Ber.*, **101**, 3329 (1968).
[113] (a) S. Quartieri, L. Benedetti, R. Andreoli, A. Rastelli, *J. Electroanal. Chem., Interfacial Electrochem.*, **122**, 247 (1981); (b) L. Benedetti, R. Andreoli, G. B. Gavioli, G. Grandi, *J. Electroanal. Chem., ibid.*, **68**, 243 (1976); (c) R. Andreoli, G. B. Gavioli, G. Grandi, L. Benedetti, A. Rastelli, *ibid.*, **108**, 77 (1980); (d) A. Rastelli, R. Andreoli, G. B. Gavioli, L. Benedetti, *ibid.*, **89**, 207 (1978); (e) L. Horner, R.-J. Singer, *Ann. Chem.*, **723**, 1 (1969).
[114] R. Kossai, B. Emir, J. Simonet, G. Mousset, *J. Electroanal. Chem., Interfacial Electrochem.*, **270**, 253 (1989).
[115] (a) V. G. Mairanovsky, *Angew. Chem. Int. Ed. Engl.*, **15**, 281 (1976); (b) M. M. Baizer, H. Lund, (eds.), *Organic Electrochemistry*, 2nd ed., Mercel Dekker, New York (1983).
[116] R. C. Roemmele, H. Rapoport, *J. Org. Chem.*, **53**, 2367 (1988).
[117] B. Nyasse, L. Grehn, U. Ragnarsson, H. L. S. Maia, L. S. Monteiro, I. Leito, I. Koppel, J. Koppel, *J. Chem. Soc., Perkin Trans. I*, **1995**, 2025.

[118] V. G. Mairanovskii, N. F. Loginova, *Zh. Obshch. Khim.*, **41**, 2581 (1971); *J. Gen. Chem. USSR*, **41**, 2615 (1971).

[119] A. Lebouc, P. Martigny, R. Carlier, J. Simonet, *Tetrahedron*, **41**, 1251 (1985).

[120] (a) T. Iwasaki, K. Matsumoto, M. Matsuoka, T. Takahashi, K. Okumura, *Bull. Chem. Soc. Jpn.*, **46**, 852 (1973); (b) K. Okumura, T. Iwasaki, M. Matsuoka, K. Matsumoto, *Chem. Ind.* (London), **1971**, 929; (c) M. A. Casadei, A. Gessner, A. Inesi, W. Jugelt, F. M. Moracci, *J. Chem. Soc., Perkin Trans. I*, **1992**, 2001.

[121] M. V. B. Zanoni, N. R. Stradiotto, *J. Electroanal. Chem., Interfacial Electrochem.*, **312**, 141 (1991).

[122] K. Oda, T. Ohnuma, Y. Ban, *J. Org. Chem.*, **49**, 953 (1984).

[123] C. G.-Dubois, A. Guggisberg, M. Hesse, *J. Org. Chem.*, **60**, 5969 (1995).

[124] D. Knittel, B. Kastening, *J. Appl. Electrochem.*, **3**, 291 (1973).

[125] H. J. Wille, B. Kastening, D. Knittel, *J. Electroanal. Chem., Interfacial Electrochem.*, **214**, 221 (1986).

[126] D. Knittel, *J. Electroanal. Chem., Interfacial Electrochem.*, **195**, 345 (1985).

[127] H. J. Wille, D. Knittel, B. Kastening, J. Mergel, *J. Appl. Electrochem.*, **10**, 489 (1980).

[128] D. Knittel, *Monatsh. Chem.*, **113**, 37 (1982).

[129] D. Knittel, B. Kastening, *Ber. Bunsenges. Phys. Chem.*, **77**, 833 (1973).

[130] D. Knittel, *Monatsh. Chem.*, **117**, 359 (1986).

[131] (a) P. Jeroschewski, W. Ruth, B. Strübing, H. Berge, *J. Prakt. Chem.*, **324**, 787 (1982); (b) G. Merkel, H. Berge, P. Jeroschewski, *ibid.*, **326**, 467 (1984).

[132] B. Gautheron, C. Degrand, *J. Electroanal. Chem., Interfacial Electrochem.*, **163**, 415 (1984).

[133] (a) A. J. Bard, *Anal. Chem.*, **35**, 1125 (1963); (b) T. F. Connors, J. F. Rusling, *Chemosphere*, **13**, 415 (1984).

[134] B. Gautheron, G. Tainturier, C. Degrand, *J. Am. Chem. Soc.*, **107**, 5579 (1985).

[135] T. Röthling, D. Creuzburg, G. Merkel, H. Berge, A. Hornuf, P. Jeroschewski, W. Kochmann, K. Naumann, L. Kranz, H. Patzer, *Ger (East)*, DD 222,355 (1985).

[136] (a) C. Degrand, *J. Org. Chem.*, **52**, 1421 (1987); (b) C. Degrand, R. Prest, *J. Electroanal. Chem., Interfacial Electrochem.*, **282**, 281 (1990).

[137] M. Genesty, O. Merle, C. Degrand, M. Nour, P. L. Compagnon, J. P. Lemaitre, *Denki Kagaku*, **62**, 1158 (1994).

[138] J. L. Poncet, R. Guilard, P. Friant, C. Goulon-Ginet, J. Goulon, *Nouv. J. Chim.*, **8**, 583 (1984) and references cited therein.

[139] G. Tainturier, B. Gautheron, C. Degrand, *Organometallics*, **5**, 942 (1986).

[140] Y. Liftman, M. Albeck, *Electrochim. Acta*, **28**, 1841 (1983).

[141] M. A. Abramov, M. E. Niyazymbetov, M. L. Petrov, *Zh. Obshch. Khim.*, **62**, 1914 (1992).

[142] C. Degrand, M. Nour, *J. Electroanal. Chem., Interfacial Electrochem.*, **190**, 213 (1985).

[143] (a) F. Fagioli, F. Pulidori, C. Bighi, A. De Battisti, *Gazz. Chim. Ital.*, **104**, 639 (1974); (b) G. Paliani, M. L. Cataliotti, *Z. Naturforsch., B: Anorg. Chem., Org. Chem.*, **29b**, 376 (1974).

[144] J. R. Bradbury, A. F. Masters, A. C. McDonell, A. A. Brunette, A. M. Bond, A. G. Wedd, *J. Am. Chem. Soc.*, **103**, 1959 (1981).

[145] S. Torii, T. Inokuchi, G. Asanuma, N. Sayo, H. Tanaka, *Chem. Lett.*, **1980**, 867.

[146] C. Degrand, R. Prest, P.-L. Compagnon, *J. Org. Chem.*, **52**, 5229 (1987).

[147] S. Torii, T. Inokuchi, N. Hasegawa, *Chem. Lett.*, **1980**, 639.

[148] (a) T. Inokuchi, M. Kusumoto, S. Torii, *J. Org. Chem.*, **55**, 1548 (1990); (b) T. Inokuchi, M. Kusumoto, T. Sugimoto, H. Tanaka, S. Torii, *Phosphorus, Sulfur Silicon Relat. Elem.*, **67**, 271 (1992).

[149] (a) T. Inokuchi, M. Kusumoto, S. Torii, *Electroorg. Synth.*, **1991**, 233; (b) T. Inokuchi, T. Sugimoto, M. Kusumoto, S. Torii, *Bull. Chem. Soc. Jpn.*, **65**, 3200 (1992).

[150] M. Leonor, S. Rivas, R. Rozas, *J. Electroanal. Chem., Interfacial Electrochem.*, **177**, 299 (1984).

[151] (a) G. le Guillanton, Q. T. Do, J. Simonet, *Stud. Org. Chem.*, **30**, 257 (1987); (b) *Idem*, *Tetrahedron Lett.*, **27**, 2261 (1986).

[152] S. Singh, R. D. Verma, *Indian J. Chem., Sect. A.*, **25A**, 51 (1986).

[153] M. Kijima, Y. Nambu, T. Endo, *J. Org. Chem.*, **50**, 2522 (1985).

[154] W. Kaim, P. Hänel, U. Lechner-Knoblauch, H. Bock, *Chem. Ber.*, **115**, 1265 (1982).

[155] M. Benaichouche, G. Bosser, J. Paris, V. Plichon, *J. Chem. Soc., Perkin Trans. II*, **1991**, 817.

7. Electroreduction of Halogenated Compounds

7.1 Electroreduction of Alkyl Halides

The aim of making carbon-carbon and carbon-hetero atom bonds has been achieved to some extent by the electroreduction of organic halides [1,2]. Improved methods using mediators or transition-metal catalysts have been reviewed in the literature [3~7]. Novel proposals for organic synthesis *via* the electroreduction of organic halides in the presence of various electrophiles using sacrificial metallic anodes have been well documented [8a].

Especially, a remarkable effect of additives such as $SmCl_3$ has been discussed with a consumable magnesium anode [8b].

7.1.1 Electroreductive Dehalogenation

Direct electroreduction of alkyl and aryl halides is mostly performed at potenitals which are more negative than –2 V (SCE) [9a~c].

The electroreduction of organic halides (R-X) **1** is known to be rate controlled by the uptake of the first electron and associated with the cleavage of the carbon-halogen bond [9d, 10, 11]. The ultimate electron transfer intermediate is usually the carbanion (R$^-$) **2**, which is formed as shown in eqs. (7.1) and (7.2).

$$R\text{--}X + e^- \longrightarrow [R\text{--}X]^{-\bullet} \longrightarrow R^\bullet + X^- \tag{7.1}$$
$$\mathbf{1}$$

$$R^\bullet + e^- \longrightarrow R^- \tag{7.2}$$
$$\mathbf{2}$$

The electrochemical alkylation of unactivated ketones has been attained by the electro-generated carbanion **2** (EG base) [12]. Alkylation of activated ketones by electroreduction of alkyl halides have been well investigated [13, 14]. A recent investigation on the electroreduction of alkyl halides demonstrates a concerted electron transfer-bond cleavage mechanism [15].

The pathway for the electroreduction of primary alkyl halides [16] at a mercury electrode can be exemplified by taking 1-iododecane as follows: The electroreduction of 1-iododecane in a DMF-Me$_4$NClO$_4$-(Hg) system is shown to be influenced by the potential of the cathode and the amount of water in an electrolysis medium. An ordinary dc polarogram for the reduction of 1-iododecane exhibits two waves (–0.96 V and –1.37 V (SCE)) [17]; an anomalous current maximum appears on the rising portion of the second wave. At potentials positive with respect to the polarographic maximum, the alkyl iodide accepts one electron to yield a decyl radical which becomes an adsorbed decylmercury radical. The adsorbed organomercury radicals disproportionate to form didecylmercury, which is the only product produced at these potentials [18] (see Section 7.6.2). However, at potentials negative with respect to the polarographic maximum, transfer of two-electrons to 1-iododecane is predominant, giving decyl carbanions as a precursor of decane. In the course of the electrolysis, EG bases produce 1-decene (E2 elimination) together with 1-decanol (S$_N$2 displacement). Large-scale electrolysis of 1-iododecane at a reticulated vitreous carbon in a DMF-Me$_4$NClO$_4$ system affords decane (86%) and 1-decene (2%), respectively [19]. The related results from the reduction of halides are listed in Table 7.1 [18~25]. The *n* value for the electroreduction of 1-iododecane depends on electrolysis time [17].

Table 7.1a Electroreductive Dehalogenation of Alkyl Halides at Mercury Cathodes

$$R^1CH(CH_2)_nCHR^2 \longrightarrow R^1CH(CH_2)_{n+1}R^2 + \text{Olefin} + Z_2Hg^a + \text{Others}$$

with X, Y (A) on left and X (B) on right.

R¹	n	R²	X	Y	Electrolysis Conditions	V (SCE)	$R^1CH(CH_2)_{n+1}R^2$ X	Olefin	Z_2Hg	Others	Ring	Ref.
							Yield, %					
Pr	2	H	H	I	DMF-Me₄NClO₄(0.1M)-(Cd/Hg) Diethyl Malonate	-1.87	79 (X = H)	-	-	4	-	[18]
Pr	2	H	H	I	DMF-Bu₄NClO₄(0.1M)-(Cd/Hg) Diethyl Malonate	-1.89	90 (X = H)	-	-	6	-	[18]
CH₃(CH₂)₅	2	H	H	I	DMF-Bu₄NClO₄(0.1M)-(Hg) N-Methylformamide(0.002%)	-1.1	1 (X = H)	-	99	-	-	[20] [21]
CH₃(CH₂)₅	2	H	H	I	DMF-Bu₄NClO₄(0.1M)-(Hg) N-Methylformamide(0.002%)	-1.7	71 (X = H)	4	3	-	-	[20]
CH₃(CH₂)₅	2	H	H	I	DMF-Me₄NClO₄(0.1M)-(C) Hexafluoroisopropanol	-1.6	86 (X = H)	2	-	8	-	[19]
CH₃(CH₂)₅	2	H	H	Br	DMF-Bu₄NClO₄(0.1M)-(Hg) N-Methylformamide(0.002%)	-1.9	53 (X = H)	21	-	5	-	[21]
CH₃(CH₂)₅	2	H	H	Br	DMF-Et₄NClO₄(0.1M)-(C)	-1.8	74 (X = H)	4	-	-	-	[19]

ᵃ $Z = R_1CH(CH_2)_nCHR^2$

Table 7.1b Electroreductive Dehalogenation of Alkyl Halides at Mercury Cathodes

R¹	n	R²	X	Y	Electrolysis Conditions	V (SCE)	$R^1CH(CH_2)_{n+1}R^2$ X	Olefin	Z_2Hg^a	Others	Ring	Ref.
H	3	H	Br	Br	DMF-Me$_4$NClO$_4$(0.05M)-(C),	-2.10	26 (X = H)	29	11	–	29	[22]
H	3	H	I	I	DMF-Me$_4$NClO$_4$(0.05M)-(C),	-1.70	22 (X = H)	26	14	–	26	[22]
H	3	H	Br	Cl	DMF-Me$_4$NClO$_4$(0.05M)-(C),	-2.00	77	–	2	–	18	[22]
H	3	H	Cl	I	DMF-Me$_4$NClO$_4$(0.05M)-(C),	-1.70	76	–	12	–	13	[22]
H	4	H	Br	Cl	DMF-Me$_4$NClO$_4$(0.1M)-(C), CF$_3$CHOHCF$_3$(50mM)	-1.95	93	2	–	–	–	[23]
H	4	H	Br	Br	DMF-Me$_4$NClO$_4$(0.1M)-(C), CF$_3$CHOHCF$_3$(50mM)	-1.95	85 (X = H)	5	–	–	–	[23]
H	2	CO$_2$Et	Br	Br	DMF-Et$_4$NClO$_4$	-1.5	2 (X = H) 6	–	–	–	72	[24]
H	3	CO$_2$Et	Br	Br	DMF-Et$_4$NClO$_4$	-1.5	5 (X = H) 33	–	–	–	34	[24]
Ph	0	H	H	Br	MeOH-Bu$_4$NClO$_4$(0.1M)-(Zn),	-1.54	55	–	–	25	–	[25]
Ph	1	H	H	Br	MeOH-Bu$_4$NClO$_4$(0.1M)-(Zn),	-1.56	32	17	–	6	10	[25]
Ph	1	Me	H	Cl	MeOH-Ba(ClO$_4$)$_2$(0.1M)-(Zn),	-1.41	36	20	–	–	18	[25]

ᵃ $Z = R_1CH(CH_2)_nCHR^2$

The electroreduction of 1,10-dihalodecanes in a DMF-Me$_4$NClO$_4$-(Hg) system has been investigated [26].   Quantitative comparison of 1-iodo- and 1-bromodecane suggests that 1,10-diiododecane is reduced in a pair of two-electron steps whereas 1,10-dibromodecane undergoes a single four electron reduction.   The preparative-scale electrolysis of 1,10-di-iododecane at a potential on its first reduction wave brings about an organomercury poly-mer as the sole product.

The electroreduction of ethyl iodides 3 is influenced by the choice of cathode materials to give either dimer 4 or olefin 6 together with hydrogenerated product 5 (eq. (7.3)) [27]. The formation of an absorption complex with the metal may cause a different set of reac-tions.   The electroreductive removal of chlorine atom from ω-chloroalkyl naphthyl sulfides and sulfoxides occurs in an MeOH-Ba(ClO$_4$)-(Pt/Zn) system to yield the corresponding alkyl (25~29%), alkenyl (23~26%), and thiols (10~20%), respectively [28].

$$
I-CH_2CH_2-Y \quad \xrightarrow[\substack{(Pt/Cu\ or\ Pb,\ Hg)\\ divided\ cell}]{H_2O-NaOH(or\ H_2SO_4)} \quad \begin{cases} (CH_2)_4Y & \mathbf{4} \\ + \\ C_2H_5-Y & \mathbf{5} \\ + \\ CH_2=CH_2 & \mathbf{6} \end{cases}
$$

(7.3)

3

Y	Electrode	Electrolyte	4	5	6
H	Cu	NaOH	50.0	45.5	–
H	Pb	NaOH	27.7	42.3	–
H	Hg	NaOH	–	71.0	–
OH	Pb	H$_2$SO$_4$	–	–	81.9
Cl	Pb	NaOH	–	5.5	86.5

Electroreduction of 8-bromobornan-2-one 7 in a DMF-Bu$_4$NBr-(Hg) system at –2.6 V (SCE) affords dihydrocarvone 8 in an approximately quantitative yield (eq. (7.4)) [29].

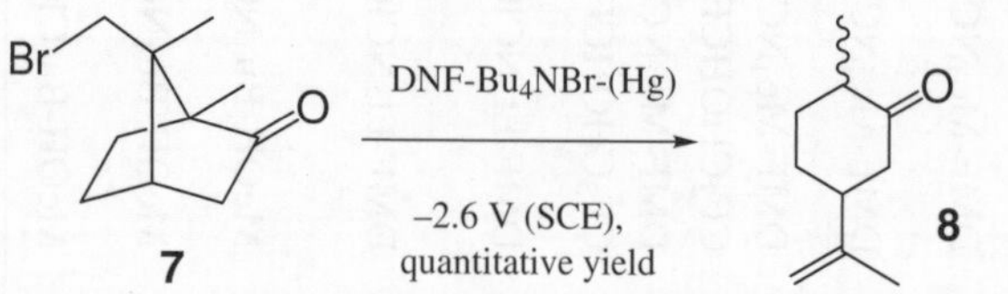

(7.4)

A cyclic voltammogram for the reduction of *tert*-butyl bromide exhibits two waves, in-dicating stepwise generation of the *tert*-butyl radical and carbanions.   The electroreduction of *tert*-butyl bromide 9 in a DMF-Me$_4$NClO$_4$-(Hg) system at the potential corresponding to the first wave produces isobutane 10 (40~43%) (eq. (7.5)), isobutylene 11 (43~45%) (eq. (7.6)), and 2,3-dimethylhexane 12 (9~10%) (eq. (7.7)).   The first two products 10 and 11 arise *via* disproportionation of *tert*-butyl radicals, whereas the third compound 12 is formed by coupling of radicals [30].

$$
\begin{array}{c}
\text{Me} \\
| \\
\text{Me—}\overset{\displaystyle |}{\underset{\displaystyle |}{\text{C}}}\text{—Br} \\
| \\
\text{Me} \\
\mathbf{9}
\end{array}
\quad\xrightarrow{\text{DMF-Me}_4\text{NClO}_4\text{-(Hg)}}\quad
\left\{
\begin{array}{l}
(\text{Me})_3\text{C—H} \quad \mathbf{10} \;\; 40\text{~}43\% \\[4pt]
+ \\[2pt]
(\text{Me})_2\text{C=CH}_2\,\mathbf{11} \;\; 43\text{~}45\% \\[4pt]
+ \\[2pt]
(\textit{tert-}\text{Bu})_2 \quad \mathbf{12} \;\; 9\text{~}10\%
\end{array}
\right.
$$

(7.5)

(7.6)

(7.7)

The investigation of the kinetics of the reductive cleavage of the carbon-halogen bond in a series of *n*-, *sec*-, and *tert*-butyl halides demonstrates that the reductive cleavage (when varying the halogen, from Cl to Br and to I) is both thermodynamically easier and kinetically faster at the standard state [31].   The aliphatic halides undergo a true dissociative electron transfer, *i.e.*, the breaking of the carbon-halogen bond works in concert with an electron-transfer process [32].

The electroreduction of α-bromo-α-methylpropionitrile in a DMF-Et$_4$NClO$_4$-(C or Hg) system at -1.8 V (SCE) undergoes a one-electron transfer reduction, and the yield of products depend on the electrode material [33a].   The reduction on Hg pool yields debrominated nitriles, dimeric, and polymeric materials.   In contrast, the reduction at the vitreous carbon electrode yields debrominated products such as unsaturated nitriles.   The electrolysis of 1-bromo-1-methylpropiononitrile **13** gives 1-methylpropiononitrile **14** in 85% yield (eq. (7.8)) [33].   The results obtained by a similar reduction of other bromonitriles and bromoesters are listed in Table 7.2 [33b, 34].

$$
\begin{array}{c}
\text{Me} \\
\diagdown \\
\text{Me}\diagup\overset{\displaystyle}{\underset{\displaystyle \text{Br}}{\text{C}}}\text{—CN} \\
\mathbf{13}
\end{array}
\quad\xrightarrow[-1.8\text{ V (SCE), 85\% yield}]{\text{DMF-Et}_4\text{NClO}_4(0.1\text{M})\text{-(Hg)}}\quad
\begin{array}{c}
\text{Me} \\
\diagdown \\
\text{Me}\diagup\text{CH—CN} \\
\mathbf{14}
\end{array}
$$

(7.8)

The electroreductive cleavage of the carbon-halogen bond of bromotrinitromethane **15** has been attained in an aqueous MeOH-H$_2$SO$_4$-(Hg) system to give the carbanion $[\text{C(NO}_2)_3]^-$ **17**, whereas the similar reduction of chlorotrinitromethane **15** undergoes loss of one nitro group to form the carbanion $[\text{ClC(NO}_2)_2]^-$ **16** exclusively (eq. (7.9)) [35].

$$
[\text{ClC(NO}_2)_2]^- \xleftarrow{\;-\text{NO}_2^-\;} \text{XC(NO}_2)_3 \xrightarrow{\;-\text{Br}^-\;} [\text{C(NO}_2)_3]^-
$$

(7.9)

$$
\begin{array}{ccc}
\mathbf{16} & \mathbf{15} & \mathbf{17} \\
X = \text{Cl} & X = \text{Br} & \\
90\% \text{ yield} & 96\% \text{ yield} &
\end{array}
$$

The kinetics of the electroreductive cleavage of arylmethyl halides in MeCN (or DMF) has been investigated in terms of a function of the energy of the π^* orbital liable to accept the incoming electron [36, 37].

The controlled potential electrolysis of benzyl chloride for direct and catalytic homogeneous reduction on mercury affords toluene as a major product in both cases [38].   The electrolysis of benzyl chloride **18** in a DMF-Me$_4$NBF$_4$(0.1M)-(Hg) system gives toluene **19** in 78% yield (eq. (7.10)) [39].   The electrogenerated anthracene anion radicals and an

$$
\text{Ph—CH}_2\text{Cl} \quad\xrightarrow[78\% \text{ yield}]{\text{DMF-Me}_4\text{NBF}_4\text{-(Hg)}}\quad \text{Ph—CH}_3
$$

(7.10)

$$
\begin{array}{ccc}
\mathbf{18} & & \mathbf{19}
\end{array}
$$

Table 7.2 Electroreductive Dehalogenation of Bromonitriles and Bromoesters

| Br(CH2)n-Y | | Potential | | Yield, % | Product Distribution, rel. % | | | | | Ref. |
n	Y	V (SCE)	Conditions		H(CH2)nCO2Et	Olefin	Ring	Dimer	Others	
2	CO2Et	-1.85	DMF-Et4NClO4-(Hg) diphenylanthracene	96	26	63	–	–	11	[34]
2	CO2Et	-2.50	DMF-Et4NClO4-(Hg)	91	20	50	–	–	30	[34]
3	CO2Et	-1.85	DMF-Et4NClO4-(Hg) diphenylanthracene	91	15	–	76[a]	–	9	[34]
3	CO2Et	-2.50	DMF-Et4NClO4-(Hg)	88	51	–	49[a]	–	–	[34]
4	CO2Et	-1.85	DMF-Et4NClO4-(Hg) diphenylanthracene	90	11	83	–	–	6	[34]
4	CO2Et	-2.50	DMF-Et4NClO4-(Hg)	72	94	6	–	–	–	[34]
2	CN	-2.25	DMF-Et4NClO4-(Hg)	96	90	–	–	–	–	[33b]
2	CN	-2.5	DMF-Et4NClO4-(Hg)	91	91	–	–	–	–	[33b]
3	CN	-2.07	DMF-Et4NClO4-(Hg)	91	48	–	32[b]	–	–	[33b]
3	CN	-2.5	DMF-Et4NClO4-(Hg)	88	94	–	–	–	–	[33b]
4	CN	-2.3	DMF-Et4NClO4-(Hg)	90	40	–	16[c]	–	–	[33b]
4	CN	-2.5	DMF-Et4NClO4-(Hg)	72	65	–	–	–	–	[33b]

[a] Ethyl Cyclopropanecarboxylate [b] Cyclopropanecarbonitrile [c] Cyclobutanecarbonitrile

anion, 1,4-dihydro-4-methoxycarbonyl-1-methylpyridine anion, have been used as mediators for the indirect electroreduction of benzyl chlorides [39]. The aromatic anion radical-mediated electroreduction of butyl chloride and bromide has been discussed in terms of a homogeneous redox catalyzed process on competing with partial destruction of the catalyst [40]. A remarkable autocatalysis process for the electroreductive cleavage of the carbon-halogen bond in benzylic type chlorides involving the carbanion as an electron transfer catalyst has been demonstrated [41].

The electroreduction of arylalkyl halides in an MeOH-Ba(ClO$_4$)$_2$(0.1M)-(Zn/Hg) system affords arylalkanes (32~55%) as a major product together with the corresponding olefins (17~24%) and dimers (5~24%), respectively [25]. Pentabromobenzyl bromide can be completely reduced electrochemically into toluene in a DMF-Bu$_4$NClO$_4$(0.01M)-(Hg) system *via* the initial formation of organomercury compounds at the surface of the electrode [42].

The cathodic peak potentials of 1-bromoadamantane **20** and 1-bromonorbornane **21** are observed as follows: **20**, –2.67 V; **21**, –3.0 V (SCE) [43]. Electrogenerated anion radicals of nitrogen-containing heteroaromatic compound **22** react with 1-bromoadamantane **20** in a DMF-Bu$_4$NI-(Hg) system to give the adamantylated product **23** in 20% yield (eq. (7.11)) [44].

$$\text{(7.11)}$$

Anion radicals of aromatic and heteroaromatic compounds have been shown to react with alkyl halides through an initial transfer of a single electron [45]. The single electron transfer rate constant (K_{SET}) from electrochemically generated anion radicals to a number of alkyl halides in DMF has been investigated [46]. The kinetic results suggest that the single-electron transfer process is the rate-determining step in these nucleophilic substitutions.

Triphenylmethyl cation derived from the reaction of triphenylmethyl chloride **24** with aluminum chloride is reduced in a one-electron process to a triphenylmethyl radical. The formation of radical species in the AlCl$_3$-melt is confirmed by ESR spectroscopy. Dimerization of the radical intermediate affords 1-(diphenylmethylene)-4-(triphenyl-methyl)-2,5-cyclohexadiene **25**, precursor of **26** (eq. (7.12)) [47].

4-Bromosydnone has been reduced with electrogenerated triphenylmethyl anion to give sydnone in 39% yield [48]. Redio- and stereoselective partial-dechlorination of oligo-cyclic insecticides performs in an MeOH-Et$_4$NBr-(Pb) system [49].

$$\text{(7.12)}$$

7.1.2   Electroreductive Alkyl-Alkyl Couplings

Preparative-scale catalytic reductions of iodoethane performed at reticulated vitreous carbon electrodes coated with polymeric nickel(II) salen in an MeCN-Me$_4$NBF$_4$(0.05M) system at a potential more negative than the reversible redox potential for the Ni(II)/Ni(I) couple in the film yields butane as a major product (90% product distribution) [50].   Butane probably arises *via* coupling of a pair of ethyl radicals and the rate for ethyl-radical coupling is approximately 6~8 times greater than the rate of disproportionation of ethyl radicals, giving ethane.   The heterogenous iron complex catalyst, Fe(acac)$_3$, can be used to couple alkyl bromides to give the corresponding homo-coupling molecules as a major product [51].   The electrolysis of alkyl bromide **27** (R = C$_6$H$_{13}$) in a DMF-Et$_4$NBr-(Al) system in the presence of Fe(acac)$_3$ affords a mixture of the dimer **28** (59%), octane **29** (23.8%), and 1-octene **30** (10.9%), respectively (eq. (7.13)).

$$
RCH_2CH_2Br \xrightarrow[\substack{Fe(acac)_3 \\ -0.9 \text{ V } vs. \text{ Cd(Hg)}}]{DMF\text{-}Et_4NBr\text{-}(Al)}
\begin{cases}
(RCH_2CH_2)_2 \\
\quad + \quad \textbf{28} \;\; 59\% \\
RCH_2CH_3 \\
\quad + \quad \textbf{29} \;\; 23.8\% \\
RCH=CH_2 \\
\qquad\qquad \textbf{30} \;\; 10.9\%
\end{cases}
\tag{7.13}
$$

27   R = alkyl

Electroreductive dimerization of α-trimethylsilylbenzyl bromide in an MeCN-LiClO$_4$-(C) system at –1.2 V (Ag/Ag$^+$) at 0 °C *dl*-1,2-bis(trimethylsilyl)-1,2-diphenylethane in 39% yield [52].

The electroreduction of β-chloropropiophenone **31** in a DMF-KClO$_4$-(Pt) system produces the dimer **32** in 23% yield together with a mixture of propiophenone **33** and acrylophenone **34** in 44% yield (eq. (7.14, 7.15)) [53].   The electrolysis of **31** on a mercury cathode yields a [3.3.0]bicyclooctane derivative [53a].

$$
Ph\text{-}COCH_2CH_2Cl \xrightarrow{DMF\text{-}KClO_4(0.1M)\text{-}(Pt)}
\begin{cases}
\begin{array}{l} Ph\text{-}COCH_2CH_2 \\ \; | \\ Ph\text{-}COCH_2CH_2 \end{array} \\
\qquad + \quad \textbf{32} \;\; 23\% \tag{7.14} \\
Ph\text{-}COCH_2CH_3 \\
\qquad + \quad \textbf{33} \\
Ph\text{-}COCH=CH_2
\end{cases}
$$

31

$$\left.\begin{array}{}\\ \end{array}\right\} 44\% \tag{7.15}$$

34

The electroreduction of 2,4-dibromo-2,4-dimethyl-3-pentanone **35** in an MeOH-Et$_4$NBr system affords tetramethylcyclopropanone methyl hemiacetal **36** in more than 90% yield (eq. (7.16)) [54~56].

$$
\underset{\textbf{35}}{Br\text{-}\!\!\!\diagdown\!\!\!\underset{O}{\overset{\|}{\diagup\!\!\!\diagdown}}\!\!\!\diagup\text{-}Br} \xrightarrow[-1.0 \text{ V (SCE)}, \; >90\% \text{ yield}]{MeOH\text{-}Et_4NBr\text{-}(?)} \underset{\textbf{36}}{\overset{MeO\quad OH}{\diagdown\diagup\!\!\!\times\!\!\!\diagdown\diagup}}
\tag{7.16}
$$

Bicyclobutanes, cyclopropane, cyclobutane, and spiropentane have been electrosynthesized by the reduction of appropriately substituted halides.   The method of electroreductive cyclization is of general utility in the synthesis of small-ring compounds.   For example, 1,3-dimethylbicyclobutane **38** is prepared from 1,3-dibromo-1,3-dimethylcyclobutane **37** in over 90% yield by electroreduction in a DMF-LiBr-(Hg) system (eq. (7.17)) [57~59].

$$(7.17)$$

The concerted mechanism of the electroreductive cyclization of 1,3-dihalides to cyclo-propanes has been proposed [58~60].   A stepwise process involving a highly efficient in-tramolecular cyclization of an intermediate bromocarbanion, $[CH_3CHCH_2CH(Br)CH_3]^-$ is also suggested [61].

The 1,5-dibromopentanes **39** bearing an electron-withdrawing group at the terminal po-sition can be converted into cyclopentane derivatives **40** in good yield [62].   The electroly-sis is carried out in a THF(or DMF)-Bu₄NClO₄(or Et₄NOTs)-(C) system in an undivided cell to give the corresponding cyclopentanones **40** in 57~87% yields (eq. (7.18)).

$$(7.18)$$

$$R^1 = H, Me, Allyl, OAc ; R^2 = H, CO_2Et$$
$$R^3 = H, Me, CH_2CO_2Et ; W = CO_2Et, CN$$

The cathodic reduction of phenacyl bromides semicarbazones **41** can lead to 1,4-diaryl-1,4-butanedione disemicarbazones **42**, which give the corresponding 3,6-diarylpyridazines **43** directly by heating (eq. (7.19)) [63].

$$(7.19)$$

41   Ar = Ph, p-BrC₆H₄,
         p-PhC₆H₄ etc.

The electroreduction of benzyl halides was formerly performed using a mercury elec-trode on which organomercury derivatives were formed.   Recently, the benzyl-benzyl cou-pling reaction has been undertaken on a platinum electrode [64].   Some results of the elec-troreductive dimerization of benzyl halides are collected in Table 7.3 [51a, 64~68].

The electroreduction of nitrobenzyl chlorides has been well investigated in an MeCN-Et₄NClO₄-(Pt) system.   The one-electron reduction proceeds initially to give an anion radi-cal.   In the case of anion radicals of o- or p-nitrobenzyl halides, halide ion is rapidly lost to give a neutral nitrobenzyl radical.   The subsequent radical dimerization takes place as a

Table 7.3 Electroreductive Coupling of Benzyl Halides

$$2 \; Ar\text{-}CHY\text{-}X \longrightarrow \underset{\textbf{A}}{(Ar\text{-}CHY)_{2}} + \underset{\textbf{B}}{Ar\text{-}CH_2Y}$$

Ar-CHY-X			Electrolysis Conditions	Yield, %		Ref.
Ar	Y	X		$(Ar\text{-}CHY)_{2}$	$Ar\text{-}CH_2Y$	
Ph	H	Br	MeCN-Bu₄NClO₄-(Pt)	70	–	[67]
Ph	H	Br	DMF-Et₄N(0.15M)-(Cu/Ni) Ni(acac)₃, Ph₃P	85		[51a]
MeC₆H₄	H	Br	MeCN-Bu₄NClO₄-(Pt)	64	2	[67]
Ph	H	Cl	THF/HMPA-LiClO₄(or Bu₄NClO₄)-(Ni) NiCl₂	95	–	[66]
Ph	H	Cl	DMF-Et₄N(0.15M)-(Cu/Ni) Ni(acac)₃, Ph₃P	87	–	[51a]
Ph	H	Cl	THF/HMPA(30%)-LiClO₄(0.5M)-(?) NiCl₂(0.5mmol), Ph₃P(1.25mmol)	86		[68]
o-NO₂C₆H₄	H	Cl	MeCN-Et₄NClO₄-(Pt)	92.3	14.9	[64]
p-NO₂C₆H₄	H	Cl	MeCN-Et₄NClO₄-(Pt)	96.3	4.5	[64]
o-NO₂C₆H₄	H	Br	MeCN-Et₄NClO₄-(Pt)	90.3	13.1	[64]
p-NO₂C₆H₄	H	Br	MeCN-Et₄NClO₄-(Pt)	96.5	3.7	[64]
Ph	Ph	Br	DMF-Et₄NClO₄-(Hg)	51	38	[65]

principal reaction together with the minor side reaction of an hydrogen atom abstraction. The electroreduction of *m*-nitrobenzyl chloride occurs more slowly than that of *ortho* and *para* isomers and the major reaction is shown to be a hydrogen atom abstraction.

Controlled potential electroreduction of benzhydryl bromide in a DMF-Et$_4$NClO$_4$-(Hg) system reveals that at the potential of -0.60 V *vs.* SCE, the reduction gives diphenyl-methane (89%) as the major product, whereas at a negative potential of -1.40 V, the inter-molecular coupling reaction favorably proceeds to give *sym*-tetraphenylethane (50%) to-gether with diphenylmethane (38%) [65].

The electroreduction of *meso*- and *dl*-2,4-dibromopentane **44** in a DMSO-Et$_4$NBr-(Hg) system affords roughly equal amounts of *cis*- and *trans*-1,2-dimethylcyclopropanes **45** and **46** as major products together with olefin **47** and pentane **48** (eq. (7.20)) [69].

$$\text{(eq. 7.20)}$$

The electroreductive cyclization of 1,3-dihaloalkanes leading to the corresponding cy-clopropanes are listed in Table 7.4 [69, 70a, 71~73] In a similar fashion, the cyclobutane **50** has been made in 90% abundance by the electroreduction of 1,4-dibromobutane **49** in a DMF-Et$_4$NBr-(Pt) system at the most negative potential (eq. (7.21)) [70a]. The controlled-potential electrolysis of 1,4-dibromo- and diiodobutanes in a DMF-Me$_4$NClO$_4$-(C) system even if in the presence of proton donors tends to give the corresponding cyclobutane in 35~36% yields [70b].

$$\text{(eq. 7.21)}$$

Spiropentane **54** has been electrosynthesized by the double reduction of 1,3-dibromo-2,2-bis(bromomethyl)propane **52** in a DMF-Bu$_4$NClO$_4$-(Pb) system *via* the intermediate **53** (eq. (7.22)) [71]. However, in the case of 2-vinyl-1,1-bis(bromomethyl)cyclopropane, the electroreduction undergoes preferentially the ring opening reaction of the cyclopropane ring along with the formation of vinylspiropentane as a minor product [74].

$$\text{(eq. 7.22)}$$

Diiodoacetylene **55** can be cathodically electropolymerized into the polymer **56** in a DMF-Bu$_4$NI-(Pt) system in the presence of diiodo[1,2-bis(diphenylphosphino)ethane]nick-el(II) under anaerobic conditions (eq. (7.23)) [75]. The electroreduction of (*E*,*E*)-1,12-di-bromo-2,10-dodecadiene **57** in a THF-Bu$_4$NClO$_4$-(Pt) system affords the corresponding twelve-membered (*E*,*E*)-1,5-cyclododecadienes **58** and (*E*,*Z*)-isomer **59** in low yields to-gether with **60** and **61** (eq. (7.24)) [76].

$$\text{(eq. 7.23)}$$

Table 7.4 Electroreductive Cyclization of Cyclopropane Compounds

$$X\text{-CH}_2\text{-CHal}^1\text{Hal}^2\text{-Y (A)} \longrightarrow X\triangle Y \text{ (B)} + \text{Olefin} + \text{Alkane} + \text{Others}$$

X	Y	Hal¹	Hal²	Electrolysis Conditions	V (SCE)	B	Olefin	Alkane	Ref.
								Yield, %	
H	H	Br	Br	DMF-Et₄NBr-(Pt)	−2.65	91	9	–	[70a]
H	H	Br	Br	DMF-Bu₄NClO₄(0.1M)-(C)	−2.56	86~90	4~7	0.1	[72]
H	H	I	I	DMF-Bu₄NClO₄(0.1M)-(C)	−2.23	91~94	1	0.1	[72]
H	H	Br	Cl	DMF-Bu₄NClO₄(0.1M)-(C)	−2.75	67~89	6~31	–	[72]
H	H	Cl	I	DMF-Bu₄NClO₄(0.1M)-(C)	−2.31	89~96	2	–	[72]
H	H	Br	I	DMF-Bu₄NClO₄(0.1M)-(C)	−2.36	91~92	2~3	0.1	[72]
Me	Me	Br	Br	DMSO-Et₄NBr-(Hg)	−2.2	83	13.5	3	[69]
CH₂Br	CH₂Br	Br	Br	DMF-Bu₄NBr-(Hg)	–	47~58	–	–	[71a]
CH₂Br	CH₂Br	Br	Br	DMF-Bu₄NClO₄-(Pb)	−1.8	78	–	–	[71b]
Ph	H	Br	Br	DMF-Et₄NBF₄(0.2M)-(Hg)	−2.10	100	–	–	[73]
Ph	Ph	Br	Br	DMF-Et₄NBF₄(0.2M)-(Hg)	−2.10	91	–	–	[73]

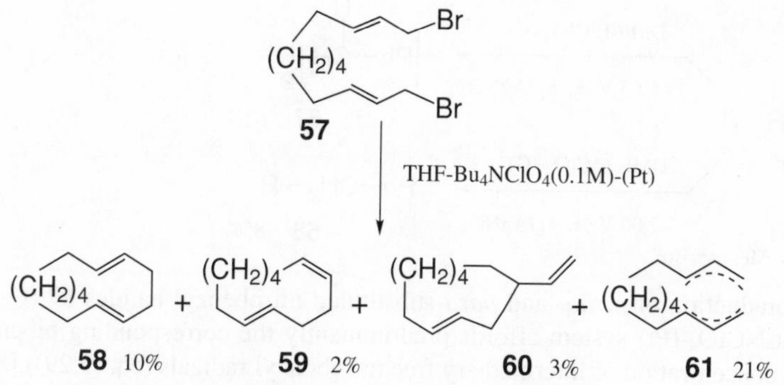

$$THF\text{-}Bu_4NClO_4(0.1M)\text{-}(Pt) \tag{7.24}$$

57

58 10%   +   **59** 2%   +   **60** 3%   +   **61** 21%

The electroreductive coupling of α-iodinated dioxolanes **62** can lead to the preparation of a series of symmetrical and non-symmetrical γ-bisdioxolanes **63** in good yield (eq. (7.25)) [77].

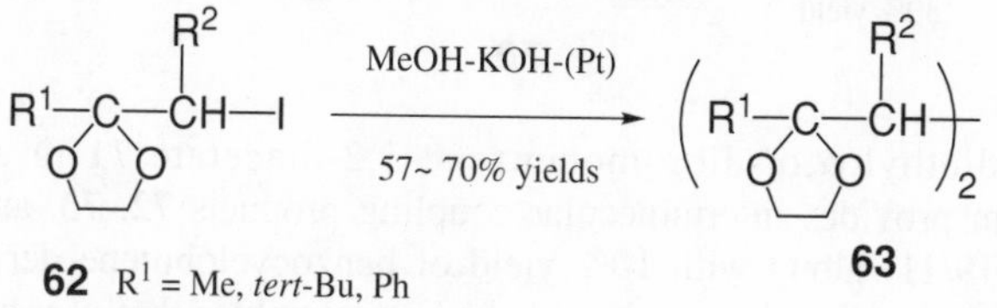

$$\xrightarrow[\text{57~70\% yields}]{MeOH\text{-}KOH\text{-}(Pt)} \tag{7.25}$$

62 R^1 = Me, *tert*-Bu, Ph
      R^2 = H, alkyl

63

Electroformylation of organic halides has been performed in a DMF-Bu$_4$NBr-(Mg/Coated Zn) system in an undivided cell to give the aldehydes in 50~80% yields [78].

Dimerization of alkyl bromides **64** can be performed by use of Ni(bpy)$_2$ complexes in a DMF(or NMP)-Bu$_4$NBF$_4$(0.1M)-(Pt) system to give the dimer **65** in good yields (eq. (7.26)) [79].

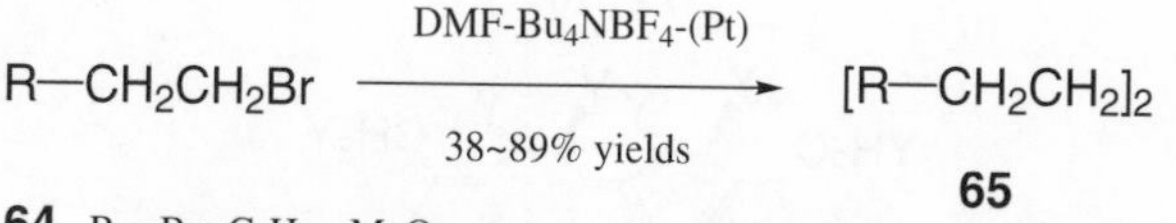

$$R\text{—}CH_2CH_2Br \xrightarrow[\text{38~89\% yields}]{DMF\text{-}Bu_4NBF_4\text{-}(Pt)} [R\text{—}CH_2CH_2]_2 \tag{7.26}$$

65

64 R = Bu, C$_5$H$_{11}$, MeO,
       MeOC(O)(CH$_2$)$_2$, Cl(CH$_2$)$_2$

The low-valent nickel-phosphine complex catalyzes reduction and homo-coupling reactions of benzyl chloride in ethanol, affording 1,2-diphenylethane together with toluene in half and half [80].

Products from the electroreduction of (1-bromo-2,2-dimethylpropyl)benzene **66** (R = *tert*-Bu) in a DMF-LiClO$_4$-(C) system are found to depend on the electrolysis potential. At relatively positive potentials, the products **67** are formed primarily from the coupling of two benzylic radicals (eq. (7.27)), whereas at more negative potentials, the products **68** are obtained from the corresponding carbanions (eq. (7.28)) [81a]. Electrosynthesis of 1,2-diarenyl diethyl succinates has been performed in a DMF-Et$_4$NClO$_4$-(C) system [81b].

$$\text{Ph–CH–R} \quad \underset{-1.65 \text{ V } vs. \text{ Ag/AgNO}_3}{\overset{\text{DMF-LiClO}_4\text{-(C)}}{\nearrow}} \quad \left[\text{Ph–}\overset{\overset{\text{R}}{|}}{\text{CH}} \right]_2 \qquad (7.27)$$

67 62%

$$\underset{\text{Br}}{|} \quad \underset{-2.00 \text{ V } vs. \text{ Ag/AgNO}_3}{\overset{\text{DMF-LiClO}_4\text{-(C)}}{\searrow}} \quad \text{Ph–CH}_2\text{–R} \qquad (7.28)$$

66   **68** 85%

R = Me, *tert*-Bu

The electroreduction of *ortho*- and *para*-substituted nitrobenzyl halides **69** (*O*-isomer) in an MeCN-Et$_4$NClO$_4$-(Pt) system affords predominantly the corresponding bibenzyls **70** due to the high concentration of intermediary free nitrobenzyl radicals (eq. (7.29)) [82].

$$\underset{\text{69}}{\overset{\text{NO}_2}{\underset{\text{–CH}_2\text{Br}}{\bigcirc}}} \quad \underset{80\% \text{ yield}}{\overset{\text{MeCN-Et}_4\text{NClO}_4\text{-(Pt)}}{\longrightarrow}} \quad \underset{\text{70}}{\overset{\text{NO}_2 \quad \text{O}_2\text{N}}{\underset{\text{–CH}_2\text{CH}_2\text{–}}{\bigcirc \quad \bigcirc}}} \qquad (7.29)$$

Electroreduction of diethyl α,α'-dibromobenzene-1,2-diacetate **71** in a DMF-Et$_4$NClO$_4$(0.1M)-(C) system provides intermolecular coupling products **72, 73**, and **74** in 46% yield (**72/ 73/ 74**:16/19/11) along with 10% yield of benzocyclobutene derivatives. α,α'-Bis(ethoxycarbonyl)-*o*-quinodimethane, whose behavior resembles that of a biradical, is proposed as a key step leading to the products (eq. (7.30)) [83].

71 Y = CO$_2$Et   **72** 16%

73 a: *meso* 16%
     b: *dl* 3%

74 11%

$$(7.30)$$

The electroreductive homo-coupling of allyl chloride **75** has been found to be catalyzed in the presence of 2,2'-bipyridine cobalt complexes to give 1,5-hexadiene **76** in 41~53% yields (eq. (7.31)) [84].   Similar coupling reactions of allyl bromide in a DMF-Et$_4$NClO$_4$(0.1M)-(Pt) system in the presence of a Cu(acac)$_2$ complex affords the homo-coupling product in 83% yield [85].

$$2 \quad \diagup\!\!\!\diagdown\!\!\!\text{Cl} \quad \underset{41~53\% \text{ yields}}{\overset{\overset{\text{MeCN-Et}_4\text{NClO}_4\text{-(Pt)}}{\text{Co(bpy)}_3(\text{ClO}_4)_2}}{\longrightarrow}} \quad \diagup\!\!\!\diagdown\!\!\!\diagup\!\!\!\diagdown \qquad (7.31)$$

75   **76**

The electroreductive intramolecular coupling of 1,3-dichloroalkanes **77** preferentially proceeds in DMSO rather than DMF, MeCN, or HMPA.   The electrolysis of **77** in a DMSO-Et$_4$NOTs-(C) system in an undivided cell to give the cyclopropane products **78** in 51~83% yields (eq. (7.32)) [86].

$$
ClCH_2-CH_2-C\underset{R^2}{\overset{Cl\ \ R^1}{\diagdown}} \quad \xrightarrow[51\sim83\% \text{ yields}]{\text{DMSO-Et}_4\text{NOTs-(C)}} \quad \triangleright\!\!<\!\!\overset{R^1}{\underset{R^2}{}} \tag{7.32}
$$

77   R^1 = H, Me, CH$_2$(CO$_2$Et)
         R^2 = CN, CO$_2$Et, Ac

                                                   78

The highly alkylated α,α'-dibromo ketones **79** can be converted into the corresponding cyclopropanone hemiacetals **81** in high current yields.   The electrolysis of **79** is carried out in an MeCN-Et$_4$NCl-(Hg) system to give the hemiacetals **81** in 40~85% yields *via* the cyclopropanones **80** (eq. (7.33)) [87].

$$
R^1-\underset{Br}{\overset{R^2}{\underset{|}{C}}}-\overset{O}{\overset{\|}{C}}-\underset{Br}{\overset{R^3}{\underset{|}{C}}}-R^4 \xrightarrow[40\sim85\% \text{ current yields}]{\text{MeCN-Et}_4\text{NCl-(Hg)}} \quad \overset{R^3\ \ R^4}{\triangle}\!=\!O \xrightarrow{\text{HNu}} \overset{R^3\ \ R^4}{\triangle}\!\!\overset{OH}{\underset{Nu}{}} \tag{7.33}
$$

79   R^1, R^2, R^3, R^4 = H, Me   **80**   **81**
Nu: OMe, NMe$_2$, NHMe, NHPh

1,3-Dibromo-1,3-diphenylpropane **82** can be cyclyzed by electroreduction in a DMF-Et$_4$NBF$_4$-(Hg) system to give the corresponding cyclopropane **83** (*cis/trans*: 22.5/74.5) in 91% yield (eq. (7.34)) [88].

$$
\underset{Br\ \ \ \ Br}{Ph\diagup\!\!\diagdown\!\!\diagup\!\!Ph} \xrightarrow[91\% \text{ yield}]{\text{DMF-Et}_4\text{NBF}_4\text{-(Hg)}} \quad \overset{H}{\underset{Ph}{}}\!\!\triangle\!\!\overset{Ph}{\underset{H}{}} \tag{7.34}
$$

         82   **83**

The intramolecular coupling of 3,7-dibromo-3,7-dinitro-bicyclo[3.3.1]nonane **84** affords the cyclized product **85** in good yields (eq. (7.35)) [89].   Electrosynthesis of other highly strained hydrocarbons has also been attempted [60].

$$
\xrightarrow[75\sim80\% \text{ yields}]{\text{H}_2\text{O/Me}_2\text{CO-LiClO}_4\text{-(Pb or C)}} \tag{7.35}
$$

         84   **85**

Electroreductive synthesis of [2.2.2]propellane has been attempted by the reduction of 1,4-dibromobicyclo[2.2.2]octane [90].

The electroreductive annelation forming indole and isoindole alkaloides **87** has been performed in a DMF-(Pt/Pb) system. The reduction of the immonium salts **86** affords the annelated products **87** in 41~60% yields (eq. (7.36)) [91].

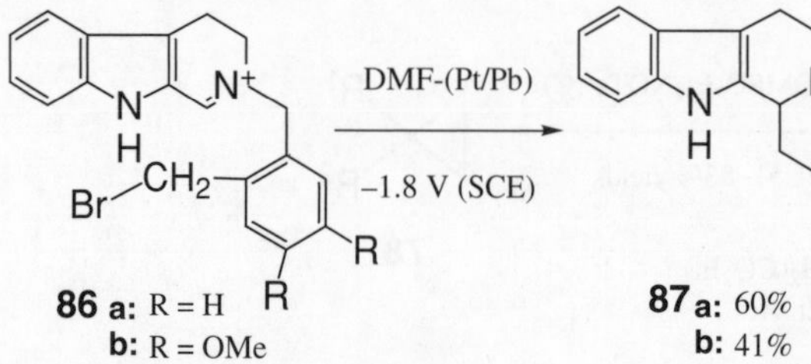

$$\text{(7.36)}$$

86 a: R = H
b: R = OMe

87 a: 60%
b: 41%

7.1.3   Electroreductive Nucleophilic Substitutions

The electroreduction of methyl iodide **88** in a DMF-Bu$_4$NClO$_4$-(Sn) system affords exclusively tetramethyltin **89** in good yield (eq. (7.37)), whereas ethyl, propyl, and butyl iodides

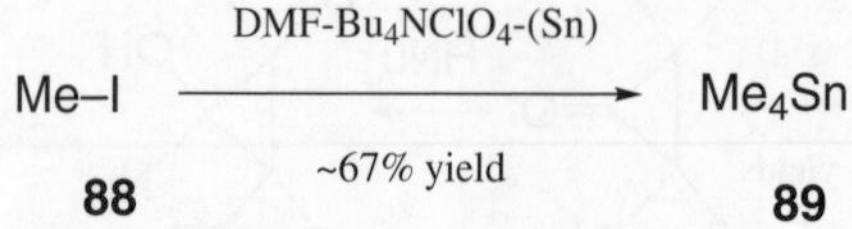

$$\text{Me–I} \xrightarrow[\text{~67\% yield}]{\text{DMF-Bu}_4\text{NClO}_4\text{-(Sn)}} \text{Me}_4\text{Sn} \qquad \text{(7.37)}$$

88   **89**

under similar conditions provides the corresponding dimers such as butane, hexane, and octane, respectively [92]. The electroreductive coupling of 2-bromonorbornane **90** and the Shiff base **91**, giving N-α-(norbornan-2-ylbenzyl)aniline **92**, has been performed in a DMF-Bu$_4$NI-(Hg) system in 55% yield (eq. (7.38)) [11].

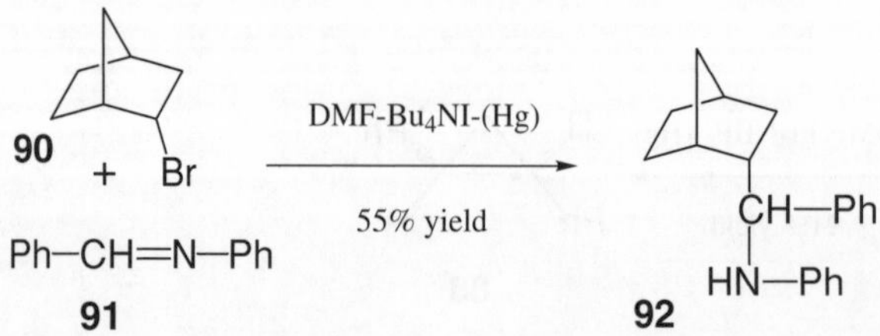

$$\text{(7.38)}$$

90

+   Br

Ph–CH=N–Ph

91   **92**

Controlled potential electroreduction of organic halides in the presence of carbon dioxide generally gives low yields of the corresponding carboxylates [93, 94]. Sacrificial anodes have been used for the carboxylation of halides, and major advantages include electrolysis performance without the use of a diaphragm in cells and the separable nature of the carboxylate salts by a simple workup procedure. The electrocarboxylation of benzylic and allylic chlorides has been found to proceed in a THF/HMPA-LiClO$_4$-(Pt/Hg) system in the presence of a catalytic amount of a cobalt Schiff base complex Co(Salen) [95a]. Electrosynthesis of ketones from organic halides and carbon monoxide has been attained in a DMF-Bu$_4$NBF$_4$/FeCl$_2$-(Ni or SUS) system in the presence of nickel bipyridine complex [95b]. The electrosynthesis of 2-arylpropionic acids **94** has been attained in 76~81% yields by electrocarboxylation of ArCH(Cl)Me in the presence of a catalytic amount of Ni(II)(dppp) complexes (eq. (7.39)) [96].

$$Ar\!-\!\underset{\underset{Me}{|}}{\overset{\overset{H}{|}}{C}}\!-\!Cl + CO_2 \xrightarrow[\text{NiCl}_2\text{(dppp), 76~81\% yields}]{\begin{array}{c}\text{THF/HMPA(2/1)-Bu}_4\text{NBF}_4\\(0.1\text{M})\text{-(Pt/C)}\end{array}} Ar\!-\!\underset{\underset{Me}{|}}{\overset{\overset{H}{|}}{C}}\!-\!CO_2H \qquad (7.39)$$

93   **94**

Ar: MeO—[naphthyl]— , Isopr—[phenyl]— , [phenyl]—O—[phenyl]

The electroreductive coupling of γ-substituted allyl bromides **95** with carbon dioxide has been attained in a DMF-Bu$_4$NClO$_4$(0.1M)-(Mg/Pt) system to yield the carboxylic acids **96** and **97** in 45~71% yields (eq. (7.40)) [97].   The electroreduction of ethyl alanine

$$\text{95} \quad X = Br \xrightarrow[\text{45~71\% yields}]{\begin{array}{c}CO_2,\\DMF\text{-Et}_4\text{NClO}_4(0.1\text{M})\text{-(Mg/Pt)}\end{array}} \text{96} \quad CO_2H \quad + \quad \text{97} \quad CO_2H \qquad (7.40)$$

96/97 = 9/1

N-methylammonium bromide **98** in a DMF-Pr$_4$NI(0.1M)-(Hg) system in the presence of carbon dioxide and subsequent treatment with ethyl bromide yields diethyl methylmalonate **99** in 76% yield (eq. (7.41)) [98].

$$\underset{\underset{+NMe_3}{|}}{Me\!-\!CHCO_2Et} \xrightarrow[\text{CO}_2\text{, EtBr;   76\% yield}]{\text{DMF-Pr}_4\text{NI(0.1M)-(Hg)}} Me\!-\!CH(CO_2Et)_2 \qquad (7.41)$$

98   **99**

The technical feasibility of the scale-up of the electrocarboxylation reactions with a sacrificial anode has recently been reported [99,100].   The results by the electrocarboxylation of halides are listed in Table 7.5 [95a, 100~104].

The electroreduction of benzyl chloride **100** in a DMF-Bu$_4$NBF$_4$-(Hg) system in the presence of anthracene as a mediator together with water as a proton source affords the benzyl ether **101** in 57% yield (eq. (7.42)) [39].   Electroreduction of allyl halides in the presence of a silylating reagent (Me$_3$SiCl, HMeSiCl, and PhMe$_2$SiCl) in a DMF-Et$_4$NOTs-(Pt)

$$PhCH_2Cl \xrightarrow[\text{57\% yield}]{\text{DMF-Bu}_4\text{NBF}_4\text{-(Hg)}} Ph\!-\!CH_2OCH_2\!-\!Ph \qquad (7.42)$$

100   **101**

system affords the corresponding organosilicon compounds [105, 106].   The trimethylsilylation of benzyl chloride proceeds in a TMU-Et$_4$NBF$_4$-(Mg) system in the presence of trimethylchlorosilane giving phenyltrimethylsilane in 70% yield [107].   The addition of the electrogenerated carbanions derived from substituted allyl halides **102** to ketones has been attained in hexamethylphosphoric triamide (HMPA).   For instance, the allylation of acetone **103** is performed in an HMPA-Bu$_4$NClO$_4$-(Pt) system in an undivided cell to afford the adduct **104** in 53% yield (eq. (7.43)) [108].

Table 7.5 Electroreductive Carboxylation of Halides

$$R{-}X \xrightarrow{\ CO_2\ } R{-}CO_2H$$
$$\quad A \qquad\qquad B$$

R–X		Electrolysis Conditions	P_{CO_2} atm	Yield, % R–CO₂H	Ref.
R	X				
CH₃(CH₂)	Cl	DMF-Bu₄NBr-(Al/C)	1	12.5	[101]
CH₃(CH₂)₃	Cl	DMF-Bu₄NBr-(Al/C)	1	9	[101]
CH₃CH=CHCH₂	Cl	THF/HMPA(4/6)-LiClO₄(0.3M)/Co(Salen)*¹(0.01M)-(Hg)	2	63	[95a]
CH₂=C(Me)CH₂	Cl	THF/HMPA(4/6)-LiClO₄(0.3M)/Co(Salen)*¹(0.01M)-(Hg)	2	62	[95a]
PhCH₂	Cl	DMF-Bu₄NI-(Mg)	1	80–90	[102]
	Cl	THF/HMPA(4/6)-LiClO₄(0.3M)/Co(Salen)*¹(0.01M)-(Hg)	2	62	[95a]
	Cl	DMF-Bu₄NClO₄-(Mg/SUS)	1	99	[103]
	Cl	THF/HMPT-Bu₄NBF₆-(Hg)*²	1	99	[104]
	Cl	DMF-Bu₄NBr-(Al/C)	1	82.5	[101]
PhCHMe	Cl	DMF-Bu₄NClO₄-(Mg/SUS)	1	80	[103]
PhCH=CH	Br	DMF-Bu₄NClO₄-(Mg/SUS)	1	80	[103]
PhCH=CHCH₂	Cl	DMF-Bu₄NClO₄-(Mg/SUS)	1	80	[103]
m-PhOC₆H₄CHMe	Cl	DMF-Bu₄NClO₄-(Mg/SUS)	1	80	[103]
i-C₆H₄CHMe	Cl	DMF-Bu₄NBr-(Zn)	1	95	[100]
(Ph)₂CH	Cl	DMF-Bu₄NBr-(Zn)	1	73	[100]
NaphtylCH₂	Cl	DMF-Bu₄NBr-(Zn)	1	79	[100]

*¹ Co(salen) = Cobalt-bis-salicylidene iminate
*² HMPT = hexamethylphosphorotriamide

$$\text{(7.43)}$$

The electrosynthetic process using a sacrificial zinc anode and catalytic amounts of NiBr$_2$(bpy) complex (bpy = 2,2'-bipyridine) affords homoallylic alcohols with good to high yields, from methally chloride and a variety of carbonyl compounds [109~111]. The zinc anode and that of a nickel catalyst, both of which prove to be essential in the coupling reaction.

The electroreduction of a catalytic amount of ZnBr$_2$ in acetonitrile provides an active Zn* able to promote the reductive coupling of allyl bromides and chlorides with carbonyl compounds in high regioselectivity to lead to the corresponding homoallylic alcohols, favoring the branched alcohol. In all cases, the ratio of branched to linear alcohols is 95/5 or higher [112]. For example, the reaction of crotyl bromide **105** and cyclohexanone **106** in an MeCN-ZnBr$_2$-(Zn) system affords the branched alcohol **107** in 85% yield together with **108** (6.5%) (eq. (7.44)) [112].

$$\text{(7.44)}$$

Electroreductive Barbier-type allylation of carbonyl compounds has been investigated in nickel catalyzed processes [109,111] and tin redox systems [113, 114]. The SmCl$_3$-catalyzed electroreductive allylation of keones has been performed in a DMF-Bu$_4$NBF$_4$-(Mg/Ni) system in the presence of a catalytic amount of samarium trichloride in an undivided cell. The reaction of methally chloride and cyclohexanone gives the corresponding adduct in 74% yield [115]. The regioselectivity in the electrochemical allylation of alkanones, alkanals, and aromatic aldehydes with allylic halides has been investigated. If substituted allyl halide **95** (X = Cl) or **109** is used in the allylations, the C-C bond formation may take place at two different sites of the allylic moiety to form two types of regioisomers, **110** and **111** (eq. (7.45)) [116, 117]. A decreasing current density in the allylation of acetone leads to the enhancement in the product ratio of the compound **111** [116, 117].

$$\text{(7.45)}$$

The electrolysis of a mixture of prenyl chloride **95** (X = Cl, $Ep = -2.2$ V) and 3-furancar-baldehyde **112** ($Ep = -1.9$ V) in an HMPA-Et$_4$NClO$_4$-(Pt) system in the presence of chlorotrimethylsilane gives **113** as a major product together with **114** (eq. (7.46)) [118].

$$
\begin{array}{c}
\text{112} + \text{95 (X = Cl)} \xrightarrow{\;2\,e^{-*}\;} \text{113 50\%} + \text{114 15\%}
\end{array}
\tag{7.46}
$$

* HMPA-Et$_4$NClO$_4$(0.1M)/TMSCl-(Pt)

The electroreduction of 5-bromodecalone **115** in a DMF-Bu$_4$NBr-(Pt/Hg) system at -2.6 V (SCE) affords a mixture of the ketones **116**, **117**, and **118** (eq. (7.47)) [119].

$$
\text{115} \xrightarrow[-2.6 \text{ V (SCE)}]{\text{DMF-Bu}_4\text{NBr-(Hg)}} \text{116 40\%} + \text{117 35\%} + \text{118 15\%}
\tag{7.47}
$$

The electroreductive addition of allyl groups in allyl halides to α,β-enoates has been attempted [120]. The electrolysis of allyl bromide **120** and diethyl fumarate **119** in a DMF-Et$_4$NOTs(0.2M)-(Pt) system in an undivided cell yields the adduct **121** in 70% yield (eq. (7.48)) [121]. Recently, the one-step electrosynthesis of γ-lactones has been attained by the electroreductive cross-coupling of 3-chloroesters **123** with carbonyl compounds **122** in the presence of a catalytic amount of samarium trichloride (eq. (7.49)) [122].

$$
\text{119} + \text{120} \xrightarrow[\text{70\% yield}]{\text{DMF-Et}_4\text{NOTs(0.2M)-(Pt)}} \text{121}
\tag{7.48}
$$

$$
\text{122} + \text{123} \xrightarrow[\text{50\textasciitilde76\% yields}]{\text{DMF-SmCl}_3\text{-(Mg/SUS)}} \text{124}
\tag{7.49}
$$

123 a: $R^1, R^2 = -[CH_2]_5-$,   **b:** $R^1, R^2 = -[CH_2]_4-$
c: $R^1 = n\text{-}C_6H_{13}$, $R^2 = Me$,   **d:** $R^1, R^2 = -[CH_2]_5-$

The electroreduction of 1-iodo-5-decyne **125** at a mercury cathode results in 5-decynyl radicals that adsorb into and interact with the mercury electrode to give 5-decynylmercury radicals which disproportionate into di-5-decynylmercury **126** efficiently (eq. (7.50)) and, partly, pentylidenecyclopentane **127** (eq. (7.51)) [123].   The similar electroreduction of **125** at a vitreous carbon electrode yields pentylidenecyclopentane **127** as a major product (eq. (7.51)) along with 5-decyne, 1-decen-5-yne, and a small amount of 5-decyn-1-ol, as shown in Table 7.6 [123, 124].   An improved procedure for the radical cyclization of 2-bromoethyl 2-alkynyl ethers is developed by the combined use of chloro (pyridine)cobaloxime(III) and a zinc plate as a sacrificial anode in an undivided cell [125].

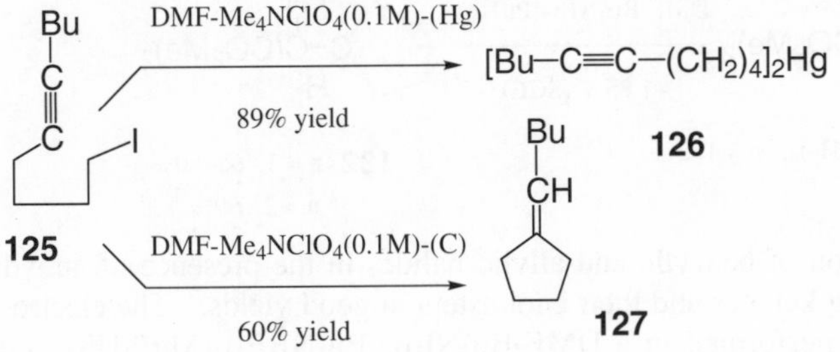

$$[Bu-C\equiv C-(CH_2)_4]_2Hg \qquad (7.50)$$

89% yield

126

125

DMF-Me$_4$NClO$_4$(0.1M)-(C)

60% yield

127

$$(7.51)$$

Table 7.6 Product Distributions for Electroreduction of 1-Halo-5-decynes in DMF-Me$_4$NClO$_4$(0.1M) System

1-Halo-5-decyne	Electrode	Product, Yield, %					Ref.
		1-Butyl-cyclohexene	Pentylidene-cyclopentane	5-Decyne	1-Decen-5-yne	Di-5-decynyl-mercury	
I	Hg pool	–	4	–	–	89	[123b]
Br	Hg pool	–	–	–	–	65	[123b]
I	Glassy Carbon	–	60	31	1	–	[124]
Br	Glassy Carbon	–	5	74	1	–	[124]
Br	Hg pool	38	–	59	–	–	[123a]

In the case of phenyl-conjugated acetylenic halides, *e.g.*, 6-iodo-1-phenyl-1-hexyne **128**, the electroreduction in an HFIP-Me$_4$NClO$_4$(0.1M)-(C) system affords a mixture of benzylidenecyclopentane **129** and 1-phenyl-1-hexyne **130** (eq. (7.52)) [126].   The use of a mercury electrode tends to give the corresponding diorganomercury compounds together with 1-phenyl-1-hexyne as major products [127, 128].

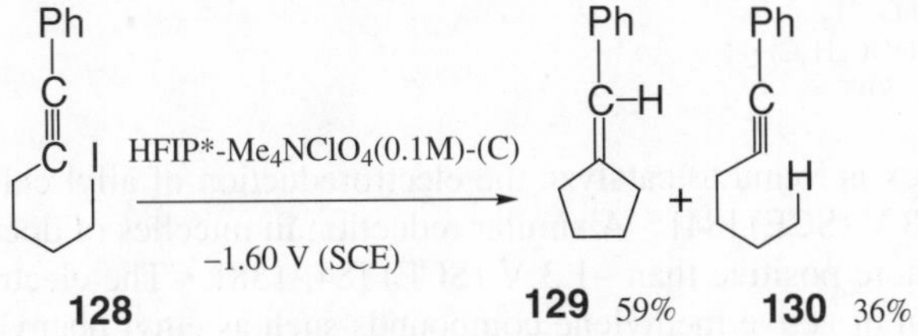

HFIP*-Me$_4$NClO$_4$(0.1M)-(C)

−1.60 V (SCE)

$$(7.52)$$

128

129 59%   **130** 36%

*HFIP = 1,1,1,3,3,3-hexafluoroisopropyl alcohol

The conjugated addition of ω-bromoalkylidenemalonate **131** by the electroreductive hydrocyclization has been discussed in terms of "concerted reduction-cyclization" during the transfer of the first electron in the potential-determining step [59, 61, 69, 129~132]. Although the reductive cyclization of **131** undergoes an overall two-electron process, the possibility of a concerted one-electron transfer and a ring closure step may not be ruled out. The reductive cyclization of **131** is carried out in a DMF-Bu$_4$NBr-(Pt/Hg) system at −1.85 V (SCE) to give the cyclic products **132** in good yields (eq. (7.53)) [133, 134].

$$Z\overset{\displaystyle CH=C(CO_2Me)_2}{\underset{\displaystyle Br}{\bigg\langle}} \xrightarrow[-1.85\ \text{V (SCE)}]{\text{DMF-Bu}_4\text{NBr-(Pt/Hg)}} Z\overset{\displaystyle \overset{H}{\underset{H}{C}}-\overset{H}{C}(CO_2Me)_2}{\bigg\langle} \tag{7.53}$$

131 Z = (CH$_2$)$_n$; n = 1, 2   **132** n = 1, 65~80%
                                                    n = 2, 60%

The electroreduction of benzylic and allylic halides in the presence of anhydrides affords the corresponding ketones and their enol esters in good yields. The electrosynthesis of the ketones **135** is performed in a DMF-Bu$_4$NI(or Bu$_4$NBF$_4$)-(Mg/SUS) system (eq. (7.54)) [135], whereas the enol esters **136** are electrosynthesized in a DMF-Et$_4$NOTs-(Pb) system (eq. (7.55)) [136]. An efficient electroreductive conversion of 3'-chloro-cephalosporins **137** and their analogues into the corresponding 3-methylenecephams **138** has been performed in an aqueous THF-LiClO$_4$-(Pb) and/or aqueous MeCN/EtOH-LiClO$_4$/NH$_4$ClO$_4$-(Pb) system in good yields (eq. (7.56)) [137].

$$\text{PhCH}_2\text{Cl} \xrightarrow[\text{80\% yield}]{\text{DMF-Bu}_4\text{NI-(Mg/SUS)}} \text{PhCH}_2-\text{COMe} \tag{7.54}$$

+ 133   **135**

(MeCO)$_2$O

134

$$\xrightarrow[\text{71\% yield}]{\text{DMF-Et}_4\text{NOTs-(Pb)}} \overset{\displaystyle Ph}{\underset{\displaystyle H}{}}C=C\overset{\displaystyle Me}{\underset{\displaystyle OCOMe}{}} \tag{7.55}$$

136

$$\tag{7.56}$$

137 R^1 = PhCH$_2$, PhOCH$_2$
        R^2 = PhCH$_2$, MeOC$_6$H$_4$CH$_2$
        X = Cl, I, OAc, SBT   **138** CO$_2$R^2

The Co(II)(bpy)$_3$ complex is found to catalyze the electroreduction of allyl chloride in an MeCN-(Pt) system at −2.8 V (SCE) [84]. A similar reduction in micelles of docecylsulfate proceeds at potentials more positive than −1.3 V (SCE) [84, 138]. The electroreductive benzylation or alkylation of active methylene compounds such as ethyl phenylacetate, phenylacetone, phenylacetonitrile and fluorene leads selectively to the corresponding monoalkylated products in good yields.   The alkylations are performed in a DMF-Bu_4NBF_4(0.5M)-(Mg/SUS) system in an undivided cell [139].   The reaction mechanism is proposed as shown in eqs. (7.57~60) (R^1 = alkyl; R^2 = phenyl, alkyl).   When R^1X is used in large excess relative to R^2CH_2A (R^2 = Ph; A = CO_2Et), the yield of alkylated product decreases in favor of the formation of the dimer R^1-R^1 in the absence of any other reagent.

$$\text{Anode}: \qquad Mg \longrightarrow Mg^2 + 2\,e^- \qquad\qquad \text{-------- (7.57)}$$

Cathode :

$$R^1\text{--}X + 2\,e^- \longrightarrow R^{1-} + X^- \qquad\qquad \text{-------- (7.58)}$$
$$\underset{\textbf{139}}{\phantom{R^1\text{--}X}}$$

$$\underset{\textbf{140}}{R^{1-}} + R^2\text{--}CH_2\text{--}A \rightleftharpoons RH + \underset{\textbf{141}}{R^2\text{--}CH^-\text{--}A} \qquad \text{---- (7.59)}$$

$$R^2\text{--}CH^-\text{--}A + \underset{\textbf{139}}{R^1\text{--}X} \longrightarrow \underset{\textbf{142}}{R^2\text{--}CH(R^1)\text{--}A} + X^- \qquad \text{---- (7.60)}$$

The electroreduction of benzyl chloride **143** in a DMF-Et_4NBr(0.1M)-(Mg/Fe) system in the presence of cyclopentadiene **144** undergoes the benzylation of cyclopentadiene to give **145** in 72% yield (eq. (7.61)) [140, 141].   The cathodic process in the electroreduction

$$PhCH_2Cl + \underset{\textbf{144}}{\bigcirc} \xrightarrow[\substack{-2.1\text{ V (Ag/AgCl)},\\ 72\% \text{ yield}}]{DMF\text{-}Et_4NBr(0.1M)\text{-}(Mg/Fe)} PhCH_2\text{--}\underset{\textbf{145}}{\bigcirc} \qquad (7.61)$$
$$\underset{\textbf{143}}{}$$

of benzyl chloride **143** may form a benzyl anion, $PhCH_2^-$, which acts as a strong EG base and reacts with cyclopentadiene with the formation of the cyclopentadienyl anion $C_5H_5^-$ on attacking the unchanged **143**.   Electroreductive benzylation of morpholine **147** proceeds in an EtOH-Et_4NClO_4(0.1M)-(Pt) system to afford the N-benzylmorpholine **148** in good yield (eq. (7.62)) [142].   In the course of the electroreduction of p-nitrobenzyl chloride **146**, the anion radical intermediate has been investigated by simultaneous electrochemical electron paramagnetic resonance [143].

$$O_2N\text{--}\underset{\textbf{146}}{\bigcirc}\text{--}CH_2Cl$$
$$+$$
$$\underset{\textbf{147}}{HN\bigcirc O} \xrightarrow[\sim 80\% \text{ yields}]{EtOH\text{-}Et_4NClO_4(0.1M)\text{-}(Pt)} O_2N\text{--}\bigcirc\text{--}CH_2\text{--}N\bigcirc O \qquad (7.62)$$
$$\underset{\textbf{148}}{}$$

Organozinc compounds are of great interest in organic synthesis since they undergo efficient and selective carbon-carbon bond formation.   Recently, a simple and mild method for zinc activation using an electrochemical process has been reported [144].   A catalytic amount of electrogenerated zinc deposit readily induces the activation of zinc metal toward allylic halides and α-bromoesters and the subsequent coupling of organozincs proceeds with various electrophilic compounds [112, 145].   For example, the benzylation to aromatic aldehydes (**150** → **151**) occurs in an MeCN-ZnBr$_2$/Me$_3$SiCl-(Zn) system in the presence of one equivalent of chlorotrimethylsilane in an undivided cell (eq. (7.63)) [146].

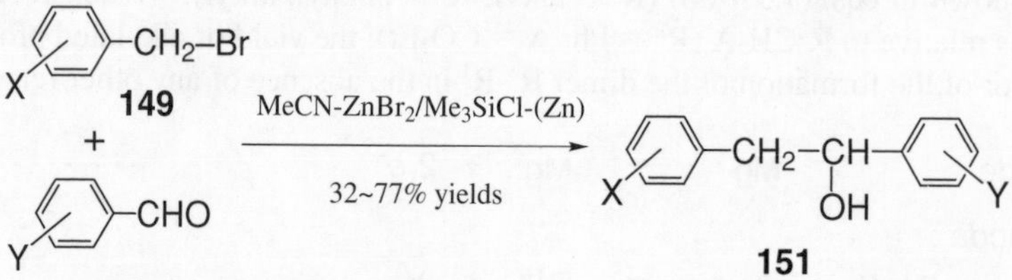

(7.63)

150   X = H, CN, Me, CF$_3$
Y = Me, OMe, CHO, SMe, Br,

Electroreductive intramolecular coupling of cyclic α-(bromomethyl) β-ketoesters **152** proceeds in the presence of trimethylsilyl chloride, and the one-carbon ring-enlarged products **153** are produced in fairly good yields (eq. (7.64)) [147].

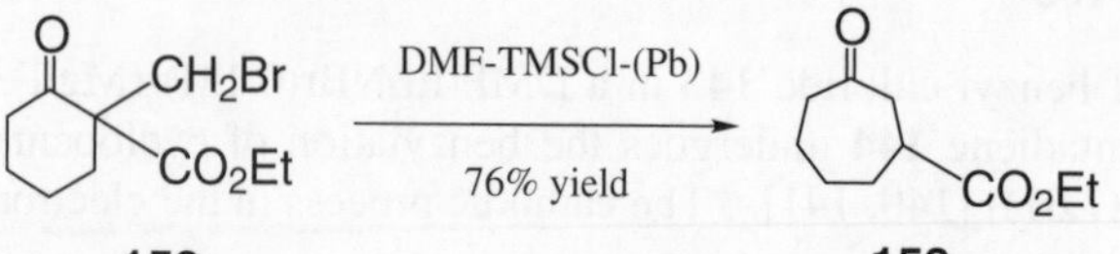

(7.64)

152   **153**

The electroreduction of 2-(1-bromo-1-methylethyl)benzofurans **154** in an MeCN-Et$_4$NOTs-(Hg) system undergoes the cleavage of a carbon-bromine bond followed by a ring expansion to afford the corresponding 2,2-dimethylchromene **155** in 87% yield (eq. (7.65)) [148].

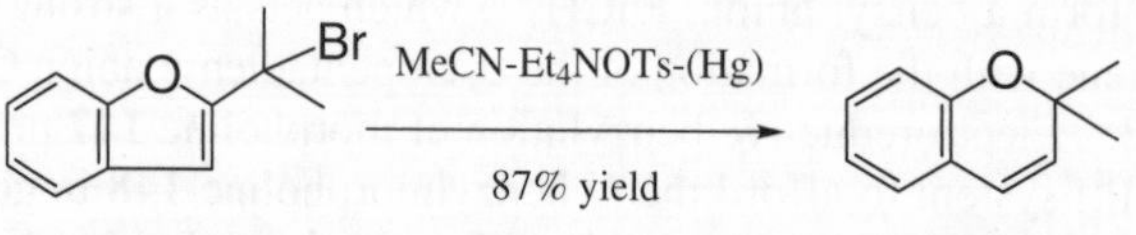

(7.65)

154   **155**

The conjugated additions of organometallic nucleophiles to α,β-enones can be used for the formation of fused and spiro carbo-cyclic rings. The electroreductive synthesis of bicyclic ketones **157** mediated by cobalt or nickel complexes has been attained by intramolecular conjugated addition (**156** → **157**) of the alkyl bromide moiety to 2-cyclohexen-1-one system using the square-planar cobalt or nickel complexes as mediators. The electrocyclization is carried out in a DMF-Et$_4$NClO$_4$(0.1M)-(C) system in the presence of Ni(cyclam)(ClO$_4$)$_2$ to give the bicyclo ketones **157** in 33~70% yields (eq. (7.66)) [149a]. Similar intramolecular cyclization is performed by use of a conductive microemulsion method in the presence of Co(I)L complex in good yields [149b].

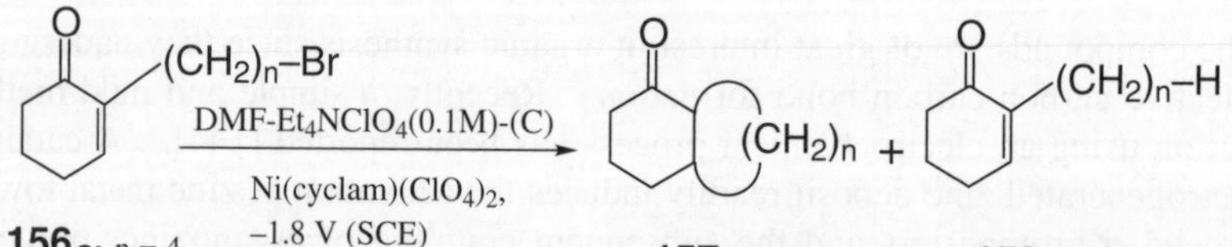

(7.66)

156 a: n = 4   DMF-Et$_4$NClO$_4$(0.1M)-(C)   **157** a: 70%   **158** a: 6%
b: n = 5   Ni(cyclam)(ClO$_4$)$_2$,   b: 33%   b: 18%
−1.8 V (SCE)

ω-Bromo-N-tosylalkanamides can be cyclized under electroreduction conditions to form the corresponding lactams.   For example, the electroreduction of **159** (n = 2) is performed in a DMF-Et$_4$NClO$_4$-(Hg) system to yield γ-lactam **160** (n = 2) in 95% yield (eq. (7.67)) [150].

$$\text{BrCH}_2(\text{CH}_2)_n\text{CONHTs} \xrightarrow[\text{93~95\% yields}]{\text{DMF-Et}_4\text{NClO}_4\text{-(Hg)}} \quad \underset{\text{O}}{\overset{(\text{CH}_2)_n}{\big|}}\!\!-\text{NH} \tag{7.67}$$

159 n = 2, 3   **160**

Electroreductive cyclization of 4-(phenylsulfonylthio)azetidin-2-ones derived from penicillin has been achieved by a Bi(0)-Bi(III) redox system to lead the cyclized products. The electrolysis of the chloride **161** is carried out in a DMF/Py(5/1)-BiCl$_3$-(Sn) system in an undivided cell to give 3-hydroxycephem **162** in 67% yield (eq. (7.68)) [151].

$$\tag{7.68}$$

161   **162**

The electroreductive formylation of 1-halogenopentane **163** has been carried out in an MeCN-Et$_4$NBr(or Et$_4$NOTs)-(Pt/SUS) system in the presence of [Fe(CO)$_5$] complexes **164** in a divided cell to give the corresponding aldehyde **165** in good yield (eq. (7.69)) [152].

$$\begin{array}{c}\text{C}_5\text{H}_{13}\text{—X}\\ \textbf{163}\\ +\\ [\text{Fe(CO)}_5]\\ \textbf{164}\ \ \text{X = Br, I}\end{array} \xrightarrow[\text{61~71\% yields}]{\text{MeCN-Et}_4\text{NBr(or Et}_4\text{NOTs)-(Pt/SUS)}} [\text{C}_5\text{H}_{13}\text{COFe(CO)}_4]^- \tag{7.69}$$

$$\downarrow \text{H}^+$$
$$\text{C}_5\text{H}_{13}\text{CHO}$$
165

The electroreductive silylation of allyl, vinyl, and aryl halides affords the corresponding organosilicon compounds.   The electrolysis of **166** in a DMF-Et$_4$NOTs-(Pt) system in the presence of chlorotrimethylsilane **167** gives the silylated product **168** in 70% yield (eq. (7.70)) [153].

$$\begin{array}{c}\text{PhCH=CHCH}_2\text{Cl}\\ \textbf{166}\\ +\\ \text{Me}_3\text{SiCl}\\ \textbf{167}\end{array} \xrightarrow[\text{70\% yield}]{\text{DMF-Et}_4\text{NOTs-(Pt)}} \begin{array}{c}\text{PhCH=CHCH}_2\text{SiMe}_3\\ \textbf{168}\end{array} \tag{7.70}$$

Studies on homogeneous rate constant for the coupling between electrogenerated alkyl radicals with aromatic anion radicals in DMF demonstrate that the redox potential of the aromatic anion radical has only a minor influence on the rate of coupling with alkyl radicals [154].

7.1.4 Double Bond Formations

Electroreduction of vicinal dihalides to olefins has been intensively investigated. Much attention has especially been paid to the steric course of the olefin-forming reactions. Upon reduction of vicinal dibromides, a single polarographic wave is commonly obtained at an electrode potential more positive than the $E_{1/2}$ of the corresponding monobromides, indicating that elimination of both bromide ions occurs either as a concerted process or as two events in fast cleavage. The cathodic elimination of dl-2,3-dibromobutane tends to give only cis-butene, and the $meso$-compound affords exclusively $trans$-butene [155]. The mechanism of the direct and indirect electroreduction of vicinal dibromoalkanes has been investigated in detail in a DMF-Bu$_4$NBF$_4$(0.1M)-(Mg/C) system in the presence (or absence) of aromatic compound at 20 °C [156]. Analysis of the kinetic data showed that aromatic anion radicals react with vicinal dibromoalkanes according to an electron-transfer mechanism in which the rate-determining step is a concerted electron-transfer bond-breaking reaction leading to the β-bromoalkyl radical [156]. The latter is then reduced very rapidly as a second step most probably along with an another concerted electron-transfer bond-breaking path leading directly to the olefin in the heterogeneous case and through halogen atom expulsion in the homogeneous case. In the absence of steric constraints, the reduction goes through an antiperiplanar conformer [156]. It has been shown that the electroreduction of $trans$-1,2-dibromocyclohexane proceeds preferentially through the $anti$-(axial, axial) conformer so long as equilibrium with the less easily reduced $gauche$ (equatorial, equatorial) conformer is maintained [157].

Electrochemical reductions of both $meso$- and dl-isomers of 3,4-dibromohexane **169** ($R^1 = R^2 = $ Et) and 2,5-dimethyl-3,4-dibromohexane **169** ($R^1 = R^2 = i$-Pr) in a DMF-Bu$_4$NBF$_4$-(Hg) system has been investigated (eq. (7.71)) [158]. Distribution of products, cis- and $trans$-hexenes, **170** in relation to the conversion yields against electrolysis potentials is demonstrated in Fig. 7.1 [158]. The reduction of 1,2-dibromobutane and 1,2-dibromoethane by electrogenerated cob(I)alamin in an aqueous MeCN(50%)-H$_3$PO$_4$/NaH$_2$PO$_4$-(C) system undergoes complete debromination to yield the corresponding alkenes [159].

$$
\underset{\substack{\textbf{169}\quad R^1, R^2 = \text{Et} \\ R^1, R^2 = i\text{-Pr}}}{R^1\text{–CHBr–CHBr–}R^2} \xrightarrow[\substack{50\sim70\% \text{ yields} \\ (cis, trans)}]{\text{DMF-Bu}_4\text{NBF}_4\text{-(Hg)}} \underset{\textbf{170}}{R^1\text{–CH=CH–}R^2} \tag{7.71}
$$

The steric course of dehalogenation reactions of the dihalides **171**, giving **172**, has been studied with configurationally uniform $erythro$- and $threo$-5,6-dihalodecanes **171** (eq. (7.72)) and the results are indicated in Table 7.7 [160]. The table shows that a gradual shift

$$
\underset{\substack{\textbf{171}\quad X, Y = \text{F, Cl, Br or I}}}{\text{Bu–CH(X)–CH(Y)–Bu}} \xrightarrow{\text{DMF-Bu}_4\text{NClO}_4\text{-(Hg)}} \underset{\textbf{172}}{\text{BuHC=CHBu}} \tag{7.72}
$$

from a clean $anti$-elimination to a prevalent syn-elimination is induced on increasing strength of the C-X and C-Y bonds in the $threo$-dihalide series. In the $erythro$-series, the stereochemical changes are analogous but less pronounced [160]. The electroreduction of citraconic acid dibromide in a phosphate buffer aqueous H$_3$PO$_4$/KOH/KCl-(Hg) system affords citraconic acid (pH 3.5) and mesaconic acid (pH 6.1), respectively [161]. By the re-

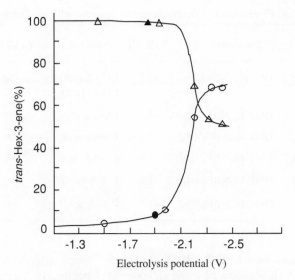

Fig. 7.1 Distribution of products, *cis*- and trans-hexene, obtained at
mercury pool (filled symbols) and platinum sheet (empty
symbols) cathodes in a DMF-0.4M-TBAF system during
controlled-potential electrolyses of 3,4-dibromohexane
($\triangle$, $\blacktriangle$ *meso*-isomer; $\bigcirc$, $\bullet$ dl-isomer).

Table 7.7 Double Bond Formation from Vicinal Dihalides

Entry	Dihalides		*erythro*, %		*threo*, %	
	X	Y	*cis*[a] (syn)	*trans*[a] (anti)	*cis*[a] (syn)	*trans*[a] (anti)
Ia	Br	I	0	100	>98	<2
Ib	Cl	I	0	100	>98	<2
Ic	F	I	32	68	48	52
Id	Br	Br	0	100	>98	<2
Ie	Cl	Br	0	100	96	4
If	F	Br	31	69	41	59
Ig	Cl	Cl	0	100	75	25
Ih	F	Cl	33	67	40	60

[a] Corrected values; for impurities present in some of the reactants.
[Reproduced with permission from J. Zavada et al., *Collect. Czech.
Chem. Commun.*, **48**, 3552 (1983)]

duction of mesaconic acid dibromide, the formation of mesaconic acid occurs at the more
positive potential than *ca.* −0.4 V (SCE) and at the more negative potential, the produced
mesaconic acid is reduced to methylsuccinic acid [161]. The electroreductive double bond
formation from other vicinal halides are shown in Table 7.8 [24, 162].

Alkyl vicinal dihalides can be reduced to alkenes at high rates by electrocatalytic reduction using the cobalt corrin vitamin B_{12} as a mediator [159, 163, 164]. The dehalogena-

Table 7.8 Electroreductive Double Bond Formation from Vicinal Halides

Substrates	Conditions	V (SCE)	Products	Yield, %	Ref.
BrCH$_2$CHBrCO$_2$Et	DMF-Et$_4$NClO$_4$	−1.5	CH$_2$=CHCO$_2$Et	70	[24]
			CH$_3$CH$_2$CO$_2$Et	2	
Dibromooleic acid	DMF-Et$_4$NBF$_4$-(Hg)	−1.4	Oleic acid	97.5	[162]
Dibromoelaidic acid	DMF-Et$_4$NBF$_4$-(Hg)	−1.4	Elaidic acid	100	[162]
Dibromolinoleic acid	DMF-Et$_4$NBF$_4$-(Hg)	−1.4	Linoleic acid	98	[162]
4,5-Dibromoheptanol	DMF-Et$_4$NBF$_4$-(Hg)	−1.4	E-4-Heptenol	96	[162]
3,4-Dibromohexanol	DMF-Et$_4$NBF$_4$-(Hg)	−1.4	Z-3-Hexenol	100	[162]

tion of vicinal dihalides with the B$_{12}$Co(II)/B$_{12}$Co(I) redox couple affords alkenes at the potential at which B$_{12}$Co(I) is produced at the cathode (see Chapter 11). Recently, the electrocatalytic reductions of dihalides has been carried out by ZnPc, CoPc and FePc systems in a bicontinuous microemulsion of DDAB/Docenane/Water (DDAB: didodecyldimethylammonium bromide) [165].

The controlled potential electroreduction of 1-bromo-4-chlorobicyclo[2.2.0]hexane **173** in a DMF-Et$_4$NBr-(Hg) system affords $\Delta^{1,4}$-bicyclo[2.2.0]hexane **174** as a sole product, identified as Diels-Alder adduct **175** with cyclopentadiene (eq. (7.73)) [166].

meso- and *dl-*1,2-Dibromo-1,2-diphenylethanes **176** are electrochemically reduced in a DMF-Et$_4$NClO$_4$(0.1M)-(Hg) system to the *trans*-stilbene **177** in quantitative yield [167]. The relative yields of *cis-* and *trans-*stilbenes in the electroreduction of *dl-*1,2-dibromo-1,2-diphenylethane **176** (*dl-*isomers) in a DMF-(Hg) system depend on the type and concentration of the supporting electrolyte [168]. The *cis/trans* ratio increases with size and concentration of the cation species. The electroreduction of **176** in an MeCN-[Et$_4$N]$_2$Oxalate-(Pt/Pb) system affords stilbenes (*cis/trans* = 38/62) in 86% yield (eq. (7.74)) [169]. The electrocatalytic reduction of 1,2-dibromo-1,2-diphenylethane to stilbene on platinum electrodes modified by coating with poly-*p*-nitrostyrene has been investigated [170]. The efficiency of catalyst increases to a limited value as the amount of chargeable polymer increases.

A polypyrrole film containing a viologen system has been used for the electroreduction of 1,2-dibromo-1,2-diphenylethane [171].

The electroreduction of a diastereomeric mixture (*dl*/*meso* = 2/1) at -0.8 V (Ag/Ag$^+$) of α,α'-dibromobenzylsulfone **178** in a DMF-Et$_4$NClO$_4$(0.1M)-(C) system yields *trans-* and *cis*-stilbenes **179** (1/1 ratio) in 84% yield together with dibenzylsulfone **180** (5%) (eq. (7.75)) [172]. In a similar manner, α,α'-dibromobenzylsulfoxide also gives the stilbenes **179** in 85% yield.

$$\underset{\textbf{178}}{\underset{\overset{|}{Br}\quad\overset{|}{Br}}{Ph-CHSO_2CH-Ph}} \xrightarrow[\text{1.8~2.1 F/mol, }-0.8\,V]{\text{DMF-Et}_4\text{NClO}_4(0.1M)\text{-(C)}} PhCH=CHPh + (PhCH_2)SO_2 \quad (7.75)$$

$$\underset{\textbf{179}\ \ 84\%}{} \qquad \underset{\textbf{180}\ \ 5\%}{}$$

The product distributions of the electroreduction of *erythro-* and *threo*-1-bromo-2-chloro-1,2-diphenylethanes have been investigated in detail in an MeCN-Bu$_4$NClO$_4$(0.1M)-(Hg) system [173, 174]. The *erythro* isomer is reduced exclusively to (E)-stilbene. Mixtures of (E)- and (Z)-stilbene in 90/10 ratio (E/Z) are formed from *threo* isomer. Electroreduction of polyhalogenated ethanes has been intensively investigated. The dehalogenative destruction of chlorofluorocarbons in particular becomes an urgent problem with respect to preservation of the ozone layer in the stratosphere. The electrochemical dehalogenation of 1,2-difluoro-1,1,2,2-tetrachloroethane **181** has been carried out in an HMPA-Bu$_4$NBF$_4$(0.1M)-(Hg) system at the potential of -3.2 V (SCE) to give 1,2-difluoro-1,2-dichloroethene **182** in 86% yield (eq. (7.76)) [175].

$$\underset{\textbf{181}}{CFCl_2CFCl_2} \xrightarrow[\text{86\% yield}]{\text{HMPA-Bu}_4\text{NBF}_4(0.1M)\text{-(Hg)}} \underset{\textbf{182}}{CFCl=CFCl} \qquad (7.76)$$

The electrochemical dechlorination of trichloroethylene is performed in an MeCN-Et$_4$NClO$_4$(0.1M)-(C) system at the potential of -2.8 V (Ag/Ag$^+$) to yield acetylene in 96% yield [176]. A variety of polyhalogenated ethanes has been subjected to electrochemical dehalogenation and the results are listed in Table 7.9 [175~183].

Electroreductive debromination of *erythro-dl*-1-aryl-1,2-dibromo-2-nitropropanes **183** can be performed in a DMSO-LiClO$_4$(0.1M)-(Pt) system to yield (E)-1-aryl-2-nitro-propenes **184** in 80~85% yields (eq. (7.77)) [184]. Dehalogenation of vicinal dibromides

$$\underset{\substack{\textbf{183}\ \ X = H, p\text{-MeO}, m\text{-Br}, p\text{-NO}_2}}{\underset{\overset{|}{Br}\ \ \overset{|}{Br}}{X-\text{C}_6\text{H}_4-\overset{\overset{\displaystyle Me}{|}}{CH}-\overset{|}{C}-NO_2}} \xrightarrow[\text{80~85\% yields}]{\text{DMSO-LiClO}_4(0.1M)\text{-(Pt)}} \underset{\textbf{184}}{X-\text{C}_6\text{H}_4-CH=\overset{\overset{\displaystyle Me}{|}}{C}-NO_2} \qquad (7.77)$$

by electrogenerated polysulfide anions, S$_6$$^{2-}$, S$_4$$^{2-}$, or S$_3$$^{2-}$, in a DMF-Et$_4$NClO$_4$-(Au) system proceeds in the presence of small amounts of sulfur as a catalyst [185]. The electrolysis of diethyl *meso*-2,3-dibromosuccinate gives diethyl fumarate in 74% yield as the sole product. Similarly, *dl*-1,2-dibromo-1,2-diphenylethane yields a mixture of Z/E-stilbene (Z/E = 75/25) in 80% yield.

Table 7.9 Electroreduction of Polyhalogenated Ethanes

Substrates	Conditions	Products, Yield, %	Ref.
$CCl_2=CHCl$	MeCN-Et$_4$NClO$_4$(0.1M)-(C)	HC≡CH 96%	[176]
CHBr$_2$CHBr	aq. MeOH(80%)-Me$_4$NI-(Pb)	HC≡CH 75.4%	[182]
CCl$_3$CF$_2$Cl	HMPA-Bu$_4$NBF(0.1M)-(Hg)	CCl$_2$=CF$_2$ 74%	[175]
CFCl$_2$CFCl$_2$	HMPA-Bu$_4$NBF(0.1M)-(Hg)	CFCl=CFCl 86%	[175]
CF$_2$ClCFCl$_2$	i-PrOH-KCl/KOH(0.5M)-(Pb)	CF$_2$=CFCl 86.3%	[179] [180]
CFBr$_2$CHBr$_2$	aq. MeOH(80%)-Me$_4$NI-(Pb)	CFBr=CHBr 28.2% CF≡CH 66.8%	[182]
CF$_2$BrCH$_2$Br	aq. EtOH(80%)-Me$_4$NBr-(Hg)	CF$_2$=CH$_2$ 75~89%	[178]
CF$_3$CClBrH		F$_2$Cl=CClH 70%	[181]
CF$_2$BrCF$_2$Br		CF$_2$=CF$_2$ 60~65% CHF$_2$CBrF$_2$ 35~40%	[177] [183]

The controlled potential electrolysis of *N*-tetrachloroethylamide **185** in a DMF-Et$_4$NClO$_4$-(Hg) system affords the corresponding *N*-(2,2-dichlorovinyl)amide **186** in 68% yield (eq. (7.78)) [186].

$$ (7.78) $$

185   **186**

The electroreductive dehalogenation of 1,2-dihalogenoadamantanes **187** proceeds *via* adamantene to give the corresponding dimers **188** and **189** in 87% yield (eq. (7.79)) [187]. In the presence of butadiene, the corresponding adduct is isolated in 50% yield.

$$ (7.79) $$

187 X , Y = Cl, Br, I   **188**   **189**

The electroreduction of α,α'-dibromo-1,2-dialkylbenzenes **190** in a DMF-Et_4NBr(or PF_6)-(Hg) system in the presence of dienophiles **192** (*e.g.*, maleic anhydride derivatives) yields Diels-Alder adducts of *o*-quinodimethanes **193** and **194** in 12~87% yields (eq. (7.80)) [188].

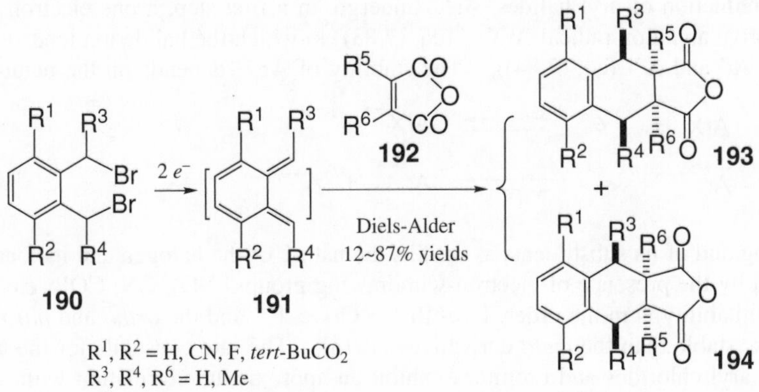

190 **191** **192** **193** **194** (7.80)

R^1, R^2 = H, CN, F, *tert*-$BuCO_2$
R^3, R^4, R^6 = H, Me
R^5 = H, Me, Ph

The constant potential electrolysis of tetra-*O*-acetyl-α-D-glucopyranosyl bromide **195** in an MeCN-Et_4NBr(0.1M)-(Hg) system affords 3,4,6-triacetylglucole **196** as the major product together with **197** (minor product) (eq. (7.81)) [189].

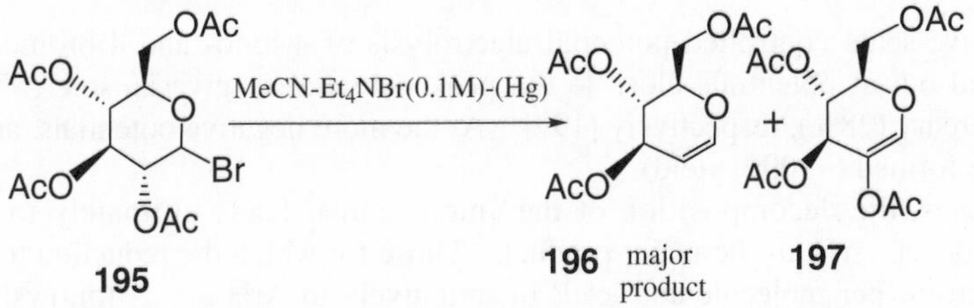

195 **196** major product **197** (7.81)

The electroreductive deacetoxybromination of **198** is carried out in a flow cell in an MeCN-Et_4NOTs system to give the corresponding olefin **199** in 42% yield (eq. (7.82)) [190a]. A large scale electrosynthesis of the olefinic compound **199** has been performed in an MeOH-AcONa-(Hg) system [190b].

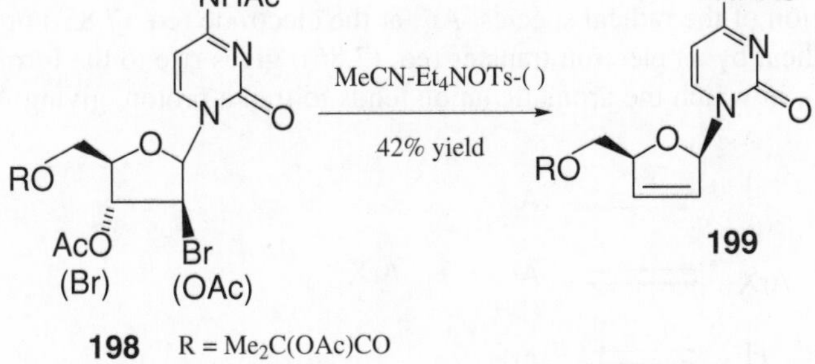

198 R = $Me_2C(OAc)CO$ **199** (7.82)

7.2   Electroreduction of Aryl Halides

7.2.1   Electroreductive Dehalogenations

The electroreduction of aryl halides, ArX, undergo, in a first step, a one-electron transfer reaction to give an anion radical ArX$^{-\bullet}$ (eq. (7.83)) in which the halide ion tends to cleave off to form Ar$^{\bullet}$ and X^{-} (eq. (7.84)).   The stability of Ar$^{-\bullet}$ depends on the nature of the

$$ArX \; + \; e^{-} \; \rightleftharpoons \; ArX^{-\bullet} \tag{7.83}$$

$$Ar^{-\bullet} \; \rightleftharpoons \; Ar^{\bullet} \; + \; X^{-} \tag{7.84}$$

aromatic ring and of its substituents as well as the nature of the halogen and its location on the ring and by the presence of electron-withdrawing groups (NO$_2$, CN, COR, etc.).   The leaving group ability is in the order: I^{-} > Br^{-} > Cl^{-} > F^{-} and the *ortho* and *para* derivatives are less stable than the *meta* derivatives [191a].   The rate constants for the cathodic cleavage of aryl chlorides and bromides exhibit an approximate correlation with the standard potentials of the starting molecule/anion radical couple [192].   In several cases, the life-time of the anion radical, Ar$^{-\bullet}$, is long enough, being on the order of minutes, to allow the recording of its ESR spectrum and the determination of the standard potential of the ArX/Ar$^{-\bullet}$ couple by techniques such as conventional polarography and slow sweep cyclic voltammetry.   This occurs with 3- and 4-fluoro- and 3-chlorobenzophenones [193], 2-fluoro, chloro- and 2-bromofluorenes[193,194], and chloro-[195] and bromonitrobenzenes [196, 197, 198].

In preparative-scale controlled potential electrolysis of 4-iodo- and 4-bromoanisoles has been carried out at potentials close to the peak potential to give anisole (63%) and bis(*p*-anisyl)mercury (28%), respectively [199].   At the more negative potentials, anisole is the only species formed (~100% yield).

In other cases, the decomposition of the anion radical leads ultimately to the hydrogenolysis product, ArH, as the major product.   Those for which the reduction totally involves two electrons per molecule and leads quantitatively to ArH are as follows: chloro- and bromobenzene [200], chlorobiphenyls [201], chloro-, bromo-, and indobenzonitriles [202], fluorobenzonitrile [203], and 1-chloro-, bromo-, and iodonaphthalenes [204].   Those for which the number of electrons per molecule is less than two and the yield of ArH is significantly less than 100% are as follows: 2-, 3-, and 4-iodo-, 2-bromonitrobenzene [191b,196~198,200], 3- and 4-bromobenzophenones [193], 6-chloroquinoline, 6-bromo- and 6-iodoquinolines, 2-iodophenazine [195], chloro-, bromo, and iodoacetophenones [205].

The further reduction of the radical species, Ar$^{\bullet}$, at the electrode (eq. (7.85)) or in solution with the anion radical by an electron transfer (eq. (7.86)) gives rise to the formation of the anion species, Ar^{-}, in which the aromatic anion tends to trap a proton, giving ArH (eq. (7.87)).

$$Ar^{\bullet} \; + \; e^{-} \; \rightleftharpoons \; Ar^{-} \tag{7.85}$$

$$Ar^{\bullet} \; + \; ArX^{-\bullet} \; \rightleftharpoons \; Ar^{-} \; + \; ArX \tag{7.86}$$

$$Ar^{-} \; + \; H^{+} \; \rightleftharpoons \; ArH \tag{7.87}$$

The proton donor in non-aqueous solvents such as dimethylformamide (DMF), acetonitrile (MeCN), and dimethylsulfoxide (DMSO) is considered to be the residual water which is much more acidic than the solvent itself [191a].   However, it cannot exclude the possibility that the reduction of Ar• into ArH occurs through H-atom transfer from the solvent, SH, leading simultaneously to S⁻ in an overall two-electron process [191a].

The nature of the solvent as an H-atom donor in the course of electrolysis has been discussed in detail [206].

The electrolysis of bromobenzene in an anode cell in a Py-NaI(0.5M)-(Mg/Pt) system undergoes "anodic reduction" to afford benzene (58.6%), 4-phenylpyridine (28.4%), 3-phenylpyridine (2.3%), and 2-phenylpyridine (11.8%), respectively [207].   Magnesium works as the reductant in this reaction.

The macroscale electrolysis of iodobenzene and bromobenzene in a DMF-Et₄NClO₄(0.1M)-(Hg) system affords benzene in 87% and 84% yields, respectively [208].   Chlorobenzene can be converted to cyclohexane in 70~80% yields [209].   The electroreductive defluorination of fluorobenzene has been carried out in a Diglyme-Bu₄NBF₄-(Pt/Hg) system to yield benzene in 76% yield [210].

The electroreduction of halonitrobenzene in a liquid NH₃-KI(0.1M)-(Pt) system at −40 °C reveals that a stable anion radical is observed when *ortho*, *meta*, and *para* isomers are reduced, and unstable dianions are observed upon reduction of anion radicals derived from *ortho* and *para* isomers.   The electrolysis of *para* isomer of anion radicals results in quantitative conversion to nitrobenzene.   The rate at which the halonitrobenzenes anion radicals are dehalogenated during electrochemical reduction is in the order: *m*-chloro < *p*-chloro < *m*-bromo < *o*-chloro < *m*-iodo~*p*-iodo < *o*-iodo [211].   Recently, an investigation of halonitrobenzenes in DMF as a function of temperature revealed that the stability of anion radicals of *m*- and *p*-iodonitrobenzene can be obtained at half-lives greater than 5 min at −40 °C [212].

Aryl radical induced H-atom transfer oxidation of primary and secondary alcoholates **200** into aldehydes and ketones **201** in liquid ammonia has been found (eq. (7.88a)) [213].   The H-atom transfer from the alcoholate **200** to the aryl radical Ar• derived from aryl halide ArX **203** by electroreduction is the key step of the oxidation process (eq. (7.88b)).   The ketyl anion radical thus formed can be oxidized into the corresponding carbonyl compound.   The reactions competing with H-atom transfer are electron transfers to the aryl radical which occur at the electrode surface and/or in the solution.   The latter reaction will play the role of termination steps for the chain reaction systems involving homogeneous initiation of the reaction.   Liquid ammonia is used as a solvent, due to its poor H-atom donor ability [214~216].   For example, diphenylcarbinolate **200** can be oxidized in a liquid NH₃-KBr(0.1M)-(Pt) system in the presence of 2(or 6)-chloroquinoline at the first wave potential (*ca.* −1.5 V *vs.* Ag/Ag⁺) to give benzophenone **201** in a quantitative yield.

The H-atom donating ability of the solvent can play an important role in the course of electrochemical reactions [217].

The kinetics of cleavage reactions of 4-halonitrobenzene anion radical in DMF and MeCN are shown as a function of temperature and a supporting electrolyte.   Rate constants ranging from 1 to 8 s⁻¹ with activation energies (Ea) in the range 17~20 kcal/mol (*e.g.*, 4-bromonitrobenzene anion radical Ea: 20.9 kcal/mol).   The rate constant in the presence of Bu₄N⁺ is found to be about 25% greater than that when Me₄N⁺ is the supporting electrolyte cation [218].

$$\underset{\textbf{200}}{\overset{\displaystyle Ph}{\underset{\displaystyle Ph}{>}}CH-OK} \xrightarrow[\substack{\text{2 (or 6)-chloroquinoline}\\ \text{quantitative yield}}]{\text{liq. NH}_3\text{-KBr-(Pt)}} \underset{\textbf{201}}{\overset{\displaystyle Ph}{\underset{\displaystyle Ph}{>}}C=O} \quad\text{--------(a)}$$

(7.88)

$$\underset{\textbf{200}}{\overset{\displaystyle Ph}{\underset{\displaystyle Ph}{>}}CH-OK}$$

ArH **202**

$$\underset{\textbf{203}}{ArX} \quad \xrightarrow{\text{liq. NH}_3\text{-KBr-(Pt)}} \quad \text{------(b)}$$

$$\underset{\textbf{201}}{\overset{\displaystyle Ph}{\underset{\displaystyle Ph}{>}}C=O}$$

Ar° ArX⁻°

X⁻

Ar = 2-chloroquinoline;

6-chloroquinoline;

On irradiation with light of wavelength 330 nm, corresponding to an adsorption band in the anion radical of *p*-bromonitrobenzene, the anion radical of the nibrobenzene is formed [219]. Photoelectroreduction of *p*-halonitrobenzenes has also been studied [220, 221].

The electroreduction of 9-bromoanthracene in a DMF-Bu$_4$NBF$_4$(0.2M)-(Hg) system at –1.65 V (SCE) affords a mixture of anthracene (73%) and di-9-anthrylmercury (10%) [222a]. The electrodechlorination of 9-chloroanthracene is attempted in an adsorbed film of cationic surfactant on an electrode [222b].

The anion of 9,10-dichloroanthracene in methylene chloride is found to be unstable at room temperature. The anion, however, increases its stability at lower temperatures, and below *ca.* –40 °C it is essentially stable [223].

The kinetics and activation parameters for the cleavage of halide ion from 9-halo- and 9,10-dihaloanthracene anion radicals have been investigated as a function of temperature [224]. Activation energies of 15.2 kcal/mol have been found for 9-chloro- and 9,10-dichlorosubstituted anion radicals, while a value of 4.0 kcal/mol is observed for bromo substituted ions [224].

The electroreductive dechlorination of 3-chloro- and 3,8-dichlorobenzofurans **204** and **205** has been performed in an MeOH-Et$_4$NBr(0.25M)-(Pt/Pb) system to yield the dechlorinated products **206** and **207** in good yields (eq. (7.89)) [225a]. 2,4-Dichlorophenoxyacetic acid can be dechlorinated by electroreduction in an MeCN-Bu$_4$NBr-(C) system to give phenoxyacetic acid in 60% yield [225b]. Selective cleavage of the carbon-bromine bond in *α*-bromo aromatic and aliphatic ethers has been attained in an MeCN-Et$_4$NClO$_4$(0.1M)-(Hg) system to give the corresponding debrominated compounds in good yields [226].

$$(7.89)$$

The direct electroreduction of polyhalogenated biphenyls has been attempted in a DMF-Et$_4$NI-(Hg) system [227]. Nearly quantitative yields of biphenyl 209 have been obtained in 25-h electrolysis of PCB mixtures 208 when using aqueous didodecyldimethylammonium bromide (DDAB) with zinc phthalocyanine (ZnPc) as a mediator. The insoluble suspension of DDAB provides remarkable co-adsorption of nonpolar reactants in an electrocatalytic cycle onto carbon felt cathodes [228]. The electrolysis of PCB has been carried out in surfactant solutions containing ZnPc, and Et$_4$NBr and a solution of perchloric acid is added after 30-min electrolysis to give biphenyl 209 in 99% yield (eq. (7.90)) [228]. The electroreduction of polychlorinated biphenyls (including PCB) proceeds also in an MeOH-Et$_4$NBr-(Pb) system to yield biphenyl in more than 95% yield [229].

$$(7.90)$$

It has been found that excitation of electrogenerated anion radicals derived from anthracene and 9,10-diphenylanthracene with visible light (> 470 nm) gives greater than 10-fold increases in the ratios of reductive dechlorination of 4-chlorobiphenyl [230].

The reduction of arylhalides by photoexcited anion radicals undergoes a photoinducing electron-transfer reactions from anthraquinone anion radical to aryl halides [231].

2,3,4,5,6-Pentafluorobenzoic acid 210 can be reduced to either 2,3,5,6-tetrafluorobenzaldehyde 212 or 2,3,5,6-tetrafluorobenzyl alcohol 211 in an aqueous H$_2$SO$_4$/Et$_4$NOTs-(Pb) system (eq. (7.91)) [232a]. A selective defluorination of p-fluoroanilines has been attained

$$(7.91)$$

by electroreduction in an aqueous KF/HF-(Ni) system [232b]. The electroreduction of pentachloropyridine 213 in a DMF-Et$_4$NBF$_4$(0.16M)-(Hg) system affords 2,3,5,6-tetrachloropyridine 214 in 57% yield (eq. (7.92)) [233]. In contrast, a similar reduction of pentafluoropyridine gives rise to the formation of the corresponding dimer at the 4-position in a 49% current efficiency [233, 234].

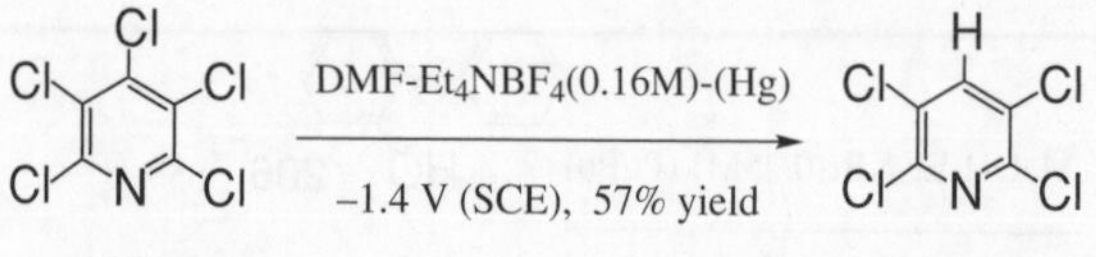

$$(7.92)$$

213   **214**

The electroreductive dechlorination of 2-chloropyridine derivative **215** in an aqueous EtOH(50%)-H$_2$SO$_4$(0.1M)-(C/Pb) system in a divided cell gives the dechlorinated product **217** in 33.3% yield (eq. (7.93)) [235].   The electroreductive dechlorination of 3,4,5,6-tetra-chloro-2-pyridine carboxylic acid **218** is performed in an aqueous HCl(50%)-(Ag) system to yield 3,6-dichloro-2-pyridine carboxylic acid **219** in a quantitative yield (eq. (7.94)) [236].

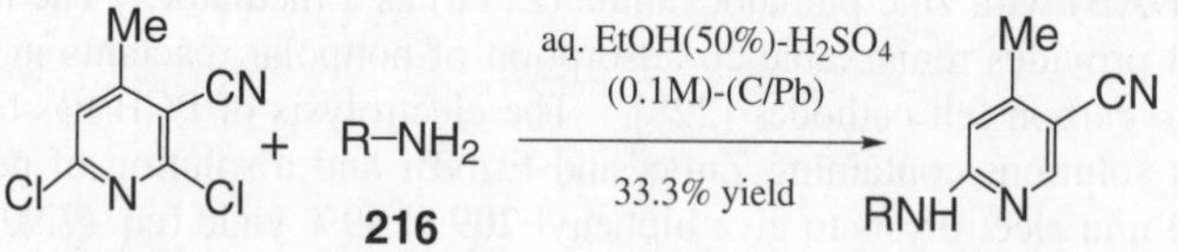

$$(7.93)$$

215   R = HO(CH$_2$)$_2$

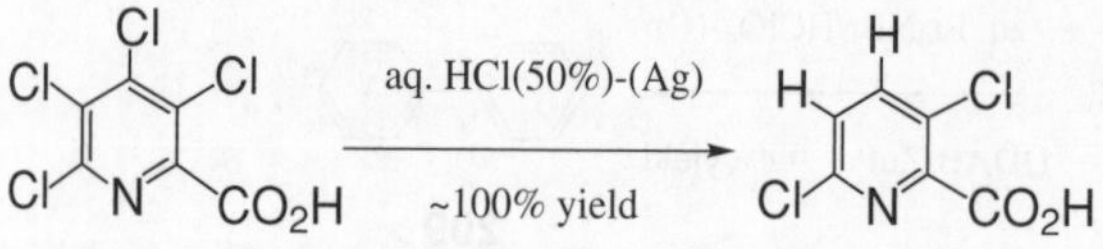

$$(7.94)$$

218   **219**

The cleavage rate (kc) of anion radicals from halogenated N-hetoroaromatic compounds is dependent on temperature (6-chloroquinoline: kc/S^{-1} 8.2 × 10^2, 27.3 °T/K).   A decrease of 20 °C in temperature lowers the cleavage rate by *ca.* 10 (kc/S^{-1}).   The electroreduction of 2-chloroquinoline **220** in a DMF-Bu$_4$NI-(C) system in the presence of ethyl iodide under the saturation with carbon dioxide gives a mixture of quinoline **221** (50%) and 2-ethoxycarbonyl quinoline **222** (20%) (eq. (7.95)) [237].   In a similar manner, the electroreductive dechlorination of **223** affords **224** in 90% yield (eq. (7.96)).

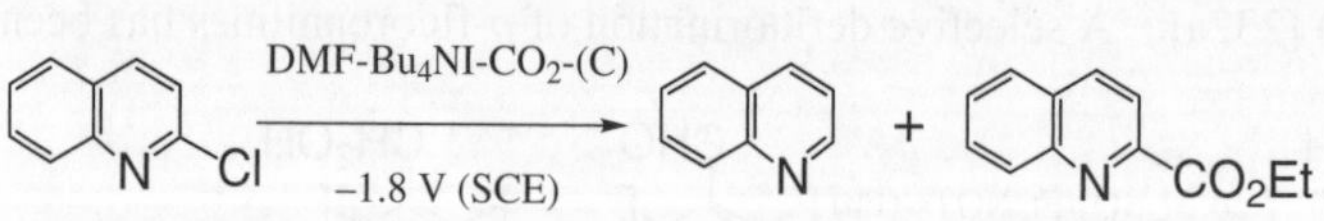

$$(7.95)$$

220   **221** 50%   **222** 20%

$$(7.96)$$

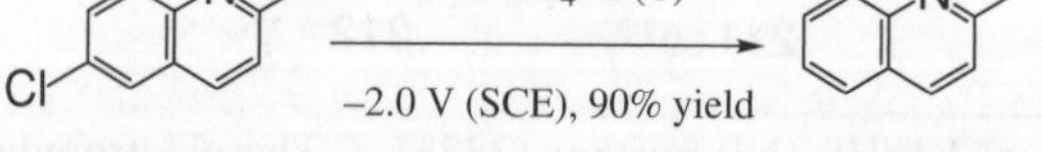

223   **224**

The electroreduction of 6,9-dichloroacridines and 6-chloroacridone in a DMF/EtOH(4/1)-Et$_4$NI-(C) system affords acridan in 90% yield after complete electrolysis [238].

2-Aminopurine has been electrosynthesized by the reduction of 2-amino-6-chloropurine **225** in an aqueous HCl(0.25M)-(Hg) system at -0.75 V (SCE) to give the product **226** in 85% yield (eq. (7.97)) [239].

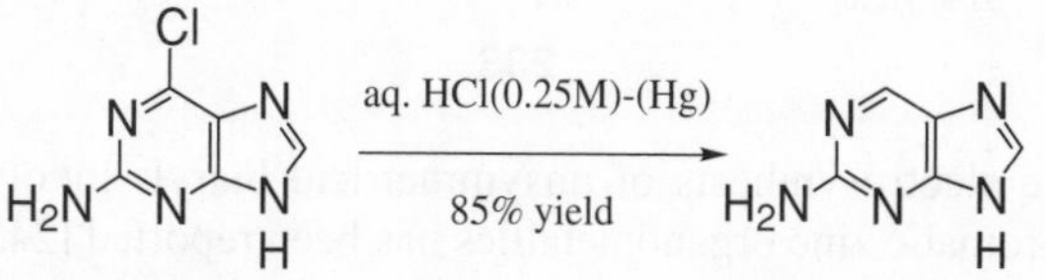

(7.97)

The electroreductive dechlorination of benzothiophene derivative **228** is performed in a THF/MeOH-Et$_4$NCl-(Pt/Cd) system to give the corresponding dechlorinated product **228** in 85% yield (eq. (7.98)) [240]. The electroreductive debromination of tetrabromothiophene

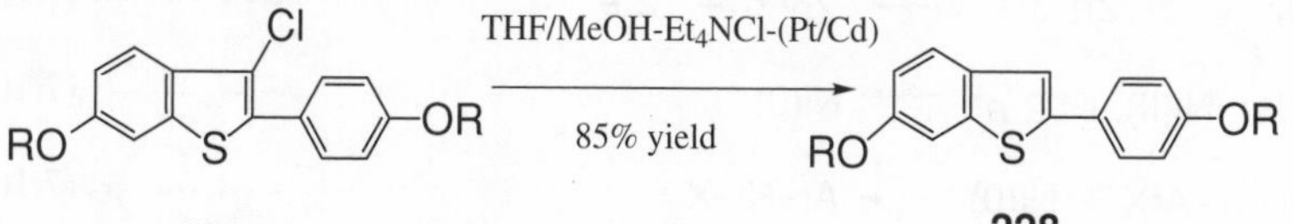

(7.98)

R = alkyl, cycloalkyl, PhCH$_2$

229 is carried out in a DMF-Et$_4$NBr(0.25M)-(Hg) system in an undivided cell [241]. The product-selective debromination can be performed by adjusting the reduction potential as follows: 2,3,4-tribromothiophene produces **230** at -0.90 V (SCE) (eq. (7.99)) and 3,4-dibromothiophene **231** can be obtained by the electrolysis at -1.20 V (eq. (7.100)).

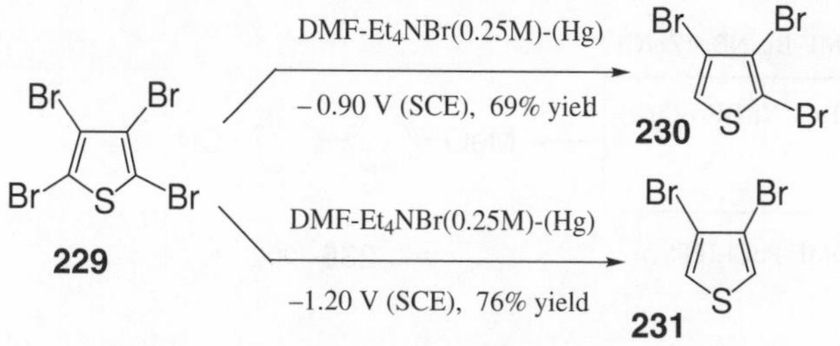

(7.99)

(7.100)

7.2.2 Electroreductive Aryl-Aryl Couplings

The electroreduction of benzyl halides has been shown to be initially a one-electron reduction which gives the corresponding anion radical as a reactive intermediate. Dimerization or hydrogen abstraction is shown to follow loss of a halide ion from the anion radical.

Reductive dimerization of benzyl halides by an electrogenerated zerovalent nickel-phosphine complex as a recycling catalyst is discussed in Chapter 10 [66, 68].

Intermolecular coupling reaction of aryl halides with use of zerovalent nickel complex by electroreduction has been carried out in a DMF-Bu$_4$NBr-(Pt/Pb) system in the presence of NiCl$_2$(PPh$_3$)$_4$ and PPh$_3$. The reduction of methyl p-chlorobenzoate **232** under a potential of -1.9 V affords the corresponding 4,4'-methoxycarbonylbiphenyl **233** in 51.1% yield (eq. (7.101)) [242]. More improved dimerization of aryl halides has been carried out in an undivided cell in N-methylpyrrolidone to yield the dimer in 70~90% yields [243, 244].

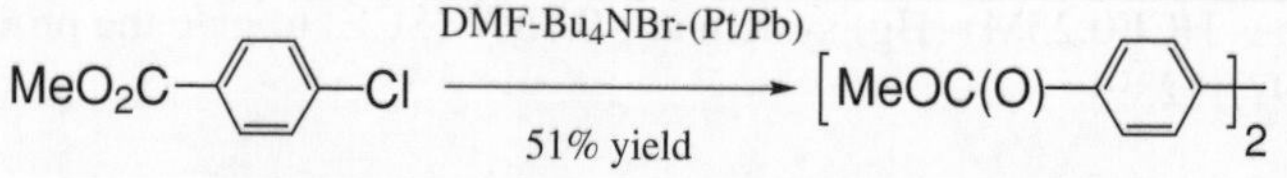

$$\text{MeO}_2\text{C}\overset{}{\underset{\textbf{232}}{\bigcirc}}\text{Cl} \xrightarrow[\text{51\% yield}]{\text{DMF-Bu}_4\text{NBr-(Pt/Pb)}} \left[\text{MeOC(O)}\overset{}{\bigcirc}\right]_2 \qquad (7.101)$$

$$\textbf{232} \qquad\qquad\qquad\qquad\qquad\qquad \textbf{233}$$

An efficient method for the electrosynthesis of unsymmetrical biaryls involving the electrochemical preparation of aromatic zinc organometallics has been reported [245]. The procedure used for the formation of the arylzinc reagent requires both the use of a catalytic amount of nickel-bipyridine complex and a sacrificial anode zinc in the following reaction sequence (eqs. (7.102~7.105)) [245]. The electrolysis is carried out in a DMF-Bu$_4$NBr-(Zn/C) system in the presence of ZnBr$_2$, p-methoxyphenyl bromide **234**, Ni(BF$_4$)$_2$(bpy)$_3$,

Anode: $\text{Zn} \longrightarrow \text{Zn}^{2+} + 2 e^-$ ---------------- (7.102)

Cathode: $\text{Ni(II)} + 2 e^- \longrightarrow \text{Ni(0)}$ ---------------- (7.103)

in solution $\text{ArX} + \text{Ni(0)} \longrightarrow \text{Ar–Ni–X}$ ---------------- (7.104)

$\text{Ar–Ni–X} + 2 \text{Zn}^{2+} \longrightarrow \text{Ar–Zn–X} + \text{Ni(II)}$ ---------------- (7.105)

and 2,2'-bipyridine, and the electrolytes are submitted to electrolysis for several hours, and then the electrolytic solution is slowly syringed into a solution of p-cyanophenyl bromide **235** and PdCl$_2$(PPh$_3$)$_2$ in DMF to give the desired coupling product **236** in 90% yield (eq. (7.106)) [245].

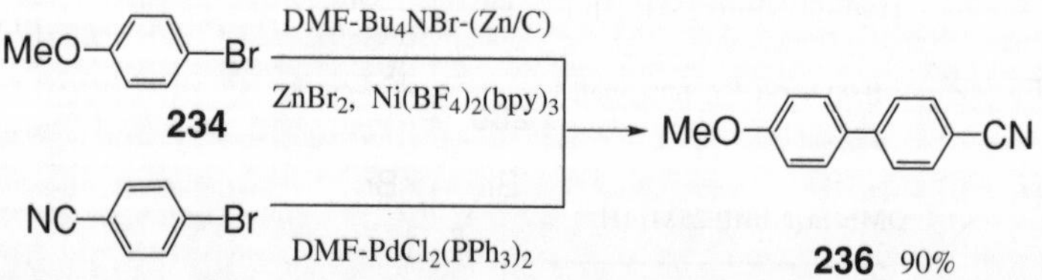

$$\text{MeO}\overset{}{\bigcirc}\text{Br} \quad \textbf{234}$$
$$\xrightarrow[\text{ZnBr}_2,\ \text{Ni(BF}_4)_2\text{(bpy)}_3]{\text{DMF-Bu}_4\text{NBr-(Zn/C)}}$$
$$\text{NC}\overset{}{\bigcirc}\text{Br} \quad \textbf{235} \qquad \text{DMF-PdCl}_2\text{(PPh}_3)_2$$
$$\longrightarrow \text{MeO}\overset{}{\bigcirc}\overset{}{\bigcirc}\text{CN} \qquad \textbf{236}\ \ 90\% \qquad (7.106)$$

Nickel complexes have been found to dramatically improve the yield and selectivity of cross-coupling reactions between Grignard reagents and aryl halides. By electrochemically recycling a nickel catalyst, it is possible to homo-couple both chlorobenzene [246] and bromobenzene [247] to biphenyl in 80 and 85% yields, respectively [246, 247].

(1,2-Bis(di-2-propylphosphino)benzene)nickel(0) complex has been found to be an efficient electrocatalyst for the coupling of aryl halides [248]. The preparative scale electrolysis of aryl chlorides in the presence of 2 mol% of the catalyst affords the corresponding biphenyls in as high as 96% yield.

Constant potential electrolysis of pentafluoronitrobenzene radical-anion **237** in an MeCN-Bu$_4$NClO$_4$-(Pt) system affords three kinds of dimerized products, **238a**, **238b** and **238c** in good yields (eq. (7.107)) [249].

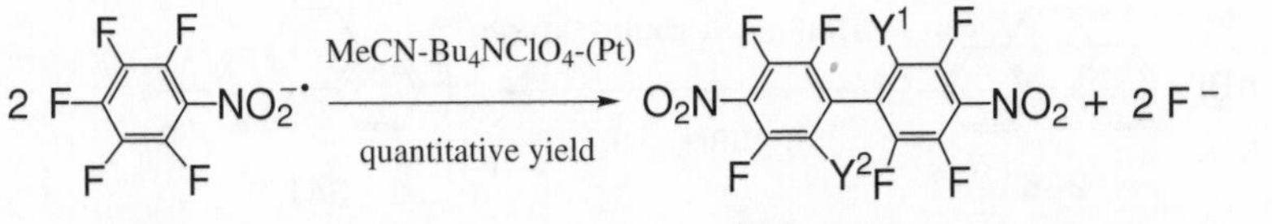

$$2 \; F{-}\!\!\!\!\bigcirc\!\!\!\!{-}NO_2^{-\bullet} \xrightarrow[\text{quantitative yield}]{\text{MeCN-Bu}_4\text{NClO}_4\text{-(Pt)}} O_2N{-}\!\!\!\!\bigcirc\!\!\!\!{-}\!\!\!\!\bigcirc\!\!\!\!{-}NO_2 + 2 \; F^- \qquad (7.107)$$

237

238 **a:** $Y^1, Y^2 = F$
b: $Y^1 = OH, \quad Y^2 = F$
c: $Y^1, Y^2 = OH$

In the case of 9-cyano-10-chloroanthracene **239** (X = CN, Y = Cl), the cleavage of chloride ion from the parent anion radical **240** is found to be a minor reaction pathway, as shown in eq. (7.108). The major product from the anion radical **240** is 9,9'-bianthryl-10,10'-dicarbonitrile **242**, which forms by a higher order reaction [250].

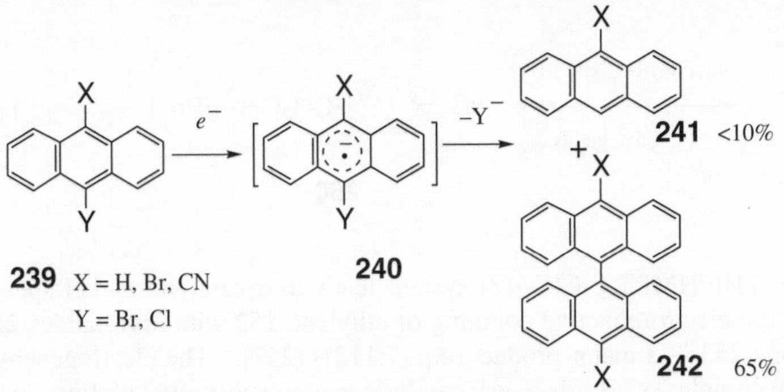

$$(7.108)$$

239 X = H, Br, CN
Y = Br, Cl

240

241 <10%

242 65%

Aryl radicals can undergo rapid intramolecular substitution onto an adjacent benzene ring, leading to the formation of a new six-membered ring. The electroreductive intramolecular coupling of aromatic nuclei of 1-(4-fluorophenyl)-5-(2-halogenophenyl)tetrazoles **243** proceeds in a DMF-Pr$_4$NBF$_4$(0.1M)-(Hg) system to give a mixture of the cyclized products **244** and **245** in 66%~85% total yields (eq. (7.109)) [251, 252].

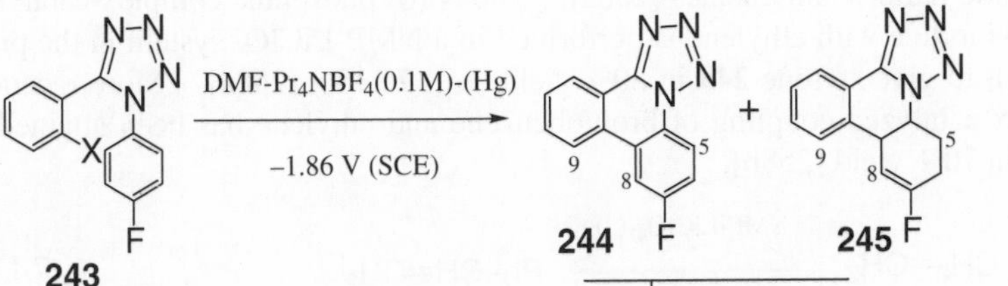

$$\xrightarrow[-1.86 \text{ V (SCE)}]{\text{DMF-Pr}_4\text{NBF}_4(0.1\text{M})\text{-(Hg)}} \qquad (7.109)$$

243

244 **245**

	Yield, %	
X	**244**	**245**
Cl	63	22
Br	50	9
I	66	0

The electroreductive synthesis of poly(*para*-phenylene) films has been performed by the reduction of dibromobiphenyl in a THF/HMPA(7/3)-LiClO$_4$-(ITO) system in the presence of a Ni(0)(diphos)$_2$ catalyst [253]. The electroreductive polymerization of 2,7-dibromo-9,10-dihydrophenanthrene **246** is performed in a DMF-Et$_4$NClO$_4$(0.25M)-(Pt) system in the presence of NiBr$_2$(bpy)$_3$ complex at −1.7 V (Ag/Ag$^+$) at 60 °C to yield a reddish yellow polymer film **247** formed on the working electrode (eq. (7.110)) [254].

$$n \; Br-\text{[phenanthrene]}-Br \xrightarrow[\text{NiBr}_2\text{(bpy)}_3]{\text{DMF-Et}_4\text{NClO}_4(0.25\text{M})\text{-(Pt)}} -(\text{[phenanthrene]})_n \qquad (7.110)$$

$$\textbf{246} \qquad\qquad\qquad\qquad\qquad\qquad\qquad \textbf{247}$$

7.2.3 Electroreductive Substitutions

The aryl radicals generated by the electroreduction of aryl halides in aprotic solvent (liquid NH_3, $PhMe_2SO$, MeCN, DMF) have been shown to react with olefins. The arylated products are obtained in good yield in the latter solvents. The electroreductive coupling of the aryl halide **248** with olefin **249** is performed in a liquid NH_3-Bu_4NBF_4-(Pt) system to yield the adduct **250** in 82~85% yields (eq. (7.111)) [255, 256]. The electroreduction of

$$NC-\text{[benzene]}-Cl$$
$$\textbf{248}$$
$$+$$
$$Ph-CH=CH_2$$
$$\textbf{249}$$

$$\xrightarrow[\text{82~85\% yields}]{\text{liq. NH}_3\text{-Bu}_4\text{NBF}_4\text{-(Pt)}} \quad NC-\text{[benzene]}-CH_2CH_2-Ph \qquad (7.111)$$
$$\textbf{250}$$

nickel(0) bromide in a THF/HMPT-$LiClO_4$-(?) system leads to a catalytic nickel species which can be used for the electroreductive coupling of ethylene **252** with aryl halides **251**, yielding 1,1-diarylethane **253** as a major product (eq. (7.112)) [257]. The electrogeneration of Ni(0) phosphine complexes affords a new catalytic way to substituted olefins *via* the

$$2 \; PhBr \; + \; CH_2=CH_2 \xrightarrow[\text{65\% yield}]{\substack{\text{THF/HMPT-LiClO}_4/ \\ \text{NiBr}_2\text{-(?)}}} (Ph)_2CH-CH_3 \; + \; 2 \; Br^- \qquad (7.112)$$
$$\textbf{251} \qquad\quad \textbf{252} \qquad\qquad\qquad\qquad\qquad \textbf{253} \qquad\quad \textbf{254}$$

coupling of aromatic halides and alkenes [258a]. The Ni(0) phosphine complex-catalyzed coupling of phenyl iodide with ethylene is performed in a NMP-$LiClO_4$ system in the presence of $NiI_2(pph_3)_2$ to give styrene **249** in 80% yield (eq. (7.113)) [258a]. More recently, the Ni(0) complex catalyzed coupling of bromobenzene and ethylene has been attained to give styrene **249** in 70% yield [258b].

$$PhI \; + \; CH_2=CH_2 \xrightarrow[\text{NiI}_2\text{(pph}_3)_2]{\substack{\text{NMP-LiClO}_4\text{-(?)} \\ \\ \text{80\% yield}}} Ph-CH=CH_2 \qquad (7.113)$$
$$\textbf{255} \qquad\quad \textbf{252} \qquad\qquad\qquad\qquad \textbf{249}$$

The electroreductive radical-cyclization of *o*-(3-butenyl)bromobenzene **256** proceeds in a DMF-Bu_4NClO_4-(Pt) system at −2.65 V (SCE) to give 1-methylindan **257** as a major product (eq. (7.114)) [259]. The more improved result (53% yield) is obtained in a DMF-Bu_4NBF_4(0.1M)-(Pb) system at 22 °C [260].

$$\text{[o-(3-butenyl)bromobenzene]} \xrightarrow[\text{41~45\% yields}]{\text{DMF-Bu}_4\text{NClO}_4\text{-(Pt)}} \text{[1-methylindan]} \qquad (7.114)$$
$$\textbf{256} \qquad\qquad\qquad\qquad\qquad\qquad\qquad \textbf{257}$$

The chemical nature of aryl anions arising from aryl halides by electroreduction has been investigated. Such aryl anions can be trapped by a variety of electrophiles. Aromatic carboxylic acids may form by the reaction of the aryl anions with carbon dioxide. The reaction is not affected by the kind of halide atom, but does not proceed in protic solvents. Electroreductions of monohalogenated benzenes in a DMF-Bu$_4$NClO$_4$-(Pt) system in the presence of carbon dioxide affords benzoic acid in 35~38% yields [261]. Many efficient methods for the electroreductive carboxylation of aryl halides, including the use of metal complexes, are discussed in Chapters 2 and 11. The electroreductive carboxylation of p-iodoanisole 258 has been performed at mercury pool cathodes in a DMF-Me$_4$NClO$_4$ system in the presence of carbon dioxide to give p-methoxybenzoic acid 259 in 50% yield together with anisole 260 (40%) (eq. (7.115)) [262]. The reduction at mercury exhibits

$$\text{258} \xrightarrow[\substack{-1.05 \text{ V (SCE)}}]{\text{DMF-Me}_4\text{NClO}_4(0.1\text{M})\text{-(Hg)}} \text{259} + \text{Anisole} \quad \text{260} \; 40\% \tag{7.115}$$

258 (I–C$_6$H$_4$–OMe) → 259 (CO$_2$H–C$_6$H$_4$–OMe) 50%

two waves in cyclic voltammograms and the electrocleavage of the carbon-iodine bond is responsible for the first wave, giving radical and further reduction to anion intermediates for the second wave, which would cause production of p-anisic acid, anisole, and bis(p-anisyl)mercury, respectively [262, 263]. The use of zerovalent nickel complex Ni(0)Ln (L = triphenylphosphine) prepared by electroreduction of the phosphinated Ni(II) complexes has been found to improve the carboxylation of aryl halides, yielding ArCO$_2^-$ in good yields [264]. The electroreductive carboxylation of phenyl halide derivatives are listed in Table 7.10 [95a, 96a, 100~103, 262~266]. Other halides are shown in Table 7.11 [96a, 100, 101, 103]. Halogenated N-heteroaromatic compounds are electroreductively coupled with carbon dioxide in an undivided cell using a sacrificial anode [267]. For example, the electrolysis of 3-bromopyridine 261 in a DMF-Bu$_4$NBF$_4$-(Mg/SUS) system under bubbling carbon dioxide yields nicotinic acid 262 in 75% yield (eq. (7.116)).

$$\text{261} \xrightarrow[\substack{75\% \text{ yield}}]{\text{CO}_2, \text{ DMF-Bu}_4\text{NBF}_4\text{-(Mg/SUS)}} \text{262} \tag{7.116}$$

The electroreductive Bouveault type formylation of aryl halides proceeds in a DMF-Bu$_4$NBr-(Mg/SUS) system by use of a modified cathode coated with zinc or cadmium in good yields [78, 268]. The electroreductive nucleophilic substitutions of aryl halides are listed in Table 7.12 [134, 269~278].

The mechanism of S$_{RN}$1 aromatic substitution reactions of aryl halides is assumed to proceed as a chain process: Ar–X (203) + ArNu$^{-\bullet}$ → Ar$^-$Nu (263) + ArX$^{-\bullet}$. The process involves three intermediate radical species: ArX$^{-\bullet}$, Ar$^\bullet$, and ArNu$^{-\bullet}$ (eq. (7.117)) [279a]. The radical intermediate Ar$^\bullet$ mainly undergoes dimerization in competition with attack of the nucleophite [279b]. The rate constant of the key step, addition of nucleophile

Table 7.10a Electroreductive Carboxylation of Aryl Halides

R¹	R²	R³	R⁴	R⁵	X	Electrolysis Conditions	P_{CO_2} atm	B	C	Ref.
H	H	H	H	H C	I	THF/HMPA(3/1)-LiClO$_4$(0.3M), NiBr(PPh$_3$)$_2$(0.02M)/PPh$_3$(0.1M)	1	87	–	[264]
H	H	H	H	H D	Cl	MF-Bu$_4$NClO$_4$-(Mg/Pt)	1	85	–	[103]
H	H	H	H	H	Cl	DMF-Bu$_4$NBr-(Mg/SUS)	1	85	–	[265]
H	H	H	H	H	Br	THF/HMPA(3/1)-LiClO$_4$(0.3M), NiBr(PPh$_3$)$_2$(0.02M)/PPh$_3$(0.1M)	1	84	–	[264]
H	H	H	H	H	Br	DMF-Et$_4$NClO$_4$(0.1M)-(Hg), −2.3 V(SCE)	1	39	11	[263]
H	H	H	H	H	Br	DMF-Bu$_4$NClO$_4$-(Mg/Pt)	1	80	–	[103]
H	H	H	H	H	I	THF/HMPA(3/1)-LiClO$_4$(0.3M), NiBr(PPh$_3$)$_2$(0.02M)/PPh$_3$(0.1M)	1	64	–	[264]
H	H	H	H	H	I	DMF-Et$_4$NClO$_4$(0.1M)-(Hg), −1.8 V(SCE)	1	0	7	[263]
H	H	H	H	H	CH$_2$Cl	DMF-Bu$_4$NBr-(Al/C)	1	82.5	–	[101]
H	H	H	H	H	CH$_2$Cl	DMF-Bu$_4$NI-(Mg/SUS)	1	80-90	–	[102]

Table 7.10b Electroreductive Carboxylation of Aryl Halides

R¹	R²	R³	R⁴	R⁵	X	Electrolysis Conditions	P_{CO_2} atm	Products, Conv. Yield, % B	C	Ref.
H	H	H	H	H	CH$_2$Cl	DMF-Bu$_4$NClO$_4$-(Mg/Pt)	1	90	–	[103]
H	H	H	H	H	CH$_2$Cl	THF/HMPA(4/6)-LiClO$_4$(0.3M) /Co(Salen)*(0.01M)-(Hg)	2	62	–	[95a]
H	H	H	H	H	CH(Cl)Me	DMF-Bu$_4$NBr-(Al/C)	1	88	–	[101]
H	H	H	H	H	CH(Cl)Me	DMF-Bu$_4$NClO$_4$-(Mg/Pt)	1	80	–	[103]
H	H	H	H	H	CH(Cl)Me	THF/HMPA(4/6)-LiClO$_4$(0.3M) /Co(Salen)*(0.01M)-(Hg)	2	65	–	[95a]
H	H	H	H	H	CH(Cl)Me	THF/HMPA(2/1)-Bu$_4$NBF$_4$(0.1M)-(C)	1	89	–	[96a]
H	H	H	H	H	CH=CHBr	DMF-Bu$_4$NClO$_4$-(Mg/Pt)	1	80	–	[103]
H	H	H	H	H	CH=CHCH$_2$Cl	THF/HMPA-Bu$_4$NClO$_4$-(Mg/Pt)	1	80	–	[103]
H	H	H	H	H	CH=CHCH$_2$Cl	THF/HMPA(4/6)-LiClO$_4$(0.3M) /Co(Salen)*(0.01M)-(Hg)	2	97	–	[95a]
H	H	H	H	H	CH(Ph)Cl	DMF-Bu$_4$NBr-(Al/C)	1	73	–	[100]
H	H	F	H	H	Cl	DMF-Bu$_4$NBr-(Mg/SUS)	1	80	–	[265]
H	H	F	H	H	Br	THF/HMPT-LiClO$_4$(0.3M)-(Hg) p-BrC$_6$H$_4$OMe	1	94	–	[266]
H	H	F	H	H	Br	DMF-Bu$_4$NClO$_4$-(Mg/Pt)	1	80	–	[103]

Table 7.10c Electroreductive Carboxylation of Aryl Halides

R¹	R²	R³	R⁴	R⁵	X	Electrolysis Conditions	P_{CO_2} atm	Conv. Yield, % B	C	Ref.
H	Cl	H	H	H	Cl	DMF-Bu₄NBr-(Al/C)	1	70	–	[101]
H	H	Cl	H	H	Cl	DMF-Bu₄NBr-(Al/C)	1	78	–	[101]
Cl	H	H	H	H	Cl	DMF-Bu₄NBr-(Mg/SUS)	1	62	–	[265]
H	Cl	H	H	H	Cl	DMF-Bu₄NBr-(Mg/SUS)	1	64	–	[265]
H	H	Cl	H	H	Cl	DMF-Bu₄NBr-(Mg/SUS)	1	62	–	[265]
H	Cl	H	H	H	CH(Cl)Me	DMF-Bu₄NBr-(Al/C)	1	61.5	–	[101]
H	H	Cl	H	H	CH(Cl)Me	DMF-Bu₄NBr-(Al/C)	1	88	–	[101]
H	H	Br	H	H	Cl	DMF-Bu₄NBr-(Mg/SUS)	1	68	–	[265]
H	H	Me	H	H	Cl	THF/HMPA(3/1)-LiClO₄(0.3M), NiBr(PPh₃)₂(0.02M)/PPh₃(0.1M)	1	87	–	[264]
H	H	Me	H	H	Cl	DMF-Bu₄NBr-(Mg/SUS)	1	78	–	[265]
H	H	Me	H	H	Br	THF/HMPA(3/1)-LiClO₄(0.3M), NiBr(PPh₃)₂(0.02M)/PPh₃(0.1M)	1	56	–	[264]
CF₃	H	H	H	H	Cl	DMF-Bu₄NBr-(Mg/SUS)	1	80	–	[265]
CF₃	H	H	H	H	Cl	DMF-Bu₄NBr-(Al/C)	1	67	–	[100]
H	CF₃	H	H	H	Cl	DMF-Bu₄NBr-(Mg/SUS)	1	85	–	[265]
H	H	CF₃	H	H	Cl	DMF-Bu₄NBr-(Mg/SUS)	1	80	–	[265]

Table 7.10d Electroreductive Carboxylation of Aryl Halides

R^1	R^2	R^3	R^4	R^5	X	Electrolysis Conditions	P_{CO_2} atm	Products, Conv. Yield, % B	C	Ref.
H	H	CF_3	H	H	Br	THF/HMPT-Bu_4NBF_4(0.3M)-(Hg) $NiCl_2(PPh_3)_2$+PPh_3	1	76	–	[266]
H	H	MeO	H	H	Cl	DMF-Bu_4NBr-(Mg/SUS)	1	75	–	[265]
H	H	MeO	H	H	Br	THF/HMPA(3/1)-$LiClO_4$(0.3M), $NiBr(PPh_3)_2$(0.02M)/PPh_3(0.1M)	1	40	–	[264]
H	H	MeO	H	H	I	DMF-Me_4NClO_4(0.1M)-(Hg), –1.05 V(SCE)	1	50	40	[262]
H	H	Ac	H	H	Cl	DMF-Bu_4NBr-(Mg/SUS)	1	50	–	[265]
H	H	Ac	H	H	Br	DMF-Bu_4NClO_4-(Mg/Pt)	1	82	–	[103]
CN	H	H	H	H	Cl	DMF-Bu_4NBr-(Mg/SUS)	1	65	–	[265]
H	H	Ph	H	H	Cl	THF/HMPA(3/1)-$LiClO_4$(0.3M), $NiBr(PPh_3)_2$(0.02M)/PPh_3(0.1M)	1	47	–	[264]
H	H	PhO	H	H	Br	THF/HMPT-$LiClO_4$(0.3M)-(Hg) p-BrC_6H_4OMe	1	52	–	[266]
H	H	PhO	H	H	Br	THF/HMPT-Bu_4NBF_4(0.3M)-(Hg) $NiCl_2(PPh_3)_2$+PPh_3	1	86	–	[266]
H	H	PhO	H	H	Br	THF/HMPT-$LiClO_4$(0.3M)-(Hg) p-BrC_6H_4OMe	1	52	–	[266]
H	H	PhO	H	H	Br	THF/HMPT-Bu_4NBF_4(0.3M)-(Hg) $NiCl_2(PPh_3)_2$+PPh_3	1	86	–	[266]

Table 7.10e Electroreductive Carboxylation of Aryl Halides

R^1	R^2	R^3	R^4	R^5	X	Electrolysis Conditions	P_{CO_2} atm	Products, Conv. Yield, % B	C	Ref.
H	PhO	H	H	H	CH(Cl)Me	DMF-Bu$_4$NClO$_4$-(Mg/Pt)	1	80	–	[103]
H	PhO	H	H	H	CH(Cl)Me	THF/HMPA(2/1)-Bu$_4$NBF$_4$(0.1M)-(C)	1	76	–	[96a]
H	H	Iso-Pr	H	H	CH(Cl)Me	DMF-Bu$_4$NBr-(Al/C)	1	83	–	[101]
H	H	Iso-Pr	H	H	CH(Cl)Me	THF/HMPA(2/1)-Bu$_4$NBF$_4$(0.1M)-(C)	1	80	–	[96a]
H	iso-Bu	H	H	H	CH(Cl)Me	DMF-Bu$_4$NClO$_4$-(Mg/Pt)	1	85	–	[103]
H	H	iso-Bu	H	H	CH(Cl)Me	DMF-Bu$_4$NBr-(Al/C)	1	95	–	[100]
Cl	Cl	H	H	H	Cl	DMF-Bu$_4$NBr-(Mg/SUS)	1	74*	–	[265]
H	Cl	H	Cl	H	Cl	DMF-Bu$_4$NBr-(Mg/SUS)	1	48	–	[265]
Cl	H	Cl	H	H	Cl	DMF-Bu$_4$NBr-(Mg/SUS)	1	57	–	[265]
H	Cl	Me	H	H	Cl	DMF-Bu$_4$NBr-(Mg/SUS)	1	53	–	[265]
Cl	MeO	H	H	H	Cl	DMF-Bu$_4$NBr-(Mg/SUS)	1	64	–	[265]
Cl	Cl	MeO	H	H	Cl	DMF-Bu$_4$NBr-(Mg/SUS)	1	36	–	[265]
Cl	MeO	Cl	H	H	Cl	DMF-Bu$_4$NBr-(Mg/SUS)	1	56	–	[265]
H	Cl	Ac	H	H	Cl	DMF-Bu$_4$NBr-(Mg/SUS)	1	56	–	[265]
Cl	H	Ac	H	H	Cl	DMF-Bu$_4$NBr-(Mg/SUS)	1	42	–	[265]

Table 7.11 Electroreductive Carboxylation of Halides

$$Ar—X \xrightarrow{\ CO_2\ } Ar—CO_2H$$

$$\quad\quad A \quad\quad\quad\quad\quad\quad\quad\quad\quad B$$

| Ar—X | | | | Yield, % | |
Ar	X	Electrolysis Conditions	P_{CO_2} atm	$Ar—CO_2H$	Ref.
(naphthalene-CH₂)	Cl (α)	DMF-Bu$_4$NBr-(Al/C)	1	57	[101]
(naphthalene)	Cl	DMF-Bu$_4$NBr-(Al/C)	1	79	[100]
(thiophene, S)	Cl (α)	THF/HMPA-Bu$_4$NClO$_4$-(Mg/Pt)	1	80	[103]
(furan, O)	Br (β)	THF/HMPA-Bu$_4$NClO$_4$-(Mg/Pt)	1	78	[103]
MeO(naphthalene)CHMe	Cl	DMF-Bu$_4$NClO$_4$-(Mg/Pt)	1	80	[103]
		THF/HMPA(2/1)-Bu$_4$NBF$_4$(0.1M)-(C)	1	81	[96a]

(7.117)

on the aryl radical, has been determined [280].   The rate constants of the substitution reaction of cyanide ion with aryl radicals in liquid ammonia at −40 °C have been surveyed [281].   It is suggested that the reaction between solvated electrons and substrates Ar − X **203** yielding fast-cleaving anion radicals ArX⁻˙ or giving rise to concerted bond cleavage occurs during the mixing of the two reactants in the diffusion layer [282].   The origin of leaving group effects in radical reactions triggered by solvated electron reduction is also discussed.

A mechanistic survey of electrochemical induction of aromatic nucleophilic substitutions has been discussed in terms of an electron transfer and a bond breaking by chemical reactions on electrodes [276, 280, 283~288].   It has been demonstrated that when the reduction potential of the substrate is positive to the standard potential of the ArNu (**263**)/ArNu⁻˙ couple, *e.g.*, with Nu = PhS, **263** is obtained directly.   In the reverse case, *e.g.*, with the 4-bromobenzophenone-cyanide system, the substitution product was partly

Table 7.12a Electroreductive Nucleophilic of Substitutions of Aryl Halides

R¹	R²	R³	R⁴	R⁵	X	Nu	Electrolysis Conditions	Conv. Yield, %	Ref.
H	H	H	H	H	Br	PhS⁻	DMSO/PhCN-Bu₄NClO₄(0.01M)/Bu₄NSPh-(Pt)	67	[269]
H	Me	Cl	H	H	Cl	⁻CH(CN)CO₂Et	liq. NH₃-KOH-(Pt)	48	[270]
H	Me	H	H	H	Cl	CN⁻	DMF-Bu₄NClO₄(0.1M)-(Ni) Et₄NCN/Ni(PPh₃)₄	54	[271]
H	H	Me	H	H	Cl	CN⁻	DMF-Bu₄NClO₄(0.1M)-(Ni) Et₄NCN/Ni(PPh₃)₄	25	[271]
H	CF₃	H	H	H	Cl	CN⁻	DMF-Bu₄NClO₄(0.1M)-(Ni) Et₄NCN/Ni(PPh₃)₄	55	[273]
CN	H	H	H	H	Cl	ArOH*	liq. NH₃-KOH-(Pt)	89	[272]
H	CN	H	H	H	Cl	ArOH*	liq. NH₃-KOH-(Pt)	68	[272]
H	H	CN	H	H	Cl	ArOH*	liq. NH₃-KOH-(Pt)	79	[272]
H	H	CN	H	H	Cl	(PhSe)₂	MeCN-NaClO₄-(C)	73	[277c]
H	H	H	H	H	Br	(PhSe)₂	MeCN-Bu₄NPF₆(0.1M)-(Pt/C)	36	[273]
CN	H	H	H	H	Br	(PhSe)₂	MeCN-Bu₄NPF₆(0.1M)-(Pt/C)	42	[273]
H	H	CN	H	H	Br	(PhSe)₂	MeCN-Bu₄NPF₆(0.1M)-(Pt/C)	58	[273]

Table 7.12b Electroreductive Nucleophilic Substitutions of Aryl Halides

R^1	R^2	R^3	R^4	R^5	X	Nu	Electrolysis Conditions	Products, Conv. Yield, %	Ref.
H	H	CN	H	H	Br	$(PhTe)_2$	$MeCN-Bu_4NPF_6(0.1M)-(Pt/C)$	42	[274]
H	H	PhCO	H	H	Cl	CN^-	$DMF-Bu_4NClO_4(0.1M)-(Ni)$ $Et_4NCN/Ni(PPh_3)_4$	84	[271]
H	H	PhCO	H	H	Br	^-CN	$DMSO-LiClO_4-(Hg)$	95	[134]
H	H	PhCO	H	H	Br	MeS	$DMSO-LiClO_4-(Hg)$	5	[275]
H	PhCO	H	H	H	Br	$\overline{C}H(Ac)_2$	$DMSO-Et_4NClO_4(0.1M)-(?)$	51	[270]
						$\overline{C}H(CN)CO_2Et$	$DMSO-Et_4NClO_4(0.1M)-(?)$	94	[270]
H	H	PhCO	H	H	Br	$PhSNBu_4$	$MeCN-Et_4NCN-(Hg)$	95	[276]
H	H	PhCO	H	H	Br	$PhSeNEt_4$	$MeCN-Bu_4NPF_6(0.1M)-(C)$	86	[277a] [277b]
PhCO	H	H	H	H	Br	$PhSeNEt_4$	$MeCN-Bu_4NPF_6(0.1M)-(C)$	75	[277a] [277b]
H	H	PhCO	H	H	Br	PhS^-	$DMSO-LiClO_4-(Hg)$	95	[275]
H	H	PhCO	H	H	Br	$t-BuS^-$	aq. $MeCN-Et_4NClO_4(0.1M)-(?)$	80	[278]
							$DMSO-LiClO_4-(Hg)$	60	[275]

obtained in the form of ArNu⁻·, which had to be independently reoxidized. The mechanistic reconsideration in which aromatic anion radicals containing leaving groups react with nucleophiles by a bimolecular displacement process has been discussed [289]. The role of polar organic solvents, *e.g.*, benzonitrile as an electron carrier on electroreduction of aromatic halides in redox-catalysis systems, has been pointed out [290].

The electroreductive cross-coupling of bromotrifluoromethane **265** with aromatic or heteroaromatic iodides **264** (X = I) and bromides **264** (X = Br) has been attained in an undivided cell fitted with a sacrificial copper anode (eq. (7.118)) [291].

$$\text{264} \quad \xrightarrow[\text{CF}_3\text{Br, 32~98\% yields}]{\text{DMF-Bu}_4\text{NBr(1M)/Ph}_3\text{P-(Cu)}} \quad \text{265} \tag{7.118}$$

264 X = Br, I
Y, Z = MeO, CN, Cl,
CO₂Et, CO₂Me

265

An unusual reaction resulting from the addition of butyl vinyl ether of trifluoromethyl radicals derived from the electron transfer between electrogenerated aromatic anion radicals and trifluoromethyl bromide has been reported [292]. The electrolysis of telephthalonitrile **266** is performed in a DMF-Bu₄NBF₄(0.1M)-(C) system in the presence of trifluoromethyl bromide and in the addition of butyl vinyl ether **267** to yield the coupling product **268** in 95% yield (eq. (7.119)).

$$\text{266} + \text{267} \quad \xrightarrow[\text{CF}_3\text{Br, -CN}^-, \text{ 95\% yield}]{\text{DMF-Bu}_4\text{NBF}_4(0.1M)-(C)} \quad \text{268} \tag{7.119}$$

NC—⟨ ⟩—CN
266

CH₂=CH–OBu
267

NC—⟨ ⟩—CH(CH₂CF₃)(OBu)
268

2-Substituted indole derivatives **271** are electrosynthesized by the reductive coupling of 2-iodoaniline **269** with keto enolates **270** in a liquid NH₃-KBr-(C) system at −33 °C (eq. (7.120)) [293].

$$\text{269} + \text{270} \quad \xrightarrow{\text{liq. NH}_3\text{-KBr-(C/C)}} \quad \text{271} + \text{Aniline} \tag{7.120}$$

269   **270**   **271**   **272**

RCOCH₂⁻K⁺ R	Yield, %	
	271	272
H	93	–
Me	87	13
Ispro	75	22

The stable potassium and ammonium salts of *p*-cyanophenyl malononitrile have been electrosynthesized as a one-pot reaction with very high yield [294]. The electrolysis of a mixture of *p*-chlorobenzonitrile and malononitrile is performed in a liquid NH₃-KNH₂-(Pt/Au) system in the presence of 4,4'-bipyridine.

The electrochemically induced $S_{RN}1$ substitution reactions with chalcogenophenoxide anions proceeds preferentially in dimethyl sulfoxide (DMSO) rather than acetonitrile (MeCN) [295].   The electrosynthesis of 1-(phenylthio)naphthalene can be obtained in a quantitative yield in DMSO and in 32% yield in MeCN.

The competition between mono- and disubstitution of dihalobenzenes by a series of aromatic sulfanions *via* the $S_{RN}1$ reaction involving two radical chains has been discussed [296].   The electrolysis of *p*-iodochlorobenzene **273** in a liquid NH$_4$-KBr(0.1M)-(Au) system in the presence of the nucelophile **274** affords mono-substituted product **275** in 87% yield (eq. (7.121)).   The electrocatalyzed substitution of 2-chloroquinoline by benzenethiolate in liquid ammonia has been investigated as a model of the application of the theoretical

$$ (7.121) $$

relationship [276, 297a].   The product distribution in preparative-scale electrolysis of the aromatic nucleophilic substitution of **276** in a liquid NH$_3$-PhSH-(Pt) system proceeds to give the substituted product **277** in good yield (eq. (7.122)) [297a].   Nucleophilic substitution of nitroaromatic halides by electrogenerated polysulfide ions in DMF is studied [297b]. Sodium benzenesulfinate reacts with *p*-halonitrobenzenes in a *SnAr*-reaction manner to give the corresponding substituted products [297c].

$$ (7.122) $$

Electron transfer to the aryl-radical resulting from the initial reductive cleavage is found to be essentially the only reaction competing with the substitution.

2,6-Di-*tert*-butyl-(4-trifluoromethyl-2-pyridyl)phenol **280** has been electrosynthesized *via* $S_{RN}1$ reaction of 2-chloro-4-(trifluoromethyl)pyridines **278** with 2,6-di-*tert*-butylphenol **279** [298].   The electrolysis of **278** is carried out in a liquid NH$_3$/DMF-KBr-(Pt/Mg) system in the presence of phthalonitrile as a mediator to give the coupling product **280** in 50% yield (eq. (7.123)).   The redox mediator activates the electron transfer initiating radical chain reaction [299].

$$ (7.123) $$

1,4-Dichlorobenzene and dichloropyridines undergo $S_{RN}1$ reactions with various nucleophiles upon electrochemical initiation.   Selective formation of monosubstitution products is observed when the nucleophiles bear electro-donating groups on the aromatic ring. The electroreductive substitution of 3,5-dichloropyridine **281** with 2,6-di-*tert*-butylphenol **282** proceeds in a liquid NH₃-*tert*-BuOK/KBr-(Pt) system at –38 °C in the presence of 4,4'-bipyridine as a mediator to give the coupling product **283** in 58% yield (eq. (7.124)) [300].

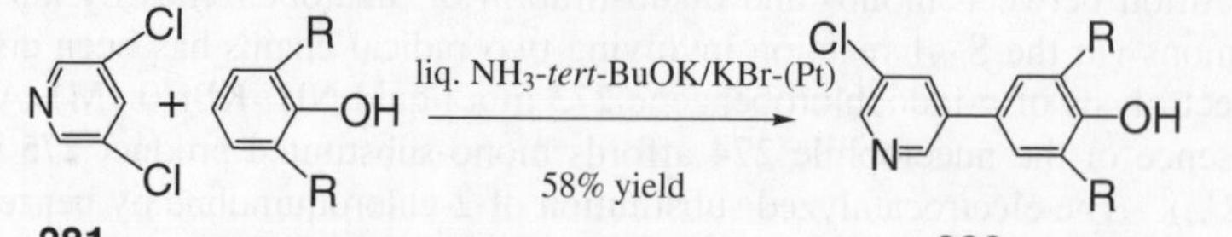

$$\text{(7.124)}$$

281

282   R = *tert*-Bu

283

The electroreduction of *o*-dibromobenzene **284** in a DMF-Bu₄NClO₄(0.1M)-(Pt) system in the presence of *N*-ethylpiperidine as a benzyne **285** acceptor affords the addition product **286** in 10~21% yields together with benzene and bromobenzene (eq. (7.125)) [301, 302].

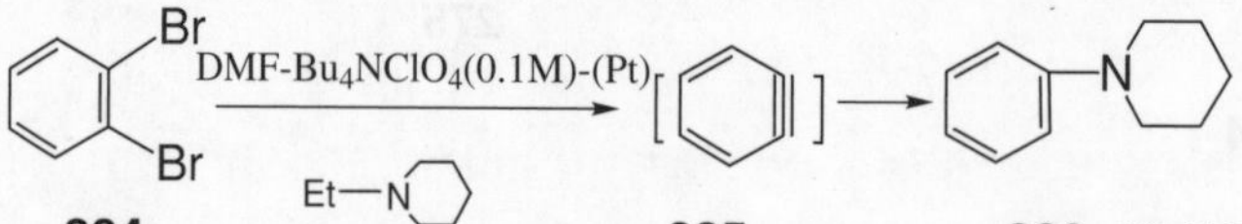

$$\text{(7.125)}$$

284

285

286   10~21% yields

Concerning the reaction of 2-chlorobenzonitrile with benzenethiolate, the yield in ArNu **263** depends on the magnitude of the rate constant, K_1 of the cleavage of the anion radical ArX⁻˙ [303].

The aryl σ-radical formed by electroreduction of aryl halides in an aprotic solvent has been shown to undergo an intramolecular 1-radical substitution on an adjacent aromatic ring. Electroreductive cyclization of the benzimidazole **287** in a DMF-Pr₄NBF₄-(Pt/Hg) system under a potential of –2.1 V gives benzimidazo[1,2-f]phenanthridine **288** in 69% yield (eq. (7.126)) [304]. Electroreductive cyclization of 2-bromo-*N*-methyl-*N*-(2-naph-thyl)benzamide **289** in a DMF-Pr₄NClO₄-(Pt/Hg) system under a potential of –1.9 V is also investigated. The relative yields of benzophenanthridones **290** is given as 54% (eq. (7.127)) [305]. The results obtained by the electrochemically induced aromatic substitutions of aryl halides with nucleophiles are listed in Table 7.13 [277a, 277b, 278, 306, 307].

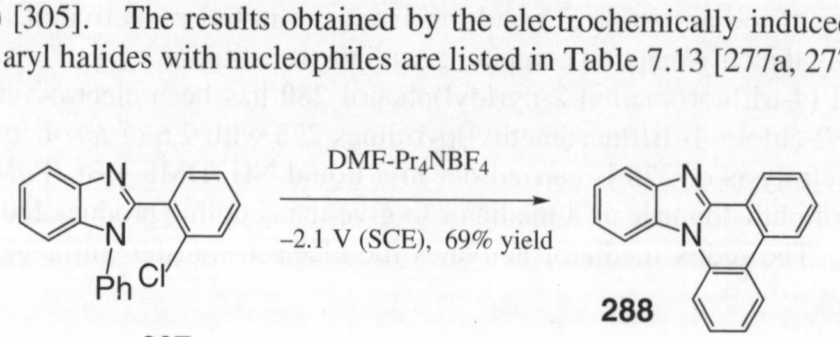

$$\text{(7.126)}$$

DMF-Pr₄NBF₄

–2.1 V (SCE), 69% yield

287

288

$$\text{(7.127)}$$

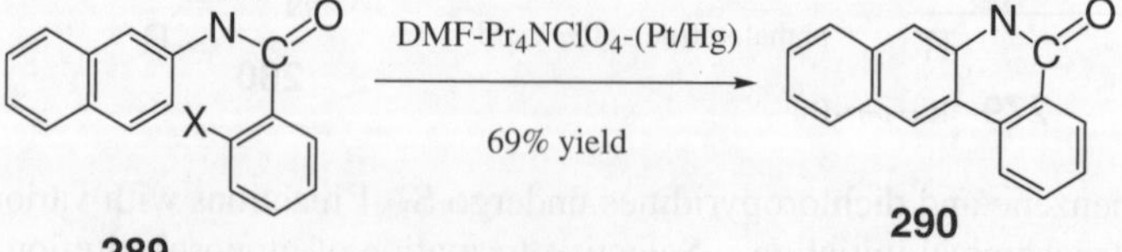

DMF-Pr₄NClO₄-(Pt/Hg)

69% yield

289   X = Br

290

Table 7.13 Electrochemically Induced Aromatic Substitution

Substrate	Nucleophile	Electrolysis Conditions	Products	Yield, %	Ref.
p-Br-C$_6$H$_4$COPh	Na$^+$MeC$^-$NO$_2$	Me$_2$SO-Bu$_4$NBF$_4$-(Hg•Au)	PhCOC$_6$H$_4$CHMe$_2$ PhCOPh	50 26	[306]
p-Br-C$_6$H$_4$COPh	Na$^+$OP(OEt)$_2^-$	Me$_2$SO-Bu$_4$NBF$_4$-(Hg•Au)	PhCOC$_6$H$_4$P(O)(OEt)$_2$ PhCOPh	10 42	[306]
p-Br-C$_6$H$_4$COPh	NaSPh	Me$_2$SO-Bu$_4$NBF$_4$-(Hg•Au)	PhCOC$_6$H$_4$SPh PhCOPh	80 10	[306]
p-Br-C$_6$H$_4$COPh	Bu$_4$NSPh	MeCN$^-$	PhCOC$_6$H$_4$SPh PhCOPh	95 3	[278]
p-Br-C$_6$H$_4$COPh	Bu$_4$NCN	MeCN$^-$	PhCOC$_6$H$_4$CN PhCOPh	95 3	[307]
p-Br-C$_{10}$H$_9$	Bu$_4$NSPh	Me$_2$SO$^-$	4-PhSC$_{10}$H$_9$	100	[307]
p-Br-C$_6$H$_4$COMe	Bu$_4$NSPh	MeCN$^-$	p-PhSC$_6$H$_4$COMe	95	[307]
2-Cl-quinoline	(EtO)$_2$POK	Me$_2$SO$^-$	quinoline N-OP(OEt)$_2$	100	[307]
Br-anthracene	PhSeNBu$_2$	MeCN-Bu$_4$NPF$_6$(0.1M)-(C)	SePh-anthracene	74	[277a][277b]
Br-anthracene	PhTeLi	MeCN-Bu$_4$NPF$_6$(0.1M)-(C)	TePh-anthracene	49	[277a][277b]

The trimethylsilylation of *o*-chlorotoluene **291** proceeds in a THF/HMPA(4/1)-Et$_4$NBF$_4$-(Al/C) system in the presence of trimethylsilyl chloride to give *o*-trimethylsilyl-toluene **292** in 70% yield (after passage of 2.2 F/mol) (eq. (7.128)), whereas, after passage of 4.4 F/mol of *cis*- and *trans*-tris(trimethylsilyl)cyclohexa-1,3 (or1,4)-dienes **293** and **294** are produced in 84% yield (eq. (7.129)) [308].   A versatile and stepwise procedure for the

$$(7.128)$$

$$(7.129)$$

* THF/HMPA(4/1)-Et$_4$NBF$_4$-(Al/C)

creation of the C-Si bond on *o*-dichlorobenzene has been developed [309].   The electro-chemical trimethylsilylation is performed in an undivided cell using a sacrificial anode, un-der passage of a constant current density.   Actually, the one-pot syntheses of *o*-(trimethylsilyl)chlorobenzene **296**, *o*-bis(trimethylsilyl)benzene **297**, 2,3,5,6-tetrakis(trimethylsilyl)cyclohexa-1,3-diene **298**, and 1,2,3,4,5,6-hexakis(trimethylsilyl)cy-clohexene **299** are performed in high yields in a THF/HMPA(4/1)-Me$_3$SiCl-(Al/SUS) sys-tem at desirable temperatures (eqs. (7.130~133)) [309].

$$(7.130)$$

$$(7.131)$$

$$(7.132)$$

$$(7.133)$$

* THF/HMPA(4/1)-Me$_3$SiCl-(Al/SUS)

7.3 Electroreduction of Polyhalogenated Compounds

7.3.1 Electroreduction of *gem*-Halogeno Compounds

The electroreduction of haloethane **300** undergoes two-electron reduction at pH 13 to produce 1-chloro-2,2,2-trifluoroethane **301** together with bromide ion (eq. (7.134)) [310, 311].

$$CHClBrCF_3 \xrightarrow{\text{aq. KOH(0.1M)-(Pt)}} CH_2ClCF_3 + Br^- + OH^- \qquad (7.134)$$

300   **301**

The first electron transfer is found to be the slowest electrochemical step. Electroreductive silylations of 7,7-dibromonorcarane exhibit dramatic enhancement of current efficiency and stereoselectivity under sonication conditions [312, 313]. Ni(II) and Co(II) ion-mediated electroreduction of 2-carbomethoxy-2-methyl-1,1-dichlorocyclopropane has been carried out in a DMF-Et₄NI(0.1M)-(Hg) system to yield 2-carbomethoxy-2-methyl-1-chlorocyclopropane in 70% yield [314].

The electroreductive cleavage of the carbon-bromine bond on 2,2-disubstituted 1,1-dibromocyclopropanes undergoes two-electron reduction to give a mixture of diastereomers of mono-bromides [315]. The electroreduction of 2-substituted 1,1-dibromo-2-methylcyclopropanes **302** in an MeOH-LiClO₄-(Cu/Hg) system affords the corresponding monobromo-2-methylcyclopropanes **303** and **304** in good yields (eq. (7.135)) [316]. The effects of current density and solvent are discussed in light of the *cis/trans* ratio of stereoisomers.

$$\qquad (7.135)$$

302   R = Me, CH=CH₂,   **303** *cis*   **304** *trans*
          CMe=CH₂, CO₂Me, CN

The electroreduction of 2-phenyl-1,1-dibromocyclopropane affords the *cis* isomer in 84% yield. The *cis* isomers are always preferentially formed when the substituents at the C₂ position are either a pair of phenyl and hydroxycarbonyl (92%) or a pair of ethoxycarbonyl and methyl (99%) groups, respectively. The electroreductive asymmetric reduction of prochiral 1,1-dibromo-2,2-diphenylcyclopropane on the poly-L-valine-coated graphite electrode has been attempted [317]. The asymmetric yield of 1-bromo-2,2-diphenylcyclopropane is 16.6% calculated with reference to optical rotating power. The electroreductive debromination of 6,6-dibromopenicillanic acid **305** proceeds in an MeOH-Et₄NOTs-(Pb) system to afford the debrominated product **306** in 75% yield (eq. (7.136)) [318].

$$\qquad (7.136)$$

305   **306**

The electroreduction of 2,2-dichloronorbornane and 2-*exo*-bromo-2-*endo*-chloro- and/or 2-*exo*-chloro-2-*endo*-bromonorbornane has been investigated in a DMF-LiBr(or

Et$_4$NBr)-(Hg) system [319].   The halides give the same mixture of *endo*-norbornyl chloride and nortricyclene.   It is well known that 7,7-dichloronorcarane is hard to reduce electrochemically.   However, the formation of 7-chloronorcarane has been observed when the electrolysis is carried out in diglyme in the presence of tetrabutylammonium salts [320]. The preparative electrolysis of the dichlorocarane in a Diglyme/EtOH-Bu$_4$NBF$_4$(0.1M)- (Hg) system affords the corresponding hydrocarbon [321].   Copper, pyrolytic graphite, and lead electrodes can be used for this reduction in a Diglyme-Bu$_4$NBF$_4$ system [322].

The electroreduction of benzal chloride **307** in the presence of a catalytic amount of Co(II)(salen) affords stilbenes **308** (*cis/trans* = 1.1/1) in good yield (eq. (7.137)) [323, 324].

$$\text{Ph--CHCl}_2 \xrightarrow{\text{Co(II)(salen)}} \text{Ph--CH=CH--Ph} + \text{Ph--CH}_2\text{CH}_2\text{--Ph} \qquad (7.137)$$

307   **308** 87%   **309** 5%

Salen: *bis*-salicylideneethylenediamine

The reductive carboxylation of benzal chloride **307** has been attained by electrolysis in a DMF-Bu$_4$NBr-(Al/Zn) system to give chlorophenylacetic acid **310** (52%) and phenylacetic acid **311** (29%) (eq. (7.138)) [325].   The cathodic reduction of phthalyl chloride **312** in a DMF-NaClO$_4$(0.4M)-(Hg) system yields phthalide **313** in 78% yield (eq. (7.139)) [326].

$$\text{Ph--CHCl}_2 \xrightarrow[\text{CO}_2]{\text{DMF-Bu}_4\text{NBr-(Al/Zn)}} \underset{\overset{|}{\text{Cl}}}{\text{Ph--CHCO}_2\text{H}} + \text{Ph--CH}_2\text{CO}_2\text{H} \qquad (7.138)$$

307   **310** 52%   **311** 29%

$$(7.139)$$

DMF/AcOH-NaClO$_4$-(Hg)

−1.2V (SCE), 78% yield

312   **313**

The electrochemical cyclopropanation of alkenes with *gem*-dibromoalkanes has been achieved in an undivided cell fitted with a sacrificial zinc anode.   The cyclopropanation is carried out in a CH$_2$Cl$_2$/MeCN(9/1)-Bu$_4$NBr(1.3mM)/Bu$_4$NI(0.5mM)-(Zn/C) system in the presence of alkene **315** and the dihalo compound **314** to give **316** in 75% yield (eq. (7.140)) [327].   The cyclopropane formation reaction of the electrogenerated monohalomalonate

$$(7.140)$$

CH$_2$Cl$_2$/MeCN(9/1)-Bu$_4$NBr/

Bu$_4$NI-(Zn/C)

~75% yields

314   X = Br, I$_2$

R, R' = alkyl

315

316

carbanions derived from ethyl dihalomalonates **317** with Michael acceptors **318** proceeds in a CH$_2$Cl$_2$/MeCN(1/1)-Bu$_4$NBr(0.1M)-(Hg) system in an undivided cell to give the cyclo-condensation products **319** in excellent yields (eq. (7.141)) [328].   The reactivity of the bromomalonate carbanion strongly depends on the nature of the cation species of the supporting electrolyte.   The chloromalonate carbanion appears to be more reactive than the bromo carbanion.

$$\text{(7.141)}$$

R^1, R^2 = H, Ph ;   X = Cl, Br
Y, W = H, Ac, CN, CO$_2$Et

A carbene intermediate has been detected in the course of the electroreduction of dichloromalononitrile in acetonitrile [329].

The electroreduction of the tetrachlorospirobicyclobutane is highly dependent on the working potential and the solvent in the preparative scale electrolysis.   The electroreduction of **320** in an MeCN-Bu$_4$NI system gives 1,2-dichloro-1,2-(dicyclohexylidene)ethane **321** in 60% yield (eq. (7.142)) [330], whereas the electrolysis in a THF-Bu$_4$NI system affords bicyclo[1.1.0]butane derivative **322** in 72% yield (eq. (7.143)).

$$\text{(7.142)}$$

$$\text{(7.143)}$$

The electroreduction of tetrachlorocyclopropene **323** undergoes a one-step Barbier-type homo-coupling silylation coupling reaction to yield hexakis(trimethylsilyl)-3,3'-bicyclo-propenyl **324** [331, 332].   The electrosynthesis of **324** is carried out in a THF-HMPA (10/1)-Bu$_4$NBr-(Al/SUS) system in the presence of chlorotrimethylsilane in an undivided cell (eq. (7.144)).

$$\text{(7.144)}$$

Polycarbosilanes have been electrosynthesized by the reductive polymerization of the corresponding dichlorocarbosilanes and their analogues. The electrolysis is performed in a DME-Bu₄NCl-(Al) system by alternating polarity in an undivided cell [333].

7.3.2   Electroreduction of Polyhalogenated Compounds

Organic halides are generally reduced in a bielectronic process [334~341]. In most cases, a two-electron wave is observed in a linear sweep voltammetry. However, some allenic and acetylenic polyhalogeno compounds exhibit a process of two separate one-electron steps [342].

The electroreductive cleavage of the carbon-fluorine bond in perfluoroalkanes has been demonstrated by performing it in a DMF-Bu₄NClO₄(0.1M)-(C) system in a divided cell [343]. The electroreduction of perfluorohexyl iodide, C_6F_{13}-I **325**, in a DMF-LiClO₄(0.1M)-(Hg) system at −0.75 V (SCE) affords the organomercuric compound, $(C_6F_{13})_2Hg$ **327**, in a quantitative yield (eq. (7.145)) [344]. The reaction proceeds through the formation of $C_6F_{13}HgI$ **326** as an intermediate. When the electrolysis is carried out at −1.7 V, electroreductive cleavage of the carbon-iodine bond occurs, giving the protonated compound, C_6F_{13}-H **328**, in 60% yield (eq. (7.146)).

$$\text{DMF-LiClO}_4(0.1M)\text{-(Hg)}$$

$$C_6F_{13}\text{-I} \xrightarrow[-0.75 \text{ V (SCE), quantitative yield}]{} C_6F_{13}\text{-HgI} \xrightarrow[-0.85V]{e^-} (C_6F_{13})_2Hg \qquad (7.145)$$

$$\mathbf{325} \qquad \qquad \mathbf{326} \qquad \qquad \mathbf{327}$$

$$\xrightarrow[-1.7 \text{ V (SCE), 60\% yield}]{\text{DMF-LiClO}_4(0.1M)\text{-(Hg)}} C_6F_{13}\text{-H} \qquad (7.146)$$

$$\mathbf{328}$$

The electrosynthesis of perfluoroalkanecarboxylic acids has been performed by the reaction of electrogenerated superoxide, $O_2^{-\cdot}$, with perfluoroalkyl iodides [345]. The electroreduction of dioxygen has been carried out in a DMF-LiCl/Bu₄NBF₄-(C) system in the presence of perfluoroalkyl iodide **329** under bubbling oxgen gas to give the corresponding carboxylic acid **330** in 43~58% yields (eq. (7.147)).

$$R_F\text{-CF}_2\text{-I} + O_2 \xrightarrow[43\sim58\% \text{ yields}]{\text{DMF-LiCl/Bu}_4\text{NBF}_4} R_F\text{-CO}_2^- + I^- + 2F^- \qquad (7.147)$$

$$\mathbf{329} \qquad \qquad \mathbf{330}$$

$$R_F = C_{n-1}F_{2n-1}$$

The electroreduction of 1,4-diiodo perfluorobutane, I-(CF₂)₄-I, in a DMF-LiClO₄-(Hg) system affords the corresponding symmetrical mercurial compound, [H-(CF₂)₄]₂Hg [346].

Addition of perfluoroalkyl iodides to allyl alcohol has been performed by electroreduction in an aqueous DMF-KCl-(C) system to yield the corresponding iodohydrins, the precursor of epoxides [347].

The electroreduction of carbon tetrachloride in an MeCN-Bu₄NBr-(Hg) system undergoes two-electron reduction to form a trichloromethyl anion as an initial intermediate and subsequent elimination of chloride ion gives rise to a carbene intermediate which can be trapped with olefin to produce cyclopropane derivatives [348, 349].

An electroreductively induced chain reaction catalyzed by an EG base in the reaction of trichloromethyl anion to vinyl acetate **331** or aldehydes **333** has been reported (eqs. (7.148) and 7.149)) [350, 351].   Further reduction of the trichloromethyl moiety of the alcohol products has been attained by the electroreduction in an aqueous MeOH-NH$_4$NO$_3$-(Pt) system as shown in eqs. (7.150a and 7.150b) [352].   The reduction of **334** in an MeOH-H$_2$SO$_4$-(C/Hg) system affords 1,1-dichloro-1-enes **338** in 48~86% yields (eq. (7.150c)) [353, 354].   Secondary trichloromethyl carbinols **334** and their corresponding ethers can be reduced by electrochemically generated chromium(II) chloride to form Z-monochlorovinyl compounds **339** in one step (eq. (7.151)) [355].   The electrolysis of the trichloromethyl carbinol **334** is carried out in an aqueous DMF-HCl/CaCl$_2$-(Pt/Pb) system in the presence of CrCl$_3$•6H$_2$O as a reducing mediator to give the monochlorovinyl compound **339** in 86% yield (eq. (7.151)) [355].

$$CH_2=CH-OAc \xrightarrow[\text{71\% yield}]{\text{DMF/CHCl}_3\text{-Et}_4\text{NOTs-(C)}} CH_3-\underset{\underset{\textbf{332}}{OAc}}{\overset{}{CH}}-CCl_3 \qquad (7.148)$$

331

$$R-CHO + CCl_4 \xrightarrow[\text{56~90\% yields}]{\text{DMF/CHCl}_3\text{-Et}_4\text{NOTs-(C)}} R-\underset{OH}{\overset{}{CH}}-CCl_3 \qquad (7.149)$$

333   R = alkyl, aryl   **334**

$$R-\underset{\underset{\textbf{335}}{OMe}}{\overset{}{CH}}-CCl_3 \xrightarrow[\text{50~84\% yields}]{\text{aq. MeOH-NH}_4\text{NO}_3\text{-(Pt)}} R-\underset{OMe}{\overset{}{CH}}-CHCl_2 ---(a) \quad \textbf{336}$$

$$\xrightarrow[\text{77~86\% yields}]{\text{aq. MeOH-NH}_4\text{NO}_3\text{-(Pt)}} R-\underset{OH}{\overset{}{CH}}-CHCl_2 -----(b) \quad \textbf{337} \qquad (7.150)$$

$$R-\underset{\underset{\textbf{334}}{OH}}{\overset{}{CH}}-CCl_3 \xrightarrow[\substack{\text{48~86\% yields}\\ \text{R = allyl, methallyl, 2-butenyl}}]{\text{MeOH-H}_2\text{SO}_4\text{-(C/Hg)}} R-CH=CCl_2 \quad -------(c) \quad \textbf{338}$$

$$\xrightarrow[\text{86\% yield}]{\text{DMF-HCl/CaCl}_2\text{/CrCl}_2\text{-(Pt/Pb)}} \underset{\textbf{339}}{\overset{R}{\underset{H}{>}}C=C\overset{Cl}{\underset{H}{<}}} + \underset{\textbf{340}}{\overset{R}{\underset{H}{>}}C=C\overset{Cl}{\underset{Cl}{<}}}$$

R =

The electroreduction of 5,5,5-trichloro-4-hydroxy-2-methyl-1-pentene **341** in an aqueous MeOH(70%)-HCl(3%)-(Pb) system affords 1,1-dichloro-4-methyl-1,4-pentadiene **342** as a major product in good yields (eq. (7.152)) [356, 357].

$$
\underset{\substack{\\ \textbf{341}}}{\underset{OH}{\overset{Me}{CH_2=CCH_2-CH-CCl_3}}} \xrightarrow[\substack{93\sim75\%\ yields}]{\substack{aq.\ MeOH(ca.70\%) \\ -HCl(3\%)-(Pb)}} \underset{\textbf{342}}{\overset{Me}{CH_2=CCH_2-CH=CCl_2}} \qquad (7.152)
$$

The electroreductive trichloromethylation of the chlorinated carbonate **343** has been accomplished in a THF/NMP-Bu₄NBF₄/Bu₄NI-(Zn/SUS) system in the presence of carbon tetrachloride, giving the cross-coupling product **344** in 98% yield (eq. (7.153)) [358]. The effect of ferrous ion for the electroreduction of carbon tetrachloride has been investigated [359]. The kinetics of the electroreduction of carbon tetrachloride is discussed in terms of the calculation using Marcus theory [360].

$$
\underset{\substack{\\ \textbf{343}}}{\underset{Me}{\overset{OH}{Cl-C-CO_2Et}}} \xrightarrow[\substack{CCl_4,\ 98\%\ yield}]{\substack{THF/NMP-Bu_4NBF_4 \\ /Bu_4NI-(Zn/SUS)}} \underset{\substack{\\ \textbf{344}}}{\underset{Me}{\overset{OH}{Cl_3C-C-CO_2Et}}} \qquad (7.153)
$$

The electroreduction of dibromodifluoromethane **346** (X = Y = Br) in the presence of diphenyldiselenide **347** and olefins **345** in a DMF-Et₄NOTs(0.1M)-(Al/Pt) system affords adducts of the *in situ* generated bromodifluoromethyl radical and the phenylselenyl group to olefins, giving **348** in 45~80% yields (eq. (7.154)) [361].

$$
\begin{array}{c}
\overset{\textbf{345}}{\diagup\!\!\!\diagdown}\!\!\!{\diagup}^{R} \\
\textbf{345}\quad + \\
CF_2XY \\
\textbf{346}\quad + \\
(SePh)_2 \\
\textbf{347}\ \ X = Br, \\
Y = Br,\ CO_2Et
\end{array}
\xrightarrow[\substack{45\sim80\%\ yields \\ R = \ O\diagup\!\!\!\diagdown\!\!\!\diagup}]{DMF-Et_4NOTs(0.1M)-(Al/Pt)}
\underset{\textbf{348}}{YF_2C\diagdown\!\!\!\diagup\overset{SePh}{\underset{R}{\diagdown}}}
\qquad (7.154)
$$

The electroreductive preparation of dichloroacetic acid from trichloroacetic acid in 75~84% yields has been performed in an aqueous NH₄OH(25%)-NH₄Cl-(Hg) system at −1.15 V (Ag/AgCl) (eq. (7.155)) [362]. Surfactant-intercalated clay films as a catalyst for

$$
Cl_3CCO_2H \xrightarrow[\substack{-1.15\ V\ (Ag/AgCl),\ 75\sim84\%\ yields}]{aq.\ NH_4OH(25\%)-NH_4Cl-(Hg)} Cl_2CHCO_2H \qquad (7.155)
$$

electroreduction of trichloroacetic acid have been devised [363, 364]. 2-Methoxy-2-trichloromethyl-4-methyltetrahydrofuran **351** is electrosynthesized by the reduction of 1,1,1-trichloro-4-methylpent-4-en-2-ol **349** in an MeOH-LiCl(0.1M)-(Hg) system in the presence of sodium bromide (eq. (7.156)) [365]. The reactions proceed first electroreduction of the trichloromethyl group followed by electrooxidative cyclization of **350** [365].

$$
\underset{\textbf{349}}{Cl_3C-\underset{\underset{OH}{|}}{CH}-CH_2-\underset{\underset{Me}{|}}{C}=CH_2} \xrightarrow[\text{59.4\% yield}]{\text{MeOH-LiCl(0.1 M)-(Hg)}} \underset{\textbf{350}}{Cl_2HC-\underset{\underset{OH}{|}}{CH}-CH_2-\underset{\underset{Me}{|}}{C}=CH_2}
$$

$$
Cl_2HC\diagdown\underset{O}{} \quad \textbf{351}
$$

(7.156)

The diastereoselective addition of the trichloromethyl and dichloro(methoxycarbonyl)methyl anions to α-branching aldehydes has also been investigated [366]. The addition of trichloromethyl anion to 2,2-dimethyl-1,3-dioxane-4-carboxaldehyde **352** is carried out in a DMF-Et$_4$NOTs-(C) system in the presence of a CCl$_4$/CHCl$_3$(1/10) mixture (or methyl trichloroacetate (Y = CO$_2$Me) to yield *syn*- and *anti*-isomers **353** in 52~89% yields (eq. (7.157)).

352 CHO + CCl$_3$Y $\xrightarrow[\text{52~89\% yields}]{\text{DMF-Et}_4\text{NOTs-(C)}}$ **353a** *syn* + **353b** *anti*

Y = Cl
Y = CO$_2$Me

(7.157)

The electroreductive cross-coupling of alkyl halides (R^1 − X) with carbon tetrachloride **354** (R = X = Cl) or substituted trichloromethanes **354** (R^2 = CO$_2$Me, CF$_3$, MeC$_6$H$_4$; X = Cl) has been performed in a THF/TMU-Bu$_4$NBF$_4$-(Zn/SUS) system to yield the cross-coupling products **355** in good yields (eq. (7.158)) [367].

$$
\underset{\textbf{354}}{R^1\!-\!X + R^2\!-\!CX_3} \xrightarrow[\text{30~70\% yields}]{\substack{\text{THF/TMU(1/1)-} \\ \text{Bu}_4\text{NBF}_4\text{-(Zn/SUS)}}} \underset{\textbf{355}}{R^1\!-\!CX_2\!-\!R^2} + 2X^-
$$

(7.158)

X = Br, Cl
R^1 = C$_{10}$H$_{21}$, allyl, PhCH$_2$, MeO$_2$CCH$_2$, EtO$_2$CHCCH(Et)
R^2 = Cl, Ph, CO$_2$Me, CF$_3$
TMU = tetramethyl urea

The electrosynthesis of diisopropyl α-chlorocyclopentylphosophonate **358** has been carried out by electrolysis of a mixture of trichlorophosphates **356** and dibromobutane in a DMF-Et$_4$NBr-(Mg/C) system in an undivided cell to give the ring compounds **357** (n = 4) in 72% yield (eq. (7.159)) [368]. Diisopropyl cyclopentylphosphonate **358** is prepared by addition of methanol in the electrolysis mixture after the complete consumption of the dibromide and the electrolysis is continued until the complete consumption of the mono chloride **357**, yielding **358** (n = 4) in 58% yield.

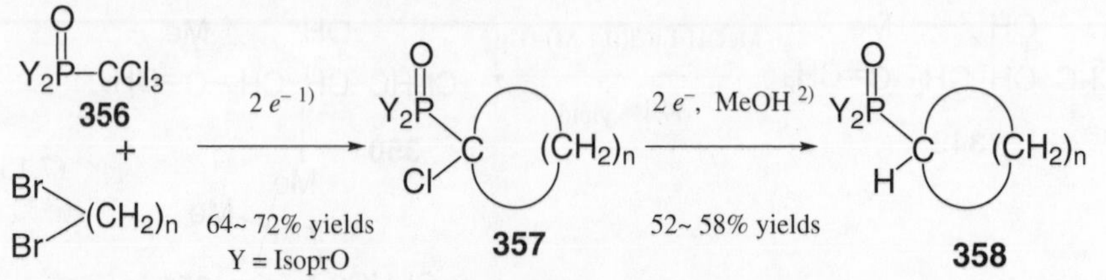

(7.159)

1) DMF-Et$_4$NBr-(Mg/C), undivided cell
2) DMF-Et$_4$NBr-(Mg/C), addition of MeOH after
   the complete consumption of **357**.

The electrolysis of bromoform in MeCN-Et$_4$NBr(0.22M)-(C/Fe) in the presence of chlorotrimethylsilane affords trimethyl(tribromomethyl)silane (11.3%) together with hexabromoethane (34.1%) [369].

The first successful attempt of the electrochemical preparation of difluoromethyl carbene has been carried out by electroreduction of dibromodifluoromethane in a CH$_2$Cl$_2$-Bu$_4$NBr(0.35M)-(Pt) system in the presence of 2-phenylpropene **359** in a divided cell at 0 °C to yield 1,1-difluoro-2-phenyl-2-methylcyclopropane **360** in 57% yield (eq. (7.160)) [370]. The trifluoromethylation of organic compounds by electrochemical methods has

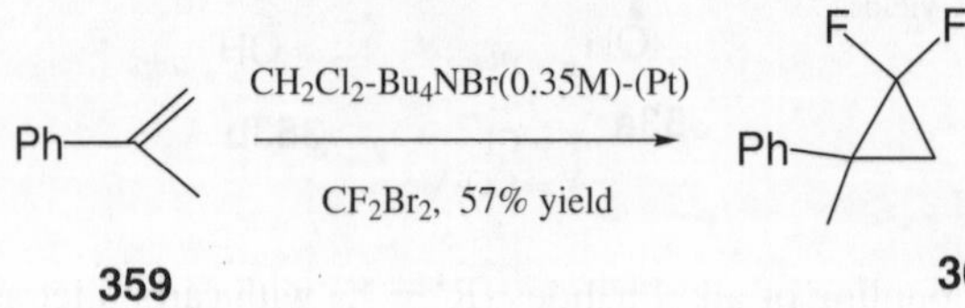

(7.160)

recently been reviewed [371]. The trifluoromethylation of aryl aldehydes has been attained in a DMF-Bu$_4$NBF$_4$/KBr-(Zn/SUS) system in which the use of zinc as a sacrificial anode is essential for the activation of the carbon-bromine bonds [372a]. The electroreductive coupling of p-chlorobenzaldehyde **361** with trifluorobromomethane **362** affords the corresponding trifluoromethylated alcohol **363** in 90% yield (eq. (7.161)).

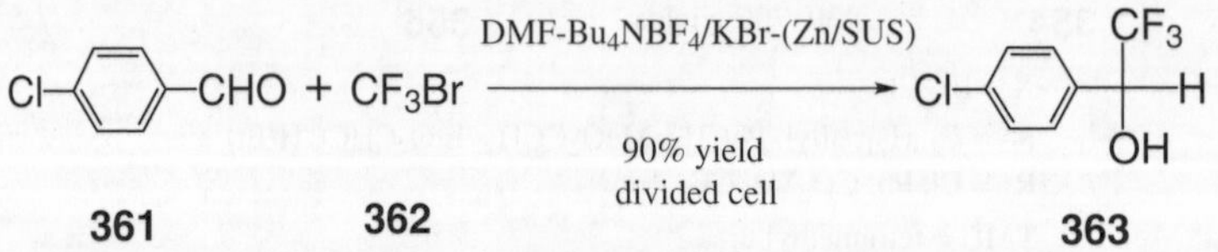

(7.161)

The electrosynthesis of trifluoroacetaldehyde, isolated as CF$_3$CH(OAc)$_2$ **365**, has been performed in a DMF-Bu$_4$NBr-(Al/Ni) system in the presence of AlCl$_3$ (or BF$_3$•Et$_2$O) under bubbling CF$_3$Br **362** and subsequent treatment of the electrolyte with an Ac$_2$O-HCl/Py system, giving **365** in 74~77.5% yields (eq. (7.162)) [372b].

$$CF_3Br \ + \ Me_2NCHO \ \xrightarrow[\substack{2) \ Ac_2O\text{-}HCl/Py \\ 74\sim77.5\% \ yields}]{\substack{1) \ DMF\text{-}Bu_4NBr\text{-}(Al/Ni)^*}} \ CF_3CH(OAc)_2 \qquad (7.162)$$

362   **364**   **365**

*Additives: Me$_3$SiCl, BF$_3$OEt$_2$, AlCl$_3$
Best Choice: Al-AlCl$_3$, Al-BF$_3$OEt$_2$, Al-Me$_3$SiCl

Quantitative analysis of the electroreduction kinetics of CF_3Br, CF_3I, $C_6F_{13}I$, $C_8F_{17}I$, and of radicals (R_F') reveals that $R_F X$ (X = Br, I) most likely undergoes a concerted electron-transfer bond-breaking reduction [373]. The R_F' radicals are strongly stabilized by interaction with the solvent. Unlike alkyl radicals, but similar to aryl radicals, R_F' radicals are good hydrogen atom scavengers.

The dissociative electron transfer between the anion radical of telephthalonitrile and trifluoromethyl bromide in the presence of styrene results in the formation of 1,1,1,4,4,4-hexafluoro-2,3-diphenylbutane in 27~40% yields [292].

Electrosynthesis of trifluoromethanesulfinic acid **366** from **362** and sulfur dioxide has been attained in a DMF-Bu$_4$NBF$_4$ system in an undivided cell with a sacrificial anode (Mg or Zn) in good yields (eq. (7.163)) [374]. A plausible mechanism for the abstraction of the bromine atom with SO_2^{-}' has been suggested [375].

$$CF_3\text{-}Br \ + \ SO_2 \ \xrightarrow[60\sim72\% \ yields]{DMF\text{-}Bu_4NBF_4\text{-}(Mg \ or \ Zn)} \ CF_3\text{-}SO_2^{-} \qquad (7.163)$$

362   **366**

The electroreductive silylation of carbon tetrachloride **367**, chloroform or methylene dichloride selectively proceeds in a DMF-Et$_4$NBF$_4$-(Zn/C) system to give the corresponding silylated products, *e.g.*, **368** from **367** in 94% yield (eq. (7.164)) [376],

$$CCl_4 \ + \ Me_3SiCl \ \xrightarrow[94\% \ yield]{DMF\text{-}Et_4NBF_4\text{-}(Zn/C)} \ Cl_3C\text{-}SiMe_3 \qquad (7.164)$$

367   **368**

(Trifluoromethyl)trimethylsilane **369** as a nucleophilic trifluoromethylating agent has been electrochemically synthesized in an HMPA/PhOMe-Bu$_4$NPF$_6$-(Al/Ni) system (eq. (7.165)) [377]. The electrogenerated CF_3^{-} species can be readily trapped by an effective

$$CF_3\text{-}Br \ + \ Me_3SiCl \ \xrightarrow[73\% \ yield]{HMPA/PhOMe\text{-}Bu_4NPF_6\text{-}(Al/Ni)} \ F_3C\text{-}SiMe_3 \qquad (7.165)$$

362   **369**

electrophile (Me$_3$SiCl) to give **369** in 73% yield. Electrosynthesis of *gem*-dichloroalkenes has been attained by electrochemical Wittig reaction [378a]. The two-electron reduction of trichloromethyl tris(dimethylamino)phosphonium tetrafluoroborate **370** leads to dichloromethylene tris(dimethylamino)phosphorane **371**, which is reacted with aldehydes to give gem-dichloroalkenes in 63~93% yields (eq. (7.166)). The trifluoromethylation of

$$(Me_2N)_3\overset{+}{P}CCl_3 \quad \xrightarrow{\text{MeCN-Bu}_4\text{NBr-(Hg)}} \quad (Me_2N)_3P=CCl_2 \qquad (7.166)$$
$$BF_4^-$$

371

370 R^1 = alkyl, aryl

R^2 = H. Me, Ph, CF_3,

63~93% yields $\Big\downarrow$ $R^1R^2C=O$

$$R^1R^2C=CCl_2$$

aldehydes **333** (R^1 = alkyl, aryl) and ketones **373** has been performed in a DMF-Bu₄NBr-(Zn) system using a sacrificial zinc anode, yielding the corresponding trifluoromethylated alcohols **372a** and **372b**, respectively, together with CF₃ZnBr and (CF₃)₂Zn (eqs. (7.167a) and 7.167b)) [378b, 379]. The addition of tetramethylethylenediamine allows improvement in the yields of the alcohols derived from ketones. The electroreduction of trifluoromethylarenes completely proceeds to give the corresponding toluene derivatives

333

$$R^1\!\!-\!\!CHO, \text{ DMF-Bu}_4\text{NBr-(Zn/Ni)}$$

$$CF_3\!\!-\!\!\underset{\underset{\textstyle OH}{|}}{CH}\!\!-\!\!R^1 \quad \text{-----(a)}$$

50~95% yields   **372a**

CF₃–Br

362

$$\text{DMF-Bu}_4\text{NBr-(Zn/Ni)}$$ (7.167)

$$R^1\!\!-\!\!CO\!\!-\!\!R^2, \text{ 5~57\% yields}$$

$$CF_3\!\!-\!\!\underset{\underset{\textstyle OH}{|}}{C}\!\!\overset{R^1}{\underset{R^2}{<}} \quad \text{-----(b)}$$

373   **372b**

R^1, R^2 = alkyl, aryl

(58~60% yields) [380a]. The electroreductive coupling of trifluoromethylarenes **374** with carbon dioxide, acetone, and DMF, giving **375**, **376**, and **377**, has been achieved in good yields by electrolysis in a DMF-Bu₄NBr-(Mg/SUS) system containing trifluoromethylarenes (eqs. (7.168, 7.169, and 7.170)) [380b]. The preparative scale electroreduction of

$$CO_2$$
$$Ar\!\!-\!\!CF_2\!\!-\!\!CO_2Me \qquad (7.168)$$
65~70% yields   **375**

$$Me_2CO$$

Ar–CX₃ $Ar\!\!-\!\!CF_2\!\!-\!\!CMe_2OH \qquad (7.169)$
40~80% yields   **376**

374

$$DMF$$
$$Ar\!\!-\!\!CF_2\!\!-\!\!CH(OAc)_2 \qquad (7.170)$$
58~62% yields   **377**

(X = F, Ar = Ph, p-FC₆H₄)

trichloromethylbenzene **374** in a DMF-Et₄NClO₄-(Hg) system at −1.3 V (SCE) affords benzal chloride **378** in 85% yield (eq. (7.171a)), whereas at −0.7 V, the major product is the dimer, 1,2-diphenyl-1,1,2,2-tetrachloroethane (63%) (eq. (7.171b)) [381a].

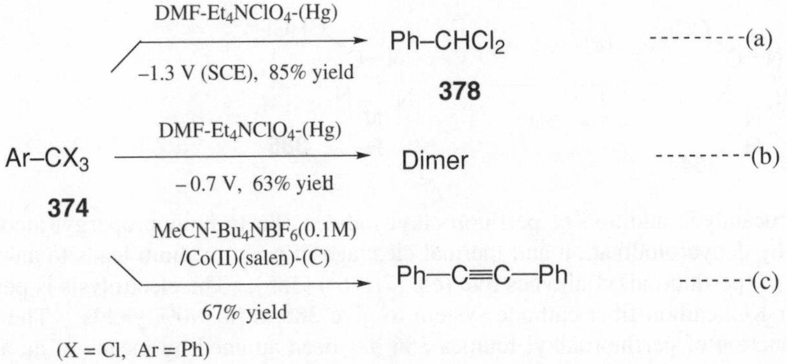

$$Ar-CX_3 \quad \begin{cases} \xrightarrow[\substack{DMF-Et_4NClO_4-(Hg)\\ -1.3\ V\ (SCE),\ 85\%\ yield}]{} & Ph-CHCl_2 \qquad \text{--------(a)} \\[2mm] \textbf{378} \\[4mm] \xrightarrow[\substack{DMF-Et_4NClO_4-(Hg)\\ -0.7\ V,\ 63\%\ yield}]{} & \text{Dimer} \qquad \text{--------(b)} \\[4mm] \xrightarrow[\substack{MeCN-Bu_4NBF_6(0.1M)\\ /Co(II)(salen)-(C)\\ 67\%\ yield}]{} & Ph-C\equiv C-Ph \qquad \text{------(c)} \end{cases} \qquad (7.171)$$

374

$(X = Cl,\ Ar = Ph)$

Electroreduction of a mixture of cobalt(II)(salen) and benzotrichloride **374** results in stilbene as a major product (eq. (7.171c)) [381b]. Electroreduction of *exo*-5-acetyl-*endo*-6-trichloromethyl-bicyclo[2.2.1]hept-2-ene **379** affords either the corresponding dichloromethyl or monochloromethyl derivatives, depending on the selection of supporting electrolyte [382]. For example, the electrolysis of **379** in a DMF-Bu$_4$NBr-(Hg) system gives the dichlorinated product **380** in 80% yield (eq. (7.172)), whereas in an MeOH-Me$_4$NCl-(Hg) system yields the monochlorinated product **381** in 94% yield (eq. (7.173)) [382]. Electroreduction of trichloromethylbenzene in a DMF-Bu$_4$NBF$_4$-(Hg) system in the

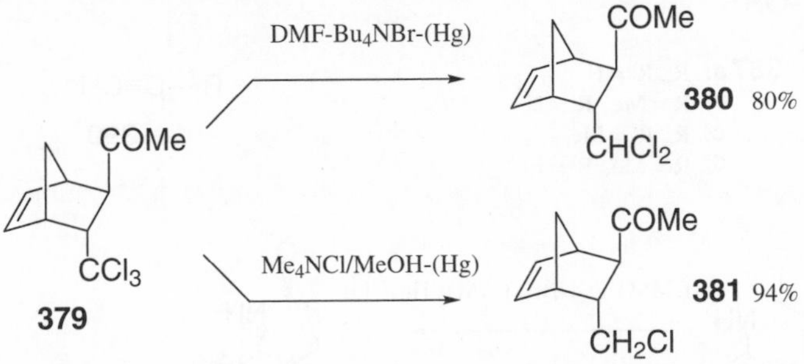

presence of acetic anhydride gives 1,2-diacetoxy-1-phenylpropene as a major product [383a]. The electroreductive coupling of trichloromethylbenzene **374** with ketones **382** proceeds in a THF-LiClO$_4$(0.5M)-(Pt/Hg) system in a divided cell to give the corresponding α,β-enones **383** in 48~65% yields together with dehalogenated product (16~29%) (eq. (7.174)) [383b]. The electroreduction of 5-trichloromethyltetrazole **384** in an aqueous HCl(2M)-(Pb) system yields 5-dichloromethyltetrazole **385** in 70% yield (eq. (7.175)) [384)].

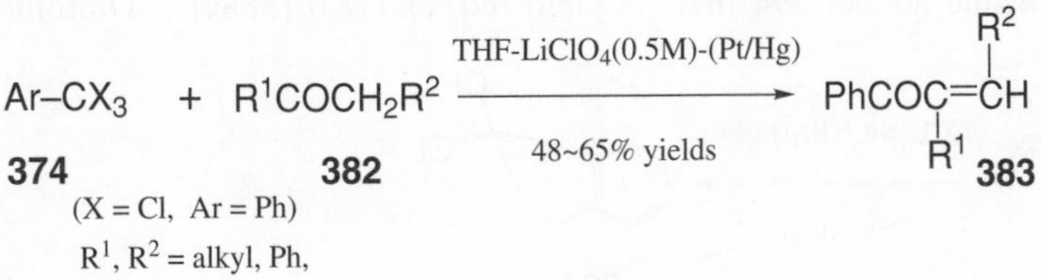

$$Ar-CX_3 \quad + \quad R^1COCH_2R^2 \quad \xrightarrow[\substack{THF-LiClO_4(0.5M)-(Pt/Hg)\\ 48\sim65\%\ yields}]{} \quad PhCOC\overset{R^2}{\underset{R^1}{=}}CH \qquad (7.174)$$

374   **382**   **383**

$(X = Cl,\ Ar = Ph)$

$R^1, R^2 = alkyl, Ph,$

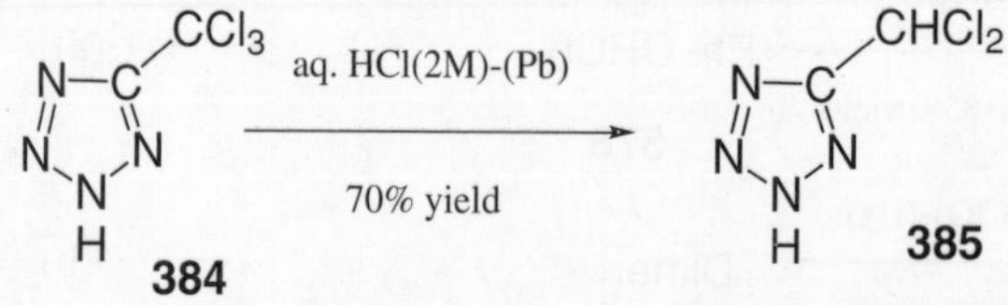

$$\text{(7.175)}$$

Electrocatalytic addition of perfluoroalkyl iodides **386** to homopropargyl alcohol **387** followed by dehydroiodination and thermal cleavage in basic medium leads to an efficient synthesis of (perfluoroalkyl)alkynes **390** (eq. (7.176)) [385]. The electrolysis is performed in a water-KCl-carbon fiber cathode system to give **388** in 80~90% yields. The indirect electroreduction of perfluoroalkyl iodides **386** has been attained by means of an aromatic anion mediator [386]. The electrolysis of a mixture of the halide **386** and uracil **391** is carried out in a DMSO-Et$_4$NBF$_4$(0.1M)-(Hg) system in the presence of nitrobenzene as a mediator to give the S$_{RN}$1 substituted product **392** in 65% yield (eq. (7.177)).

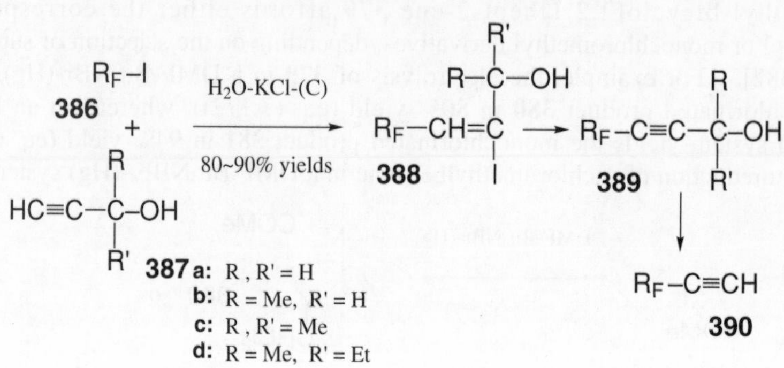

$$\text{(7.176)}$$

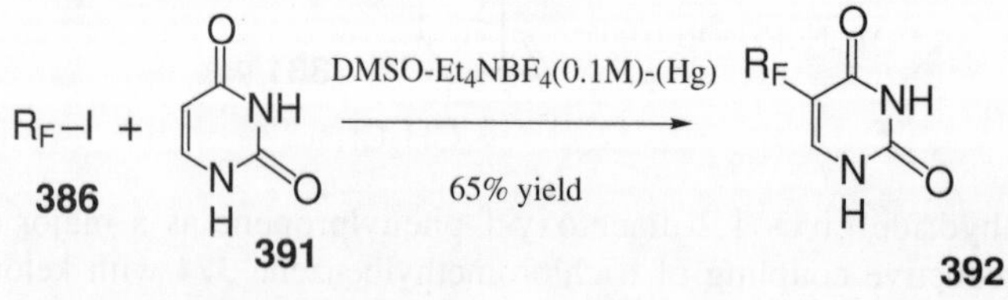

387 **a:** R, R' = H
b: R = Me, R' = H
c: R, R' = Me
d: R = Me, R' = Et

$$\text{(7.177)}$$

R$_F$ = perfluoroalkyl

The cathodic generation of dichlorocarbene has been improved by the reduction of carbon tetrachloride and chloroform in a CHCl$_3$-Bu$_3$NBr-(Pb) system at −5 °C [387]. The electrochemically generated dichlorocarbene reacts with indene **393** to give the 1,1-dichlorocyclopropane adduct **394** in 75% yield (eq. (7.178)) [388a]. Dibromocarbene

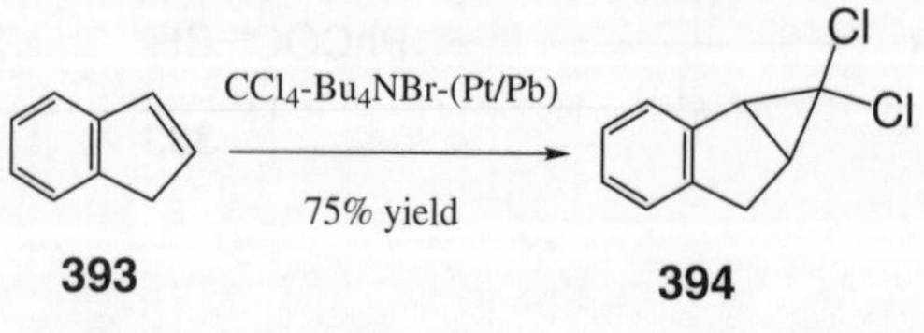

$$\text{(7.178)}$$

produced by the electroreduction of carbon tetrabromide has been shown to react with cyclohexene to give dibromocyclopropane derivative in 88% yield [388b]. The reaction of the electrogenerated dichlorocarbene with methylindoles **395** has been achieved in a $CHCl_3$-Bu_4NBr(0.1M)-(Pb) system to give a mixture of **396** and **397** in the presence of styrene as a bromine scavenger at the anode (eq. (7.179)) [389]. The electroreductive

(7.179)

dechlorination of 1,1,2-trichlorotrifluoroethane **398** leading to chlorotrifluoroethene **399** has been attained in an aqueous MeOH(50~90%)-NH_4Cl-(Zn/Cu) system in 87~90% yields (eq. (7.180)) [390, 391]. The electroreduction of 1,1,2-trichlorotrifluoroethane (TCTFE) **398** in an aqueous MeOH(90%)-NH_4Cl(0.2M)-(Zn) system has been attempted [392]. At 25 °C, when the current density is greater than 300 A/m^2, the indirect electroreduction of TCTFE **398** predominantly proceeds by zinc mediation.

$$Cl_2FC-CF_2Cl \xrightarrow[\text{87~90% yields}]{\text{aq. MeOH(50~90%)-NH}_4\text{Cl-(Zn/Cu)}} ClFC=CF_2 \qquad (7.180)$$

398   **399**

The electroreductive cross-coupling of 1,1,1-trichloro-2,2,2-trifluoroethane **400** with benzaldehyde **401** proceeds in a DMF-Et_4NOTs-(Pt) system in the presence of chlorotrimethylsilane to give the cross-coupling product **402** in 87% yield (eq. (7.181)) [393]. The coupling product **402** can be converted into a variety of compounds, **403 ~ 412**, by electrolysis under the conditions described in equations (7.182~7.188).

$$CCl_3CF_3 + Ph-CHO \xrightarrow[\text{87% yield}]{\text{DMF-Et}_4\text{NOTs/Me}_3\text{SiCl-(Pt)}} \underset{\substack{| \\ OH}}{Ph-CHCCl_2CF_3} \qquad (7.181)$$

400   **401**   **402**

1) DHP, TsOH
/benzene
2) tert-BuOK/THF
3) 10% H₂SO₄

DMF-Et₄NOTs/MsOH

Pb cathode

$$Ph \underset{Cl}{\overset{OH \quad F}{=}} F \qquad \textbf{403} >95\% \qquad (7.182)$$

MeOH-NH₄NO₃

Pb cathode

$$Ph \overset{OH}{\underset{Cl}{\bigwedge}} CF_3 \; + \; \textbf{403} \; 50\% \qquad (7.183)$$

404 35%

$$Ph \underset{\textbf{402}}{\overset{OH}{\bigwedge}} CCl_2CF_3$$

MsCl, Et₃N
/CH₂Cl₂

MeOH-Me₄NCl/Et₃N

Pb cathode

$$Ph \underset{\textbf{405} >95\%}{\overset{OH \quad F}{=}} F \qquad (7.184)$$

$$Ph \underset{\textbf{406} \; 96\%}{\overset{OMs}{\bigwedge}} CCl_2CF_3$$

MeOH-NH₄NO₃

Pb cathode

$$Ph \underset{Cl}{\bigwedge} CF_3 \qquad \textbf{407} \; 82\% \qquad (7.185)$$

MeOH-Me₄NCl

Pb cathode

$$Ph \bigwedge CF_3 + Ph \bigwedge CF_3 \qquad (7.186)$$

408   **409**

F/mol	408, %	409, %
6.5	50	19
11	trace	51

DMF-Et₄NOTs/MsOH

Pb cathode

$$Ph \overset{O}{\bigwedge} CF_3 \qquad \textbf{411} \; 90\% \qquad (7.187)$$

MeOH-NH₄NO₃

Pb cathode

$$Ph \overset{OH}{\bigwedge} CF_3 \qquad (7.188)$$

412 55%

$$Ph \underset{\underset{Cl}{|}}{\overset{O}{\bigwedge}} CF_3$$

410 92%

The electroreductive, nickel-catalyzed Reformatsky reaction with methyl chlorodifluoroacetate **413** has been carried out in a CH₂Cl₂/DMF-ZnBr₂-(Zn/C) system in the presence of NiBr₂(bpy) complexes [394]. The coupling reaction of **413** with **401** affords the adduct **414** in 79% yield (eq. (7.189)). The electroreductive coupling of ethyl trichloroacetate

$$\begin{array}{c} \text{Ph–CHO} \\ + \; \textbf{401} \\ \\ \text{ClCF}_2\text{CO}_2\text{Me} \\ \textbf{413} \end{array} \xrightarrow[\text{79\% yield}]{\text{CH}_2\text{Cl}_2/\text{DMF-ZnBr}_2\text{-(Zn/C)}} \begin{array}{c} \text{Ph–CH–CF}_2\text{CO}_2\text{Me} \\ \underset{\text{OH}}{|} \\ \textbf{414} \end{array} \qquad (7.189)$$

with acrylonitrile **415** proceeds in a DMF-LiCl(0.5M)-(Hg) system at −0.84 V (SCE) to afford the cyclopropane derivative **417** in 80% current yield (eq. (7.190)) [93].

$$\begin{array}{c} \text{CH}_2\text{=CH–CN} \\ + \; \textbf{415} \\ \\ \text{Cl}_3\text{C—CO}_2\text{Et} \\ \textbf{416} \end{array} \xrightarrow[\text{80\% current yield}]{\text{DMF-LiCl(0.5M)-(Hg)}} \begin{array}{c} \text{Cl} \\ \text{EtO}_2\text{C} \overset{\triangle}{\longleftarrow} \text{CN} \\ \textbf{417} \end{array} \qquad (7.190)$$

The electroreductive cyclization of the monocyclic adduct **418** is carried out in a DMF-Bu$_4$NBr-(Zn/SUS) system to give the bicylo compound **419** in 69% yield (eq. (7.191)) [395].

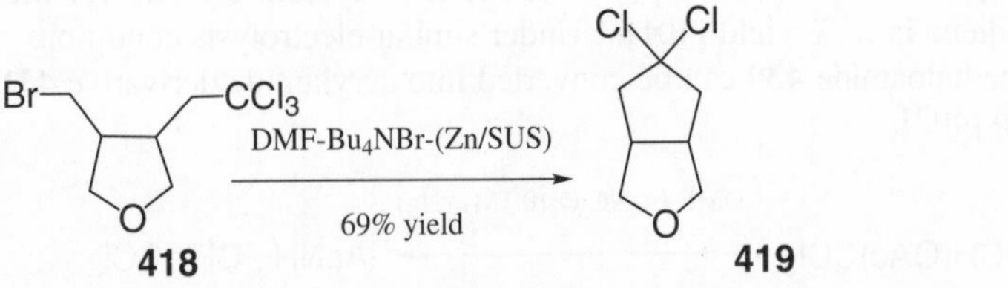

Br—⟨ ⟩—CCl$_3$ $\xrightarrow[\text{69\% yield}]{\text{DMF-Bu}_4\text{NBr-(Zn/SUS)}}$

418   **419**   (7.191)

The addition of electrogenerated dichloro(methoxycarbonyl)methyl anion, derived from **421**, to the α-aminoaldehyde **420** has been performed in a DMF-Et$_4$NOTs-(C) system in a divided cell to give the corresponding amino alcohol **422** in 44% yield (eq. (7.192)) [396].

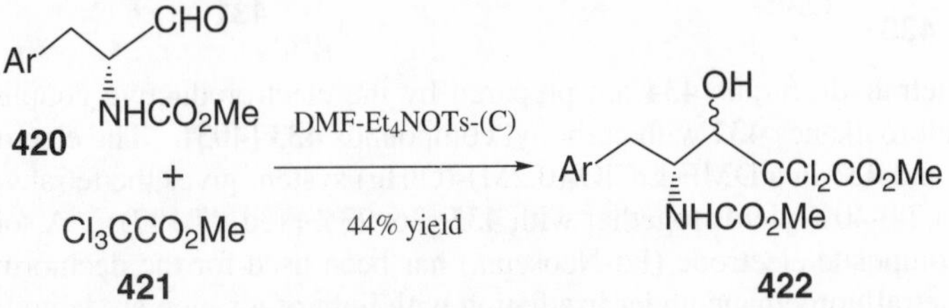

Ar—CHO
    NHCO$_2$Me
420
  +
Cl$_3$CCO$_2$Me
421

$\xrightarrow[\text{44\% yield}]{\text{DMF-Et}_4\text{NOTs-(C)}}$

OH
Ar—   CCl$_2$CO$_2$Me
    NHCO$_2$Me
422   (7.192)

The electroreductive aromatization of 4-methyl-4-trichloromethyl-p-quinone-(1)-arylimines **423** has been attained by electrolysis in an MeCN-Et$_4$NOTs-(C) system in the presence of acetic acid [397]. The imines undergo the cathodic elimination of trichloromethyl group to give the corresponding N-tolylarylamines **424** in 78~89% yields (eq. (7.193)).

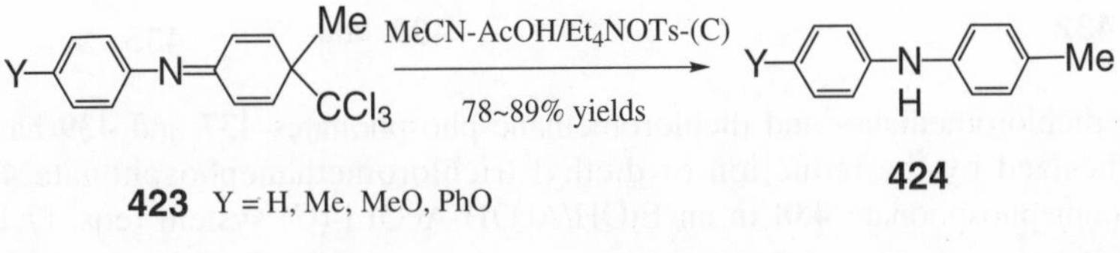

Y—⟨ ⟩—N=⟨ ⟩ Me CCl$_3$ $\xrightarrow[\text{78\textasciitilde89\% yields}]{\text{MeCN-AcOH/Et}_4\text{NOTs-(C)}}$ Y—⟨ ⟩—N(H)—⟨ ⟩—Me

423 Y = H, Me, MeO, PhO   **424**   (7.193)

The electroreduction of o-trichloroacetylamide **425** in an MeCN-LiClO$_4$(0.02M)-(Hg) system at −1.1 V (SCE) undergoes two kinds of dehalogenation reactions to give a mixture of dehalogenated amide **426** (38~46%) together with 2-aryl-4H-3,1-benzoxazin-4-one **427** (40~45%) (eq. (7.194)) [398].

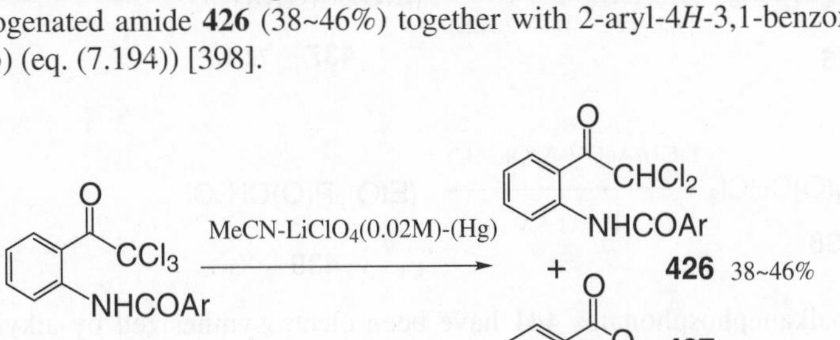

⟨ ⟩ C(=O)CCl$_3$ NHCOAr $\xrightarrow{\text{MeCN-LiClO}_4\text{(0.02M)-(Hg)}}$

425

⟨ ⟩ C(=O)CHCl$_2$ NHCOAr   **426** 38~46%
  +
⟨ ⟩ O N—Ar   **427** 40~45%   (7.194)

The electroreduction of N-(2,2,2-trichloro-1-acetoxyethyl)acetamide **428** in a DMF-Et_4NClO_4(0.1M)-(Hg) system at −1.5 V (SCE) affords the corresponding N-dichloroviny-lacetamide **429** in 94% yield (eq. (7.195)) [399, 400]. The electroreduction of 4-methyl-1,1,1-trichloro-4-penten-2-ol in a IsoprOH-TsOH-(Pb) system affords 1,1-dichloro-4-methyl-1,4-pentadiene in 82% yield [401]. Under similar electrolysis conditions, the electroreduction of the haloamide **430** can be converted into acrylamide derivative **431** in good yield (eq. (7.196)) [402].

$$AcNH\text{—}CH(OAc)CCl_3 \xrightarrow[\text{−1.5 V (SCE), 94\% yield}]{\text{DMF-Et}_4\text{NClO}_4\text{(0.1M)-(Hg)}} AcNH\text{—}CH\text{=}CCl_2 \qquad (7.195)$$

$$\underset{\textbf{428}}{} \qquad\qquad\qquad\qquad\qquad \underset{\textbf{429}}{}$$

$$\underset{\substack{| \\ OAc \\ \textbf{430}}}{Br(CH_2)_2CONHCHCCl_3} \xrightarrow[\text{70\% yield}]{\text{DMF-Et}_4\text{NClO}_4\text{-(Hg)}} \underset{\textbf{431}}{CH_2\text{=}CHCONHCH\text{=}CCl_2} \quad (7.196)$$

3,3-Dichlorotetrahydrofurans **434** are prepared by the electroreductive coupling of 3-bromo-1,1,1-trichloroalkanes **432** with carbonyl compounds **433** [403]. The electrolysis of a mixture of **432** and **433** in a DMF-LiClO$_4$(0.2M)-(C/Hg) system gives the tetrahydrofuran derivatives **434** in 24~40% yields together with **435** (26~73%) (eq. (7.197)). A solid polymer electrolyte composite electrode (Pd-Neosepta) has been used for the dechlorination of 2-chloro-1,1,1,2-tetrafluoroethane under irradiation with light of a xenon arc lamp [404].

$$(7.197)$$

432   **434** 40%   **435** 26%

Diethyl dichloromethane- and dichloromethane-phosphonates **437** and **439** have been electrosynthesized by the reduction of diethyl trichloromethanephosphonate **436** and dichloromethanephosphonate **438** in an EtOH/AcOH-AcOLi-(C) system (eqs. (7.198 and 7.199)) [405].

$$(EtO_2)P(O)CCl_3 \xrightarrow[\text{− Cl, H}^+, \text{ −1.0 V (SCE)}]{\text{EtOH/AcOH-AcOLi-(C)}} (EtO)_2P(O)CHCl_2 \qquad (7.198)$$

$$\underset{\textbf{436}}{} \qquad\qquad\qquad\qquad\qquad \underset{\textbf{437}\ \ 72\,\%}{}$$

$$(EtO)P_2(O)CHCl_2 \xrightarrow[\text{− Cl, H}^+, \text{ −1.5 V (SCE)}]{\text{EtOH/AcOH-AcOLi-(C)}} (EtO)_2P(O)CH_2Cl \qquad (7.199)$$

$$\underset{\textbf{438}}{} \qquad\qquad\qquad\qquad\qquad\quad \underset{\textbf{439}\ \ 85\,\%}{}$$

1,1-Dichloroalkanephosphonates **441** have been electrosynthesized by alkylation of **440** under electrolysis in a DMF-Et$_4$NBr(0.02M)-(Mg/C) system in the presence of alkylating reagents (eq. (7.200)) [406].

$$(i\text{-PrO})_2\underset{\underset{O}{\|}}{P}CCl_3 \xrightarrow[\text{65~78\% yields}]{\text{DMF-Et}_4\text{NBr(0.02M)-(Mg/C)}} (i\text{-PrO})_2\underset{\underset{O}{\|}}{P}CCl_2R \qquad (7.200)$$

440 R = Pr, isobutyl, C_6H_{13}, allyl,
benzyl, TMS, PhOCH$_2$

441

The electroreductive Wittig-Horner reaction with an anion of diethyl (dichloro-methane)phosphonate **442** has been studied [407].   The reaction of diethyl trichloromethanephosphonate **442** and benzaldehyde **401** in a DMF-Et$_4$NCl-(Al/C) system at −0.85 V (E$_{1/2}$, SCE) gives 2,2-dichloro-1-phenylethane **444** in 70% yield together with **443** (eq. (7.201)) [407].   The electroreduction of **442** undergoes the two-electron reduction in a stepwise manner in aqueous media, leading to diethyl dichloromethanephosphonate anion [408].

$$(EtO)_2P(O)-CCl_3 \xrightarrow[\text{70\% yield}]{\overset{\text{PhCHO}}{\overset{\textbf{401}}{}}} (EtO)_2P(O)O^- + PhCH{=}CCl_2 \qquad (7.201)$$

442   **443**   **444**

The electroreductive removal of the β-haloethyl moiety as a protecting group of β-haloethane diethanephosphates **445** has been attained by electrolysis in an MeCN-LiClO$_4$-(Hg) system (eq. (7.202)) [409].

$$(PhO)_2-\underset{\underset{O}{\|}}{P}-OCH_2-\underset{\underset{X}{|}}{\overset{\overset{X}{|}}{C}}-Y \xrightarrow[\text{60~96\% yields}]{\text{MeCN-LiClO}_4\text{-(Hg)}} (PhO)_2-\underset{\underset{O}{\|}}{P}-OH \qquad (7.202)$$

445   X = H, Cl, Br
Y = H, Cl, Br, I

446

Electrochemically induced free radical addition of 1-(N,N-dimethylamino)-4-(chlorodi-fluoroacetyl)-2-methylnaphthalene **447** with butyl vinyl ether **448** followed by intramolecu-lar cyclization of the resulting α,α-difluoroalkyl radical affords the cyclized product **449** in 50% yield (eq. (7.203)) [410a].   The similar reaction with 2,3-dihydrofuran has been at-tained in 32~71% yields [410b].

$$(7.203)$$

447   **448**   **449**

Electrochemically induced nucleophilic substitution of perfluoroalkyl halides under-goes a slightly modified version of the classical S$_{RN}$1 mechanism where the reaction is trig-gered by dissociative electron transfer.   Direct electroinduction is possible with the iodides but not with the bromides because the reduction potentials of the substrate and the perfluo-roalkyl radicals are fairly close to each other in the latter case.   Thiolates react at the sulfur atom whereas phenoxide as well as imidazolate ions react at ring carbons rather than at the negatively charged heteroatom.   For instance, the nucleophilic substitution of 4-nitroimidazole anion **451** with perfluoroalkyl iodide **450** has been performed in a DMF-Bu$_4$NBF$_4$-(C) system to give the product **452** in 65% yield (eq. (7.204)) [373, 411, 412].

$$
\text{C}_6\text{F}_{13}\text{–I} \quad + \quad \underset{\textbf{451}}{\overset{\text{O}_2\text{N}}{\text{H}}} \quad \xrightarrow[\text{51\% yield}]{\text{MeCN-Bu}_4\text{NBF}_4(0.1\text{M})\text{-(C)}} \quad \underset{\textbf{452}}{\overset{\text{O}_2\text{N}}{\text{C}_6\text{F}_{13}}} \tag{7.204}
$$

The preparative electroreduction of polyfluoroalkyl iodides in the presence of lithium *p*-chlorothiophenolate yields the corresponding polyfluoroalkyl sulfides in 48~72% yields [413].   The electrochemically induced substitution of 4-bromobenzophenone by cyanide ions proceeds in a non-catalytic fashion [414].   Tetrafluoroethane-1,2-disulfonic acid has been found to have the best combination of properties for a fuel cell electrolyte [415].

7.4   Electroreduction of α-Halogeno Carbonyl Compounds and Related Halides

7.4.1   Reductive Dehalogenation of α-Halogeno Carbonyl Compounds

The electroreductive conversion of dichloroacetic acid **453** in an aqueous HCl(0.125M)-(C) system in the presence of lead ions proceeds to give monochloroacetic acid **454** (eq. (7.205)) [416].   Lead, copper, and compact pyrographite are shown to be the most applicable cathode materials for the dechlorination of α,α-dichloropropionic acid to monochloropropionic acid [417].   A double role of lead ions has been mentioned: they form a multilayer deposit on the graphite electrode, inhibiting the undesired hydrogen evolution and accelerating the dichloroacetic acid reduction.   The selectivity of the electroreduction of dichloroacetic acid increases with rising temperature [418].

$$
\underset{\textbf{453}}{\text{Cl}_2\text{CHCO}_2\text{H}} \quad \xrightarrow[\text{Pb}^{2+}]{\text{aq. HCl(0.125M)-(C)}} \quad \underset{\textbf{454}}{\text{ClCH}_2\text{CO}_2\text{H}} \tag{7.205}
$$

The electroreductive removal of chlorine atom on 2-chloro-4,4-dimethyl-2-cyclohexenone **455** has been carried out in an MeCN-Bu$_4$NBF$_4$(0.5M)-(Hg) system to give the dechlorinated cyclohexenone **456** in 66% yield (eq. (7.206)) [419].

$$
\underset{\textbf{455}}{\text{[structure with Cl]}} \quad \xrightarrow[\text{66\% yield}]{\text{MeCN-Bu}_4\text{NBF}_4(0.5\text{M})\text{-(Hg)}} \quad \underset{\textbf{456}}{\text{[structure]}} \tag{7.206}
$$

The electroreduction of 2-bromo-*N*-phenylisobutyramide **457** in a DMF-Et$_4$NClO$_4$-(Hg) system in the presence of phenol as a proton donor affords 2-methyl-*N*-phenylpropanamide **458** in a quantitative yield (eq. (7.207)) [420, 421].   A reduction mechanism by which electrogenerated carbanions derived from 2-bromoisobutyramides **457** and their self-protonation by the starting material gives rise to producing debrominated product **458** has been proposed [422].

$$
\underset{\textbf{457}}{\text{(Br, NHPh)}} \xrightarrow[\text{PhOH, 100\% yield}]{\text{DMF-Et}_4\text{NClO}_4\text{-(Hg)}} \underset{\textbf{458}}{\text{(H, NHPh)}} \tag{7.207}
$$

The electroreduction of the diastereoisomers of α,α'-dibromosuccinic acids has been investigated at a mercury cathode over the pH range of 0.4 to 8.5.   Fumaric acid is produced quantitatively from the *meso* form at all pH levels.   The maleic acid is formed by the reduction of the *dl*-isomer at pH 4.0 [423].

Electroreductive defluorination proceeds by the electrolysis of α,α,α-trifluoroac-etanilide **459** in an EtOH-Et$_4$NCl-(Hg) system to give monofluorocetamilide **460** in 65% yield (eq. (7.208)) [424].

$$
\underset{\textbf{459}}{\text{NHCOCF}_3} \xrightarrow[\text{65\% yield}]{\text{EtOH-Et}_4\text{NCl-(Hg)}} \underset{\textbf{460}}{\text{NHCOCH}_2\text{F}} \tag{7.208}
$$

The selective electrocatalytic reduction of hexachloroacetone has been attained using a viologen polymer modified electrode in aqueous media [425].   The corresponding pentachloro derivative can be obtained in 92% yield in an aqueous LiClO$_4$(0.1M)-(C) system in which the modification of the cathode material has been performed on growing poly(pyrrole)-viologen films by electroreduction of the methylvilogen monomer in an MeCN-Bu$_4$NClO$_4$(0.1M) system on a glassy carbon disk electrode.

Electroreductive cleavage of the carbon-bromine bond of *p*-nitrophenacyl bromide in an EtOH/H$_2$O(1/1)-Et$_4$NBr-(Hg) system has been attempted [426].   The electroreduction of 9-anthryl bromoacetate in a DMF-Et$_4$NClO$_4$-(C) system at -1.1 V (SCE) affords 9,10-anthraquinone as a major product due to the subsequent oxidation of an intermediary radical species by dioxygen [427].   The electroreductive debromination of 2,2-dibromo-1,3-diphenyl-propane-1,3-dione **461** in a DMF-LiClO$_4$-(Hg or Pt) system affords the corresponding mono-bromo compound **462** in a quantitative yield (eq. (7.209)) [428].

$$
\underset{\textbf{461}}{\underset{\text{Br}}{\overset{\text{Br}}{\text{Ph-CO-C-CO-Ph}}}} \xrightarrow[\text{quantitative yield}]{\text{DMF-LiClO}_4\text{-(Hg or Pt)}} \underset{\textbf{462}}{\underset{\text{Br}}{\overset{\text{H}}{\text{Ph-CO-C-CO-Ph}}}} \tag{7.209}
$$

The charge consumption of a platinum electrode is double of that on a mercury electrode.   The electroreduction of α,α'-dibromo ketones in acetic acid affords α-acetoxy ketones as major products when the α-carbons bear at least three alkyl substituents.   The electroreductive mono-acetoxylation of α,α'-dibromo ketones **463** proceeds in an AcOH-AcONa(0.1M)-(Hg) system to give the products **464** and **465** (64/15 ratio) in 79% yield (eq. (7.210)) [56, 429].   Electroreduction of α, α'-dichloroketones has been studied in a

DMF-Et$_4$NClO$_4$-(Hg or C) system [430].   The electroreduction of alicyclic α-haloketones in a DMF-Et$_4$NClO$_4$(0.1M)-(Hg) system in the presence of phenol as a proton donor affords the corresponding dehalogenated ketones in 84~92% yields [431].

$$
\underset{\substack{\textbf{463}}}{\underset{\text{R, R}^1\text{, R}^2 = \text{H, Me, Ph}}{\underset{\substack{\text{Br} \quad \text{Br}}}{\text{Ph}\diagdown\underset{\text{R}}{\diagup}\overset{\text{O}}{\diagdown}\underset{\text{R}^2}{\diagup}\text{R}^1}}} \xrightarrow{\text{AcOH-AcONa(0.1M)-(Hg)}} \underset{\substack{\textbf{464} \ 64\%}}{\underset{\substack{\text{OAc}}}{\text{Ph}\diagdown\underset{\text{R}}{\diagup}\overset{\text{O}}{\diagdown}\underset{\text{H}}{\underset{\text{R}^2}{\diagup}}\text{R}^1}} + \underset{\substack{\textbf{465} \ 15\%}}{\underset{\substack{\text{R} \quad \text{OAc}}}{\text{Ph}\diagdown\overset{\text{O}}{\diagdown}\text{R}^1\,\text{R}^2}}
$$

(7.210)

The electroreduction of bis(α-bromocyclopropyl) ketone **466** in a DMF-Bu$_4$NBr(0.1M)-(Hg) system at –0.80 V (SCE) affords α-bromocyclopropyl cyclopropyl ketone **467** in 90% yield (eq. (7.211)) [432].   A DMF/AcOH(9/1)-AcONa system gives a similar result.

$$
\underset{\substack{\textbf{466}}}{\underset{\substack{\text{Br} \quad \text{Br}}}{\diagup\!\!\triangle\!\!\overset{\text{O}}{\diagdown}\!\!\triangle}} \xrightarrow[\text{90\% yield}]{\text{DMF-Bu}_4\text{NBr(0.1M)-(Hg)}} \underset{\substack{\textbf{467}}}{\underset{\substack{\text{Br} \quad \text{H}}}{\triangle\overset{\text{O}}{\diagdown}\triangle}}
$$

(7.211)

The electroreductive cyclization of ethyl α,γ-dibromobutyrate **468** in a DMF-Et$_4$NClO$_4$(0.1M)-(Hg) system yields the corresponding cyclopropane-carboxylate **469** in 89% yield (eq. (7.212)) [24, 433].   At more negative potentials, an electrocyclization via an electrode-assisted ionization has been suggested.   The prominent nature of ammonium ions for the electroreductive cleavage of the carbon-bromine bond on dimethyl 1-bromocyclopropane 1,2-dicarboxylate **470** has been found [434].

$$
\underset{\textbf{468}}{\underset{\substack{\text{Br} \quad\quad \text{Br}}}{\text{CH}_2\text{CH}_2\!-\!\text{CHCO}_2\text{Et}}} \xrightarrow[\text{–2.5 V (SCE), 89\% yield}]{\text{DMF-Et}_4\text{NClO}_4\text{(0.1M)-(Hg)}} \underset{\textbf{469}}{\triangleright\!-\!\text{CO}_2\text{Et}}
$$

(7.212)

The electroreduction of the *trans* isomer with NH$_4^+$ ions gives rise to a 90% retention, while 90% inversion is attained when Bu$_4$N$^+$ cations are present (eq. (7.213)) [434, 435]. For the reduction of the *cis* isomer, 100% inversion of the configuration has been achieved in the presence of Bu$_4$N$^+$ ions.

$$
\underset{\textbf{470}}{\underset{\substack{\text{H} \quad\quad \text{CO}_2\text{Me}}}{\overset{\text{MeO}_2\text{C} \quad\quad \text{Br}}{\triangle^*}}} \xrightarrow[\text{H}^+]{2\,e^-,\ -\text{Br}^-} \underset{\textbf{471}}{\underset{\substack{\text{MeO}_2\text{C} \quad\quad \text{H}}}{\overset{\text{H} \quad\quad \text{CO}_2\text{Me}}{\triangle^*}}} + \underset{\textbf{472}}{\underset{\substack{\text{H} \quad\quad \text{H}}}{\overset{\text{MeO}_2\text{C} \quad\quad \text{CO}_2\text{Me}}{\triangle^*}}}
$$

(7.213)

Electrogenerated carbanions are derived from 2-bromoalkanamides in a DMF-Et₄NClO₄(0.1M)-(Hg) system and their self-protonation leads to the corresponding de-brominated amides [422, 436].   The electrochemical behavior of 3-halogeno-1,4-diphenyl-β-lactams has been investigated in dry DMF.   The electrolysis of 3-bromo-β-lactam **473** (X = Br) carried out in a DMF-Et₄NClO₄(0.1M)-(Hg) system in the presence of acetic acid as a proton donor yields the reduced product **474** (Y = H) in a quantitative yield (eq. (7.214)) [437].   In the presence of electrophiles (CO₂), the electroreduction of **473**

$$\text{473} \xrightarrow[\text{X = Cl, CO}_2]{\text{DMF-Et}_4\text{NClO}_4\text{(0.1M)-(Hg)}} \text{474} \tag{7.214}$$

X	V (SCE)	Y	Yield, %
Br	−1.3	H	100
Cl	−2.0	CO₂H	95

(X = Cl) gives rise to the carboxylated product **474** (Y = CO₂H) in 95% yield.   The electroreduction of 3,3-dichloro-β-lactam in a DMF-Et₄NClO₄(0.1M)-(Hg) system in the presence of acetic anhydride affords 3-acetyl-β-lactam in 55% yield [438].   The electrosynthesis of sulbactam by electroreductive debromination of 6,6-dibromopenicillanic acid sulfone **475** has been performed in an aqueous H₂SO₄/NaHCO₃-(Cu/C) system to give the dehalogenated product **476** in 63~67% yields (eq. (7.215)) [318, 439, 440].

$$\text{475} \xrightarrow[\text{63~67\% yields}]{\text{Aq. H}_2\text{SO}_4\text{/NaHCO}_3\text{-(Cu/C)}} \text{476} \tag{7.215}$$

The electroreduction of 3,3-dichlorosuccinimides **477** in an aqueous EtOH/NH₃-Bu₄NOAc-(Hg) system affords a mixture of two monochloro derivatives **478** (eq. (7.216)) [441, 442].   Whatever the starting compound, preferential formation of the *trans* isomer is observed when electrolyses are carried out in a basic medium (ammoniacal buffer).   In acidic medium (acetic buffer or sulfuric acid), *N*-alkyl-succinimides give rise to an excess of the *cis* isomer when reductions are carried out at very cathodic potentials.   An asymmetric synthesis has been attempted by reduction of *N*-methyl-3,3-dichloro-4,4-diphenylsuccinimide **477** (R = Me, R¹ = Ph) at a mercury pool in the presence of a strongly adsorbed alkaloid [442].   Preferential formation of the (-)-enantiomer of **478** has been observed, the best optical yield being given by strychnine.

$$\text{477} \xrightarrow[\substack{\text{70~80\% yields} \\ \text{(trans)}}]{\text{aq. EtOH/NH}_3\text{-Bu}_4\text{NOAc-(Hg)}} \text{478} \tag{7.216}$$

477   R = H, Me, Ph; R¹ = Et, Me, Ph

The electroreduction of α-bromopropiophenone **479** in a DMF-LiClO$_4$-(Hg) system at −1.0 V (SCE) produces 1,4-diphenyl-2,3-dimethyl-1,4-butanedione **480** in 65% yield (eq. (7.217)) [443a].   Electroreduction of 1,2-dibenzoylchloroethane offers the corresponding dimer, which suffers further cyclization reaction to give the cyclopentanol derivatives in 41~43% yields [443b].   The reduction process is rationalized *via* anionic intermediates.

$$
\underset{\textbf{479}}{\underset{\underset{Br}{|}}{Ph-\overset{\overset{O}{\|}}{C}-CHMe}} \quad \xrightarrow[\text{65\% yield}]{\text{DMF-LiClO}_4\text{-(Hg)}} \quad \underset{\textbf{480}}{\underset{\underset{Me}{|}\;\underset{Me}{|}}{Ph-\overset{\overset{O}{\|}}{C}-CH-CH-\overset{\overset{O}{\|}}{C}-Ph}}
\tag{7.217}
$$

The electroreduction of 2,2-dibromo-1,3-diketones **481** in the presence of olefins **345** undergoes [3 + 2] cycloaddition to give 2,3-dihydrofuran derivatives **482** in 37~94% yields (eq. (7.218)) [444].

$$
\tag{7.218}
$$

7.4.2   **Reaction of α-Halogeno Carbonyl Compounds with Nucleophiles**

It is well known that arylnickel compounds can be coupled with various electrophiles such as carbon dioxide as usual chemical reactions.   The *in situ* electroreductive generation of Ni(0) complexes in the presence of both the aromatic halide and the nucleophile is an interesting alternative.   The electroreductive coupling of α-chloroketones **484** and arylhalides **483** in a DMF-Bu$_4$NBF$_4$/NiBr$_2$(bpy)-(Zn/C) system proceeds to give the coupling product **485** in 23~79% yields (eq. (7.219)) [445].   The use of a Al- or Zn-sacrificial anode and a

$$
\underset{\textbf{483}}{Ar-X} \;+\; \underset{\textbf{484}}{ClCH_2COMe} \quad \xrightarrow[\text{23~79\% yields}]{\substack{\text{DMF-Bu}_4\text{NBF}_4(0.6M) \\ /\text{NiBr}_2(\text{bpy})\text{-(Zn/C)}}} \quad \underset{\textbf{485}}{Ar-CH_2-COMe}
\tag{7.219}
$$

X = I, Br

catalytic amount of a nickel complex is essential.   The nickel-catalyzed electroreductive coupling of α-halogenoesters **487** with aryl or vinyl halides has been attained in a DMF-Bu$_4$NBr-(Al/C) system in the presence of a catalytic amount of nickel bromide-2,2'-bipyridine complex (eq. (7.220)) [446a].   The electrolysis of methyl α-chloropropionate **487** in the presence of 1-naphthyl iodide **486** affords methyl α-naphthylpropionate **488** in 80% yield (eq. (7.220)).   The electrochemically induced Reformatsky reaction of aliphatic and aromatic ketones with ethyl α-bromoisobutyrate at a sacrificial indium anode has been investigated in comparison with the cases of zinc anode [446b, 446c].

$$ \text{(7.220)} $$

The preparative electrolysis of ethyl α-bromo-p-chlorophenylacetate **489** in a DMF-Et$_4$NClO$_4$(0.1M)-(C) system undergoes the electroreductive dimerization to give the corresponding *dl* and *meso* succinates **490** in 56% yield together with debrominated product (21%) (eq. (7.221)) [447, 448].   Ethyl α-bromophenylacetate provides *meso-* and *dl*-diethyl

$$ \text{(7.221)} $$

dl	39%
meso	17%

2,3-diphenylsuccinate in *ca.* 50% yield [449].   Methyl 2-bromo-2-phenylpropanoate **491** dimerizes under similar conditions to give the corresponding dimer **492** in 60% yield together with miner products **493** and **494** (eq. (7.222)) [450].

$$ \text{(7.222)} $$

The electrosynthesis of diastereoisomeric diethyl 2,3-bis(dihalogenophenyl)succinates is also performed at −1.6 ~ −1.8 V (SCE), giving the dimers in moderate yields [451].

The electroreduction of α-bromo-α-fluorophenylacetate under similar conditions preferentially affords the substituted *dl*-succinates as major products [452].

The electroreductive cyclization of dimethyl α,α'-dibromoalkanedioates **495** has been attained in a DMF-Et$_4$NClO$_4$(0.1M)-(Cu/Pt) system to afford the cyclized products **496** in 30~77% yields (eq. (7.223)) [453].

$$ \text{(7.223)} $$

Cyclopropane *gem*-dicarboxylates **499** are electrosynthesized by the reaction of ethyl halomalonate **497** with Michael acceptor **498** (eq. (7.224)) [454]. The electrolysis of **497** is carried out in an MeCN/CH$_2$Cl$_2$(1/1)-Bu$_4$NBr(0.1M)-(Pt) system in the presence of **498** to yield the products **499** in 70~88% yields. A similar cycloaddition with dihalomalonate under electroreduction conditions has also been attempted, forming cyclopropane dicarboxylates in good yields [328].

$$\text{(7.224)}$$

497   **498** $R^1, R^2 = H, Ar$
W = Ac, CN, CO$_2$Me
Y = H, Ac, CN

The electroreductive reaction of cyclohexanone **500** with the Reformatsky reagent derived from 2-bromo-2-methylpropanoate **501** by using a zinc anode in an electrolysis cell proceeds to give the crystalline β-lactone **502** in 80% yield together with **503** (4%) (eq. (7.225)) [455]. With this method, the reaction of succinic anhydride with 2-bromoalkanoates affords 2-alkyl substituted 3-oxohexanedioic acid monoesters in good yields [456]. The use of other metals as a sacrificial anode shows that indium gives even better yields than zinc [457].

$$\text{(7.225)}$$

500   **502** 80%   **503** 4%

An electrochemically activated zinc metal-assisted condensation of α-bromoesters **504** with nitriles **505** has been developed [458]. The electro-coupling is carried out in an MeCN-Bu$_4$NBF$_4$/ZnBr$_2$-(Zn/Au) system in an undivided cell to give the coupling product **506** in 33~84% yields (eq. (7.226)) [458].

$$\text{(7.226)}$$

504   MeCN-Bu$_4$NBF$_4$/ZnBr$_2$-(Zn/Au)
+
R^2—CN   33~84% yields
505

R^1 = methyl, Ph
R^2 = alkyl, vinyl, halogenoalkyl
X = Br or I

The electroreduction of α-halogeno carbonyl compounds has been widely investigated. The cathodic reduction of phenacyl bromide yields different products, depending on the combination of the solvent and supporting electrolytes [459].   First, the electrosynthesis of 2,4-diarylfurans **508** has been attained by electrolysis of phenacyl bromides **507** in a DMF-LiClO$_4$-(Hg) system at the potential of –1.0 V (SCE) in 67~80% yields (eq. (7.227)) [460]. However, when the electrolysis is carried out in methanol solution, the major products are acetophenone (53%) and 2-bromo-1,3-diphenyl-3,4-epoxybutan-1-one **509** (37%) (eq. (7.228)) [461].   The structure of the latter product has been proven to be 4-bromo-1,3-diphenyl-2,3-epoxybutan-1-one **510** (eq. (7.229)) [462], which can be converted into 2,4-diphenylfuran under refluxing in DMF.   Under the high diluted conditions, the initial inter-mediate has been isolated [462].   4-Aryl-2-methylfurans have been synthesized by electrol-ysis of phenacyl bromide in an acetone-LiClO$_4$-(Hg) system in 44~84% yields [459, 463, 464].

$$\text{(7.227)}$$

$$\text{(7.228)}$$

$$\text{(7.229)}$$

The enolate intermediates derived from phenacyl bromides **511** have been trapped as **513** in a CH$_2$Cl$_2$-Bu$_4$NBF$_4$-(Hg) system in the presence of ethyl chloroformate **512** (eq. (7.230)) [465a].

$$\text{(7.230)}$$

R = H, Cl, Br, Me, MeO, Ph

A new carbene route for the electroreduction of phenacyl bromide under high dilution conditions created by the slow addition of the bromide has been found [465b].   The elec-trolysis is performed in a DMF-LiClO$_4$(10 mmol)-(Hg) system in the presence of phenacyl bromide **514** at –0.6 V (SCE) to yield 2-(3*H*)-3,5-diphenyl furanone **515** (44%) and 1,c-2,t-3-tribenzoylcyclopropane **516** (24%) (eq. (7.231)) [465b].   The direct preparation of

$$\text{Ph-CO-CH}_2\text{-Br} \xrightarrow{\text{DMF-LiClO}_4\text{-(Hg)}} \textbf{515} \; 44\% \quad + \quad \textbf{516} \; 24\%$$

(7.231)

514

hetero-cyclic compounds from the phenacyl bromide semicarbazones has been attained by electroreduction. The semicarbazone **517** is converted into 3,7-diphenyl-2*H*-imidazo[2,1-*b*][1,3,4]oxadiazine **518** by electrolysis in a DMF-LiClO$_4$-(Pt/Hg) system in 70% yield (eq. (7.232)) [466].

$$2 \; \text{Ph-C-CH}_2\text{-Br} \xrightarrow[\text{75\% yield}]{\text{DMF-LiClO}_4\text{-(Pt/Hg)}} \textbf{518}$$

(7.232)

517

The electroreduction of 2-bromo-2,2-diphenylacetyl bromide in a CH$_2$Cl$_2$-Et$_4$NBr-(C) system in the presence of sodium thiosulfate affords 2-benzhydrylidene-4,4-diphenyl-1,3-oxathiolan-5-one in 21% yield [467].

Electrochemical aldol condensation is found to proceed depending on the efficient mediating effect of trivalent lanthanide ions. The lanthanide ion assisted electrochemically initiated aldol condensation of α-bromopropiophenone **519** (R = Me) with benzaldehyde **520** (R^2 = Ph) is carried out in a THF-LiClO$_4$/LaBr-(C) system to give the corresponding aldol **521** in 80% yield (eq. (7.233)) [468, 469].

$$\text{Ph-CO-C}\overset{R^1}{\underset{Br}{\overset{|}{\underset{|}{\text{C}}}}}\text{H} \; + \; \text{R}^2\text{-CHO} \xrightarrow[\text{41~93\% yields}]{\text{THF-LiClO}_4/\text{LaBr-(C)}} \text{Ph-}\overset{O}{\overset{||}{\text{C}}}\text{-}\overset{R^1}{\underset{H}{\overset{|}{\text{C}}}}\text{-}\overset{}{\underset{OH}{\overset{|}{\text{CH}}}}\text{-R}^2$$

(7.233)

519 R^1 = Me, Et, *tert*-Bu
         R^2 = *i*-Br, Ph

521

Electrosynthesis of β-lactams from readily available substituted haloacetamides has been attempted. The electroreductively produced anion sites of **522** promote cyclization by intramolecular nucleophilic substitution. The electroreduction of **523** is performed in a DMF-Et$_4$NClO$_4$(O.1M)-(Hg) system to give lactams **524** in 67~93% yields (eq. (7.234)) [470]. The condensation of *N*-phenyl bromoacetamide, BrCH$_2$CONHPh with electrogenerated diethyl bromoalonate anion yields *N*-phenyl-4,4-diethoxycarbonyl-β-lactam in 52% yield [471]. The electrochemical method can be extended to the synthesis of γ- and δ-lactams [472].

$$\underset{\textbf{522}}{X^1-\overset{\overset{\displaystyle X^2}{|}}{CH}-\overset{\overset{\displaystyle }{||}}{\underset{\underset{\displaystyle R^3}{|}}{C}}-\overset{\overset{\displaystyle R^1}{|}}{N}-\overset{\overset{\displaystyle }{|}}{\underset{\underset{\displaystyle CO_2R^2}{|}}{CH}}} \quad \xrightarrow[\text{67~93\% yields}]{\text{DMF-Et}_4\text{NClO}_4\text{-(Hg)}}$$

$$\underset{\textbf{524}}{\text{(structure)}} \qquad (7.234)$$

$$\underset{\textbf{523}}{Br-CH_2-\overset{\overset{\displaystyle }{||}}{\underset{\underset{\displaystyle R^3}{|}}{C}}-\overset{\overset{\displaystyle R^1}{|}}{\underset{\underset{\displaystyle CO_2R^2}{|}}{C}}-Br} \quad \xrightarrow[\text{85\% yield}]{\text{DMF-Et}_4\text{NClO}_4\text{-(Hg)}}$$

523 $X^1 = Br, Cl;$ $R^1 = H, CO_2Et$
$X^2 = H, Cl;$ $R^2 = H, Et$
$R^3 = CH_2CO_2Et$ (or Me), p-MeOC$_6$H$_4$

7.4.3 Electroreduction of Acyl Halides

The electroreductive acylation of cyclohexene **525** is performed in a CH_2Cl_2-$Et_4NClO_4(0.1M)$-(Al/Cu) system in the presence of acetylchloride **526** to give the acylated product **527** in 70~80% yields (eq. (7.235)) [473].

$$\underset{\textbf{525}}{\text{(structure)}} \quad \xrightarrow[\text{MeCOCl}]{\text{CH}_2\text{Cl}_2\text{-Et}_4\text{NClO}_4(0.1M)\text{-(Al/Cu)}} \quad \underset{\textbf{527}\ \ 70\sim80\%}{\text{(structure)}} + \underset{\textbf{528}\ \ 5\sim7\%}{\text{(structure)}} \qquad (7.235)$$

The preparative-scale synthesis of fluorinated α-diketones from perfluoroacyl halides **529** is performed in a $CH_2Cl_2/MeCN(2/1)$-$Bu_4NBF_4(0.1M)$-(C) system, giving the α-diketones **530** in 15~20% yields (eq. (7.236)) [474a].   Electrogenerated bis(trimethylsilylcyclopentadienyl)niobium(III) complexes (η^5-Me$_3$SiClSiC$_5$H$_4$)$_2$NbX$_2{}^-$ (X = Cl, Br, and I) have been shown to be effective catalysts for the electroreduction of acyl chlorides $R^1R^2CHCOCl$ ($R^1 = R^2 = Ph$ and $R^2 = Me$, $R^2 = Ph$).   Indirect electroreduction yields α-diketones (R^1R^2CHCO)$_2$, ketones ($R^1R^2CHCOCHR^1R^2$), and alkanes (R^1R^2CH)$_2$ [474b].

$$\underset{\textbf{529}}{R_F-\overset{\overset{\displaystyle O}{||}}{C}\diagdown_X} \quad \xrightarrow[\text{15~20\% yields}]{\text{CH}_2\text{Cl}_2/\text{MeCN}(2/1)\text{-Bu}_4\text{NBF}_4(0.1M)\text{-(C)}} \quad \underset{\textbf{530}}{\underset{R_F-C\diagdown_O}{\overset{R_F-C\diagup^O}{|}}} \qquad (7.236)$$

R_F = perfluoroalkyl
X = Cl, Br, I

Cyclic voltammograms of heptanoyl chloride in an $MeCN$-Et_4NClO_4-(Hg) system exhibit two waves; the first wave is attributed to cleavage of the carbon-chlorine bond, whereas the second wave arises from the reduction of electrogenerated heptaldehyde and hydrolytically formed heptanoic anhydride [475a].   Actually, the electrolysis of heptanoyl chloride at -1.50 V (SCE) affords heptanol (59%) together with heptanoic anhydride (40%) [475b].   The electroreduction of glutaryl dichloride gives 5-chlorovalerolactone (14%), valerolactone (18%), and glutaric anhydride (34%), respectively [476].   The electroreduction of cyclohexanecarbonyl chloride **531** at a potential slightly negative of its first cathodic wave generates cyclohexanecarboxaldehyde **532** in 40~45% yields (eq. (7.237)) [475a, 477a]. The reduction of 1-adamantanecarbonyl chloride in an MeCN-Et$_4$NClO$_4$-(C) system at −0.84 V (SCE) affords the corresponding aldehyde in quantitative yield [477b].

$$\underset{\textbf{531}}{\overset{\text{COCl}}{\bigcirc}} \xrightarrow[-1.51 \text{ V (SCE)}]{\text{MeCN-Et}_4\text{NClO}_4(0.1\text{M})\text{-(Hg)}} \underset{\textbf{532} \ 40\text{~}45\%}{\overset{\text{CHO}}{\bigcirc}} + \underset{\textbf{533} \ 43\%}{\text{Acid Anhydride}} \tag{7.237}$$

The direct electrochemical carboxylation of acetyl and benzoyl chlorides has been attained by using sacrificial zinc anode [478]. The electrolysis of acetyl chloride **526** in a DMF-Bu$_4$NBF$_4$(0.1M)-(Zn/Pt) system under bubbling CO$_2$ gas affords pyruvic acid **534** in 60% yield (eq. (7.238)).

$$\underset{\textbf{526}}{\text{MeCOCl}} + \text{CO}_2 \xrightarrow[60\% \text{ yield}]{\text{DMF-Bu}_4\text{NBF}_4(0.1\text{M})\text{-(Zn/Pt)}} \underset{\textbf{534}}{\text{MeCOCO}_2\text{H}} \tag{7.238}$$

The electroreduction of trimethylacetyl chloride in an MeCN-Et$_4$NClO$_4$-(C or Hg) system affords a mixture of the corresponding aldehyde and anhydride [479a]. The controlled-potential electrolysis of 2,4,6-trimethylbenzoyl chloride **535** in an MeCN-Et$_4$NClO$_4$(0.1M)-(C) system affords a mixture of the aldehyde **536** (58%) and the 3-substituted acrylonitrile **537** (11%) (eq. (7.239)) [479b].

$$\underset{\textbf{535}}{\text{Ar–COCl}} \xrightarrow[-1.76 \text{ V (SCE)}]{\text{MeCN-Et}_4\text{NClO}_4(0.1\text{M})\text{-(C)}} \underset{\textbf{536} \ 58\%}{\text{Ar–CHO}} + \underset{\textbf{537} \ 11\%}{\text{Ar–CH}=\text{CHCN}} \tag{7.239}$$

$$\text{Ar} = \underset{\text{Me}}{\overset{\text{Me}}{\underset{|}{\overset{|}{\bigcirc}}}}\text{—Me}$$

The polarographical survey of cathodic reductions on acyl halides suggests the formation of free radicals in these reductions [480]. The preparative-scale electroreduction of aroyl chlorides **535** leads to the corresponding 1,2-diaroyl-1,2-ethenediol diaroylates **538** and **539** in good yields (eq. (7.240)) [481, 482].

$$4 \ \underset{\textbf{535}}{\text{Ar–COCl}} \xrightarrow[\text{divided cell, 18 °C}]{\text{Acetone-LiClO}_4/\text{Na}_2\text{CO}_3\text{-(Hg)}} \underset{\textbf{538}}{\overset{\text{W} \diagup \text{Ar}}{\underset{\text{Ar} \diagup \text{W}}{\bigg\rangle\!=\!\bigg\langle}}} + \underset{\textbf{539}}{\overset{\text{W} \diagup \text{Ar}}{\underset{\text{W} \diagup \text{Ar}}{\bigg\rangle\!=\!\bigg\langle}}} \tag{7.240}$$

W = ArCO$_2$

Ar	E/2, V (SCE)	Yield, %	538/539 (trans/cis)
Ph	−1.35	95	68/32
Naphtyl	−1.20	93	100/0

The electroreduction of 2-furoyl chloride **535** (Ar = furyl) in an MeCN-Et$_4$NClO$_4$ (0.1M)-(Hg) system generates the tetramer, 1,2-bis(2-furyl)ethene-1,2-diol bis(2-furoate) **540**, as a major product (eq. (7.241)) [483a]. Cyclic voltammograms for the reduction of **535** at a hanging mercury drop electrode under the above conditions exhibit six cathodic waves. The electrolytic cleavage of the carbon-chlorine bond is responsible for the first wave (−1.36 V *vs.* SCE). Electroreductive coupling of aroyl or arylacetyl chlorides is performed in an undivided cell in an MeCN-Bu$_4$NBF$_4$-(SUS/Ni) system, giving the corresponding symmetrical ketone derivatives in 30~80% yields [483b]. The controlled-potential electrolysis of 2-thiophenecarbonyl chloride in an MeCN-Bu$_4$NClO$_4$-(C) system affords 1,2-di(2-thienyl)ethene-1,2-diol di-(2-thiophenecarboxylate) in 95% yield together with 2-thiophenecarxaldehyde (5%) [483c].

$$\text{Ar–COCl} \xrightarrow[\text{major product}]{\text{MeCN-Et}_4\text{NClO}_4(0.1\text{M})\text{-(Hg)}} \begin{array}{c} \text{ArCO}_2 \quad\quad \text{Ar} \\ \diagdown\;\;\diagup \\ \text{C=C} \\ \diagup\;\;\diagdown \\ \text{Ar} \quad\quad \text{O}_2\text{CAr} \end{array} \qquad (7.241)$$

535 **540**

$$\text{Ar} = \text{(furyl)}$$

The electroreductive acylation of benzylchlorides **541** has been attempted to yield alkyl benzyl ketones **543** [484]. The electrolysis of **541** is carried out in an MeCN-Et$_4$NOTs-(C) system in the presence of acyl chlorides **542** in 27~71% yields (eq. (7.242)).

$$\underset{\textbf{541}}{\overset{\text{R}^1}{\text{Ph–CH–Cl}}} + \underset{\textbf{542}}{\text{R}^2\text{–COCl}} \xrightarrow[\text{27~71% yields}]{\text{MeCN-Et}_4\text{NOTs-(C)}} \underset{\textbf{543}}{\overset{\text{R}^1}{\text{Ph–CH–CO–R}^2}} \qquad (7.242)$$

$$\text{R}^1 = \text{H}; \quad \text{R}^2 = \text{Me}$$

The electroreduction of 2-chloro-2-phenylacetyl chloride **544** in a CH$_2$Cl$_2$-Et$_4$NCl-(Hg) system can lead to the formation of a mixture of α- and γ-6-benzyl-3,5-diphenylhydrox-ypyranones **545a** and **545b** in 62% total yield (eq. (7.243a)) [485a]. The formation of 2-benzhydrylidene-4,4-diphenyl-1,3-oxathiolan-5-one **547** has been observed by the electrolysis of **546** in a CH$_2$Cl$_2$-Et$_4$NBr-(C) system in the presence of sodium thiosulphate (eq. (7.243b)) [467]. The pathway of the cyclization involves the reaction of an electrogenerated diphenylketene, 2-bromo-2,2-diphenylacetylbromide **546**, and thiosulfate anion [485b]. The mechanism for the trimerization of a phenylketene intermediate has been rationalized.

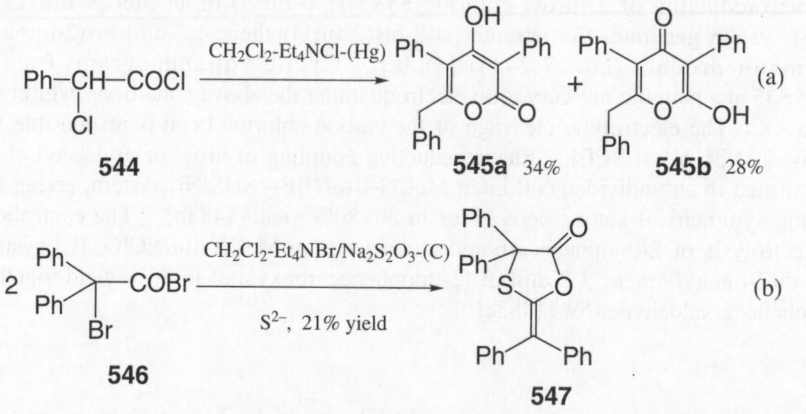

$$\text{Ph–CH–COCl} \quad \xrightarrow{\text{CH}_2\text{Cl}_2\text{-Et}_4\text{NCl-(Hg)}} \quad \text{545a} \; 34\% \quad + \quad \text{545b} \; 28\% \tag{a}$$

544

(7.243)

$$2 \quad \text{546} \quad \xrightarrow[\text{S}^{2-}, \; 21\% \; \text{yield}]{\text{CH}_2\text{Cl}_2\text{-Et}_4\text{NBr/Na}_2\text{S}_2\text{O}_3\text{-(C)}} \quad \text{547} \tag{b}$$

The electroreduction of cinnamoyl chloride **548** in an acetone-LiClO$_4$-(Pt/Hg) system affords 2,4-dibenzyl-1,3-cyclobutanedione **549** (eq. (7.244)) [486].   A mechanism for the dimerization of an aldoketene intermediate has been proposed.

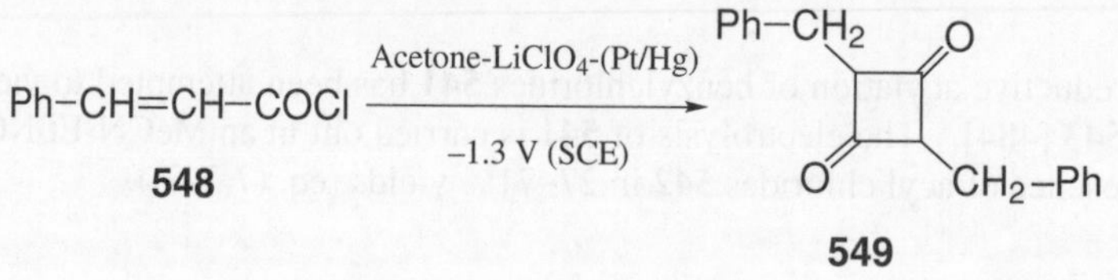

$$\text{Ph–CH=CH–COCl} \quad \xrightarrow[-1.3 \; \text{V (SCE)}]{\text{Acetone-LiClO}_4\text{-(Pt/Hg)}} \quad \text{549} \tag{7.244}$$

548

A stable anion radical species, characterized as 1,4-dihydroxybenzene dibenzoate ester lithium anion radical **552**, has been prepared [487].   The electroreduction of 1,4-benzoquinone **550** in an MeCN-LiClO$_4$-(Pt/Hg) system in the presence of benzoyl chloride **551** at -1.9 V (SCE) affords the anion radical **552** (eq. (7.245)).

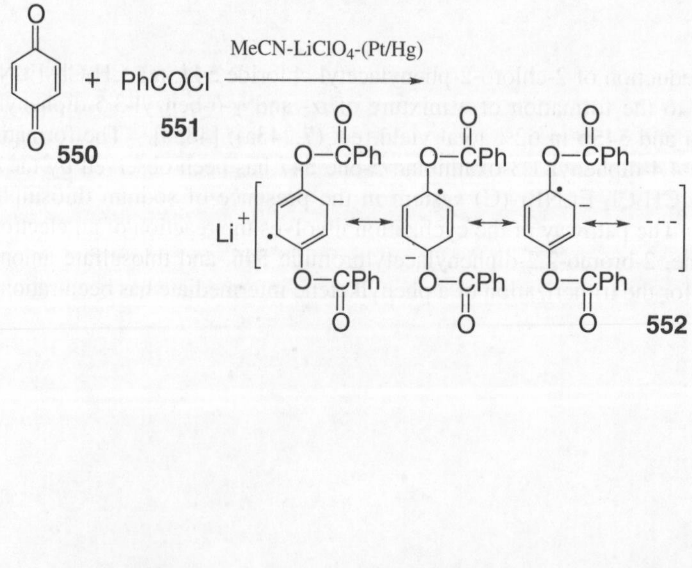

$$\text{550} \quad + \quad \text{PhCOCl} \quad \xrightarrow{\text{MeCN-LiClO}_4\text{-(Pt/Hg)}} \quad \text{552} \tag{7.245}$$

551

7.5 Electroreduction of Vinyl Halides

7.5.1 Electroreductive Dehalogenations

The stepwise electroreductive dechlorination of tetrachloroethylene **553** has been performed in an HMPA/H$_2$O(10%)-Bu$_4$NBF$_4$(0.1M)-(Hg) system at –3.2 V (SCE) to give a mixture of **554** and **555** in good yields (eq. (7.246)) [488]. The electrochemical properties of halogenated allenes have been investigated by analysis of polarographic waves [489].

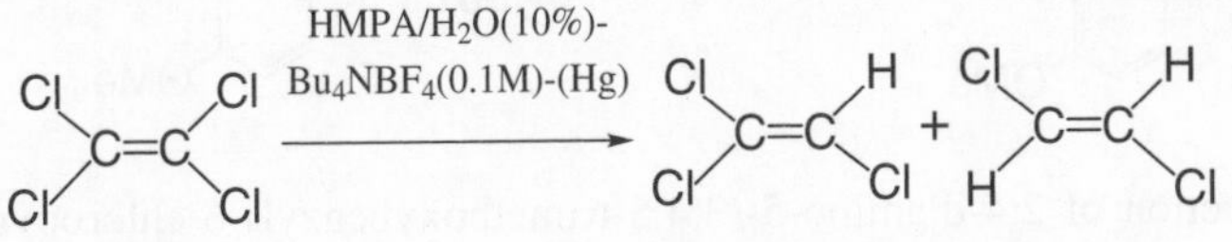

$$\text{(7.246)}$$

553   **554** 61%   **555** 27%

The electroreductive dechlorination of 2-chloro-2-cyclohexenones proceeds at a mercury cathode in acetonitrile to yield cyclohexenones [419]. The electroreduction of aryl 2-chlorovinyl ketones (Ar-CO-CH=CHCl) has been shown to give the corresponding diaroyl-butenes and diaroylbutadienes [490]. 2-Substituted 3-chloro-2-cyclopentenones **556** can be reduced in an MeOH-H$_2$SO$_4$-(Hg or Pb) system to afford the corresponding 2-cyclopentenones **557** in good yields (eq. (7.247)) [491]. Electroreductive dechlorinations of

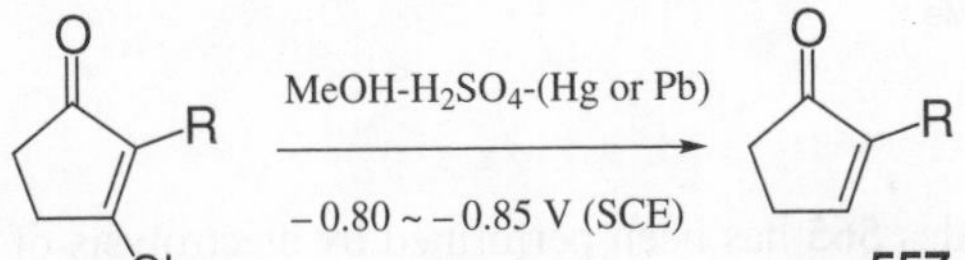

$$\text{(7.247)}$$

556 R = alkyl

Cathode	Yield, %
Hg	76~96
Pb	78~95

4-chloro-3-formyl-(2H)-benzopyran **558** and β-chlorocinnamaldehyde have been attempted in a DMF-Bu$_4$NClO$_4$(0.1M)-(Hg) system to give a gummy brown material [492]. However, the electrolysis of **558** in an aqueous EtOH(60%)-Buffer(pH 7.0)-(Hg) system at –1.40 V affords the dimer **559** in 68% yield (eq. (7.248)) [493].

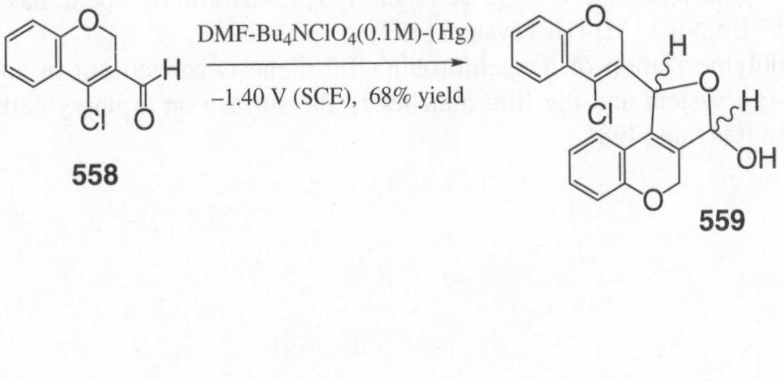

$$\text{(7.248)}$$

558   **559**

The electrochemical behavior of β-chlorovinylimines in aprotic media has been investigated. The macroscale electrolysis of the substituted β-chlorovinylimines **560** in a DMF-Bu$_4$NClO$_4$(0.1M)-(Hg) system yields the corresponding unsubstituted imines **561** in good yields (eq. (7.249)) [494].

$$\text{(7.249)}$$

The electroreduction of 2,4-diamino-5-(3,4,5-trimethoxybenzyl)-6-chloropyrimidine **562** in an aqueous HCl(5%)-(Pb/Hg) system at 1.00 V (SCE) undergoes dechlorination reaction to give pyrimidine derivatives **563** in 81% yield (eq. (7.250)) [495].

$$\text{(7.250)}$$

The electrosynthesis of isocyanides **565** has been performed by electrolysis of carbonimidoyl dichloride **564** in a DMF-LiClO$_4$(0.2M)-(Hg) system in a divided cell. The electrolysis of **564** (R = Ph) affords phenyl isocyanide **565** (R = Ph) in 91% yield (eq. (7.251)) [496].

$$\text{(7.251)}$$

R = Ph, 4-ClC$_6$H$_4$, 2-Cl-4-MeC$_6$H$_3$
benzyl, cyclohexyl

The polarographic reduction of 1-(1-adamantyl)-1,3,3-tribromopropene has been attempted in a DMF-Bu$_4$NI(0.1M)-(Hg) system [497].

The electropolymerization of hexachlorobuta-1,3-diene is carried out in an MeCN-Bu$_4$NBF$_4$(0.1M)-(C) system and the film attaches to the surface on a glassy carbon electrode tightly when it is thin [498].

7.5.2   Electroreductive Coupling with Electrophiles

The vinyl radical induced intramolecular cyclization of vinyl bromides with double bonds has been attained by a nickel(II) complex catalyzed electroreduction (See Chap.11.1.1.1) [499]. The electroreduction of 3-chloro-2-phenylpropenenitrile **566** in an MeCN-Et$_4$NClO$_4$(0.1M)-(C/S) system yields the sulfur substituted product **567** in 55% yield (eq. (7.252)) [500]. The role of a carbon-sulfur electrode is discussed.

$$
\begin{array}{c}
\text{Ph} \\
\diagdown \\
\text{C=C} \\
\diagup \quad \diagdown \\
\text{NC} \quad \text{Cl}
\end{array}
\quad
\begin{array}{c}
\text{H} \\
\xrightarrow[\substack{-1.90\text{ V (SCE)},\\ 55\% \text{ yield}}]{\text{MeCN-Et}_4\text{NClO}_4(0.1\text{M})\text{-(C/S)}}
\end{array}
\quad
\begin{array}{c}
\text{Ph} \quad\quad\quad\quad\quad \text{Ph} \\
\diagdown \quad\quad\quad\quad\quad \diagup \\
\text{C=CH–S–S–CH=C} \\
\diagup \quad\quad\quad\quad\quad \diagdown \\
\text{NC} \quad\quad\quad\quad\quad \text{CN}
\end{array}
\quad (7.252)
$$

566   **567**

α-Chloro-β-acylvinylsulfides **568** are reduced in an MeCN-Et$_4$NBr(0.1M)-(Al/Pt) system in the presence of carbon dioxide to give the corresponding acids **569** in 30~53% yields (eq. (7.253)) [501].

$$
\begin{array}{c}
\text{SR}^2 \\
\diagup \\
\text{R}^1\text{–CO–CH=C} \\
\diagdown \\
\text{Cl}
\end{array}
\quad
\xrightarrow[\text{CO}_2, \ 30\text{~}53\% \text{ yields}]{\text{MeCN-Et}_4\text{NBr}(0.1\text{M})\text{-(Al/Pt)}}
\quad
\begin{array}{c}
\text{SR}^2 \\
\diagup \\
\text{R}^1\text{–CO–CH=C} \\
\diagdown \\
\text{CO}_2\text{H}
\end{array}
\quad (7.253)
$$

568 R^1 = Me, i-Pr, Ph
R^2 = Bu, Ph

569

7.6   Electroreduction of Halogenated Hetero Bonds

7.6.1   Electroreduction of Halogen-Nitrogen and Halogen- Sulfur Bonds

Tetramethylammonium hydroxide **571** has been prepared by the electroreduction of quaternary ammonium chloride **570** in an electrolytic cell equipped with a cation-exchange membrane as a diaphragm (eq. (7.254)) [502]. The semiconductor industry requires higher purity of ammonium hydroxide **571**. There is a particular demand for the complete removal of small amounts of chloride ions. An industrial process for the production of a chloride-free solution of **571** by using tetramethylammonium formate has been developed [503]. Electrolysis of the ammonium formate is carried out in an aqueous H$_2$SO$_4$(1.13mol)-(Ti/IrO$_2$-SUS) system with monitoring by the titration of the catholyte with aqueous HCl (0.1 M) solution using phenolphthalein as an indicator [504]. The elec

$$
\text{Me}_4\text{N}^+\text{Cl}^- + \text{H}_2\text{O} \quad \xrightarrow[-1/2\ \text{Cl}_2,\ -1/2\ \text{H}_2]{\text{aq. H}_2\text{SO}_4(1.13\text{M})\text{-(Ti/IrO}_2\text{–SUS)}} \quad \text{Me}_4\text{N}^+\text{OH}^- \quad (7.254)
$$

570   **571**

troreductive preparation of tetramethylammonium hydroxide from tetramethylammonium chloride 570 has been attained in an aqueous Me_4NCl-(SUS) system on a pilot plant scale [505]. The methanol solution of tetramethylammonium methoxide ($Me_4N^+OMe^-$) in various concentrations can be prepared by electrolyzing a methanol solution of tetramethylammonium chloride in a divided cell [506].

Electrophilic fluorinating agents containing nitrogen-fluorine bonds are known as an extremely useful agent for introducing fluorine into organic compounds. The reduction potential of compounds bearing nitrogen-fluorine bonds is highly affected by the substituents of the nitrogen atom. The nitrogen-fluorine bond activated by electron-withdrawing substituents possesses a fairly high oxidizing power which induces electron-transfer reactions with a variety of nucleophites [507]. The electroreductive cleavage of the nitrogen-fluorine bond causes loss of the fluoride anion. The reductive cleavage of N-fluoro-, N-chloro-, and N-bromosaccharin sultams and of their 4-nitro-substituted analogues has been investigated in an $MeCN$-Et_4NBF_4(0.1M)-(C) system [508]. The controlling factors of stepwise versus concerted reductive cleavage are discussed.

The formation of a complex between N-bromosuccinimide and anions of succinimide 573 has been suggested in the course of the electroreduction of N-bromosuccinimide 572 in an $MeCN$-Bu_4NBF_4(0.1M)-(Pt) system (eq. (7.255)) [509]. N-Halophthalimides 574 can also be reduced into the corresponding phtalimide 575 in good yields (eq. (7.256)). The succinimide anion generated by a two-electron reduction of N-chlorosuccinimide 576 in an $MeCN$-Me_4NBF_4-(Pt) system can be trapped with methyl tosylate to afford N-methylsuccimide 577 in 85% yield (eq. (7.257)) [510].

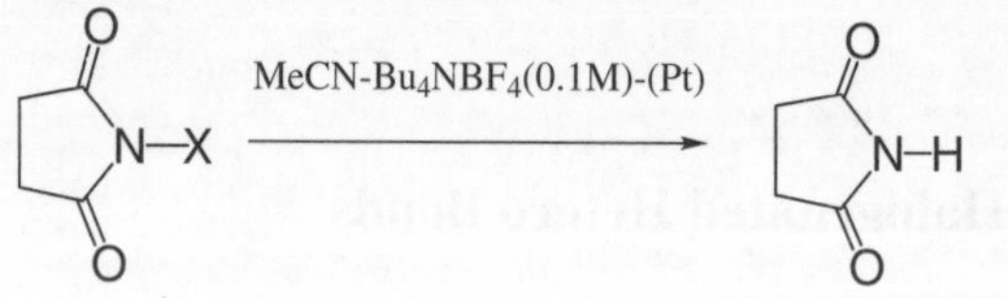

$$MeCN\text{-}Bu_4NBF_4(0.1M)\text{-}(Pt) \quad (7.255)$$

572   X = Br; 92%
      I; 93%

573

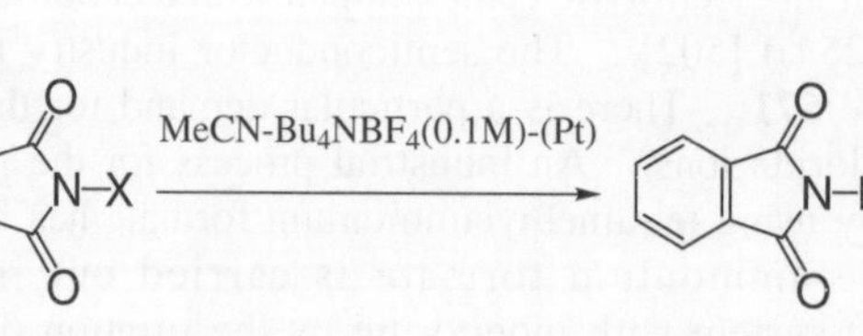

$$MeCN\text{-}Bu_4NBF_4(0.1M)\text{-}(Pt) \quad (7.256)$$

574   X = Br; 84%
      I; 86%

575

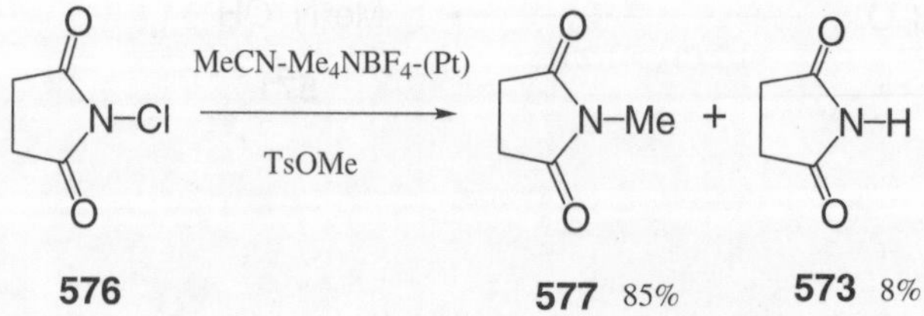

$$MeCN\text{-}Me_4NBF_4\text{-}(Pt) \quad TsOMe \quad (7.257)$$

576

577 85%   573 8%

The electroreductive defluorination of *N*-fluorourethans **578** undergoes a hydrogen-abstraction with an EG base (urethane anion, etc.) to generate EtOCONF$^-$, which loses F$^-$ to give carbethoxynitrene as an intermediate, giving **579** and **580** (eq. (7.258)) [511]. The α-elimination of F$^-$ from the anionic intermediate [EtOCONF]$^-$ proceeds much easier than that of Cl$^-$ from [EtOCONCl]$^-$.

$$
\underset{\textbf{578}}{\overset{\text{EtOCO}-\underset{|}{\text{N}}\text{Me}}{\underset{\text{F}}{}}} \xrightarrow[\text{$-$F}^{\ominus}]{\text{MeCN-Et}_4\text{NClO}_4\text{-(Hg)}} \underset{\textbf{579} \ 26\%}{\text{EtOCO}-\underset{\text{H}}{\overset{|}{\text{N}}}\text{Me}} + \underset{\textbf{580} \ 25\%}{\text{EtOCO}-\text{NH}_2} \qquad (7.258)
$$

The electrochemical generation of toluene-*p*-sulfonylnitrene by electroreduction of *N,N*-dichlorotoluene-*p*-sulfonamide **581** in an MeCN/dioxane(3/7)-LiClO$_4$•3H$_2$O-(Pt) system has been suggested by the formation of 2-(*p*-tolylsulfonylamino)-1,4-dioxane **582** in 32% yield (eq. (7.259)) [512].

$$
\underset{\textbf{581}}{\text{TsNCl}_2} \xrightarrow{\substack{\text{MeCN/Dioxane(3/7)-} \\ \text{LiClO}_4\text{•3H}_2\text{O-(Pt)}}} \underset{\substack{\textbf{582} \ 32\% \\ + \\ p\text{-MeC}_6\text{H}_4\text{SO}_2\text{NH}_2 \\ \textbf{583} \ 52\%}}{\text{TsNH}-\text{(dioxane)}} \qquad (7.259)
$$

The electroreduction of arenediazonium tetrafluoroborates in the presence of unsaturated compounds tends to give the polymer of the unsaturated compounds [513]. However, the electroreduction of benzenediazonium chloride **584** in the presence of isoprene **585** affords the aryl-chlorination product **586** in 45% yield (eq. (7.260)) [514].

$$
\underset{\textbf{584}}{\text{PhN}_2^+\text{Cl}^-} + \underset{\textbf{585}}{\text{(isoprene)}} \xrightarrow[\text{45\% yield}]{\text{H}_2\text{O/Acetone-(Fe/Pt)}} \underset{\textbf{586}}{\text{Ph}\diagdown\diagup\diagdown\text{Cl}} \qquad (7.260)
$$

The electroreductive cross-coupling of quaternary ammonium salts of 3-aminoalkanoic acid esters **587** with aldehyde **588** affords the corresponding γ-lactones **589** in one step (eq. (7.261)) [515].

$$
\underset{\textbf{587}}{\overset{\text{Me}\diagdown\overset{+}{\underset{|}{\text{N}}}\diagup\overset{I^-}{\underset{\text{CO}_2\text{Me}}{\text{R}^1}}}{\underset{R^2}{}}} + \underset{\textbf{588}}{\text{C}_3\text{H}_7\text{CHO}} \xrightarrow{\text{DMF-Et}_4\text{NOTs-(Pb)}} \underset{\textbf{589}}{\overset{O}{\underset{R^2 \quad \text{C}_3\text{H}_7}{\text{(lactone)}}}} \qquad (7.261)
$$

a: R^1 = Et; R^2 = H
b: R^1, R^1 = –(CH$_2$)$_4$–; R^2 = Me

The electroreductive coupling of 2-methyl-3,4-dihydro-6,7-dimethoxyisoquinolium iodide **590** and 3-bromomeconine **591** has been attained in a DMF-Et$_4$NOTs-(Pt) system to give the adducts **592** in 77% yield (eq. (7.262)) [516].

$$(7.262)$$

590   **591**   **592**

R^1, R^2, R^3, R^4 = OMe

The electroreductive cleavage behavior of unsymmetrically substituted diphenyliodonium salts has been investigated at a mercury cathode in DMF [517].   The cleavage site of carbon-iodine bonds depends on the electronic character of the substituent at the aromatic ring.   The electrolysis of phenyl p-methoxyphenyl iodonium salt **593** in a DMF-Me$_4$NClO$_4$(0.1M)-(Hg) system at 0.3 V (SCE) yields p-iodoanisole **594** (77%) and iodobenzene **595** (16%) (eq. (7.263)) [517].   The controlled potential electroreduction of phenyliodonio-bis(acyloxy)methanides (PhI$^+$ $-$ C$^-$(COR)$_2$) has been investigated in an MeOH-LiClO$_4$-(Hg) system [518].

$$(7.263)$$

593   **594** 77%   **595** 16%

The preparative electroreduction of phenyl(heptafluoropropyl)iodonium tetrafluoroborate in the presence of sodium p-chlorophenylsulfinate yields the corresponding sulfone and perfluoropropionic acid [519].   Electroreduction of p-toluene-N,N-dichlorosulfonamide **596** proceeds in an MeCN-LiClO$_4$-(Pt) system to give the dechlorinated product **597** in 84% yield (eq. (7.264)) [520].   The electroreduction of substituted aromatic sulfonyl fluorides, ArSO$_2$F, has been investigated in an MeOH-Et$_4$NI(0.1M)-(Hg) system [521].   An examination of changes in the transfer coefficient in the electroreduction of a series of sulfonyl fluorides has been made.

$$p-MeC_6H_4-SO_2NCl_2 \xrightarrow[\text{84\% yield}]{\text{MeCN-LiClO}_4\text{-(Pt)}} p-MeC_6H_4-SO_2NH_2 \qquad (7.264)$$

596   **597**

7.6.2   Electroreductive Formation of Carbon-Metal Bonds

Cathodic alkylation is one of the promising methods for the preparation of metal alkyls. Tetramethylstanane **601** has been electrosynthesized as a major product (59~86%) in an aqueous MeCN(90%)-Et$_4$NBr-(Sn) system, giving hexamethyldistanane **600** as a minor product *via* an unstable intermediate **599** (eq. (7.265)) [522].

$$3 \text{ MeX } + \text{ Sn} \xrightarrow[\text{-Et}_4\text{NBr-(Sn)}]{\text{MeCN/H}_2\text{O}(7/1)} [\text{ Me}_3\text{Sn }]$$

598 X = I, Br

599 e^-
Unstable
intermediate
on electrode

1/2 Me₃SnSnMe₃
600 8~16% yields

MeX
598

Me₄Sn
601 59~86% yields

(7.265)

The ethylation of lead metal under electroreduction conditions can be carried out in an aqueous quaternary ammonium salt solution by vigorously circulating the aqueous electrolyte solution in order to maintain ethyl bromide in an emulsified state [523]. The electrosynthesis of tetraethyllead at a lead trickle-bed electrode has been attained as a large-scale manufacturing method (65% current yield) [524].

The electroreduction of ethyl iodide in an EtOH-Bu₄NI-(Hg) system gives rise to an oxidizable intermediate adsorbed on mercury cathode. Diethylmurcury is found to be the major product of electrolysis at not very negative (–1.6 V (SCE)) potentials with 99% yield [525]. The identification and distribution of products of a large scale electrolysis have been performed at different potentials in a solution of iododecane in DMF, indicating that alkyl radicals are produced in the first step of reduction and then combine in the form of alkylmercury radicals to give dialkylmercury [17, 20]. It has been shown that on mercury, the alkyl iodide is reduced in a one-electron process to form an adsorbed radical, [RHg]˙ (ads), which may be influenced by dimerization of [RHg]˙ followed by disproportionation to form R₂Hg [526].

7.6.3 Miscellaneous

The electroreductive removal of bromine atom from 4-bromosydnone **602** has been attained in a DMF-Bu₄NBF₄(0.3M)-(Pt) system (eq. (7.266)) [527]. The electrogenerated trityl anion from **603** plays a role as a reducing agent by electron transfer to **602** leading to **604**.

$$\text{X} \underset{}{\overset{}{\bigcirc}} \underset{\text{N}}{\overset{\text{Br}}{-\text{N}-\text{C}}} + \text{Ph}_3\text{C}^{-+}\text{NBu}_4 \xrightarrow[\text{33~45\% yields}]{\text{DMF-Bu}_4\text{NBF}_4\text{-(Pt)}} \text{X} \bigcirc -\text{N}-\text{C}$$

603

602 X = H, Br

604

(7.266)

The electroreduction of hypervalent iodine compounds involving I(III) or I(V) compounds has been performed in an aqueous HClO₄(0.5M)-(Pt) system [528]. All compounds undergo a reductive cleavage of their I – O bonds, yielding the corresponding I(I) derivatives. For example, the electrolysis of **605** yields iodobenzene **595** and sulfonic acid **606** smoothly (eq. (7.267)) [528].

$$\bigcirc\overset{\text{OH}}{\underset{\text{OSO}_2\text{R}}{-\text{I}}} \xrightarrow[e^-, 2\text{H}^+]{\text{aq. HClO}_4(0.5\text{M})\text{-(Pt)}} \bigcirc -\text{I} + \text{RSO}_3\text{H} + \text{H}_2\text{O}$$

605

595 **606**

(7.267)

R = 10-camphoryl,
    p-tolyl, 2-naphthyl

Oxybis(iodoniumbenzene)bis(tetrafluoroborate) **607** can be reduced different ways in an aqueous $H_2SO_4(0.5M)$-(Pt) system, depending on the electrolysis conditions. At the first stage, the products are iodosylbenzene **609** and iodobenzene **608** (eq. (7.268)), and the latter compound is formed from the former as a result of further reduction (eq. (7.269)) [529]. In the final stage, iodide ions **610** are formed (eq. (7.270)).

$$
\begin{array}{l}
\mathrm{BF_4^-} \\
\mathrm{(Ph\!-\!I^+\!-\!O\!-\!I^+\!-\!Ph)_{ads}} \\
\mathrm{BF_4^-} \\
\qquad \textbf{607}
\end{array}
$$

$$2\,H^+ + 4\,e^- \longrightarrow (PhI)_{ads} + H_2O \qquad \text{------ (7.268)}$$
$$\textbf{608}$$
$$2\,e^- \longrightarrow (PhI\!=\!O)_{ads} + (PhI)_{ads} \qquad \text{------ (7.269)}$$
$$\textbf{609} \qquad\qquad \textbf{608}$$
$$4\,H^+ + 8\,e^- \longrightarrow 2\,I^-_{ads} + 2\,C_6H_6 + H_2O \qquad \text{---- (7.270)}$$
$$\textbf{610}$$

The electroreductive cleavage of carbon-halogen bonds of haloferrocenes FeX (X = Cl, Br, I) has been attained in a DME-Bu_4NClO_4(0.2M)-(Pt) system (DME = ethylene glycol dimethyl ether) in good yield [530].

References

[1]   T. Shono, *Electroorganic Chemistry as a New Tool in Organic Synthesis*, Springer-Verlag, Berlin (1984).

[2]   M. M. Baizer, *Tetrahedron*, **40**, 944 (1984).

[3]   J. Simonet, G. Le Guillanton, *Bull. Soc. Chim. Fr.*, **1985**, 180.

[4]   H. Lund, *J. Mol. Catalysis*, **38**, 203 (1986).

[5]   E. Steckhan, *Angew. Chem., Int. Ed. Engl.*, **25**, 683 (1986).

[6]   E. Steckhan, *Topics Curr. Chem.*, **142**, 1 (1987).

[7]   S. Torii, *Synthesis*, **1986**, 873.

[8]   (a) J. Chaussard, Jean-Claude Folest, J.-Y. Nédelec, J. Périchon, S. Sibille, M. Troupel, *Synthesis*, **1990**, 369; (b) H. Hebri, E. Duñach, J. Périchon, *Synth. Commun.*, **21**, 2377 (1991).

[9]   (a) M. Baizer, *Organic Electrochemistry* (eds. M Baizer, H. Lund), 2nd Ed., Marcel Dekker, New York (1983); (b) J. Simonet, *Organic Electrochemistry* M. Baizer, H. Lund (eds.), 2nd Ed., pp 843, Marcel Dekker, New York (1983); (c) O. N. Efimov, V. V. Strelets, *Coord. Chem. Rev.*, **99**, 15 (1990); (d) M. D. Hawley, *Encyclopedia of Electrochemistry of the Elements* A. J. Bard, H. Lund (eds.), Vol. XIV, Marcel Dekker, New York (1980).

[10] J. Y. Becker, *Chemistry of Functional Groups, Supplement D* (eds. S. Patai, J. Rappoport), Chapter 6, pp 203-285, John Wiley & Sons, Inc., New York (1983).

[11] U. Hess, S. Czapla, *Z. Chem.*, **25**, 334 (1985).

[12] A. Curulli, A. Inesi, E. Zeuli, *Electrochim. Acta*, **32**, 1117 (1987).

[13] L. H. Kristensen, H. Lund, *Acta Chem. Scand.*, **B33**, 735 (1979).

[14] A. O. Moing, J. Delaunay, A. Lebouc, J. Simonet, *Tetrahedron*, **41**, 4483 (1985).

[15] F. Maran, E. Vianello, *Tetrahedron Lett.*, **31**, 5803 (1990).

[16] J. A. Dougherty, A. J. Diefenderfer, *J. Electroanal. Chem., Interfacial Electrochem.*, **21**, 531 (1969).

[17] D. M. La Perriere, B. C. Willett, W. F. Carroll, Jr., E. C. Torp, D. G. Peters, *J. Am. Chem. Soc.*, **100**, 6293 (1978).

[18] M. S. Mubarak, D. G. Peters, *J. Org. Chem.*, **47**, 3397 (1982).

[19] J. A. Cleary, M. S. Mubarak, K. L. Vieira, M. R. Anderson, D. G. Peters, *J. Electroanal. Chem., Interfacial Electrochem.*, **198**, 107 (1986).

[20] D. M. La Perriere, W. F. Carroll, Jr., B. C. Willett, E. C. Torp, D. G. Peters, *J. Am. Chem. Soc.*, **101**, 7564 (1979).

[21] G. M. McNamee, B. C. Willett, D. M. La Perriere, D. G. Peters, *J. Am. Chem. Soc.*, **99**, 1831 (1977).

[22] W. A. Pritts, D. G. Peters, *J. Electrochem. Soc.*, **141**, 3318 (1994).

[23] M. S. Mubarak, D. G. Peters, *J. Org. Chem.*, **60**, 681 (1995).

[24] C. Giomini, A. Inesi, E. Zeuli, *J. Chem. Res. (S)*, **1983**, 280.

[25] M. T. Ismail, A. A. Abdel-Wahab, O. S. Mohamed, A. A. Khalaf, *Bull. Soc. Chim. Fr.*, **1985**, 1174.

[26] J. C. Bart, D. G. Peters, *J. Electroanal. Chem., Interfacial Electrochem.*, **280**, 129 (1990).

[27] L. G. Feoktistov, A. P. Tomilov, Yu. D. Smirnov, M. M. Gol'din, *Elektrokhimiya*, **1**, 887 (1965).

[28] M. T. Ismail, M. F. El-Zohry, *J. Chem. Tech. Biotechnol.*, **56**, 135 (1993).

[29] D. P. G. Hamon, K. R. Richards, *Aust. J. Chem.*, **36**, 109 (1983).

[30] (a) K. L. Vieira, M. S. Mubarak, D. G. Peters, *J. Am. Chem. Soc.*, **106**, 5372 (1984); (b) K. L. Vieira, D. G. Peters, *J. Electroanal. Chem., Interfacial Electrochem.*, **196**, 93 (1985); (c) K. L. Vieira, D. G. Peters, *J. Org. Chem.*, **51**, 1231 (1986).

[31] C. P. Andriux, I. Gallardo, J.-M. Savéant, K. B. Su, *J. Am. Chem. Soc.*, **108**, 638 (1986).

[32] C. P. Andrieux, J.-M. Savéant, K. B. Su, *J. Phys. Chem.*, **90**, 3815 (1986).

[33] (a) I. Carelli, A. Curulli, A. Inesi, *Electrochim. Acta*, **30**, 941 (1985); (b) A. Curulli, I. Carelli, A. Inesi, *J. Electroanal. Chem., Interfacial Electrochem.*, **235**, 209 (1987).

[34] A. Inesi, E. Zeuli, *J. Electroanal. Chem., Interfacial Electrochem.*, **195**, 129 (1985).

[35] V. A. Petrosyan, V. I. Slovetskii, A. A. Fainzil'berg, *Izv. Akad. Nauk SSSR*, **1973**, 2027.

[36] C. P. Andrieux, A. Le Gorande, J.-M. Savéant, *J. Am. Chem. Soc.*, **114**, 6892 (1992).

[37] J.-M. Savéant, *J. Am. Chem. Soc.*, **114**, 10595 (1992).

[38] L. A. Avaca, E. R. González, E. A. Ticianelli, *Electrochim. Acta*, **28**, 1473 (1983).

[39] T. Lund, H. Lund, *Acta Chem. Scand.*, **B41**, 93 (1987).

[40] L. Nadjo, J.-M. Savéant, K. B. Su, *J. Electroanal. Chem., Interfacial Electrochem.*, **196**, 23 (1985).

[41] C. P. Andrieux, A. Merz, J.-M. Savéant, *J. Am. Chem. Soc.*, **107**, 6097 (1985).

[42] K. P. Butin, A. A. Ivkina, O. A. Reutov, *Izv. Akad. Nauk SSSR, Ser. Khim.*, **1985**, 546.

[43] J.-E. Dubois, P. Bauer, B. Kaddani, *Tetrahedron Lett.*, **26**, 57 (1985).

[44] U. Hess, D. Huhn, H. Lund, *Acta Chem. Scand.*, **B34**, 413 (1980).

[45] N. Kornblum, *Angew. Chem.*, **87**, 797 (1975).

[46] I. Lund, H. Lund, *Tetrahedron Lett.*, **27**, 95 (1986).

[47] G. D. Luer, D. E. Bartak, *J. Org. Chem.*, **47**, 1238 (1982).

[48] A. Konno, T. Fuchigami, K. Suzuki, T. Nonaka, H.-J. Tien, *Denki Kagaku*, **62**, 260 (1994).

[49] J. Gassmann, J. Voss, G. Adiwidjaja, *Z. Naturforsch., B: Chem. Sci.*, **50**, 953 (1995).

[50] (a) M. S. Mubarak, D. G. Peters, *J. Electroanal. Chem.*, **332**, 127 (1992); (b) C. E. Dahm, D. G. Peters, *Anal. Chem.*, **66**, 3117 (1994).

[51] (a) P. W. Jennings, D. G. Pillsbury, J. L. Hall, V. T. Brice, *J. Org. Chem.*, **41**, 719 (1976); (b) J. L. Hall, R. D. Geer, P. W. Jennings, *J. Org. Chem.*, **43**, 4364 (1978).

[52] A. J. Fry, J. M. Porter, P. F. Fry, *J. Org. Chem.*, **61**, 3191 (1996).

[53] (a) V. G. Garcia, V. Montiel, J.-M. Feliu, A. Aldaz, M. Feliz, *J. Chem. Res. (S)*, **1990**, 144; *ibid., (M)*, **1990**, 1101; (b) V. G. Garcia, V. Montiel, A. Aldaz, *J. Chem. Res. (S)*, **1992**, 58; *ibid., (M)*, **1992**, 54.

[54] P. A. Leermakers, G. F. Vesley, N. J. Turro, D. C. Neckers, *J. Am. Chem. Soc.*, **86**, 4213 (1964).

[55] A. J. Fry, R. Scoggins, *Tetrahedron Lett.*, **1972**, 4079.

[56] A. J. Fry, J. J. O'Dea, *J. Org. Chem.*, **40**, 3625 (1975).

[57] K. Griesbaum, W. Naegele, G. G. Wanless, *J. Am. Chem. Soc.*, **87**, 3151 (1965).

[58] M. R. Rifi, *J. Am. Chem. Soc.*, **89**, 4442 (1967).

[59] M. R. Rifi, *Tetrahedron Lett.*, **1969**, 1043.

[60] M. R. Rifi, *Coll. Czech. Chem. Commun.*, **36**, 932 (1971).

[61] A. J. Fry, W. E. Britton, *Tetrahedron Lett.*, **1971**, 4363.

[62] M. Mitani, H. Takeuchi, K. Koyama, *Chem. Lett.*, **1986**, 2125.

[63] F. Barba, M. D. Velasco, A. Guirado, N. Moreno, *Synth. Commun.*, **15**, 939 (1985).

[64] J. G. Lawless, D. E. Bartak, M. D. Hawley, *J. Am. Chem. Soc.*, **91**, 7121 (1969).

[65] Y. Matsui, T. Soga, Y. Date, *Bull. Chem. Soc. Jpn.*, **44**, 513 (1971).

[66] Y. Rollin, M. Troupel, J. Périchon, J. F. Fauvarque, *J. Chem. Res. (M)*, **1981**, 3801.

[67] D. Bérubé, R. N. Renaud, G. Pierre, *Electrochim. Acta*, **28**, 1367 (1983).

[68] M. Troupel, Y. Rollin, S. Sibille, J. F. Fauvarque, J. Périchon, *J. Chem. Res. (S)*, **1980**, 26.

[69] A. J. Fry, W. E. Britton, *J. Org. Chem.*, **38**, 4016 (1973).

[70] (a) K. B. Wiberg, G. A. Epling, *Tetrahedron Lett.*, **1974**, 1119; (b) W. A. Pritts, D. G. Peters, *J. Electroanal. Chem.*, **380**, 147 (1995).

[71] (a) M. R. Rifi, *Org. Synth.*, **52**, 22 (1972); (b) M. R. Rifi, *J. Org. Chem.*, **36**, 2017 (1971).

[72] W. A. Pritts, D. G. Peters, *J. Electrochem. Soc.*, **141**, 990 (1994).

[73] J. Casanova, H. R. Rogers, J. Murray, R. Ahmed, O. Rasmy, M. Tarle, *Croat. Chem. Acta*, **63**, 225 (1990).

[74] N. S. Zefirov, S. I. Kozhushkov, T. S. Kuznetsova, I. M. Sosonkin, A. M. Domarev, V. N. Leibzon, T. I. Egorova, *Zh. Org. Khim.*, **23**, 2109 (1987).

[75] M. Kijima, Y. Sakai, H. Shirakawa, *Chem. Lett.*, **1994**, 2011.

[76] M. Tokuda, K. Satoh, H. Suginome, *Bull. Chem. Soc. Jpn.*, **60**, 2429 (1987).

[77] C. Daubié, C. B.-Einhorn, D. Lelandais, *Can. J. Chem.*, **62**, 1552 (1984).

[78] C. Saboureau, M. Troupel, S. Sibille, E. d'Incan, J. Périchon, *J. Chem. Soc., Chem. Commun.*, **1989**, 895.

[79] S. Mabrouk, S. Pellegrini, J.-C. Folest, Y. Rollin, J. Périchon, *J. Organomet. Chem.*, **301**, 391 (1986).

[80] S. Sibille, J.-C. Folest, J. Coulombeix, M. Troupel, J. F. Fauvarque, J. Périchon, *J. Chem. Res. (S)*, **1980**, 268.

[81] (a) A. J. Fry, T. A. Powers, *J. Org. Chem.*, **52**, 2498 (1987); (b) L. Mattiello, U. Rampazzo, G. Sotgiu, *J. Chem. Res. (M)*, **1992**, 2732.

[82] P. Peterson, A. K. Carpenter, R. F. Nelson, *J. Electroanal. Chem., Interfacial Electrochem.*, **27**, 1 (1970).

[83] C. De Luca, A. Inesi and L. Rampazzo, *J. Chem. Soc., Perkin Trans. II*, **1983**, 1821.

[84] S. Margel, F. C. Anson, *J. Electrochem. Soc.*, **125**, 1232 (1978).

[85] M. Tokuda, K. Satoh, H. Suginome, *Chem. Lett.*, **1984**, 1035.

[86] M. Mitani, Y. Yamamoto, K. Koyama, *J. Chem. Soc., Chem. Commun.*, **1983**, 1446.

[87] W. J. M. Van Tilborg, R. Plomp, R. de Ruiter, C. J. Smit, *Ann. Soc. Sci. Brux. Pays-Bas.*, **99**, 206 (1980).

[88] M. Tarle, O. Rasmy, H. Rogers, J. Casanova, *Croat. Chem. Acta*, **63**, 239 (1990).

[89] V. N. Leibzon, A. S. Mendkovich, T. A. Klimova, M. M. Krayushkin, S. G. Mairanovskii, S. S. Novikov, V. V. Sevast'yanova, *Elektrokhimiya*, **11**, 349 (1975).

[90] K. B. Wiberg, G. A. Epling, M. Jason, *J. Am. Chem. Soc.*, **96**, 912 (1974).

[91] T. Shono, K. Yoshida, K. Ando, Y. Usui, H. Hamaguchi, *Tetrahedron Lett.*, **1978**, 4819.

[92] M. Fleischmann, G. Mengoli, D. Pletcher, *Electrochim. Acta*, **18**, 231 (1973).

[93] M. Baizer, J. Shruma, *J. Org. Chem.*, **37**, 1951 (1972).

[94] D. Tyssee, J. Shruma, *J. Org. Chem.*, **39**, 2819 (1974).

[95] J.-C. Folest, J.-M. Duprilot, J. Périchon, Y. Robin, J. Devynck, *Tetrahedron Lett.*, **26**, 2633 (1985); (b) M. Ocafrain, M. Devaud, M. Troupel, J. Périchon, *J. Chem. Soc., Chem. Commun.*, **1995**, 2331.

[96] (a) J. F. Fauvarque, A. Jutand, M. Francois, *Nouv. J. Chim.*, **10**, 119 (1986); (b) J. F. Fauvarque, A. Jutand, M. Francois, *J. Appl. Electrochem.*, **18**, 109 (1988).

[97] M. Tokuda, T. Kabuki, Y. Katoh, H. Suginome, *Tetrahedron Lett.*, **36**, 3345 (1995).

[98] J. H. Wagenkecht, *J. Electroanal. Chem., Interfacial Electrochem.*, **52**, 489 (1974).

[99] G. Silvestri, S. Gambino, G. Filardo, *Tetrahedron Lett.*, **27**, 3429 (1986).

[100] J. Chaussard, M. Troupel, Y. Robin, G. Jacob, J. P. Juhasz, *J. Appl. Electrochem.*, **19**, 345 (1989).

[101] G. Silvestri, S. Gambino, G. Filardo, A. Gulotta, *Angew. Chem., Int. Ed. Engl.*, **23**, 979 (1984)

[102] J. Pouliquen, M. Heintz, O. Sock, M. Troupel, *J. Chem. Educ.*, **63**, 1013 (1986)

[103] O. Sock, M. Troupel, J. Périchon, *Tetrahedron Lett.*, **26**, 1509 (1985).

[104] J. Périchon, O. Sock, M. Troupel, *French Pat.*, 8,409,787 (1984).

[105] T. Shono, Y. Matsumura, S. Katoh, N. Kise, *Chem. Lett.*, **1985**, 463.

[106] J. Yoshida, K. Muraki, H. Funahashi, N. Kawabata, *J. Org. Chem.*, **51**, 3996 (1986).

[107] P. Pons, C. Biran, M. Bordeau, J. Dunogués, S. Sibille, J. Périchon, *J. Organomet. Chem.*, **321**, 27 (1987).

[108] S. Satoh, H. Suginome, M. Tokuda, *Bull. Chem. Soc. Jpn.*, **56**, 1791 (1983).

[109] S. Sibille, E. d'Incan, L. Leport, J. Périchon, *Tetrahedron Lett.*, **27**, 3129 (1986).

[110] S. Sibille, E. d'Incan, L. Leport, M.-C. Massebiau, J. Périchon, *Tetrahedron Lett.*, **28**, 55 (1987).

[111] S. Durandetti, S. Sibille, J. Périchon, *J. Org. Chem.*, **54**, 2198 (1989).

[112] Y. Rollin, S. Derien, E. Duñach, C. Gebehenne, J. Périchon, *Tetrahedron*, **49**, 7723 (1993).

[113] K. Uneyama, H. Matsuda, S. Torii, *Tetrahedron Lett.*, **25**, 6017 (1984).

[114] M. Ninato, J. Tsuji, *Chem. Lett.*, **1988**, 2049.

[115] H. Hebri, E. Duñach, J. Périchon, *Tetrahedron Lett.*, **34**, 1475 (1993).

[116] M. Tokuda, S. Satoh, H. Suginome, *J. Org. Chem.*, **54**, 5608 (1989).

[117] M. Tokuda, M. Uchida, Y. Katoh, H. Suginome, *Chem. Lett.*, **1990**, 461.

[118] M. Tokuda, S. Satoh, Y. Katoh, H. Suginome, *Electroorg. Synth.*, **1991**, 83.

[119] D. P. G. Hamon, K. R. Richards, *Aust. J. Chem.*, **36**, 2243 (1983).

[120] S. Satoh, H. Suginome, M. Tokuda, *Tetrahedron Lett.*, **22**, 1895 (1981).

[121] S. Satoh, H. Suginome, M. Tokuda, *Bull. Chem. Soc. Jpn.*, **54**, 3456 (1981).

[122] H. Hebri, E. Duñach, J. Périchon, *J. Chem. Soc., Chem. Commun.*, **1993**, 499.

[123] (a) M. Michael, D. G. Peters, *Tetrahedron Lett.*, **1972**, 453; (b) R.-L. Shao, J. A. Cleary, D. M. La Perriere, D. G. Peters, *J. Org. Chem.*, **48**, 3289 (1983).

[124] R.-L.Shao, D. G. Peters, *J. Org. Chem.*, **52**, 652 (1987).

[125] T. Inokuchi, H. Kawafuchi, K. Aoki, A. Yoshida, S. Torii, *Bull. Chem. Soc. Jpn.*, **67**, 595 (1994).

[126] M. S. Mubarak, D. D. Nguyen, D. G. Peters, *J. Org. Chem.*, **55**, 2648 (1990).

[127] W. M. Moore, A. Salajegheh, D. G. Peters, *J. Am. Chem. Soc.*, **97**, 4954 (1975).

[128] B. C. Willett, W. M. Moore, A. Salajegheh, D. G. Peters, *J. Am. Chem. Soc.*, **101**, 1162 (1979).

[129] J. P. Petrovich, J. D. Anderson, M. M. Baizer, *J. Org. Chem.*, **31**, 3897 (1966).

[130] J. D. Anderson, M. M. Baizer, J. P. Petrovich, *J. Org. Chem.*, **31**, 3890 (1966).

[131] A. J. Fry, *Synthetic Organic Electrochemistry*, pp 250, Harper & Row, New York (1972).
[132] A. J. Bard, L. R. Faulkner, *Electrochemical Methods*, Chapter 3, John Wiley & Sons, Inc., New York (1980).
[133] P. G. Gassman, O. M. Rasmy, T. O. Murdock, K. Saito, *J. Org. Chem.*, **46**, 5455 (1981).
[134] S. T. Nugent, M. M. Baizer, R. D. Little, *Tetrahedron Lett.*, **23**, 1339 (1982).
[135] E. d'Incan, S. Sibille, J. Périchon, M.-O. Moingeon, J. Chaussard, *Tetrahedron Lett.*, **27**, 4175 (1986).
[136] I. Nishiguchi, T. Oki, T. Hirashima, J. Shiokawa, *Chem. Lett.*, **1991**, 2005.
[137] S. Torii, H. Tanaka, T. Ohshima, M. Sasaoka, *Bull. Chem. Soc. Jpn.*, **59**, 3975 (1986).
[138] J. F. Rusling, G. N. Kamau, *J. Electroanal. Chem., Interfacial Electrochem.*, **187**, 355 (1985).
[139] J. C. Folest, S. Guibe, J. Y. Nédélec, J. Périchon, *J. Chem. Res. (S)*, **1990**, 258.
[140] E. A. Chernyshev, A. V. Bukhtiarov, L. V. Evstifeev, S. A. Zinov'eva, B. K. Kabanov, O. V. Kuz'min, A. P. Tomilov, *Elektrokhimiya*, **19**, 1003 (1983).
[141] A. V. Bukhtiarov, E. A. Chernyshev, O. V. Kuz'min, B. K. Kabanov, S. A. Zinov'eva, A. P. Tomilov, *Zh. Obshch. Khim.*, **55**, 395 (1985).
[142] D. L. Kirkpatrick, K. E. Johnson, A. C. Sartorelli, *J. Med. Chem.*, **29**, 2048 (1986).
[143] I. Barelmann, J. K. Blum, C. H. Hamann, *J. Chem. Soc., Faraday Trans. I*, **86**, 3233 (1990).
[144] N. Zylber, J. Zylber, Y. Rollin, E. Duñach, J. Périchon, *J. Organomet. Chem.*, **444**, 1 (1993).
[145] Y. Rollin, C. Gebehenne, S. Derien, E. Duñach, J. Périchon, *J. Organomet. Chem.*, **461**, 9 (1993).
[146] C. Gosmini, Y. Rollin, C. Gbehenne, E. Lojou, V. Ratovelomanana, J. Périchon, *Tetrahedron Lett.*, **35**, 5637 (1994).
[147] T. Shono, N. Kise, N. Uematsu, S. Morimoto, E. Okazaki, *J. Org. Chem.*, **55**, 5037 (1990).
[148] M. Tsukayama, H. Utsumi, A. Kunugi, *J. Chem. Soc., Chem. Commun.*, **1995**, 615.
[149] (a) S. Ozaki, T. Nakanishi, M. Sugiyama, C. Miyamoto, H. Ohmori, *Chem. Pharm. Bull.*, **39**(1), 31 (1991); (b) J. Gao, J. F. Rusling, De-ling Zhou, *J. Org. Chem.*, **61**, 5972 (1996).
[150] A. Inesi, M. A. Casadei, F. M. Moracci, W. Jugelt, *Gazz. Chim. Ital.*, **124**, 81 (1994).
[151] (a) H. Tanaka, Y. Kameyama, D. Nonen, S. Torii, *Chem. Express*, **7**, 885 (1992); (b) S. Torii, H. Tanaka, M. Taniguchi, M. Sasaoka, T. Shiroi, Y. Kameyama, R. Kikuchi, *Pict Int. Appl.* Wo 92/16532 (1992).
[152] K. Yoshida, M. Kobayashi, S. Amano, *J. Chem. Soc., Perkin Trans. I*, **1992**, 1127.
[153] J. Yoshida, K. Muraki, H. Funahashi, N. Kawabata, *J. Organomet. Chem.*, **284**, C33 (1985).
[154] S. U. Pedersen, T. Lund, *Acta Chem. Scand.*, **45**, 397 (1991).
[155] J. Casanova, H. R. Rogers, *J. Org. Chem.*, **39**, 2408 (1974).
[156] (a) D. Lexa, J.-M. Savéant, H. J. Schäfer, Khac-Binh Su, B. Vering, D. Li Wang, *J. Am. Chem. Soc.*, **112**, 6162 (1990); (b) C. P. Andrieux, A. Le Gorande, J.-M. Savéant, *J. Electroanal. Chem.*, **371**, 191 (1994).
[157] K. M. O'Cornnell, D. H. Evans, *J. Am. Chem. Soc.*, **105**, 1473 (1983).
[158] O. R. Brown, P. H. Middleton, *J. Chem. Soc., Perkin Trans. II*, **1984**, 955.
[159] T. F. Connors, J. V. Arena, J. F. Rusling, *J. Phys. Chem.*, **92**, 2810 (1988).
[160] J. Zavada, J. Krupicka, O. Kocian, M. Pankova, *Collect. Czech. Chem. Commun.*, **48**, 3552 (1983).
[161] I. Spirevska, V. Rekalic, *Glas. Hem. Drus. Beograd*, **49**, 683 (1984).
[162] U. Husstedt, H. J. Schäfer, *Tetrahedron Lett.*, **22**, 623 (1981).
[163] A. Owlia, Z. Wang, J. F. Rusling, *J. Am. Chem. Soc.*, **111**, 5091 (1989).
[164] J. F. Rusling, T. F. Connors, A. Owlia, *Anal. Chem.*, **59**, 2123 (1987).
[165] M. O. Iwunze, N. Hu, J. F. Rusling, *J. Electroanal. Chem.*, **333**, 353 (1992).
[166] J. Casanova, H. R. Rogers, *J. Org. Chem.*, **39**, 3803 (1974).
[167] A. Inesi, L. Rampazzo, *J. Electroanal. Chem., Interfacial Electrochem.*, **54**, 289 (1974).
[168] H. Lund, E. Hobolth, *Acta Chem. Scand.*, **B30**, 895 (1976).
[169] R. Engels, C. J. Smit, W. J. M. Van Tilborg, *Angew. Chem., Int. Ed. Engl.*, **22**, 492 (1983).
[170] J. B. Kerr, L. L. Miller, M. R. Van De Mark, *J. Am. Chem. Soc.*, **102**, 3383 (1980).
[171] L. Coche, A. Deronzier, J.-C. Moutet, *J. Electroanal. Chem., Interfacial Electrochem.*, **198**, 187 (1986).
[172] A. J. Fry, K. Ankner, V. Handa, *Tetrahedron Lett.*, **22**, 1791 (1981).
[173] P. Fawell, J. Avraamides, G. Hefter, *Aust. J. Chem.*, **43**, 1421 (1990).
[174] P. Fawell, J. Avraamides, G. Hefter, *Aust. J. Chem.*, **44**, 791 (1991).
[175] M. Tezuka, M. Iwasaki, *Denki Kagaku*, **62**, 1230 (1994).
[176] T. Nagaoka, J. Yamashita, M. Kaneda, K. Ogura, *J. Electroanal. Chem.*, **335**, 187 (1992).
[177] L. G. Feoktistov, M. M. Gol'din, *Zh. Obshch. Khim.*, **43**, 515 (1973).
[178] L. G. Feoktistov, M. M. Gol'din, *Zh. Obshch. Khim.*, **43**, 520 (1973).
[179] K. M. Smirnov, A. P. Tomilov, L. G. Feoktistov, M. M. Gol'din, *Zh. Prinkl. Khim.*, **51**, 701 (1978).
[180] K. M. Smirnov, A. P. Tomilov, L. G. Geoktistov, M. M. Gol'din, *Zh. Prinkl. Khim.*, **51**, 703 (1978).
[181] L. G. Geoktistov, A. P. Tomilov, M. M. Gol'din, *Izv. Acad. Nauk SSSR, Ser. Khim.*, **1963**, 1352.
[182] K. M. Smirnov, A. P. Tomilov, *Elektrokhimiya*, **11**, 784 (1975).
[183] M. M. Gol'din, V. R. Polishchuk, N. S. Stepanova, L. G. Feoktistove, *Zh. Obshch. Khim.*, **43**, 525 (1973).
[184] B. R. Cho, E. J. Cho, S. J. Lee, W. S. Chae, *J. Org. Chem.*, **60**, 2077 (1995).

[185] G. Bosser, J. Paris, *J. Chem. Soc., Perkin Trans. II*, **1992**, 2057.

[186] M. A. Casadei, S. Cesa, F. M. Moracci, A. Inesi, *Gazz. Chim. Ital.*, **123**, 457 (1993).

[187] D. Lenoir, W. Kornrumpf, H. P. Fritz, *Chem. Ber.*, **1983**, 2390.

[188] E. Eru, G. E. Hawkes, J. H. P. Utley, P. B. Wyatt, *Tetrahedron*, **51**, 3033 (1995).

[189] S. Rondinini, P. R. Mussini, V. Ferzetti, D. Monti, *Electrochim. Acta*, **36**, 1095 (1991).

[190] (a) P. S. Manchand, P. S. Belica, M. J. Holman, T,-N. Huang, H. Maehr, S. Y.-K. Tam, R. T. Yang, *J. Org. Chem.*, **57**, 3473 (1992); (b) O. Johansen, S. M. Marcuccio, A. W.-H. Mau, *Aust. J. Chem.*, **47**, 1843 (1994).

[191] (a) F. M'Halla, J. Pinson, J.-M. Savéant, *J. Electroanal. Chem., Interfacial Electrochem.*, **89**, 347 (1978); (b) B. J. Coté, D. Despres, R. Labrecque, J. Lamothe, J.-M. Chapuzet, J. Lessard, *J. Electroanal. Chem.*, **355**, 219 (1993).

[192] C. P. Andrieux, J.-M. Savéant, D. Zann, *Nouv. J. Chim.*, **8**, 107 (1984).

[193] L. Nadjo, J.-M. Savéant, *J. Electroanal. Chem., Interfacial Electrochem.*, **30**, 41 (1971);

[194] J. Grimshaw, J. T.-Grimshaw, *J. Electroanal. Chem., Interfacial Electrochem.*, **56**, 443 (1974).

[195] K. Alwair, J. Grimshaw, *J. Chem. Soc., Perkin Trans. II*, **1973**, 1811.

[196] J. G. Lawless, M. D. Hawley, *J. Electroanal. Chem., Interfacial Electrochem.*, **21**, 365 (1969).

[197] R. P. Van Duyne, C. N. Reilley, *Anal. Chem.*, **44**, 158 (1972).

[198] R. F. Nelson, A. K. Carpenter, E. T. Seo, *J. Electrochem. Soc.*, **120**, 206 (1973).

[199] M. S. Mubarak, L. L. Karras, N. S. Murcia, J. C. Bart, J. Z. Stemple, D. G. Peters, *J. Org. Chem.*, **55**, 1065 (1990).

[200] (a) S. O. Farwell, F. A. Beland, R. D. Geer, *J. Electroanal. Chem., Interfacial Electrochem.*, **61**, 303 (1975); (b) K. V. Plekhanov, A. I. Tsyganok, S. M. Kulikov, *Izv. Akad. Nauk SSSR, Ser. Khim.*, **1995**, 1129.

[201] S. O. Farwell, F. A. Beland, R. D. Geer, *J. Electroanal. Chem., Interfacial Electrochem.*, **61**, 315 (1975).

[202] D. E. Bartak, K. J. Houser, B. C. Rudy, M. D. Hawley, *J. Am. Chem. Soc.*, **94**, 7526 (1972).

[203] K. J. Houser, D. E. Bartak, M. D. Hawley, *J. Am. Chem. Soc.*, **95**, 6033 (1973).

[204] R. N. Renaud, *Can. J. Chem.*, **52**, 376 (1974).

[205] G. J. Gores, C. E. Koeppe, D. E. Bartak, *J. Org. Chem.*, **44**, 380 (1979).

[206] F. M'Halla, J. Pinson, J.-M. Savéant, *J. Am. Chem. Soc.*, **102**, 4120 (1980).

[207] T. T. Tsai, W. E. McEwen, J. Kleinberg, *J. Org. Chem.*, **26**, 318 (1960).

[208] J. Casado, I. Gallardo, *Electrochim. Acta*, **32**, 1145 (1987).

[209] V. N. Andreev, G. Horanyi, V. F. Stenin, *Elektrokhimiya*, **21**, 968 (1985).

[210] E. K.-Miller, Z. Vajtner, *J. Org. Chem.*, **50**, 1394 (1985).

[211] T. Teherani, A. J. Bard, *Acta Chem. Scand.*, **B37**, 413 (1983).

[212] A. Demortier, A. J. Bard, *J. Am. Chem. Soc.*, **95**, 3495 (1973).

[213] C. Amatore, J. B.-Lambling, C. B.-Huyghes, J. Pinson, J.-M. Savéant, A. Thiébault, *J. Am. Chem. Soc.*, **104**, 1979 (1982).

[214] J. F. Bunnett, B. F. Gloor, *J. Org. Chem.*, **39**, 382 (1974).

[215] J. F. Bunnett, R. G. Scamehorn, R. P. Traber, *J. Org. Chem.*, **41**, 3677 (1976).

[216] J. Belloni, *Actions Chim. Biol. Radiat.*, **15**, 47 (1971).

[217] C. P. Andrieux, J. B.-Lambling, C. Combellas, D. Lacombe, J.-M. Savéant, A Thiébault, D. Zann, *J. Am. Chem. Soc.*, **109**, 1518 (1987).

[218] V. D. Parker, *Acta Chem. Scand.*, **B35**, 655 (1981).

[219] R. G. Compton, R. A. W. Dryfe, *J. Electroanal. Chem.*, **375**, 247 (1994).

[220] R. G. Compton, R. A. W. Dryfe, A. C. Fisher, *J. Electroanal. Chem.*, **361**, 275 (1993).

[221] R. G. Compton, R. A. W. Dryfe, A. C. Fisher, *J. Chem. Soc., Perkin Trans. II*, **1994**, 1581.

[222] (a) S. Wawzonek, S. M. Heilmann, *J. Electrochem. Soc.*, **121**, 516 (1974); (b) A. Sucheta, I. Ul Haque, J. F. Rusling, *Langmuir*, **8**, 1633 (1992).

[223] T. Matsumoto, M. Sato, S. Hirayama, *Bull. Chem. Soc. Jpn.*, **46**, 369 (1973).

[224] V. D. Parker, *Acta Chem. Scand.*, **B35**, 595 (1981).

[225] (a) J. Voss, M. Altrogge, H. Wilkes, W. Francke, *Z. Naturforsch., B; Chem. Sci.*, **46**, 400 (1991). (b) A. Tsyganok, K. Otsuka, I. Yamanaka, V. Plekhanov, S. Kulikov, *Chem. Lett.*, **1996**, 261.

[226] N. Bhuvaneswari, C. S. Venkatachalam, K. K. Balasubramanian, *Tetrahedron Lett.*, **33**, 1499 (1992).

[227] J. F. Rusling, J. V. Arena, *J. Electroanal. Chem., Interfacial Electrochem.*, **186**, 225 (1985).

[228] M. O. Iwunze, J. F. Rusling, *J. Electroanal. Chem., Interfacial Electrochem.*, **226**, 197 (1989).

[229] D. Petersen, M. Lemmrich, M. Altrogge, J. Voss, *Z. Naturforsch., B; Chem. Sci.*, **45**, 1105 (1990).

[230] S. S. Shukla, J. F. Rusling, *J. Phys. Chem.*, **89**, 3353 (1985).

[231] P. Nelleborg, H. Lund, J. Eriksen, *Tetrahedron Lett.*, **26**, 1773 (1985).

[232] (a) N. Sato, A. Yoshiyama, P.-C. Cheng, T. Nonaka, M. Sasaki, *J. Appl. Electrochem.*, **22**, 1082 (1992); (b) J. S. Moilliet, I. K. Jones, *Eur. Pat. Appl.*, EP 508,578 (1992).

[233] R. D. Chambers, W. K. R. Musgrave, C. R. Sargent, F. G. Drakesmith, *Tetrahedron*, **37**, 591 (1981).

[234] C. K. Bon, A. J. Kamp, T. J. Sobieralski, *U. S. Pat.*, 4,592,810 (1986).

[235] R. Skowronski, W. Strzyzewski, *Pol. J. Chem.*, **65**, 883 (1991).

[236] D. Kyriacou, D. N. Brattesani, *U. S. Pat.*, 4,533,454 (1985).
[237] P. Fuchs, U. Hess, H. H. Holst, H. Lund, *Acta Chem. Scand.*, **B35**, 185 (1981).
[238] A. N. Gaidukevich, T. V. Zhukova, A. A. Kravchnko, *Zh. Obshch. Khim.*, **56**, 1887 (1986).
[239] J. T. Kusmierek, B. Czochralska, N. G. Johansson, D. Shugar, *Acta Chem. Scand.*, **B41**, 701 (1987).
[240] R. M. Justice, Jr., D. A. Hall, *U. S. Pat.* 4,588,484 (1986).
[241] R. N. Gedye, Y. N. Sadana, R. Leger, *Can. J. Chem.*, **63**, 2669 (1995).
[242] M. Mori, Y. Hashimoto, Y. Ban, *Tetrahedron Lett.*, **21**, 631 (1980).
[243] Y. Rollin, M. Troupel, D. G. Tuck, J. Périchon, *J. Organomet. Chem.*, **303**, 131 (1986).
[244] G. Meyer, M. Troupel, J. Périchon, *J. Organomet. Chem.*, **393**, 137 (1990).
[245] S. Sibille, V. Ratovelomanana, J.-Y. Nédélec, J. Périchon, *Synlett.*, **1993**, 425.
[246] M. Troupel, Y. Rollin, S. Sibille, J. F. Fauvarque, J. Périchon, *J. Organomet. Chem.*, **202**, 435 (1980).
[247] C. Amatore, A. Jutand, *Organometallics*, **7**, 2203 (1988).
[248] M. A. Fox, D. A. Chandler, C. Lee, *J. Org. Chem.*, **56**, 3246 (1991).
[249] G. A. Selivanova, V. F. Starichenko, V. D. Shteingarts, *Izv. Akad. Nauk SSSR, Ser. Khim.*, **1988**, 1155.
[250] O. Hammerich, V. D. Parker, *Acta Chem. Scand.*, **B37**, 851 (1983).
[251] S. Donnelly, J. Grimshaw, J. T.-Grimshaw, *J. Chem. Soc., Perkin Trans. I*, **1993**, 1557.
[252] S. Donnelly, J. Grimshaw, J. T.-Grimshaw, *Denki Kagaku*, **62**, 1125 (1994).
[253] G. Froyer, G. Ollivier, C. Chevrot, A. Siove, *J. Electroanal. Chem.*, **327**, 159 (1992).
[254] N. Saito, T. Kanbara, T. Sato, T. Yamamoto, *Polym. Bull.*, **30**, 285 (1993).
[255] Z. Chami, M. Gareil, J. Pinson, J.-M. Savéant, A. Thiébault, *Tetrahedron Lett.*, **29**, 639 (1988).
[256] Z. Chami, M. Gareil, J. Pinson, J.-M. Savéant, A. Thiébault, *J. Org. Chem.*, **56**, 586 (1991).
[257] Y. Rollin, G. Meyer, M. Troupel, J. F. Fauvarque, *Tetrahedron Lett.*, **23**, 3573 (1982).
[258] (a) Y. Rollin, G. Meyer, M. Troupel, J. F. Fauvarque, J. Périchon, *J. Chem. Soc., Chem. Commun.*, **1983**, 793; (b) G. Meyer, Y. Rollin, *C. R. Acad. Sci. Ser. II*, **302**, 303 (1986).
[259] M. D. Koppang, G. A. Ross, N. F. Woolsey, D. E. Bartak, *J. Am. Chem. Soc.*, **108**, 1441 (1986).
[260] D. M. Loffredo, J. E. Swartz, E. Kariv-Miller, *J. Org. Chem.*, **54**, 5953 (1989).
[261] F. Barba, A. Guirado, A. Zapata, *Electrochim. Acta*, **27**, 1335 (1982).
[262] N. S. Murcia, D. G. Peters, *J. Electroanal. Chem.*, **326**, 69 (1992).
[263] T. Matsue, S. Kitahara, T. Osa, *Denki Kagaku*, **50**, 732 (1982).
[264] M. Troupel, Y. Rollin, J. Périchon, J. F. Fauvarque, *Nouv. J. Chim.*, **5**, 621 (1981).
[265] M. Heintz, O. Sock, C. Saboureau, J. Périchon, *Tetrahedron*, **44**, 1631 (1988).
[266] J. F. Fauvarque, C. Chevrot, A. Jutand, M. François, J. Périchon, *J. Organomet. Chem.*, **264**, 273 (1984).
[267] N. Zylber, A. Druilhe, J. Zylber, J. Périchon, *New. J. Chem.*, **13**, 535 (1989).
[268] E. Lojou, M. Devaud, M. Heintz, M. Troupel, J. Périchon, *Electrochim. Acta*, **38**, 613 (1993).
[269] J. E. Swartz, T. T. Stenzel, *J. Am. Chem. Soc.*, **106**, 2520 (1984).
[270] M. A. Oturan, J. Pinson, J.-M. Savéant, A. Thiébault, *Tetrahedron Lett.*, **30**, 1373 (1989).
[271] J. B. Davison, P. J. Peerce-Landers, R. J. Jashinski, *J. Electrochem. Soc.*, **130**, 1862 (1983).
[272] N. Alam, C. Amatore, C. Combellas, A. Thiébault, J. N. Verpeaux, *Tetrahedron Lett.*, **28**, 6171 (1987).
[273] C. Degrand, *J. Org. Chem.*, **52**, 1421 (1987).
[274] C. Degrand, *J. Chem. Soc., Chem. Commun.*, **1986**, 1113.
[275] C. Amatore, J. Pinson, J.-M. Savéant, A. Thiébault, *J. Am. Chem. Soc.*, **104**, 817 (1982).
[276] C. Amatore, J. Chaussard, J. Pinson, J.-M. Savéant, A. Thiébault, *J. Am. Chem. Soc.*, **101**, 6012 (1979).
[277] (a) C. Degrand, R. Prest, *J. Org. Chem.*, **55**, 5242 (1990); (b) C. Degrand, *Tetrahedron*, **46**, 5237 (1990); (c) M. Genesty, C. Thobie, A. Gautier, C. Degrand, *J. Appl. Electrochem.*, **23**, 1125 (1993).
[278] J. Pinson, J.-M. Savéant, *J. C. S., Chem. Commun.*, **1974**, 933.
[279] (a) C. Amatore, J. Pinson, J.-M. Savéant, A. Thiébault, *J. Am. Chem. Soc.*, **103**, 6930 (1981); (b) R. Ettayeb, J.-M. Savéant, A. Thiébault, *J. Am. Chem. Soc.*, **114**, 10990 (1992).
[280] J.-M. Savéant, *Bull. Soc. Chim. Fr.*, **1988**, 225.
[281] C. Amatore, C. Combellas, S. Robveille, J.-M. Savéant, A. Thiébault, *J. Am. Chem. Soc.*, **108**, 4754 (1986).
[282] C. P. Andrieux, J.-M. Savéant, *J. Am. Chem. Soc.*, **115**, 8044 (1993).
[283] J.-M. Savéant, *Acc. Chem. Res.*, **13**, 323 (1980).
[284] C. Amatore, J. Pinson, J.-M. Savéant, A. Thiébault, *J. Electroanal. Chem., Interfacial Electrochem.*, **107**, 75 (1980).
[285] W. J. M. Van Tilborg, C. J. Smit, J. J. Scheele, *Tetrahedron Lett.*, **1977**, 2113.
[286] W. J. M. Van Tilborg, C. J. Smit, J. J. Scheele, *Tetrahedron Lett.*, **1978**, 776.
[287] C. Amatore, J.-M. Savéant, A. Thiébault, *J. Electroanal. Chem., Interfacial Electrochem.*, **103**, 303 (1979).
[288] C. Amatore, J. Pinson, J.-M. Savéant, A. Thiébault, *J. Electroanal. Chem., Interfacial Electrochem.*, **107**, 59 (1980).
[289] D. B. Denney, D. Z. Denney, *Tetrahedron*, **47**, 6577 (1991).
[290] J. Simonet, M. C. el Badre, G. Mabon, *J. Electroanal. Chem., Interfacial Electrochem.*, **281**, 289 (1990).
[291] J.-M. Paratian, S. Sibille, J. Périchon, *J. Chem. Soc., Chem. Commun.*, **1992**, 53.

[292] C. P. Andrieux, L. Gelis, J.-M. Savéant, *Tetrahedron Lett.*, **30**, 4961 (1989).

[293] K. Boujlel, J. Simonet, G Roussi, R. Beugelmans, *Tetrahedron Lett.*, **23**, 173 (1982).

[294] C. Combellas, M. Lequan, R. M. Lequan, J. Simon, A. Thiébault, *J. Chem. Soc., Chem. Commun.*, **1990**, 542.

[295] C. T.-Gautier, M. Genesty, C. Degrand, *J. Org. Chem.*, **56**, 3452 (1991).

[296] C. Amatore, R. Beugelmans, M. B.-Choussy, C. Combellas, A. Thiébault, *J. Org. Chem.*, **54**, 5688 (1989).

[297] (a) C. Amatore, J. Pinson, J.-M. Savéant, *J. Electroanal. Chem., Interfacial Electrochem.*, **123**, 231 (1981); (b) M. Benaichouche, G. Bosser, J. Paris, V. Plichon, *J. Chem. Soc., Perkin Trans. II*, **1991**, 817; (c) H. Balslev, H. Lund, *Tetrahedron*, **50**, 7889 (1994).

[298] P. Boy, C. Combellas, A. Thiébault, *Synlett.*, **1991**, 923.

[299] N. Alam, C. Amatore, C. Combellas, A. Thiébault, J. N. Verpeaux, *J. Org. Chem.*, **55**, 6347 (1990).

[300] C. Amatore, C. Combellas, N.-E. Lebbar, A. Thiébault, J. N. Verpeaux, *J. Org. Chem.*, **60**, 18 (1995).

[301] N. Egashira, I. Sakurai, F. Hori, *Denki Kagaku*, 54, 282 (1986).

[302] N. Egashira, J. Takenakga, F. Hori, *Bull. Chem. Soc. Jpn.*, **60**, 2671 (1987).

[303] C. Amatore, M. A. Oturan, J. Pinson, J.-M. Savéant, A. Thiébault, *J. Am. Chem. Soc.*, **106**, 6318 (1984).

[304] J. Grimshaw, R. Hamilton, J. T.-Grimshaw, *J. Chem. Soc., Perkin Trans. I*, **1982**, 229.

[305] J. Grimshaw, R. J. Haslett, *J. Chem. Soc., Perkin Trans. I*, **1980**, 657.

[306] C. Amatore, M. Gareil, M. A. Oturan, J. Pinson, J.-M. Savéant, A. Thiébault, *J. Org. Chem.*, **51**, 3757 (1986).

[307] J. Pinson, J.-M. Savéant, *J. Am. Chem. Soc.*, **100**, 1506 (1978).

[308] M. Bordeau, C. Biran, P. Pons, M.-P. Léger-Lambert, J. Dunogués, *J. Org. Chem.*, **57**, 4705 (1992).

[309] D. Deffieux, M. Bordeau, C. Biran, J. Dunogués, *Organometallics*, **13**, 2415 (1994).

[310] P. Tebbutt, C. E. W. Hahn, *J. Electroanal. Chem., Interfacial Electrochem.*, **261**, 205 (1989).

[311] A. R. Mount, M. S. Appleton, W. J. Albery, D. Clark, C. E. W. Hahn, *J. Electroanal. Chem.*, **334**, 155 (1992).

[312] J. Touster, *Diss. Abstr. Int. B*, **52**, 2564 (1991).

[313] A. J. Fry, J. Touster, U. N. Sirisoma, B. Raimundo, *Electroorg. Synth.*, **1991**, 99.

[314] V. V. Yanilkin, N. I. Maksimyuk, E. I. Gritsenko, Yu. M. Kargin, B. M. Garifullin, *Izv. Akad. Nauk SSSR, Ser. Khim.*, **1992**, 292.

[315] R. Hazard, S. Jaquannet, A. Tallec, *Electrochim. Acta*, **28**, 1095 (1983).

[316] Yu. M. Kargin, E. I. Gritsenko, V. V. Yanilkin, V. V. Piemenkov, L. K. Dubovik, N. I. Maksimyuk, B. M. Garifullin, Sh. K. Letypov, A. V. II'yasov, *Izv. Akad. Nauk SSSR, Ser. Khim.*, **1992**, 2023.

[317] S. Abe, T. Fuchigami, T. Nonaka, *Chem. Lett.*, **1983**, 1033.

[318] S. Ikeda, G. Tsukamoto, I. Nishiguchi, T. Hirashima, *Chem. Pharm. Bull.*, **36**, 1976 (1988).

[319] A. J. Fry, R. G. Reed, *J. Am. Chem. Soc.*, **94**, 8475 (1972).

[320] L. G. Feoktistov, G. P. Girina, V. A. Kokorekina, N. M. Alpatova, E. V. Ovsyannikova, *J. Electroanal. Chem., Interfacial Electrochem.*, **180**, 67 (1984).

[321] G. P. Girina, V. A. Kokorekina, L. G. Feoktistov, N. M. Alpatova, E. V. Ovsyannikova, A. I. D'yachenko, *Elektrokhimiya*, **20**, 778 (1984).

[322] G. P. Girina, V. A. Kokorekina, L. G. Feoktistov, N. M. Alpatova, *Elektrokhimiya*, **23**, 696 (1987).

[323] N. Egashira, S. Tomishige, Y. Takita, F. Hori, *Denki Kagaku*, **51**, 338 (1983).

[324] (a) A. J. Fry, U. N. Sirisoma, A. S. Lee, *Tetrahedron Lett.*, **34**, 809 (1993); (b) A. J. Fry, A. H. Singh, *J. Org. Chem.*, **59**, 8172 (1994).

[325] G. Silvestri, S. Gambino, G. Filardo, G. Greco, A. Gulotta, *Tetrahedron Lett.*, **25**, 4307 (1984).

[326] A. Guirado, F. Barba, M. B. Hursthouse, A. Martinez, A. Arcas, *Tetrahedron Lett.*, **27**, 4063 (1986).

[327] S. Durandetti, S. Sibille, J. Périchon, *J. Org. Chem.*, **56**, 3255 (1991).

[328] J. C. Le Menn, J. Sarrazin, A. Tallec, *Electrochim. Acta*, **35**, 563 (1990).

[329] R. C. Duty, B. V. Pepich, *J. Electrochem. Soc.*, **127**, 1261 (1980).

[330] T. Strelow, J. Boss, G. Adiwidjaja, *J. Chem. Res. (S)*, **1989**, 136.

[331] G. K. S. Prakash, S. Quaiser, H. A. Buchhollz, J. Casanova, G. A. Olah, *Synlett.*, **1994**, 113.

[332] G. K. S. Prakash, H. A. Buchholz, D. Deffieux, G. A. Olah, *J. Org. Chem.*, **59**, 7532 (1994).

[333] M. Umezawa, M. Kojima, H. Ichikawa, T. Ishikawa, T. Nonaka, *Electrochim. Acta*, **38**, 529 (1993).

[334] M. M. Baizer, *Organic Electrochemistry*, Marcel Dekker, New York (1973).

[335] C. K. Mann, K. K. Barnes, *Electrochemical Reactions in Non Aqueous Systems*, pp 201, Marcel Dekker, New York (1970).

[336] M. R. Rifi, F. H. Covitz, *Introduction to Organic Electrochemistry*, p. 194, Marcel Dekker, New York (1974).

[337] F. L. Lambert, K. Kobayashi, *J. Am. Chem. Soc.*, **82**, 5324 (1960).

[338] R. Annino, R. E. Erickson, J. Michalovic, B. McKay, *J. Am. Chem. Soc.*, **88**, 4424 (1966).

[339] J. W. Sease, F. G. Burton, S. L. Nickol, *J. Am. Chem. Soc.*, **90**, 2595 (1968).

[340] P. Zuman, *Chem. Listy*, **56**, 219 (1962).

[341] J. L. Webb, C. K. Mann, H. M. Walborsky, *J. Am. Chem. Soc.*, **92**, 2042 (1970).
[342] J. Simonet, H. Doupeux, P. Martinet, D. Bretelle, *Bull. Soc. Chim. Fr.*, **1970**, 3930.
[343] A. A. Pud, L. L. Gervits, G. S. Shapoval, O. É. Mikulina, *Elektrokhimiya*, **28**, 825 (1992).
[344] P. Calas, P. Moreau, A. Commeyras, *J. Electroanal. Chem., Interfacial Electrochem.*, **78**, 271 (1977).
[345] C. D.-Avignon, P. Calas, A. Commeyras, C. Amatore, *J. Fluorine Chem.*, **51**, 357 (1991).
[346] P. Calas, P. Moreau, A. Commeyras, *J. Fluorine Chem.*, **12**, 67 (1978).
[347] P. Calas, A. Commeyras, *J. Fluorine Chem.*, **16**, 553 (1980).
[348] S. Wawzonek, R. C. Duty, *J. Electrochem. Soc.*, **1961**, 1135.
[349] G. P. Girina, V. A. Kokorekina, L. G. Feoktistov, V. A. Petrosyan, *Elektrokhimiya*, **28**, 517 (1992).
[350] T. Shono, H. Ohmizu, S. Kawakami, S. Nakano, N. Kise, *Tetrahedron Lett.*, **22**, 871 (1981).
[351] T. Shono, N. Kise, M. Masuda, T. Suzumoto, *J. Org. Chem.*, **50**, 2527 (1985).
[352] T. Shono, N. Kise, A. Yamazaki, H. Ohmizu, *Tetrahedron Lett.*, **23**, 1609 (1982).
[353] R. Voigtlander, H. Matschiner, C. Krzeminski, H. Biering, *J. Prakt. Chem.*, **327**, 649 (1985).
[354] U. Fechtel, H. Matschiner, *J. Prakt. Chem.*, **331**, 545 (1989).
[355] L. Wolf, E. Steckhan, *J. Chem. Soc., Perkin Trans. I*, **1986**, 733.
[356] L. F. Filimonova, S. M. Makarochkina, I. N. Chernykh, V. G. Soldatov, G. V. Motsak, A. P. Tomilov, *Elektrokhimiya*, **28**, 725 (1992).
[357] H. Biering, *Z. Chem.*, **20**, 217 (1980).
[358] J.-Y. Nédélec, H. A. Haddou, J.-C. Folest, J. Périchon, *J. Org. Chem.*, **53**, 4720 (1988).
[359] R.-A. Doong, S.-C. Wu, *Chemosphere*, **24**, 1063 (1992).
[360] L. Eberson, M. Ekström, T. Lund, H. Lund, *Acta Chem. Scand.*, **43**, 101 (1989).
[361] H. Asai, K. Uneyama, *Chem. Lett.*, **1995**, 1123.
[362] P. E. Iversen, *J. Chem. Educ.*, **48**, 136 (1971).
[363] N. Hu, J. Rusling, *Anal. Chem.*, **63**, 2163 (1991).
[364] I. A. Avrutskaya, T. A. Arkhipova, G. V. Itov, N. E. Krasnoshchekova, *Elektrokhimiya*, **29**, 926 (1993).
[365] U. Fechtel, K. Westphal, H. Matschiner, *J. Prakt. Chem.*, **332**, 394 (1990).
[366] (a) T. Shono, N. Kise, T. Suzumoto, *J. Am. Chem. Soc.*, **106**, 259 (1984); (b) T. Shono, H. Ohmizu, N. Kise, *Tetrahedron Lett.*, **23**, 4801 (1982).
[367] J.-Y. Nédélec, H. A.-H.-Mouloud, J. C. Folest, J. Périchon, *Tetrahedron Lett.*, **29**, 1699 (1988).
[368] P. Jubault, C. Feasson, N. Collignon, *Tetrahedron Lett.*, **37**, 3679 (1996).
[369] A. V. Bukhtiarov, V. N. Golyshin, A. P. Tomilov, O. V. Kuz'min, *Dokl. Akad. Nauk SSSR*, **294**, 875 (1987).
[370] H. P. Fritz, W. Kornrumpf, *J. Electroanal. Chem., Interfacial Electrochem.*, **100**, 217 (1979).
[371] M. A. McClinton, D. A. McClinton, *Tetrahedron*, **48**, 6555 (1992).
[372] (a) J.-M. Paratian, E. Labbe, S. Sibille, J. Y. Nédélec, J. Périchon, *Denki Kagaku*, **62**, 1129 (1994); (b) S. Sibille, J. Périchon, J. Chaussard, *Synth. Commun.*, **19**, 2449 (1989).
[373] C. P. Andrieux, L. Gélis, M. Médebielle, J. Pinson, J.-M. Savéant, *J. Am. Chem. Soc.*, **112**, 3509 (1990).
[374] J.-C. Folest, J.-Y. Nédélec, J. Périchon, *Synth. Commun.*, **18**, 1491 (1988).
[375] C. P. Andrieux, L. Gelis, J.-M. Savéant, *J. Am. Chem. Soc.*, **112**, 786 (1990).
[376] P. Pons, C. Biran, M. Bordeau, J. Dunogués, *J. Organomet. Chem.*, **358**, 31 (1988).
[377] G. K. S. Prakash, D. Deffieux, A. K. Yudin, G. A. Olah, *Synlett*, **1994**, 1057.
[378] (a) P. Jubault, C. Feasson, N. Collignon, *Bull. Soc. Chim. Fr.*, **131**, 1001 (1994); (b) S. Sibille, S. Mcharek, J. Périchon, *Tetrahedron*, **45**, 1423 (1989).
[379] H. J. Schäfer, H. Clause, *Stud. Org. Chem.*, **1987**, 227.
[380] (a) J. P. Coleman, H. G. Gilde, J. H. P. Utley, B. C. L. Weedon, *J. C. S., Chem. Commun.*, **1970**, 738; (b) C. Saboureau, M. Troupel, S. Sibille, J. Périchon, *J. Chem. Soc., Chem. Commun.*, **1989**, 1138.
[381] (a) J.-P. Gisselbrecht, H. Lund, *Acta Chem. Scand.*, **B39**, 773 (1985); (b) S. A. Kaufman, T. Phanijphand, A. J. Fry, *Tetrahedron Lett.*, **37**, 8105 (1996).
[382] S. Tsuboi, Y. Ishiguro, A. Takeda, *Bull. Chem. Soc. Jpn.*, **60**, 830 (1987).
[383] (a) J.-P. Gisselbrecht, H. Lund, *Acta Chem. Scand.*, **B39**, 823 (1985); (b) O. Schneider, M. Hanack, *Angew. Chem., Int. Ed. Engl.*, **21**, 79 (1982).
[384] M. E. Niyazymbetov, V. A. Petrosyan, V. V. Kozlov, M. S. Pevzner, *J. Electroanal. Chem.*, **338**, 239 (1992).
[385] P. Calas, P. Moreau, A. Commeyras, *J. Chem. Soc., Chem. Commun.*, **1982**, 433.
[386] M. Médebielle, J. Pinson, J.-M. Savéant, *Tetrahedron Lett.*, **33**, 7351 (1992).
[387] H. P. Fritz, W. Kornrumpf, *Liebigs Ann. Chem.*, **1978**, 1416.
[388] (a) V. A. Petrosyan, M. E. Niyazymbetov, T. K. Baryshnikova, *Izv. Akad. Nauk SSSR, Ser. Khim.*, **1988**, 91; (b) M. E. Niyazymbetov, T. Yu. Rudashevskaya, L. V. Adaevskaya, V. A. Petrosyan, *Izv. Akad. Nauk SSSR, Ser. Khim.*, **1990**, 1802.
[389] F. De Angelis, A. Inesi, M. Feroci, R. Nicoletti, *J. Org. Chem.*, **60**, 445 (1995).
[390] V. L. Kornienko, N. V. Kalinichenko, I. A. Kedrinskii, Yu. G. Chirkov, *Zh. Prikl. Khim.*, **59**, 1179 (1986).
[391] A. Savall, R. Abdelhedi, S. Dalbera, M. L. Bouguerra, *Electrochim. Acta*, **35**, 1727 (1990).

[392] A. Savall, S. Dalbéra, R. Abdelhedi, M. L. Bouguerra, *J. Appl. Electrochem.*, **20**, 1045 (1990).

[393] T. Shono, N. Kise, H. Oka, *Tetrahedron Lett.*, **32**, 6567 (1991).

[394] (a) S. Mcharek, S. Sibille, J.-Y. Nédélec, J. Périchon, *J. Organomet. Chem.*, **401**, 211 (1991); (b) A. Conan, S. Sibille, J. Périchon, *J. Org. Chem.*, **56**, 2018 (1991).

[395] I. Lachaise, K. Nohair, M. Hakiki, J.-Y. Nédélec, *Synth. Commun.*, **25**, 3529 (1995).

[396] T. Shono, N. Kise, *Chem. Lett.*, **1987**, 697.

[397] A. Kunugi, T. Takahashi, K. Abe, *Denki Kagaku*, **61**, 864 (1993).

[398] P. Molina, C. Conesa, M. D. Velasco, *Tetrahedron Lett.*, **34**, 175 (1993).

[399] I. Carelli, A. Inesi, M. A. Casadei, B. Di Rienzo, F. M. Moracci, *J. Chem. Soc., Perkin Trans. II*, **1985**, 179.

[400] (a) M. A. Casadei, F. M. Moracci, D. Occhialini, A. Inesi, *J. Chem. Soc., Perkin Trans. II*, **1987**, 1887; (b) M. A. Casadei, F. M. Moracci, C. Giomini, A. Inesi, *Bull. Soc. Chim. Fr.*, **1989**, 63.

[401] Yu. D. Smirnov, A. P. Tomilov, *Elektrokhimiya*, **31**, 713 (1995).

[402] M. A. Casadei, A. Inesi, *Bull. Soc. Chim. Fr.*, **127**, 54 (1991).

[403] H. Claus, H. J. Schäfer, *Tetrahedron Lett.*, **26**, 4899 (1985).

[404] M. Inaba, K. Sawai, Z. Ogumi, Z. Takehara, *Chem. Lett.*, **1995**, 471.

[405] B. Tue Bi, M. Devaud, *Tetrahedron Lett.*, **28**, 3799 (1987).

[406] P. Jubault, C. Feasson, N. Collignon, *Tetrahedron Lett.*, **36**, 7073 (1995).

[407] M. Devaud, F. Azzouzi, B. T. Bi, *J. Chem. Res. (S)*, **1991**, 120; *ibid.,(M)*, **1991**, 1052.

[408] J.-C. Le Menn, A. Tallec, J. Sarrazin, *J. Chem. Educ.*, **68**, 513 (1991).

[409] J. Engels, *Leibigs Ann. Chem.*, **1980**, 557.

[410] (a) M. Médebielle, *Tetrahedron Lett.*, **36**, 2071 (1995); (b) M. Médebielle, *Tetrahedron Lett.*, **37**, 5119 (1996).

[411] M. Médebiell, J. Pinson, J.-M. Savéant, *Tetrahedron Lett.*, **31**, 1279 (1990).

[412] M. Médebielle, J. Pinson, J.-M. Savéant, *J. Am. Chem. Soc.*, **113**, 6872 (1991).

[413] S. D. Datsenko, N. V. Ignat'ev, L. M. Yagupol'skii, *Elektrokhimiya*, **27**, 1674 (1991).

[414] C. Amatore, J.-M. Savéant, *J. Electroanal. Chem., Interfacial Electrochem.*, **184**, 25 (1985).

[415] H. Saffarian, P. Ross, F. Behr, G. Gard, *J. Electrochem. Soc.*, **139**, 2391 (1992).

[416] (a) R. Holze, U. Fette, *J. Electroanal. Chem.*, **339**, 247 (1992); (b) R. Holze, U. Fette, *Dechema-Monogr.*, **125**, 769 (1992).

[417] G. V. Itov, I. A. Avrutskaya, *Elektrokhimiya*, **31**, 1245 (1995).

[418] A. N. Zhuravlev, S. D. Shamshinov, I. A. Avrutskaya, *Elektrokhimiya*, **28**, 874 (1992).

[419] P. Tissot, P. Margaretha, *Electrochim. Acta*, **23**, 1049 (1978).

[420] F. Maran, S. Roffia, M. G. Severin, E. Vianello, *Electrochim. Acta*, **35**, 81 (1990).

[421] F. Maran, E. Vianello, G. Cavicchioni, F. D'Angeli, *J. Chem. Soc., Chem. Commun.*, **1985**, 660.

[422] F. Maran, E. Vianello, *Bull. Electrochem.*, **6**, 276 (1990).

[423] P. J. Elving, I. Rosenthal, A. J. Martin, *J. Am. Chem. Soc.*, **77**, 5218 (1955).

[424] J. Kopilov, D. H. Evans, *J. Electroanal. Chem., Interfacial Electrochem.*, **280**, 435 (1990).

[425] L. Coche, J.-C. Moutet, *J. Electroanal. Chem., Interfacial Electrochem.*, **245**, 313 (1988).

[426] A. S. Reddy, S. J. Reddy, *Trans. SAEST*, **19**, 270 (1984).

[427] C. De Luca, A. Inesi, L. Rampazzo, *J. Electroanal. Chem., Interfacial Electrochem.*, **198**, 369 (1986).

[428] F. Barba, M. D. Velasco, N. Moreno, A. Aldaz, *Electrochim. Acta*, **32**, 1507 (1987).

[429] (a) A. J. Fry, J. P. Bujanauskas, *J. Org. Chem.*, **43**, 3157 (1978); (b) A. J. Fry, A. T. Lefor, *J. Org. Chem.*, **44**, 1270 (1979).

[430] I. Chiarotoo, M. Feroci, C. Giomini, A. Inesi, *Bull. Soc. Chim. Fr.*, **133**, 167 (1996).

[431] (a) I. Carelli, A. Curulli, A. Inesi, E. Zeuli, *J. Chem. Res. (S)*, **1990**, 74; *ibid., (M)*, **1990**, 620; (b) F. De Angelis, M. Feroci, A. Inesi, *Bull. Soc. Chim. Fr.*, **130**, 712 (1993).

[432] A. J. Fry, J. T. Andersson, *J. Org. Chem.*, **46**, 1490 (1981).

[433] C. Giomini, A. Inesi, E. Zeuli, *Electrochim. Acta*, **29**, 1107 (1984).

[434] R. Hazard, S. Jaouannet, A. Tallec, *Bull. Soc. Chim. Belg.*, **94**, 199 (1985).

[435] R. Hazard, S. Jaouannet, E. Raoult, A. Tallec, *Nouv. J. Chim.*, **6**, 325 (1982).

[436] F. Maran, M. G. Severin, E. Vianello, F. D'Angeli, *J. Electroanal. Chem.*, **352**, 43 (1993).

[437] M. A. Casadei, F. M. Moracci, A. Inesi, *J. Chem. Soc., Perkin Trans. II*, **1986**, 419.

[438] M. A. Casadei, A. Inesi, F. M. Moracci, D. Occhialini, *Tetrahedron*, **45**, 6885 (1989).

[439] S. Ikeda, M. Moriyama, G. Tsukamoto, T. Hirashima, I. Nishiguchi, *Japan Kokai Tokkyo Koho,* 61-63683 (1986).

[440] B. K. V.-Lukanova, L. S. Changov, N. N. Agappova, I. A. Pesheva, *Pharmazie*, **45**, 63 (1990).

[441] M. F. C.-Pommeret, S. Jaquannet, A. Lebouc, A. Tallec, *Electrochim. Acta*, **29**, 1287 (1984).

[442] A. Tallec, R. Hazard, A. Le Bouc, J. Grimshaw, *J. Chem. Res. (S)*, **1986**, 342.

[443] (a) F. Barba, M. D. Velasco, A. Guirado, I. Barba, A. Aldaz, *Electrochim. Acta*, **30**, 1119 (1985); (b) F. Barba, J. L. de la Fuente, *J. Org. Chem.*, **58**, 7685 (1993).

[444] J. Yoshida, M. Yamamoto, N. Kawabata, *Tetrahedron Lett.*, **26**, 6217 (1985).

[445] M. Durandetti, S. Sibille, J.-Y. Nédélec, J. Périchon, *Synth. Commun.*, **24**, 145 (1994).

[446] (a) A. Conan, S. Sibille, E. d'Incan, J. Périchon, *J. Chem. Soc., Chem. Commun.*, **1990**, 48; (b) H. Shick, R. Ludwig, K.-H. Schwarz, K. Kleiner, A. Kunath, *Angew. Chem.*, **105**, 1218 (1993); (c) H. Schick, R. Ludwig, K.-H. Schwarz, K. Kleiner, A. Kunath, *J. Org. Chem.*, **59**, 3161 (1994).

[447] C. De Luca, A. Inesi, L. Rampazzo, *J. Chem. Soc., Perkin Trans. II*, **1985**, 209.

[448] C. De luca, A. Inesi, L. Rampazzo, *J. Chem. Soc., Perkin Trans. II*, **1987**, 847.

[449] L. Rampazzo, A. Inesi, *J. Electrochem. Soc.*, **127**, 2388 (1980).

[450] C. De Luca, A. Inesi, L. Rampazzo, *J. Chem. Soc., Perkin Trans. II*, **1982**, 1403.

[451] L. Mattiello, C. De Luca, *J. Chem. Soc., Perkin Trans. II*, **1990**, 1041.

[452] L. Mattiello, L. Rampazzo, G. Sotgiu, *J. Chem. Res. (S)*, **1992**, 321.

[453] M. Tokuda, A. Hayashi, H. Suginome, *Bull. Chem. Soc. Jpn.*, **64**, 2590 (1991).

[454] J.-C. Le Menn, A. Tallec, *Can. J. Chem.*, **69**, 761 (1991).

[455] H. Schick, R. Ludwig, K.-H. Schwarz, K. Kleiner, A. Kunath, *Angew. Chem., Int. Ed. Engl.*, **32**, 1191 (1993).

[456] K.-H. Schwarz, K. Kleiner, R. Ludwig, H. Schick, *J. Org. Chem.*, **57**, 4013 (1992).

[457] K.-H. Schwarz, K. Kleiner, R. Ludwig, H. Schick, *Chem. Ber.*, **126**, 1247 (1993).

[458] N. Zylber, J. Zylber, Y. Rollin, E. Duñach, J. Périchon, *J. Organomet. Chem.*, **444**, 1 (1993).

[459] I. Carelli, A. Curulli, A. Inesi, E. Zeuli, *J. Chem. Res. (S)*, **1988**, 154.

[460] F. Barba, M. D. Velasco, A. Guirado, *Synthesis*, **1981**, 625.

[461] F. Barba, M. D. Velasco, A. Guirado, *Electrochim. Acta*, **28**, 259 (1983).

[462] F. Barba, M. D. Velasco, M. I. Lopez, A. Zapata, A. Aldaz, *J. Chem. Res. (S)*, **1988**, 44.

[463] F. Barba, J. L. de la Fuente, *Port. Electrochim. Acta*, **9**, 145 (1991).

[464] F. Barba, J. L. de la Fuente, *Tetrahedron Lett.*, **33**, 3911 (1992).

[465] (a)F. Barba, M. G. Quintanilla, G. Montero, *Synthesis*, **1992**, 1215; (b) G. Montero, M. G. Quintanilla, F. Barba, *J. Electroanal. Chem.*, **345**, 457 (1993).

[466] F. Barba, B. Batanero, *J. Org. Chem.*, **58**, 6889 (1993).

[467] J. I. Lozano, F. Barba, *Tetrahedron Lett.*, **35**, 9623 (1994).

[468] (a) A. J. Fry, M. Susla, M. Weltz, *J. Org. Chem.*, **52**, 2496 (1987).

[469] A. J. Fry, M. Susla, *J. Am. Chem. Soc.*, **111**, 3225 (1989).

[470] I. Carelli, A. Inesi, V. Carelli, M. A. Casadei, F. Liberatore, F. M. Moracci, *Synthesis*, **1986**, 591.

[471] (a) M. A. Casadei, B. Di Rienzo, A. Inesi, F. M. Moracci, *J. Chem. Soc., Perkin Trans. I*, **1992**, 379; (b) M. A. Casadei, B. Di Rienzo, A. Inesi, F. M. Moracci, *J. Chem. Soc., Perkin Trans. I*, **1992**, 375; (c) F. Maran, *J. Am. Chem. Soc.*, **115**, 6557 (1993).

[472] M. A. Casadei, A. Gessner, A. Inesi, W. Jugelt, H. Liebezeit, F. M. Moracci, *Bull. Soc. Chim. Fr.*, **1989**, 650.

[473] R. Vukicevic, S. Konstantinovic, L. Joksovic, G. Ponticelli, M. L. Mihailovic, *Chem. Lett.*, **1995**, 275.

[474] (a) E. A. Smertenko, S. D. Datsenko, N. V. Ingnat'ev, L. E. Deev, I. K. Bil'dinov, P. V. Podsevalov, *Elektrokhimiya*, **30**, 1284 (1994); (b) D. Lucas, Y. Mugnier, A. Antinolo, A. Otero, M. Fajardo, C. Lopez-Mardomingo, A. Fakhr, J. Mofidi, *New. J. Chem.*, **18**, 817 (1994).

[475] (a) G. A. Urove, *Diss. Abstr. Int. B.*, **53**, 6270 (1993); (b) G. A. Urove, D. G. Peters, M. S. Mubarak, *J. Org. Chem.*, **57**, 786 (1992).

[476] G. A. Urove, D. G. Peters, *Tetrahedron Lett.*, **34**, 1271 (1993).

[477] (a) G. A. Urove, D. G. Peters, *J. Electrochem. Soc.*, **140**, 932 (1993); (b) M. S. Mubarak, D. G. Peters, *J. Electrochem. Soc.*, **142**, 713 (1995).

[478] V. D. Pokhodenko, V. G. Koshechko, V. E. Titov, V. A. Lopushanskaja, *Tetrahedron Lett.*, **36**, 3277 (1995).

[479] (a) G. A. Urove, D. G. Peters, *J. Org. Chem.*, **58**, 1620 (1993); (b) G. A. Urove, D. G. Peters, *Electrochim. Acta*, **39**, 1441 (1994).

[480] P. Arthur, H. Lyons, *Anal. Chem.*, **24**, 1422 (1952).

[481] (a) A. Guirado, F. Barba, C. Manzanera, M. D. Velasco, *J. Org. Chem.*, **47**, 142 (1982); (b) C. Polo, M. G. Quintanilla, F. Barba, *Synth. Commun.*, **24**, 907 (1994).

[482] (a) G. T. Cheek, P. A. Horine, *J. Electrochem. Soc.*, **131**, 1796 (1984); (b) M. S. Mubarak, G. A. Urove, D. G. Peters, *J. Electroanal. Chem.*, **350**, 205 (1993).

[483] (a) G. A. Urove, D. G. Peters, *J. Electroanal. Chem.*, **365**, 221 (1994); (b) J.-C. Folest, E. Pereira-Martins, M. Troupel, J. Périchon, *Tetrahedron Lett.*, **34**, 7571 (1993); (c) M. S. Mubarak, *J. Electroanal. Chem.*, **394**, 239 (1995).

[484] T. Shono, I. Nishiguchi, H. Ohmizu, *Chem. Lett.*, **1977**, 1021.

[485] (a)J. I. Lozano, F. Barba, *Heterocycles*, **38**, 1339 (1994); (b) J. I. Lozano, F. Barba, *Tetrahedron*, **52**, 1259 (1996).

[486] A. Guirado, F. Barba, J. Martin, *Electrochim. Acta*, **29**, 587 (1984).

[487] A. Guirado, F. Barba, J. M. Cuadrado, *Electrochim. Acta*, **28**, 761 (1983).

[488] M. Tezuka, T. Yajima, *Denki Kagaku*, **59**, 517 (1991).

[489] H. Doupeux, P. Martinet, J. Simonet, *Bull. Soc. Chim. Fr.*, **1971**, 2299.
[490] H. Matchiner, R. Voigtländer, R. Liesenberg, G. W. Fischer, *Electrochim. Acta*, **24**, 331 (1979).
[491] H. Matshiner, R. Voightländer, H. Schick, H. Lund, *Acta Chem. Scand.*, **B34**, 136 (1980).
[492] R. Saiganesh, K. K. Balasubramanian, C. S. Venkatachalam, *Proc. Indian Acad. Sci. (Chem. Sci.)*, **101**, 473 (1989).
[493] R. Saiganesh, K. K. Balasubramanian, C. S. Venkatachalam, *J. Electroanal. Chem., Interfacial Electrochem.*, **262**, 221 (1989).
[494] R. Saiganesh, K. K. Balasubramaniank, C. S. Venkatachalam, *Bull. Electrochem.*, **6**, 515 (1990).
[495] Z. Vajtner, B. Lovrecek, *J. Electroanal. Chem., Interfacial Electrochem.*, **213**, 111 (1986).
[496] A. Guirado, A. Zapata, M. Fenor, *Tetrahedron Lett.*, **33**, 4779 (1992).
[497] G. S. Shapoval, V. A. Bagrii, S. I. Vdovenko, V. F. Baklan, V. P. Kukhar, A. A. Kisilenko, *Zh. Obshch. Khim.*, **55**, 1850 (1985).
[498] H. Nishihara, H. Harada, S. Kaneko, M. Tateishi, K. Aramaki, *J. Chem. Soc., Chem. Commun.*, **1990**, 26 ( ).
[499] S. Ozaki, I. Horiguchi, H. Matsushita, H. Ohmori, *Tetrahedron Lett.*, **35**, 725 (1994).
[500] G. Le Guillanton, Q. Tho Do, J. Simonet, *Bull. Soc. Chim. Fr.*, **1989**, 433.
[501] H.-H. Ruttinger, H. Matschiner, W. D. Gollnow, *J. Prakt. Chem.*, **328**, 539 (1986).
[502] H. Harada, *Japan Kokai Tokkyo Koho*, 57-181385 (1982).
[503] Yagi, S. Shimizu, *Nippon Kagaku Kaishi*, **1993**, 291.
[504] (a) O. Yagi, S. Shimizu, *Chem. Lett.*, **1993**, 2041; (b) O. Yagi, S. Shimizu, *Chem. Lett.*, **1994**, 1683.
[505] M. Kashiwase, H. Harada, K. Tomiie, *Stud. Org. Chem.*, **1987**, 467.
[506] S. Hashiba, T. Fuchigami, T. Nonaka, *Bull. Chem. Soc. Jpn.*, **62**, 2424 (1989).
[507] E. Differding, P. M. Bersier, *Tetrahedron*, **48**, 1595 (1992).
[508] C. P. Andrieux, E. Differding, M. Robert, J.-M. Savéant, *J. Am. Chem. Soc.*, **115**, 6592 (1993).
[509] J. E. Barry, M. Finkelstein, W. M. Moore, S. D. Ross, L. Eberson, *Tetrahedron Lett.*, **25**, 2847 (1984).
[510] J. E. Barry, M. Finkelstein, W. M. Moore, S. D. Ross, L. Eberson, *J. Org. Chem.*, **50**, 528 (1985).
[511] J. Lessard, D. Bérubé, *Can. J. Chem.*, **62**, 768 (1984).
[512] T. Fuchigami, T. Nonaka, K. Iwata, *J. C. S., Chem. Commun.*, **1976**, 951.
[513] E. P. Koval'chuk, N. I. Ganushchak, V. I. Kopylets, I. N. Krupak, N. D. Obushak, *Zh. Obshch. Khim.*, **52**, 2540 (1982).
[514] N. I. Ganushchak, N. D. Obushak, E. P. Koval'chuk, G. V. Trifonova, *Zh. Obshch. Khim.*, **54**, 2334 (1984).
[515] T. Shono, Y. Matsumura, S. Kashimura, *J. Chem. Res. (S)*, **1984**, 216; *ibid., (M)*, **1984**, 1922.
[516] T. Shono, Y. Usui, H. Hamaguchi, *Tetrahedron Lett.*, **21**, 1351 (1980).
[517] M. S. Mubarak, D. G. Peters, *J. Org. Chem.*, **50**, 673 (1985).
[518] G. Kokkinidis, D. Sazou, E. Hatzigrigoriou, A. Varvoglis, *Electrochim. Acta*, **35**, 455 (1990).
[519] N. V. Ingat'ev, L. A. Nechitailo, A. A. Mironova, I. I. Maletina, V. V. Orda, *Elektrokhimiya*, **28**, 502 (1992).
[520] T. Fuchigami, T. Nonaka, *Denki Kagaku*, **53**, 582 (1985).
[521] P. Sanecki, *Pol. J. Chem.*, **66**, 101 (1992).
[522] M. Atobe, K. Matsuda, T. Nonaka, *Denki Kagaku*, **62**, 1298 (1994).
[523] *Soviet Pat.*, No. 833,976; *Byull. Izobret.*, No. 20 (1981).
[524] S. M. Makarochkina, Yu. I. Rozin, K. M. Samarin, V. F. Pavlichenko, L. V. Zhitarena, A. P. Tomilov, *Elektrokhimiya*, **21**, 1617 (1985).
[525] O. R. Brown, K. Taylor, *J. Electroanal. Chem., Interfacial Electrochem.*, **50**, 211 (1974).
[526] R. Bilewicz, J. Osteryoung, *J. Electroanal. Chem., Interfacial Electrochem.*, **226**, 27 (1987).
[527] A. Konno, T. Fuchigami, K. Suzuki, T. Nonaka, H. Tien, *Denki Kagaku*, **62**, 260 (1994).
[528] G. Kokkinidis, E. Hatzigrigoriou, D. Sazou, A. Varvoglis, *Electrochim. Acta*, **36**, 1391 (1991).
[529] S. A. Kuliev, A. M. Magerramov, A. I. Ismiev, R. G. Guseinova, Ya. A. Yaraliev, V. V. Zhdankin, A. S. Koz'min, N. S. Zefirov, *Zh. Obshch. Khim.*, **62**, 748 (1992).
[530] (a) N. Ito, T. Saji, S. Aoyagui, *J. Organomet. Chem.*, **247**, 301 (1983); (b) T. Saji, N. Ito, *Bull. Chem. Soc. Jpn.*, **58**, 3375 (1985)

Abbreviations and Symbols

Electrodes

DME	Dropping mercury electrode
GC	Glassy carbon
HMDE	Hanging mercury drop electrode
NHE	Normal hydrogen electrode
OTE	Optically transparent electrode
RDE	Rotating disc electrode
PRDE	Rotating ring-disc electrode
SCE	Saturated calomel electrode
SHE	Standard hydrogen electrode
SUS	Stainless steel electrode

Techniques

CP	Constant potential
CV	Cyclic voltammetry
DPSC	Double potential step chronoamperometry
LSV	Linear sweep voltammetry
dc pol, ac pol	dc, ac polarography

Solvents

AcOH	Acetic acid
Ac$_2$O	Acetic anhydride
aq.	aqueous
BN	Benzonitrile
CH$_2$CL$_2$	Dichloromethane
DME	Dimethoxyethane
DMF	Dimethylformamide
DMSO	Dimethylsulfoxide
EtOH	Ethanol
Glyme	1,2-Dimethoxyethane
HMPA(HMPT)	Hexamethylphosphoramide
MeCN(AN)	Acetonitrile
Me$_2$CO	Acetone
MeOH	Methanol
NMP	N-Methylpyrrolidone
PC	Propylene carbonate
Py	Pyridine
TFA	Trifluoroacetic acid
THF	Tetrahydrofuran
TMP	Trimethyl phosphate

Chemicals, Ligands, Radicals

Ac	Acetyl
acac	Acetylacetone
aib	Tripeptide of a-aminoisobutyric acid
Ant	Anthracene
Ar	Aryl
bipy(bpy)	2,2'-Bipyridine
Boc	t-Butoxycarbonyl
Bu	Butyl
Bn	Benzyl
Bu_4NBF_4	Tetrabutylammonium tetrafluoroborate
Bu_4NBr	Tetrabutylammonium bromide
Bu_4NClO_4	Tetrabutylammonium perchloate
Bu_4NI	Tetrabutylammonium iodide
Bu_4NOTs	Tetrabutylammonium tosylate
Bu_4NPF_6	Tetrabutylammonium hexafluorophosphate
CbZ	Benzyloxycarbonyl
COT	Cyclooctatetraene
Cp	Cyclopentadiene
Cyclam	1,4,8,11-tetraazacyclotetradecane
Dim	Diimine
$DMPBF_4$	Dimethylpyrrolidinium tetrafluoroborate
dppe	1,2-bis(diphenylphosphino)ethane, $(Ph_2PCH_2)_2$ dppe 1,2-bis(diphenylphosphino)propane, $(Ph_2PCH_2)_2CH_2$
E^+	Electrophile
Et	Ethyl
Et_4NClO_4	Tetraethylammomium perchlorate
Et_4NI	Tetraethylammonium iodide
Et_4NOTs	Tetraethylammonium tosylate
EWG	Electron withdrawing group
Fc	Ferrocenyl
H_2P	Porphyrin
H_2OP	Octaethylporphyrin
KCl	Potassium Chloride
KI	Potassium iodide
$KClO_4$	Potassium perchlorate
LiBr	Lithium bromide
LiCl	Lithium chloride
$LiClO_4$	Lithium perchorate
Me	Methyl
Me_4NBF_4	Tetramethylammonium tetrafluoroborate
Me_4NClO_4	Tetramethylammonium perchlorate
$Me_3SiCl(TMSCl)$	Trimethylchlorosilane
MP	Metalloporphyrin
MsOH	Methanesulfonic acid

MTPP	Tetraphenylmetalloporphyrine
NaClO$_4$	Sodium perchlorate
Naph	Naphthalene
Nu	Nucleophile
Pa	Polyacetylene
Pc	Phthalocyanine
Ph	Phenyl
phen	1,10-Phenanthroline
PPh$_3$	Triphenylphosphine
Pr	Propyl
Pr$_4$NClO$_4$	Tetrapropylammonium perchlorate
R	Alkyl
Sal	Salcylaldehyde
salen	*N,N'*-ethylenebis(salycylideneamine)
Bu$_4$NClO$_4$	Tetrabutylammonium perchlorate
Et$_4$NClO$_4$	Tetrabutylammonium perchlorate
THP	Tetrahydropyran
TMA	Tetramethylammonium
TPA	Tetrapropylammonium
sec(s)	Secondary
tert(t)	Tertiary

Index